Linear Electric Machines, Drives, and MAGLEVs Handbook

Linear Electric Machines, Drives, and MAGLEVs Handbook

ION BOLDEA

CRC Press is an imprint of the
Taylor & Francis Group, an informa business

CRC Press
Taylor & Francis Group
6000 Broken Sound Parkway NW, Suite 300
Boca Raton, FL 33487-2742

First issued in paperback 2017

Version Date: 20130311

ISBN 13: 978-1-138-07633-4 (pbk)
ISBN 13: 978-1-4398-4514-1 (hbk)

Library of Congress Cataloging-in-Publication Data

Boldea, I.
Linear electric machines, drives, and maglevs handbook / Ion Boldea.
pages cm
Includes bibliographical references and index.
ISBN 978-1-4398-4514-1 (alk. paper)
1. Electric motors, Linear. I. Title.

TK2537.B648 2013
621.46--dc23 2012044178

Visit the Taylor & Francis Web site at
http://www.taylorandfrancis.com

and the CRC Press Web site at
http://www.crcpress.com

Contents

Preface

According to Aristotle's *Politics* written in 360 BC, a society is worth existing if it provides freedom and prosperity to its people. While engineering is arguably one of the main agents of prosperity, the shortage of energy is a major concern on the path to a better quality of life. Electric energy represents about 40% of all energy used today as it enhances industrial productivity and causes less pollution. Electric motion control in industry is ultimately related to electric energy conversion and its flow control.

Linear electric (electromagnetic) machines (LEMs) realize the conversion of electrical energy to linear motion mechanical energy (or vice versa) directly through electromagnetic forces. Linear motion is very common in industry. LEMs were invented in the nineteenth century; however, they gained prominence at the industrial level only in 1960 due to the necessity of using power electronics for control (in the absence of any mechanical transmission).

Typical LEM applications in industry are as follows:

- Magnetically levitated vehicles (e.g., Shanghai's "Transrapid")
- Urban people movers (e.g., Dallas-Forth Worth's Airline Commuter)
- Linear electric motor–driven refrigeration compressors
- X–Y planar motion industrial platforms
- Espresso coffee steamer drivers
- Microphones and loudspeakers in cellular phones
- Digital camera zoomers
- Linear electric generators for deep space missions
- Hotel locker solenoids
- Electric power switch solenoids

LEMs are characterized by

- Low initial cost
- Higher reliability
- Adhesion—less propulsion (lighter vehicles: lower Wh/passenger/km)
- Better position tracking (no backlash)

However, to perform better than rotary electric motors with mechanical transmission, LEMs need power electronics for linear position, speed, and/or force control. The steady improvement of LEMs since 1960 is related to new topologies, materials, modeling, design, and testing and control methods; however, the enhanced dynamics of LEMs since 2000 prompted us to write this new, rather comprehensive and practical book.

This book is dedicated mainly to R&D and decision-making engineers in industry and to senior undergraduate and graduate students in electric power, mechanical, robotics, power electronics, and control engineering. It is based mainly on the author's vast experience in the field (40 years), which includes five books in English published in 1976, 1985, 1987, 1996, and 2001, but also draws heavily from recent contributions to the field worldwide. The book integrates tutorial and monograph attributes.

Consequently, it deals with a wide range of subjects from simple to complex, from classifications to practical topologies, to modeling, design, and control and provides numerous case studies, examples, and sample results based on an up-to-date survey of the field.

The contents of the book, covering 22 chapters, clearly reflect this ambivalent (textbook and monograph) nature.

Heartfelt thanks are due to my associates Sorin Agarlita, Ana Moldovan, Ana Maria Ungurean, Mircea Baba, and Dragos Ursu, who understood the contents and edited the text with all its equations and drawings, incorporating my corrections into the manuscript. Thanks are also due to the staff at CRC Press/Taylor & Francis Group for their competence and patient assistance during the preparation of this book. As with any other book, this book reflects the possibilities as well the limitations of the field. I hope it presents a balanced and in-depth, but practical, treatment of subjects, drawing heavily from R&D efforts worldwide but also giving due attention to personal contributions. I welcome any feedback from the readers.

Prof. Ion Boldea
IEEE Life Fellow
Timisoara, Romania, EU

1 Fields, Forces, and Materials for LEMs

Linear electric machines (LEMs) are electromagnetic, electrostatic, piezoelectric, and magnetostriction force devices capable of producing directly progressive or oscillatory translational (linear) motion.

They transform electric energy to linear motion mechanical energy via magnetic, electrostatic, etc., energy storage. Just as rotary electric machines, they may operate as motors (from electric to mechanical energy) or as generators (from mechanical to electric energy, Figure 1.1).

LEMs [1–11] may be considered as counterparts of rotary electric machines (as visible again in Chapter 2).

However, the absence of mechanical transmission in LEMs opens up the possibility of inventing a myriad of new topologies tied to the application. This demands solid fundamental knowledge on fields, forces, materials, and methods of approach to start with.

Our attention in this book will be devoted to electromagnetic devices as they have higher thrust density (N/kg, N/m^3) and lower loss/thrust (W/N). We call thrust the force along the direction of progressive or oscillatory linear motion, where most of the mechanical work is performed.

Electromagnetic fields [12], a rather abstract concept, but a form of matter, play a key role in explaining electric and magnetic phenomena with deep implications in linear (rotary as well) electric machines.

1.1 REVIEW OF ELECTROMAGNETIC FIELD THEORY

The basic laws of electricity are expressed by a set of equations called Maxwell's equations. The presence of moving parts (media) in electric machines makes the direct application of Maxwell's equations to energy conversion rather involved [13]. We hereby restrict ourselves to a review of Maxwell's equations, with a few LEM applications.

When placed in electric and magnetic fields, charged particles experience forces. In particular, the force $\widehat{F}$ on a charge q moving with a speed U in an electric field E and magnetic field B is given by the Lorenz force equation:

$$\widehat{F} = q \cdot \left(\overline{E} + \overline{U} \times \overline{B}\right) = \overline{F}_E + \overline{F}_B \tag{1.1}$$

The electric field intensity $\overline{E}$ in (V/m) is defined as

$$\overline{E} = \frac{\overline{F}_E}{\Delta q} \tag{1.2}$$

where Δq is an infinitesimal electric charge.

The magnetic flux density $\overline{B}$ in (V s/m^2) can also be defined as the force $\overline{F}_B$ on a unit charge moving at speed $\overline{U}$ at right angle to the direction of $\overline{B}$. The unit of $\overline{B}$ (V s/m^2) is also called Wb/m^2 or Tesla, after Nikola Tesla, the inventor of induction machines.

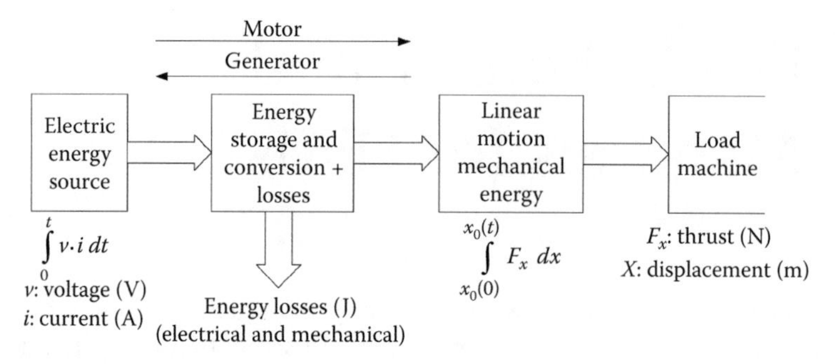

FIGURE 1.1 LEMs: energy conversion.

The electric scalar potential difference dV between two points separated by the distance $\overline{dl}$ is

$$\Delta V = -\overline{E} \cdot \overline{dl} \tag{1.3}$$

Using the vector operator ∇ (gradient), Equation 1.3 becomes

$$\overline{E} = -\nabla V = -\frac{\partial V}{\partial x} \cdot \overline{u}_x - \frac{\partial V}{\partial y} \cdot \overline{u}_y - \frac{\partial V}{\partial z} \cdot \overline{u}_z \tag{1.4}$$

where $\overline{u}_x$, $\overline{u}_y$, $\overline{u}_z$ are the unit vectors along x, y, z orthogonal coordinates. For the magnetic flux density, $\overline{B}$, the total magnetic flux ϕ through a surface S is

$$\phi = \iint_S \overline{B} \cdot \overline{dS} \tag{1.5}$$

The key law of electromagnetism is, however, the Faraday's law. This law states that an electromotive force (*emf*)—measured in volts—is induced in a closed circuit when the magnetic flux linking that circuit changes in time. For an N-turn coil closed circuit,

$$emf = -N \cdot \frac{d\phi}{dt} \tag{1.6}$$

The negative sign was introduced by Lenz to provide consistence with the positive sense of circulation about a path with respect to the positive direction of flux flow through the surface (Figure 1.2). When (1.6) is expressed in integral form, we obtain

$$emf = \int \overline{E} \cdot \overline{dl} \tag{1.7}$$

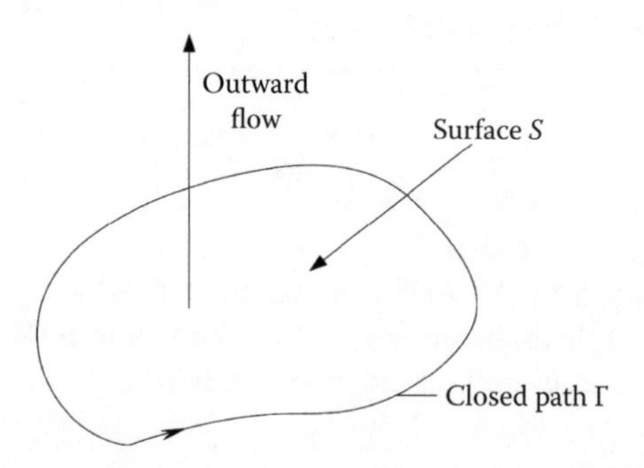

FIGURE 1.2 Positive circulation on closed path Γ with positive flow through the surface S.

From (1.5) through (1.7) for $N=1$ (one turn),

$$\int \bar{E}\cdot \bar{dl} = -\frac{\partial}{\partial t}\iint_{S} \bar{B}\cdot \bar{dS} \tag{1.8}$$

The partial derivative in Equation 1.8 refers to time as $\bar{B}$ may be dependent on space also. Equation 1.8 expresses Faraday's law in integral form for bodies at standstill.

By using Stokes theorem for vector fields,

$$\int_{\Gamma} \bar{E}\cdot \bar{dl} = \iint_{S} \left(\nabla\times\bar{E}\right)\cdot \bar{dS} \tag{1.9}$$

The surface $\bar{S}$ is enclosed by the closed path Γ. The so-called rotor of field $\bar{E}(\nabla\times\bar{E})$ is in orthogonal (Cartesian) coordinates:

$$\nabla\times\bar{E} = rot\,\bar{E} = \left(\frac{\partial E_z}{\partial y} - \frac{\partial E_y}{\partial z}\right)\cdot\bar{u}_x + \left(\frac{\partial E_x}{\partial z} - \frac{\partial E_z}{\partial x}\right)\cdot\bar{u}_y + \left(\frac{\partial E_y}{\partial x} - \frac{\partial E_x}{\partial y}\right)\cdot\bar{u}_z \tag{1.10}$$

$$\text{with} \quad \bar{E} = E_x\cdot\bar{u}_x + E_y\cdot\bar{u}_y + E_z\cdot\bar{u}_z \tag{1.11}$$

Using (1.9) in Equation 1.8 yields

$$\nabla\times\bar{E} = -\frac{\partial\bar{B}}{\partial t} \tag{1.12}$$

Vector algebra shows that the divergence of a rotor (curl) field is zero:

$$\nabla\cdot\left(\nabla\times\bar{E}\right) = 0 \tag{1.13}$$

And thus, $div\,\bar{B}=0$

$$div\,\bar{B} = \frac{\partial B_x}{\partial x} + \frac{\partial B_y}{\partial y} + \frac{\partial B_z}{\partial z} \tag{1.14}$$

$$\bar{B} = B_x\cdot\bar{u}_x + B_y\cdot\bar{u}_y + B_z\cdot\bar{u}_z \tag{1.15}$$

So far, we referred to the case of no relative motion between the closed circuit and the magnetic field, with $\bar{E}$ and $\bar{B}$ in stationary coordinates in Equation 1.1. If the speed of charge $\bar{U}$ is considered to be $\bar{U}'$ with respect to a moving reference at $\bar{U}$, the charge will move with speed $\bar{U}+\bar{U}'$ with reference to a stationary reference system and thus (1.1) becomes simply

$$F = q\cdot\left[\bar{E} + \bar{u}\times\bar{B} + \bar{u}'\times\bar{B}\right] \tag{1.16}$$

$\bar{E}' = \bar{E} + \bar{u}' \times \bar{B}$ is the electric field $\bar{E}$ in moving coordinates, with $rot\,\bar{E} = -\partial\bar{B}/\partial t$. Consequently, for an electric field $\bar{E}$ induced in a circuit moving with a speed $\bar{U}$ in a time varying field $\bar{B}$, we have

$$rot\,\bar{E} = -\frac{\partial B}{\partial t} + rot\left(\bar{u} \times \bar{B}\right) \tag{1.17}$$

The first term in (1.17) corresponds to transformer *emf*, while the second one represents the motional *emf*. The total *emf* does not depend on the reference frame speed but its components do. In LEMs, there is a mover and the presence of a motional *emf* is mandatory for force production.

So far, we considered only how an electric field $\bar{E}$ is produced by a magnetic field $\bar{B}$. The Ampere's low considers the relationship between existing (given) electric currents and the resulting magnetic fields $\bar{H}$:

$$\int \bar{H} \cdot d\bar{l} = I \tag{1.18}$$

$\bar{H}$ is the magnetic field intensity (in A/m), and I is the electric current enclosed by the closed path of integration:

$$I = \iint_S \bar{J}\, d\bar{S} \tag{1.19}$$

$\bar{J}$ is the surface current density in (A/m^2) through the surface $\bar{S}$. Again, Stokes' theorem (1.18) and (1.19) leads to

$$rot\,\bar{H} = \bar{J} \tag{1.20}$$

But the surface current density $\bar{J}$ is related to the volume charge density ρ by the continuity equation:

$$\nabla \cdot \bar{J} = -\frac{\partial \rho}{\partial t} \tag{1.21}$$

As there is an inconsistency between (1.20) and (1.21) ($\nabla \cdot \nabla \times H = 0$), an extra term $\partial\bar{D}/\partial t$ is added to the right term of (1.20) to yield

$$rot\,H = \bar{J} + \frac{\partial \bar{D}}{\partial t} \tag{1.22}$$

provided $\nabla \cdot \bar{D} = \rho$, which is, in fact, Gauss' law; $\bar{D}$ is the electric flux density, and $\partial\bar{D}/\partial t$ represents the displacement current density. We may now summarize Maxwell's equations as

$$\nabla \times \bar{E} = -\frac{\partial \bar{B}}{\partial t} + \nabla \times \left(\bar{U} \times \bar{B}\right) \tag{1.23}$$

$$\nabla \times \bar{H} = \bar{J} + \frac{\partial \bar{D}}{\partial t}$$

$$\nabla \cdot \bar{B} = 0; \quad \nabla \cdot D = \rho$$

To complete the picture, we add a few material constitutive laws such as

Ohm's law, for electric conductors of electrical conductivity σ (S/m),

$$\bar{J} = \sigma \cdot \bar{E} \tag{1.24}$$

Permittivity (ε) law:

$$\bar{D} = \varepsilon \cdot \bar{E} \tag{1.25}$$

ε-material permittivity (F/m) for insulation materials.

Permeability (μ) law:

$$\bar{B} = \mu \cdot \bar{H} \tag{1.26}$$

Electric conductivity σ, permittivity ε, and permeability μ (H/m) are scalars in isotropic materials, but they are tensors in anisotropic materials.

Finally, let us notice that in the majority of electromagnetic energy conversion devices (LEMs in our case) there are no free charges; so the Lorenz' force (1.11) becomes

$$\bar{F} = \bar{J} \times \bar{B}; \quad \frac{\text{N}}{\text{m}^3} \tag{1.27}$$

since the motion of charges constitutes the flow of electric current.

With $\bar{J}$ in A/m^2 and $\bar{B}$ is V s/m^2, the force $\bar{F}$ refers to the force on unit volume rather than the total force in an electromagnetic mechanic energy conversion device.

1.1.1 Poisson, Laplace, and Helmholtz Equations

Let us consider a magnetostatic (solenoidal) field, described by Equations 1.14, 1.20, and 1.26:

$$div\,\bar{B} = 0; \quad rot\,\bar{H} = \bar{J}; \quad \bar{B} = \mu \cdot \bar{H}; \quad \frac{\partial \bar{D}}{\partial t} = 0 \tag{1.28}$$

We may denote $\bar{B} = rot\,\bar{A}$ as the divergence of a rotor field is zero.

Consequently,

$$rot\,\bar{H} = rot\frac{\bar{B}}{\mu} = rot\left(\frac{1}{\mu}rot\,\bar{A}\right) = \bar{J} \tag{1.29}$$

So

$$\nabla^2\bar{A} = rot\left(rot\,\bar{A}\right) = +\mu \cdot \bar{J} \tag{1.30}$$

is the Poisson equation.

The Laplacian of $\bar{A}$, $\nabla^2\bar{A}$ is

$$\nabla^2\bar{A} = \left(\frac{\partial^2 A_x}{\partial x^2} + \frac{\partial^2 A_x}{\partial y^2} + \frac{\partial^2 A_x}{\partial z^2}\right) \cdot \bar{u}_x + \left(\frac{\partial^2 A_y}{\partial x^2} + \frac{\partial^2 A_y}{\partial y^2} + \frac{\partial^2 A_y}{\partial z^2}\right) \cdot \bar{u}_y + \left(\frac{\partial^2 A_z}{\partial x^2} + \frac{\partial^2 A_z}{\partial y^2} + \frac{\partial^2 A_z}{\partial z^2}\right) \cdot \bar{u}_z \tag{1.31}$$

If the current density vector $\bar{J}$ falls only along one direction (x or y or z), Equation 1.31 greatly simplifies.

Equation 1.30 is called Poisson's equation and $\bar{A}$ is called the magnetic vector potential. When the current density $\bar{J}$ is zero in a domain, (1.30) becomes

$$\nabla^2 \bar{A} = 0 \tag{1.32}$$

This is called the Laplace's equation.

If $\bar{H}$ and $\bar{J}$ vary sinusoidally in time in a uniform medium characterized by constant μ and, respectively, σ, the total current density $\bar{J}$ is the sum of source current density $\bar{J}_S$ and the induced current density $\bar{J}_i$ with

$$\bar{J}_i = \sigma \cdot \bar{E}_i = -\sigma \cdot \frac{\partial \bar{A}}{\partial t} = -j \cdot \omega \cdot \sigma \cdot \bar{A}; \quad J = \bar{J}_i + \bar{J}_S \tag{1.33}$$

Corroborating (1.33), we obtain

$$\nabla^2 \bar{A} - j \cdot \omega \cdot \sigma \cdot \bar{A} = +\mu \cdot \bar{J}_S \tag{1.34}$$

This is the Helmholtz equation.

Poisson, Laplace, and Helmholtz equations represent particular cases of electromagnetic field distributions, very instrumental for analytical or finite element (numerical) solution by a general equation:

$$L(\phi(P,t)) = f(P,t) \tag{1.35}$$

where

- L is a differential (linear) operator
- ϕ is an unknown function to be determined (magnetic vector potential $\bar{A}$ for magnetostatic fields in Poisson, Laplace, or Helmholtz equations); boundary conditions are required also, besides position in space $P(x, y, z)$

For electrostatic fields, which are irrotational, $rot\,\bar{E} = 0$, ϕ is a scalar potential V, but this time from (1.23) and (1.25)

$$\bar{E} = -grad\,V; \quad -div(\varepsilon\, grad\,V) = \rho \tag{1.36}$$

$$\text{Or} \quad -\left(\frac{\partial^2 V}{\partial x^2} + \frac{\partial^2 V}{\partial y^2} + \frac{\partial^2 V}{\partial z^2}\right) = \rho \tag{1.37}$$

in Cartesian coordinates.

1.1.2 Boundary Conditions

The behavior of $\phi(P, t)$ on the boundary Γ is described as boundary conditions. Among the most used boundary conditions, we mention here

$$\phi = 0 \quad \text{or} \quad \phi = \phi_n \quad \text{on} \quad \Gamma_1 \tag{1.38}$$

which is Dirichlet condition.

$$\frac{\partial \phi}{\partial n} = 0 \quad \text{or} \quad \frac{\partial \phi}{\partial n} + k \cdot \phi = 0 \quad \text{on} \quad \Gamma_2 \tag{1.39}$$

This is Newmann condition ($\bar{n}$ is the normal to the boundary Γ).

1.1.3 Energy Relations

Once the field distribution is obtained through equations such as (1.30), the energy stored in the field is determined as

$$W_E = \frac{1}{2} \cdot \iiint_V \left(\bar{D} \cdot \bar{E} \right) dV \, (J); \quad V \text{ is the volume} \tag{1.40}$$

for electrostatic fields and

$$W_m = \frac{1}{2} \cdot \iiint_V \left(\bar{B} \cdot \bar{H} \right) dV \, (J) \tag{1.41}$$

for magnetic fields.

Also, the power density is given by the Poynting vector P_e (W/m^2):

$$P_e = \bar{E} \times \bar{H} \tag{1.42}$$

P_e is used mainly for the steady state, while W_E and W_m are used to calculate the thrust. P_E is also used to calculate the losses in LEMs.

1.1.4 Resistor, Inductor, and Capacitor

Let us consider a conductor of electrical conductivity σ_e (Ω^{-1}m^{-1}) length l_c and cross section ac and a dc voltage V_{dc} applied to it to produce a current I_{dc} (Figure 1.3a).

From (1.3) with (1.14)

$$\frac{V_{dc}}{I_{dc}} = R_{dc}; \quad R_{dc} = \frac{1}{\sigma_e} \cdot \frac{l_c}{A_c} \, (\Omega) \tag{1.43}$$

This is Ohm's law in dc The total flux linkage λ in a coil made of N turns is

$$\lambda = N \cdot \iint_S \bar{B} \cdot d\bar{S} = N \cdot B \cdot S \tag{1.44}$$

for a uniform distribution of B on area S. But according to Ampere's law (with a constant value of H along the flux path around the coil l_H),

$$H \cdot l_H = N \cdot I \tag{1.45}$$

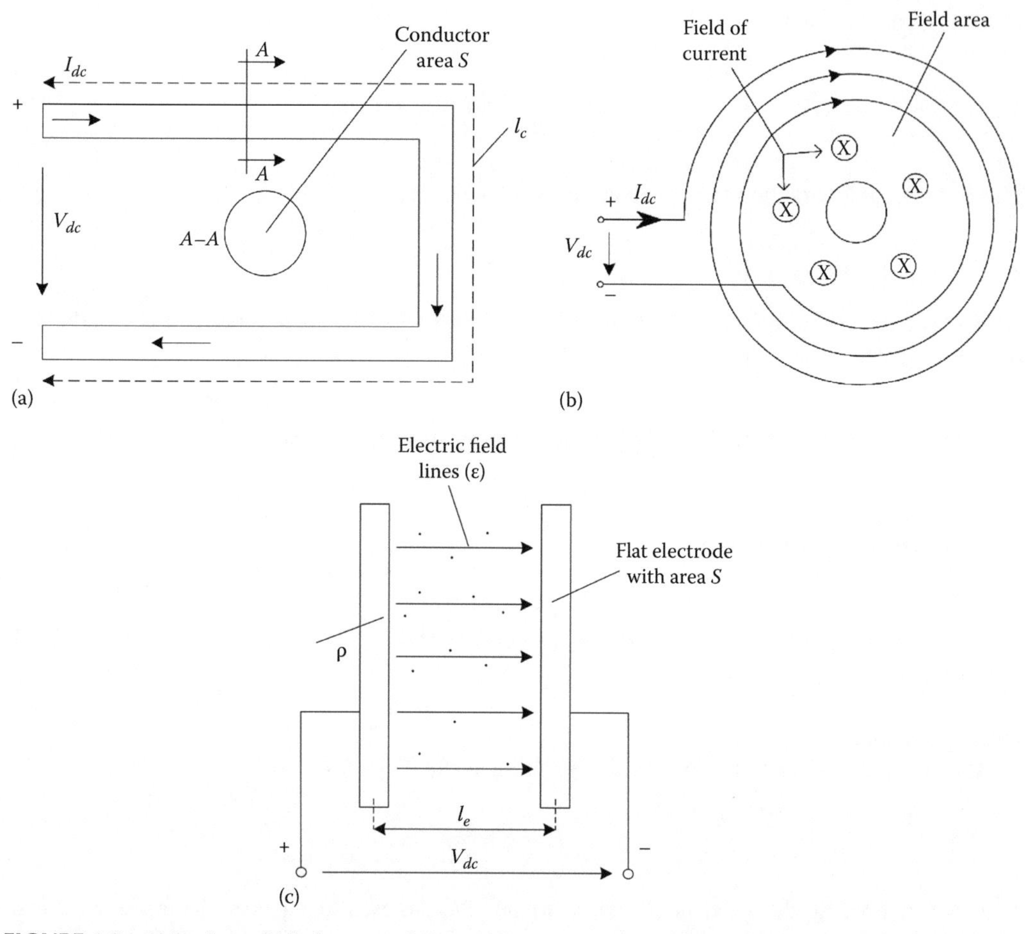

FIGURE 1.3 (a) Resistor, (b) inductor, and (c) capacitor.

As $B = \mu \cdot H$ for the coil's magnetic core, we get the definition of the inductance L (in Henry)—Figure 1.3b:

$$L = \frac{\lambda(\text{Wb})}{I(\text{A})} = \frac{\mu \cdot S \cdot N^2}{l_n} = \frac{N^2}{R_m}; \quad R_m = \frac{1}{\mu} \cdot \frac{l_n}{S} \tag{1.46}$$

R_m is called magnetic reluctance and is very similar to electric resistance R; so, there is a parallel between electric and magnetic circuits.

Similarly, the electric charge ρ, stored in a capacitor, is related to the voltage V between the capacitor plates, V, parted by an insulator of permittivity ε, by what we call the capacitance C:

$$C = \frac{\rho}{V} \tag{1.47}$$

In a homogenous insulator, the permittivity is constant throughout the insulator volume and, for a planar capacitor, the electric field lines are straight and parallel:

$$V = \int_l E\,dl = E \cdot l_e; \quad D = \varepsilon \cdot E \tag{1.48}$$

The electric flux law yields

$$\phi_e = \int D\,dS = \rho \tag{1.49}$$

So, finally

$$C = \frac{\rho}{V} = \frac{\varepsilon \cdot S}{l_e}; \text{ (Farad = Coulomb/volt)} \tag{1.50}$$

l_e is the distance between the metallic electrodes of the flat capacitor (Figure 1.3c).

As we can see, the capacitance formula for a flat electrode capacitor is very similar to the one of inductance and resistance, and they all are derived from Maxwell (field) equations.

So there is a correspondence between field variables *H*, *B*, *E*, *D* and circuit variables *V*, *emf*, *I*, *R*, *L*, *C*. They all are used to characterize LEMs as well. They make the interface between field and circuit theory of LEMs.

In a constant permeability magnetic medium (μ) with the stored magnetic energy W_m, (1.41) may be written as

$$W_m = \int_0^i \lambda\,di = \int_0^i (L \cdot i)\,di = \frac{L \cdot i^2}{2} \tag{1.51}$$

Similarly, for a flat electrode capacitor, the electric stored energy (1.41) may be written as

$$W_e = \int_0^V \rho\,dV = \int_0^V (C \cdot V)\,dV = \frac{C \cdot V^2}{2} \tag{1.52}$$

1.2 FORCES IN ELECTROMAGNETIC FIELDS OF PRIMITIVE LEMs

Forces are required in LEMs to produce progressive or oscillatory translational motion. They may be obtained from the energy conservation principle or directly from Maxwell's tensor (from Equation 1.1 onward) or from the even more general Hamiltonian principle through Lagrange equations [14].

We will illustrate here only the energy methods as they are used currently in both analytical and numerical field circuit methods, to characterize LEMs.

Let us consider here in parallel the plunger solenoid LEM, a capacitor LEM, and a voice-coil LEM (Figure 1.4a through c).

We consider first the plunger solenoid and apply to it the energy conservation law for zero losses in the system:

$$\text{Input electric energy} = \text{Mechanical work done} + \text{increase in stored energy} \tag{1.53}$$

Or

$$V \cdot i\,dt = F_{em}dx + dW_m \tag{1.54}$$

According to Faraday's law, $V = d\lambda/dt$ and thus

$$F_{em}dx = -dW_m + i \cdot d\lambda \tag{1.55}$$

In such a system, i and x (mover displacement) or λ and x may be chosen as independent variables:

$$d\lambda = \frac{\partial \lambda}{\partial i} \cdot di + \frac{\partial \lambda}{\partial x} \cdot dx$$

$$dW_m = \frac{\partial W_m}{\partial i} \cdot di + \frac{\partial W_m}{\partial x} \cdot dx \tag{1.56}$$

With (1.56) in (1.55), and observing that the force F_{em} has to be independent of di, we need

$$\frac{\partial W_m}{\partial \lambda} = i$$

And finally,

$$F_{em} = \frac{-dW_m}{dx}(\lambda, x) \tag{1.57}$$

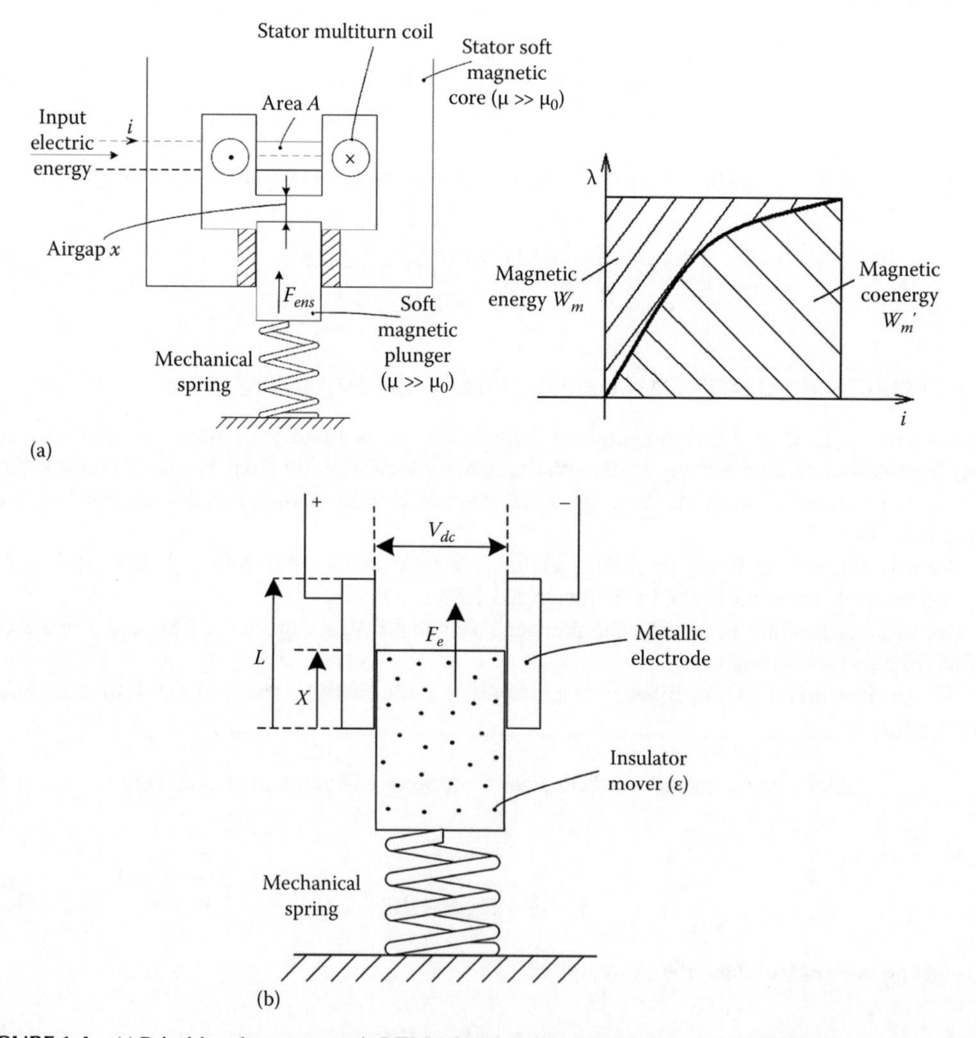

FIGURE 1.4 (a) Primitive electromagnetic LEM (plunger solenoid), (b) electrostatic LEM (plunger capacitor).

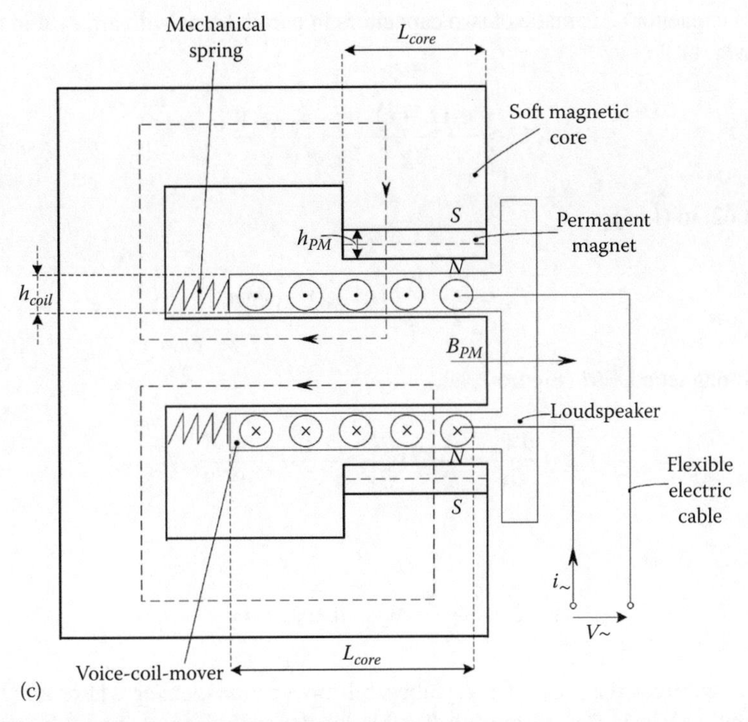

FIGURE 1.4 (continued) (c) voice-coil-mover loudspeaker.

The stored energy W_m, in the absence of motion, is equal to input energy and thus

$$W_m = \int_0^\lambda i d\lambda' = i \cdot \lambda - \int_0^i \lambda' di \tag{1.58}$$

The term $\int_0^i \lambda' di$ is called magnetic coenergy W_m':

$$W_m + W_m' = i \cdot \lambda \tag{1.59}$$

If the magnetic circuit is nonlinear ($\mu \neq$ const), then $W_m \neq W_m'$. For a linear system, $W_m = W_m'$ (this explains Equations 1.51 and 1.52, which are valid only for linear systems [μ = const, ε = const]).

When we use the coenergy function W_m', the force F_{em} is

$$F_{em} = \frac{\partial W_m'}{\partial x}(i,x) \tag{1.60}$$

A similar rationale is valid for the capacitor with insulation mover (Figure 1.4b):

$$F_{es} = \frac{\partial W_{es}}{\partial x}(V,x) \tag{1.61}$$

For a constant permittivity ε of insulation mover,

$$F_{es} = -\frac{V^2}{2} \cdot \frac{\partial C_e}{\partial x} \tag{1.62}$$

The equivalent capacitor C_e is made of two capacitors in parallel (one with air, ε_0, and the other with insulation mover, ε):

$$C_e = \frac{\varepsilon \cdot (L - x) \cdot W}{g} + \frac{\varepsilon_0 \cdot x \cdot W}{g} \tag{1.63}$$

So F_{es} from (1.62) to (1.63) is

$$F_{es} = \frac{V^2}{2} \cdot \frac{W}{g} \cdot (\varepsilon - \varepsilon_0) = \text{const.} \tag{1.64}$$

For the electromagnetic LEM (Figure 1.4a),

$$F_{em} = \frac{\partial W_m'}{\partial x} = \frac{1}{2} \cdot i^2 \cdot \frac{\partial L}{\partial x} = -\frac{1}{2} \cdot i^2 \cdot \frac{\mu_0 \cdot S}{X^2} \cdot N_C^2 \tag{1.65}$$

where

$$L = \frac{\mu_0 \cdot S}{X} \cdot N_C^2, \quad \text{if } \mu_{iron} = \infty \tag{1.66}$$

In Figure 1.4c, we treat the case of the voice-coil-mover loudspeaker where a cylindrical coil in a nonmagnetic resin is fed, through a flexible electric cable, from a power source in ac, in general.

The magnetic circuit comprising soft iron parts contains permanent magnets that produce a flux density B_{PM} in the magnetic airgap $g_t = 2 * g + h_{PM} + h_{coil}$, where g is the mechanical radial airgap g, h_{PM} is the PM radial thickness, and h_{coil} is the coil radial thickness.

The Lorenz force density expression in (1.27) is applied directly for the active part of the coil (situated under the PMs whose characteristics will be discussed later in this chapter):

$$\bar{F}_{em} = \iiint \left(j \times \bar{B} \right) dV = B_{PM} \cdot I \cdot l \cdot N_{turns/coil} \frac{L_{core}}{L_{coil}} \tag{1.67}$$

$$l = \pi \cdot D_{av\_coil}; \quad D_{av\_coil} = \text{Average coil diameter} \tag{1.68}$$

Ideally, with constant B_{PM} and constant current in the coil, the force $\bar{F}_{em}$ along the direction of motion is independent of mover position (as for the capacitor electrostatic LEM, but in contrast to the plunger solenoid).

When the polarity of electric current I changes through voltage polarity $V_{\sim}$ (Figure 1.4c), the direction of force changes and so does the motion direction. This is in contrast to the plunger solenoid and capacitor electrostatic LEM, where the direction of force does not change with current (respectively, voltage) polarity.

Example 1.1: The Plunger Solenoid Circuit Equations and Force

Let us consider a simplified structure of a plunger solenoid as in Figure 1.5, with an infinite permeability magnetic core.

For $N_C \cdot I = 1000$ A turns/coil (these are two twin coils), after writing the circuit and motion (force) equations, calculate the force for $x = 2$ mm and $x = 4$ mm if the core area $A_{core} = 100$ mm × mm and the air permeability $\mu_0 = 4 \cdot \pi \cdot 10^{-7}$ H/m.

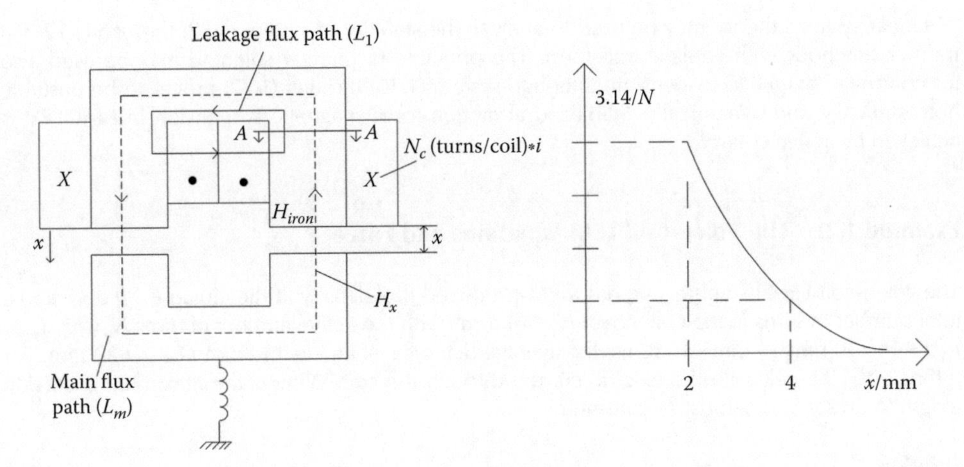

FIGURE 1.5 A simplified flat plunger solenoid and its force versus proven position for constant current.

Solution

Let us first notice that we can build a rather simple electric circuit for the plunger solenoid as a resistor R and inductance $L_t = L_l + L_m$ (Figure 1.5).

The force equation (1.65) may be used where L_l=const (leakage inductance) does not produce force:

$$F_{em} = -\frac{1}{2}\cdot i^2\cdot(2\cdot N_C)^2\cdot\frac{\mu_0\cdot A_{core}}{2\cdot X^2};\quad L_m = \frac{\mu_0\cdot A_{core}\cdot(2N_c)^2}{2\cdot X} \tag{1.69}$$

The voltage equation is simply

$$\underbrace{V(t)}_{\text{Source voltage}} = \underbrace{R\cdot i(t)}_{\text{Resistive voltage drop}} + \underbrace{(L_l + L_m)\cdot\frac{di(t)}{dt}}_{\text{Transformer emf}} + \underbrace{\frac{\partial L_m}{\partial x}i(t)\cdot\frac{dx}{dt}}_{\text{Motion emf}} \tag{1.70}$$

The motion equation writes simply

$$m_m\frac{d\bar{U}}{dt} = F_{em} + m\cdot g + k\cdot x;\quad F_{em} < 0 \tag{1.71}$$

$$\frac{dx}{dt} = U \tag{1.72}$$

Using (1.69),

$$(F_{em})_{x=2\,mm} = -\frac{1}{2}\cdot 4\cdot\frac{1000^2\cdot 4\cdot\pi\cdot 10^{-7}\cdot 100\cdot 10^{-6}}{2\cdot(2\cdot 10^{-3})^2} = 3.14\text{N} \tag{1.73}$$

Similarly for x=4 mm, the force is four times smaller, that is 3.14/4 N.

In reality, the state equations (1.70) through (1.72) should be solved together to calculate the current i, the mover position x, and its speed U evolution in time, for a given input voltage function $U(t)$.

This simple case has an analytical solution, but as long as magnetic core permeability μ_{iron} is not ∞, but it is variable with the magnetic flux density $B_{iron} = \mu_{iron}(B_{iron})H_{iron}$, numerical methods are required.

Linear systems theory may be used to analyze the stability of system (1.70) through (1.72) via transfer functions with Laplace transform. The principle of plunger solenoid may be used also for controlled magnetic suspension, although system (1.70) through (1.72) proves to be unstable both statically and dynamically. Stabilized attraction force magnetic suspension in MAGLEV is achieved by active control.

Example 1.2: The Voice-Coil LEM Equations and Force

The voice-coil LEM in Figure 1.4c has a PM-produced flux density in the airgap $B_{PM}=0.5T$ and a total number of turns in the coil mover $N_c=14$ turns with the active number of turns $N_a=N_c \cdot L_{core}/L_{coil}=14\times 10$ mm/14 mm $=10$ turns; the average diameter of an electric turn $D_{av\,coil}=20$ mm. For a thrust of 1 N, calculate the electric current through the coil. What is the maximum excursion length to preserve the thrust as constant?

Solution

We simply use (1.67)

$$F_{em} = B_{PM}\cdot i\cdot \pi\cdot D_{av\,coil}\cdot N_c\cdot \frac{L_{core}}{L_{coil}} = 0.5x\cdot i\cdot \pi\cdot 0.02\cdot 14\cdot \frac{10}{14} = 1\text{N};\quad i = 3.18\text{ A}$$

The maximum excursion length for constant thrust, l_{stroke}, is the difference between total coil length l_{coil} and core length $l_{core}-l_{stroke}=14-10=4$ mm. So the maximum motion amplitude would be ±2 mm in the presence of ±3.18 A in the coil. Again, the thrust is bidirectional, according to the current polarity.

Example 1.3: The Capacitor (Electrostatic) LEM

The air permittivity $\varepsilon_0 = 1/(4\cdot\pi\cdot 9\cdot 10^9)\ (\varepsilon_0\cdot\mu_0 = (1/2)U_{light})$, while that of the insulator $\varepsilon=5\cdot\varepsilon_0$; the electrode height $L=10$ cm and its depth $W=10$ cm; the distance between them $g=2$ mm. Calculate the force on the insulation mover as dependent on its position, x. The applied voltage $V_{dc}=1000$ V.

Solution

According to (1.64), the vertical force, which tends to keep the insulation mover inside the electrodes, is constant:

$$F_{es} = -\frac{V^2}{2}\cdot\frac{W}{g}\cdot(\varepsilon-\varepsilon_0) = \frac{10^6}{2}\cdot\frac{10^{-1}}{2\cdot 10^{-3}}\cdot\frac{(5-1)}{4\cdot\pi\cdot 9\cdot 10^9} = 0.0008846\text{ N}$$

As seen the force is insignificant even at $V_{dc}=1$ kV, which explains why electrostatic LEMs are not used for large devices; in the micrometer size, however, they become competitive.

Example 1.4: The Linear Induction Motor

A primitive linear induction motor is presented in Figure 1.6.

The slots house a primitive three-phase winding with symmetric currents:

$$i_{ABC} = I\cdot\sqrt{2}\cdot\cos\left(\omega t-(i-1)\cdot\frac{2\pi}{3}\right) \tag{1.74}$$

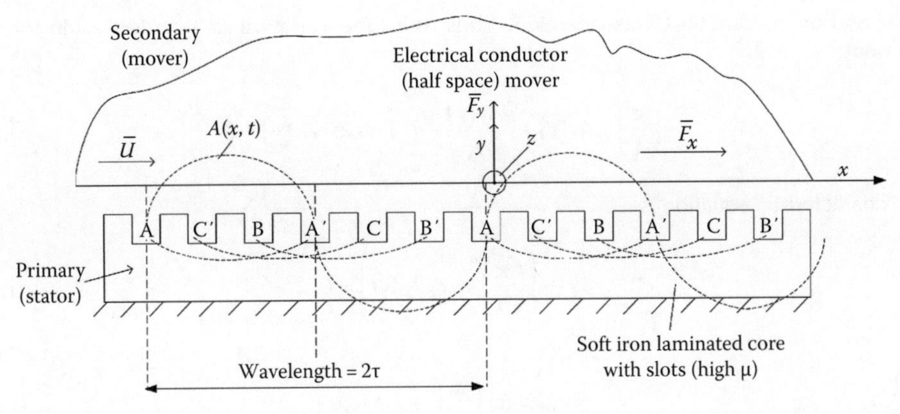

FIGURE 1.6 Linear induction motor and its forces F_x and F_y versus slip S with a conducting half space secondary.

They produce a traveling magnetomotive force:

$$F(x,t) = F_m \cdot e^{j\cdot\left(\omega t - \frac{\pi}{\tau}x\right)} \tag{1.75}$$

The speed U_S of this traveling mmf is obtained by making the time derivative of its argument zero:

$$U_S = 2\cdot\tau\cdot f; \quad \omega = 2\cdot\pi\cdot f \tag{1.76}$$

Calculate the forces F_x and F_y that are exerted on the conducting half space.

Solution

This is a traveling wave field diffusion problem. As long as $U \neq U_S$ the traveling wave $A(x, t)$ produces eddy currents in the conducting half space. Their reaction field in turn reduces them along "y" direction.

Let us consider the traveling mmf in mover coordinates (at speed U):

$$F(x,t) = F_m \cdot e^{j\cdot\left(S\cdot\omega t - \frac{\pi}{\tau}x\right)} \tag{1.77}$$

with $S = 1 - U/U_s$, known as slip in electric machinery. We may now consider that there is no motion between primary and secondary, but the magnetic field variables vary at frequency $S\cdot\omega_0$. Moreover, the source current density $\bar{J}_{sz}$ in the stator varies along x direction, while that induced in the mover, $\bar{J}_z$, varies with x, y; both are considered only along z direction.

Now we can directly apply Helmholtz equation (1.34):

$$\frac{\partial^2 A_z}{\partial x^2} + \frac{\partial^2 A_z}{\partial y^2} - j\cdot S\cdot\omega\cdot\sigma\cdot\bar{A} = -\mu_0\cdot J_{sz}(x,t) = -\mu_0\cdot J_{sz}\cdot e^{j\cdot\left(S\cdot\omega t - \frac{\pi}{\tau}\cdot x\right)} \tag{1.78}$$

The vector potential A_z will vary with x and t as $J_{sz}(x, t)$ does:

$$A_z(x,y,t) = e^{j\left(S\cdot\omega t - \frac{\pi}{\tau}\cdot x\right)}\cdot A_z(y) \tag{1.79}$$

A separation of variables was operated. Thus, the time variation can be left aside for the moment:

$$-\left(\frac{\pi}{\tau}\right)^2 \cdot A_z(y) + \frac{\partial A_z(y)}{\partial y^2} - j \cdot S \cdot \omega \cdot \sigma \cdot \bar{A}_z(y) = 0 \tag{1.80}$$

The characteristic equation is

$$\underline{\gamma}^2 - \left(\left(\frac{\pi}{\tau}\right)^2 + j \cdot S \cdot \omega\sigma\right) = 0 \tag{1.81}$$

$$\underline{\gamma} = \pm sqrt\left(\left(\frac{\pi}{\tau}\right)^2 + j \cdot S \cdot \omega\sigma\right) \tag{1.82}$$

$A_z(x, y, t)$ in the mover is thus

$$A_z(x, y, t) = \left(c_1 \cdot e^{\underline{\gamma} y} + c_2 \cdot e^{-\underline{\gamma} y}\right) \cdot e^{j \cdot \left(S \cdot \omega t - \frac{\pi}{\tau} \cdot x\right)} \tag{1.83}$$

The first boundary condition is simply

$$A_z(x, y, t)_{y=\infty} = 0, \text{ which means } c_1 = 0 \tag{1.84}$$

The second one is based on the Ampere's law, along a contour on the mover surface, which states that

$$\frac{\partial F(x, t)}{\partial x} = (H_x)_{y=0}; \quad H_x = \frac{B_x}{\mu_0} \tag{1.85}$$

But as

$$\bar{B} = rot\, A_z \bar{u}_z \tag{1.86}$$

$$B_x = \frac{\partial A_z}{\partial y}; \quad B_y = -\frac{\partial A_z}{\partial x} \tag{1.87}$$

So

$$c_2 = +j \cdot \frac{\pi}{\tau \cdot \underline{\gamma}} \cdot F_{xm} \tag{1.88}$$

Further on, the current density $\bar{J}_z$ in the mover is

$$\bar{J}_z = rot\, \bar{H} = \frac{\partial H_y}{\partial x} - \frac{\partial H_x}{\partial y} \tag{1.89}$$

The field distribution in the secondary is thus practically solved. We still have to calculate the force components F_x and F_y:

$$F_x + j \cdot F_y = \iiint \bar{j} \times \bar{B}\, dv = L \cdot R_e \cdot \left(-\int_0^{2\tau}\int_0^{\infty} J_z \cdot B_y dx dy + \int_0^{2\tau}\int_0^{\infty} J_z \cdot B_x dx dy\right) \tag{1.90}$$

Finally, we obtain the forces per wave length 2τ:

$$\left\langle F_{y(2\tau)}\right\rangle = -\mu_0 \cdot \frac{l \cdot F_{xm}^2}{4}\left(\frac{\left(1+S'^2\right)^{\frac{1}{2}}-1}{\left(1+S'^2\right)^{\frac{1}{2}}}\right); \quad S' = \mu_0 \cdot \sigma \cdot S \cdot \omega_1 \cdot \frac{\tau^2}{\pi^2} \tag{1.91}$$

$$\left\langle F_{x(2\tau)}\right\rangle = \mu_0 \cdot \frac{l \cdot F_{xm}^2}{4}\left(\frac{S'}{\left(1+S'^2\right)^{\frac{1}{2}}}\right) \tag{1.92}$$

l is the laminated core length along axis z.

It was assumed that the current density in the secondary has only a J_z component like if the secondary conductor would have been short-circuited at ends by a superconductor slab (or that the core is very long (along z direction) with respect to pole pitch τ; not so in reality as seen in forthcoming chapters).

A few remarks are in order in the following:

- At zero slip S, there is no propulsion force F_x, neither a repulsive normal force F_y (for $S=0$, $S'=0$): this was to be expected as at $S=0$ the conducting half space does not experience any magnetic flux change. So, $(J_z)_{S=0}=0$.
- In reality, the primary length is finite and the secondary is longer than the primary; thus, new positions of secondary experience magnetic flux change when they enter or exit the stator mmf zone. Additional currents are induced in the secondary even at zero slip. This is the so-called dynamic longitudinal end effect that reduces, in general, the thrust F_x, the efficiency, and the power factor of real LIMs.
- The LIM may be used rather with a fix ac mmf (not traveling) to stir melted metals for better homogeneity (for quality steel, etc.).
- The thrust (F_x) is positive (motoring) for positive slip ($S>0$ or $U<U_s$); $F_x<0$ also for $S>1$ ($U_s>0$ and $U<0$)—typical braking.
- Alternatively if the LIM primary is single-phase dc fed and is moving above a conductive sheet at speed U, for $U \neq 0$ a repulsive force (F_y) together with a braking force ($F_x<0$) are produced as in repulsive levitation systems.

1.3 MAGNETIC, ELECTRIC, AND INSULATION MATERIALS FOR LEMs

We have dealt so far with primitive topologies of LEMs and applied the electromagnetic field theory to calculate forces and thus uncover basic principles of LEMs. We have discussed soft magnetic materials ($\mu \gg \mu_0$), permanent magnets, electrical conductors, and insulations but did not elaborate on their characteristics. It is time to dwell a little on them.

1.3.1 Soft Magnetic Materials

Soft magnetic or ferromagnetic materials ($\mu_m \gg \mu_0$) are used as magnetic cores to flow the flux lines of magnetic fields in LEMs at the "cost" of reasonable mmf (Ampere turns) and core losses, considering that any LEM has an airgap zone to be traveled too by the closed flux paths ($div\,\bar{B}=0$).

Soft magnetic materials are characterized by

- The magnetization curve $B_m(H_m)$
- The magnetic permeability $\mu_m=B_m/H_m$; μ_m is a scalar for homogenous and a tensor for nonhomogenous materials

- The hysteresis cycle
- Core loss (W/kg) as a function of B_m for given frequency
- Electric conductivity

In general, soft magnetic materials made in thin sheets (to reduce core loss), which are used in LEMs, include alloys of iron, nickel, cobalt, and silicon, with $\mu_{rel}=\mu_m/\mu_0 \gg 1$. For silicon steel laminations, $\mu_{rel}>2000$, for $B_m=1$ T.

For frequencies above 500 Hz, compressed or injected soft powder materials that contain iron particles suspended in an epoxy or plastic matrix are also used, though their permeability is so far only moderate $\mu_{mrel}=500-700$ at $B_m=1.5$ T (Somaloy 500, 700).

However, they are easier to fabricate in intricate topologies.

Figure 1.7 presents the magnetization curve and the hysteresis cycle for standard silicon (3.5%) steel M19.

The initial magnetization process implies the gradual orientation of micro magnetic dipoles in the material by an external magnetic field (mmf). For monotonous rising and falling of mmf (positive and negative), the hysteresis cycle is obtained.

So the magnetization curve is obtained by connecting the tips of amplitude-decreasing hysteresis cycles. The hysteresis cycle is traveled every second many times (at frequency rate) in ac LEMs and leads to hysteresis losses.

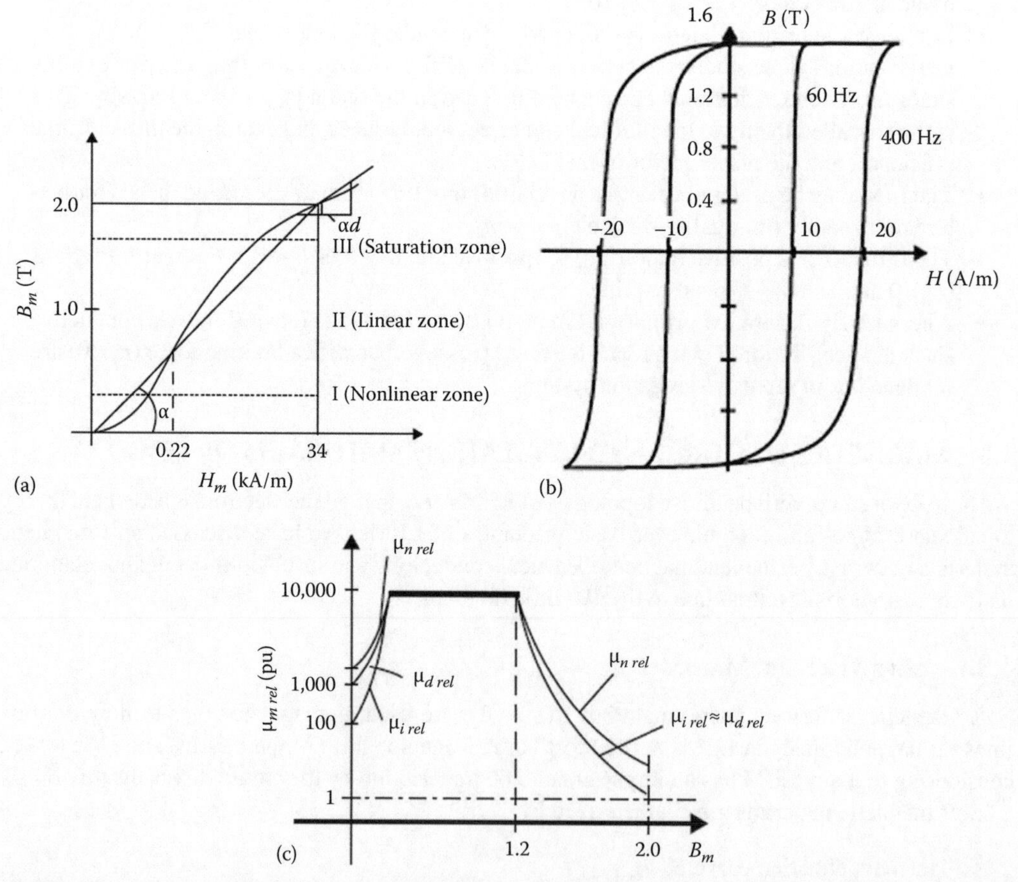

FIGURE 1.7 (a) Magnetization curve and (b) hysteresis cycle of delta max tape-wound core 0.5 mm strip, (c) permeabilities: $\mu_{n\,rel}$, $\mu_{diff\,rel}$, $\mu_{i\,rel}$.

The presence of magnetic saturation (B_m leveling) implies a variable permeability $\mu_m(B_m)$. In fact, we may define three permeabilities, all used in LEMs:

- Normal (steady-state) permeability μ_n:

$$\mu_n = \frac{B_m}{H_m}; \quad \mu_{nrel} = \frac{\mu_n}{\mu_0} \tag{1.93}$$

- Differential permeability μ_{diff}:

$$\mu_{diff} = \frac{dB_m}{dH_m}; \quad \mu_{diff\,rel} = \frac{\mu_{diff}}{\mu_0} \tag{1.94}$$

- Incremental permeability μ_i:

$$\mu_i = \frac{\Delta B_m}{\Delta H_m}; \quad \mu_{irel} = \frac{\mu_i}{\mu_0} \tag{1.95}$$

Let us note that

- Below 1.2 T in silicon mild steel tapes (laminations), the three permeabilities do not differ much:

  Above 1.2 T $\mu_n > \mu_{diff}$

- As in LEMs ac and dc magnetization coexist, care must be exercised when the normal differential or incremental inductance is calculated under magnetic saturation conditions.
- Although not shown here, the magnetization curve varies above 500 Hz and pertinent tests are needed.
- Hysteresis losses per cycle also increase with frequency.

1.3.2 Permanent Magnets

Permanent magnets [15] are solid hard magnetic materials with an extremely large (wide) hysteresis cycle and a recoil permeability $\mu_{rec} \approx (1.05 - 1.3) \cdot \mu_0$ (Figure 1.8).

Only the second quadrant of the PM hysteresis cycle is given with the "knee" demagnetization point ($K1$, $K2$, $K3$) in the third quadrant.

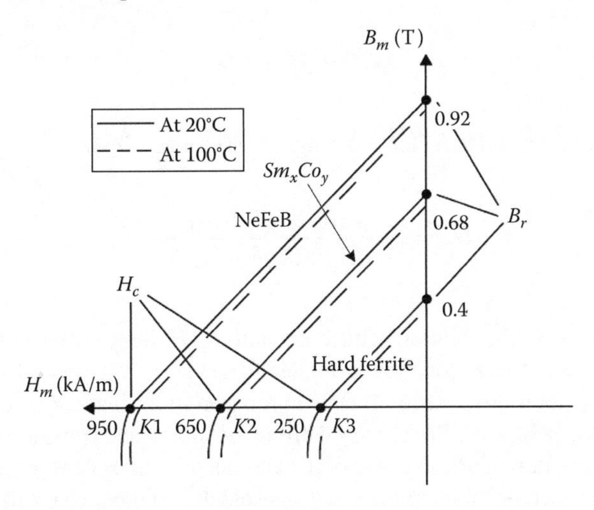

FIGURE 1.8 PMs characteristic in the second quadrant.

The remanent flux density $B_r(H_m=0)$ and the coercive field $H_c(B_m=0)$ are also key characteristics of PMs.

Temperature increases move the $B_m(H_m)$ curve downward for sintered and bonded NeFeB and Sm_xCo_y material but upward for some hard ferrites.

Instead of $B_m(H_m)$ curve, the PMs may be characterized by

$$M(H_m) = B_m - \mu_0 \cdot H_m \tag{1.96}$$

M is called PM magnetization and the $M(H_m)$ curve is called intrinsic demagnetization curve, which is, in reality, calculated from $B_m(H_m)$ curve. PMs have to be magnetized first and then they can hold the stored energy for years, thus providing the magnetic field at low cost in compact volumes for LEMs.

Linear demagnetization curve (modern) PMs (Figure 1.8) at 20°C can be magnetized a priori by putting them in special soft magnetic circuits without air paths and feeding them through a few millisecond long high mmf field signals by discharging a capacitor through the magnetization special coil. In general, the magnetization device should produce a strong magnetic field ($2B_r$ and at least $2H_c \cdot h_{PM}$ mmf) to secure sufficient magnetization.

The zero airgap short circuit corresponds to $B_r(H_m=0)$ point on the $B_m(H_m)$ curve while, for the open magnetic circuit situation (PM in air), the $H_c(B_m=0)$ situation corresponds.

In reality, in an LEM, the magnetic circuit contains some soft magnetic cores and an airgap and thus the operation point on the demagnetization curve is between B_m and H_c, say, at A_0, in the absence of any mmf (current) in the LEM (Figure 1.9).

To calculate the position of no-load operation point ($A_0(i=0)$), we may consider here approximately that μ_{rec}=const and thus

$$B_m \approx B_r + \mu_m \cdot H_m \tag{1.97}$$

Ampere's law provides

$$H_m \cdot h_{PM} + H_g \cdot (2g + h_{coil}) = N_c \cdot \left(1 - \frac{x}{l_{coil}}\right) \cdot i \tag{1.98}$$

where $0 < x < l_{core} < l_{coil}$.

Neglecting the leakage flux in the airgap, Gauss flux law yields ideally (no flux fringing).

$$B_m = B_g = \mu_0 \cdot H_g \tag{1.99}$$

In the absence of current ($i=0$), from (1.97) through (1.98) we get finally

$$H_{m0} = \frac{-B_m \cdot (2g + h_{coil})}{\mu_e \cdot h_{PM} + (2g + h_{coil}) \cdot \mu_{rec}} \tag{1.100}$$

Now, when the current $i \neq 0$, the solution is different and B_m, H_m vary with the position x in the PM.

For $i<0$, at point A' the PM flux reduction is smaller than in point A'' (Figure 1.9). For $i>0$ a PM flux increase takes place and total flux density in B' is lower than in B''. Therefore, the influence of mmf makes the total magnetic field in the PM to vary with current, for a given position of the mover. This is the known transverse armature reaction effect of dc commutator and synchronous rotating machines. Now, on top of that, the current varies in time, say sinusoidally, to produce oscillatory motion (voice-coil LEM), and thus the total flux density (in the PM) varies in time.

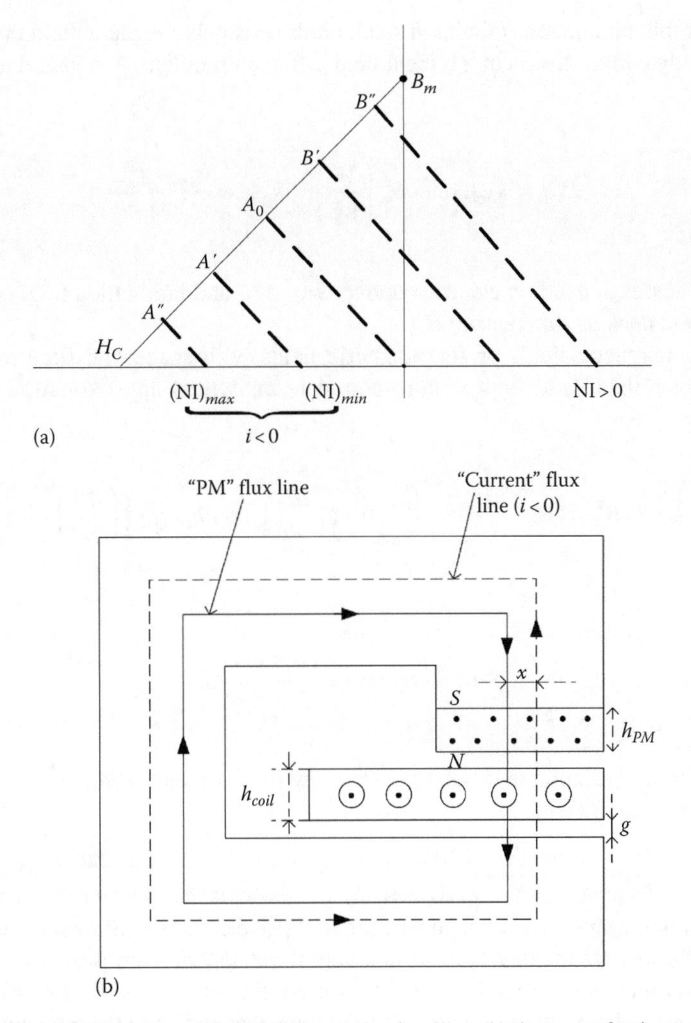

FIGURE 1.9 (a) Operation points on the PM demagnetization line, (b) the case of voice-coil LEM.

Local magnetic saturation in the soft magnetic core may occur, and eddy current losses will be induced in the PM solid materials; this in turn would deteriorate the performance and eventually lead to PM demagnetization, by reaching points $K1$, $K2$, $K3$ (Figure 1.8), which, at high temperatures, move in the third quadrant.

Avoiding PM demagnetization becomes thus a major design problem in PM-LEMs.

1.3.3 Magnetic Core Losses

In general, soft magnetic material losses have been divided into hysteresis losses P_h (W/kg) and eddy current losses P_{eddy} (W/kg).

For standard materials, they have been measured and then curve fitted. Very simplified formulae are used, for sinusoidal time variations:

$$P_h \approx k_h \cdot f_1 \cdot B_m^2 \left[\frac{\text{W}}{\text{kg}}\right] \tag{1.101}$$

B_m, peak ac flux density; f_1, frequency; and k_h, a coefficient that accounts for hysteresis cycle contour and frequency (see more on hysteresis cycle models in Ref. [16]).

Similarly, for thin laminations of silicon steel, unidimensional magnetic field theory was used to determine the eddy current losses in a typical field diffusion problem. A standard approximation of this method is

$$P_{eddy} = k_{eddy} \cdot \omega_1^2 \cdot B_m^2 \left[\frac{\text{W}}{\text{kg}}\right]; \quad k_{eddy} \approx \frac{\sigma_{iron} \cdot d_{iron}^2}{24} \tag{1.102}$$

So eddy current losses depend on electric conductivity σ_{iron} and lamination thickness d_{iron}, besides frequency (ω_1) and peak ac flux density B_m.

For traveling magnetic fields or for magnetic fields with space and time harmonics, these formulae are not reliable and thus a more complete analytical approximation is getting wide acceptance [17]:

$$P_{iron} = k_h \cdot f \cdot B_m^2 \cdot k(B_m) + \frac{\sigma_{iron}}{12} \cdot \frac{d_{iron}^2 \cdot f}{\gamma_{iron}} \cdot \int_{\frac{1}{f}} \left(\frac{dB}{dt}\right)^2 dt + k_{ex} \cdot f \cdot \int_{\frac{1}{f}} \left(\frac{dB}{dt}\right)^{1.5} dt; \left[\frac{\text{W}}{\text{kg}}\right] \tag{1.103}$$

$$\text{With } k(B_m) = 1 + \frac{0.65}{B_m} \sum_{1}^{n} \Delta B_i \tag{1.104}$$

ΔB_i is the flux density variation during integration step (for various harmonics)
k_{ex} is the excess loss coefficient

The formulae (1.102) through (1.103) are valid only if the field penetration depth $\delta_{iron} > d/2 \left(\delta_{iron} = \sqrt{2/\omega_1 \cdot \mu_{iron} \cdot \sigma_{iron}}\right)$. Excess losses accounts for nonsinusoidal time fields, and k_h, k_{ex}, k have to be always checked against proper experiments as core losses depend heavily on mechanical machining technology used to fabricate the magnetic core provided, in general, with slots. Even when finite element method (FEM) is used, the latter provides the dB/dt and ΔB_i, but then still (1.104) is used, eventually for all FEM elements that add up to the total core losses of magnetic cores (to be considered in critical operation modes of LEM).

Tables 1.1 and 1.2 give the magnetization curve of 3.5% silicon steel 0.5 mm thick sheets and, respectively, the core losses for a few frequencies for 0.5 mm thick M19 fully processed CRNO laminations.

TABLE 1.1
***B–H* Curve for Silicon (3.5%) Steel (0.5 mm Thick) at 50 Hz**

B (T)	0.05	0.1	0.15	0.2	0.25	0.3	0.35	0.4	0.45	0.5
H (A/m)	22.8	35	45	49	57	65	70	76	83	90
B (T)	0.55	0.6	0.65	0.7	0.75	0.8	0.85	0.9	0.95	1
H (A/m)	98	106	115	124	135	148	162	177	198	220
B (T)	1.05	1.1	1.15	1.2	1.25	1.3	1.35	1.4	1.45	1.5
H (A/m)	237	273	310	356	417	482	585	760	1050	1340
B (T)	1.55	1.6	1.65	1.7	1.75	1.8	1.85	1.9	1.95	2.0
H (A/m)	1760	2460	3460	4800	6160	8270	11170	15220	22000	34000

TABLE 1.2
Typical Core Loss W/lb of As-Sheared 29 Gage M19 Fully Processed CRNO at Various Frequencies

	Frequencies										
[kG]	50 Hz	60 Hz	100 Hz	150 Hz	200 Hz	300 Hz	400 Hz	600 Hz	1000 Hz	1500 Hz	2000 Hz
1.0	0.008	0.009	0.017	0.029	0.042	0.074	0.112	0.205	0.465	0.900	1.451
2.0	0.031	0.039	0.072	0.119	0.173	0.300	0.451	0.812	1.786	3.370	5.318
4.0	0.109	0.134	0.252	0.424	0.621	1.085	1.635	2.960	6.340	11.834	18.523
7.0	0.273	0.340	0.647	1.106	1.640	2.920	4.450	8.180	17.753	33.720	53.971
10.0	0.494	0.617	1.182	2.040	3.060	5.530	8.590	16.180	36.303	71.529	116.702
12.0	0.687	0.858	1.648	2.860	4.290	7.830	12.203	23.500	54.258	108.995	179.321
13.0	0.812	1.014	1.942	3.360	5.060	9.230	14.409	27.810	65.100	131.918	
14.0	0.969	1.209	2.310	4.000	6.000	10.920	17.000				
15.0	1.161	1.447	2.770	4.760	7.150	13.000	20.144				
15.5	1.256	1.559	2.990	5.150	7.710	13.942	21.619				
16.0	1.342	1.667	3.179	5.466	8.189						
16.5	1.420	1.763	3.375	5.788	8.674						
17.0	1.492	1.852	3.540	6.089	9.129						

1.4 ELECTRIC CONDUCTORS AND THEIR SKIN EFFECTS

Electric conductors in LEMs flow in high electric conductivity wires (cables) properly insulated from each other (by an insulation coating) and with respect to the magnetic core via insulation sheets/liners.

Pure (electrolytic) copper is mostly used for primary windings in LEMs, while aluminum is used in the secondary (track) as sheets on solid iron or as ladders in slotted cores (similar to the rotary induction motor rotor) in linear induction machines (LIMs).

The conductor electric resistivity $\rho_{co(Al)}$ increases rather linearly with temperature, approximately as

$$\rho_{co(Al)} = 1.8(3.0)\cdot 10^{-8}\cdot\left(1+\frac{1}{273}\cdot\left(T-20^{\circ}\right)\right) \tag{1.105}$$

The distribution of the current density in a conductor cross section is uniform only if the current is constant in time (dc).

Round conductors are built up to 3 mm in diameter in standardized gauges, while, above 6 mm^2 total cross section, rectangular cross section conductors are used, in general, to increase the slot filling factor. For direct current (dc), the copper losses are

$$\left(P_{co}\right)_{dc} = R_{dc}\cdot I_{dc}^{2};\quad R_{dc} = \rho_{co}\cdot\frac{l_{cond}}{A_{cond}} \tag{1.106}$$

But dc current is hardly used in LEMs with the exception of the dc field winding of long stator linear synchronous motor MAGLEV (Transrapid system).

AC currents are predominant in LEMs, which are, in general, brushless.

An ac copper conductor in air has a penetration depth of its own field of

$$\delta_{co} = \sqrt{\frac{2}{\mu_0 \cdot \omega_1 \cdot \sigma_{copper}}}; \quad \sigma_{copper} = 5.55 \cdot 10^7 \, \Omega^{-1}\mathrm{m}^{-1} \tag{1.107}$$

For $f_1 = 60$ Hz, $\delta_{co} = 6.16 \cdot 10^{-3}$ m. So, if alone in air, all round copper conductors should have $\delta_{co} \gg diam$ and thus a rather uniform current distribution would be an acceptable assumption even in ac up to 60 Hz. Unfortunately, in most LEMs, the multiple turn coils in the primary are placed in the slots of the laminated core in quite a few layers or as single bar coils (Figure 1.10).

The vicinity of magnetic core changes the picture. Essentially, the current density in the single conductor (bar) in Figure 1.10b decreases in amplitude (and changes phase) along slot depth. Consequently, both the bar in the slot resistance and its leakage inductance (reactance for given frequency) vary with frequency ω: the resistance increases and the inductance decreases:

$$R_{ac} = R_{dc} \cdot k_{Rac}; \quad k_{Rac} \geq 1 \tag{1.108}$$

$$L_{eac} = L_{ldc} \cdot k_{xac}; \quad k_{xac} \leq 1$$

For the open (rectangular) slot shape in Figure 1.10b, after using Faraday's law and Ampere's law for the magnetic field produced by the current in the slot ($\mu_{iron} \approx \infty$), the following equations yield:

$$\frac{\partial H_y}{\partial x} = J_z; \quad \frac{1}{\sigma} \cdot \frac{\partial J_z}{\partial x} = \mu_0 \cdot \frac{\partial H_y}{\partial t}; \quad \sigma \tilde{E}_z = J_z \tag{1.109}$$

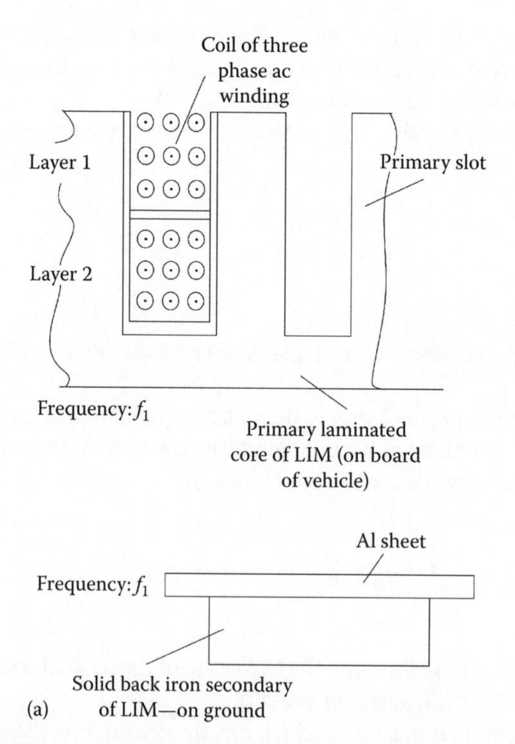

FIGURE 1.10 Typical windings of LEMs: (a) LIM with passive guideway.

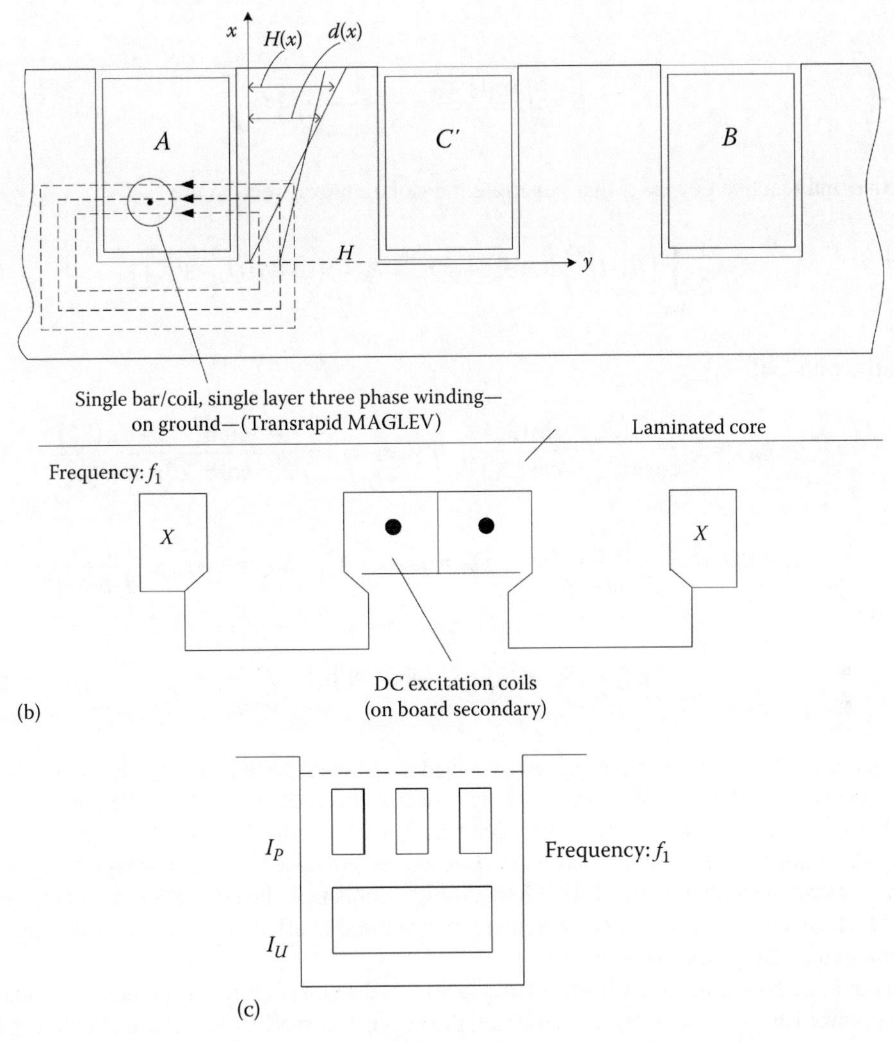

FIGURE 1.10 (continued) Typical windings of LEMs: (b) linear synchronous motor with active guideway (c) multilayer conductor in a slot.

In addition, separating the variables

$$H_y = H_y(x) \cdot e^{j\omega t} \tag{1.110}$$

with

$$(H_y)_{x=0} = 0 \quad \text{and} \quad b_{slot} \cdot \int_0^{h\,slot} J_z \, dx = I \cdot \sqrt{2} \tag{1.111}$$

we get:

$$\frac{\partial^2 H_y(x)}{\partial x^2} = \underline{\gamma}^2 \cdot H_y(x); \quad \underline{\gamma} = \pm\beta \cdot (1 + j) \tag{1.112}$$

and

$$\beta = \sqrt{\frac{\omega \cdot \mu_0 \cdot \sigma}{2}} = \frac{1}{\delta_{penetration}} \tag{1.113}$$

The active and reactive powers $\underline{S}$ that penetrate the slot along x direction are

$$S = \frac{1}{2} \int_{Amp} \left(\bar{E}_z \cdot \bar{H}_y^* \right) dt = P_{ac} + jQ_{ac} = P_{dc} \cdot \varphi(\zeta) + j \cdot Q_{dc} \cdot \Psi(\zeta) \tag{1.114}$$

Finally, with $\zeta = \beta \cdot h_{slot}$

$$\varphi(\zeta) = k_{Rac} = \zeta \cdot \frac{\sinh(2\zeta) + \sin(2\zeta)}{\cosh(2\zeta) - \cos(2\zeta)}; \quad \Psi(\zeta) = k_{Xac} = \frac{3}{2\zeta} \cdot \frac{\sinh(2\zeta) - \sin(2\zeta)}{\cosh(2\zeta) - \cos(2\zeta)} \tag{1.115}$$

$$P_{dc} = R_{dc} \cdot I^2 = \frac{l_{stack}}{\sigma \cdot h_{slot} \cdot b_s} \cdot I^2; \quad Q_{dc} = \omega \cdot L_{ldc} \cdot I^2; \quad L_{ldc} = \mu_0 \cdot l_{stack} \cdot \frac{h_{slot}}{3 \cdot b_s} \tag{1.116}$$

$$\text{For } \zeta > 2.5; \quad \varphi(\zeta) \approx \zeta \quad \text{and} \quad \Psi(\zeta) \approx \frac{3}{(2\zeta)} \tag{1.117}$$

The frequency or (and) slot height h_{slot} increases lead to ζ increase and thus give rise to notable skin effect (increased resistance and reduced leakage (slot) inductance in ac). Now the bar coil is used typically for high currents and, even if the frequency is not high, the skin effect is important: So the bar is made of quite a few conductor layers transposed such that each of them occupies all positions in the slot, along machine length (Roebel bar). At high frequency, the conductor is broken into thin wires, which are again transposed (Litz-wire) to reduce the skin effect to values typical to the single elementary conductor standing alone.

In general, we have quite a few layers of conductors in the slot (Figure 1.10c) such as in small and medium-power LEMs. With m layers in the slot, the average skin effect resistance coefficient K_{Rm} is

$$K_{Rm} = \varphi(\zeta) + \frac{m^2 - 1}{3} \cdot \Psi'(\zeta); \quad \zeta = \frac{h_{slot}}{\delta_{pen}} \tag{1.118}$$

$$\text{with } \Psi'(\zeta) = \frac{\sinh(\zeta) + \sin(\zeta)}{\zeta \cdot (\cosh(\zeta) + \cos(\zeta))} \tag{1.119}$$

In (1.118), all conductors in the slot have been considered in series and no transposition applied.

For $m > 1$ layers $K_{Rm} > \varphi(\zeta)$, characteristic for a single conductor, so care must be taken to check the skin effect in multiple turn coil LEMs. Even here, limited transposition (in the end connection zone)—twisting of conductors—will help in reducing K_{Rm} further. There is an optimum number of layers for a given slot height to obtain minimum K_{Rm}, for given frequency.

Finally, with a few elementary conductors in parallel to make up for the cable cross section demanded by the current level, there are additional currents induced by adjacent conductors in the other (proximity effect); this additional effect has to be also added to K_{Rm} of (1.118).

In general, the total skin effect resistance coefficient K_{Rm} should not be over 1.1–1.15 to secure good efficiency of LEM.

There is severe skin effect (even at the moderate slip frequency $S \cdot f_1$) in the solid back iron of the LIM secondary. As the aluminum sheet (Figure 1.10a) of LIM secondary is less than 6–8 mm thick, the skin effect is moderate (at $S \cdot f_1$), but it still has to be considered in a realistic design approach.

LEMs convert mechanical energy to electrical energy and are thus prone to losses: winding (copper) losses and core losses, besides mechanical losses (if any, as LEM is attached to the load machine tightly).

The losses increase the temperatures of windings (conductors) and of cores. These temperatures depend on load changes and on the specific cooling system of LEMs.

Now the conductors flowed by electric currents have to be insulated from each other and from the magnetic cores, except for the aluminum sheet on iron LEM secondary.

1.5 INSULATION MATERIALS FOR LEMs

Insulation materials in electrical engineering are divided (IEC 85, 1984 standard, etc.) into a few classes out of which we describe the following:

Class E: up to 120°C: phenol formaldehyde, melamine formaldehyde moldings, and laminated with cellulosic additions; polyvinyl formal, polyurethane, epoxy resins and varnishes, oil modified alkyd, etc.
Class F: up to 150°C: inorganic fibrous (mica, glass asbestos fibers, and fabrics) bounded and impregnated with alkyd epoxy, silicon-alloyed, etc.
Class H: up to 180°C: class F, but with silicon resins and silicon rubber.
Class C: above 180°C: mica, asbestos, glass, polyamide, polytetrafluoroethylene, etc.

As insulation materials are characterized not only by temperature but also by leaning, bending strength, abrasion, resistance, etc., there are specialized insulation materials for conductors (enamels), slot-insulation/slot liner (insulating papers); multilayer (composite) materials (Nomex 410+polyester+Nomex 410 (Class F), glass-fabric+kapton 410+glass-fabric (Class C), with adhesives); and preimpregnated sheet materials (to impregnate preformed coils, etc).

1.5.1 Piezoelectric and Magnetostriction Effect Materials in LEMs

Direct piezoelectric effect (1880) means the electrical polarization by mechanical stress (generator mode); inverse piezoelectric effect means that a solid crystal becomes strained when an electric field is applied.

Quite a few piezoelectric materials with better and better piezoelectric coefficients have been discovered since 1880: quartz, Rochelle salt crystals, $BaTiO_3$ (a ferroelectric material), and $LiNbO_3$.

From the conversion of electrical energy (electrostatic) to mechanical energy, the road was short to introduce rotary and linear piezoelectric motors [18], especially with two-phase ac traveling wave (Figure 1.11).

In essence, phenomenologically, the piezoelectric effect is described by the four basic electromechanical variables [19,20] D (C/m^2)—electric displacement vector (3×1), E (V/m)—electric field vector (3×1), S (m/m) relative mechanical deformation (6×1), and T (N/m^2) the mechanical strain (6×1).

In an electric solid, a strain T produces a deformation such as

$$S = T \cdot S_y;\quad S_y = \left(Y^{-1}\right);\quad Y \text{ is the young modulus (N/m}^2\text{)};\quad S_y \text{ is the slimness} \tag{1.120}$$

Also

$$D = \varepsilon \cdot E \tag{1.121}$$

ε (C m/V) is the electric permittivity.

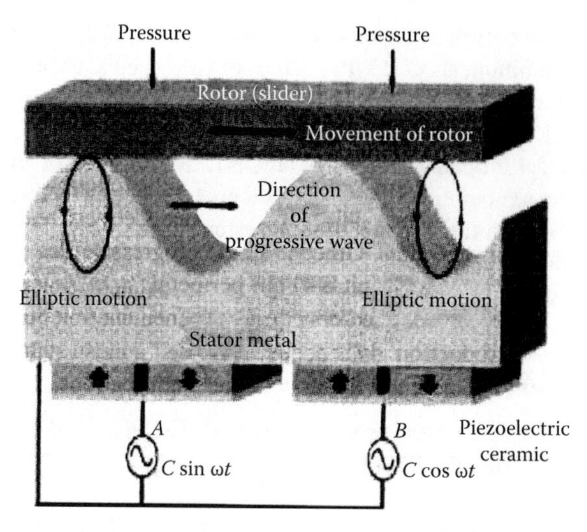

FIGURE 1.11 Principle of piezoelectric linear motor with traveling wave. (After Bullo, M., Modeling and control of a traveling wave piezoelectric motor, PhD thesis (in French), École Polytechnique Fédérale de Lausanne, Vaud, Switzerland, 2005.)

In a piezoelectric material, the mechanical and electrical effects are mutually coupled ($\partial D/\partial T = 0$, $\partial S/\partial \overline{t} \neq 0$).

Consequently,

$$S = S_y^{\varepsilon} \cdot T + d^T \cdot E; \quad D = d \cdot T + \varepsilon^T \cdot E \tag{1.122}$$

The first linear piezoelectric motor (1982) is shown in Figure 1.12.

The extremities of a long metal bar are welded to form a closed rail (stator). A carriage (mover) is pressed against the rail. The two piezoelectric-ceramic blocks form the two phase excitation parts placed on the unused part of the rail.

The electric progressive field of the two ceramic parts produces a wave that propagates through the continuous bar to finally move the chariot.

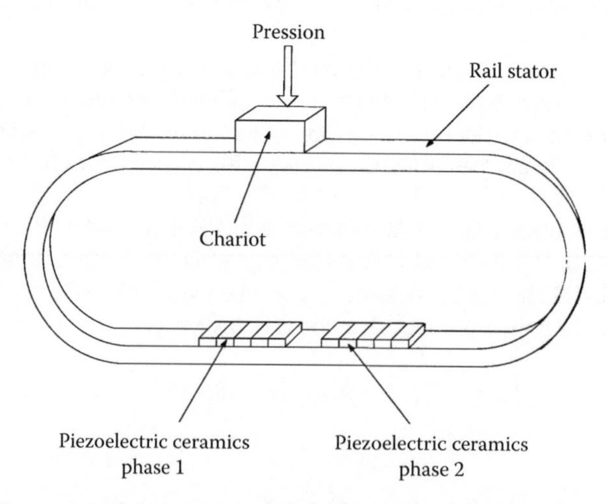

FIGURE 1.12 First linear piezoelectric motor with traveling field (1982). (After Bullo, M., Modeling and control of a traveling wave piezoelectric motor, PhD thesis (in French), École Polytechnique Fédérale de Lausanne, Vaud, Switzerland, 2005.)

There are many other more efficient configurations of piezoelectric linear and rotary motors for applications up to 5 kW (at 5 kg of weight), 886 N m, motors for aviation [20]. High force (torque) density and good force (torque)/W of losses at very low speeds are characteristic for such motors.

1.6 MAGNETOSTRICTION EFFECT LEMs

So far, the effect of magnetization on the elastic properties of solid materials was not mentioned. However, there is a myriad of effects of magnetization and strain in ferromagnetic solids. These phenomena and their reciprocal (changes in magnetization $\bar{M}$ due to stress or strain) are called magnetostriction [21].

Spontaneous magnetization is, at the domain level, caused by the alignment of electron spins in the 3D subshell of Fe, Ni, and Co, and thus small macroscopic change in the shape of such materials is feasible. There are the reversible magnetostriction effects that we are interested in, to build LEMs on these principles.

The relationship between relative deformation $\varepsilon = \Delta L/L$ produced by a longitudinal magnetic field H is shown in Figure 1.13a for three different materials, and the magnetization M dependence of H for zero positive and negative longitudinal stress is visible in Figure 1.13b.

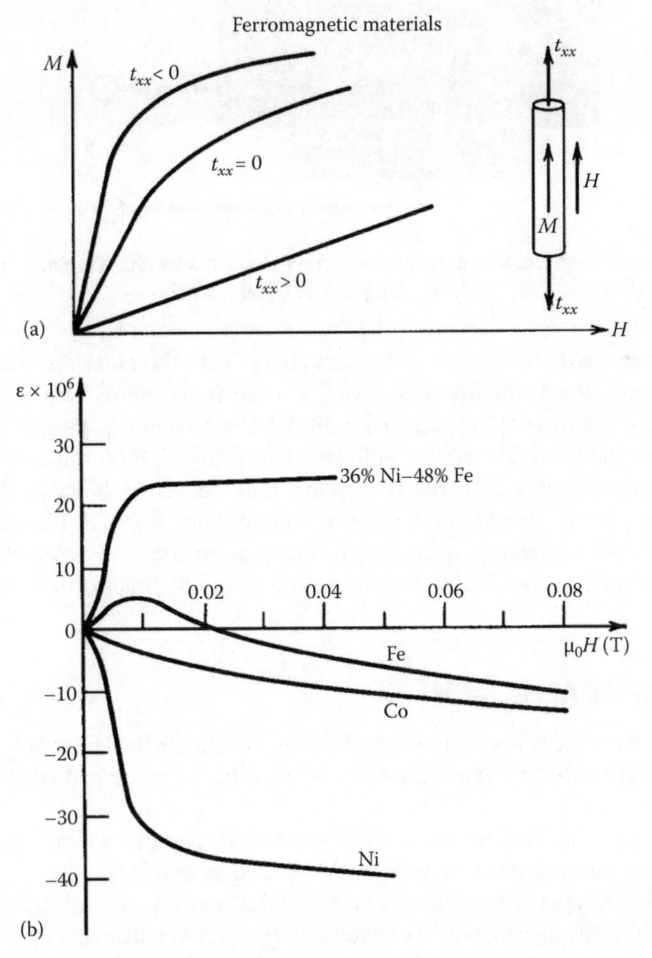

FIGURE 1.13 Magnetostrictive effects (a) p.u. deformation versus $\mu_0 \cdot H$; (b) $M(H)$ for ± longitudinal stress t_{xx}. (After Moon, F., *Magnetosolid Mechanics*, John Wiley & Sons, New York, 1984.)

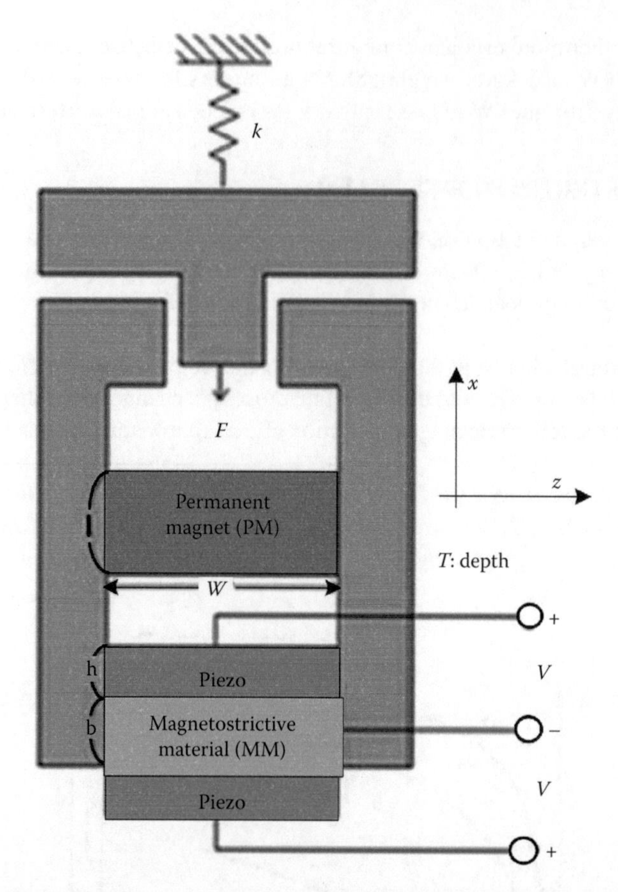

FIGURE 1.14 Linear PM-piezoelectric-magnetostriction plunger solenoid. (Galfenol, Terfenol-D, or other.) (After Ueno, T. and Higuchi, T., *IEEE Trans. Magn.*, 43(6), 2594, 2007.)

Combining piezoelectric and magnetostriction effects with the presence of a PM, a controlled magnetic force linear device with low losses, which is coil-free (Figure 1.14) [22], is obtained.

The magnetostrictive material is bounded with two Piezos and it is aligned along 0*z* direction such that its magnetization varies largely with the compressive stress (produced by the Piezos) due to inversed magnetostriction effect. But this means that the permeability of the shunt magnetic path of magnetostriction material to PM changes and thus more or less PM flux travels through the main airgap to produce vertical plunger motion. The plunger motion is controlled by the electric field intensity produced in the Piezos, which in turn modifies (finally) the PM flux through the plunger airgap.

1.7 METHODS OF APPROACH

All LEMs are based on forces in electromagnetic fields, so first the field distribution has to be found. It may be found by magnetic circuit [24]—analytical—or by numerical solutions of Maxwell's equations [25].

So far we just alluded to the force calculation by magnetic energy methods but Maxwell tensors' method and the Lagrange equation methods could be used as well [14].

The 2D and 3D FEM is by now standard in the calculations of numerical field, forces, and losses.

In addition, from fields distribution via stored energy to circuit elements *R*, *L*, *C*, *emfs* and force expressions could be obtained, and thus the steady-state and transient performance of various LEMs could be approached.

FEM itself has now commercial versions of coupled field–circuit type, where the circuit model is obtained directly through FEM field distribution.

Analytical models for studying the heat transmission in LEMs have been developed. Once the location of main losses and their time variation is known, FEM thermal models could be used to investigate steady-state and transient thermal behavior of LEMs. A similar path is valid for FEM investigation of mechanical stresses in LEM. Consequently, the picture of the LEM design, steady-state and transient performance and control, with thermal and mechanical stress verifications, can be completed by combining electromagnetic, thermal, and mechanical models: analytical and/or by FEM.

1.8 SUMMARY

- LEMs are electromagn etic, electrostatic, piezoelectric, or magnetostriction force devices capable of producing directly progressive or oscillatory translation motion.
- They may transform electrical energy to mechanical work (motor mode) or vice versa (generator mode).
- LEMs may be considered counterparts of rotary electric machines obtained by radial cutting and unfolding in a plane (flat LEMs) or, further, rolling them around the direction of motion (tubular LEMs). Even LEM counterparts of dc brush motors have been proposed [1,26].
- The absence of mechanical transmission characterizes LEMs and this should lead to better precision in motion control in the absence of backlash, etc.
- LEMs may be characterized by the electromagnetic field theory and methods synthesized in the so-called Maxwell's equations.
- The main variables in the electromagnetic field theory are magnetic flux density $\bar{B}$ (V s/m^2), magnetic field $\bar{H}$ (A/m), electrostatic field $\bar{E}$ (V/m), and $\bar{D}$ (C/m^2), the electric flux density.
- The two fundamental Maxwell equations for the bodies in motion are $rot\,\bar{H} = \bar{J} + \left(\partial\bar{D}/\partial t\right)$ Ampere's law and $rot\,\bar{E} = -\left(\partial\bar{B}/\partial t\right) + rot\left(\bar{U}\times\bar{B}\right)$; $\bar{U}$ is the speed.
- These two key laws show the interdependence of electric ($\bar{E}$, $\bar{D}$) fields and magnetic (B, H) fields.
- The basic material laws are $\bar{B}=\mu\cdot\bar{H}$ in magnetic materials (except for permanent magnets) and $\bar{D}=\varepsilon\cdot\bar{E}$ in insulation materials; μ, magnetic permeability (Henry/m) and ε, electric permittivity of insulation materials (F/m). Also, Ohm's law is valid in electric conductors $\bar{J}=\sigma\cdot\bar{E}$; σ, electric conductivity (S/m or Ω^{-1}m^{-1}).
- In magnetostatic fields ($\partial\bar{D}/\partial t=0$), with $div\ \bar{B}=0$ (Gauss flux law), $\bar{B}=\mu\cdot\bar{H}$ and thus $rot\ \bar{H}=\bar{J}$, we may express the flux density $\bar{B}$ as a rotor (curl) of a magnetic vector potential $\bar{A}$: $\bar{B}=rot\,\bar{A}$ as the divergence of a rotor is always zero. This way, for constant source current density $\bar{J}$, we may obtain $-rot(rot)\bar{A}=\nabla^2 A=-\mu\cdot\bar{J}_S$.

This is Poisson's equation, which, when solved, will yield B, H distributions.

- When $\bar{J}_S=0$, $\nabla^2 A=0$, Laplace equation, typical in the airgap and back iron of LEMs magnetic circuits, is obtained.
- Finally, for a sinusoidal source current $\bar{J}_S=\bar{J}_S(x, y, z)\cdot e^{j\cdot\omega\cdot t}$ we get $\nabla^2 A - j\cdot\omega\cdot\sigma\cdot\bar{A}=-\mu\cdot\bar{J}_S$ or Helmholtz equation.
- All these equations are used to get the field distribution in various LEMs, provided proper boundary conditions are set between different media in LEMs.
- Typical boundary conditions have been introduced by Dirichlet ($A=0$ or $A=A_n$ on the border) and Newmann ($\partial A/\partial n=0$ or $(\partial A/\partial n)+k\cdot A=A_g$).
- Once the electromagnetic field equations are solved by using Maxwell equations and the LEM geometry containing coils, magnetic materials, airgap, and permanent magnets are known, the forces developed in LEMs may be calculated, based on energy relations in electromagnetic fields.

- The electrostatic W_e and magnetic W_m stored energies are defined as

$$W_e = \frac{1}{2} \cdot \iiint_V (\bar{D} \cdot \bar{E}) dV$$

$$W_m = \frac{1}{2} \cdot \iiint_V (\bar{B} \cdot \bar{H}) dV$$

- Also, the surface power density P_e (Poynting vector) is $\bar{P}_e = \bar{E} \times \bar{H}$ (W/m^2).
- P_e is used mainly for steady state and is instrumental in calculating losses in continuous magnetic cores or conductors (in slots).
- The electric circuit elements R, L, C are defined based on energy relations: $R = V_{dc}/I_{dc}$, $L = \lambda/I$, $C = \rho/V$.
- Based on energy conservation law, the force in an electromagnetic field, F_{em}, is

$$F_{em} = \frac{-\partial W_m}{\partial x}(\lambda, x) = \frac{\partial W_m'}{\partial x}(i, x) = \frac{1}{2} \cdot L^2 \cdot \frac{\partial L}{\partial x},$$

where λ is the flux linkage of the coil (winding) and W_m' is the coenergy: $W_m + W_m' = \lambda \cdot I$.
- In a capacitor, the electrostatic force on the movable electrode is

$$F_{em} = \frac{-\partial W_e}{\partial x}(V, x) = \frac{V^2}{2} \cdot \frac{\partial C_e}{\partial x}.$$

- Forces are developed only when the LEM circuit storage element L (or C) varies with position. It could be a few coupled circuits for more complex LEMs.
- A plunger solenoid, a capacitor solenoid, and a voice-coil linear motor are investigated as basic LEM case studies, and their forces are calculated. The capacitor linear motor has a very small force density in mm, cm, m, length, and geometry range but may be acceptable in the μm range, when electromagnetic LEMs are even worse.
- The plunger solenoid and the voice-coil linear motion circuit and motion equations are then developed as they are used to calculate steady-state and dynamic performance (and design).
- The traveling field (at speed U_s) three-phase winding linear induction motor (LIM) case is investigated in the presence of half space conducting secondary, and the propulsion (breaking) force and the normal (repulsive) force are calculated.

For positive slip $S = (U_s - U)/U_s < 1$, the ideal LIM is motoring, while for $S < 0$, it is generating.

- The finite longitudinal length of LIM primary leads to entry and exit additional induced currents in the conducting secondary that die slowly from entry to exit. Their adverse effect on losses, thrust, efficiency, and power factor is called dynamic longitudinal end effect, to be tackled in the LIM chapters.
- LEMs use magnetic, electric, and insulation materials.
- Soft (ferromagnetic $\mu \gg \mu_0$) materials are used for LEMs magnetic circuits; they are laminated in sheets to reduce their eddy current losses in ac (or traveling) magnetic fields; hysteresis losses also occur in mild (soft) magnetic materials used as paths for magnetic fields in LEMs; soft magnetic materials have a nonlinear $B_m(H_m)$ dependence, with μ decreasing with B.
- Permanent magnets are used to produce a magnetic field in LEMs which, in interaction with currents, produce Lorenz force $F_x = \bar{j} \times \bar{B}$ (N/m^3) and thus motoring or generating effects.

- PMs are hard magnetic materials with a huge hysteresis cycle [large B_r ($H=0$) and H_c ($B=0$)]; $\mu_{re\ coil} \approx (1.05 - 1.3) \cdot \mu_0$.
- PMs are magnetized in large, short-duration, magnetic field millisecond pulses ($B_{max}>2B_r$, $NI>2H_c h_{PM}$). As the PM magnetization process duration is short but lasts for years, the total losses with PMs is much smaller than with dc coils for flux density production in LEMs; efficiency is thus better in PM LEMs.
- Electric conductors (mainly copper) in slots experience ac skin effects, which increase the ac resistance, and thus additional copper losses occur. The skin effect losses have to be limited to 10%–15% to secure good performance.
- Piezoelectric effects in dielectrics (mechanical stress produces electric polarization (electric field) or vice versa) and magnetostriction (mechanical stress modifies magnetic polarization M in some ferromagnetic materials) are also used to produce mechanical work in special LEMs characterized by high force density for low-speed applications.
- The chapter ends by recapitulating the main stages and methods in LEM characterization by field models, then circuit models for LEM performance assessment, design, and control, all to be detailed throughout the book.

REFERENCES

1. E. Laithwaite, *Induction Machines for Special Purposes*, John Newness, London, U.K., 1966.
2. S. Yamamura, *Theory of Linear Induction Motors*, John Wiley & Sons, New York, 1972.
3. S.A. Nasar and I. Boldea, *Linear Motion Electric Machines*, John Wiley & Sons, New York, 1976.
4. P.K. Budig, *A.C. Linear Motors* (in German), Veb Verlag Technik, Berlin, Germany, 1978.
5. M. Poloujadoff, *Theory of Linear Induction Motors*, Oxford University Press, Oxford, U.K., 1980.
6. I. Boldea and S.A. Nasar, *Linear Motion Electromagnetic Systems*, John Wiley & Sons, New York, 1985.
7. S.A. Nasar and I. Boldea, *Linear Electric Motors*, Prentice Hall, Inc., Englewood Cliffs, NJ, 1987.
8. J.F. Gieras, *Linear Induction Drives*, Oxford University Press, Oxford, U.K., 1994.
9. I. Boldea and S.A. Nasar, *Linear Electric Actuators and Generators*, Cambridge University Press, London, U.K., 1997.
10. J.F. Gieras and Z.J. Piech, *Linear Synchronous Motors*, 2nd edn., CRC Press, Boca Raton, FL, 2000.
11. I. Boldea and S.A. Nasar, *Linear Motion Electromagnetic Devices*, Taylor & Francis, New York, 2001.
12. P. Silvester, *Modern Electromagnetic Fields*, Prentice Hall, Englewood Cliffs, NJ, 1968.
13. N. Bianchi, *Electrical Machine Analysis Using Finite Elements*, CRC Press, Boca Raton, FL, 2005.
14. I. Boldea and S.A. Nasar, *Electric Machine Dynamics*, Chapter 1, MacMillan Publishing Company, Oxford, U.K., 1986.
15. P. Campbell, *Permanent Magnet Materials and Their Applications*, Cambridge University Press, Cambridge, MA, 1994.
16. I.D. Mayergoyz, *Mathematical Models of Hysteresis*, Springer Verlag, New York, 1991.
17. M.A. Mueller, Calculation of iron losses from the stepped FE models of cage induction machines, *International Conference on EMD*, IEE Conference Publication 412, Boulder, CO, 1995.
18. T. Sashida, A prototype ultrasonic motor—Principles and experimental investigations, *Appl. Phys.*, 51, 713–733, 1982.
19. T. Ikeda, *Fundamentals of Piezoelectricity*, Oxford University Press, New York, 1996.
20. M. Bullo, Modeling and control of a traveling wave piezoelectric motor, PhD thesis (in French), École Polytechnique Fédérale de Lausanne, Vaud, Switzerland, 2005.
21. F. Moon, *Magnetosolid Mechanics*, John Wiley & Sons, New York, 1984.
22. D.H. Sul, Y.W. Park, and H.J. Park, Lumped modeling of magnetic actuator using the inverse magnetostriction effect, *IEEE Trans.*, MAG-43(6), 2007, 2591–2593.
23. T. Ueno and T. Higuchi, Magnetic circuit for stress—Based magnetic force control using iron – gallium alloy, *IEEE Trans. Magn.*, 43(6), 2007, 2594–2596.
24. V. Ostovic, *Dynamics of Saturated Electric Machines*, John Wiley, New York, 1989.
25. S. Rotnajeevan and H. Hooke, *Computer Aided Analysis and Design of Electromagnetic Devices*, Elsevier, New York, 1989.
26. A. Basak, *Permanent Magnet DC Linear Motors*, Oxford University Press, Oxford, U.K., 1996.

2 Classifications and Applications of LEMs

Linear electromagnetic machines (LEMs) develop electromagnetic forces based on Faraday's and Ampere's laws, as described in Chapter 1, and produce directly linear motion. Linear motion may be either progressive (Figure 2.1a) or oscillatory (Figure 2.1b) [1–4]. Linear progressive motion even when experiencing back and forth, but nonperiodic, operation modes leads to LEMs whose topology differ (in general) from that of linear oscillatory machines (LOMs). The linear oscillatory motion takes place in general at resonance—when mechanical eigenfrequency equals the electrical frequency—to secure high efficiency in the presence of a strong springlike force (mechanical or even magnetic) [5].

In general, the progressive motion LEMs are three-phase ac devices that operate in brushless configurations, while LOMs are typically single-phase ac devices.

The progressive motion LEMs operate at variable voltage and frequency to vary speed at high efficiency for wide speed control ranges. On the contrary, LOMs operate in general at resonant (fixed) frequency and variable voltage. Close-loop position control may be applied for both LEMs and LOMs. A slight variation of frequency in LOMs may be needed to adjust the resonance frequency with the mechanical (magnetic) springs rigidity variation due to temperature and (or) aging and thus secure high frequency over the entire device life (10 years or more). Linear progressive motion machines may be classified by principle into

- Linear induction machines (with sinusoidal current control)
- Linear synchronous (brushless ac) machines (with sinusoidal current control)
- Linear brushless dc machines (with trapezoidal (block) current control)
- Linear dc brush machines

2.1 LINEAR INDUCTION MACHINES

Linear induction machines (LIMs) resemble the principle and the topologies of rotary induction machines by the "cutting" and unrolling principle to obtain flat LIMs (Figure 2.2).

When the flat LIM is rerolled around the motion axis, the tubular LIM is obtained (Figure 2.3).

Three-phase ac windings are placed in the slots of the primary. They are obtained by the same "cut" and "unrolling" routine from those of rotary induction machines.

A typical simplified single-layer full pitch four-pole three-phase winding is shown in Figure 2.4a and a five-pole two-layer one in Figure 2.4b.

As expected, the limited length of primary—along the direction of motion—leads to winding construction peculiarities (limits) such as the half-wound end poles in the two-layer five-pole winding (Figure 2.4b), which also originates from a four-pole two-layer rotary machine winding.

The mechanical airgap is rather large: 1 mm for up to 2 m travel, 6–8 mm for urban vehicles (up to 20 m/s), and 10 mm above 20 m/s speed, in general.

The large mechanical airgap is augmented (magnetically) by a rather notable 1.5–6 mm thick aluminum plate in the secondary and a 20–30 mm thick solid mild steel back iron placed

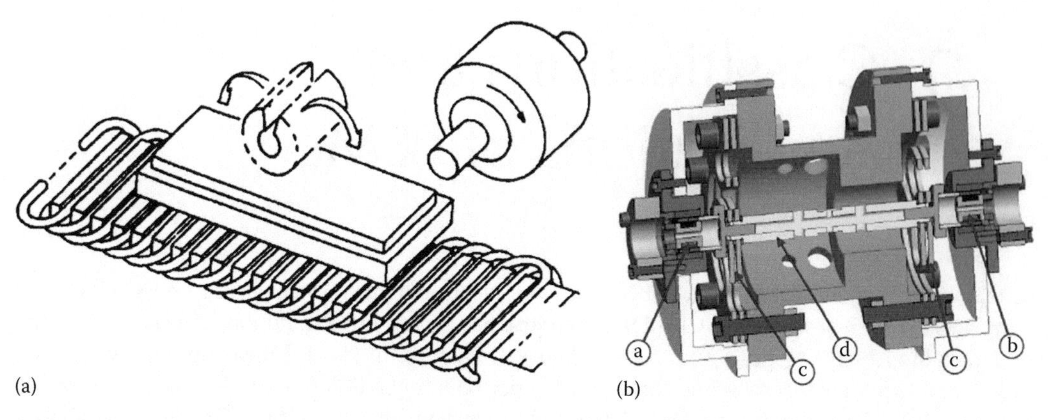

FIGURE 2.1 Linear electric machines (a) with progressive motion (LEMs) (After Boldea, I. and Nasar, S.A., *Linear Motion Electromagnetic Devices*, Taylor & Francis, New York, 2001); (b) with oscillatory resonant motion (LOMs): motor plus generator (a. linear motor, b. linear generator, c. resonant springs (features), d. coupling shaft). (After Pompermaier, C. et al., *IEEE Trans. IE*, 58, 2011.)

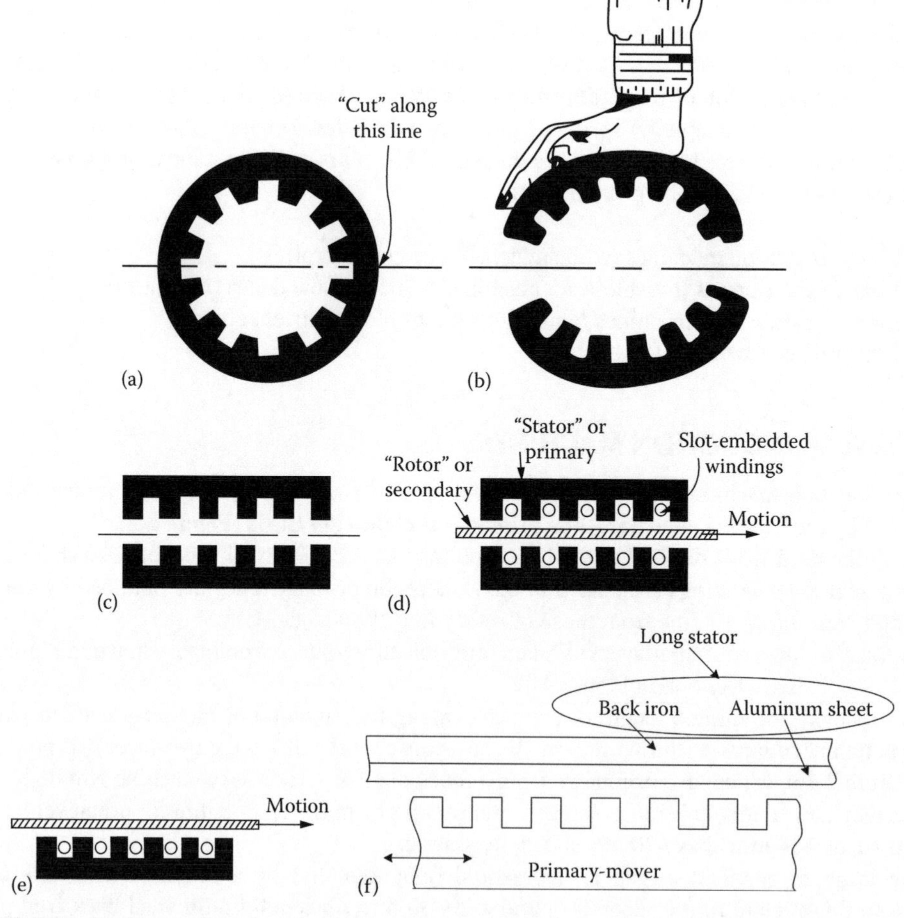

FIGURE 2.2 Double- and single-sided LIMs: (a) round primary, (b) unrolling of primary (after cutting), (c) flat double-sided LIM primary with slots, (d) double-sided LIM primary plus Aluminum (copper) sheet secondary, (e) single-sided LIM (primary) plus Al (copper) sheet, and (f) with back iron in the secondary. (After Boldea, I. and Nasar, S.A., *Linear Motion Electromagnetic Devices*, Taylor & Francis, New York, 2001.)

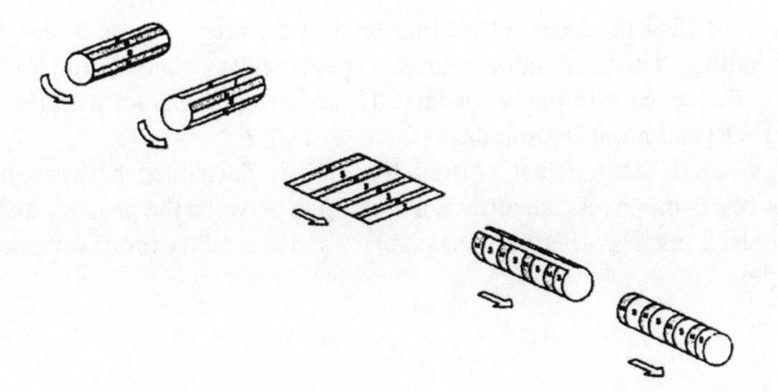

FIGURE 2.3 Flat and tubular LIMs. (After Boldea, I. and Nasar, S.A., *Linear Motion Electromagnetic Devices*, Taylor & Francis, New York, 2001.)

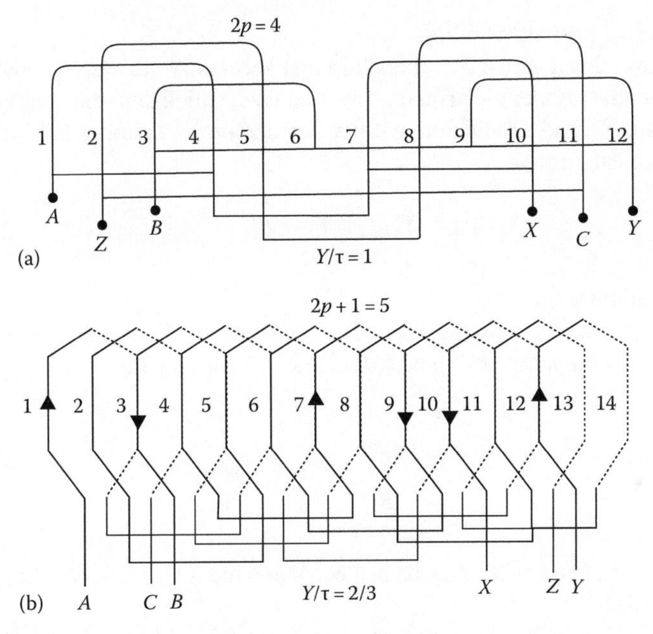

FIGURE 2.4 Typical three-phase LIM windings: (a) with four poles (single layer) and (b) with five poles (double layer). (After Boldea, I. and Nasar, S.A., *Linear Motion Electromagnetic Devices*, Taylor & Francis, New York, 2001.)

behind the aluminum sheet in single-sided LIMs for transportation. All these lead to a moderate goodness factor G:

$$G = \frac{\omega_1 \cdot L_m}{R_2^{'}}; \quad \omega_1 = 2 \cdot \pi \cdot f_1 \tag{2.1}$$

where

ω_1 is the primary frequency
L_m is the magnetization inductance
$R_2^{'}$ is the equivalent secondary resistance

The short primary of LIM produces a traveling field in the airgap by its three-phase ac currents through the ac windings. This field induces emfs in the secondary plate on iron (or ladder type), and thus secondary currents occur in the secondary. These currents interact with the primary current magnetic field ($\bar{j} \times \bar{B}$) and produce propulsion force (thrust), F_{xc}.

Also, in single-sided LIMs, a net normal force F_n is developed between the primary and secondary; it has two components: an attraction one F_{na}—between the primary and secondary iron cores—and a repulsive one F_{nr}—between the primary and secondary mmfs (magnetomotive forces: ampere turns/pole):

$$F_n = F_{na} - F_{nr} \tag{2.2}$$

The conventional synchronous (field) speed of LIM is, as for the rotary IM, U_s:

$$U_s = 2 \cdot \tau \cdot f_1 \tag{2.3}$$

with τ the pole pitch of primary winding.

Besides the conventional thrust F_{nc}, the additional secondary currents, produced mainly at the entrance of the secondary under the primary traveling field, called dynamic longitudinal effect, lead to an end effect force F_{xe} and a reduction in efficiency and power factor of LIM (this is accentuated by the large mechanical airgap):

$$F_{x\,total} = F_{xc} + F_{xe} \tag{2.4}$$

where F_{xc} is the ideal thrust.

The attenuation length of longitudinal end effect currents in the secondary is related to the slip S, goodness factor G, and the number of poles of primary ($2p$ or $2p+1$).

Basically if

$$2 \cdot \tau \cdot f_1 \cdot \frac{L_m}{R_2'} = \frac{\tau}{\pi} \cdot G < \frac{L_{primary}}{10} \tag{2.5}$$

the dynamic longitudinal end effect may be neglected and the rotary IM modeling may be extended to such low-speed LIMs.

In contrast, for high-speed LIM condition (2.5) is not fulfilled at full speed (f_1), and thus the dynamic longitudinal end effects have to be considered.

Clever adaption in the circuit model of rotary IM have been introduced to handle high-speed LIMs modeling.

A high goodness factor increases conventional performance but also increases the adverse influences of dynamic longitudinal end effects on performance.

So, for high-speed LIMs there is an optimal goodness factor G_0 that leads to zero end effect thrust $(F_{xe})_{S=0}=0$ at zero slip ($S=0$); G_0 increases linearly with the number of poles as shown in a later chapter dedicated to LIMs.

For low-speed LIMs, in general, the design is approached around the condition

$$S_n \cdot G = 1 \tag{2.6}$$

This implies maximum thrust per primary copper losses. The low-speed LIM has found good usage with a ladder-type secondary, for limited travel (a few meters) and an airgap around 1 mm, for moving machining tables in industry in a harsh environment (Figure 2.5).

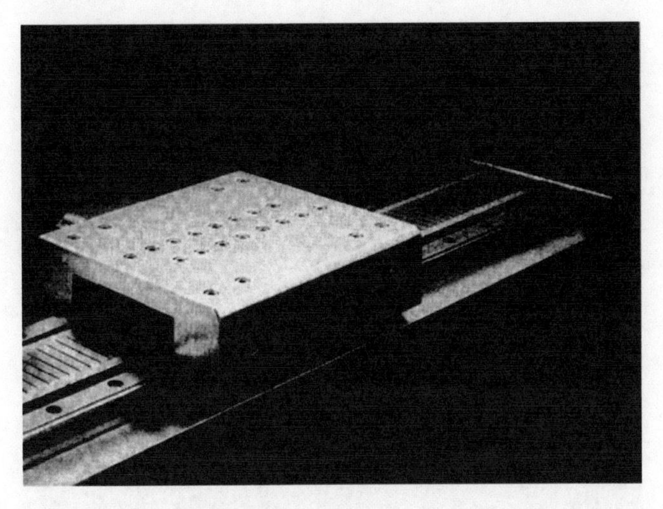

FIGURE 2.5 LIM for a machining table. (After Gieras, J.F., *Linear Induction Drives*, Oxford University Press, Oxford, U.K., 1994.)

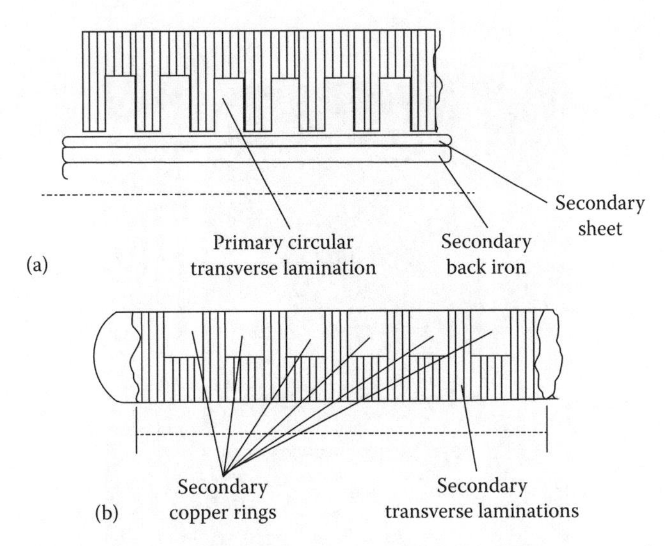

FIGURE 2.6 Tubular LIM with (a) disk shape laminations and (b) secondary copper rings.

Tubular LIMs for short travel (less than 1 m) may be built for robotic applications (Figure 2.6).

In medium (suburban) speed range (30 m/s), the single-sided LIM has found applications both in wheeled (Figure 2.7) and MAGLEV vehicles (Figure. 2.8), after [9].

2.2 LINEAR SYNCHRONOUS MOTORS FOR TRANSPORTATION

We mean here, by "transportation," people movers for urban/interurban purposes.

A first classification of linear synchronous motors (LSMs) may be

- Active guideway (long primary) LSMs
- Passive guideway (short primary-on board) LSMs

FIGURE 2.7 JFK–New York AirTrain on wheels propulsed by single-sided LIMs. (From http://en.wikipedia.org/wiki/File:JFK_AirTrain.agr.jpg)

FIGURE 2.8 Linimo MAGLEV (HSST) in Japan, with single-sided LIM propulsion. (From http://en.wikipedia.org/wiki/File:Linimo_approaching_Banpaku_Kinen_Koen,_towards_Fujigaoka_Station.jpg)

The active guideway LSMs may be built with

- DC heteropolar excitation on board (without or with PM assistance) and active guideway with laminated iron core and three-phase windings in slots (Figure 2.9)
- DC superconducting heteropolar excitation on board with air-core three-phase cable winding guideway (Figure 2.10)

The peak speed of Transrapid, now commercial in Shanghai, China, is 450 km/h with a flight at 10 ± 2.5 mm height while JR–MAGLEV reached more than 560 km/h with a flight height of about 150 ± 30 mm.

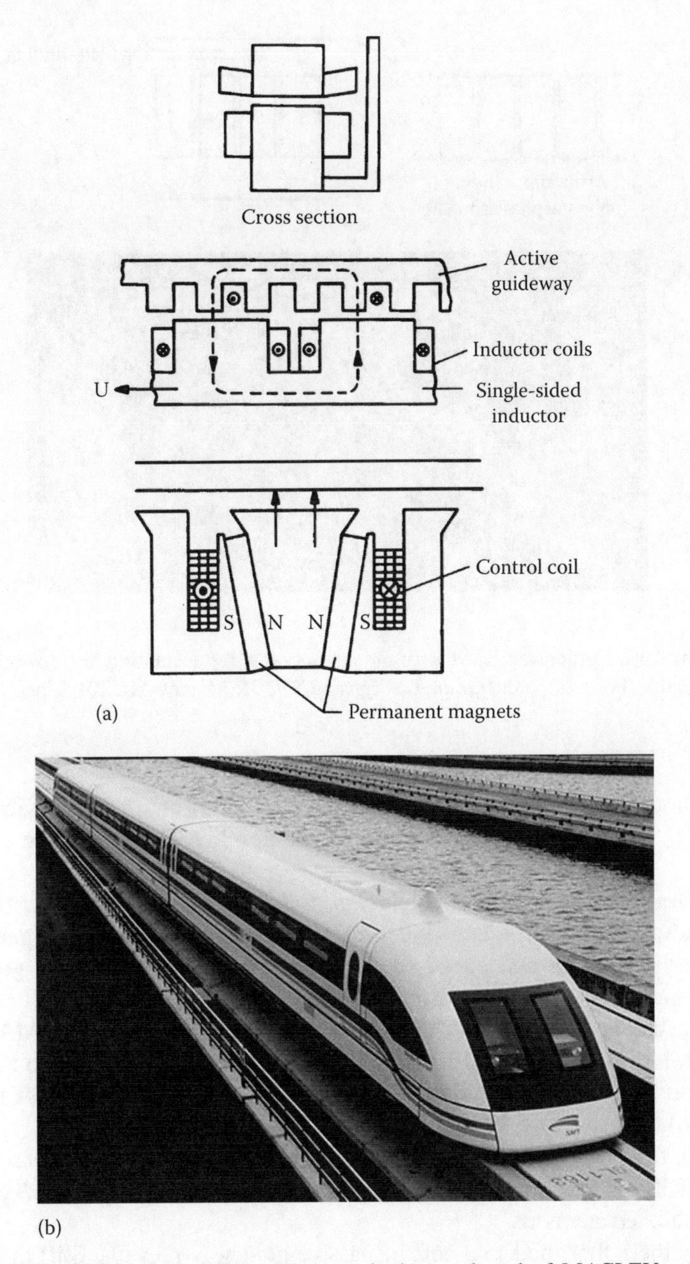

FIGURE 2.9 Active guideway with LSM with dc excitation on board of MAGLEV vehicle (Transrapid, Germany). (a) LSM for propulsation and magnetic levitation and (b) the first commercial Transrapid. (From http://en.wikipedia.org/wiki/File:A_maglev_train_coming_out,_Pudong_International_Airport,_Shanghai.jpg)

The active guideway (stator with three-phase cable ac windings) of both solutions leads to rather expensive initial costs of guideway, especially if we add the 1–3 km sections to be supplied from other power stations on ground, 10–20 km away from each, provided with large power PWM converters and numerous power on/off switches.

But the vehicle is claimed to be less heavy as no large (full) power energy collection on board is required. Some power on board is required to feed the dc excitation coils and auxiliaries that control levitation and guidance, respectively, and to replenish the superconducting coils with current daily, etc. Power on board is transferred, by electromagnetic induction from

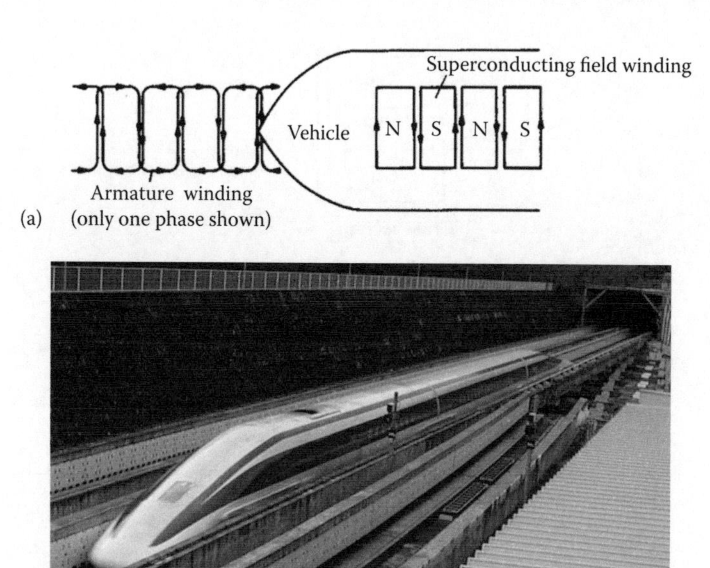

FIGURE 2.10 (a) Active guideway LSM with dc superconducting excitation and (b) on board of vehicle (JR-MAGLEV, Japan). (From http://en.wikipedia.org/wiki/File:JR-Maglev-MLX01-2.jpg)

the long stator mmf space harmonic fields, into dedicated on-board linear generator coils and a backup battery. The major (propulsion) control of such MAGLEVs is done from the ground stations [6].

As with all linear-motor-propulsed vehicles, the direct electromagnetic thrust allows to build lower deadweight vehicles; consequently, lower energy consumption per passenger and kilometer for given commercial speed is obtained. The dedicated active guideway system means notable initial costs in comparison with wheeled high-speed trains (TAGV or similar). But, so far, for urban people movers, the passive guideway LIM-propulsed MAGLEVs (HSST in Japan and Rotem in S. Korea (Figure 2.11)) with separate dc-controlled electromagnets for levitation and guidance control have been introduced in LIM-MAGLEVs with full power transfer on board.

A vehicle with full power PWM converters on board for propulsion control is required.

However, as IGBT PWM converters can be built today at less than 1 kVA/kg, the additional weight is not prohibited anymore.

Consequently, the rather moderate cost of passive guideway prevails. Still the LIM moderate performance (70%–75% efficiency and (0.5–0.6) power factor) makes the search for better passive guideway solutions very welcome.

A homopolar LSM was introduced for integrated propulsion levitation of passive guideway (made of solid iron segment poles and air interpoles) higher performance MAGLEVs (Figure 2.12).

Here, both the dc and ac windings are on board and so are the PWM power converters. Full power transfer to the vehicle is required. But the track is made of solid iron segments spaced with 2τ period.

Experiments have demonstrated 80% propulsion plus levitation efficiency at 0.8 power factor for less than 2 kVA/ton for levitation at 8–10 m/s [7], with the H-LSM providing 10/1 levitation/own weight ratio, for natural cooling.

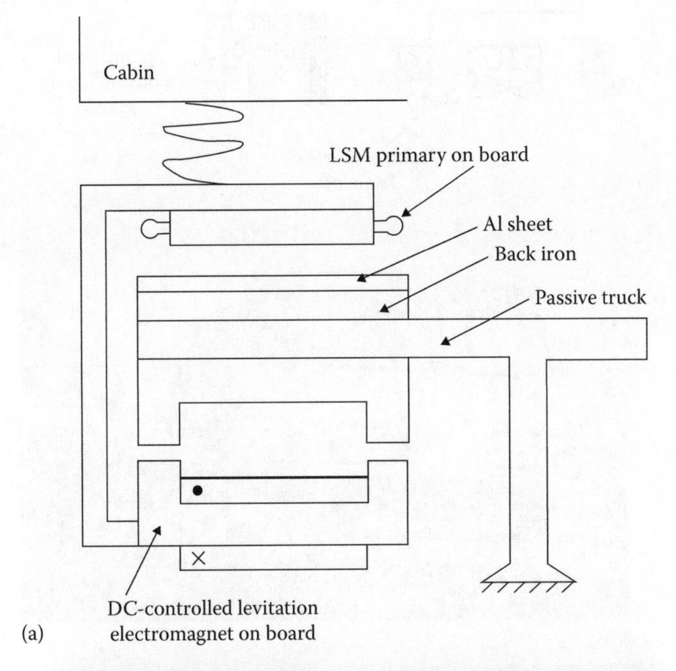

FIGURE 2.11 The urban Korean MAGLEV with LIM propulsion and additional dc-controlled levitation and guidance electromagnets: (a) generic structure and (b) general view.

Other passive guideway LSMs have been recently investigated for transportation in industrial rooms or for people movers.

Some of them such as the reluctance LSMs (with three phases), the transverse-flux PM LSMs, or the six-phase brushless two-level bipolar current controlled reluctance linear motors (with two field phases and four thrust phases at any time) and switched reluctance LSMs will all be treated in dedicated chapters of this book.

They all share a passive variable reluctance laminated guideway with short primary on board of vehicle. The simplicity and low cost of the passive guideway are attractive features of such LSMs, despite full power transfer and larger kVA rating PWM converters on board of vehicles.

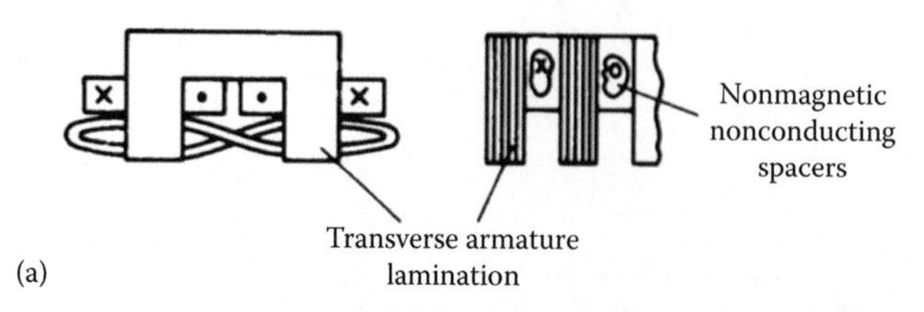

(b)

FIGURE 2.12 Homopolar linear synchronous motor (H-LSM), as an integrated propulsion-levitation system for a passive guideway MAGLEV (Magnibus 01). (a) Homopolar linear synchronous motor and (b) general view. (From Boldea, I. et al., *IEEE Trans.*, VT-37(4), 213, 1988.)

2.3 INDUSTRIAL USAGE LINEAR SYNCHRONOUS MACHINES

By industrial usage, we mean various industrial applications with limited travel (less than 2–3 m) and wheels or bearings suspension and LSM propulsion, with magnetic suspension and LSM propulsion, or with magnetic suspension for single axis or *X–Y* linear motion precise positioning. There is a myriad of topologies proposed so far; here we will refer to a series of such devices fabricated by a few manufacturers, to give a coherent up to date status of the field.

LSMs provide linear motion high position speed control performance advantages, such as

- High accuracy: down to 2.5 μm/300 mm travel
- High repeatability: resolution to 0.1 μm
- No backlash due to direct driving
- Faster acceleration (1–10 g) than rotary motors with rotary to linear transmission solutions
- Higher linear speeds (up to 8 m/s) for faster travel to excursion end
- High reliability for long life due to simplicity
- Low maintenance costs
- Compatibility with clean rooms

Typical industrial applications include

- Bottle labeling, baggage handling, electronic assembly, food processing, laser cutting machines, laser surgery equipment, machine tools, mail sorting, material handling, packaging devices, part transfer systems, PCB assembly/inspection/drilling, precision grinding, printing applications, robotics, surface mount assembly, wafer etch machines, and vision inspection

Ball-screw or timing belt rotary (stepper) drives have been drastically surpassed in positioning performance by LSMs of various types (Table 2.1).

TABLE 2.1
Performance Comparison: Ball-Screw/Timing Belt Rotary Steppers versus PM-LSMs

	Close Loop		Open Loop	
	Rotary	LSM	Rotary	LSM
Maximum speed (m/s)	1	10+	2.5	10+
Maximum acceleration (m/s^2)	20	98+	20	98+
Repeatability (μm)	50	1	250	10

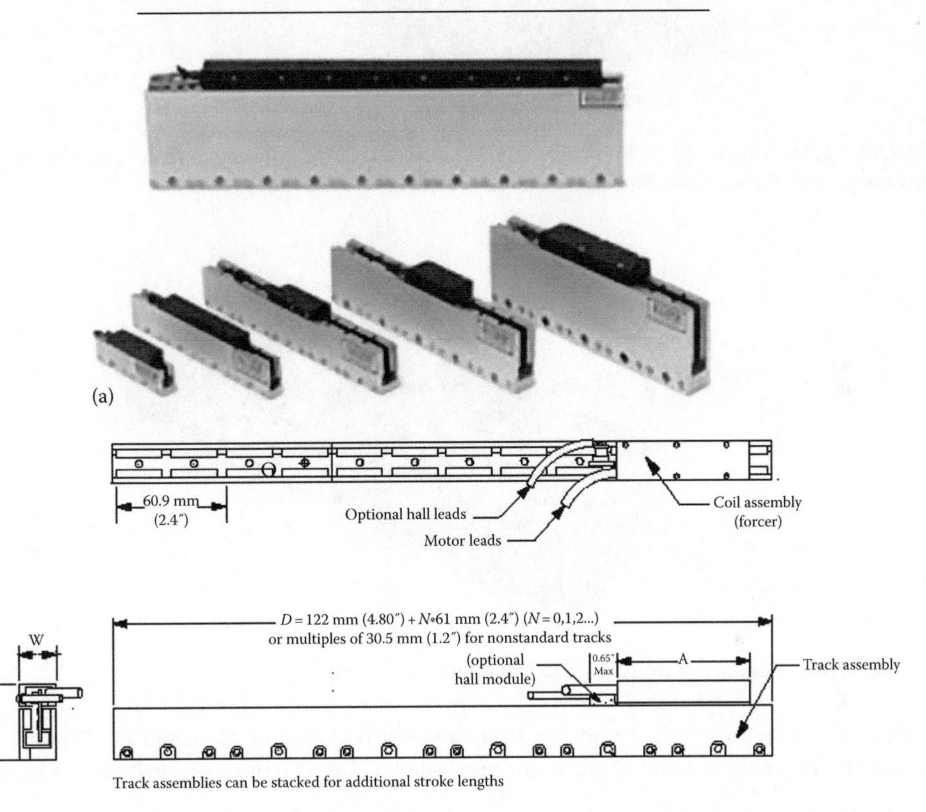

FIGURE 2.13 Commercial iron-core double-sided PM-LSM (a) general view and (b) detailed geometry. (From http://www.baldormotion.com/products/linearproducts/lmcf.asp)

Two representative LSMs are built with PM tracks and three-phase ac windings on the mover, which is fed through a flexible power cable:

- Air-core (cogging force free) PM-LSM, Figure 2.13
- Iron-core PM-LSM, Figure 2.14

The double-sided track uses a U-shape solid iron core and a row of NSNS PMs along the travel length. The short primary with a three-phase ac winding (similar to those of LIMs, but also of nonoverlapping coil type as shown later in the book) is fixed in an insulation adhesive or core, respectively, in a laminated core.

When superprecision in positioning and superfast response are required with very low vibrations, the air-core PM-LSM is used despite its larger PM weight and lower efficiency, for same thrust and speed, than iron-core PM-LSMs. Linear stepper motors in single- and dual-axis motion

FIGURE 2.14 Commercial iron-core double-sided PM-LSM. (From http://www.baldormotion.com/products/linearproducts/lmic.asp)

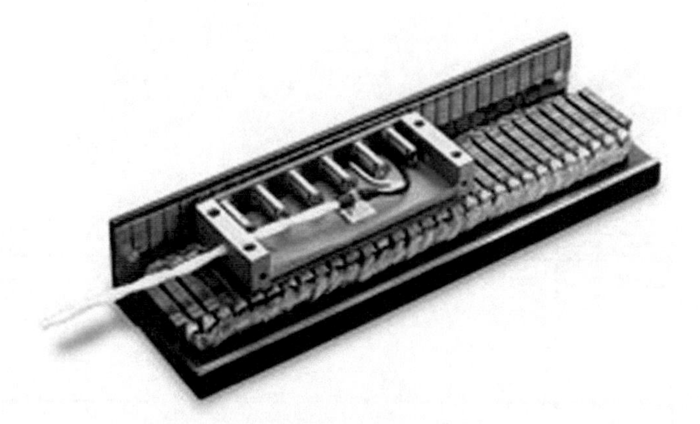

FIGURE 2.15 Commercial brush-dc PM linear motors.

configurations are also fabricated due to their simplified control. To avoid the limitation of the flexible power cable in some applications brush-dc PM linear motors are still used (Figure 2.15).

2.4 SOLENOIDS AND LINEAR OSCILLATORY MACHINES

Plunger solenoids produce thrust over a limited length (Figure 2.16 [8,9]), but also up to even 0.3 m in special PM configurations in single-phase (coil) movers with PM tracks [9].

The coil current is modified to control the airgap as constant (as for magnetic suspension in MAGLEVs) or as variable, according to a planned "trip" (as in valve actuators for internal combustion engines or in high-power electromagnetic power-switch drives).

With numerous applications, such as door lockers in hotels, plunger solenoids are a growing industry (Figure 2.17), and they will be discussed in a chapter in the book.

The principle of the microphone/speaker and of a linear vibrator is illustrated in Figure 2.18a and b, while a long stroke (up to 0.3 m) moving coil PM actuator is illustrated in Figure 2.18c.

A homopolar PM piece (cylindrical or flat) produces the magnetic field—which interacts with current (i) in an air-core coil—mover to produce thrust F_x, $(\bar{j} \times \bar{B})$:

$$F_x = B \cdot i \cdot l_{coil} \cdot n_{turns} \tag{2.7}$$

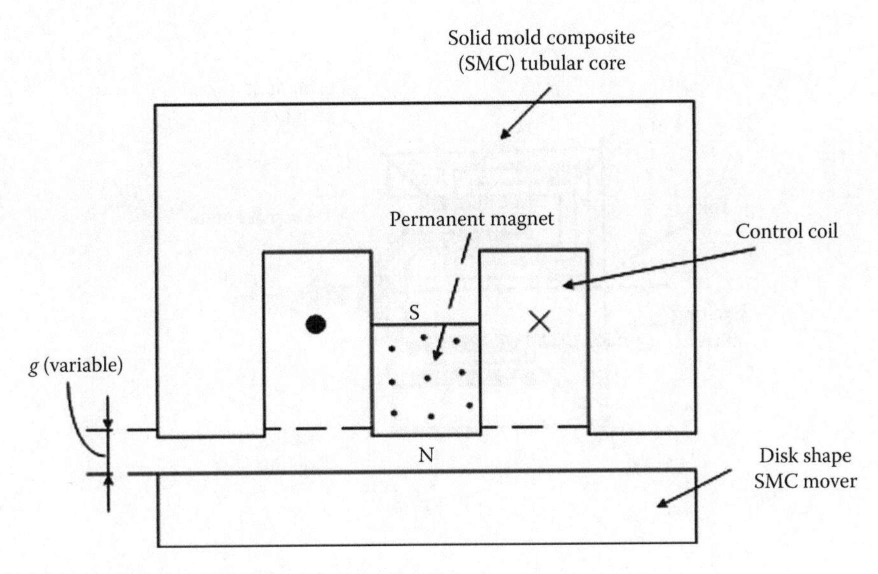

FIGURE 2.16 PM-plunger solenoid with control coil.

	Tubular solenoids	Low-profile solenoids	Soft shift solenoids	DC open frame	AC open frame	Laminates
Solenoid power	Average power consumption; moderate force output	Average power consumption; high force output	Average power consumption; moderate force output	Higher power consumption; moderate force output	Higher power consumption; moderate force output	Higher power consumption; highest force output
Solenoid life	25 + million actuations for STA models	1–5 million cycles	10 million cycles	50,000–100,000 cycles	50,000–100,000 cycles	50,000–100,000 cycles
Solenoid operation	Push/pull engagement: well suited to lock/latch operations	Push/pull engagement: well suited to lock/latch operations	Quiet operation with 3–5 times the starting force of standard solenoids	Pull-in engagement	Pull-in engagement	Pull-in engagement
Force	Up to 97.86 N	Up to 355.86 N	Up to 133.45 N	Up to 133.45 N	Up to 155.7 N	Up to 133.45 N
Stroke	Up to 63.5 mm	Up to 9525 mm (4.76 mm nominal)	Up to 9525 mm	Up to 25.4 mm	Up to 25.4 mm	Up to 31.75 mm

FIGURE 2.17 Sample commercial linear solenoids. (After http://www.solenoids.com/linear_solenoids.html)

A typical handheld commercial speaker appears in Figure 2.19a, while a waterproof speaker/microphone is shown in Figure 2.19b.

Resonant linear oscillatory motors and generators also use PMs and single-/multiple-coil single-phase currents (Figure 2.20) for strokes up to ±25 mm (Figure 2.21a and b) or in applications such as refrigerator-compressor drives (Figure 2.21a) or Stirling engine-driven linear electric generators (Figure 2.21b).

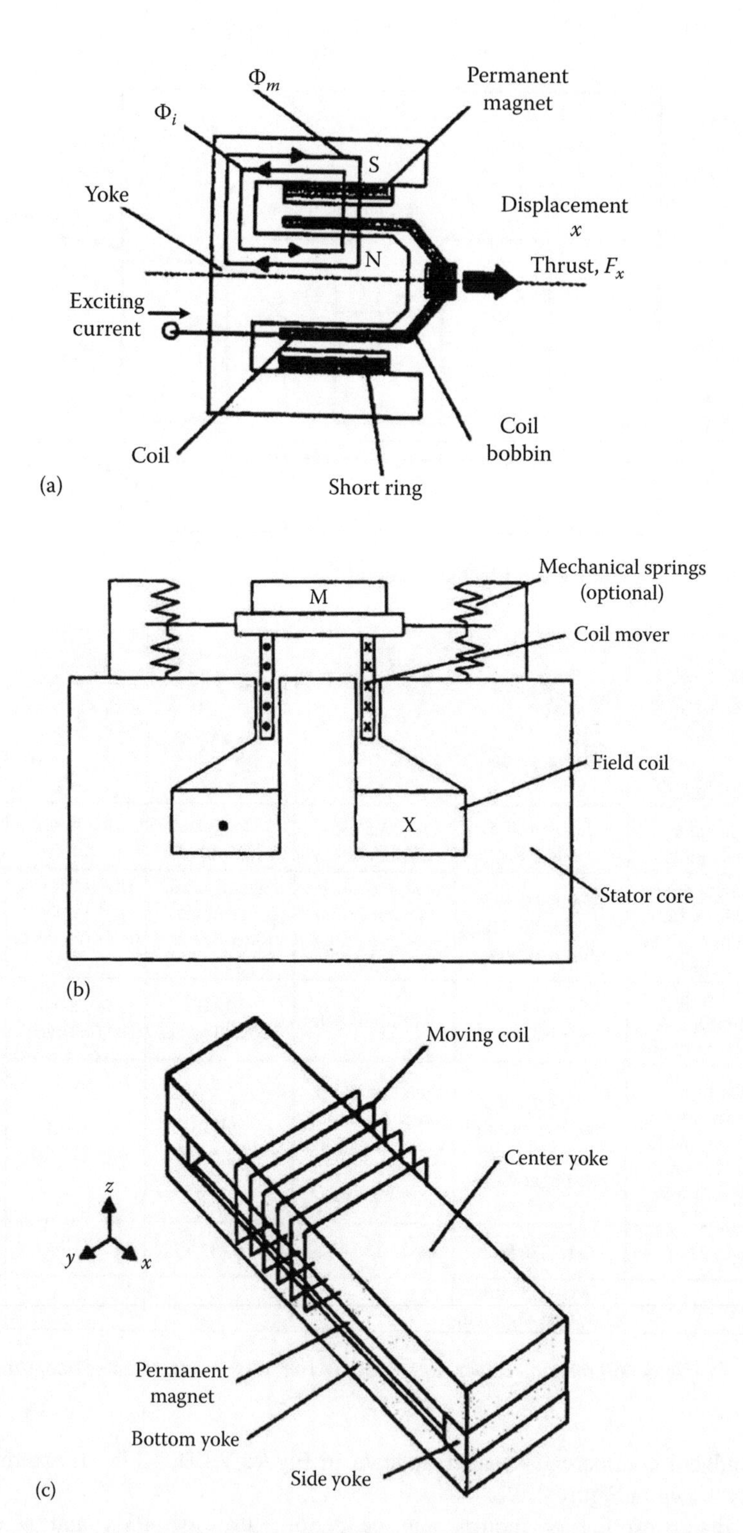

FIGURE 2.18 (a) Speaker/microphone, (b) linear vibrator, and (c) moving coil PM actuator. (After Boldea, I. and Nasar, S.A., *Linear Motion Electromagnetic Devices*, Taylor & Francis, New York, 2001.)

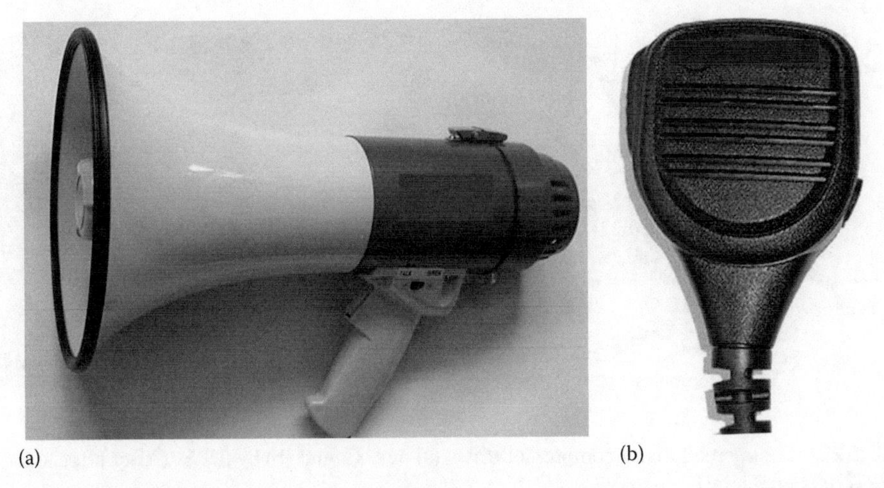

FIGURE 2.19 Sample commercial speaker/microphone: (a) handheld (After http://www.alibaba.com/product-gs/667141691/professional_wireless_microphone_for_recording.html) and (b) waterproof (After http://i00.i.aliimg.com/photo/v1/424513460/Handheld_speaker_mic_Water_resistant_IP54_for.jpg).

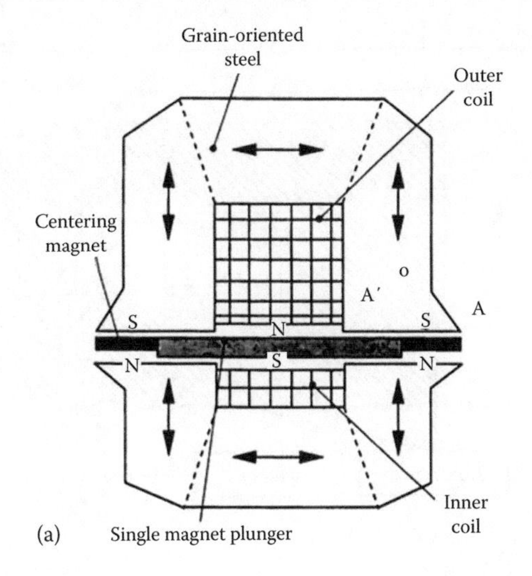

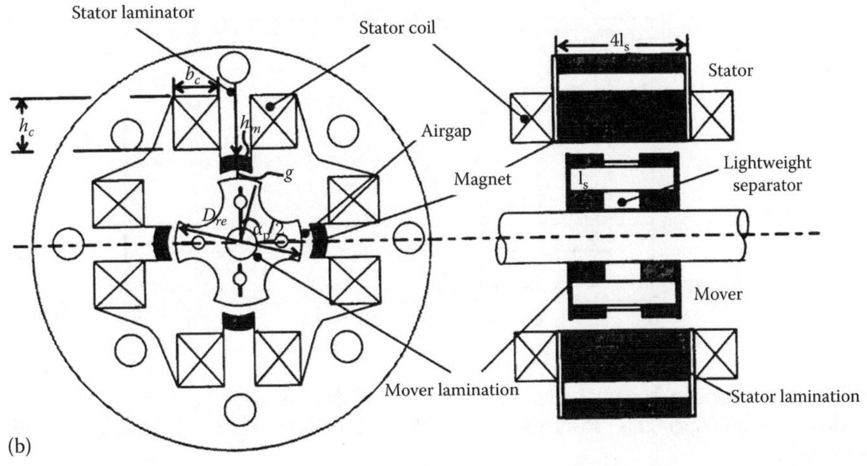

FIGURE 2.20 Resonant linear oscillator machines with: (a) PM mover and (b) iron mover. (From Boldea, I. and Nasar, S.A., *Linear Electric Actuators and Generators*, Cambridge University Press, London, U.K., 1997.)

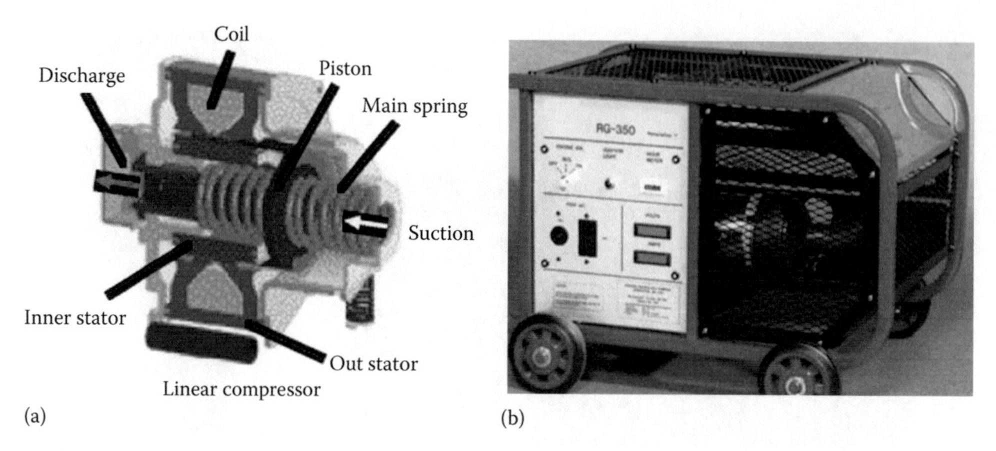

FIGURE 2.21 Commercial linear compressor drive: (a) by L.G. and (b) by S.T.C. (After http://spinoff.nasa.gov/spinoff2002/er_9.html)

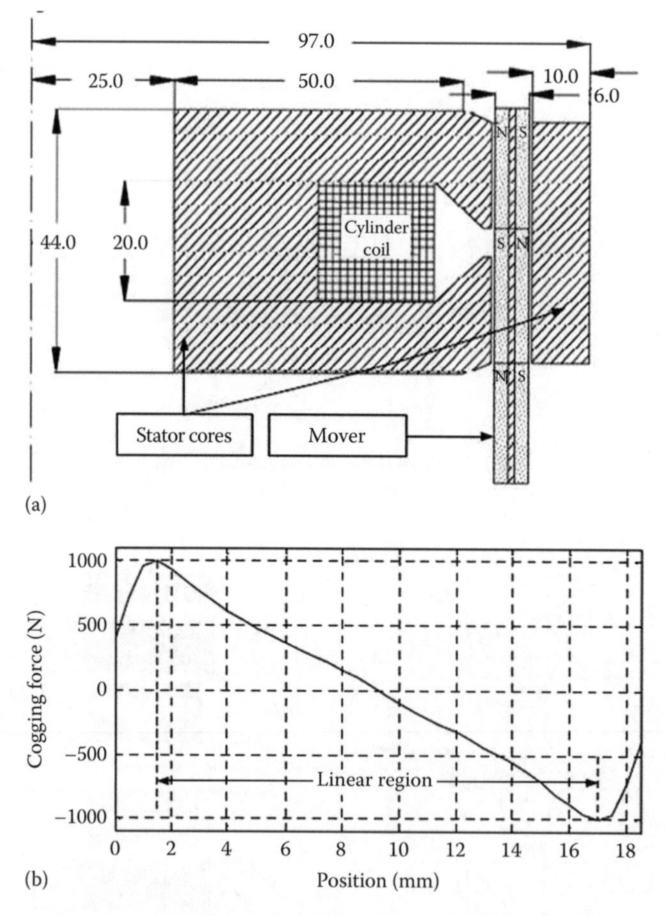

FIGURE 2.22 Springless resonant LOM: (a) cross section and (b) cogging force ("magnetic spring"). (After Boldea, I. et al., Springless resonant linear PM oscillomotors, *Record of LDIA-2011*, Eindhoven, the Netherlands, 2011.)

The PM mover device (Figure 2.20a) leads to a lower weight mover (lower rigidity mechanical spring), while the core-mover device protects better (thermally and mechanically) the PMs placed on the stator (Figure 2.20b). Commercial embodiments of LOMs are shown in Figure 2.21.

Efficiency above 0.9 from 50 W upward has been obtained with such resonant LOMs with PMs both in motoring and generating. As the mechanical spring has to be designed with its rigidity k corresponding to the electric frequency f_1,

$$f_1 = \frac{1}{2\pi}\cdot\sqrt{\frac{k}{m_t}} = f_m;\quad m_t\text{, mover mass} \tag{2.8}$$

Extreme precision is required in manufacturing to match (2.8), unless the LOM is single-phase inverter-fed, when the electrical frequency f_1 is adjusted to the mechanical eigen frequency f_m.

In an effort to replace the mechanical spring entirely—for cost and volume reduction reasons—particular designs of LOM capable of large enough and linear and centralizing cogging (zero current, PM) force have been proposed recently (Figure 2.22) for a 1.9 kW, 2 kg mover, ±7.5 mm motion, at 41 Hz at above 0.9 efficiency for 10 kg of active materials [10].

2.5 SUMMARY

- Linear electric machines are counterparts of rotary electric machines, and thus there are LIMs, LSMs, and dc-brush linear motors.
- Linear electric machines are reversible and thus transit easily from motoring to generating modes.
- In applications, linear progressive or linear oscillatory motion control is needed.
- In general, for linear progressive motion, even with frequent returns, three (two)-phase ac or dc brushless linear motors are used.
- In contrast, for linear oscillatory motion at controlled frequency for small strokes (50 mm or less), single-phase (coil) linear PM synchronous motors/generators are used.
- Linear electric machine (LEM) topologies are flat (single or double sided) or tubular—only for short travels (max. 1–1.5 m).
- Three-phase ac windings of LEMs are similar to those of rotary machine counterparts, but, for double-layer topologies, they have $2p+1$ poles (with half-filled end-pole slots).
- Linear induction motors (LIMs)—in single-sided flat configuration with short primary and long (fixed) secondary are used for low speed (machine table movers) or high speed (urban/interurban people movers).
- To reduce costs, the long fixed secondary of LIMs is made of an aluminum sheet fixed by bolts to a solid back iron slab but, to increase efficiency and power factor, a ladder-type secondary with a stator laminated core is preferable (for short travels: a few meters).
- Due to its ruggedness, LIMs via variable voltage and frequency converters may adapt to both people movers and industrial transport uses; such applications have become an industry by offering lower cost good performance transport.
- The rather large airgap of LIMs in transportation led to lower than desired efficiency and power factor (0.8 and 0.6, respectively) penalizing the kVA ratings of the PWM converter used for speed control. This is how linear synchronous motors (LSMs) came into play.
- LSMs may be first classified as
  - Active guideway (primary)
  - Passive guideway (secondary)
- In active guideway LSMs, the primary laminated core with its three-phase ac windings is spread along the entire travel length and is fed through on-ground converters in sections, to reduce energy consumption.

- The long primary is wound with a cable winding for a $q=1$ slot/pole/phase single-layer winding. A $q=\frac{1}{2}$ nonoverlapping winding is also feasible to reduce guideway initial costs.
- On board of mover there is a dc excitation source:
  - With dc excitation coils (without or with PM assistance)
  - With dc superconducting excitation
- The active guideway LSMs have found applications in superhigh-speed trains with magnetic levitation. The LSM provides integrated levitation and propulsion or, in addition, guidance control (in superconducting MAGLEV vehicles).
- Passive guideway LSMs include
  - Homopolar LSMs (H-LSM)
  - Transverse-flux PM-LSMs
  - Reluctance LSMs
  - Switched reluctance LSMs
  - Multiple-phase bipolar two-level current reluctance brushless linear machines
- Only the H-LSM has been proved on a prototype for an urban MAGLEV vehicle, but all the other types are candidates in both people movers and for indoors transporters.
- Besides people movers, industrial linear position (speed) control is today realized on many occasions by LSMs or dc-brush PM linear motors with higher quickness, precision, and repeatability than ball-screw rotary motor actuators.
- Both air-core and iron-core primary PM-LSMs in flat and tubular topologies are commercial or proposed for single- or multiple-axis linear motion control in various applications, from movable tables to robot arms.
- Short stroke linear oscillatory motors, mostly in single-phase (coil) PM linear synchronous embodiments, are in commercial use as loudspeakers/microphones, gas-compressor drives, or Stirling or internal combustion engine-driven linear generators [11].
- Such linear oscillomachines (LOMs) work at resonance and thus need mechanical well-calibrated springs. To eliminate the mechanical springs, but keep the resonance conditions, to secure very good efficiency, special designs with large, linear, and centralizing cogging force in PM LOMs have been recently proposed.
- Finally, solenoids (electromagnets)—dc or ac fed—without or with PM assistance, with simple or sophisticated position control, have constituted for quite some time an industry with applications from hotel door lockers to electromechanical power-switch drives.
- Quite a few applications of LEMs or LOMs have been named or illustrated in this chapter.

Subsequent chapters will analyze most LEM and LOM topologies in terms of modeling, design, and control for specific applications.

REFERENCES

1. S.A. Nasar and I. Boldea, *Linear Motion Electric Machines*, John Wiley Interscience, New York, 1976.
2. I. Boldea and S.A. Nasar, *Linear Motion Electromagnetic Systems*, John Wiley & Sons, New York, 1985.
3. J.F. Gieras, *Linear Induction Drives*, Oxford University Press, Oxford, U.K., 1994.
4. J.F. Gieras, *Linear Synchronous Motors*, 2nd edn., CRC Press, Boca Raton, FL, 2000.
5. C. Pompermaier, F.J.H. Kalluf, A. Zambonetti, M.V. Ferreira, and I. Boldea, Small linear PM oscillatory motor magnetic circuit modeling corrected by asymmetric FEM and experimental characterization, *IEEE Trans. IE*, 58, 2012, 1389–1396.
6. H.-W. Lee, K.-C. Kim, and J. Lee, Review of MAGLEV train technologies, *IEEE Trans.*, MAG—42(7), 2006, 1917–1925.
7. I. Boldea, I. Trica, G. Papusoiu, and S.A. Nasar, Field tests on a Maglev with passive guideway linear inductor motor transportation system, *IEEE Trans.*, VT-37(4), 1988, 213–219.
8. I. Boldea and S.A. Nasar, *Linear Electric Actuators and Generators*, Cambridge University Press, London, U.K., 1997.

9. I. Boldea and S.A. Nasar, *Linear Motion Electromagnetic Devices*, Taylor & Francis, New York, 2001.
10. I. Boldea, S. Agarlita, and L. Tutelea, Springless resonant linear PM oscillomotors, *Record of LDIA-2011*, Eindhoven, the Netherlands, 2011.
11. R.Z. Unger, Linear compressors for clean and specialty gases, *Record of 1998 International Compressor Engineering Conference*, Purdue University, West Lafayette, IN, 1998.
12. http://en.wikipedia.org/wiki/File:JFK_AirTrain.agr.jpg
13. http://en.wikipedia.org/wiki/File:Linimo_approaching_Banpaku_Kinen_Koen,_towards_Fujigaoka_Station.jpg
14. http://en.wikipedia.org/wiki/File:A_maglev_train_coming_out,_Pudong_International_Airport,_Shanghai.jpg
15. http://en.wikipedia.org/wiki/File:JR-Maglev-MLX01-2.jpg
16. http://www.baldormotion.com/products/linearproducts/lmcf.asp
17. http://www.baldormotion.com/products/linearproducts/lmic.asp
18. http://www.baldor.com/support/Literature/Load.ashx/BR1202-G?LitNumber=BR1202-G
19. http://www.solenoids.com/linear_solenoids.html
20. http://spinoff.nasa.gov/spinoff2002/er_9.html
21. http://powertime.en.alibaba.com/viewimg/picture.html?picture = http://i00.i.aliimg.com/photo/v1/424513460/Handheld_speaker_mic_Water_resistant_IP54_for.jpg

3 Linear Induction Motors
Topologies, Fields, Forces, and Powers Including Edge, End, and Skin Effects

This chapter deals with flat linear induction motor (LIM) topologies, double sided and single sided, with their windings and field technical theories accounting for edge, longitudinal end, and skin effects, with the scope of providing a deep understanding of flat LIMs performance and limitations and, also, preparing analytical expressions for equivalent circuit models of LIMs to be developed and then used to investigate the dynamics and control of LIMs (Chapter 4).

3.1 TOPOLOGIES OF PRACTICAL INTEREST

As seen in Chapter 2, LIMs are counterparts of rotary induction motors. They may be obtained by "cutting" and unrolling the rotary induction machines to yield flat, single-sided topologies (Figure 3.1a), where the cage secondary may be used as such or replaced by an aluminum sheet placed between two primaries to make the double-sided LIM (Figure 3.1b).

The passive long secondary (track) is fixed for short-primary-mover configurations (Figure 3.1) and secures a reasonable low cost track/km, while it imposes transfer of full propulsion electric power on board of vehicle, together with the static power converters and their control. However, the vehicle becomes independent, and a faulty vehicle may be put away to let the track available for the healthy vehicles.

Though the research on LIMs has started with double-sided configurations, it ended up so far in using single-sided moving primary topologies for transportation (on wheels or on magnetic suspension) in commercial applications. For long tracks, an aluminum sheet on solid mild iron (made of 3–4 "thick" laminations) secondary (Figure 3.1c) may be preferable on account of lower cost, at the price of lower LIM efficiency and power factor, with corresponding consequences in increased inverter kVA ratings and costs.

The ladder secondary allows for a lower (up to 10–12 mm) magnetic airgap (in transportation) than the aluminum sheet on iron secondary (16–18 mm). The double-sided LIM implies for transportation applications a total magnetic airgap of 26–40 mm, which, unless the pole pitch is high, leads to small power factors (less than 0.4).

Besides propulsion force (thrust) F_x, normal force F_n is exerted on the LIM primary. But for double-sided LIM (Figure 3.1b), the net normal force on secondary is zero for symmetrical position of the latter in the airgap. On the contrary, the single-sided LIM is characterized by a net nonzero normal force (on secondary and primary)—$F_n/F_x \approx$ 3–4—which may be used for active magnetic suspension by adequate control in MAGLEVs. The normal force is notably higher for ladder secondary single-sided LIMs ($F_n/F_x \approx$ 8–10), providing perhaps all necessary magnetic suspension for an acceleration of 1 m/s in vehicles where the LIMs represent 10% of loaded vehicle weight.

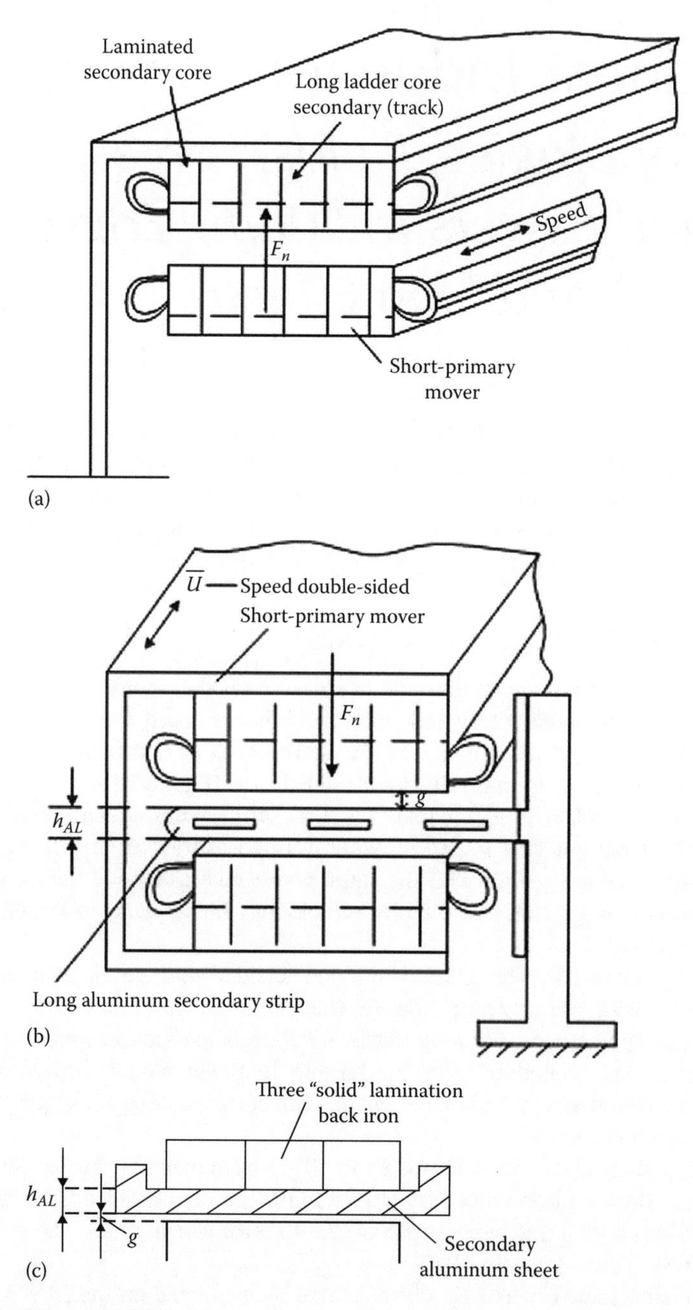

FIGURE 3.1 LIM flat topologies with short-primary mover: (a) single-sided with cage-core secondary (track), (b) double-sided with aluminum-strip secondary (track), and (c) single-sided with aluminum sheet on iron secondary.

For the double-sided LIM (DLIM), with close to zero normal force, separate dc-fed controlled electromagnets acting on a separate solid iron track are required for active magnetic suspension in LIM-MAGLEVs.

There are also tubular LIMs—as introduced in Chapter 2—but they are destined to low excursion (lower than 1.5–2 m) and will be treated separately in Chapters 4 and 5 as low-speed LIMs.

Long-primary active track with wound-secondary short movers have also been proposed [1], but the development of synchronous long-primary (active guideway) linear motors with superior performance (due to dc mover magnetization) has rendered them less practical.

Considering flat, double-sided, and single-sided short-primary LIMs, we should notice that their cores are to be laminated (even coarsely) as in rotary IMs. The LIM short-primary cores are open magnetic circuits along the direction of motion, and thus the three-phase ac windings located in their uniform slots have special peculiarities, though their purpose is the same as in rotary IMs: to produce traveling magnetomotive forces (mmfs) and thus close to traveling airgap magnetic flux density. Basically, there are two main types of LIM practical windings:

- Single-layer $2p$ (even) pole windings (Figure 3.2a)
- Double-layer $2p+1$ (odd) pole windings with half-filled end slots (Figure 3.2b)

While Figure 3.2 illustrates $q_1=1$ slot/pole/phase, full-pitch-coil three-phase LIM windings, for large (transportation) LIMs, $q_1=2,3$ and short span coils (for two-layer windings) may be used to produce closer to traveling mmf, as in rotary IMs (with lower space harmonics).

As the power factor of LIMs, due to larger airgap/pole pitch ratios than in rotary IMs, tends to be lower, the magnetization current in p.u. tends to be larger—around or even higher than $I_{rated}/\sqrt{2}$—and thus, if end-turns relative length (dependent on l_{stack}/pole pitch) tends to be large in some applications, the two-layer chorded-coil windings are to be preferred as their end turns are shorter (10%–20%).

To further shorten the end-turns length, some peculiar windings such as those in Figure 3.3 may be adopted.

The three-layer winding with chorded coils (Figure 3.3a) provides a 0.867 fundamental winding factor and uses the primary slots less efficiently but shows notable shorter end coils while, however, the phase resistance and self and mutual inductances are slightly nonsymmetric. The tooth-coil winding in Figure 3.3b has $q=1/2$ slots/pole and phase and poor fundamental winding factor, but it shows very short end connections.

Finally, the tooth-coil winding in Figure 3.3c, with 12 coils, in a different connection can produce a 10-pole mmf, with a good fundamental winding factor (around 0.933), but not all 10 poles are symmetric. Two such primaries, phase shifted properly, could improve the 10-pole mmf symmetry and be satisfactory for some applications, though higher order space harmonics still show up in the mmf, in a double-sided primary LIM configuration.

The winding in Figure 3.3c is typical for linear PM synchronous motors, as it is easy to manufacture and the end coil length is low; they have a high fundamental winding factor. The fundamental winding factor is the ratio between real emf and its ideal value when the emf in all coils in series per phase would add arithmetically. Reducing the subharmonics in tooth-coil (fractionary) windings with N_{s1} and $2p$ slightly different from each other may be tried by dividing some of the slot volume into 2, 3 coils of different phases. The ac windings so far are three phase, but split-phase (or two-phase) windings may also be envisaged for small thrust (power) applications as for rotary IMs. Two-phase symmetric windings may be preferable when $2p=2, 4$ in low-speed applications to avoid phase current asymmetry.

The topologies of practical interest so far warrant remarks as follows:

- The magnetic airgap (g_m) is larger than the mechanical airgap (g) in DLIMs and in single-sided LIMs with conductive sheet on iron secondary thickness:

$$g_m = K \times g + h_{AL} \tag{3.1}$$

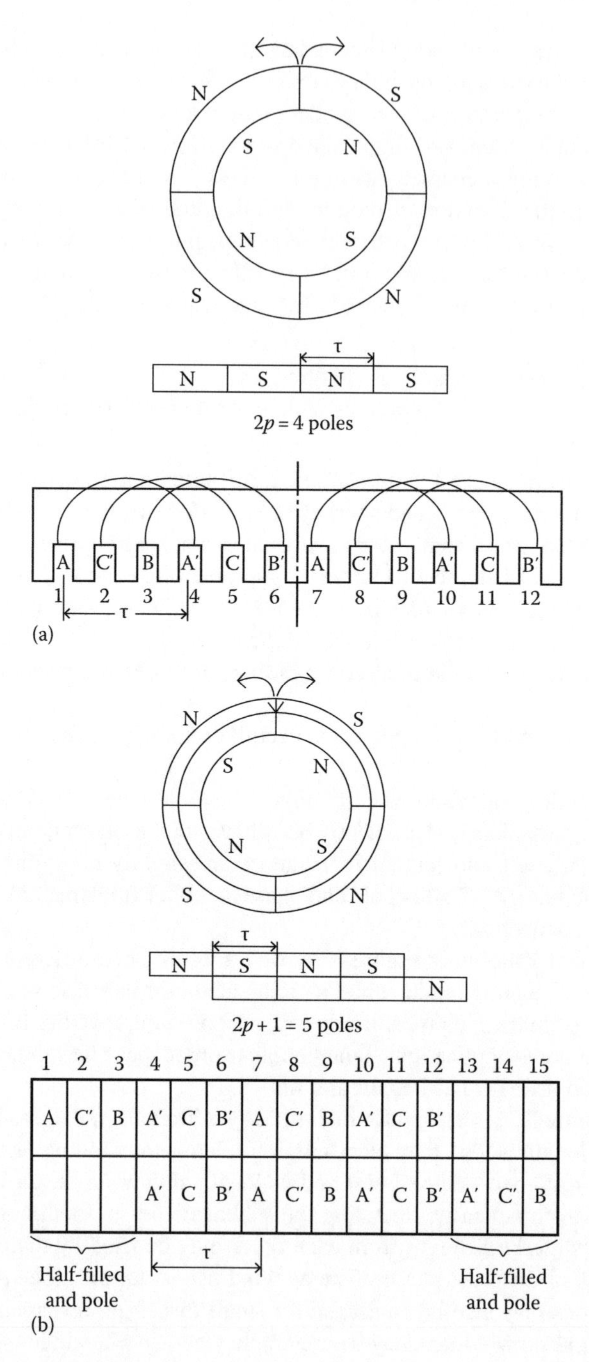

FIGURE 3.2 Single- and two-layer three-phase LIM windings: (a) $2p=4$, $q_1=1$ slot/pole/phase, single layer; (b) $2p+1=5$, $q_1=1$, double layer.

$K=1$ for single-sided LIM (SLIM) and $K=2$ for DLIM.

- The magnetic circuit of LIMs is open along the direction of motion (it has limited length); however, the Gauss flux law applies, so there may be a redistribution of airgap in the back-iron (yoke) flux densities, in contrast to rotary IMs where circumferentially the magnetic circuit is closed.

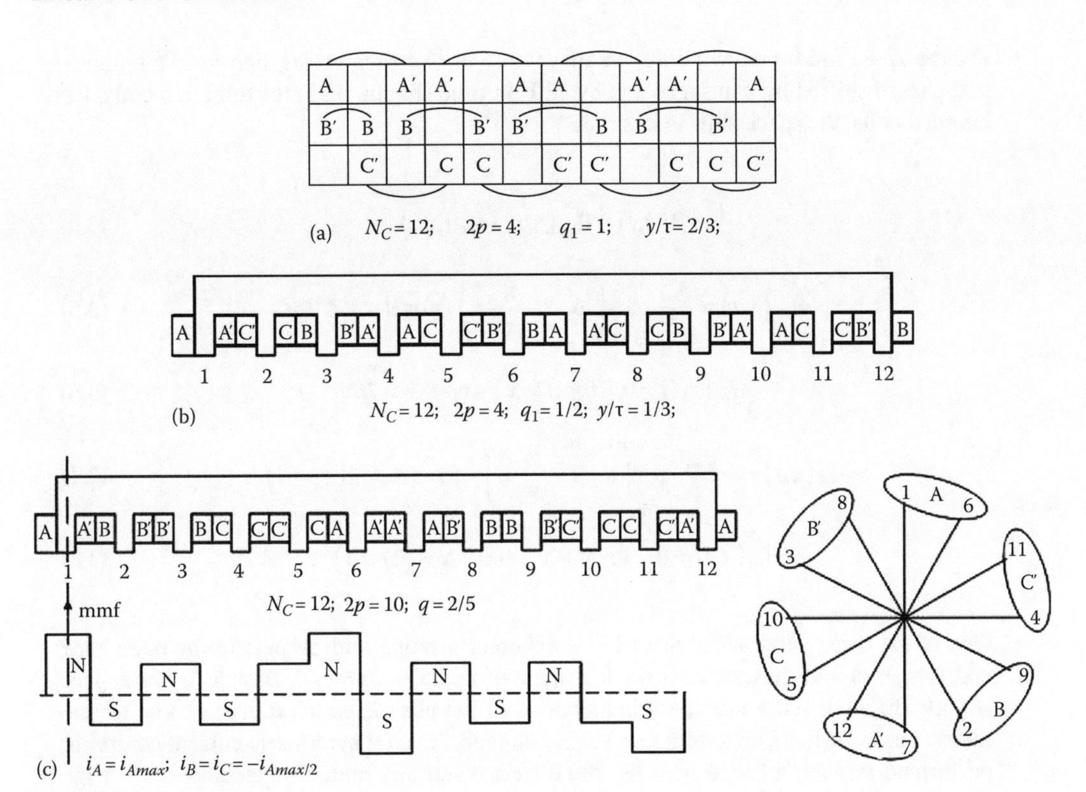

FIGURE 3.3 Short end-coil LIM windings: (a) $N_C=12$, $2p=4$, $q_1=1$, $y/\tau=2/3$, 3 layers and (b) $N_C=12$, $2p=4$, $q_1=1/2$, $y/\tau=1/3$—tooth coils; (c) $N_{s1}=12$, $2p=10$, $q=2/5$.

- The primary mmf fundamental may be described by a traveling wave, along the primary length, as in rotary IMs [2]:

$$\theta(x,t) = \theta_{m1} \cdot \cos\left(\omega_1 \cdot t - \frac{\pi}{\tau}x\right), \quad \text{for } 0 < x < 2p\tau \tag{3.2}$$

$$\theta(x,t) = 0 \quad \text{for } x < 0 \quad \text{and} \quad x > 2p\tau$$

$$F_{m1} = \frac{3\sqrt{2} \cdot W_1 \cdot I_1 \cdot k_{\omega_1}}{\pi p} \tag{3.3}$$

where

p is pole pairs
W_1 is turns per phase in series (for one current path only)
$\omega_1 = 2 \cdot \pi \cdot f_1, f_1$ is the primary frequency, τ is the pole pitch or half the period of primary mmf wave, I_1 is the rms value of primary phase current

- Formulae (3.2) through (3.3) are valid strictly for a $2p$ (even) pole-count windings (Figure 3.2a).

- For the $2p+1$ (odd) pole winding, we may consider two such waves, halved in amplitude and phased shifted by τ in space and by 180° in time (Figure 3.2b) for full pitch coils; for chorded coils, the space shift is less than τ

$$\theta(x,t) = \theta_{xu}(x,t) + \theta_{xl}(x,t) \tag{3.4}$$

$$\theta_{xu}(x,t) = \frac{F_{m1}}{2} \cdot \cos\left(\omega_1 \cdot t - \frac{\pi}{\tau} \cdot x\right) \quad \text{for } 0 < x \le 2p\tau \tag{3.5}$$

$$\theta_{xu}(x,t) = 0 \quad \text{for } 0 > x \quad \text{and} \quad x > 2p\tau \tag{3.6}$$

$$\theta_{xl}(x,t) = -\frac{\theta_{m1}}{2} \cdot \cos\left(\omega_1 \cdot t - \frac{\pi}{\tau} \cdot x\right) \quad \text{for } \tau \le x \le (2p+1) \cdot \tau \tag{3.7}$$

$$\theta_{xu}(x,t) = 0 \quad \text{for } \tau > x \quad \text{and} \quad x > (2p+1) \cdot \tau \tag{3.8}$$

- One of the three windings is placed in a different position with respect to the open core end (phase c) and thus, at least for low number of poles $2p=2$, 4, 6 or 5, 7, we expect slightly different self and mutual inductances of the phases, which translates into asymmetric phase currents for symmetric supply voltages. Even if symmetric current control is performed through a PWM inverter, the different self and mutual inductances will trigger slight thrust pulsations. This effect, present from zero speed, may be called static end effect. In a multiple unit, we may envisage three primary modules where the order of phases in slots is "rotated" and thus, for their series connection phase by phase, the phase currents will become symmetric; also, the total thrust pulsations are reduced.

  It may be argued that by adopting a symmetric two-phase winding with $2p=2$, 4, the phase currents will be symmetric and thus the static end effect is zero; true, but we need a two-phase dedicated inverter, which shows a smaller number of nonzero voltage vectors and thus higher current ripple is expected. Even so, for $2p=2$, 4 poles in small thrust applications, the two-phase winding seems the first option, which justifies building a dedicated two-phase inverter and control system.
- As is well known, the back-iron flux, in the primary and secondary of rotary IMs, is half the airgap pole flux Φ_{1p}:

$$\Phi_{ys} = \frac{\Phi_{1p}}{2} \tag{3.9}$$

For LIMs, the maximum back-iron flux becomes larger, up to almost equal to the pole flux (airgap flux per pole) due to airgap flux redistribution to suit Gauss flux law over the limited core longitudinal extension of primary:

$$B_{g1}(x,t) = \frac{\mu_0 \cdot F_{m1} \cdot \cos\left(\omega_1 \cdot t - \frac{\pi}{\tau} \cdot x\right)}{K_C \cdot g_m \cdot (1+k_s)} \tag{3.10}$$

$$B_{ys}(x,t) = \frac{1}{h_{ys}} \cdot \int_0^x B_{g1}(x,t) = \frac{B_{g1} \cdot \tau}{\pi \cdot h_{ys}} \cdot \left(\sin(\omega_1 \cdot t) - \sin\left(\omega_1 \cdot t - \frac{\pi}{\tau} \cdot c\right)\right) \tag{3.11}$$

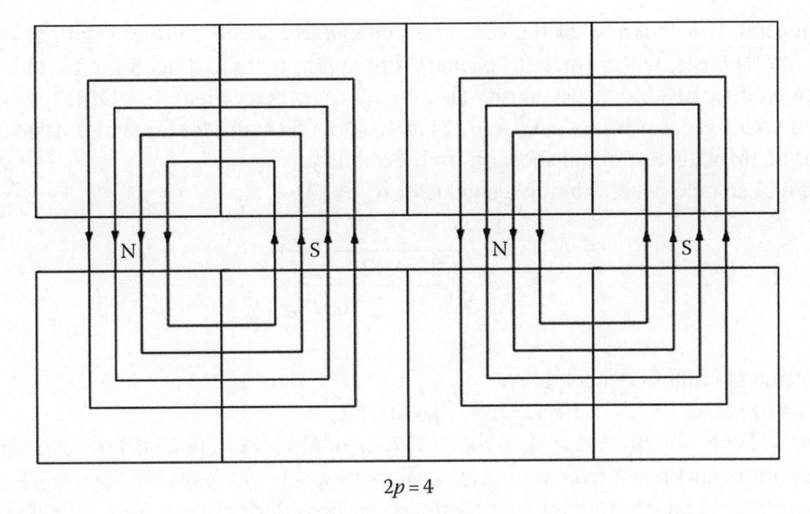

FIGURE 3.4 DLIM, zero secondary conductivity (ideal no load) flux lines produced by primary traveling mmf ($2p=4$ poles).

Even for a pure traveling airgap flux density, which, for zero secondary conductivity, fulfills Gauss flux law, the back-iron flux density B_{ys} has an additional component, and thus for $x=\tau$, 3τ, 5τ, B_{ys} is $B_{ys\,max} = (2 \cdot B_{g1} \cdot \tau)/(\pi \cdot h_{ys})$ instead of $(B_{g1} \cdot \tau)/(\pi \cdot h_{ys})$ as in rotary LMs.

Figure 3.4 illustrates the fact that the total flux per end pole is forced to flow through the back iron as its flux lines do not close through the air zones outside the primary (longitudinal core ends). A similar assertion is valid for the central poles of the $2p+1$ pole primaries.

- The almost doubling of back-iron flux density in some points in comparison with rotary LIM has notable design consequences in low-pole-count LIMs ($2p=2$, 4 or $2p+1=5$) in the sense of doubling the back-iron depth for given flux density in the primary; for the single-sided LIM, the solid back iron of the secondary has to handle also almost all pole flux, which means a much higher degree of magnetic saturation and skin effects, to be considered in any realistic design.

As the maximum yoke flux density appears at $x=3\tau$, 5τ, it suggests avoiding drawing holes in those regions for stacking the laminations or cooling, etc.

3.2 SPECIFIC LIM PHENOMENA

3.2.1 Skin Effect

The secondary frequency f_2 of the secondary emfs and their currents may be calculated as a function of slip S, as for the rotary IMs:

$$f_2 = S \cdot f_1 \tag{3.12}$$

with

$$S = \frac{(U_s - U)}{U_s}; \quad U_s = 2\tau f_1 \tag{3.13}$$

The ideal synchronous speed U_s (or the mmf wave speed) is, as for rotary IMs, $U_s=2 \cdot \tau \cdot f_1$, where U is the linear speed opposite to U_s, in short (moving) primary LIMs.

As, in general, the thickness of the secondary conductive sheet is rather small, to reduce long secondary (track) costs, with respect to primary slot depth, there is a need for a rather large slip, 5%–10%, to produce notable thrust density (1–2 N/cm^2 of primary area). In a DLIM, the aluminum secondary thickness (per primary side) $h_{AL}/2$ is up to about 6–8 mm for large high-speed LIMs and so is h_{AL} for SLIM with aluminum sheet on iron secondary.

The depth of ac field penetration in a conductor δ_{AL} is [2]

$$\delta_{AL} = \sqrt{\frac{2}{S \cdot 2\pi \cdot f_1 \cdot \mu_0 \cdot \sigma_{AL}}} \tag{3.14}$$

where σ_{AL} is the aluminum conductivity.

For $S \cdot f_1 = 10$ Hz, $\sigma_{AL} = 3.33 \times 10^7\ \Omega^{-1}\ \text{m}^{-1}$, $\delta_{AL} = 0.0276$ m.

As δ_{AL} at 10 Hz is notably larger than the thickness of aluminum plate (sheet), the skin effect is not very important (as known from rotary IMs; $K_R \approx 1 + \xi = 1 + h_{AL}/\delta_{AL}$, because $h_{AL}/\delta_{AL} \ll 1$).

The depth of field penetration may be corrected by considering the traveling field (and not the ac field characteristic to rotary IM slots) variation with x in a simplified bidimensional airgap field model where δ_{AL} would be replaced by δ'_{AL}:

$$\delta'_{AL} = \text{Re}\left(\frac{1}{\sqrt{(\pi/\tau)^2 + j \cdot S \cdot f_1 \cdot 2\pi \cdot \mu_0 \cdot \sigma_{AL}}}\right) \tag{3.15}$$

For high pole pitches ($\tau > 0.1$ m) the first term under the square root in (3.15) is rather small but it may still be considered, for a few percent better precision. Now the aluminum skin effect coefficient $K_{R\,skin}$ of rotary IMs may be applied [2]:

$$K_{R\,skin} = \xi \frac{\sinh(2\xi) + \sin(2\xi)}{\cosh(2\xi) - \cos(2\xi)} \tag{3.16}$$

$$\xi = \frac{h_{AL}}{\delta'_{AL}} \tag{3.17}$$

In essence, the current density crowds closer to the aluminum sheet surface, perpendicular to the airgap normal flux density: its phase is also shifted along the aluminum sheet depth. The h_{AL} is replaced by $h_{AL}/2$ for the DLIM in (3.16) because the magnetic field penetrates the aluminum sheet from both sides (Figure 3.5).

If the SLIM secondary back iron is solid, its field depth of penetration at $f_2 = S \cdot f_1$ is notably smaller unless the latter gets heavily saturated:

$$\delta'_{iron} = \text{Re}\left(\frac{1}{\sqrt{(\pi/\tau)^2 + j \cdot S \cdot f_1 \cdot 2\pi \cdot \mu\left(B_{t\,iron}\right) \cdot \sigma_{iron}}}\right) \tag{3.18}$$

For same $S \cdot f_1 = 10$ Hz, $\tau = 0.15$ m, and $\sigma_{iron} = 3.5 \times 10^6\ (\Omega\ \text{m})^{-1}$, even with $\mu_{iron} = 60 \cdot \mu_0$ (at $B_{t\,iron} \approx 2$ T, $H_{iron} = 26{,}540$ A/m), $\delta'_{iron} = 11$ mm.

B_t is the tangential flux density on the back-iron upper surface.

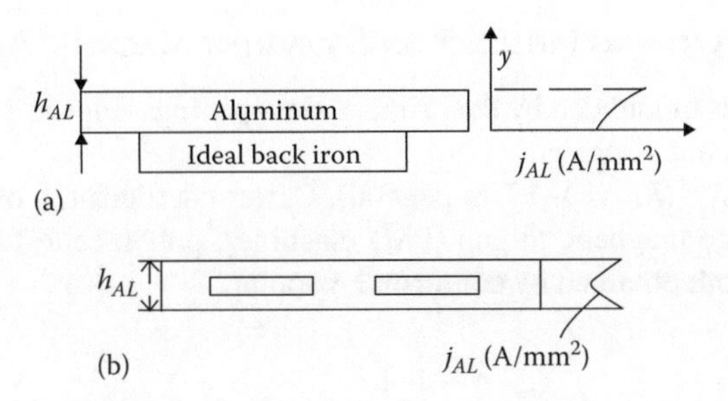

FIGURE 3.5 Current density amplitude variation in aluminum versus aluminum sheet thickness: (a) SLIM and (b) DLIM.

As the approximate back-iron flux is about the flux per pole, we have

$$B_{t\,max} \cdot \delta'_{iron}\left(1 - e^{-1}\right) \approx B_{g1} \cdot \frac{2}{\pi} \cdot \tau \tag{3.19}$$

Consequently, B_{g1}, $B_{t\,max} \cdot \delta'_{iron}\left(1 - e^{-1}\right)$, the peak airgap flux density is $B_{g1} = 0.209$ T. This is a rather small value. Consequently, the back iron should be divided into a few (3–4) sections or "laminations" (Figure 3.6) so as to reduce the effective electric conductivity of iron and thus increase the field penetration depth to 20 mm or more and to provide for a load airgap flux density of at least 0.4 T.

Note: For ladder secondary, if the secondary iron core is laminated, the skin effect has to be treated as in rotary IMs [2]. This rationale leads to the conclusion that the skin effect produces a decrease in the equivalent conductivity $\sigma_e = \sigma / K_{R\,skin}$. The secondary leakage inductance as a consequence of skin effect (due to secondary iron eddy current density phase shifting, along iron depth) may be neglected for DLIMs but not for SLIMs.

3.2.2 Large Airgap Fringing

At least in transportation applications, the magnetic airgap g_m per pole pitch τ ratio is rather large in comparison with rotary IMs. In a simplified $2 \cdot D(x, y)$ field theory, the traveling mmf varying with $e^{j \cdot \frac{\pi}{\tau} \cdot x}$ will lead to a vertical (along z) variation with $\sinh(\pi/\tau)y$ and $\cosh(\pi/\tau)y$. In essence, the equivalent airgap is approximately increased to $g_{mec} = g_m \cdot K_{Fg}$, [3–7]:

$$K_{Fg} \approx \frac{\sinh(\pi/\tau) \cdot g_m}{(\pi/\tau) \cdot g_m} \geq 1 \tag{3.20}$$

A typical 2D field distribution of airgap flux lines for a DLIM is shown in Figure 3.6, where both skin effect and large airgap fringing are visible.

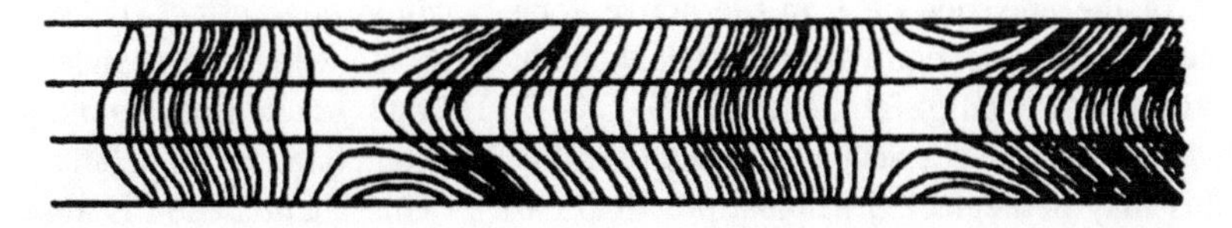

FIGURE 3.6 2D airgap field in DLIM under load (with aluminum sheet currents). (After Nasar, S.A. and Boldea, I., *Linear Motion Electromagnetic Devices*, Taylor & Francis, New York, 2001.)

3.2.3 Primary Slot Opening Influence on Equivalent Magnetic Airgap

The airgap flux density is modulated by the primary slot openings (Figure 3.7a); an equivalent effect consists of a larger constant airgap g_e.

The ratio $g_{mec}/g_m = K_C$ (K_C=1.1–1.7 in general), Carter coefficient, a one-century-old concept, revised recently for large magnetic airgap (PM) machines, puts face to face the old and a newly derived formula [5]—both obtained by conformal mapping:

$$K_c = \frac{g_{mec}}{g_m} = \frac{1}{1-\sigma_c \cdot \dfrac{b_s}{\tau_s}}; \quad \tau_s, \text{slot pitch} \tag{3.21}$$

$$\sigma_c = \frac{2}{\pi} \cdot \tan^{-1}\left(\frac{b_s}{2 \cdot g_m}\right) - \frac{1}{\pi} \cdot \frac{2 \cdot g_m}{b_s} \cdot \ln\left(1 + \left(\frac{b_s}{2 \cdot g_m}\right)^2\right) \tag{3.22}$$

and

$$K_{c\,new} = 1 + \frac{\tau_s}{2\pi g_m}\left[(1+\alpha) \cdot \log(1+\alpha) + (1-\alpha) \cdot \log(1-\alpha)\right] \tag{3.23}$$

$$\alpha = \frac{b_s}{\tau_s}; \quad b_s, \text{slot opening}$$

Reference [5] establishes regions where one or the other formula is preferable (Figure 3.7) but, in general, for large g_m/τ_s and b_s/τ_s values, the new formula (3.23) is more precise.

If the magnetic airgap $g_m = g + h_{AL}$ is large (14–30 mm), then the slots may be open, as long as $b_s < 2 \cdot g_m$, which leads to easy insertion of coils in slots, while K_c is, in general, less than 1.15. For low (0.7–1.2 mm) mechanical airgap (industrial, short-travel machine tool table applications), the ladder secondary is preferable, as it leads to satisfactory efficiency and power factor (both above 0.7) even at speeds below 3 m/s. In such cases, semiclosed slots, at least in the primary, are to be used.

3.2.4 Edge Effects

For the aluminum sheet or aluminum sheet on iron secondaries of LIM, the trajectory of induced secondary currents (by the traveling stator mmf) is closed (*div* $\bar{J} = 0$)—Figure 3.8—and thus a 2D field analysis is required to portray them; only the transverse (J_y) component produces thrust, with J_x producing, in the active zone, only additional losses. Active zone is the one between the primaries (in DLIMs) and between primary and secondary back iron (in SLIMs). The laterally outside zone corresponds to the "end-ring" zone in rotary IMs.

The effect of the presence of J_x in the active zone is called edge effect. It boils down to a decrease in conductive sheet equivalent electrical conductivity K_{tR}, which is dependent on machine geometry and secondary (slip) frequency $f_2 = S \cdot f_1$ in DLIM and, in addition, on magnetic saturation in SLIM. There is also a very small decrease in the magnetization inductance due to edge effect, but this may be neglected in most practical cases. (This phenomenon is absent in ladder secondary SLIMs.) We will rederive here in short the 2D field analysis of edge effects for DLIMs and then generalize it to SLIMs.

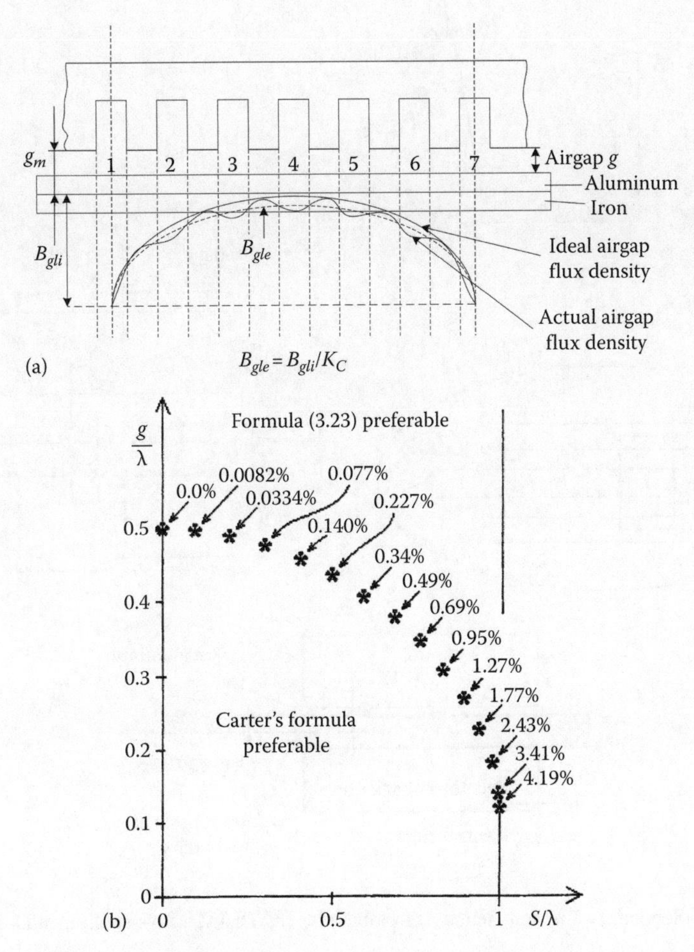

FIGURE 3.7 Slot opening influence on airgap flux density (Carter coefficient): (a) airgap flux density waveform; (b) Carter coefficient formulae: dividing regions. (After Matagne, E., *Electromotion*, 15(4), 171, 2008.)

To separate the edge effect in a 2D field analysis, let us assume that

- The two primaries slotting is replaced by an increased airgap via $K_C > 1$
- The airgap fringing is accounted for by $K_{Fg} > 1$
- So the total equivalent airgap $g_{et} = g_m \cdot K_C \cdot K_{F_g}$
- The stator mmf is represented by its fundamental traveling wave in secondary coordinates

$$\theta_1(x,t) = F_{1m} \cdot e^{-j\cdot\left(S\cdot\omega_1\cdot t - \frac{\pi}{\tau}\cdot x\right)} \tag{3.24}$$

- $S \cdot \omega_1$ is the secondary angular frequency of induced currents
- We may separate the time dependence as the secondary current density varies sinusoidally in time with secondary (slip) frequency $S \cdot \omega_1$
- The LIM length is so large along motion direction that there is no end effect (to be treated later in this chapter)

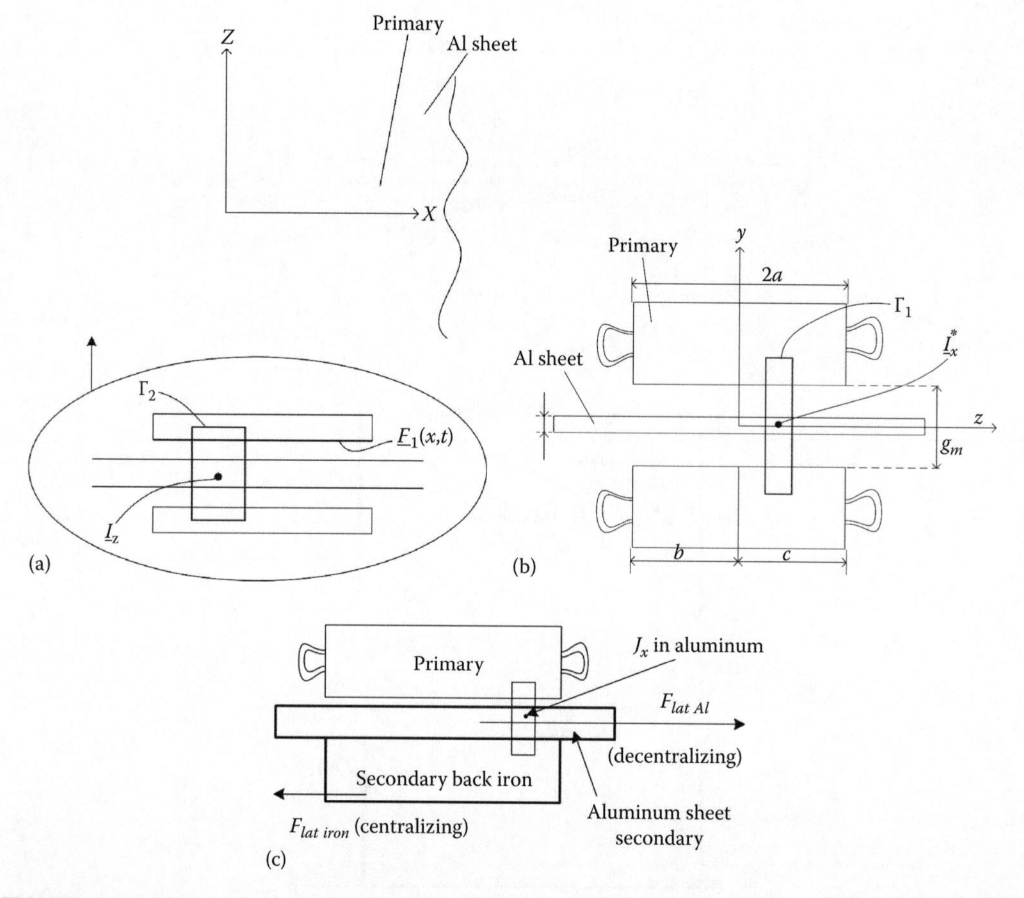

FIGURE 3.8 (a) Secondary current density components, (b) DLIM cross-section, and (c) decentralized SLIM secondary.

- Outside the active zone ($|y| > a_e$, $a_e \approx a + g_{et}$), g_{et} is introduced to account for transversal flux fringing), as $div\,\overline{J}_2 = 0$:

$$\frac{\partial J_{2x}}{\partial x} + \frac{\partial J_z}{\partial z} = 0 \tag{3.25}$$

- At $y = -b, c$, as expected, and $J_z = 0$
- Inside the active zone, we apply Ampere's law along contours Γ_1 and Γ_2 in Figure 3.8:

$$g_{et} \cdot \frac{\partial H_y}{\partial z} = J_{2x} \cdot d_{AL} \tag{3.26}$$

$$-g_{et} \cdot \frac{\partial \left(H_y + H_0\right)}{\partial x} = J_m + J_{2z} \cdot d_{AL} \tag{3.27}$$

where H_y is the only magnetic field component, produced by the secondary currents, while H_0 is the initial airgap field in the absence of secondary currents; J_m is the primary current sheet (in A turns/m):

$$\underline{J}_m = \frac{\partial \underline{\theta}_1(x,t)}{\partial x} \tag{3.28}$$

Adding Faraday's law,

$$\frac{\partial J_{2z}}{\partial z} - \frac{\partial J_{2x}}{\partial z} = -j \cdot S \cdot \omega_1 \cdot \mu_0 \cdot \sigma_{et} \cdot \left(H_y + H_0\right) \tag{3.29}$$

$\sigma_{et} = \sigma / K_{R\ skin}$, accounts for the skin effect in the secondary ((3.16), with $d_{AL}/2$ instead of d_{AL} for DLIM).

H_0 is obtained from (3.27) with $J_{2z} = 0$:

$$H_0 = \frac{j \cdot J_m \cdot e^{-j \cdot \frac{\pi}{\tau} \cdot x}}{(\pi/\tau) \cdot g_{et}} \tag{3.30}$$

Equations 3.26 through 3.29 lead to the equation of magnetic field of secondary currents (H_y):

$$\frac{\partial^2 H_y}{\partial x^2} + \frac{\partial^2 H_y}{\partial x^2} - j \cdot S \cdot \omega_1 \cdot \sigma_{et} \cdot \mu_0 \frac{d_{AL}}{g_{et}} \cdot H_y = j \cdot S \cdot \omega_1 \cdot \sigma_{et} \cdot \mu_0 \frac{d_{AL}}{g_{et}} \cdot H_0 \tag{3.31}$$

Equation 3.31 is valid for $|z| \leq a_e$, that is, in the active zone.

The solution of (3.31) is straightforward:

$$H_y = \frac{-H_0 \cdot \left(j \cdot S \cdot \omega_1 \cdot \mu_0 \cdot \sigma_{et} \cdot d_{AL}/g_{et}\right)}{\left(\left(\pi^2/\tau^2\right) + j \cdot S \cdot \omega_1 \cdot \sigma_{et} \cdot \mu_0 \left(d_{AL}/g_{et}\right)\right)} + A \cdot \cosh\left(\alpha \cdot z\right) + B \cdot \sinh\left(\alpha \cdot z\right) \tag{3.32}$$

with

$$\alpha^2 = \frac{\pi^2}{\tau^2} + j \cdot \omega_1 \cdot S \cdot \mu_0 \cdot \sigma_{et} \frac{d_{AL}}{g_{et}} = \left(\frac{\pi}{\tau}\right)^2 \cdot \left(1 + j \cdot S \cdot G_{et}\right) \tag{3.33}$$

If we define by G_{et},

$$G_{et} = \frac{\mu_0 \cdot \tau^2 \cdot \omega_1}{\pi^2} \cdot \sigma_{et} \cdot \frac{d_{AL}}{g_{et}} \tag{3.34}$$

Then,

$$H_y = \frac{-H_0 \cdot \left(j \cdot S \cdot G_{et}\right)}{\left(1 + j \cdot S \cdot G_{et}\right)} \cdot \left(A \cdot \cosh\left(\alpha \cdot z\right) + B \cdot \sinh\left(\alpha \cdot z\right)\right) \tag{3.35}$$

G_{et} is the so-called goodness factor [3,7] or the equivalent of Reynold's number in electrical terms [1,8].

The goodness factor G_{et} lumps up quite a few geometrical variables, by considering the slot opening, airgap fringing, and secondary skin effects. Thus, G_{et} is a very powerful variable to be used for design.

To continue with the edge effect analysis, let us notice that, for the overhang areas of secondary ($|z| \le a_e$), the current densities retain from (3.35) one of the two last terms:

$$J_{zr} = -C \cdot \sinh\frac{\pi}{\tau}\cdot(c - z), \quad \text{for } a_e < z < c$$

$$J_{xr} = +jC \cdot \cosh\frac{\pi}{\tau}\cdot(c - z), \quad \text{for } a_e < z < c \tag{3.36}$$

$$J_{zl} = -D \cdot \sinh\frac{\pi}{\tau}\cdot(b + z), \quad \text{for } -a_e > z > -b$$

$$J_{xl} = jD \cdot \cosh\frac{\pi}{\tau}\cdot(b + z), \quad \text{for } -a_e > z > -b$$

The r and l correspond to left and right. From the current density continuity conditions at $z=\pm a_e$, we get the constants A, B. C, D.

Also from (3.27),

$$J_{2z} = \frac{-J_m \cdot j \cdot S \cdot G_{et}}{d_{AL}\cdot(1 + j\cdot S\cdot G_{et})}\cdot e^{j\cdot\left(S\cdot\omega_1\cdot t - \frac{\pi}{\tau}\cdot x\right)}\cdot\left(A\cdot\cosh(\alpha\cdot z) + B\cdot\sinh(\alpha\cdot z)\right), \quad \text{for } |z| \le a_e \tag{3.37}$$

$$J_{2x} = \frac{-g_{et}}{d_{AL}}\cdot\alpha\cdot\left(B\cdot\cosh(\alpha\cdot z) - A\cdot\sinh(\alpha\cdot z)\right)\cdot\frac{J_m\cdot j\cdot S\cdot G_{et}}{d_{AL}\cdot(1 + j\cdot S\cdot G_{et})}\cdot e^{j\cdot\left(S\cdot\omega_1\cdot t - \frac{\pi}{\tau}\cdot x\right)} \tag{3.38}$$

As expected $J_{2x}=0$ for $z=0$ (in the center) if $b=c$ (symmetrical placement of secondary inside the primary). For a numerical example characterized by the data, $\tau=0.35$ m, $f_1=173.3$ Hz, $S=0.08$, $d_{AL}=6.25$ mm, $g_{mt}=37.5$ mm, $J_m=2.25\cdot 10^5$ A turns/m, the results in Figure 3.9 were obtained [3].

A few remarks are in order:

- The edge effect produces a nonuniform total airgap flux density along "z" direction, which may lead to a larger magnetic saturation in the primary core left and right edge sides. However, the average airgap flux density is not modified notably.
- Both current density components vary with "z" due to edge effects.
- The global consequences of edge effects can be calculated by comparing the active and reactive powers (by Poynting vector) with and without edge effect ($J_x=0$ or the overhangs are made of a fictitious superconductor, and the secondary is made of aluminum bars).

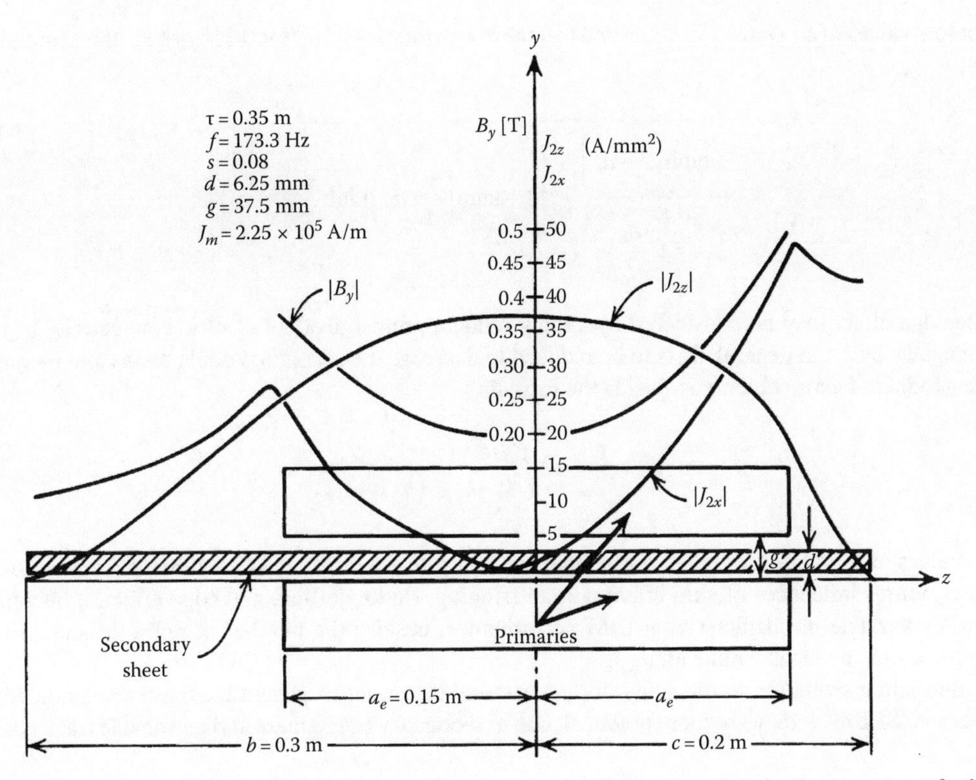

FIGURE 3.9 Airgap flux density and secondary current density components in the presence of edge effect ($b \neq c$). (After Nasar, S.A. and Boldea, I., *Linear Motion Electric Machines*, John Wiley & Sons, New York, 1976.)

Doing this, Ref. [6] has found for $b=c$ (symmetrically placed secondary), for the active power ratio, a reduction coefficient k_{tr} of equivalent conductivity σ_e of aluminum sheet:

$$\sigma_e = \frac{\sigma}{\left(k_{skin} \times k_{tr}\right)} \tag{3.39}$$

$$k_{tr} = \frac{k_x^2}{k_R} \cdot \left[\frac{1 + S \cdot G_{et}^2 \cdot k_R^2 / k_x^2}{1 + S^2 \cdot G_{et}^2}\right] \geq 1 \tag{3.40}$$

$$k_{tx} = \frac{k_R^2}{k_x} \approx 1; \quad k_R = 1 - \text{Re}\left[\left(1 - S \cdot G_{et}\right) \cdot \frac{\lambda}{\alpha \cdot a_e} \cdot \tanh\left(\alpha \cdot a_e\right)\right] \tag{3.41}$$

$$k_x = 1 + \text{Re}\left[\left(S \cdot G_{et} + j\right) \cdot \frac{S \cdot G_{et} \cdot \lambda}{\alpha \cdot a_e} \cdot \tanh\left(\alpha \cdot a_e\right)\right] \tag{3.42}$$

$$\lambda = \frac{1}{1 + \sqrt{1 + j \cdot S \cdot G_{et}} \cdot \tan\left(\alpha \cdot a_e\right) \cdot \tanh\frac{\pi}{\tau} \cdot \left(c - a_e\right)} \tag{3.43}$$

For low values of $S \cdot G_{et}(S \cdot G_{et} << 1)$ and (or) narrow primaries $2 \cdot a_e/\tau < 0.3$, k_{tr} yields the expression

$$k_{tr} = \frac{1}{1 - \dfrac{\tanh\left(\dfrac{\pi}{\tau} \cdot a_e\right)}{\dfrac{\pi}{\tau} \cdot a_e} \Bigg/ \left[1 + \tanh\left(\dfrac{\pi}{\tau} \cdot a_e\right) \tanh\dfrac{\pi}{\tau}(c - a_e)\right]} \geq 1 \quad (3.44)$$

The edge effect may be considered by simply reducing the equivalent electric conductivity of the secondary by k_{tr}; in general, k_{tr} is to be reduced to decrease the secondary Joule losses and increase the goodness factor, which now yields the formula

$$G_e = \frac{G_{et}}{k_{tr}} = \frac{\mu_0 \cdot \tau^2 \cdot \omega_1 \cdot \sigma_{AL} \cdot d_{AL}}{\pi \cdot k_{tr} \cdot k_{skin} \cdot \left(g \cdot k_c \cdot k_{Fr}\right)} \quad (3.45)$$

Values of $k_{tr} \approx 1.3$–1.5 lead to reasonable overhangs ($c - a = \tau/\pi$) and thus limit secondary costs. As G_e lumps influences of skin effect, airgap fringing, stator slotting, and edge effect, it becomes the key variable in characterizing LIM performance, besides the number of poles $2p$ and slip S, frequency ω_1, and stator mmf (or J_{1m}).

But before switching to the study of dynamic end effect, let us generalize these results on edge effects to SLIMs as they are more practical, due to secondary ruggedness and reasonable track costs.

3.2.5 Edge Effects for SLIMs

A SLIM for short excursion (a few meters) may use a ladder-laminated core secondary to obtain high efficiency (above 0.8) and good power factor (above 0.7) for mechanical gaps of less than 1.5 mm. SLIMs do not experience edge effects and should be treated as rotary IMs unless the number of poles $2p = 2, 4$, when the static end effect occurs, as already mentioned in this chapter. For excursions longer than a few meters, the cost of secondary becomes important and thus an aluminum sheet on secondary becomes important and, consequently, an aluminum sheet on iron secondary (Figure 3.10a) is the only economical solution. To increase the contribution of solid back iron of secondary to the thrust, the depth of penetration of the magnetic field in it should be increased, and making it of 3–4 pieces ("laminations") seems the way to do it (Figure 3.10b).

To account for the edge effect in SLIMs with aluminum sheet on iron, we should first notice that

- The secondary back-iron width is about equal to the "stack width" of primary, $2a$, while the aluminum sheet (flat or bent toward back-iron sides for larger ruggedness) is wider, to produce enough thrust. Also, to counteract the lateral expulsion force on the aluminum sheet by a self-centering force due to the primary and secondary iron cores, in case of off-center (laterally) placement of the secondary with respect to primary, the secondary back-iron width equals the primary stack length.
- The airgap flux per pole has to close through the back iron along (basically) the field penetration depth δ_{iron} (3.15); fortunately, the division of secondary back iron into 3–4 "laminations"—Figure 3.10b leads to a pronounced edge effect in iron, which reduces the iron equivalent electrical conductivity:

$$\sigma_{ie} = \left(\frac{\sigma_{iron}}{k_{tri}}\right) \quad (3.46)$$

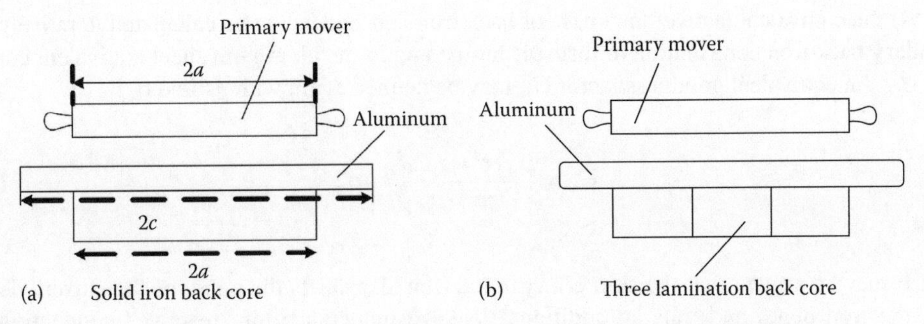

FIGURE 3.10 SLIM secondaries: (a) with solid back iron and (b) made of three "laminations."

The k_{tri} coefficient for edge effect for i back-iron laminations may be considered based on Equation 3.40 with $c=a_e$ and a_e/i instead of a_e:

$$k_{tri} \approx \frac{1}{1-\dfrac{\tanh\left(\dfrac{\pi}{\tau}\cdot\dfrac{a_e}{i}\right)}{\dfrac{\pi}{\tau}\cdot\dfrac{a_e}{i}}} \tag{3.47}$$

Now, the skin effect (3.15) in iron may be calculated with σ_{ie} and not with σ_{iron}.

Also, the back iron adds a "virtual length" to the airgap because it is very heavily saturated; the increased airgap g_e is

$$g_{se} \approx (1+k_{sat})\cdot(g+h_{AL})\cdot k_c\cdot k_{Fr} \tag{3.48}$$

The saturation coefficient k_{sat} in (3.43) stems from the reluctance ratio (of back iron and of airgap pole flux paths):

$$k_{ss} \approx \frac{2\tau^2}{\pi^2}\cdot\frac{\mu_0}{2\cdot(g+h_{AL})\cdot k_c\cdot\delta_i\cdot\mu_{iron}} \tag{3.49}$$

Let us note that δ_i (3.15) is dependent on μ_{iron}, which in turn depends on flux density B_{ti} tangential on the iron upper surface:

$$B_{ti}\cdot\delta_i \approx B_g\cdot\frac{\tau}{\pi} \tag{3.50}$$

This is an alternative calculation of B_{ti} for given B_{g1} (airgap flux density at rated (or peak) thrust) or $B_{ti}\approx 2$ T (or in general, as required).

Now, again, we may lump up the iron contribution to thrust by an increase in the equivalent conductivity of secondary σ_{se}:

$$\sigma_{se} = \frac{\sigma_{AL}}{k_{tr\,AL}\cdot k_{skin}}\cdot\left(1+\frac{\sigma_i}{\sigma_{AL}}\cdot\frac{\delta_i\cdot k_{tr\,AL}\cdot k_{skin}}{k_{ti}\cdot d_{AL}}\right)=\frac{\sigma_{AL}}{k_{tr\,AL}\cdot k_{skin}}(1+k_{is}) \tag{3.51}$$

In (3.51) the equivalent (active) thickness of back iron is δ_i and has to be calculated iteratively. The secondary back-iron contribution to thrust is lumped up in the aluminum sheet equivalent conductivity σ_{se}. An equivalent goodness factor G_{se} may be defined again with g_{se} and σ_{se}:

$$G_{se} = \frac{\mu_0 \cdot \tau^2 \cdot \omega_1}{\pi^2} \cdot \frac{d_{AL}}{g_{se}} \cdot \sigma_{se} \tag{3.52}$$

Note: It may be argued that the skin effect in the iron also shifts the phase of the current density along the iron depth and thus an additional leakage inductance (or "reactive" conductivity) is added.

3.3 DYNAMIC END-EFFECT QUASI-ONE-DIMENSIONAL FIELD THEORY

Let us consider a $2p$ pole single-layer winding short-primary long-secondary DLIM whose primary is moving at speed $U < U_s$; U_s—the traveling primary mmf. Speed ($U_s = 2 + \tau . f_1$) (Figure 3.11). The phase shifting of eddy current in secondary iron may be considered through a leakage reactance equal to its resistance: $S \cdot \omega_1 \cdot L_{2li} = R_{2i}' \cdot R_{2i}'$ is related to total secondary resistance as in (3.46) by the coefficient $k_{i\sigma}$. Also notice that the lateral force (Figure 3.8c) $j_x \times B_y$ by aluminum currents becomes decentralizing when the secondary is placed asymmetrically in the lateral direction (0y): For SLIM the solid back iron adds two lateral forces, one decentralizing (due to its eddy currents) and one centralizing (due to iron).

The contour $c_0(t_0)$ of secondary is in time in positions $c_1(t_1)$, $c_2(t_2)$, $c_3(t_3)$, $c_4(t_4)$, with respect to the moving primary.

In position c_0, there is no magnetic flux and so no induced voltage or current occurs.

When we consider contour c_2, there is a strong variation of magnetic flux in it and thus additional currents occur along the contour (secondary). As the primary moves and contour c_2 position is reached, the current induced in position c_1 has already decreased by the secondary electrical time constant (calculated as for a rotary IM) T_{e2}:

$$T_{e2} \approx \frac{L_m(g_m)}{R_2'} = \frac{G_e}{\omega_1} \tag{3.53}$$

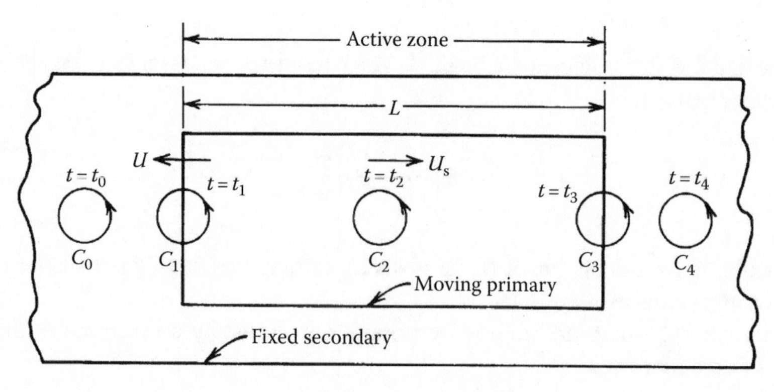

FIGURE 3.11 Dynamic end effect. (After Nasar, S.A. and Boldea, I., *Linear Motion Electric Machines*, John Wiley & Sons, New York, 1976.)

where

L_m is the magnetization inductance (the secondary leakage inductance is almost zero for aluminum sheet secondary once the airgap fringing effect has been considered).

R_2' is the secondary equivalent resistance reduced to the primary; the expressions of L_m and R_2' are calculated as if the machine were rotary with $2p$ poles but identical geometry of primary and secondary.

It is a known fact that if the speed is high, $U_{rated} > 20$ m/s (100 m/s, for a 360 km/h in high-speed trains), the mechanical and the magnetic airgap g_m is large, but still the conventional performance (without dynamic end effect) increases with higher goodness factor (G_e)—as in rotary IMs. A $G_e \geq 20$ at $f_1 = 175$ Hz ($\tau = 0.35$ m, $2p = 10$) should be typical for a high-speed application.

But in this case, $3T_{e2} = (3\times 20)/(2\cdot\pi\cdot 175) = 0.055\,\text{s}$. For a slip $S = 0.08$, the LIM speed would be

$$U = 2\tau f_1(1-S) = 2\cdot 0.35\cdot 175\cdot(1-0.08) = 112.7\,\text{m/s} \tag{3.54}$$

The length traveled during three time constants ($3T_{e2}$) would be $L_{lef} = U\times 3T_{e2} = 112.7\times 0.055 = 6.198$ m; the length of the primary $L = 2p\tau = 10\times 0.35 = 3.5$ m. Consequently, the additional (dynamic end-effect-produced) entry secondary currents would not die out before, but rather after, the LIM primary exit end. It means a notable dynamic end effect or a high-speed LIM. But in the position of exit-end c_3, new additional dynamic end-effect currents occur. However, these currents would die out with a different (much smaller) time constant introduced here as T_{exit}:

$$T_{exit} = \frac{L_{m\,exit}(\tau/\pi)}{R_2'} = \frac{G_{exit}}{R_2'};\quad g_{m\,exit} = \tau/\pi \tag{3.55}$$

The equivalent airgap behind the primary (Figure 3.11) corresponds to an average flux line length in air, which is approximately τ/π, because the current periodicity is still τ at initiation (there are two primaries in DLIMs); for a SLIM $g_{m\,exit} = \tau/\pi$, still, as half the travel of flux lines is considered to take place in the back iron.

For the same example (with $g_m = 37.5$ mm), the new time constant is $3\cdot T_{exit} = 3\cdot T_{e2}\cdot(g_m\cdot\pi)/\tau = 0.055\cdot 37.5/(2\cdot 350)\times\pi = 0.00925\,\text{s}$; the corresponding exit field "tail" length would be $L_{exit} \approx 3T_{exit}\cdot U = 0.0925\times 112.7 = 1.0425$ m.

Now these additional secondary currents are due to the motion of a short primary in front of a long secondary, and thus the dynamic end effects manifest themselves as follows:

- The existence of additional secondary current density components both within the active zone (decaying with the T_{e2} time constant) and in the primary tail ($x > 2p\tau$, decaying with $T_{exit} < 3\cdot T_{e2}$ in general).
- Consequently, considering the primary core to be infinitely long would exaggerate grossly the exit dynamic end effect, though most 1–2 DLIM field theories rely on it [4,7–9]. It may be safe to say that this exit-end part of dynamic end effect (in the tail of LIM primary) produces only additional secondary losses (besides a small reluctance braking force [3]).

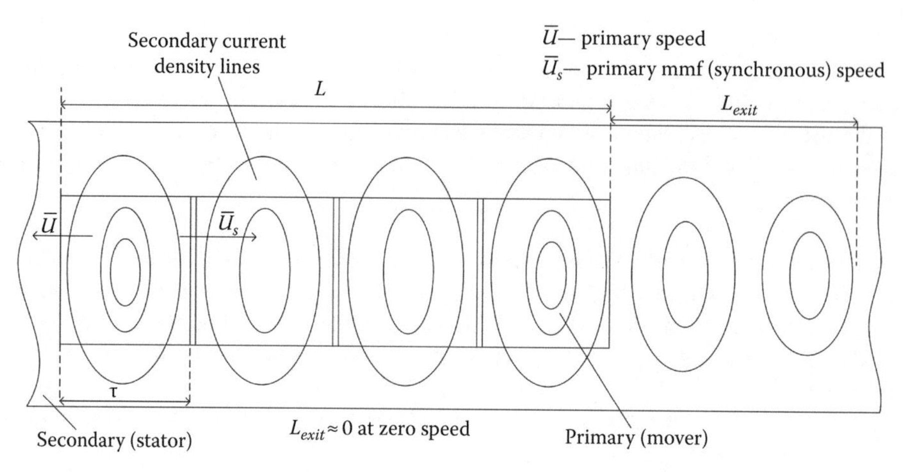

FIGURE 3.12 Secondary current density lines in a high-speed LIM ($L_{exit} >> \tau$).

- The dynamic end effect in the entry zone ($x < 0$) may be considered zero (Figure 3.11).
- The dynamic end effect in active zone ($0 \leq x \leq 2p\tau$) is the most important, and, as expected, it produces additional secondary current losses, demagnetization of the machine (a reduction of secondary power factor), and a dynamic end-effect force $F_{x\,end}$, which at zero slip may be propulsive (in lower goodness factor LIMs) or braking (for large dynamic end effect: high speed) (Figure 3.12).
- Though there are two main types of primary windings, $2p$ pole single layer and $2p+1$ double layer (with half-filled end slots); they behave very similarly with respect to dynamic end effect. The two-step increase in primary mmf in the half-filled slots of entry end pole does not reduce the dynamic end effect if its attenuation length L_{ef} is larger than a pole pitch τ. It suffices to consider, in a first approximation, not $2p+1$ poles but approximately $L=2p\tau$ if $p \geq 4$.
- Only the fundamental of primary mmf $\theta_1(x, t)$ traveling wave is considered here (or its primary linear current sheet $J_1(x, t)$); also, all variables vary sinusoidally in time.
- The other effects, typical to a sheet secondary and large airgap, are lumped into the equivalent goodness factor G_e.
- Other more subtle consequences of dynamic end effect will surface on the way of unfolding the quasi 1D field theory of LIMs in the following sections. We will call this a technical theory as it is suitable for design and for defining the circuit model parameters (with dynamic end effect) in the next chapter.

Let us consider the DLIM with the increased airgap in the primary core extended to infinity (Figure 3.13) but at larger airgap (τ/π).

Because we are dealing with a sinusoidal time and space variation of primary current sheet J_1, complex variables may be used:

$$\underline{J}_1 = J_m \cdot e^{j \cdot \left(\omega_1 \cdot t - \frac{\pi}{\tau} x \right)} \tag{3.56}$$

Equation 3.56 is written in primary coordinates. We split the infinite-length model in three regions and apply Maxwell equations with a normal (single) airgap flux density component (along $0y$) and a single variation, along the direction of motion x. The secondary current density has only one component $J_z(x)$; the sinusoidal time variations are well understood.

The active zone ($0<x<2p\tau$).

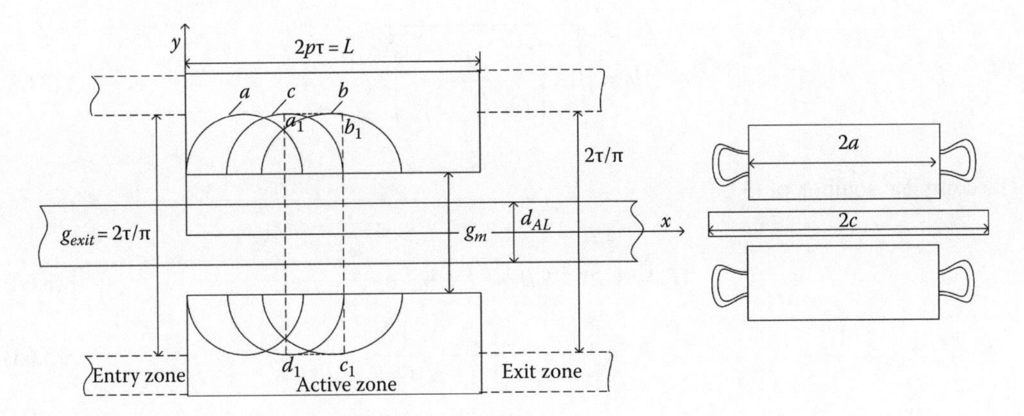

FIGURE 3.13 LIM model for technical field theory.

Ampere's law along contour $a_1b_1c_1d_1$ in Figure 3.13 leads to

$$g_e \cdot \frac{\partial H}{\partial x} = J_m \cdot e^{j\cdot\left(\omega_1 t - \frac{\pi}{\tau}x\right)} + J_2 \cdot d_{AL} \tag{3.57}$$

From Faraday's law,

$$\frac{\partial J_2}{\partial x} = j\omega_1 \cdot \mu_0 \cdot \sigma_e \cdot \frac{d_{AL}}{g_e} \cdot H_y + \mu_0 \cdot U \cdot \sigma_e \cdot \frac{d_{AL}}{g_e} \cdot \frac{\partial H_y}{\partial x} \tag{3.58}$$

where

U is the relative speed of primary with respect to secondary

H_y is the resultant magnetic field in the airgap

From (3.57) through (3.58) we obtain

$$\frac{\partial^2 H_y}{\partial x^2} - \mu_0 \cdot \sigma_e \cdot \frac{d_{AL}}{g_e} \cdot U \cdot \frac{\partial H_y}{\partial x} - j \cdot \omega_1 \cdot \mu_0 \cdot U \cdot \sigma_e \cdot \frac{d_{AL}}{g_e} = -j \cdot \frac{\pi}{\tau \cdot g_e} \cdot J_m \cdot e^{-j\frac{\pi}{\tau}x} \tag{3.59}$$

The characteristic equation of (3.59) is

$$\gamma^2 - \mu_0 \cdot \sigma_e \cdot \frac{d_{AL}}{g_e} \cdot U \cdot \gamma - j \cdot \omega_1 \cdot \mu_0 \cdot \sigma_e \cdot \frac{d_{AL}}{g_e} = 0 \tag{3.60}$$

with the roots

$$\underline{\gamma}_{1,2} = \frac{a_1}{2} \cdot \left(\pm\sqrt{\frac{b_1 + 1}{2}} + 1 \right) \pm \frac{a_1}{2} \cdot j \cdot \sqrt{\frac{b_1 - 1}{2}} = \gamma_{1,2r} \pm j \cdot \gamma_r \tag{3.61}$$

where

$$a_1 = \mu_0 \cdot \sigma_e \cdot \frac{d_{AL}}{g_e} \cdot U = \frac{\pi}{\tau} \cdot G_e \cdot (1 - S) \tag{3.62}$$

$$b_1 = \sqrt{1 + \left(\frac{4}{G_e \cdot (1-S)^2}\right)^2} \tag{3.63}$$

The complete solution of (3.60) is

$$H_a = A \cdot e^{\gamma_1 \cdot x} + B \cdot e^{\gamma_2 \cdot x} + B_m \cdot e^{-j \cdot \frac{\pi}{\tau} \cdot x} \tag{3.64}$$

$$B_n = j \cdot \frac{\tau \cdot J_m}{\pi \cdot g_e \cdot (1 + j \cdot S \cdot G_e)} \tag{3.65}$$

Entry zone ($x \le 0$) and exit zone ($x \ge 2p\tau$).

With no primary currents in these zones and an equivalent airgap $g_{exit} = \tau/\pi \left(G_{exit} << G_e\right)$,

$$H_{entry} = C \cdot e^{\gamma_{1exit}}, \quad \text{for } x \le 0 \tag{3.66}$$

$$H_{exit} = D \cdot e^{\gamma_{2exit} \cdot (x - 2 \cdot p \cdot \tau)} \quad \text{for } x \ge 2p\tau \tag{3.67}$$

$$\underline{\gamma}_{1exit} = \left(\underline{\gamma}_1\right)_{g_e = \frac{\tau}{\pi} G_{exit}}; \quad \underline{\gamma}_{2exit} = \left(\underline{\gamma}_2\right)_{g_e = \frac{\tau}{\pi} G_{exit}} \tag{3.68}$$

The continuity conditions of magnetic field and current density at entry and exit ends ($x = 0$, $2p\tau$) yield

$$\left(H_{entry}\right)_{x=0} = \left(H_a\right)_{x=0}; \quad \left(H_{exit}\right)_{x=2p\tau} = \left(H_a\right)_{x=2p\tau} \tag{3.69}$$

$$\left(\frac{\partial H_{entry}}{\partial x}\right)_{x=0} = \left(J_2\right)_{x=0}; \quad \left(\frac{\partial H_{exit}}{\partial x}\right)_{x=2p\tau} = \left(J_2\right)_{x=2p\tau}$$

We may replace the second boundary condition in (3.69) by

$$\int_{-\infty}^{\infty} H dx = 0 \tag{3.70}$$

to get more realistic results.

However, using $\underline{\gamma}_1$ in the entry zone (not $\underline{\gamma}_{1entry}$) and γ_2 in the exit zone, as done in most literature (infinite core length), we get easily A, B, C, D constants as

$$A = -j \cdot \frac{J_m}{\Delta} \cdot \left(\frac{\gamma_2}{\beta + S \cdot G_e}\right) \cdot e^{-2p \cdot \tau \cdot \gamma_1}$$

$$B = j \cdot \frac{J_m}{\Delta} \cdot \left(\frac{\gamma_1}{\beta + S \cdot G_e}\right)$$

$$\Delta = g_e \cdot (\gamma_2 - \gamma_1) \cdot (1 + j \cdot S \cdot G_e) \tag{3.71}$$

$$C = A \cdot \left(1 - e^{-2p \cdot \tau \cdot \gamma_1}\right)$$

$$D = B \cdot \left(1 - e^{2p \cdot \tau \cdot \gamma_2}\right)$$

To make a technical compromise, we may use the value of C and D from (3.71) into (3.66) through (3.67) and then calculate $J_{exit}(x, t)$ (with $J_{entry} \approx 0$) and the corresponding exit end secondary Joule losses at a more realistic value.

3.3.1 Dynamic End-Effect Waves

The airgap flux density in the airgap (active zone) B_{ag} ($\mu_0 \cdot H_a$ in Equation 3.64) has a first conventional term (typical to rotary IMs: or for LIMs with "infinite length") and two additional terms known as end-effect waves [1]: one forward wave $\mu_0 \cdot A \cdot e^{2p\tau - \gamma_1 \cdot x}$ and one backward wave ($\mu_0 \cdot B \cdot e^{\gamma_2 \cdot x}$):

$$\underline{B}_{end-} = \mu_0 \cdot A \cdot e^{\gamma_1 \cdot x} = -j \cdot \mu_0 \cdot \frac{J_{1m}}{\Delta} \cdot \left(\frac{\gamma_2}{\beta} + S \cdot G_e\right) \cdot e^{(\gamma_{1r} + j \cdot \gamma_i) \cdot (x - 2p\tau)} \tag{3.72}$$

$$\underline{B}_{end+} = \mu_0 \cdot B \cdot e^{\gamma_2 \cdot x} \tag{3.73}$$

$$\underline{B}_{end+} = \frac{j \cdot \mu_0 \cdot J_{1m}}{\Delta} \cdot \left(\frac{\gamma_1}{\beta} + j \cdot S \cdot G_e\right) \cdot e^{\gamma_{2r} \cdot x} \cdot e^{-j \cdot \gamma_i \cdot x} \tag{3.74}$$

They are also called entry end and exit end waves [1].

Both waves are characterized by the same pole pitch $\tau_e = 1/\gamma_i$, but they are attenuated along the direction of motion differently as $|\gamma_{1r}| >> \gamma_{2r}$ (Equation 3.61). Typical values of $1/\gamma_{1r} \cdot \tau$ and $1/\gamma_{2r} \cdot \tau$ for a DLIM, as a function of goodness factor for two typical slip values ($S=0$ and $S=0.08$), are given in Figure 3.14 (based on Equation 3.61).

It is obvious that the entry end wave (Figure 3.14a) is penetrating a very short length (less than 10% of a pole pitch), and thus its effects may be neglected. In contrast, the exit-end wave penetration depth (Figure 3.14b) may be large and increases almost linearly with goodness factor.

For a DLIM, at 100 m/s and $\tau=0.35$ m and $G_e \approx 24$–28, the exit end wave penetration depth would be about 6 pole pitches (2.1 m), not far away from the approximate prediction made earlier in this paragraph. For an urban LIM transport system $U \approx 30$ m/s and $\tau=0.25$–0.3 m, still $G_e \approx 15$–20 and thus the attenuation length of exit-end wave would be 4–5 pole pitches; consequently a LIM with 7–9 poles would experience a notable but perhaps still reasonable dynamic end effect.

Let us now explore the pole pitch of the exit-end wave $\tau_e/\tau = 1/\gamma_i$ at $S=0$ and 0.08 versus goodness factor (Figure 3.15, from (3.61)):

$$\frac{\tau_e}{\tau} \approx \frac{U_{se}}{U} = \frac{2}{G_e \cdot \left(\dfrac{\sqrt{1 + \dfrac{16}{G_e^2 \cdot (1-S)^2}} - 1}{2}\right)^{1/2}} \tag{3.75}$$

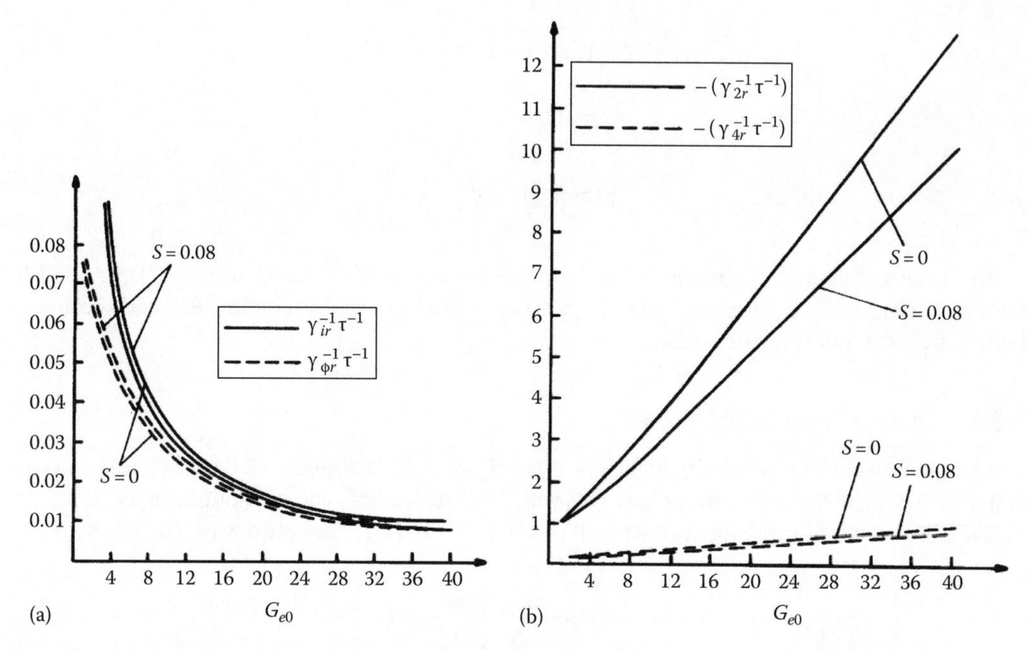

FIGURE 3.14 Relative penetration depths of end-effect waves: (a) entry end wave and (b) exit end wave.

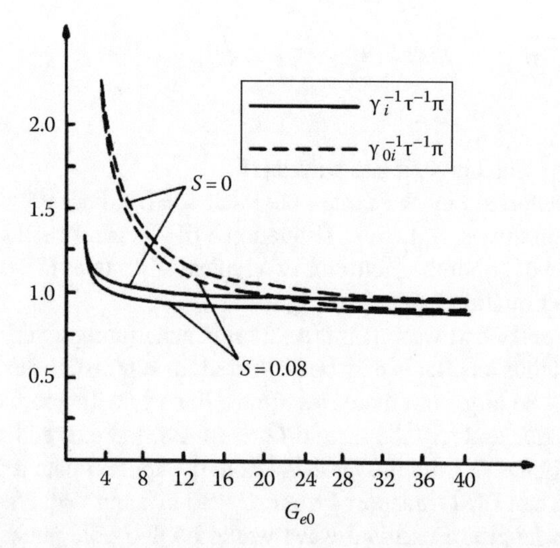

FIGURE 3.15 Relative pole pitch (τ_e/τ) of end-effect waves versus goodness factor for slip $S=0$, 0.08. (After Nasar, S.A. and Boldea, I., *Linear Motion Electromagnetic Devices*, Taylor & Francis, New York, 2001.)

It is very evident that, at small values of goodness factor (when the dynamic end (dynamic) effect is smaller anyway), the speed of end-effect waves is larger than the synchronous speed; so, as expected, at $S=0$, the end-effect force has a braking character, thus reducing the mechanical power produced by the high-speed LIM.

As the LIM length for a vehicle may not go over 2–4 m due to necessity to negotiate tight curves, it seems that with $G_e = 10 - 25$ for LIMs from 30 to 120 m/s, the dynamic end effect will always be present. To improve performance, efficiency means to reduce end effects, and (or) optimal design methods should be used, as both conventional and end effects increase with the goodness factor.

3.3.2 Dynamic End-Effect Consequences in a DLIM

The main consequences of dynamic end effect are

- Reduction of thrust
- Higher secondary Joule losses (lower efficiency)
- Higher absorbed reactive power (lower power factor)
- Nonuniform airgap flux density, thrust, normal force along DLIM length

The thrust is composed of two components (conventional F_{xc} and end-effect force F_{xe}):

$$F_{xc} = \mu_0 \cdot a \cdot J_1 \cdot \mathrm{Re}\left(\int_0^{2p\tau} \overset{*}{B_n} dx \right) \quad (3.76)$$

$$F_{xe} = \mu_0 \cdot a \cdot J_1 \cdot \mathrm{Re}\left(\int_0^{2p\tau} B^* \cdot e^{\overset{*}{\gamma_2} \cdot x} \cdot e^{-j \cdot \beta \cdot x} dx \right) \quad (3.77)$$

The ratio of the two force components f_e is

$$f_e = \frac{F_{xe}}{F_{xc}} = \frac{\mathrm{Re}\left[B^* \cdot \left(\dfrac{-1 + \exp 2p \cdot \tau \cdot \left(\overset{*}{\gamma_2} - j \cdot \beta \right)}{\overset{*}{\gamma_2} - j \cdot \beta} \right) \right]}{\mathrm{Re}\left(\overset{*}{B_n} \cdot 2p \cdot \tau \right)}$$

$$= \left(1 + S^2 \cdot G_e^2\right) \cdot \mathrm{Re}\left[\frac{j \cdot \left(\dfrac{\gamma_1}{\beta} + S \cdot G_e \right) \cdot \left(\exp\left(2p \cdot \pi \cdot \left(\dfrac{\overset{*}{\gamma_2}}{\beta} - j \right) \right) - 1 \right)}{S \cdot G_e \cdot \dfrac{1}{\beta} \cdot \left(\overset{*}{\gamma_2} - \overset{*}{\gamma_1} \right) \cdot \left(1 - j \cdot S \cdot G_e\right) \cdot 2p \cdot \pi \cdot \left(\dfrac{\overset{*}{\gamma_2}}{\beta} - j \right)} \right] \quad (3.78)$$

Equation 3.78 is fairly general in the sense that the relative influence of dynamic end effect is dependent only on the number of poles $2p$, slip S, and on the equivalent goodness factor G_e (which includes slot opening, airgap fringing, skin and edge effects, by correction coefficients). If we add the primary stack length transversally $2a$ and the primary electric current sheet J_n as variables, we see that the DLIM thrust is dependent only on five variables and only one of them, G_e, depends on secondary frequency, magnetic saturation, and other geometrical data of DLIM.

The end-effect force at zero slip switches from propulsion to braking mode with increasing G_e and decreasing number of poles (Figure 3.16).

For $2p=8$ poles, and different goodness factors, the dynamic end effect at small slips (Figure 3.17) shows how severely the end-effect braking character increases with the goodness factor G_e (implicitly with speed).

At low goodness factor values, the dynamic end-effect force changes from braking to propulsion mode once at small slip values, while for high goodness factors (high speed in general), it oscillates from propulsion to braking mode a few times (at a few slip values). As shown in a later chapter on LIM design, it is very tempting to choose as optimum goodness factor G_{e0} its values for which the dynamic

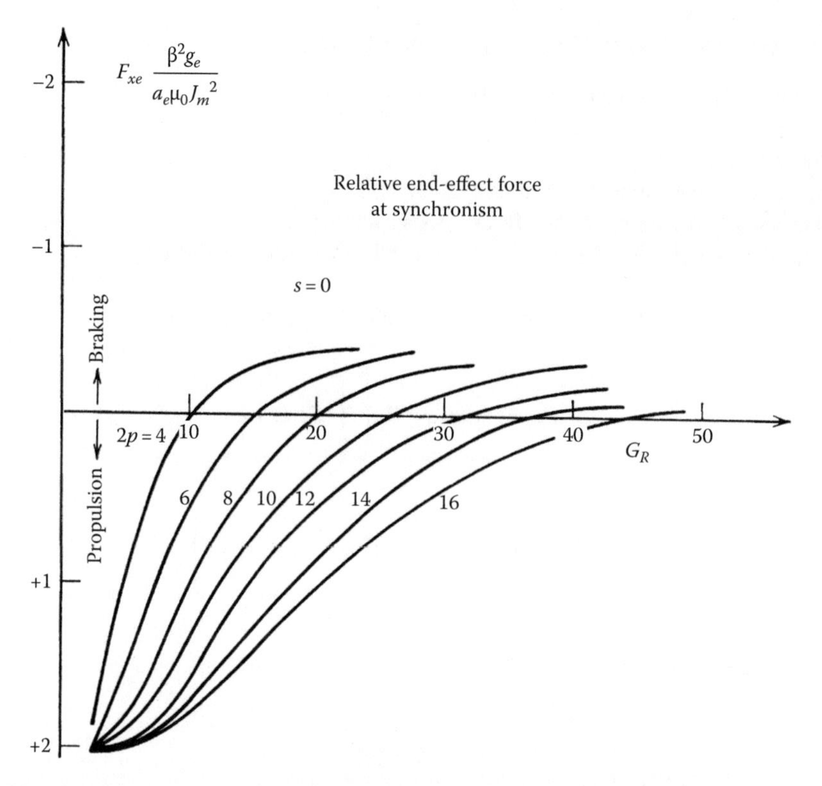

FIGURE 3.16 Normalized dynamic end-effect force versus goodness factor at zero slip. (After Nasar, S.A. and Boldea, I., *Linear Motion Electric Machines*, John Wiley & Sons, New York, 1976.)

end-effect force at zero slip is zero [10]. Reasonable performance, without dynamic end-effect compensation, may be obtained this way, even for speeds up to 120 m/s. As Equation 3.61 shows, the airgap flux density ($\mu_0 \cdot H_a$) amplitude varies along the DLIM primary length (Figure 3.18a):

$$B_a = \mu_0 \cdot B_n \cdot e^{-j\cdot\frac{\pi}{\tau}\cdot x} + \mu_0 \cdot B \cdot e^{\gamma_2 \cdot x} + \mu_0 \cdot A \cdot e^{\gamma_1 \cdot x} \tag{3.79}$$

Similarly, the secondary current density J_2 amplitude will vary along primary length:

$$J_2 = \frac{g_e}{d_{AL}} \cdot \left[\gamma_1 \cdot A \cdot e^{\gamma_1 \cdot x} + \gamma_2 \cdot B \cdot e^{\gamma_2 \cdot x} - \frac{j \cdot S \cdot G_e \cdot e^{-j\cdot\beta\cdot x} \cdot J_1}{g_e \cdot (1 + j \cdot S \cdot G_e)} \right] \tag{3.80}$$

Consequently, the thrust will vary along primary length due to dynamic end effect:

$$F_x = a \cdot \mathrm{Re}\left[\int_0^{2p\tau} \left(J_2 \cdot B_a^* \right) dx \right] \tag{3.81}$$

This is illustrated in Figure 3.18b for an unusually large pole pitch τ ($\tau = 1$ m, but for $2p = 6$) and a reasonably low goodness factor $G_e = 15$.

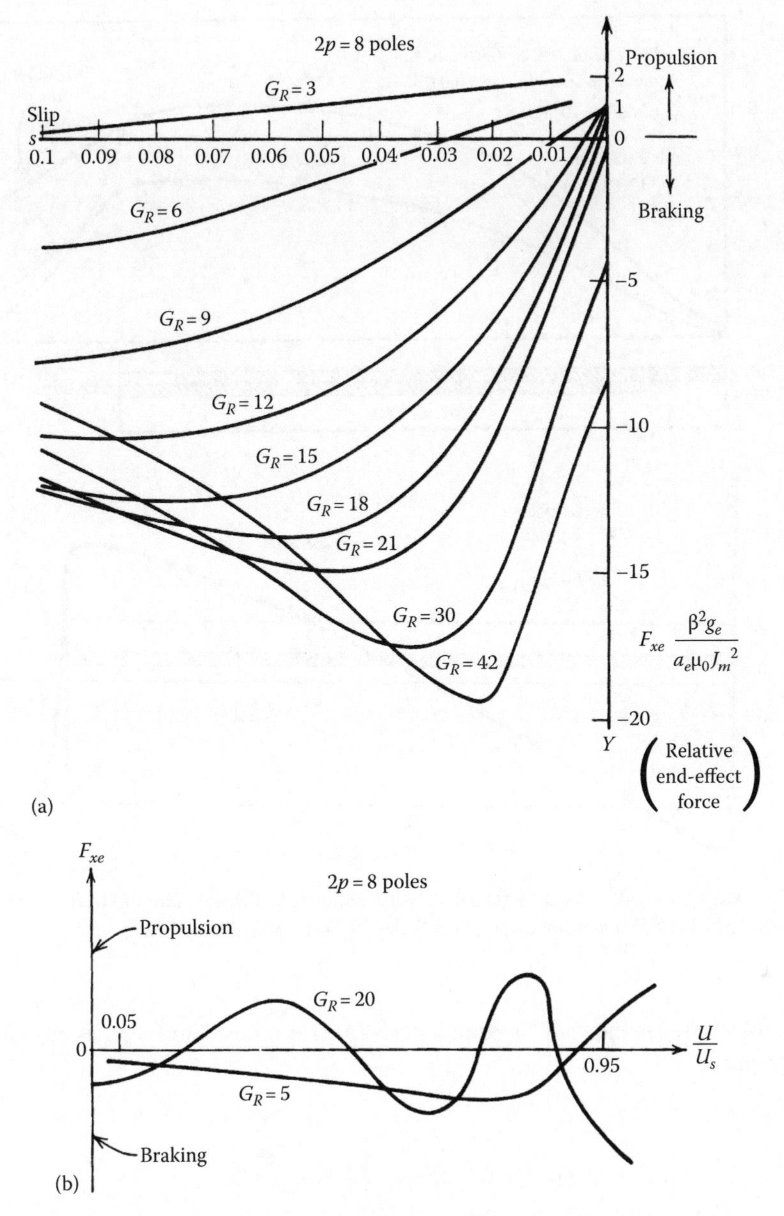

FIGURE 3.17 Normalized dynamic end-effect force versus slip for $2p=8$ poles and a few values of equivalent goodness factor values, (a), and the same force versus speed for low ($G_e=5$) and high ($G_e=20$ goodness factor and $2p=8$ poles), (b). (After Nasar, S.A. and Boldea, I., *Linear Motion Electric Machines*, John Wiley & Sons, New York, 1976.)

With an increased airgap in the primary tail, the decrease in the flux density after the end of primary core is much faster, as expected. We may now calculate the secondary power losses p_2 and the reactive power (Q_2):

$$p_2 = \frac{a \cdot d^2}{\sigma_e \cdot g_e} \cdot \int_0^{2p\tau + L_{exit}} \left(J_2 \cdot J_2^*\right) dx \tag{3.82}$$

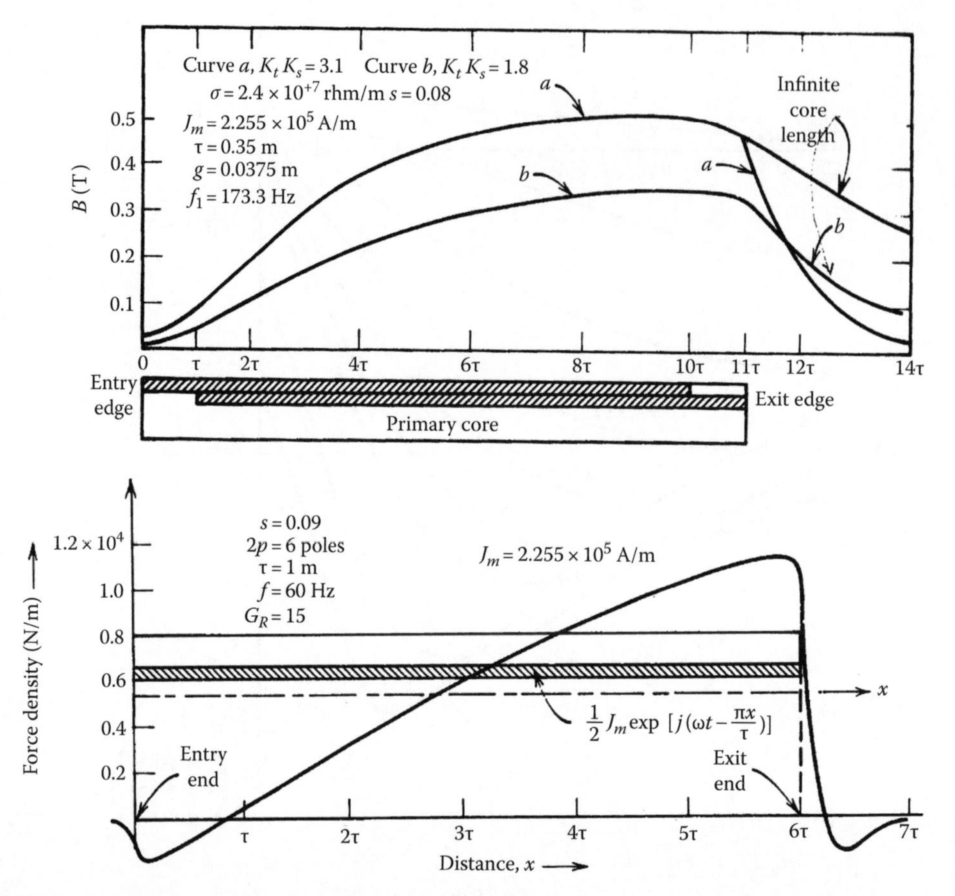

FIGURE 3.18 Airgap flux density and thrust density along the primary length. (After Nasar, S.A. and Boldea, I., *Linear Motion Electric Machines*, John Wiley & Sons, New York, 1976.)

The secondary losses in the "tail" of primary ($x > 2p \cdot \tau$) are considered here, as introduced earlier in this paragraph:

$$Q_2 \approx a \cdot \omega_1 \cdot \mu_0 \cdot g_e \cdot \int_0^{2p\tau} \left(H_a \cdot H_a^* \right) dx \tag{3.83}$$

We may define a secondary (or airgap) efficiency η_2 and power factor cos φ_2 to emphasize the dynamic end effects on them; the primary losses are added to yield the total efficiency:

$$\eta_2 = \frac{F_x \cdot U}{F_x \cdot U + p_2}; \quad \cos \varphi_2 = \frac{F_x \cdot U + p_2}{\sqrt{\left(F_x \cdot U + p_2\right)^2 + Q_2^2}} \tag{3.84}$$

In the absence of dynamic end effect, $\eta_{2i} = 1 - S$.

A typical example of η_2 and cos φ_2 results for a high-speed LIM ($U_s \approx 120$ m/s) with $2p = 12$ and $G_e = 31$ shows acceptably good performance (Figure 3.19).

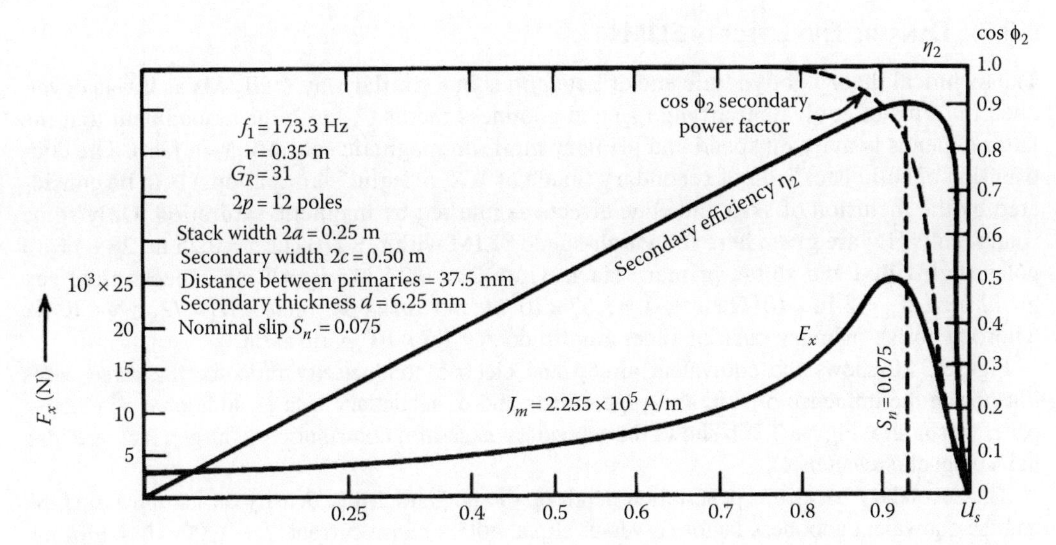

FIGURE 3.19 Secondary efficiency and power factor versus speed (p.u.) for a high-speed LIM ($U_s \approx 120$ m/s).

In addition, the dynamic end effect leads to a gradual increase in core flux density along primary length. Also, the nonuniformity of the secondary current density produces two nonuniformly distributed normal forces: an attraction force F_{na}

$$F_{na} \approx \mu_0 \cdot a \cdot \int_0^{2p\tau} |H_a|^2 dx \tag{3.85}$$

and a repulsive one between one primary and the secondary

$$F_{nr_1} \approx \mu_0 \cdot a \cdot d \cdot \mathrm{Re}\left[\int_0^{2p\tau}\int_0^{2p\tau} \frac{J_m}{2} \cdot e^{-j\beta \cdot x_1} \cdot J_2(x) dx dx_1 \cdot \frac{\Delta_1}{\Delta_1^2 + (x - x_1)^2}\right] \tag{3.86}$$

$\Delta_1, \Delta_2 = g_1, g_2$—the distances between the aluminum plate and the two primaries of a DLIM.

For the second primary,

$$F_{nr2} \approx F_{nr1} \frac{\Delta_2}{\Delta_1} \tag{3.87}$$

After some approximations,

$$F_{nr_1} \approx \frac{2p \cdot \tau}{\Delta \cdot \pi} \cdot \mu_0 \cdot a \cdot \frac{J_m}{2} \cdot \mathrm{Re}\left\{\frac{\overset{*}{\gamma_2} \cdot B^* \cdot \left[\exp\left(\overset{*}{\gamma_2} - j \cdot \beta\right) \cdot 2p \cdot \tau - 1\right]}{\overset{*}{\gamma_2} - j \cdot \beta} + \frac{j \cdot S \cdot G_e \cdot J_m \cdot 2p \cdot \tau}{g_e \cdot (1 - j \cdot S \cdot G_e)}\right\} \tag{3.88}$$

In a DLIM, the repulsive normal force acts as a spring trying to keep the secondary in the middle of the airgap ($\Delta_2 = \Delta_1$); it also has a small damping effect on this "normal" motion.

Now adding the attraction and repulsive force F_{na} and F_{nr}, we may get in a high-speed DLIM, a resultant repulsive force in the first half of primary with an attraction total force in the second part of primary length (the so-called "dolphin" effect [7]).

In a SLIM, this latter effect also exists, and in Equation 3.80 we have to use twice J_m (total primary current sheet) and g_e in place of Δ.

If used for MAGLEVs, the SLIM normal forces, altered by the dynamic end effect, have to be considered in detail.

3.3.3 Dynamic End Effect in SLIMs

The technical theory of dynamic end effect applies in a similar way to SLIMs as it was developed in terms of equivalent airgap (g_e) and goodness factor G_e, with the amendment that the latter depends heavily on speed and primary mmf (on magnetic saturation, in fact). The contribution of solid back iron of secondary (made of 1, 2, 3 "solid" laminations) is to be considered by the inclusion of skin and edge effects as marked by magnetic saturation. Only some final results [11] are given here for a high-speed SLIM with $f_n = 210$ Hz, $\tau = 0.25$ m, $2p + 1 = 13$ poles (half-filled end slots), primary stack width $2a = 0.24$ m, $d_{AL} = 4$ mm, mechanical gap $g = 12$ mm, $\sigma_{AL} = 2.16 \times 10^7\ \Omega^{-1}\text{m}^{-1}$, $\sigma_i = 3.52 \times 10^6\ \Omega^{-1}\text{m}^{-1}$ turns per phase: $W_1 = 72$, $y/\tau = 10/12$ (chorded coils), primary current sheet amplitude $J_1 = 1.5 \times 10^5$ A turns/m.

Figure 3.20 shows the equivalent airgap and electric conductivity ratio for the three cases illustrating the influence of skin and edge effects and of secondary iron as additional conductive plate, versus slip; Figure 3.20b shows the secondary back-iron contribution to airgap increase (due to its magnetic reluctance).

The secondary back-iron penetration depth δ_i, the tangential flux density on its upper surface, and the equivalent goodness factor G_e versus slip at 400 A phase current $J_1 = 1.55 \times 10^5$ A turns/m are shown in Figure 3.21.

Finally, the thrust and normal forces (F_x, F_n) versus slip are illustrated in Figure 3.22a and the secondary efficiency and power factor in Figure 3.22b.

Airgap flux density, secondary current density, and the variation along primary length as influenced by dynamic end effect are similar to those presented for DLIM [11].

The SLIM and DLIM numerical examples related to dynamic end-effect technical (1D) field theory revealed results, which can be summarized as follows:

- A quasi 1D theory of DLIM (and SLIM) may be developed simultaneously if the concept of equivalent airgap g_e and equivalent goodness factor G_e are used to lump the slot opening, airgap fringing, and skin and edge effects.
- The $2p \cdot \tau$ primary winding type of DLIM is not crucial in influencing the dynamic end-effect consequences of airgap flux, secondary current density, and force density distribution along primary length.

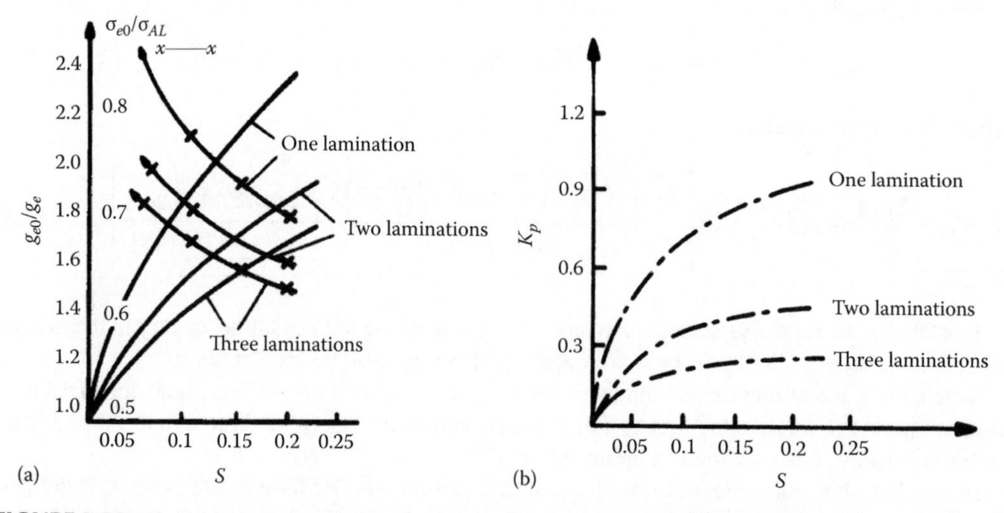

FIGURE 3.20 Equivalent airgap conductivity, (a), and p.u. secondary back-iron reluctance K_p versus slip for a high-speed SLIM with 1, 2, 3 "solid" lamination secondary back iron, (b). (After Nasar, S.A. and Boldea, I., *Linear Motion Electromagnetic Devices*, Taylor & Francis, New York, 2001.)

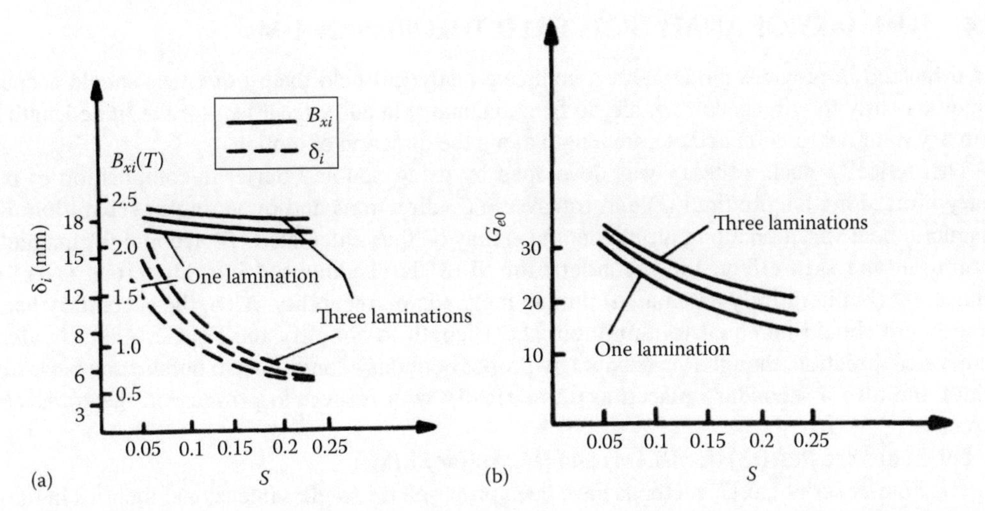

FIGURE 3.21 Secondary back-iron penetration depth (δ_i) and surface-tangential flux density (B_{xi}), (a); equivalent goodness factor G_e versus slip (at 210 Hz and $J_1 = 1.55 \times 10^5$ A turns/m), (b). (After Nasar, S.A. and Boldea, I., *Linear Motion Electromagnetic Devices*, Taylor & Francis, New York, 2001.)

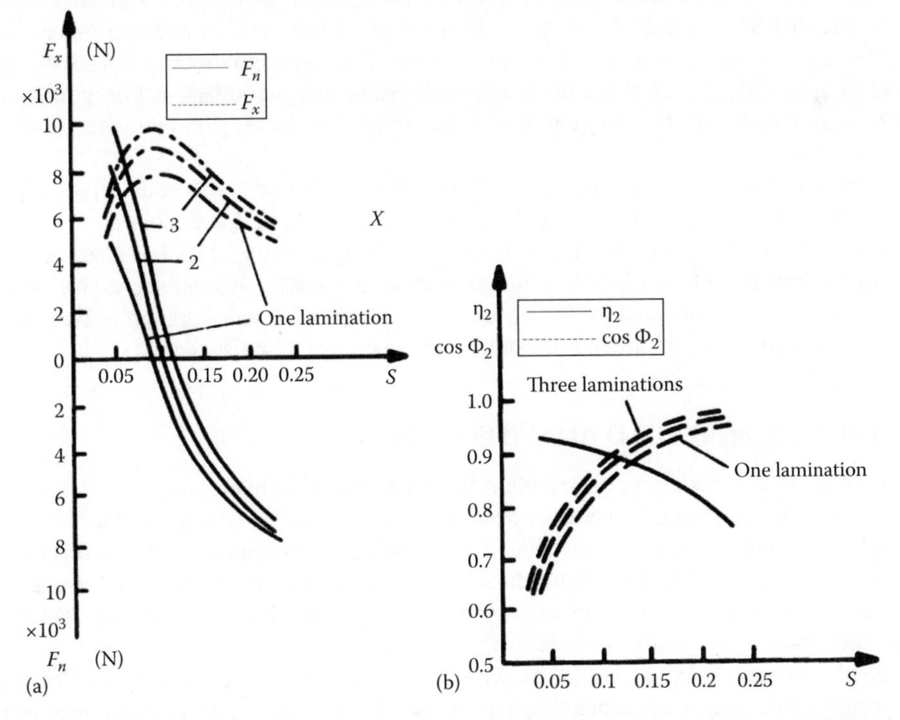

FIGURE 3.22 Secondary efficiency and power factor, (a); forces F_x, F_n versus slip for a high-speed SLIM, (b). (After Nasar, S.A. and Boldea, I., *Linear Motion Electromagnetic Devices*, Taylor & Francis, New York, 2001.)

- The dynamic end-effect force, produced by the forward (exit-end) or end-effect wave, is motoring at low G_e (low-speed) LIMs and braking in character at high G_e (high speed) at zero slip, indicating the great divide between low-speed and high-speed LIMs.
- As the technical theory uses only five variables to characterize LIMs, with given slip S and geometry (G_e, g_e, $2a$ (stack width) and primary electric current sheet J_m), it is fairly general and may be used in preliminary and optimal designs, as demonstrated indirectly in latter studies on LIMs for urban and interurban applications [12].

3.4 SUMMARY OF ANALYTICAL FIELD THEORIES OF LIMs

As discussed in previous paragraphs, a complete analytical field theory of LIMs should account simultaneously for airgap slotting, airgap fringing, and skin and edge effect for the finite length of primary mmf distribution and of core length along the direction of motion.

Theoretically, such a theory was developed by using double Fourier decomposition of primary mmf along longitudinal (x) and transverse (z) directions and by variables separation; the magnetic field variation along airgap depth (y) may be thus calculated. To account for magnetic saturation and skin effect, the secondary for SLIM is decomposed into numerous layers of equivalent (but iteratively calculated) unique magnetic permeability. Also, the secondary back-iron length should be equal to aluminum sheet length to simplify the Fourier analysis along transverse direction, though it is feasible by proper boundary conditions to handle any back-iron width and also a secondary placed asymmetrically with respect to primary in the transverse direction.

For details see Ref. [13] (for DLIM) and [14,15] (for SLIMs).

The Fourier series [16,17] methods have been proposed for single-dimensional theories in third-order models [7] to deal with longitudinal end effect in DLIMs and SLIMs (with Al sheet on iron secondary [17] and with ladder winding [18,19]).

On the other hand, Ref. [20] introduced a pole by pole unidimensional theory of dynamic end effect in DLIMs with remarkable insight into phenomena. However, for Fourier transform methods, the difficulty in calculating the self and mutual inductances and resistances of various fictitious secondary circuits (even in the absence of magnetic saturation) makes the method not easy to apply. Measuring these secondary–secondary and secondary–primary loops inductances by search coils may be a more practical way to get the circuit parameters required for the pole-by-pole method.

More recently, the wavelet analysis of LIM dynamic end effect was proposed [20] but apparently with no strong conclusion, yet, in terms of its practicality for LIM design or control.

Dynamic end effect in short-secondary (mover) and long-primary LIMs has also been studied [4,7,21] by analytical field theories but their applications are scarce. Wound-secondary doubly fed LIMs have also been investigated for short excursion (a few meters) transport applications, but their more practical theories are of circuit type and will be investigated in Chapter 4.

3.5 FINITE ELEMENT FIELD ANALYSIS OF LIMs

The 3D nature of magnetic field distribution in LIMs with aluminum sheet on iron secondary, which experiences (in general) 2D current density trajectories, seems to suggest that the way to go is 3D-FEM. True, but from the start we need to mention that a complete analysis with thrust, normal force decentralizing lateral force, losses, efficiency, power factor, etc., versus speed characteristics would need, even today, tens of hours of computer time (Table 3.1, [22]), while the 3D analytical multiple-layer theory [14] would take a few minutes.

2D-FEM studies, one accounting for transverse (along 0z) and one accounting for slot opening, airgap fringing, and skin effect (along 0y), may be used to check (or duplicate) analytical results of equivalent airgap g_e and secondary conductivities σ_e, σ_i with which a second 2D-FEM (in x,y plane) would include magnetic saturation in the primary and secondary back iron and the dynamic end effect. Such a study may reduce to hours (from tens of hours for 3D-FEM) the total computation time of main LIM curves discussed earlier.

To illustrate the FEM, as applied to LIMs in terms of computation effort, a dual flat $2p=4$ poles primary SLIM with aluminum sheet-on-iron secondary was investigated by 3D-FEM (Figure 3.23) [22]:

The computation time is very large and thus, its use is, still, advisable only for special cases (situations).

TABLE 3.1
Discretization Data and CPU Time

Analysis Method	Step-by-Step Method	Combination Method
Number of elements	468,090	
Number of nodes	82,552	
Number of edges	559,442	
Number of unknowns	557,492	
Number of time steps	108	
CPU time (h)	109	86

Computer used: Pentium III 800 MHz PC.

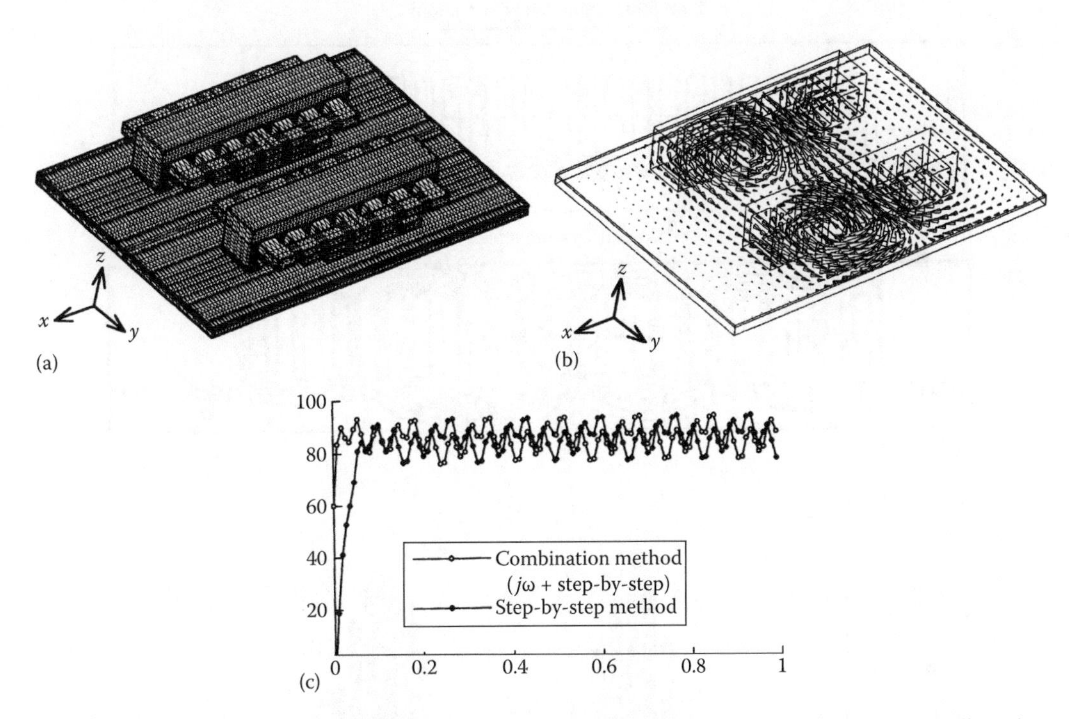

FIGURE 3.23 3D-FE mesh of LIM, (a); secondary eddy current density vectors, (b); thrust, (c); discretization data and computation time. (After Yamaguchi, T. et al., 3D FE analysis of a LIM with two armatures, *Record of LDIA-2001*, Nagano, Japan, pp. 300–303, 2001.)

A 2D-FEM investigation of a cage-secondary LIM is illustrated in Figure 3.24 [23].

As the amplitude of secondary depends on the contact resistance between the secondary bars and the sidebar (R_c), the normal force increases, the flux density distribution is perturbed, and the thrust decreases. The computation time is reasonable, and thus the 2D-FEM coupled circuit approach may be used extensively in LIM analysis.

3.6 DYNAMIC END-EFFECT COMPENSATION

Though the dynamic end effect leads to a notable reduction of efficiency and power factor, the reduction of thrust in LIMs for transportation (urban at 20–30 m/s and interurban at up to 100–120 m/s) is a strong enough indicator expressing the ratio of end-effect thrust F_{xe} to conventional thrust (F_{xe}/F_{xc}(3.78)).

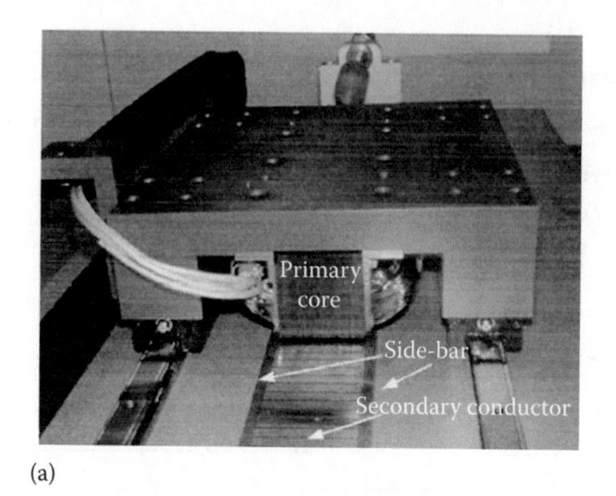

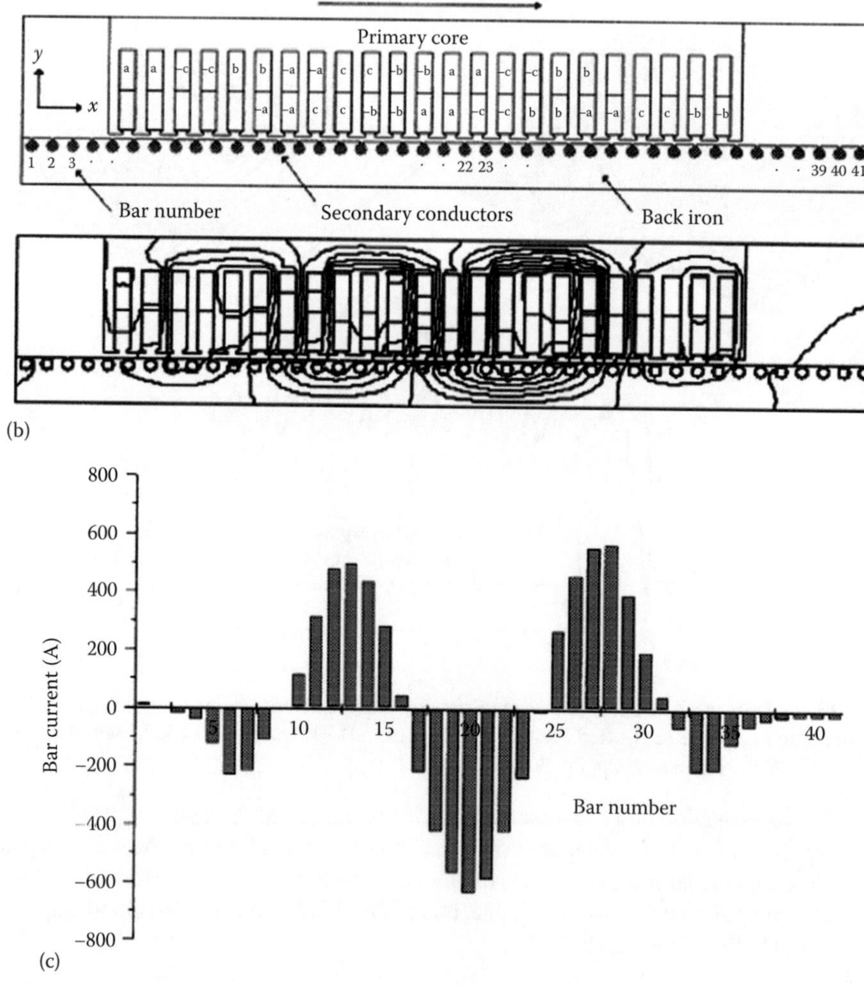

FIGURE 3.24 2D-FEM analysis of ladder cage-secondary SLIM: the lab model, (a), 2D model, (b), flux lines, (c), bar currents ($R_c/R_e = 0.5$; R_c, R_e—contact and side-bar segment resistances). (After Poloujadoff, M. and El Khashab, H., Eigen values and eigen functions: A tool to synthesize different models of LIMs, *Record of LDIA-2001*, Nagano, Japan, pp. 78–83, 2001.)

For the 30–120 m/s, at $S=0.1$ the ratio F_{xe}/F_{xc}—dependent on G_e, $2p$, S—varies in general in well-designed SLIMs in the interval of 0.2–0.6. This means a significant reduction of performance for high speed.

3.6.1 Designing at Optimum Goodness Factor

As both dynamic end effects and conventional performance grow with speed (with G_e increasing, for given $2p$ and S), it seems reasonable to assume that there is an optimum goodness factor G_{e0}. The criterion to calculate it depends on the objective function (max. thrust/primary weight, max thrust/Joule losses, max thrust per kVA at rated speed, etc.).

To simplify the problem and strike a compromise between criteria as mentioned previously, we use here the condition of zero end-effect force F_{xe} at $S=0$ [10]:

$$\left(F_{xe}\right)_{S=0}=0 \Rightarrow G_e\left(2p\right) \tag{3.89}$$

Solving Equation 3.78 iteratively, we get $G_e(2p)$, illustrated in Figure 3.25.

As expected, the optimum goodness factor G_{e0}, which includes quite a few parameters such as pole pitch (τ), primary frequency (ω_1), secondary equivalent conductivity (including skin effect and edge effect), and equivalent airgap (which includes slot opening, airgap fringing, secondary back-iron magnetic saturation), increases with the number of poles.

In general, designing at equivalent optimum goodness factor may reduce the end effect influence on total thrust at rated slip to 10%–15% at 30 m/s and 20%–25% at 120 m/s, under the condition that the maximum primary length is about 2.5 m for urban transportation and about 3.5 m for interurban vehicles, to facilitate negotiating tight curves. With pole pitches of 0.25–0.35 m, such conditions can be met.

3.6.2 LIMs in Row and Connected in Series

If further improvements are needed, a few SLIMs may be placed in row at a distance L_{exit} between each other (Figure 3.26a) [24].

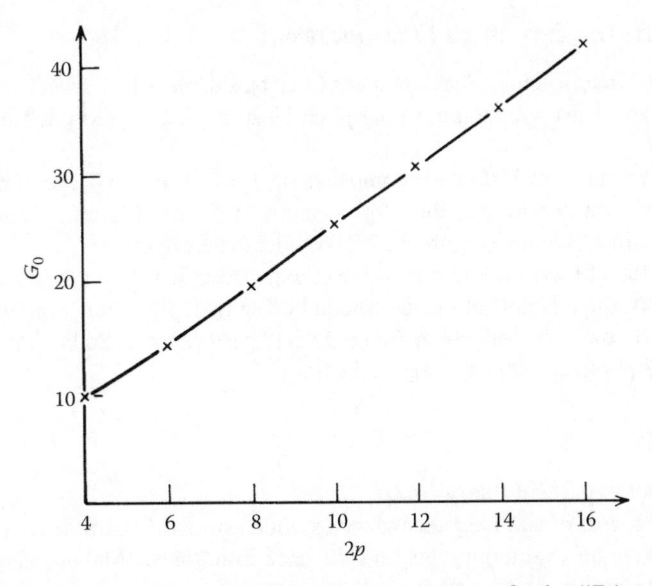

FIGURE 3.25 Optimum equivalent goodness factor G_e versus number of poles ($(F_{xe})_{S=0}=0$).

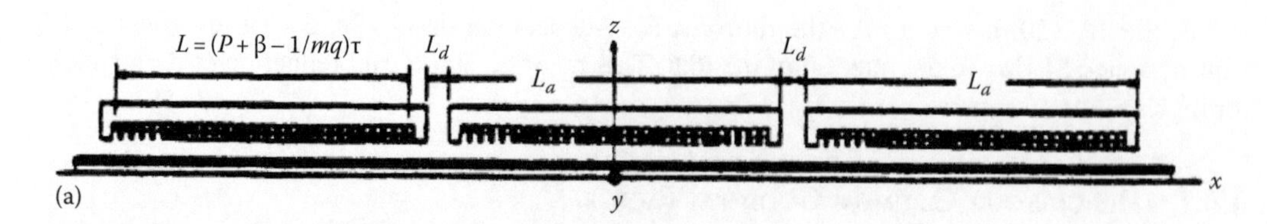

(a)

Type of Connection	Arrangement of the Primary Winding (Expression by the Phase)			Distance between Stators L_d
	First Stator	Second Stator	Third Stator	
A	$a, -c, b, -a, ..$	$c, -b, a, -c, ..$	$b, -a, c, -b, ..$	$1/3\ \tau$
B	$a, -c, b, -a, ..$	$-b, a, -c, b, ..$	$c, -b, a, -c, ..$	$2/3\ \tau$
C	$a, -c, b, -a, ..$	$a, -c, b, -a, ..$	$a, -c, b, -a, ..$	τ
D	$a, -c, b, -a, ..$	$-c, b, -a, c, ..$	$b, -a, c, -b, ..$	$4/3\ \tau$
E	$a, -c, b, -a, ..$	$b, -a, c, -b, ..$	$c, -b, a, -c, ..$	$5/3\ \tau$
F	$a, -c, b, -a, ..$	$-a, c, -b, a, ..$	$a, -c, b, -a, ..$	$2\ \tau$

(b)

FIGURE 3.26 Series connection of SLIMs (a) 3 LIMs in series, (b) border of phases in 3 SLIMs in series.

The distance L_{exit} between the SLIMs in row is such that, at rated speed (slip), the exit end-effect magnetic flux density of first SLIM reaches the entry end of the next LIM at such an amplitude and phase that it reduces drastically the end-effect airgap flux density in the second SLIM. With $L_{exit}=(1, 2, 3, 4, 5, 6)\ \tau/3$ and a certain sequence of phases (as in Figure 3.26b), such conditions can be met.

As illustrated in Ref. [24], the influence of dynamic end effect on thrust is thus reduced to, perhaps, 10% at 300 km/h for 3 SLIMs in row, 8 poles each, pole pitch $\tau=0.3$ m.

Besides these solutions, already in Refs. [1,3], compensation windings in front of LIM have been proposed to reduce the dynamic end effects. But though the thrust has been increased, with slightly lower efficiency, there is apparently no progress in terms of total thrust/weight, which is essential in transportation.

3.6.3 PM Wheel for End-Effect Compensation

Quite a different dynamic end-effect compensation based on a PM wheel rotating at a speed slightly higher than SLIM speed and a pole pitch close to SLIM pole pitch has been proposed [25]—Figure 3.27.

At a certain core radius of PM wheel compensator ($r_c=75$ mm [25]), the torque (Figure 3.27d) would be zero. In such conditions, the improvement in thrust (Figure 3.27b), in power factor (Figure 3.27c), and in efficiency (Figure 3.27c) is evident and effective.

But the cost of the PM wheels and their drive (one for each SLIM on a vehicle) is not trivial (the PMs are very thick), and it is not yet proven whether using the optimum goodness factor design and placing of SLIMs in row is not less costly for equivalent performance. So the jury is still out on this issue of dynamic end-effect compensation (reduction).

3.7 SUMMARY

- LIMs are counterparts of rotary IMs.
- To reduce the cost of the long secondary extended over the entire excursion length, the latter is made of an aluminum sheet on solid back iron (for SLIMs) or solely an aluminum sheet in DLIMs.

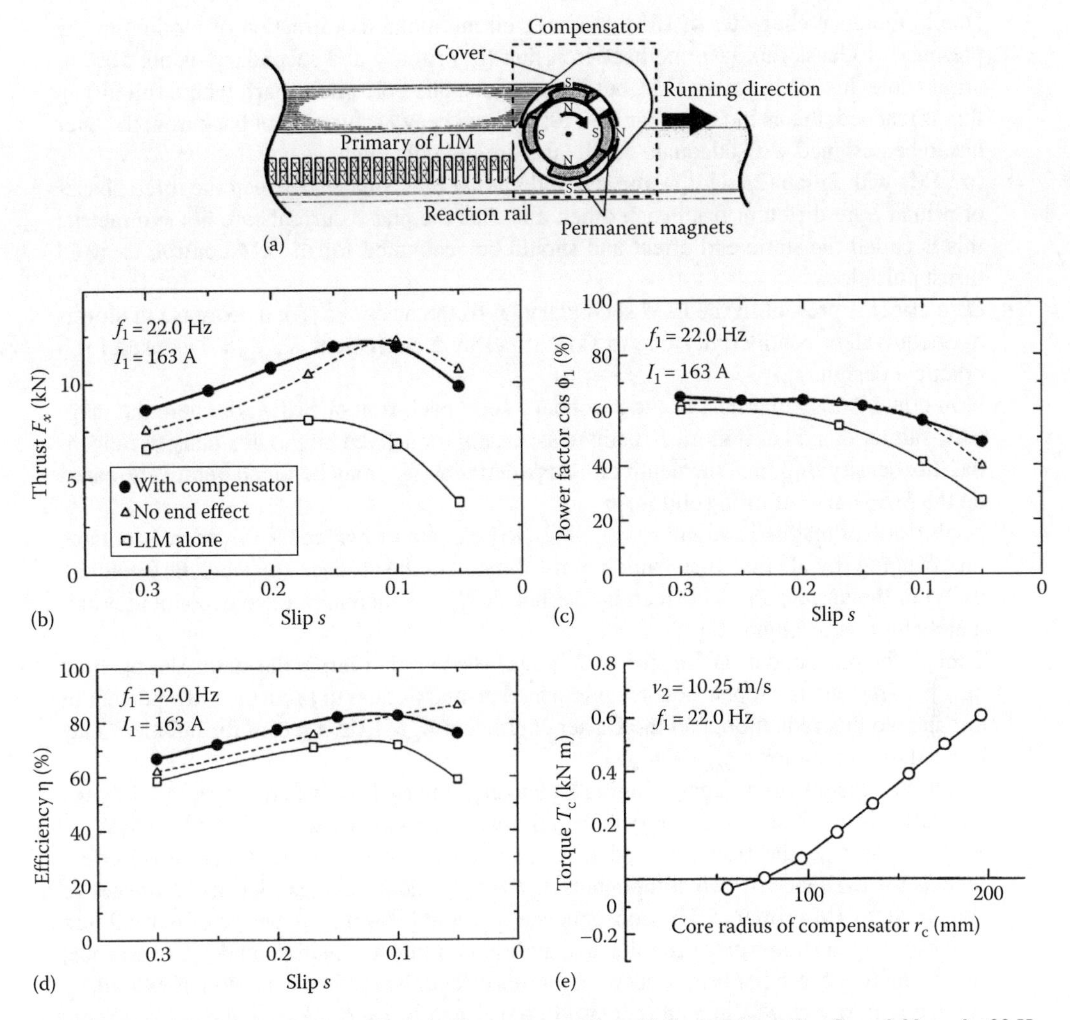

FIGURE 3.27 PM wheel for dynamic end-effect compensation; (a) the topology ($2p=8$, $\tau=0.28$ m, $f_1=22$ Hz, $S_n=0.17$, $v=10$ m/s); (b) thrust; (c) power factor; (d) efficiency; (e) torque of the PM wheel. (After Fuji, N. et al., End effect compensator based on new concept for LIM, *Record of LDIA-2001*, Nagano, Japan, pp. 102–106, 2001.)

- The SLIM is favored since the secondary is more rugged and less costly in general.
- Besides the thrust, the LIM also produces normal force, which in a SLIM is repulsive at high secondary (slip) frequency S^*f_1 and attraction in character for small slip frequency S^*f_1 values.
- In SLIMs with ladder (cage) secondary, placed in a laminated (or even solid) core, however, the resultant normal force is of attraction character and, at $B_g=0.6$ T in the airgap, it may produce a normal force $F_n \geq 6F_{xn}$ (thrust) and thus "cover" a good part of vehicle weight; a few additional controlled electromagnets would provide the rest of the normal force for a MAGLEV operation.
- Single-layer $2p$ (even) pole-count or double-layer $2p+1$ (odd) pole-count three-phase windings are the most practical solutions for flat LIMs. Tubular LIM will be discussed in a later chapter dedicated to low-speed, low-excursion length (less than 2 m) applications.
- Simplified three-phase windings may be used in some applications; among them typical 12 slot/10 pole modules (or other combinations) are good candidates for low-thrust, low-speed, low-excursion length (copper losses are reduced) applications.

- Due to the open character of LIM magnetic circuit along $0x$ (direction of motion) in the presence of Gauss flux law, the back-iron flux (in primary and secondary) is not 50% of airgap pole flux, as in rotary IMs, but it has points, one pole pitch apart, where full airgap flux is reached; this is in fact so for $2p<10$; to avoid heavy saturation of back iron, the later has to be designed with adequate depth (and limited pole pitch).
- In LIMs with $2p<6$ ($2p+1<7$), the self and mutual inductances between the three phases of primary are different from each other, and thus the phase currents are not symmetric; this is called the static end effect and should be accounted for in LIM control, to avoid thrust pulsations.
- Skin effect is present in the LIM secondary (at frequency $f_2=S^*f_1$); a decrease in aluminum equivalent conductivity due to skin effect by a coefficient $K_{R\,skin}>1$ is used for a practical design.
- Skin effect is also manifest in the secondary solid back iron of SLIM, but here the magnetic saturation makes it more difficult to ascertain; for a given airgap flux density, tangential flux density Bt_{max} and the depth of field penetration δ_{iron} may be found iteratively, based on the $B(H)$ curve of mild solid iron.
- As the ratio of magnetic airgap $g_m=g_{air}+g_{AL}$ to pole pitch τ may reach 0.1, there is notable flux fringing (by 2D field distribution in the airgap "in the absence" of secondary conductivity) in the airgap, and a correction coefficient $k_{Fg}>1$ increases g_m to account approximately for this phenomenon.
- Though the magnetic airgap g_m (or $g_m/2$ in DLIMs) is rather large, the stator slot opening may be large enough (open slots for preformed primary coils) to produce a few percent of airgap pole flux reduction, and the Carter coefficient $K_c>1$ (in a recent formulation) may be used to account for it ($g_e=K_cg_m$).
- So the slot opening and airgap fringing lead to an increased equivalent airgap $g_e=k_ck_{Fg}g_m$.
- Similarly, the equivalent secondary conductivity is decreased by skin effect $\sigma_{Alc}=\sigma_{AL}/K_{skin}$.
- With g_e and σ_{Alc}, the transverse edge effect is approached; the transverse edge effect reflects the existence of two components J_z, J_x of secondary current density in the active zone (under LIM primary). The consequence is a nonuniform distribution of normal flux density along the transverse direction and additional Joule losses in secondary. In essence, two correction coefficients may account for this effect: the further reduction of secondary conductivity $\sigma_e=\sigma_{eA}/k_{TR}$ and an increase in the airgap $g_{et}=g_e/k_x$ ($k_x\le 1$, $k_{TR}>1$).

  For SLIMs, the equivalent airgap will further increase due to magnetic saturation: $g_{ete}=g_{et}\,(1+k_s)$, and the equivalent aluminum conductivity will increase to σ_{se} (3.46), to account for secondary solid iron conductivity and magnetic reluctance, approximately. When the secondary is placed asymmetrically (in the transverse direction), a decentralizing lateral force occurs due to transverse edge effect.
- With the equivalent airgap and secondary conductivity, the dynamic end effect may be approached in a 1D field (technical) theory.
- The dynamic end effect occurs as new secondary sections enter the airgap magnetic field zones. Thus, additional secondary currents are induced at LIM primary entry.
- These additional secondary currents "born" at entry end decay toward the LIM primary end with the secondary electric time constant $T_e\approx G_e/\omega_1$, $G_e=\mu_0\left(\tau^2\omega_1/\pi^2\right)\left(\sigma_{te}d_{AL}/g_{te}\right)$.
- G_e is the goodness factor or the Reynolds magnetic number; it may be shown that G_e is equal to X_m/R_2' with X_m—magnetization reaction, R_2'—secondary resistance.
- It has been demonstrated that the influence of dynamic end effect depends essentially on G_e, S, $2p$, $2a$ (stack width). G_e is a key variable, whose influence on performance is very strong.
- The influence of dynamic end effect on airgap flux distribution (which becomes more and more nonuniform as the speed (or G_e) increases, for low slip values and a_e given and for given number of poles of primary). The end effect is basically portrayed by the so-called exit-end forward wave B_{exit}, which decays slowly (with T_e time constant), along primary length.

- The pole pitch of B_{exit}, τ_e at $S=0$ is larger than the primary mmf wave pole pitch τ for low values of G_e, and thus the end-effect thrust is motoring: for larger G_e, $\tau_e \leq \tau$ and thus the end-effect thrust is braking.
- So, in high-speed LIMs (with high dynamic end effect), the end-effect thrust at $S=0$ (or for $S<0.1$) is braking.
- Additionally, the dynamic end effect produces a reduction in power factor, efficiency, and thrust (power) in large speed (large G_e) LIMs.
- To reduce the dynamic end effect, an optimal goodness factor G_e would be a reasonable compromise because both longitudinal end-effect intensity and conventional performance increase with G_e.
- A G_e defined simply as corresponding to zero end-effect force at $S=0$ seems a reasonable compromise [10]. G_e depends solely on $2p$ (pole counts) and increases rather linearly with $2p$ ($G_e=10$ for $2p=4$ and $G_e=30$ for $2p=10$).
- Placing 2–3 LIMs in row at a precalculated distance (in between) and connected in series may produce further reduction in dynamic end effects.
- Compensation windings have been proposed at the entry end of LIM [25,26] to reduce the dynamic end effect in the main LIM; still no such solutions have been applied as they are not clearly superior in global performance/cost terms.
- A PM wheel speeding above LIM speed and placed in front of LIM primary [27] was shown to produce better overall performance at zero torque of the wheel; still, the additional cost of the PM wheel makes the solution application at best a future goal.
- As the LIM field analysis [28], a must for design, has been approached in this chapter, the circuit model, needed for dynamics and control, is the aim of the next chapter.

REFERENCES

1. S. Yamamura, *Theory of Linear Induction Motors*, John Wiley & Sons, New York, 1972.
2. I. Boldea and L. Tutelea, *Electric Machines*, CRC Press, Taylor & Francis, New York, 2009.
3. S.A. Nasar and I. Boldea, *Linear Motion Electric Machines*, John Wiley & Sons, New York, 1976, Chapter 1.
4. S.A. Nasar and I. Boldea, *Linear Motion Electromagnetic Devices*, Taylor & Francis, New York, 2001, Chapter 3.
5. E. Matagne, Slot effect on the airgap reluctance Carter-like calculation suitable for electric machines with large air-gap or surface mounted permanent magnet, *Electromotion*, 15(4), 2008, 171–176.
6. H. Bolton, Transverse edge effect in sheet rotor induction machines, *Proc. IEE*, 116, 1969, 725–739.
7. E.R. Laithwaite, *Induction Machines for Special Purposes*, G. Newness, London, U.K., 1966.
8. M. Poloujadoff, *The Theory of Linear Induction Motors*, Clarendon Press, Oxford, U.K., 1980.
9. J.F. Gieras, *Linear Induction Drives*, Oxford University Press, Oxford, U.K., 1992.
10. I. Boldea and S.A. Nasar, Optimum goodness criterion for linear induction motor design, *Proc. IEE*, 123(1), 1976, 89–92.
11. I. Boldea and S.A. Nasar, Improved performance of high speed single-sided LIMs: A theoretical study; *EMELDG International Quarterly* (now EPCS Taylor & Francis), 2(2), 1978, 155–166.
12. I. Boldea and S.A. Nasar, Optical design criterion for linear induction motors, *Proc. IEEE*, 123(1), 1976, 89–92.
13. K. Oberretl, Three dimensional analysis of the linear motor taking into account edge effects and winding distribution, *Arch. Electrotech.*, 55, 1973, 181–190.
14. I. Boldea and M. Babescu, Multilayer approach to the analysis of single-sided linear induction motors, *Proc. IEE*, 125(4), 1978, 283–285.
15. I. Boldea and M. Babescu, Multilayer approach of d.c. linear brakes with solid iron secondary, *Proc. IEE*, 123(3), 1976, 220–222.
16. K. Budig, *Linear A.C. Electric Motors*, VEB Verlag, Berlin, Germany, 1974 (in German).
17. S. Nonaka, Simplified Fourier transform method of LIM analysis based on space harmonic method, *Record of LDIA-1998*, Tokyo, Japan, 1998, pp. 187–190.

18. K. Oberretl, Single sided linear motor with ladder type secondary, *Archiv fur Electrotechnik*, 56, 1974, 305–319 (in German).
19. M. Fujii, Analysis of cage type LIM with ladder shaped pole pitch conductors, *Record of LDIA-1998*, Tokyo, Japan, 1998, pp. 191–194.
20. T. Sugixama and S. Torii, An establishment of the wavelet analysis of LIM aimed at analyzing end effect, *Record of LDIA-2001*, Nagano, Japan, 2001, pp. 310–314.
21. T. Onuki, K. Yoshihida, K. Yushi, and H. Forekmasa, An approach to a suitable secondary shape for improving the end effect of a short rotor DLIM, *LDIA-1995*, Nagasaki, Japan, 1995, pp. 369–372.
22. T. Yamaguchi, Y. Kawase, M. Yoshida, M. Nagai, Y. Ohdachi, and Y. Saito, 3D FE analysis of a LIM with two armatures, *Record of LDIA-2001*, Nagano, Japan, 2001, pp. 300–303.
23. M. Poloujadoff and H. El Khashab, Eigen values and eigen functions: A tool to synthesize different models of LIMs, *Record of LDIA-2001*, Nagano, Japan, 2001, pp. 78–83.
24. S. Nonaka and M. Fuji, The series connection of short stator LIMs for intercity transit, *Record of International Conference on MAGLEV and Linear Devices*, Las Vegas, NV, 1987, pp. 23–29.
25. N. Fuji, Y. Sakumoto, and T. Kayasuga, End effect compensator based on new concept for LIM, *Record of LDIA-2001*, Nagano, Japan, 2001, pp. 102–106.
26. T. Kaseki and R. Mano, Flux synthesis of a LIM for compensating end effect based on insight of a control engineer, *Record of LDIA-2003*, Birmingham, U.K., 2003, pp. 359–362.
27. N. Fujii, T. Hashi, and Y. Tanabe, Characteristics of two types of end effect compensators for LIM, *Record of LDIA-2003*, Birmingham, U.K., 2003, pp. 73–76.
28. J.L. Gong, F. Gillon, and P. Brochet, Linear induction modeling with 2D and 3D finite elements, *Record of ICEM 2010*, Rome, Italy, 2010.

4 Linear Induction Motors *Circuit Theories, Transients, and Control*

4.1 LOW-SPEED/HIGH-SPEED DIVIDE

Field theories presented in Chapter 3 proved static and dynamic end effects as the main differences between rotary and linear induction motors.

Other specific phenomena related to the conductive sheet (alone or one placed on solid iron) such as edge effects, besides skin effect, airgap fringing, and primary slot opening influence on performance, have been reduced to equivalent correction coefficients on airgap (g_e) and on secondary sheet electrical conductivity σ_e.

The dynamic end effect divides the LIMs into high-speed (when this effect counts) and low-speed LIMs when the dynamic end effect is negligible. It is not so easy to hammer out a border between the two cases, but a first approximation would be to consider a LIM as of low speed if the exit end wave penetration depth along motion direction $1/\gamma_{2r}$ (3.61–3.62) is less than 10% of primary length $(2 \cdot p \cdot \tau)$:

$$\frac{1}{\gamma_{2r}} < \frac{2 \cdot p \cdot \tau}{10} \tag{4.1}$$

As γ_{2r} depends on the equivalent goodness factor G_e, slip S, and pole pitch τ [1],

$$\gamma_{2r} = \frac{a_1}{2}\left(1 - \sqrt{\frac{1+b_1}{2}}\right) \tag{4.2}$$

$$a_1 = \frac{\pi}{\tau} \cdot G_e \cdot (1-S); \quad b_1 = \sqrt{1 + \frac{16}{G_e^2 \cdot (1-S)^4}} \tag{4.3}$$

Condition (4.1) becomes

$$p \cdot \pi \cdot G_e \cdot (1-S) \cdot \left(1 - \sqrt{\frac{1+b_1}{2}}\right) > 10 \tag{4.4}$$

As this condition depends only on equivalent goodness factor G_e, slip S, and number of poles $2p$, it may be simply represented on a fairly general graph (Figure 4.1).

The intersection of continuous and interrupted lines in Figure 4.1 for a given slip S suggests compliance with condition (4.1).

For example, at $S=0.1$, for $2p=4$, 6, 8, there is no low-speed solution but for $2p=10$ there is one with large G_e (and good performance): $G_e=21$.

Also for $2p=12$, $S=0.05$, there is a solution at $G_e=25$; for $S=0.1$ and for $2p=12$, a larger G_e is allowed ($G_e=35$!). Values of G_e larger than 25 correspond in general to well-designed high-speed LIMs (100–110 m/s): the optimum goodness factor G_e for $2p=12$ is $G_{e0}=27$ in Figure 3.26.

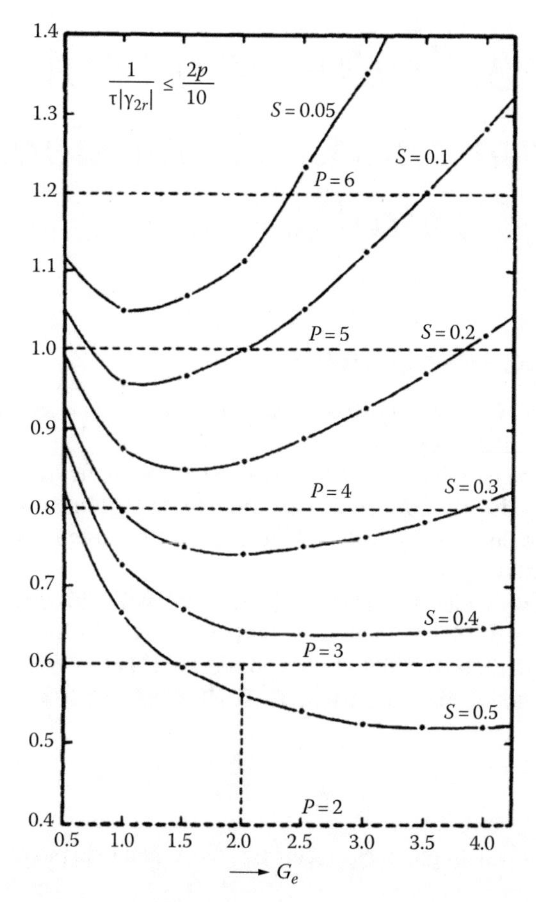

FIGURE 4.1 Exit end effect wave (interrupted lines) penetration depth per pole pitch versus G_e for given p and S (slip). (After Boldea, I. and Nasar, S.A., *Linear Motion Electromagnetic Devices*, Taylor & Francis, New York, 2001.)

Therefore, the optimal goodness factor is rather close to the condition that the penetration depth of exit end effect wave equals only 10% of primary length; this means that, grossly, the end effect influence on thrust reduction at rated slip is less than 10%.

Now we may proceed to the investigation of circuit models, transients, and control of low-speed LIMs as they resemble rotary IMs, though dedicated circuit parameter expressions for flat (and tubular) LIMs with Al-on-iron and for ladder secondary have to be derived.

4.2 LIM CIRCUIT MODELS WITHOUT DYNAMIC END EFFECT

There are a few basic topologies for low-speed LIMs:

- Flat DLIMs with Al sheet secondary
- Flat SLIMs with Al-on-iron secondary
- Flat SLIMs with ladder secondary
- Tubular SLIMs with cage secondary

We will derive the equivalent circuit model with its parameter expressions for all these configurations, one by one.

Note: Short secondary (mover) long primary (stator) LIMs are not considered here as their applications are still uncertain (due to strong competition from dc excited or PM linear synchronous machines).

4.2.1 Low-Speed Flat DLIMs

Double-sided LIMs use an aluminum sheet secondary with a thickness d_{AL} and an equivalent electrical conductivity σ_{Ale}, which accounts for transverse (edge) effect and skin effect:

$$\sigma_{Ale} = \frac{\sigma_{Al}}{k_{tr} \cdot k_{skin}} \tag{4.5}$$

In addition, the equivalent magnetic airgap g_e accounting for slot opening (by Carter coefficient k_c), airgap fringing (k_{fg}), and stator yoke magnetic saturation coefficient k_{ss} is

$$g_{eAl} = (2 \cdot g + d_{AL}) \cdot k_c \cdot k_{fg}; \quad k_c > 1, k_{fg} \geq 1 \tag{4.6}$$

The equivalent goodness factor G_e is

$$G_e(\omega_1) = \frac{\mu_0 \cdot \omega_1 \cdot \tau^2}{\pi^2} \cdot \sigma_{Ale}(S\omega_1) \cdot \frac{d_{Al}}{g_{eAl} \cdot (1 + k_{ss})} \tag{4.7}$$

as the other definition of goodness factor is

$$G_e(\omega_1) = \frac{X_m(\omega_1)}{R_2'}; \quad X_m = \omega_1 \cdot L_m \tag{4.8}$$

But in the absence of end effects, the magnetization inductance $L_m(\omega_1)$ is (as for rotary IMs)

$$L_m(\omega_1) = \frac{6 \cdot \mu_0 \cdot (2 \cdot a_e) \cdot (k_{w1} \cdot W_1)^2 \cdot \tau}{\pi^2 \cdot p \cdot g_{eAl} \cdot (1 + k_{ss})} = k_{X_m} \cdot W_1^2 \tag{4.9}$$

where

- $2a$ is the stack width
- $a_e \approx a + g_{eAl}/2$
- $2p$ is the pole count
- W_1 is the turns in series per phase
- k_{w1} is the winding factor

Consequently, the secondary phase equivalent resistance (reduced to the primary) is from (4.7) through (4.9) (Figure 4.2):

$$R_2'(s\omega_1) = \frac{X_m}{G_e} = \frac{12 \cdot a_e \cdot (k_{w1} \cdot W_1)^2}{d_{AL} \cdot \tau \cdot p \cdot \sigma_{eAl}} = k_{R_2'} \cdot W_1^2 \tag{4.10}$$

For the aluminum sheet, due to its small thickness (4–6 mm), the leakage inductance, skin [2], and transverse edge effects may be neglected for low speed (low goodness factor: $S_n G_e \ll 1$).

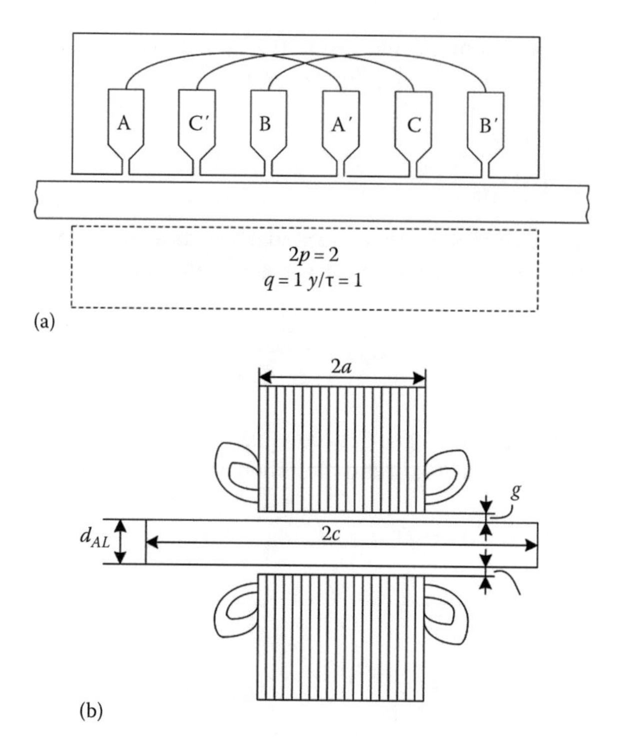

FIGURE 4.2 DLIM: (a) longitudinal cross section and (b) transverse cross section.

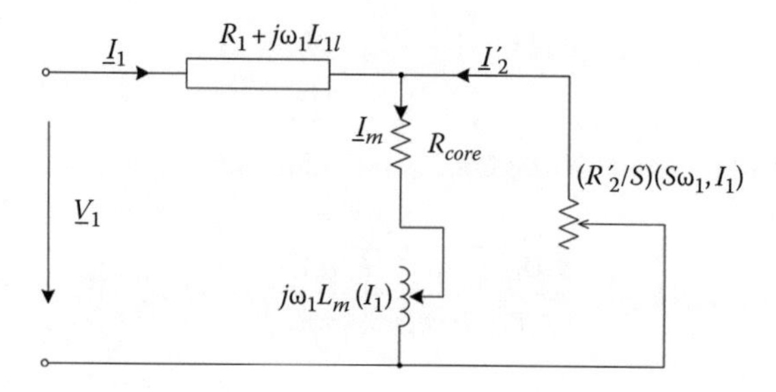

FIGURE 4.3 LIM equivalent circuit.

As expected, the equivalent circuit (Figure 4.3) will be similar to that of rotary IMs, with the observation that $X_m = \omega_1 \cdot L_m$ and R_2' vary with slip frequency $S\omega_1$, with current I_1 (through magnetic saturation), and with airgap g_{eAL}.

To complete the expressions of equivalent parameters, R_1 and L_{1l} are

$$R_1 \approx \frac{1}{\sigma_{co}} \cdot \frac{(4 \cdot a + 2 \cdot l_{ec})}{W_1 \cdot I_{1rated}} \cdot W_1^2 \cdot j_{c0rated} = k_{R_1} \cdot W_1^2 \tag{4.11}$$

where

l_{ec} is the primary coil end connection length
$j_{c0rated}$ is the rated design current density
$W_1 \cdot I_{1rated}$ is the rated phase Ampere turns per phase (RMS)
σ_{co} is the copper electrical conductivity

$$L_{1l} \approx \frac{2\cdot\mu_0}{p\cdot q}\cdot\left[\left(\lambda_{slot}+\lambda_{diff}\right)\cdot 2a+\lambda_{end}\cdot l_{ec}\right]\cdot W_1^2 = k_{1l}\cdot W_1^2 \tag{4.12}$$

where

λ_{slot} is the slot nondimensional (specific) permeance $\lambda_s = h_{1s}/(3\cdot b_{1s})$ for open slots

λ_{dif3f} is the differential leakage specific nondimensional permeance (as the airgap g_e/τ ratio is rather large, $\lambda_{diff} < 0.2 - 0.25$)

λ_{end} is the coil end connection specific permeance: $\lambda_e \approx 0.66q$, for single-layer windings (Figure 4.2), and $\lambda_e \approx 0.33q$, for double-layer windings

Note: The $2p=2$ pole primitive LIM in Figure 4.1 is characterized by asymmetric phase currents when fed from symmetric ac phase voltages, due to the fact that phase C is in a different position with respect to primary core ends. To calculate more correctly the behavior of such a LIM, the self and mutual inductances between primary phases and primary/secondary have to be calculated based on field theories. The equivalent circuit in Figure 4.2 would produce good results only for the positive symmetric component of the currents. From $2p=6$ onward, this static end effect becomes negligible.

We may advance the circuit theory by calculating the thrust F_x, efficiency (η_1), power factor ($\cos\varphi_1$), and normal force (F_n). Based on the electromagnetic power definition,

$$F_x = \frac{3\cdot(I_2^{'})^2\cdot R_2^{'}}{2\cdot\tau\cdot f_1\cdot S} = \frac{3\cdot\pi\cdot L_m\cdot I_1^2\cdot S\cdot G_e}{\tau\cdot\left(1+S^2\cdot G_e^2\right)} \tag{4.13}$$

$$\eta_1 = \frac{F_x\cdot 2\cdot\tau\cdot f_1\cdot(1-S)}{F_x\cdot 2\cdot\tau\cdot f_1+3\cdot R_1\cdot I_1^2+3\cdot R_{cores}\cdot I_m^2} \tag{4.14}$$

R_{cores} represents the primary core losses (as in rotary IMs) and is to be calculated in the design stage or measured in adequate tests:

$$\cos\varphi_1 = \frac{F_x\cdot 2\cdot\tau\cdot f_1+3\cdot R_1\cdot I_1^2+3\cdot R_{cores}\cdot I_m^2}{3\cdot V_1\cdot I_1} \tag{4.15}$$

The large airgap/pole pitch ratio (g_e/τ) leads to rather low primary power factor ($\cos\varphi_1 = 0.4$–0.55). The normal force F_n, between primaries and secondary in DLIM has two components:

- An attraction force, F_{na}, between the two primary magnetic cores
- A repulsion force, F_{nr}, between primary and secondary mmfs

In the absence of end effects, F_{na} is simply

$$F_{na} = \frac{(2\cdot a_e)\cdot B_{g1}^2\cdot(2\cdot p\cdot\tau)}{2\cdot\mu_0} \tag{4.16}$$

B_{g1} (3.65) is the airgap flux density peak value under load:

$$B_{g1} = \frac{\mu_0\cdot\tau\cdot J_{1m}}{\pi\cdot g_{eAL}\cdot(1+k_{ss})\cdot(1+j\cdot sG_e)};\quad J_{1m} = \frac{3\sqrt{2}\cdot(W_1\cdot k_{m1})\cdot I_1}{p\cdot\tau} \tag{4.17}$$

So, from (4.15) and (4.16),

$$F_{na} = \frac{a_e \cdot 2 \cdot p \cdot \tau^3 \cdot \mu_0 \cdot J_{1m}^2}{\pi^2 \cdot g_{eAL}^2 \cdot (1+k_{ss})^2 \cdot (1+S^2 \cdot G_e^2)} \tag{4.18}$$

To calculate F_{nr}, the repulsion force, we need the tangential component of airgap flux magnetic field H_x on the primary surface:

$$H_x = J_{1m}; \quad B_x = \mu_0 \cdot J_{1m} \tag{4.19}$$

$$F_{nr} \approx 4 \cdot \mu_0 \cdot \frac{a_e}{2} \cdot \tau \cdot p \cdot d_{AL} \cdot \mathrm{Re}\left[\underline{J}_2^* J_{1m}\right] \tag{4.20}$$

$\underline{J}_2$ is the secondary current density:

$$\underline{J}_2 = S \cdot \omega_1 \cdot \sigma_{eAL} \cdot \frac{\tau}{\pi} \cdot \underline{B}_{g1} \tag{4.21}$$

From (4.20) and (4.21),

$$F_{nr} \approx 2 \cdot a_e \cdot \tau \cdot S \cdot \omega_1 \cdot \sigma_{eAL} \cdot d_{AL} \cdot \left(\frac{\tau}{\pi}\right)^2 \cdot \frac{(\mu_0 \cdot J_{1m})^2 \cdot S \cdot G_e}{g_{eAL} \cdot (1+k_{ss}) \cdot (1+S^2 \cdot G_e^2)} \tag{4.22}$$

Finally,

$$F_n \approx \frac{2a_e \cdot p \cdot \mu_0 \cdot J_{1m}^2 \cdot \tau^3}{\pi^2 \cdot g_e^2 \cdot (1+k_{ss})^2 \cdot (1+S^2 \cdot G^2)} \cdot \left(1 - \frac{\pi^2 \cdot g_{eAL}^2 \cdot (1+k_{ss})^2 \cdot S^2 \cdot G_e^2}{\tau^2}\right) \tag{4.23}$$

As visible in (4.23), in the low slip region the normal force is of attraction character (positive) while it becomes repulsive (negative) at large slip values.

In low-speed LIMs, in general at rated slip ($S=0.1$–0.15), the normal force is of attraction type. The zero net normal force occurs at a rather large $S^* f_1 = f_2$ (secondary frequency), which leads to lower efficiency.

As in general the maximum thrust per given mmf (J_{1m}, linear current sheet)—I_1 is obtained from (4.13):

$$\frac{\partial F_x}{\partial S} = 0; \quad S_k \cdot G_e = 1. \tag{4.24}$$

Condition (4.24) for a low-speed LIM with low end effect (Figure 4.1) is not easy to reach and, for example, for $G_e=10$, $S_k=0.1$, it implies $2p=8$.

Now for $G_e=10$, $\tau=0.2$ m, $d_{AL}=8$ mm, $g_e=20$ mm, $\sigma_e=2.7 \cdot 10^7$ $(\Omega\text{ m})^{-1}$, and $k_{ss}=0{,}1$, it would mean f_1 (from (4.7)) is 29 Hz. This would lead to a synchronous speed $U_s=2 \cdot \tau \cdot f_1 = 2\times 0.2 \times 29 = 11.6$ m/s and the LIM speed $U=U_s \cdot (1-S_k) = 11.6 \cdot (1-0.1) = 10.49$ m/s.

As evident in this numerical example, condition (4.24) at $S=0.1$ implies a fairly large LIM (1.6 m) and a notable speed (10 m/s), to secure low (negligible) end effect. In low-speed (<3 m/s) applications, the maximum thrust/current should be placed at large S_k to fulfill condition (4.24) $S_k=(0.5$–$1)$.

As expected, both motoring and regenerative braking operation modes are feasible. As variable frequency supply is a must in LIM drives, to secure variable speed, the control of slip

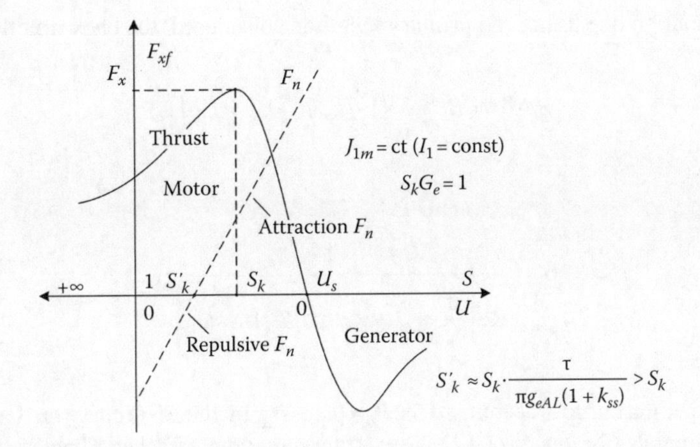

FIGURE 4.4 Repulsion F_x and normal F_n forces of DLIM.

frequency $S \cdot \omega_1$ and of the primary current offers the first option for thrust (or speed) control of LIMs (Figure 4.4).

Note: When the secondary is placed asymmetrically in the airgap (closer to one of the two primaries), the self-aligning character of repulsive force centralizing component (Chapter 3) becomes visible (for details see (3.8)). Also, when the primary is placed laterally asymmetric in the airgap, a decentralizing force is developed by the interaction of longitudinal secondary current density and the airgap flux density (Figure 3.8c).

4.3 FLAT SLIMs WITH AL-ON-IRON LONG (FIX) SECONDARY

As shown in Chapter 3, the presence of the solid back iron (made of 1 or 2 or 3 "laminations" in SLIMs (Figure 4.5)), in order to reduce eddy currents and increase the field penetration depth, may be treated as a modification of aluminum sheet electric conductivity, the occurrence of a magnetic reluctance of iron, and, eventually, of a small leakage inductance of secondary eddy currents paths. Other than that, the equivalent circuit theory of SLIM resembles that of DLIM.

Basically,

$$\sigma_{eAL} = \frac{\sigma_{AL}}{k_{TR}k_{skin}} \rightarrow \sigma_{eALi} = \sigma_{eAL} \cdot (1+k_i) \tag{4.25}$$

$$g_{ea} = g \cdot k_c \cdot k_{Fg} \rightarrow g_{eai} = g_{ea} \cdot (1+k_{ss}) \tag{4.26}$$

From (3.44) through (3.46),

$$k_{ss} = \frac{2\tau^2}{\pi^2} \cdot \frac{\mu_0}{2 \cdot (g+h_{AL}) \cdot \delta_i \cdot \mu_{iron}}; \quad k_i = \frac{\sigma_i}{\sigma_{AL}} \cdot \frac{\delta_i}{k_{ti}} \cdot k_{trAl} \tag{4.27}$$

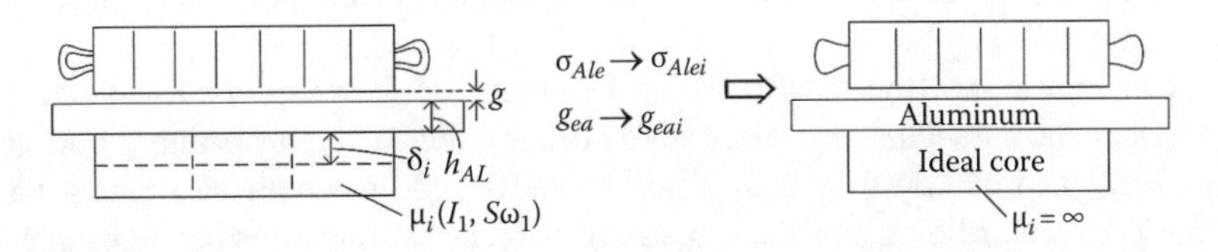

FIGURE 4.5 SLIM transverse cross section: equivalent cross section with ideal secondary back iron.

If the redistribution due to limited primary length is considered, the back-iron flux is

$$B_{ti} \cdot \delta_i = B_{g1} \cdot \frac{\tau}{\pi} \cdot (1 + k_{rd}); \quad k_{rd} = (1\text{–}1.5) \tag{4.28}$$

with

$$\delta_i \approx \frac{1}{\mathrm{Re}\left[\frac{\pi^2}{\tau^2} + j \cdot \mu_{iron} \cdot (\sigma_i / k_{ti}) \cdot S \cdot \omega_1\right]^{1/2}} \tag{4.29}$$

The iron magnetization curve is accounted for $B_{ti} = \mu_{iron} \cdot H_{ti}$, by the $B(H)$ curve on the secondary iron surface. Approximately (for low-$S \cdot G_e$ LIMs), the transverse edge effect coefficients (for Al and iron), k_{trAL} and k_{ti}, are (3.40) and (3.42):

$$k_{trAL} \approx \frac{1}{1 - \dfrac{\tanh\left(\frac{\pi}{\tau} \cdot a_e\right)}{\left(\frac{\pi}{\tau} \cdot a_e\right)} \cdot \left[1 + \tanh\left(\frac{\pi}{\tau} \cdot a_e\right) \cdot \tanh\frac{\pi}{\tau}(c - a_e)\right]^{-1}} > 1 \tag{4.30}$$

$$k_{ti} = \frac{1}{1 - \dfrac{\tanh\left(\frac{\pi}{\tau} \cdot \frac{a_e}{i}\right)}{\frac{\pi}{\tau} \cdot \frac{a_e}{i}}} > 1; \quad i = 1, 2, 3 \tag{4.31}$$

The equivalent goodness factor G_{eALi} becomes

$$G_{eALi} = \frac{\mu_0 \cdot \omega_1 \cdot \tau^2 \cdot d_{AL}}{\pi^2 \cdot g_{eai}} \cdot \sigma_{eAli} \tag{4.32}$$

For reasonably small k_{trAL}, $c = \tau/\pi$.

These expressions allow for SLIM circuit model following the same routine as for DLIM in the previous paragraph. But the secondary back-iron current density variation along penetration depth produces a leakage inductance L'_{2l}, which may be approximated as (from eddy current theories [3])

$$L'_{2l} \approx \frac{R'_{2i}}{S \cdot \omega_1} = \frac{R'_2}{|S \cdot \omega_1|} \cdot \frac{\sigma_i \cdot \delta_i}{\sigma_{AL} \cdot d_{AL}} \cdot \frac{k_{trAl}}{k_{ti}} \cdot k_{skin}; \quad |S \cdot \omega_1| > 2\pi \times 1\,[\mathrm{rad/s}] \tag{4.33}$$

So the equivalent circuit for SLIMs has to be completed with the secondary back-iron leakage inductance (Figure 4.6).

As evident in Figure 4.6, the magnetization inductance L_m, secondary resistance R'_2, and the secondary iron leakage inductance L'_{2l} (reduced to primary) depend on the primary mmf (or on linear current sheet $J_1(I_1)$) and on slip frequency $S \cdot \omega_1$. Moreover, the mechanical airgap g varies due to secondary (track) irregularities and on the primary on vehicle framing, light but inevitable, vibrations. To illustrate the magnitude of secondary solid iron influence on SLIM performance, let us consider the

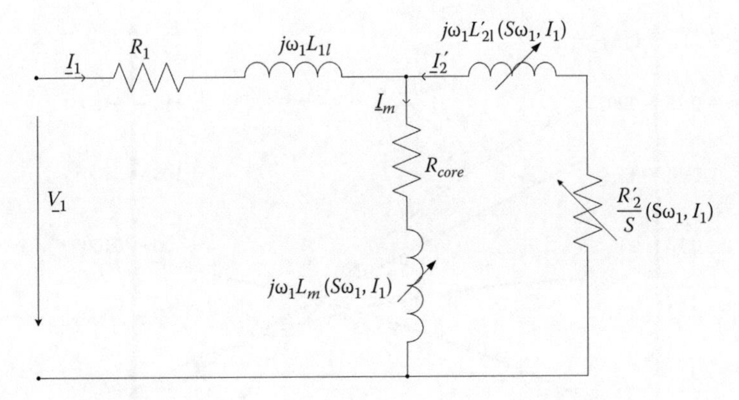

FIGURE 4.6 SLIM equivalent circuit.

numerical results for a LIM with the data: $\tau = 0.084$ m, $2a = 0.08$ m, back-iron laminations count $i = 3$, number of poles $2p = 6$, W_1(turns per phase) $= 480$, mechanical gap $g + h_{AL} = 0.012$ m, $d_{AL} = 0.06$ m, aluminum sheet overhang width $2c = 0.12$ m, $\sigma_{AL} = 3.5 \times 10^7$ $(\Omega\ \text{m})^{-1}$, $\sigma_i = 3.55 \times 10^6$ $(\Omega\ \text{m})^{-1}$; stator slot depth $h_s = 0.045$ m, open slot width $b_s = 0.009$ m, $y/\tau = 5/6$ (chorded coils).

The influence of iron on equivalent secondary conductivity accounts both for edge and for skin effects; this explains why $\sigma_{eALi}/\sigma_{AL} < 1$ and the presence of a minimum point (Figure 4.7a).

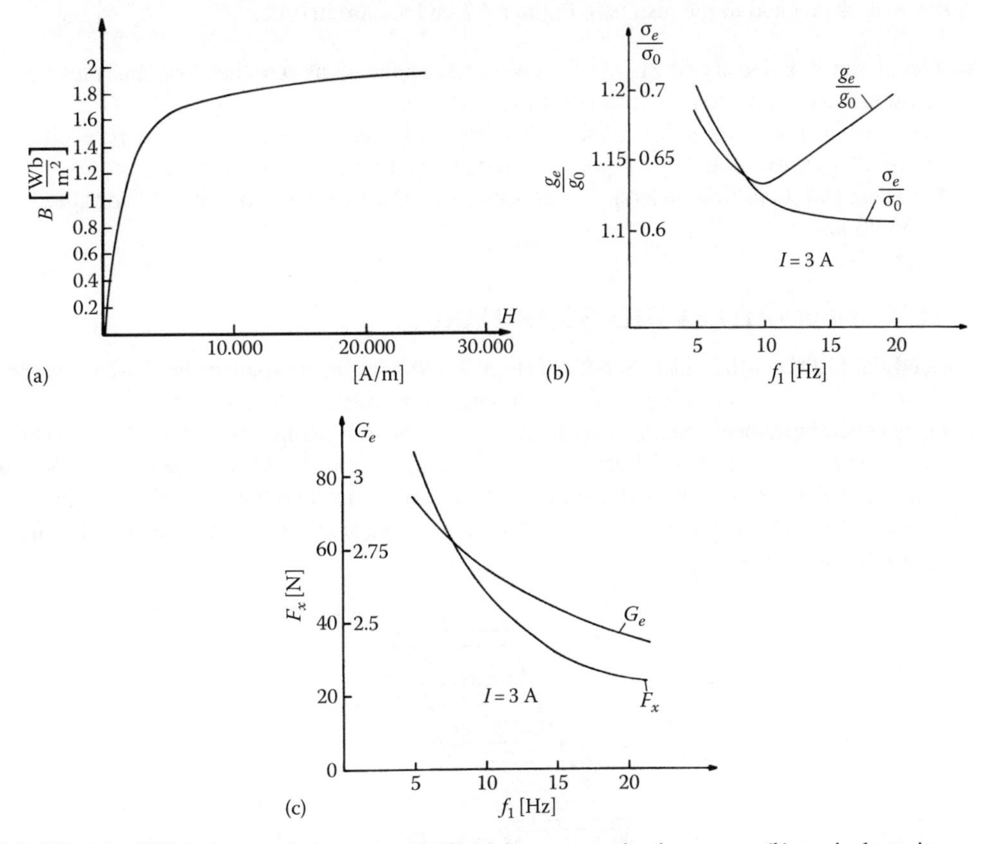

FIGURE 4.7 SLIM characteristics at standstill (a) iron magnetization curve, (b) equivalent airgap and electrical conductivity versus frequency, and (c) thrust and equivalent goodness factor versus frequency, at constant current. (After Boldea, I. and Nasar, S.A., *Linear Motion Electromagnetic Systems*, John Wiley, New York, 1985.)

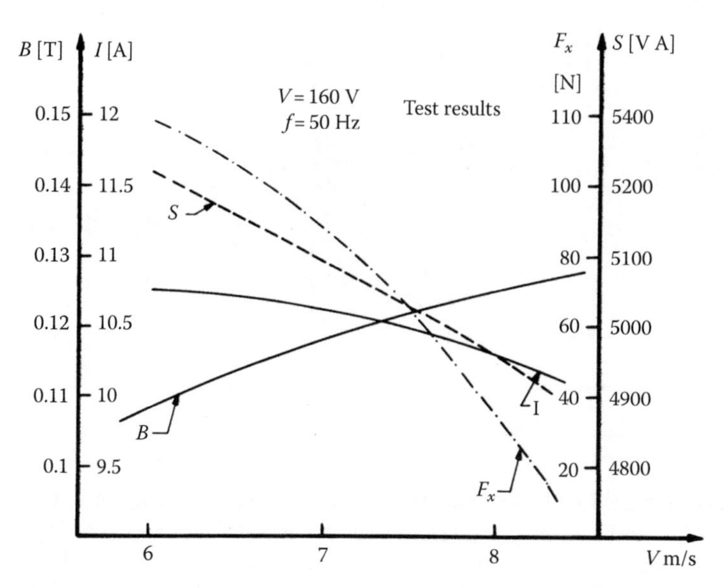

FIGURE 4.8 SLIM running test results (synchronous speed $U_S = 2 \cdot \tau \cdot f = 8.4$ m/s). (After Boldea, I. and Nasar, S.A., *Linear Motion Electromagnetic Systems*, John Wiley, New York, 1985.)

A few remarks related to the results in Figures 4.7 and 4.8 are in order:

- The airgap flux density of SLIMs for low-speed applications is rather low, and thus the thrust density is low, so are efficiency and power factor.
- The case in point suggests to use, for same synchronous speed, a higher pole pitch $\tau = 2 \cdot 0.084 = 0.168$ m and $2p = 4$ poles, to somehow alleviate the performance.
- For short travels (a few meters), ladder secondary flat SLIM should be used for speeds below 10 m/s.

4.4 FLAT SLIMs WITH LADDER SECONDARY

Low-speed flat SLIMs with ladder secondary (Figure 4.9) have been proposed (and used) for short travel (less than 3–4 m) applications (machining tables) for thrusts up to a few kN.

In such applications, the mechanical airgap may be as low as 1–2 mm. This is the whole magnetic airgap. For small airgap, at least the primary slots have to be semiclosed with openings (b_{s1}) up to 4–6 mm, while the secondary ladder may be placed in a laminated core with slot openings b_{s2} of 2–3 mm. This way the Carter coefficient, due to double slotting, K_C <1.3–1.4 and the airgap fringing coefficient $k_{Fg} \approx 1$.

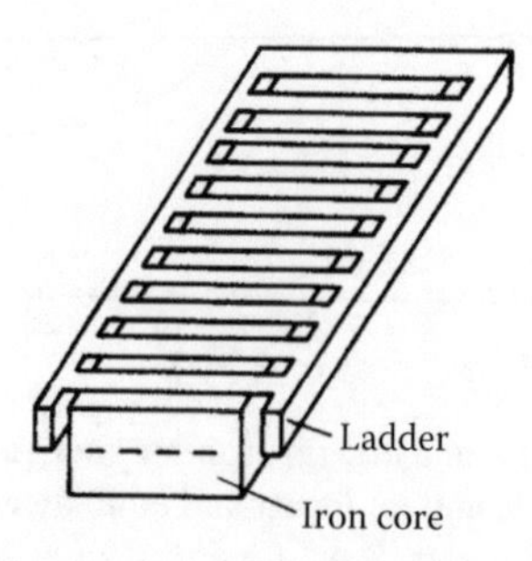

FIGURE 4.9 Ladder secondary of SLIM.

The airgap flux density at rated thrust may now be chosen as high as 0.65–0.7 T because the back iron in the secondary is laminated. The secondary ladder, made of aluminum, may be die casted in sections as for rotary IMs.

The parameters of the equivalent circuit have expressions similar to those of rotary cage rotor IMs:

$$R_2' = \frac{2\cdot a\cdot\left(W_1\cdot k_{W_1}\right)^2}{p\cdot\tau\cdot d_{AL}\cdot\frac{b_{s2}}{\tau_{s2}}\cdot\left(\sigma_{AL}/k_{skin}\right)}\cdot\left(1+k_{end}\right);\quad k_{end} = \frac{R_r}{2\cdot R_b\cdot\sin^2\left(\pi\cdot p/N_{s2}\right)} \tag{4.34}$$

k_{end} accounts for end-bars resistance in the ladder of secondary:

$$L_{2l}' = 2\cdot\mu_0\cdot 2\cdot a\cdot 12\cdot\frac{\left(W_1\cdot k_{w_1}\right)^2}{N_{s2}}\cdot\left(\lambda_{s2}+\lambda_{diff2}+\lambda_{end2}\right) \tag{4.35}$$

$$L_m = \frac{6\mu_0}{\pi^2}\cdot\frac{\left(W_1\cdot k_{w_1}\right)^2\cdot\tau\cdot 2a}{g\cdot k_c\cdot\left(1+k_{ss}\right)};\quad \lambda_{s2}\approx 1\text{–}2;\quad \lambda_{diff2}=0.3\text{–}0.4;\quad \lambda_{end}=0.7\text{–}1.5\approx 0.6\left(\frac{l_{ec}}{2a}\right) \tag{4.36}$$

where

N_{s2} is the number of secondary slots per primary length

λ_{s2}, λ_{diff2}, λ_{end2} are the geometric specific (nondimensional) permeances for secondary leakage field.

k_{ss} is the total (primary and secondary) core magnetic saturation coefficient ($k_{ss}\approx 0.3$–0.6, in general)

b_{s2}, τ_{s2} are the secondary slot opening and slot pitch ($b_{s2}/\tau_{s2}\approx 1/2$)

R_r is the ladder end-bar section resistance

R_b is the secondary in-slot bar resistance

Let us consider a numerical example with $\tau=0.1$ m, $2p=10$, $g=1.5$ mm, $d_{AL}=16$ mm, $b_{s2}/\tau_{s2}\approx 1/2$; $k_{end2}=0.3$, $2a=0.2$ m, $k_c=1.3$, $k_{skin}=1.25$, $\sigma_{AL}=3.5\times 10^7\ (\Omega\ \text{m})^{-1}$, $k_{ss}=0.1$.

The goodness factor G_e, for $U_s=4$ m/s ($f_1=20$ Hz) is

$$\left(G_e\right)_{20\,\text{Hz}} = \frac{\mu_0\cdot\omega_1\cdot\tau^2\cdot\left(\sigma_{AL}/k_{skin}\right)}{\pi^2\cdot g\cdot k_c}\cdot d_{AL}\cdot\left(\frac{\frac{b_s}{\tau_s}}{1+k_{end}}\right)$$

$$= \frac{1.256\cdot 10^{-6}\cdot 2\pi\cdot 20\cdot 0.1^2\cdot\left(3.5\cdot 10^7/1.2\right)\cdot 1.6\cdot 10^{-2}\cdot\frac{1}{2}}{\pi^2\cdot 1.5\cdot 1.3\cdot 10^{-3}\cdot 1.1\cdot 1.3} = 13.38$$

This rather high value would allow for maximum thrust ($s_kG_e=1$), the slip $s_k=0.0747$, to provide for a good power factor (above 0.75), besides a good efficiency (above 0.75); the rated slip may be even larger than S_k for $S_{rated}G_e=1.3$, for PWM vector control.

The ladder secondary thus seems capable of producing notable performance even for low-speed, low travel (3–4 m) and also for higher speeds (higher $G_e \le G_{e0}(2p)$). The value of $G_e=13.38<20$ (the optimum goodness factor for $2p=10$). Also, even at $S=0.1$, the dynamic end effect is negligible as seen in Figure 4.1.

Note: As seen here, the dynamic end effect may be present even at small speeds (see in the previous example for $2p=4, 6$) – $U=U_s*(1-S)=4*(1-0.1)=3.6$ m/s—if the goodness factor is high. In the earlier example, for $2p=6$, for example, it suffices to increase the airgap from 1.5 mm to 1.5 * 10/6 = 2.5 mm to reduce G_e to $G_e \approx 8$ and thus reduce the dynamic end effect to negligible influence, at the price of also reducing the conventional performance. In some small-speed applications, it may be thus preferable to allow for some dynamic longitudinal end effect for ($2p=4, 6$) but still secure good performance due to higher G_e.

4.5 TUBULAR SLIM WITH LADDER SECONDARY

Tubular SLIM may be favored in short length (less than 2.5 m) travel applications due to zero net radial force between primary and secondary (for symmetric mover position in the airgap) and practically zero end connections of short-primary circular coils (Figure 4.10). Also, the stator is to be made of disk-shape laminations, with open slots (inevitably so) to cut the material and manufacturing costs; to reduce eddy current iron losses in the back iron (where the magnetic field goes perpendicular to lamination plane), thin radial slits are cut in the back-iron laminations (Figure 4.10b); the same is valid for the secondary core; the secondary ladder is made of aluminum rings or short-circuited circular-shape coils in open slots; the slot opening should not surpass 3 airgap lengths in the secondary and 4–5 airgaps in the primary, to limit Carter coefficient. This way, with an airgap of 1–1.5 mm, good results are expected in terms of performance/cost, with respect to tubular linear permanent magnet synchronous motors, which represent such a strong competition that the Al-on-iron secondary was discarded here for tubular SLIMs (Figure 4.10). Only the ladder secondary is considered.

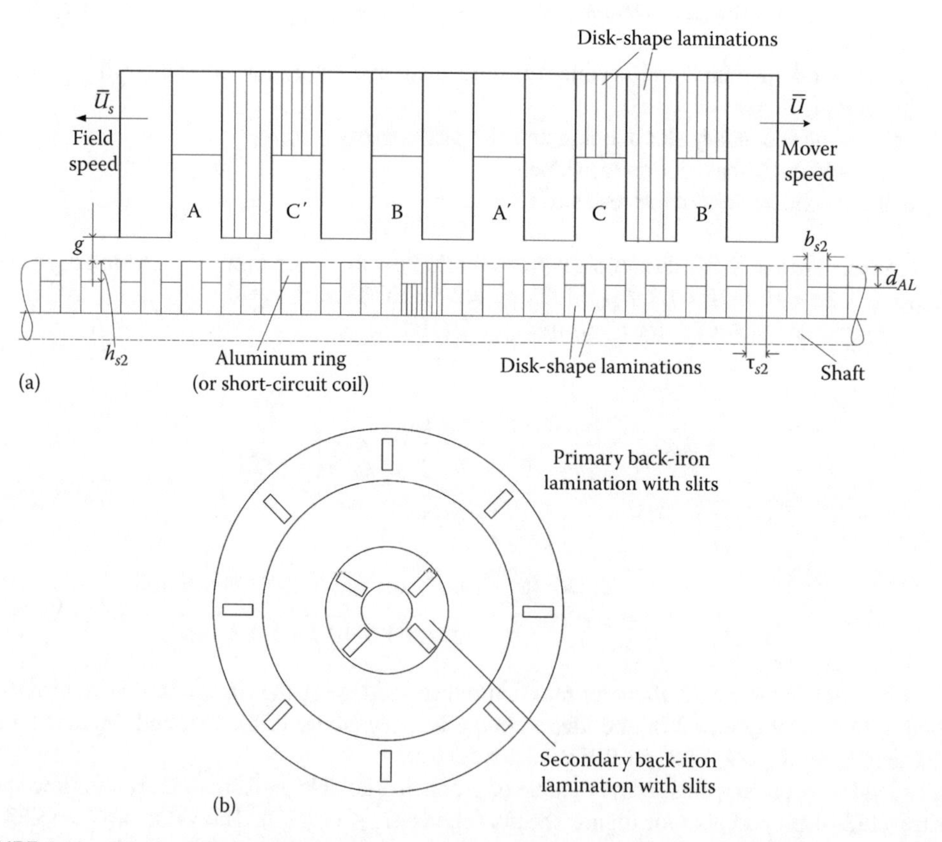

FIGURE 4.10 Tubular SLIM with ladder secondary: (a) longitudinal cross section, (b) back-iron laminations with slits.

Note: For very short travels (less than 0.4–0.5 m) and long (fix) primary with short mover secondary, to reduce the mover weight and avoid a flexible power cable, a dual airgap structure, with an aluminum cylinder with embedded Somaloy inserts (as teeth), which reduces the magnetic airgap to the mechanical dual one, is proposed. As linear PM synchronous motors represent a strong contender for such applications, this solution is not pursued here further.

The tubular SLIMs with ladder secondary parameters have expressions similar to flat SLIMs, but the primary and secondary coils are circular and only slot and differential leakage inductances exist; the stack width now becomes $\pi \cdot D_{is}$ (D_{is}, secondary outer diameter); so no secondary transverse edge effect occurs; skin effect in the secondary remains present and is treated as for rotary cage rotor IMs:

$$R_{1t} = \rho_{co} \frac{\pi \cdot D_{av1} \cdot j_{cor} \cdot W_1^2}{I_{1r} \cdot w_1} = k_{R1t} \cdot W_1^2; \quad X_{1l} = \frac{2\mu_0 \cdot \pi \cdot D_{av1}}{p \cdot q} \cdot \left(\lambda_{slot1} + \lambda_{diff1}\right) \cdot W_1^2 = k_{1lt} \cdot W_1^2$$

$$L_m = \frac{6\mu_0}{\pi^2} \cdot \frac{\left(W_1 \cdot k_{W1}\right)^2 \cdot \tau \cdot \pi \cdot D_{is}}{p \cdot g \cdot k_c}$$

$$L'_{2l} = \frac{2\mu_0 \cdot \pi \cdot D_{av2} \cdot 12 \cdot \left(W_1 \cdot k_{W1}\right)^2}{N_{s2}} \cdot \left(\lambda_{s2} + \lambda_{diff2}\right) = k_{2lt} \cdot \omega_1^2$$

$$R'_2 = \rho_{AL} \frac{\pi \cdot D_{av2} \cdot 12 \cdot \left(W_1 \cdot k_{W1}\right)^2}{A_{ring} \cdot N_{s2}} = k_{R2t} \cdot W_1^2 = k_{Lm} \cdot W_1^2; \quad A_{ring} = b_{s2} \cdot h_{s2} \cdot k_{fill2} \tag{4.37}$$

With the given rated phase mmf $W_1 \cdot I_{1n}$ and rated current density j_{cor}, all parameters of the equivalent circuit (as for flat LIMs) are proportional to the number of turns/phase squared. This peculiarity will help notably in the design process as evident in the next chapter.

Also, the reasonably small airgap (1–1.5 mm) for short travel (0.3–1.5 m), though larger than that in rotary IMs, provides for acceptable performance/cost and reliability in applications where PMs are to be avoided.

4.6 CIRCUIT MODELS OF HIGH-SPEED (HIGH GOODNESS FACTOR) SLIMs

By high-speed LIMs, we mean LIMs where dynamic end effects have to be considered; as already explained in this chapter, the dynamic end effect can occur in ladder secondary LIMs of small airgap (short travels) even at low speeds (4–5 m/s), at large goodness factor ($G_e > 8$) values, for a small number of poles ($2p = 2, 4$). So we might as well use the term high goodness factor or large dynamic end effect LIMs, instead of high-speed LIMs.

Accounting for dynamic end effect in the circuit model would mean introducing its influence on the conventional circuit parameters and adding new ones.

The first attempt [4] relied on the idea that the end effect reduces the thrust and consequently reduces the emf k_e times. This effect may be translated into an additional parallel term in the equivalent circuit Z_e:

$$\underline{Z}_e = \frac{j \cdot X_m \cdot \left(R'_2/s + j \cdot X'_{2l}\right)}{\frac{R'_2}{s} + j \cdot \left(X_m + X'_{2l}\right)} \cdot \frac{1 - k_e}{k_e} \tag{4.38}$$

with

$$E_{se} \approx E_{ms} \cdot (1 - k_e); \quad E_{ms} = X_m \cdot I_m$$

$$k_e = \frac{k_{wel}}{k_{w1}} \frac{\frac{\pi \cdot \tau_e}{\tau^2}}{\gamma_{2r}^2 + \left(\frac{\pi}{\tau_e}\right)^2} \cdot f(\delta) \cdot e^{-p \cdot \tau_e |\gamma_{2r}|} \cdot \frac{\sinh(p \cdot \tau_e \cdot \gamma_{2r})}{p \cdot \sinh(\tau_e \cdot \gamma_{2r})}; \quad f(\delta) \approx \gamma_{2r} \cdot \sin\delta + \frac{\pi}{\tau_e} \cdot \cos\delta$$

$$\delta = \delta_0 + c_1 \cdot U; \quad \delta_0 \approx \frac{3\pi}{4}$$

$$k_{wel} = \frac{\sin\left(\frac{\tau}{\tau_e} \cdot \frac{\pi}{6}\right)}{q_1 \cdot \sin\left(\frac{\tau}{\tau_e} \cdot \frac{\pi}{6 \cdot q_1}\right)} \cdot \sin\left(\frac{\tau}{\tau_e} \cdot \frac{\pi}{2} \cdot \frac{y}{\tau}\right); \quad \gamma_{2r} \text{ from (4.2)}$$

with k_{wel}, τ_e winding factor and pole pitch for the exit end wave (3.67); U is the LIM speed and c_1 is [4]

$$c_1 \approx \frac{1}{U_{s\max}} \cdot \tan^{-1}\left(\frac{\pi \cdot \tau_e}{\tau}\right)_{U = U_{s\max/2}} \tag{4.39}$$

The equivalent circuit in Figure 4.11 has been checked against two LIMs one of medium speed and one of high speed, with good agreement in terms of thrust versus speed for given phase current (Figures 4.12 and 4.13) [4].

As the verifications of active, reactive powers and secondary losses agreement of equivalent circuit in Figure 4.11 with field theory results (Chapter 3) have not been performed, subsequently, starting from the same idea of emf reduction by end effect, modifications in all components of the standard circuit model have also been added (Figure 4.14) [5].

The end effect impedance $\underline{Z}_e$ in Figure 4.14 [5] is the same as in Figure 4.11 [4] but, in addition, the coefficients k_{xe} and k_{Re} are calculated from the single dimensional field theory (see Chapter 3) by airgap reactive and active secondary power balance. The thrust and power factor agreement of this

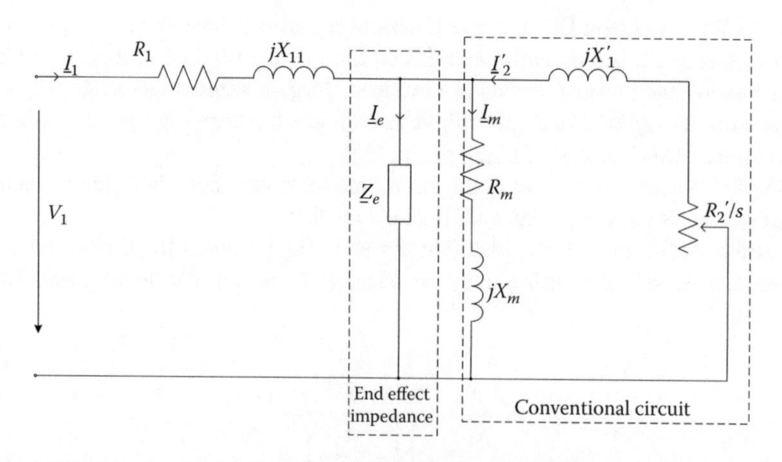

FIGURE 4.11 LIM equivalent circuit with simplified end effect impedance $\underline{Z}_e$.

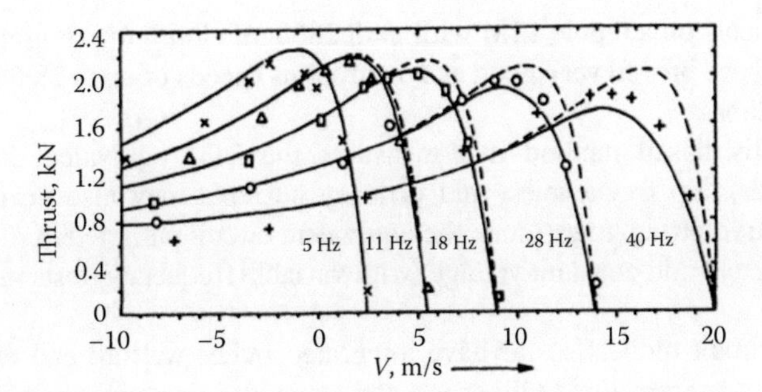

FIGURE 4.12 Thrust/speed curves for constant current: theory versus experiments—CIGGT–LIM. (After Gieras, J.F. et al., *IEEE Trans.*, EC-2(1), 152, 1987.)

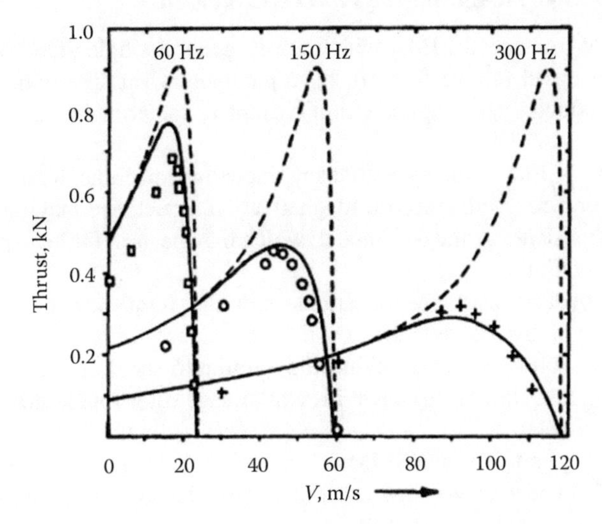

FIGURE 4.13 Thrust/speed curves for constant current: GEC–LIM. (After Gieras, J.F. et al., *IEEE Trans.*, EC-2(1), 152, 1987.)

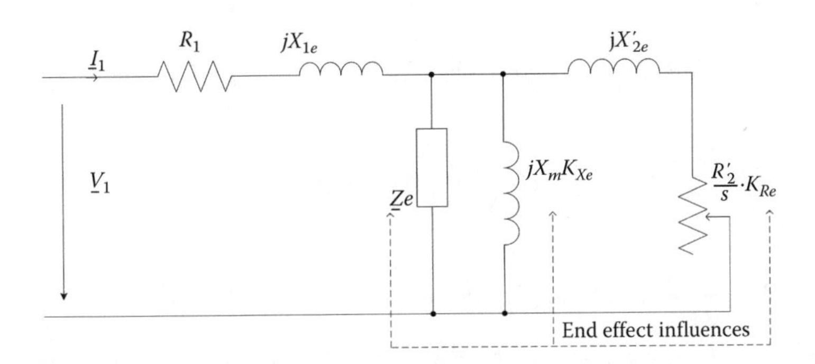

FIGURE 4.14 More complete equivalent circuit of LIM with end effect. (After Cabral, C.M.A.R., Analysis of LIMs using new longitudinal end effect factors, *Record of LDIA-2003*, Birmingham, U.K., pp. 291–294, 2003.)

circuit with experiments on a 6 pole LIM with $\tau=0.2868$ m, aluminum (6 mm) on iron secondary, with 10 mm airgap, have proved very good at synchronous speeds of up to 28.68 m/s (140 kW SLIM for urban transportation).

An experimentally based method that measures the LIM equivalent impedance at a few representative speeds, slip frequencies, and primary currents may also be used to identify the equivalent circuit curve fitting, to estimate the equivalent circuit parameters.

Both constant current and constant voltage (with variable frequency) tests may be performed for the scope.

The equivalent circuit models so far have integrated rather well all end effect influences, but at the price of variable parameters (with speed and current); in essence, only L_m and R_2' vary with speed and current; and an additional end effect impedance (resistance, even) $\underline{Z}_e(k_e)$ is added; k_e is variable with speed. Even linear variations (L_m and R_2' increase with speed) would suffice at a first approximation to be used in circuit models for transients and control.

4.7 LOW-SPEED LIM TRANSIENTS AND CONTROL

The circuit models of low-speed LIMs, which are in general single sided, with aluminum-iron secondary for longer travel (above 3–4 m), have parameters variable with slip frequency (due to skin and/or edge effects) and magnetization current I_m (or primary current) due to magnetic saturation.

In short, the LIM is like a rotary equivalent induction machine with variable parameters. Magnetic saturation or airgap influence on magnetization inductance may be (and is) handled by $L_m(I_m)$ to investigate transients by the d–q model, which may be used for low-speed (low end effect) LIMs, with small adaptations.

Moreover, the secondary resistance variations with slip frequency may be small as the slip frequency is controlled in inverter-fed (all) LIM drives.

But the presence of solid back iron behind the aluminum sheet of LIM secondary leads to a variation of R_2' with I_m also: this is unknown in regular cage rotor rotary IMs (but known to solid rotor high-speed rotary IMs).

Let us consider $R_2'(I_m)$ only in such SLIMs.

Now it becomes evident that we may use the d–q model for low-speed LIM transients and control.

4.7.1 Space–Phasor (dq) Model

The rotary IM space–phasor model can be adapted rather easily for low-speed LIMs; in general coordinates,

$$\bar{V}_1 = R_1 \cdot \bar{i}_1 + \frac{d\bar{\Psi}_1}{dt} + j\omega_b\bar{\lambda}_1; \quad \bar{V}_1 = V_d + jV_q; \quad \bar{i}_1 = i_d + ji_q; \quad \omega_b = \omega_1$$

$$\bar{V}_2' = 0 = R_2' \cdot \bar{i}_2' + \frac{d\bar{\Psi}_2'}{dt} + j\cdot(\omega_b - \omega_r)\cdot\bar{\Psi}_2'; \quad i_2' = i_D + ji_Q \tag{4.40}$$

$$F_x = \frac{-3}{2}\cdot\frac{\pi}{\tau}\cdot \mathrm{Imag}\left(\bar{\Psi}_1\bar{i}_1^*\right); \quad \bar{\Psi}_1 = L_1\cdot i_1 + L_m\cdot i_2'; \quad \bar{\Psi}_2' = L_2'\cdot i_2' + L_m\cdot i_1$$

$$L_1 = L_{1l} + L_m; \quad L_2' = L_{2l}' + L_m; \quad i_m = i_1 + i_2'; \quad \omega_1 = \frac{U_s\cdot\pi}{\tau}; \quad \omega_r = \frac{U\cdot\pi}{\tau}$$

$$F_{na} \approx \frac{3}{2} \cdot \frac{L_m \cdot i_m^2}{g_e}; \quad \theta_b = \int \omega_1 dt = -\int (S \cdot \omega_1 + \omega_r) dt \tag{4.41}$$

A few remarks are in order:

- The equations are similar to those of single-cage rotary IMs (in the absence of end effects).
- The thrust, F_x, equation has the sign (–) because the primary is the mover.
- ω_b is the speed of the reference system $\omega_b = \omega_1$ for synchronous coordinates.
- The sign (–) in the reference system angle θ_b is due to the fact that the primary is the mover.
- The normal attraction force (total normal force for cage secondary LIM) F_{na} is obtained here as the derivative of magnetic energy in the airgap with respect to airgap g_e.
- Apparently, the number of pole pairs p does not occur in thrust F_x, but, in fact, it is involved implicitly by the flux $\bar{\Psi}_1$ expression. The space–phasor diagrams for steady state ($d/dt = 0$) in synchronous coordinates ($\omega_b = \omega_1$) are shown in Figure 4.15 for motoring and generating (regenerative braking).

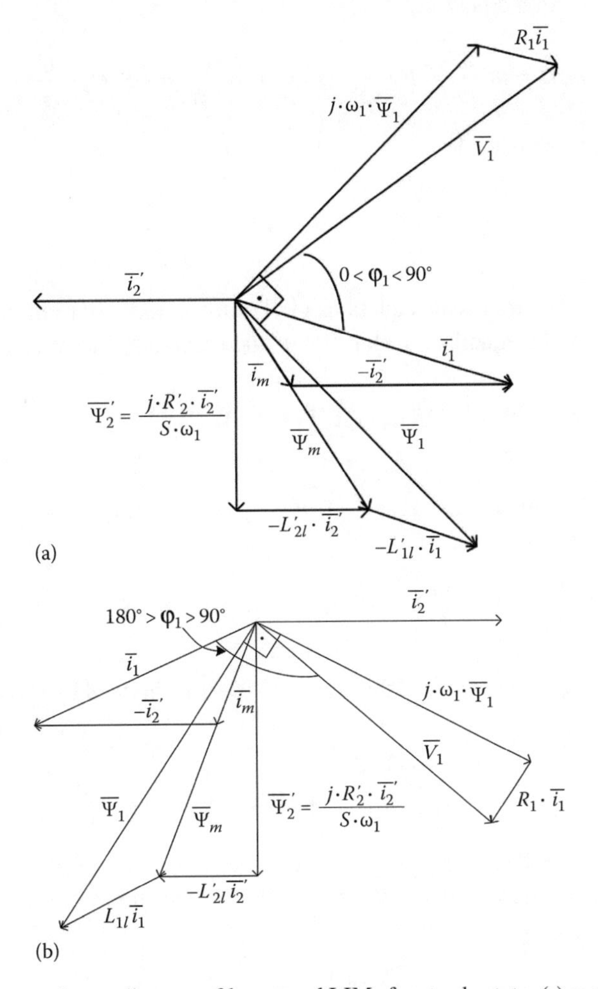

FIGURE 4.15 The space–phasor diagram of low-speed LIMs for steady state: (a) motoring, (b) generating.

The investigation of transients may be approached directly by Equation 4.40, say, in synchronous coordinates $\omega_b = \omega_1$ with the d/dt replaced by s (Laplace operator):

$$V_1(s) = R_1 \cdot i_1(s) + s \cdot \Psi_1(s) + j \cdot \omega_1 \cdot \Psi_1(s) \tag{4.42}$$

$$0 = R_2' \cdot i_2'(s) + s \cdot \Psi_2'(s) + j \cdot S \cdot \omega_1 \cdot \Psi_2'(s)$$

Apparently, we do have four variables and two equations, but the flux/current relationships reduce the number of complex variables to two. To include the wound secondary case $V_2' \neq 0$ in (4.40) and for more generality we adopt this case.

Three types of controls seem practical:

- $I_1 - f_1$ (scalar control)
- Secondary ("rotor") vector control $\left(\overline{\Psi}_2', \overline{i}_1\right)$: FOC
- Direct thrust and flux control $\left(F_x, \Psi_1\right)$: DTFC

The earlier control methods introduce constraints and thus reduce the order of the system (4.41), which still needs the motion equations:

$$msU = m\frac{dU}{dt} = F_x - F_{load}; \quad s \cdot \theta_b = \frac{d\theta_b}{dt} = -S \cdot \omega_1 - \frac{U}{\tau} \cdot \pi \tag{4.43}$$

Also, the primary voltage vector $\overline{V}_1$ is

$$\overline{V}_1 = \frac{2}{3} \cdot \left(V_a + V_b \cdot e^{j\frac{2\pi}{3}} + V_c \cdot e^{-j\frac{2\pi}{3}} \right) \cdot e^{-j\theta_b} \tag{4.44}$$

For constant speed ($U = ct$) transients equations (4.41) suffice and, after eliminating the currents $\overline{i}_1$ and $\overline{i}_2'$, we are left with two equations with primary and secondary flux vectors as variables [6]:

$$\tau_s' \cdot s \cdot \Psi_1 + \left(1 + j \cdot \omega_1 \cdot \tau_s'\right) \cdot \Psi_1 = \tau_s' \cdot V_1 + k_r \cdot \Psi_2'; \quad \tau_s = \frac{L_1}{R_1}$$

$$\tau_r' \cdot s \cdot \Psi_2' + \left(1 + j \cdot s \cdot \omega_1 \cdot \tau_r'\right) \cdot \Psi_2' = \tau_r' \cdot V_2' + k_s \cdot \Psi_1; \quad \tau_r = \frac{L_2'}{R_2'} \tag{4.45}$$

$$k_s = \frac{L_m}{L_1}; \quad k_r = \frac{L_m}{L_2'}; \quad \tau_s' = \tau_s \cdot \sigma; \quad \tau_r' = \tau_r \cdot \sigma; \quad \sigma = 1 - \frac{L_m^2}{L_2' \times L_1}$$

It is now evident that the eigen values of the system (4.45) are obtained by solving the characteristic equation:

$$\begin{vmatrix} 1 + \left(s + j \cdot \omega_1\right) \cdot \tau_s' & -k_r \\ -k_s & 1 + \left(s + j \cdot S \cdot \omega_1\right) \cdot \tau_r' \end{vmatrix} = 0 \tag{4.46}$$

It may be proved that the real part of $\gamma_{1,2}$ eigen values is negative, for the stable part of the mechanical curve $F_x(U)$, for given V_1 and $V_2' = 0$ as known from the critical slip frequency of rotary IMs:

$$\left(S \cdot \omega_1\right)_k \approx \frac{R_2'}{L_{1l} + L_{2l}'} \tag{4.47}$$

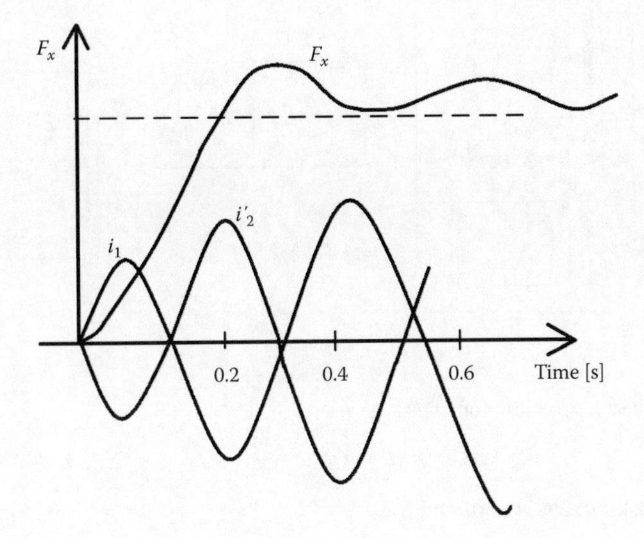

FIGURE 4.16 Typical sudden ac symmetrical input at standstill, from zero current.

In essence, however, there are two complex eigen values and two rather fast time constants for LIM transients $\tau_s^{'}$ and $\tau_r^{'}$; one inflicted by the primary $\left(\tau_s^{'}\right)$ and the other by the secondary $\left(\tau_r^{'}\right)$. As the solutions of (4.46) are analytical, the LIM transients at constant speed are straightforward. Also, as the LIM mover mass is rather large, the mechanical transients are notably slower than electromagnetic transients, so that they can be solved separately by inserting the flux and current instantaneous values in the thrust expression present in the motion equation.

A typical qualitative response, at standstill, with sudden sinusoidal voltages (of given frequency) application and zero initial current, is shown in Figure 4.16.

The sum of two currents $i_1 + i_2^{'}$ is the magnetization current i_m, and the thrust builds up after some time because it takes time to install the magnetic flux in the machine, to interact with current for thrust production.

Much faster thrust response is obtained with the flux already installed in the machine, even if a single voltage vector is preapplied to produce a dc field (with fixed position) in the LIM, before starting.

4.8 CONTROL OF LOW-SPEED LIMs

As already stated in this chapter, scalar (SC), vector (FOC), and direct torque and flux control (DTFC) seem [6] the most practical solutions.

Let us start with scalar $I_1 - Sf_1$ close-loop control.

4.8.1 Scalar $I_1 - Sf_1$ Close-Loop Control

As shown in the circuit model of LIMs, the thrust is

$$F_x = \frac{3 \cdot I_1^2 \cdot R_2^{'}}{2 \cdot \tau \cdot f_1 \cdot S \cdot \left(1 + \left(\dfrac{1}{S \cdot G_e}\right)^2\right)}; \quad G_e \approx \omega_1 \cdot \frac{L_m}{R_2^{'}} \tag{4.48}$$

with current and slip frequency as controlled variables, a simple control strategy may be developed with and without a speed control loop.

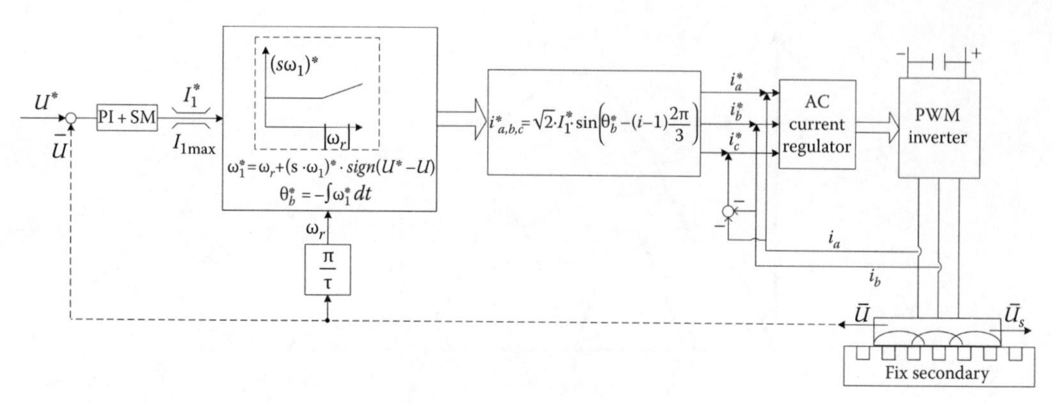

FIGURE 4.17 I–f close-loop control of LIMs.

The maximum thrust/current is obtained for

$$\left(S \cdot \omega_1\right)_k = \frac{R_2'}{L_m} \tag{4.49}$$

The slip frequency may be kept constant, up to base speed obtained at full voltage, for sustained full load and then increased a little for flux weakening.

Increasing $S \cdot \omega_1$ over $(S \cdot \omega_1)_k$ means reducing the efficiency, but still allowing higher speed/thrust envelope.

A generic block diagram is shown in Figure 4.17.

A few notes on Figure 4.17 are in order:

- Motoring and regenerative braking may be provided and the slip frequency is feed forwarded.
- The speed controller (illustrated here by a PI plus sliding mode implementation—for robustness) outputs the stator phase current RMS value which is limited, for protection.
- With measured LIM speed, the primary frequency ω_1^* is feed forwarded via the current angle θ_b^* for the three reference currents i_a^*, i_b^*, i_c^*.
- AC current regulators (individual or current vector error or predictive regulators) are used to control the PWM inverter.
- An outer position close loop may be added if position control action is targeted.
- Though called scalar control, the scheme in Figure 4.17 is almost equivalent with indirect current vector control (used for rotary IMs).
- As the parameters (L_m, R_2') may vary with current I_1 and slip frequency, $S\omega_1$, the control is not fully robust, but the PI + SM speed regulator compensates for this drawback notably, despite the apparent simplicity of the control scheme.

4.8.2 Vector Control of Low-Speed LIMs

Vector control means independent flux and thrust control in LIMs. It may be primary flux $\bar{\Psi}_1$, airgap flux $\bar{\Psi}_m$, or secondary flux $\bar{\Psi}_2'$. The implementation of vector control (FOC) relates to separate control of the primary-flux current i_M and thrust current i_T components in synchronous coordinates ($\omega_b = \omega_1$; dc in steady state).

While FOC can be performed for all the defined fluxes, only with constant secondary flux Ψ_2' amplitude the thrust/speed curves are linear; that is ideal for control purposes (as in dc brush PM motors!).

So, returning to the space–phasor model (4.32 through 4.42), we may eliminate secondary current I_2' and primary flux $\overline{\Psi}_1$ to obtain

$$\overline{V}_1 = \left(R_1 + \left(s + j\cdot\omega_1\right)\cdot L_{sc}\right)\cdot \overline{i}_1 + \left(s + j\cdot\omega_1\right)\cdot \frac{L_m}{L_2'}\cdot \Psi_2' \tag{4.50}$$

If the secondary flux is kept constant, in synchronous coordinates, it means $s\cdot\Psi_2' = 0$ in the two equations mentioned previously.

So the system becomes a first-order one:

$$0 = R_2'\cdot i_2'(s) + j\cdot S\cdot\omega_1\cdot\Psi_2'(s)$$

$$\overline{V}_1 = \left(R_1 + \left(s + j\cdot\omega_1\right)\cdot L_{sc}\right)\cdot \overline{i}_1 + j\cdot\omega_1\cdot \frac{L_m}{L_2'}\cdot \overline{\Psi}_2'$$

$$\overline{i}_s = \frac{\Psi_2'}{L_m} + j\cdot S\cdot\omega_1 \frac{\overline{\Psi}_2'}{L_m}\cdot\frac{L_2'}{R_2'} = i_M + j\cdot i_T \tag{4.51}$$

$$i_M = \frac{\overline{\Psi}_2'}{L_m};\quad i_T = i_M\cdot S\cdot\omega_1\cdot\frac{L_2'}{R_2'}$$

Also,

$$\overline{\Psi}_1 = L_1\cdot i_M + j\cdot L_{sc}\cdot i_T \tag{4.52}$$

Finally, the thrust is simply

$$F_x = \frac{-3}{2}\cdot\frac{\pi}{\tau}\cdot \mathrm{Imag}\left(\overline{\Psi}_1 * \overline{i}_1\right) = \frac{-3}{2}\cdot\frac{\pi}{\tau}\cdot\left(L_1 - L_{sc}\right)\cdot I_M\cdot I_T \tag{4.53}$$

where

$$L_{sc} = L_1 - \frac{L_m^2}{L_2'} \tag{4.54}$$

With (4.51), F_x becomes

$$F_x = \frac{-3}{2}\cdot\frac{\pi}{\tau}\cdot\left(L_1 - L_{sc}\right)\cdot\left(\frac{\overline{\Psi}_2'}{L_m}\right)^2\cdot\frac{L_2'}{R_2'}\cdot S\cdot\omega_1 = \frac{-3}{2}\cdot\frac{\pi}{\tau}\cdot\frac{\left(\Psi_2'\right)^2}{R_2'}\cdot\left(\omega_1 - \omega_r\right) \tag{4.55}$$

Equation 4.55 suggests a straight-line $F_x(U)$ curve for constant secondary flux amplitude Ψ_2'. By varying the frequency at constant Ψ_2', a family of $F_x(U)$ straight lines are obtained (Figure 4.18).

Now the voltage vector components under steady state are

$$V_M = R_1\cdot I_M - \omega_1\cdot L_{sc}\cdot I_T \tag{4.56}$$

$$V_T = R_1\cdot I_T + \omega_1\cdot L_1\cdot I_M$$

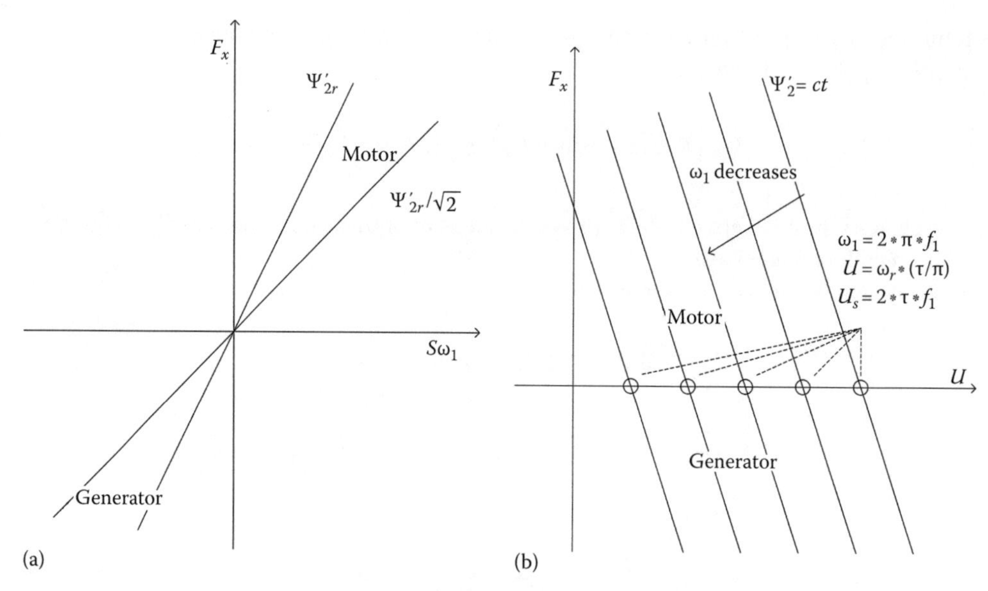

FIGURE 4.18 Mechanical characteristics at constant secondary flux Ψ'_2: (a) $F_x(S\omega_1)$, (b) $F_x(U)$.

Example 4.1

Let us consider a SLIM with the data: $R_1 = 1\ \Omega$, $R'_2 = 2\ \Omega$, $\Psi'_2 = 1\ \text{Wb}$, $L_m = 0.1$ H, $L_{1l} = 0.03$ H, $L'_{2l} = 0.01\text{H}$, $\tau = 0, 1$ m, which works at $f_2 = S \cdot f_1 = 4$ Hz, at a primary frequency of 40 Hz.

Calculate:

1. The speed
2. IM, IT, and thrust
3. VM, V_T, V_1 (RMS per phase), I_1 (RMS per phase), the efficiency and power factor (η_1, $\cos\varphi_1$)

Solution

1. The synchronous speed $U_s = 2 \cdot \tau \cdot f_1 = 2 \cdot 0.1 \cdot 40 = 8$ m/s, while the LIM speed $U = 2 \cdot \tau \cdot (f_1 - f_2) = 2 \cdot 0.1 \cdot (40 - 4) = 7.2$ m/s.
2. From (4.51), $I_M = \dfrac{\bar{\Psi}'_2}{L_m} = \dfrac{1}{0.1} = 10\text{A}$, $I_T = i_M \cdot \dfrac{L'_2}{R'_2} \cdot S \cdot \omega_1$

$$= 10 \cdot \frac{0.11}{2} \cdot 2 \cdot \pi \cdot 4 = 13.81\text{A} \left(L'_2 = L_m + L'_{2l} = 0.1 + 0.01 = 0.11\text{H}\right)$$

The short circuit inductance L_{sc} (4.54) is $L_{sc} = L_1 - L_m^2/L_2 = 0.13 - 0.1^2/0.11 = 0.0391\text{H}$.
So the thrust (4.53) is

$$F_x = \frac{-3}{2} \cdot \frac{\pi}{\tau} \cdot (L_1 - L_{sc}) \cdot I_M \cdot I_T = -\frac{\pi}{0.1} \cdot \frac{3}{2} \cdot (0.13 - 0.0391) \cdot 10 \cdot 13.816 = -591.5\text{ N}$$

3. V_M and V_T from (4.56) are

$$V_M = 1 \cdot 10 - 2 \cdot \pi \cdot 40 \cdot 0.0394 \cdot 13.816 = -126\text{ V}$$

$$V_T = 1 \cdot 13.816 + 2 \cdot \pi \cdot 40 \cdot 0.13 \cdot 10 = 340.37\text{ V}$$

$$V_1 = \sqrt{\frac{V_M^2 + V_T^2}{2}} = 256\,\text{V (RMS per phase)}$$

$$I_1 = \sqrt{\frac{I_M^2 + I_T^2}{2}} = 12.06\,\text{A (RMS per phase)}$$

The $\eta_1 \cos \varphi_1$ (with zero core losses and mechanical losses) is

$$\eta_1 \cos\varphi_1 = \frac{F_x \cdot U}{3 \cdot V_1 \cdot I_1} = \frac{591.5 \times 7.2}{3 \times 256 \times 12.06} = 0.4598$$

And efficiency η_1:

$$\eta_1 = \frac{F_x \cdot U}{F_x \cdot U + 3 \cdot R_2^{'} \cdot I_2^{'2} + 3 \cdot R_1 \cdot I_1^2}$$

$$I_2^{'}\sqrt{2} = \frac{-L_m \cdot I_T}{L_2^{'}}\left(\Psi_{2T}^{'} = 0\right) \text{ and thus } I_2^{'} = \frac{0.1}{0.11} \cdot \frac{13.816}{\sqrt{2}} = 8.907\,\text{A (RMS/phase).}$$

So

$$\eta_1 = \frac{591.5 \cdot 7.2}{591.5 \cdot 7.2 + 3 \cdot 2 \cdot 8.907^2 + 3 \cdot 1 \cdot 13.816^2} = 0.8$$

Finally, the power factor $\cos \varphi_1 = 0.4598/0.8 = 0.573$.

Note: The results show acceptable efficiency but rather low power factor; only with a cage secondary and rather small airgap (1–1.5 mm), power factor values above 0.7 can be hoped for.

As for a vector control (FOC), there are quite a few options; we will illustrate an indirect combined current and voltage vector control as it is rather robust, except for very low speeds (Figure 4.19). For another FOC of LIMs, see Ref. [7].

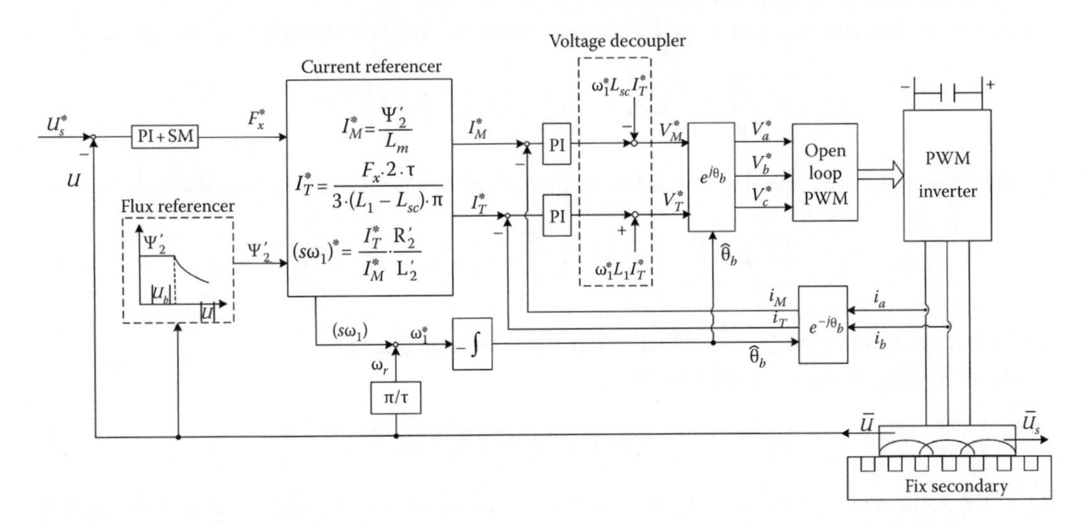

FIGURE 4.19 Field-oriented LIM control.

The control block in Figure 4.19 is characterized by

- A secondary-flux referencer where the secondary-flux amplitude stays constant up to base speed, and then, for flux weakening, it decreases with speed
- A current referencer that calculates online flux and thrust current components i_M and i_T in synchronous coordinates
- A slip frequency ($S\omega_1$) and rotation angle θ_b^* online calculator
- PI dc current controllers
- A voltage decoupler
- An open-loop PWM method that "copies" the sinusoidal symmetric reference voltages V_a^*, V_b^*, V_c^* in the inverter

4.8.3 Sensorless Direct Thrust and Flux Control of Low-Speed LIMs

In an attempt to condense the presentation of an implemented sensorless direct thrust and flux control (DTFC) system for low-speed LIMs, a motion-sensorless solution is presented [7].

DTFC is in fact the equivalent of a direct primary-flux-oriented combined vector current and voltage control system. As primary-flux orientation control is targeted, the $F_x(U)$ curves are not any more straight lines, but they show up for voltage-source supply, those known maximums (for motoring and generating). To secure stability, the slip frequency has to be limited:

$$\left(S\cdot\omega_1\right)_{\Psi_s} \le \frac{R_2'}{L_{sc}} \tag{4.57}$$

In DTFC, the primary-flux amplitude and thrust errors ($\Delta\Psi_1$, ΔF_x), corroborated with the position of the primary-flux vector in one of the six 60° wide sectors, starting with the magnetic axis of phase A, lead to a planned combination of neighboring nonzero and zero voltage vectors in the inverter. In effect, we provoke a primary-flux vector accelerating or decelerating (for motoring and generating) and the adjustment of flux amplitude.

So there should be two internal closed loops, one for primary flux and one for thrust, besides the speed control loop that outputs the reference thrust F_x^*.

The primary-flux amplitude may be constant up to base speed and then decrease with speed to allow additional speed range for maximum inverter voltage V_{max}.

DTFC presupposes, then, a primary-flux observer and a thrust estimator.

Speed sensorless control requires a speed observer additionally.

A primary-flux observer based on the primary voltage model is represented by the equation

$$\frac{d\widehat{\Psi}_1}{dt} = -R_1\cdot\overline{i}_1 + \overline{V}_1 + \left(k_p + k_i/s\right)\cdot\mathrm{sgn}\left(\overline{i}_1 - \widehat{i}_1\right) \tag{4.58}$$

$\widehat{i}_1$ is the estimated stator current vector; for its estimation, we first need an estimation of secondary flux:

$$\frac{d\widehat{\Psi}_{2r}'}{dt} + \frac{R_2'}{L_{1l}}\cdot\frac{L_1}{L_m}\cdot\widehat{\Psi}_{2r}' = \frac{R_2'}{L_{1l}}\cdot\widehat{\Psi}_{1r} + k_2\cdot\mathrm{sgn}\left(i_1 - \widehat{i}_1\right) \tag{4.59}$$

In (4.59) for simplification $L_{2l}' = 0$, which is a realistic assumption.

The secondary flux in primary coordinates $\widehat{\Psi}_2'$ is

$$\widehat{\Psi}_2' = \widehat{\Psi}_1 - L_{1l}\cdot\overline{i}_1;\quad \widehat{\theta}_{\Psi_2} = \tan^{-1}\widehat{\Psi}_{2\beta}/\widehat{\Psi}_{2\alpha} \tag{4.60}$$

$\widehat{\theta}_{\Psi_2}$ is the estimated position of the secondary flux with respect to primary. The adopted sliding mode functional $S_s = \mathrm{sgn}(i_1 - \widehat{i}_1)$.

To reduce the chattering, the sgn function is replaced by *sat*(*x*):

$$\text{sat}(x)=\begin{cases}\text{sgn}(x), & \text{for } |x|>h\\ x/h, & \text{for } |x|\le h\end{cases} \tag{4.61}$$

The block diagram for the primary- and secondary-flux observers based on these equations is shown in Figure 4.20.

The thrust estimation is

$$\hat{F}_x=-\frac{\pi}{\tau}\cdot\frac{3}{2}\cdot \text{Imag}\left(\hat{\Psi}_1 * i_1\right)=\frac{\pi}{\tau}\cdot\frac{3}{2}\cdot\hat{\Psi}_2' * \hat{i}_2' \tag{4.62}$$

From (4.62), the secondary current is estimated:

$$\left|\hat{i}_2'\right|=\frac{\left|\text{Imag}\left(\hat{\Psi}_1 * i_1\right)\right|}{\left|\hat{\Psi}_2'\right|} \tag{4.63}$$

Equation 4.63 is strictly valid for constant secondary flux.

Thus, with $\hat{\bar{\Psi}}_2'$ and $\hat{i}_2$ estimated, we may even estimate the magnetization inductance $\hat{L}_m$:

$$\hat{L}_m=\frac{\left|\hat{\Psi}_2'\right|}{\left|\hat{I}_m\right|};\quad I_m\approx\sqrt{\left|i_1^2\right|-\left|i_2^2\right|} \tag{4.64}$$

The secondary resistance estimation may also be accomplished:

$$\hat{R}_2'=R_2'+\Delta\hat{R}_2'=R_2'+k_{R_2}\cdot \text{Imag}\left[\hat{\bar{\Psi}}_2'\cdot\left(\bar{i}_1-\hat{\bar{i}}_1\right)\right] \tag{4.65}$$

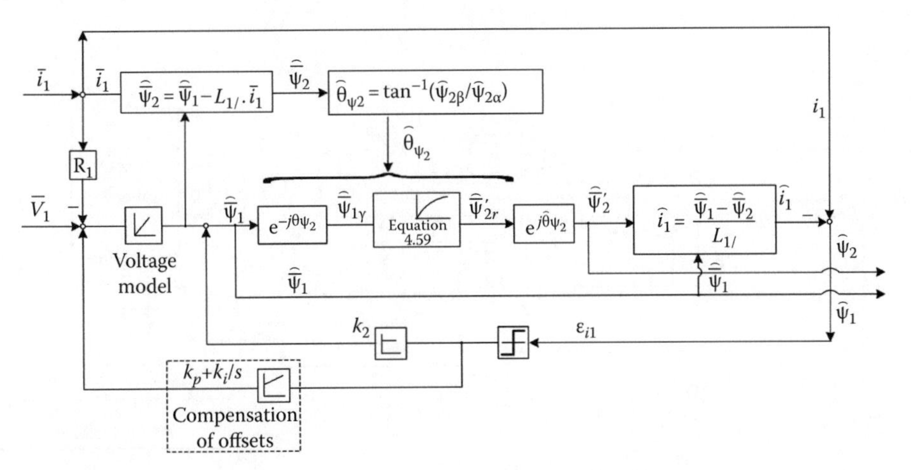

FIGURE 4.20 Observers for primary and secondary flux linkages $\hat{\Psi}_1$, $\hat{\Psi}_2'$ of LIMs.

Approximately, the secondary and primary resistance variations are proportional to each other as long as the LIM is loaded:

$$\widehat{\Delta R_1} = k_R \cdot \widehat{\Delta R_2'} \tag{4.66}$$

What still remains is to estimate on line the LIM speed $\widehat{U}$:

$$\widehat{U} \approx \frac{\pi}{\tau} \cdot \left(\widehat{\omega}_{\Psi_2'} - \left(S \cdot \widehat{\omega}_1\right)\right); \quad \widehat{\omega}_{\Psi_2} = \frac{d\widehat{\theta}_{\Psi_2}}{dt} \tag{4.67}$$

$$S \cdot \widehat{\omega}_1 = \frac{2}{3} \cdot \frac{\tau}{\pi} \cdot \frac{\widehat{F}_x \cdot \widehat{R}_2'}{\left|\widehat{\Psi}_2'\right|^2} \tag{4.68}$$

The emf compensation (voltage decoupling) is applied after the primary flux and thrust PI plus sliding mode close loops:

$$V_{1d}^* = \left(k_{p\Psi} + k_{I\Psi} \cdot \frac{1}{s}\right) \cdot \operatorname{sgn}\left(S_{\Psi_s}\right)$$

$$V_{1q}^* = \left(k_{pF} + k_{I_F} \cdot \frac{1}{s}\right) \cdot \operatorname{sgn}\left(S_{F_x}\right) + \widehat{\omega}_{\Psi_1} \cdot \widehat{\Psi}_1 \tag{4.69}$$

$$S_{\Psi_s} = \varepsilon_\Psi + c_\Psi \cdot \frac{d\varepsilon_\Psi}{dt}$$

$$S_{F_x} = \varepsilon_{F_x} + c_{F_x} \cdot \frac{d\varepsilon_{F_x}}{dt}$$

The block diagram of the whole control system is shown in Figure 4.21.

For a SLIM with the data in Tables 4.1 through 4.3, sample test results are shown in Figure 4.22: the stator current i_a, i_b, the primary flux, and the fast response of thrust [8].

The fast response of thrust, with sensorless control, in Figure 4.22, suggests robust state observers. At the same time, the notable thrust pulsations may be partly due to the asymmetry of

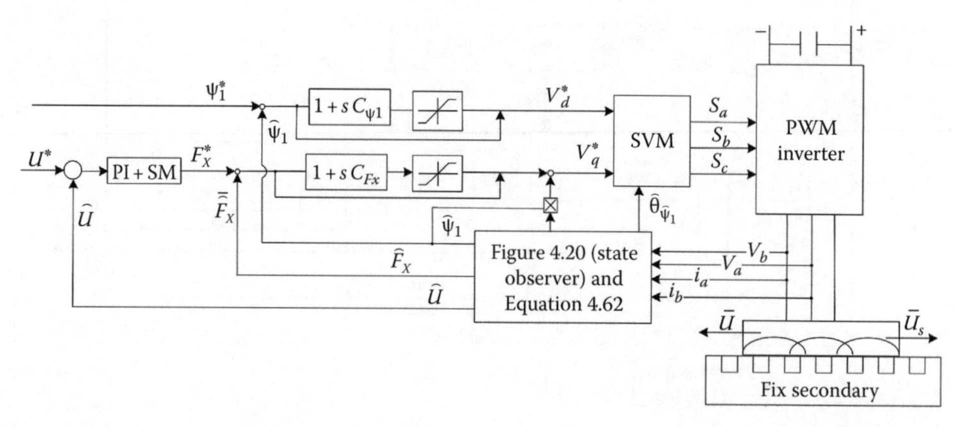

FIGURE 4.21 DTFC sensorless control of LIM.

TABLE 4.1
Motor Parameter

Rated Parameter	Value	Unit
Voltage	78	V
Frequency	25	Hz
Current	20	A
Force	80	N
Flux density	0,35	T
Current density	1488	A/cm
No. of phases	3	
Pole pitch	120	mm
Primary resistance	1,3	Ω
Slip	0,15	
Speed	5,1	m/s

TABLE 4.2
Control Parameters

Parameter	Value
$K_{p\psi}$	0.01
K_{pFx}	0.04
$K_{i\psi}$	6.00
K_{iFx}	0.10
C_{Fx}	0.001
C_{Ψ}	0.001

TABLE 4.3
Observer Parameters

Parameter	Value
K_p	$40+j0.10$
K_I	$0.005+j0.001$
K_2	$-20+j0.10$

stator currents (Figure 4.22), associated to static end effect caused by the small number of poles ($2p=4$), via the backward primary mmf wave interaction with the secondary (as in rotary IMs with asymmetric stator currents due to the asymmetric input voltages).

4.9 HIGH-SPEED LIM TRANSIENTS AND CONTROL

An exhaustive solution for high-speed LIM transients would imply using the 3D–FE coupled field/circuit method. While such an enterprise can be profitable in special situations (though it requires prohibitive computation time), it is not for extensive studies or for the design of the control system.

For such purposes, we may start from the equivalent circuit in Figure 4.14, where L_m decreases, R_2' increases with speed, and an additional resistance R_e is added, all these in relation to the dynamic end effect [9].

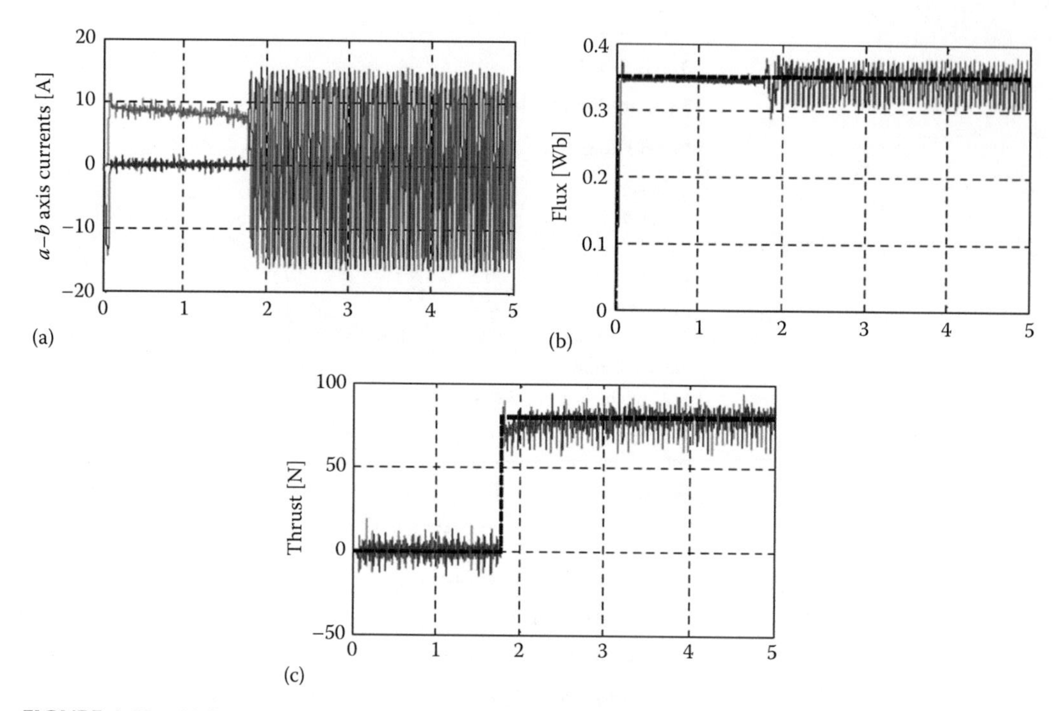

FIGURE 4.22 (a) Stator currents i_a, i_b, (b) primary flux, and (c) thrust. (After Cancelliere, P. et al., Sliding mode speed sensorless control of LIM drives, *Record of LDIA-2003*, Birmingham, U.K., 2003.)

$$L_m = L_{m0} \cdot \frac{g}{g_0} \cdot \left(1 - k_e(u)\right) \tag{4.70}$$

$$R_2' \approx R_{20}' = const$$

$$L_{2l}' = L_{2l0}' = const$$

$$R_e \approx \frac{1-k_e}{k_e} \cdot \frac{R_2'}{s}$$

These changes may be applied by including the previously mentioned variable parameters and the paralleled resistance R_e by additional equations, which have to reflect the thrust reduction due to end effect:

$$\bar{V}_1 = R_1 \cdot \bar{i}_1 + s \cdot \bar{\Psi}_1$$

$$0 = R_2' \cdot \bar{i}_2' + s \cdot \bar{\Psi}_2' + j \cdot \frac{\tau \cdot U}{\pi} \cdot \bar{\Psi}_2'$$

$$0 = R_e \cdot i_e - s \cdot \bar{\Psi}_m;\ \hat{F}_x = \frac{3}{2} \cdot \frac{\pi}{\tau} \cdot \text{Imag}\left[\left(\bar{\Psi}_1 * i_1\right) - \left(\Psi_m * i_e\right)\right] \tag{4.71}$$

$$\bar{\Psi}_1 = L_{1l} \cdot \bar{i}_1 + L_m \cdot \left(1 - k_e\right) \cdot \bar{i}_m$$

$$\bar{\Psi}_2' = L_{2l}' \cdot \bar{i}_2' + L_m \cdot (1 - k_e) \cdot \bar{i}_m$$

$$\bar{\Psi}_m = L_m \cdot (1 - k_e) \cdot \bar{i}_m$$

$$\bar{i}_e + \bar{i}_m = \bar{i}_1 + \bar{i}_2'$$

The last term in the thrust F_x expression relates to thrust reduction due to end effect, characteristic only to small slip frequency operation, a practical constraint met in PWM inverter supplied variable-speed LIMs.

An approximative function of speed, stator current, and slip frequency may be used to calculate k_e; as a first approximation, a linear increase of k_e with speed may be adopted in the presence of a robust DTFC or vector control system.

4.10 DTFC OF HIGH-SPEED LIMs

In a simplified DTFC application, we need only a state observer to yield $\hat{\Psi}_1$ and $\hat{F}_x$, in the presence of a vehicle speed sensor.

Using the voltage model in primary coordinates,

$$\hat{\Psi}_1 = \int \left(\bar{V}_1^* - R_1 \cdot \bar{i}_1 \right) dt \approx \frac{T_e}{1 + sT_e} \cdot \left(\bar{V}_1^* - R_1 \cdot \bar{i}_1 \right) \tag{4.72}$$

where V_1^* is the reference voltage vector; the low-pass filter deals with the offsets, but still, this simple and fast estimator is not suitable for low speeds (low-frequency ω_1, when an I–f starting strategy may be used). To calculate the thrust, $\hat{\bar{\Psi}}_m$ and $\hat{\bar{i}}_e$ (end effect current) have to be estimated.

From (4.72),

$$\hat{\bar{\Psi}}_m = \hat{\bar{\Psi}}_1 - L_{1l} \cdot \bar{i}_1 \tag{4.73}$$

$$\hat{\bar{i}}_e = \frac{R_2' \cdot \left(\bar{i}_s - \hat{\Psi}_m / L_m \right) - j \cdot \frac{\tau}{\pi} \cdot U \cdot \bar{\Psi}_m}{R_2' + R_e}; \quad R_e \approx \frac{1 - k_e}{k_e} \cdot \frac{R_2'}{s} \tag{4.74}$$

Also,

$$\hat{F}_x = -\frac{3}{2} \cdot \frac{\pi}{\tau} \cdot \mathrm{Imag}\left[\left(\hat{\bar{\Psi}}_1 * \bar{i}_1 \right) - \hat{\bar{\Psi}}_m * \hat{\bar{i}}_e \right] \tag{4.75}$$

For thrust and primary-flux control, a block diagram as in Figure 4.23 [9] is thus obtained.

The end effect factor k_e, which increases with speed and depends also on $S\omega_1$ and, for SLIMs, on Al-on-iron secondary, also on primary current, may be calculated by 1D analytical or 2(3)D FEM field theories of LIMs.

A linear variation of k_e with speed would be a first approximation.

Simulation results on an urban LIM with the data in Table 4.4 are given in Figure 4.24 [10].

Step increase of k_e in Figure 4.24d is due to the decrease in slip frequency when the thrust reference is reduced. In addition, Figure 4.24f reflects the necessity to consider R_e, as the estimated thrust is smaller than the actual thrust. Other pertinent LIM control systems are introduced in Refs. [10–14].

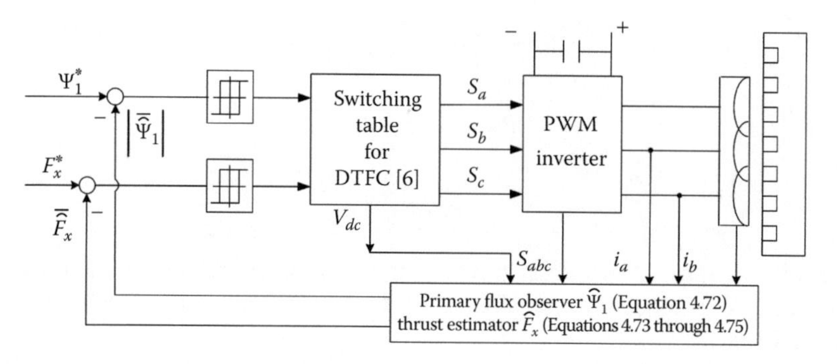

FIGURE 4.23 DTFC of high-speed LIM.

TABLE 4.4
Urban LIM Data

Parameter	Value	Units
Motor length	2.476	m
Pole pitch	280.8	mm
Number of poles	8	
Resistance of stator per phase	0.107	Ω
Resistance of rotor per phase	0.394	Ω
Self-inductance of stator	4.188	mH
Mutual inductance	22.101	mH
Self-inductance of rotor	1.33	mH

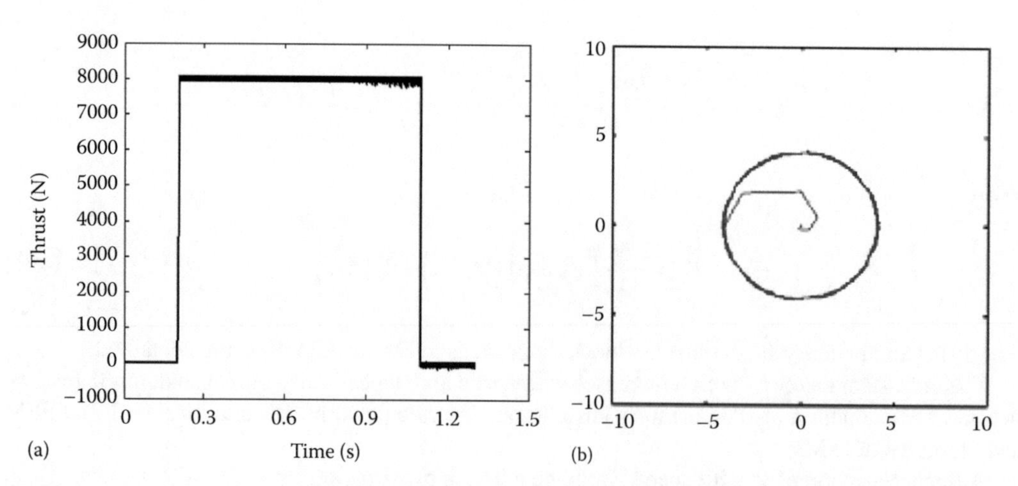

FIGURE 4.24 Urban SLIM thrust control response: (a) thrust response and (b) primary-flux response.

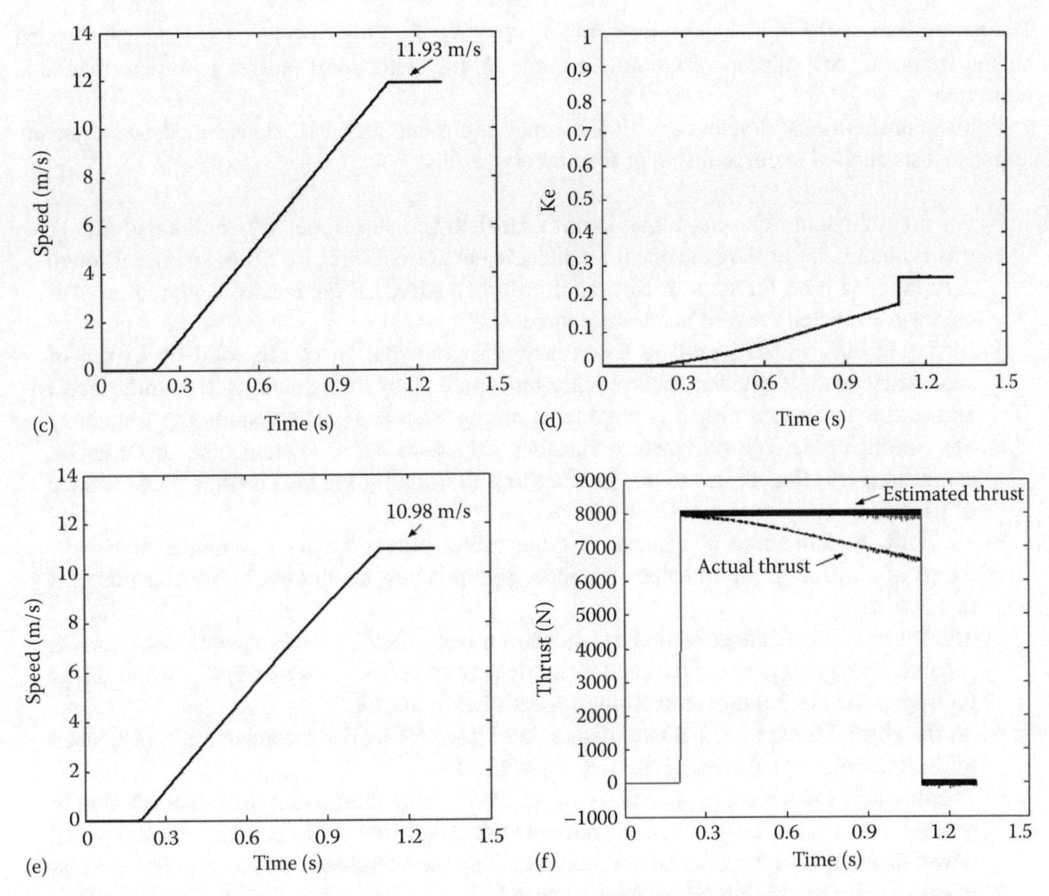

FIGURE 4.24 (continued) Urban SLIM thrust control response: (c) speed response, (d) end effect factor k_e, (e) actual speed response with end effect, and (f) estimated thrust and actual thrust with R_e neglected. (After Wang, K. et al., Direct thrust control of LIM considering end effects, *Record of LDIA-2007*, 2007; Takahashi, I. and Ide, Y., *IEEE Trans.*, IA-29(1), 161, 1993.)

4.11 SUMMARY

- Circuit models for LIMs needed to investigate transients and control and design stem from the equivalent circuit of rotary IM, with adequate modifications.
- To simplify the control of low-speed (or short travel) LIMs, they are kept in the negligible dynamic end effect conditions.
- The divide between low or high speed (low or notable dynamic end effect) is settled in this chapter by the condition that the exit end effect wave depth of penetration within LIM primary from the entry end is less than 10% of primary length. It turns out that this condition depends only on the number of poles $2p$, equivalent goodness factor G_e, and slip S. The equivalent goodness factor included edge effect, skin effect, slot opening, airgap fringing, and magnetic saturation (for SLIMs with Al-on-iron secondary)
- The low-speed LIM equivalent circuit parameters are then calculated for
    - Flat DLIM
    - Flat SLIM with Al-on-iron secondary
    - Flat SLIM with ladder secondary
    - Tubular SLIM with cage secondary

The parameters of the equivalent circuit, mainly L_m, R_2', L_{2l}' (for Al-on-iron secondary), depend on slip frequency $S\omega_1$, airgap variation (Lm), and on the stator mmf (current)—due to magnetic saturation.

Equivalent (fictitious) double cage (ladder) may represent such variations based on frequency response tests applied to curve fitting of field theory results:

- For constant stator current, a maximum thrust/current slip frequency is obtained $(S\omega_1)_k$; this is notably lower than the one for which the net normal force (in SLIM) passes through zero (an idea used for some LIM-propulsed urban MAGLEVs); consequently, lower efficiency is expected for zero net normal force.
- In flat SLIMs with Al-on-iron secondary, the contribution of the solid back iron of secondary (even if divided transversally into three slabs ("laminations")) to total thrust and to the equivalent airgap is notable. Running such LIMs at constant slip frequency $S\omega_1$ would render the parameters variable only with stator current (due to magnetic saturation) and thus easier to use in assessing thoroughly the steady-state performance of LIM.
- To limit the influence of transverse edge effect on secondary resistance, the overhangs of aluminum (or end slabs in ladder secondaries) should not be shorter than τ/π $(c - a \approx \tau/\pi)$.
- The flat SLIM with ladder secondary, for short travel (when its cost is acceptable), is shown capable, with good performance and thrust density ($\eta \geq 0.8$, $\cos\phi > 0.65$) ($B_{g1} \approx 0.6$–0.7 T) for airgaps of 1.0–2.0 mm at maximum speeds below 10 m/s.
- As the normal force in SLIM is unilateral, and large for ladder secondary, it may be used for mover (vehicle) suspension (partial or total).
- The tubular LIM with cage secondary, adequate for very short travel (less than 1.5–2 m in general) and airgap of 1–1.5 mm, is credited with even better thrust density, efficiency, and power factor, due to the absence of radial force and stator end coils and edge effect, despite lower speeds (less than 6 m/s maximum speed).
- Circuit models for high speed introduce a reduction by $(1 - k_e)$ of emf and then a new end effect impedance $\underline{Z}_e$ in parallel with $j\omega_1 L_m$: $\underline{Z}_e = ((1-k_e)/k_e)\underline{Z}_{2c}'$ is added.

$\underline{Z}_{2c}'$ is the conventional airgap plus secondary part of the equivalent circuit, typical to low-speed LIMs:

- The end effect factor k_e increases with speed, for a given high-speed LIM geometry, but it also increases when slip frequency decreases; a linear increase of k_e with speed is acceptable at given slip S.
- The equivalent circuits with end effects have been verified in experiments enquiring thrust reduction and then efficiency and power factor reduction and finally for secondary Joule losses [14].
- Identifying the equivalent circuit parameters with dynamic end effect, for a given LIM, may also be done by processing key running tests (at key speed and loads) by regressive methods of curve fitting.
- Also, simplified analytical expressions for $k_e(U)$, derived from 1D field theories, are given and checked against two representative SLIMs.
- The separate treatment of transients for low- and high-speed LIMs (low or notable dynamic end effect) by space vector models is then approached.
- The similarity of low dynamic end effect (low-speed) LIM space vector models with those of rotary IMs is striking: thrust instead of torque, linear speed versus angular speed, and no number of pole pairs p but τ (pole pitch) present in the thrust expression.

- Motoring and generating vector (space–phasor) diagrams are presented to illustrate that, in both operation modes, the amplitude of stator flux vector $|\bar{\Psi}_1|$ is greater than that of secondary flux $|\bar{\Psi}_2'|$: $\left(|\bar{\Psi}_1| > |\bar{\Psi}_2'|\right)$.
- The equations of low-speed LIMs in space phasor for constant speed may be solved analytically and reveal two complex eigen values $\underline{\gamma}_{1,2}$; the real parts of $\underline{\gamma}_{1,2}$ are negative at less than critical slip frequencies $(S\omega_1 < (S\omega_1)_k)$; $(S\omega_1)_k$ corresponds to maximum thrust.
- As the flux initiation in LIMs is not very fast, it is good to have it first "installed" even if as a single voltage vector (say $\bar{\Psi}_1$ aligned to phase a axis), before activating the speed control.
- Three control techniques, one scalar (I–Sf_1 close-loop control), two vectorial (one secondary-flux orientation control-FOC), and one direct thrust and primary-flux control DTFC.
- The simplicity and robustness of I–Sf_1 scalar control are notable, if PI plus sliding mode speed control is provided, but fast thrust transient response is not guaranteed.
- FOC of LIMs reveals the linear thrust/speed, $F_x(U)$, curves, typical for rotor flux orientation control in rotary IMs.
- A combined indirect vector current and voltage control FOC is described, to secure fast thrust response and increased stability and dynamic robustness, especially if a position sensor (with speed calculation) is available for positioning applications.
- To illustrate DTFC of low-speed LIMs, a motion-sensorless speed control system with space vector modulation and PI + SM primary- and secondary-flux observer and thrust and speed estimators has been introduced; corrections for secondary (and primary) resistances are introduced together with magnetization inductance L_m and $L_m(Im)$ curve online adjustment. L_m varies not only due to magnetic saturation but also due to airgap inevitable fluctuation; this justifies its online correction. The response in thrust and flux step variation are remarkably fast and stable in experiments.
- For high-speed (large end effect) LIMs' transients and control, simplified corrections of L_m (it becomes $(1 - k_e)\cdot L_m$) and addition of a parallel resistive circuit $R_e = ((1-k_e)/k_e)R_2'/s$ are introduced in the space–phasor model equations to model dynamic end effects. The dynamic end effect factor k_e (which decreases the emf and the thrust and causes additional secondary Joule losses) has to be calculated (or calibrated upon commissioning the LIM drive), but it may also be considered to increase linearly with speed increase and with slip decrease; an inferior limit for $|S\cdot f_1|$ might keep k_e just linearly varying with speed for a given LIM.
- An implementation of a DTFC of an existing urban LIM shows notable end effect (k_e goes up to 0.3) with a step increase for a step thrust reference reduction. Again, it is G_e, $2p$ and S that influence k_e (end effect influence, with speed implicit by frequency f_1 (in G_e) and slip, with the given pole pitch of primary winding).
- Short (mover) secondary and long (fix) primary LIMs have not been investigated in this chapter, even in the form of wound secondary on board of mover with energy "induction" on the vehicle, without power collection; it will be treated shortly only in the chapter on MAGLEV, as they did not reach markets yet and are heavily in competition with PMs or dc excited long primary stator linear synchronous motors.
- Proposed decades ago for oscillators, short secondary long primary LIMs have been since outperformed by PM linear synchronous resonant single-phase oscillators; this is the reason why they have not been treated here in any detail (see [3] for a field theory of short secondary long primary LIMs).
- Now that we have elucidated field theories, circuit models, the transients and control of LIMs, the design of LIMs will be approached separately for low-speed (low end effect) LIMs and for high-speed (notable end effect) LIMs.

REFERENCES

1. I. Boldea and S.A. Nasar, *Linear Motion Electromagnetic Devices*, Taylor & Francis, New York, 2001.
2. A. Krawczyk and J. Tegopoulous, *Numerical Modelling of Eddy Currents*, Oxford University Press, New York, 1993.
3. I. Boldea and S.A. Nasar, *Linear Motion Electromagnetic Systems*, John Wiley, New York, 1985 (Chapter 4).
4. J.F. Gieras, G.E. Dawson, and A.R. Eastham, A new longitudinal end factor for LIMs, *IEEE Trans.*, EC-2(1), 1987, 152–159.
5. C.M.A.R. Cabral, Analysis of LIMs using new longitudinal end effect factors, *Record of LDIA-2003*, Birmingham, U.K., pp. 291–294.
6. I. Boldea and S.A. Nasar, *Electric Drives*, 2nd edn., CRC Press, Taylor & Francis, New York, 2006.
7. G. Bucci, S. Meo, A. Ometo, and S. Scarano, The control of LIM by a generalization of standard vector techniques, *IEEE—IAS Annual Meeting Conference*, Bologna, Italy, 1995.
8. P. Cancelliere, V. Delli Colli, F. Marignetti, and I. Boldea, Sliding mode speed sensorless control of LIM drives, *Record of LDIA-2003*, Birmingham, U.K.
9. K. Wang, L. Shi, and Y. Li, Direct thrust control of LIM considering end effects, *Record of LDIA-2007*.
10. I. Takahashi and Y. Ide, Decoupling thrust and attractive forces of LIM using space vector control inverter, *IEEE Trans.*, IA-29(1), 1993, 161–167.
11. B.-I. Kwon, K.-I. Woo, and S. Kim, Finite element analysis of direct thrust—Controlled LIM, *IEEE Trans.*, MAG-35(3), 1999, 1306–1309.
12. M. Abbasian, Adaptive input—Output control of linear induction motor considering the end effect, *Record of ICEM*, 2010, Rome, Italy.
13. A. Gastli, Improved field oriented control of a LIM having joints in its secondary conductors, *IEEE Trans.*, EC-17(3), 2002, 349–355.
14. W. Xu, J. G. Zhu, Y. Zhang, Z. Li, Y. Wung, Y. Guo, and Y. Li, Equivalent circuits for single—Sided linear induction motors, *IEEE Trans.*, IA-46(6), 2010, 2410–2423.

5 Design of Flat and Tubular Low-Speed LIMs

5.1 INTRODUCTION

Low-speed LIMs are LIMs whose dynamic end effects can be neglected $\left[(\gamma_{2r}\tau)^{-1} < 2p/10\right]$ (as described in Chapter 3). They have to be designed, consequently, for a goodness factor less than the optimum value G_{e0} (Chapter 3), which varies linearly with the number of poles $2p$ ($G_e = 10$ for $2p = 4$ and $G_e = 40$ for $2p = 16$).

Though we are intuitively tempted to assimilate low speed with low goodness factor and thus end effect neglection, the situation is more complex, as in principle we could obtain the same equivalent goodness factor G_e and thus the same end effect (for the given pole count and slip), for up to 3/1 speed ratio [1].

So, we should choose for a low-speed application a low goodness factor (with negligible end effect) or, better, an optimal goodness factor design, which shows some end effect, but for better overall performance (thrust density, efficiency, and power factor)? The answer to this question is related to the application, as higher G_e means inevitably higher pole pitch, which means thicker back iron both in the primary and secondary of SLIMs (the favorite practical topology) and higher initial costs of SLIM.

Therefore, the "low-speed LIM" concept is just a simplified design option based on the strong similarities with rotary IM design methodologies, which constitute a valuable heritage to use for the scope.

In fact the "low-speed LIM" design concept may be the first (very preliminary) design step for any LIM (even for high speed) to get a strong feeling of magnitudes in the absence of a vast industrial experience data, as for rotary IMs.

By *design* we mean here "dimensioning" or "sizing," which is the finding of complete manufacturing data (geometry, parameters, and performance characteristics) to meet initial specifications, by using adequate LIM analytical (or numerical) field and circuit models. This operation is a synthesis while using a given geometry of LIM to calculate circuit parameters, and performance is, in fact, an analysis stage.

Optimization design means repetitive usage of design to comply best with initial specifications based on a multiobjective function (with constraints) and a mathematical algorithm to obtain convergence within reasonable computation effort (time). The design refers to three main aspects:

- Electromagnetic
- Thermal
- Mechanical

In what follows we will refer only to electromagnetic design. As for thermal and mechanical design, we will only limit current and flux densities and the shear stress components (propulsion and normal surface force densities in N/cm^2) to make sure they comply with standard material properties.

A complete thermal and mechanical design of LIMs (based on analytical and (or) on finite element multilayer model) is beyond our scope here. To save space we will unfold analytical electromagnetic design methodologies for

- Flat SLIMS with ladder secondary
- Tubular SLIMS with ladder secondary, for speeds below 10 m/s and negligible dynamic end effect conditions, and travel lengths of 1–2 m travel for tubular SLIMS in industry, and tens of meters for other, special, applications.

The aluminum sheet on iron secondary SLIM has been proven to yield poor performance (η_1 cos φ_1<0.2–0.3) [2] for low-speed applications, and thus, in the presence of strong competition of linear permanent magnet (or dc-excited) synchronous motors (LSMs), it will not be treated here further. However it will be considered for medium- and high-speed LIMs because of lower cost of the long secondary (track) for long travels, despite lower η_1 cos φ_1 (0.4–0.45) than for LSMs (η_1 cos φ_1<0.6–0.7), which implies higher KVA (and costs) PWM inverter supplies in the overall cost of the system (such as net present worth). The costs of the secondary (track) for long travels (tens and hundreds of kms) retain a very important share. To keep the presentation closer to the scope, we will present the design methodologies by numerical examples. The optimization design will be addressed only for medium to high speed (transportation LIMs in Chapter 6). We should emphasize that the LIM design methodology here revolves around the equivalent goodness factor concept. And it is based on 1D analytical field theory, which includes slot opening, magnetic saturation, and edge and skin effects by correction coefficients, as a basis to define the parameters of the standard IM equivalent circuit. The parameters expressions, however, are related to LIM topology and vary with primary mmf and slip frequency via equivalent goodness factor.

Alternative design methodologies are available elsewhere [2].

5.2 FLAT SLIM WITH LADDER LONG SECONDARY AND SHORT PRIMARY

The electromagnetic design of LIMs is related strongly to the application, due to the absence of any mechanical transmission between the latter and the load.

Let us consider here a machine-tool table application with the following general specifications:

1. Specifications
   Rated thrust: F_{xn}=500 N
   Maximum travel length: l_{travel}=3 m
   Rated speed: u_n=3 m/s
   a. Power source: three-phase ac 220 V/60 Hz line voltage, with diode rectifier and PWM voltage source IGBT inverter

Some additional data from previous experience are added as follows:

2. Additional experience-based data
   a. The mechanical airgap, with linear bearings, may be chosen as low as g=1.0 mm, especially if the host primary mover is placed below the ladder secondary (stator) such that the large normal force F_n covers most of the mover total weight (by adequate F_n control); in the latter case, the stress of linear bearings becomes notably smaller than usual.
   b. The allowance for normal force leads to the possibility of choosing a large rated airgap flux density (B_{g1n}=0.55 – 0.7 T), not larger, to avoid heavy magnetic saturation of laminated cores of primary and secondary.
   c. Slot shape factor $h_{s1}/b_s \approx$ 4–6 in the primary, lower in the secondary, with slot openings b_{s1}/g<6 and b_{s2}/g<4, to keep the leakage inductances of primary and secondary within

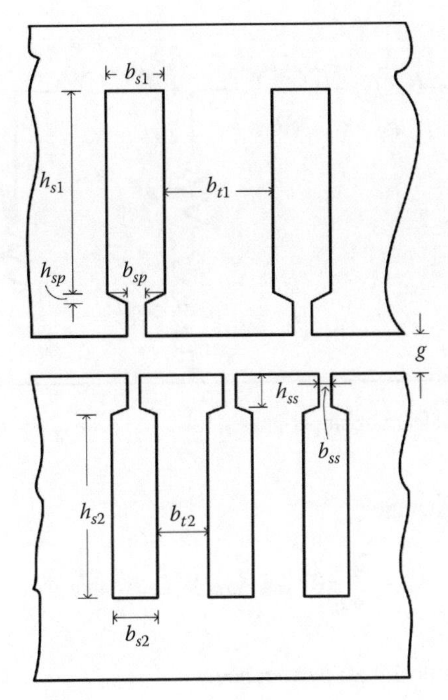

FIGURE 5.1 Flat primary and secondary slot geometries. (After Boldea, I. and Nasar, S.A., *Linear Electric Actuators and Generators*, Cambridge University Press, Cambridge, U.K., 1997.)

bounds and, at the same time, reduce Carter coefficient and surface core losses (due to slot space harmonics field); see Figure 5.1.

d. The primary winding may be single layer (with $q_1=1$ slot/pole/phase or $q_1'=2$ for lower the mmf. space harmonics), and thus lower the additional cage and core surface eddy current losses, when the number of primary slots $N_{s1}=2pq_1$; it may be also 2 layer type with $q_1=1$, 2 and (eventually) chorded coils ($x/\tau=5/6$ for $q_1=2$). For $2p+1$ poles in the primary (with half-filled end pole slots), the coils are shorter so the copper is better used, but primary core is longer (by almost one pole) and thus not completely used. We will choose here $2p$ poles single-layer, full pitch, ($q_1=2$) primary winding, with the slot pitch τ_{s1}.
e. The number of poles $2p$ is still to be chosen but $2p>4$, to avoid asymmetrical phase currents due to static end effect.
f. The slot pitch in the secondary τ_{s2} should be different from τ_{s1} (in the primary), but not far from each other (as for rotary IMs), to avoid notable thrust pulsations.
g. The key dimensioning (sizing) factor considered here is the shear secondary stress: for free air coiling $f_{xn}=F_{xn}/(primary\ area)\approx 1$ N/cm^2 and 1.5–2 N/cm^2 for forced air cooling. This value interval of f_{xn} is justified only by the rather small airgap and the ladder secondary (for Al-on-iron secondary SLIMs f_{xn} is smaller for low-speed applications).
h. As we make use of the rotary IM equivalent circuit (Figure 5.2), we gather here the expressions of circuit parameters (as developed in Chapter 4, [4,5]):
The primary resistance per phase:

$$R_1=\frac{2\rho_{Co}(l_{stack}+l_{ec})j_{cor}w_1^2}{w_1 I_{1r}} \tag{5.1}$$

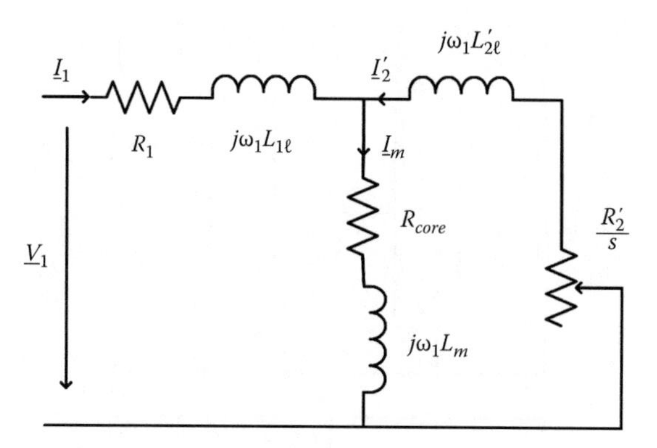

FIGURE 5.2 Low-speed equivalent circuit per phase.

Primary leakage inductance:

$$L_{1l} = \frac{2\mu_0}{pq_1}((\lambda_{s1} + \lambda_{diff1})l_{stack} + \lambda_{ec1}l_{ec})w_1 \tag{5.2}$$

Slot specific (nondimensional) permeance:

$$\lambda_{s1,2} = h_{s1,2}\frac{(1+3\beta)}{12b_{s1,2}} + \frac{h_{sp,s}}{b_{sp,s}} \tag{5.3}$$

Airgap leakage specific permeance:

$$\lambda_{diff1,2} \approx \frac{5K_c g/b_{sp,s}}{5+4(K_c g/b_{sp,s})} \tag{5.4}$$

Primary end-coil leakage specific permeance:

$$\lambda_{ec1} = 0.3(3\beta - 1)q_1 \tag{5.5}$$

The magnetization inductance:

$$L_m = \frac{6\mu_0(K_{w1}w_1)^2\tau l_{stack}}{\pi^2 K_c g p(1+K_{ss})} \tag{5.6}$$

Secondary resistance (referred to primary):

$$R_2' = 12\rho_{Al}\frac{(K_{w1}w_1)^2}{N_{s2}}\left(\frac{l_{stack}}{A_{s2}} + \frac{2l_{lad}}{A_{lad}}\right) \tag{5.7}$$

N_{s2} is the secondary slots/primary length.

Secondary leakage inductance (referred to primary):

$$L_{2l}' \approx 24\mu_0 l_{stack}(\lambda_{s2} + \lambda_{diff2})\frac{(K_{w1}w_1)^2}{N_{s2}} \times (1 + K_{ladder}) \tag{5.8}$$

with

$$A_{lad} = \frac{A_{s2}}{2\sin(\alpha_{es}/2)}; \quad \alpha_{es} \approx \frac{2\pi p}{N_{s2}}; \quad l_{lad} \approx \frac{2p\tau}{N_{s2}}; \quad A_{s2} = h_{s2} \cdot b_{s2} \tag{5.9}$$

Other symbols in (5.1) through (5.9) are as follows:

- l_{stack}—primary (and secondary) stack width
- g—mechanical gap
- K_c—Carter coefficient for dual slotting
- l_{ec}—end-coil length per side; $l_{ec} \approx \sim 0.01 + 1.5y$ (y-coil span); $\beta = y/\tau = 1$ or 5/6 etc.; τ—pole pitch

Note: Similar expressions are valid for Al-on-iron, with $g + d_{Al}$ instead of g, d_{Al}—aluminum thickness such as $N_{s2}A_{s2} = 2pd_{Al}\tau$; ρ_{Al} is augmented by the edge effect coefficient $K_{tr} > 1$ and reduced by the contribution of solid back in thrust; the magnetic saturation factor K_s accounts for secondary back iron and primary teeth and back-iron contribution to magnetization current (Chapters 3 and 4).

3. Primary sizing

For start we lump by coefficient K_{l2} the secondary leakage inductance $\left(L'_{2l}\right)$ influence, and thus the airgap flux density (3.24) is

$$B_{g1} = \mu_0 K_e = \frac{\mu_0 J_{1m}}{\frac{\pi}{\tau} g K_c (1+K_s)\sqrt{1+s^2 Ge^2}} = \frac{\mu_0 \theta_{1m}}{g K_c (1+K_s)\sqrt{1+s^2 Ge^2}} \tag{5.10}$$

with the primary mmf/pole θ_{1m} (as for rotary IMs)

$$\theta_{1m} = \frac{3\sqrt{2}(K_{w1}\omega_1)I_1}{\pi p} \tag{5.11}$$

and

$$\left(G_{ei}\right) = \omega_1 L_m/(R'_2 \cdot K_{l2}) \approx \frac{\mu_0 \omega_1 \tau^2 \sigma_{Al} h_{s2}(1 - b_{s2}/\tau_{s2})}{\pi^2 g K_c (1+K_s) K_r \cdot K_{l2}}$$
$$K_r = 1 + \frac{L_{lad}}{A_{lad}} \cdot \frac{A_{s2}}{l_{stack}}; \quad K_{l2} = \sqrt{1 + \left(\frac{\omega_2 L'_{2l}}{R'_2}\right)^2} \tag{5.12}$$

K_{l2} is assigned an initial value (1.2 ÷ 1.0).

As demonstrated in Chapter 4, the maximum thrust per primary mmf (or I_1) is obtained for $S_n G_e = 1$. Let us consider this condition fulfilled for rated thrust when $B_{g1k} = 0.7$ T:

$$B_{g1k} = \frac{\mu_0 \theta_{mn}}{g K_c \left(1+K_s\right)\sqrt{1+1^2}} = 0.7 \text{ T} \tag{5.13}$$

Assuming, realistically, that $K_c \approx 1.25$, $K_s \approx 0.4$ with $g = 1.0$ mm, the rated primary mmf θ_{1m} per pole is

$$\theta_{1m} = \frac{0.7 \times 1.0 \times 10^{-3} \times 1.25 \times 1.4 \times 1.41}{1.256 \times 10^{-6}} = 1.375 \cdot 10^3 \text{ A turns/pole} \quad (5.14)$$

For $q_1 = 2$, $y/\tau = 5/6$ the primary winding factor K_{w1} yields

$$K_{w1} = \frac{\sin(\pi/6)}{q \cdot \sin(\pi/6q)} \cdot \sin\left(\frac{\pi}{2} \cdot \frac{y}{\tau}\right) = \frac{0.5}{2 \times \sin \pi/12} \times \sin\left(\frac{5}{6} \cdot \frac{\pi}{2}\right) = 0.933 \quad (5.15)$$

The lower limit of the number of poles is $2p = 6$, to reduce static end effect (phase currents asymmetry) and preserve the low dynamic end effect condition.

Approximately

$$w_1 I_1 = \frac{\theta_{1m} \cdot \pi p}{3\sqrt{2} K_{\omega 1}} = \frac{1.375 \times 10^3 \times \pi \times 3}{3\sqrt{2} \times 0.933} = 3.28 \times 10^3 \text{ A turns/phase} \quad (5.16)$$

Let us consider $l_{stack}/\tau = 2$ to reduce end-coil length influence in copper losses, and, with a rated specific thrust $f_{xn} = 0.88$ N/cm², we obtain the primary active area A_p:

$$A_p \approx 2p\tau l_{stack} = 2p\tau^2 \left(\frac{l_{stack}}{\tau}\right) = \frac{F_{xn}}{f_{xn}} = \frac{500}{0.88 \cdot 10^4} = 0.05682 \text{ m}^2 \quad (5.17)$$

From (5.17), $\tau = 0.069$m.

The primary slot pitch $\tau_s = \tau/6 = 0.069/6 = 11.5 \times 10^{-3}$ m. To avoid excessive magnetic saturation in the primary tooth width, $b_{s1}/\tau_s = 0.55$, and thus $b_{s1} = 6.32 \times 10^{-3}$ m. Envisaging good efficiency, we may choose the secondary frequency at maximum thrust/current ($sG_e = 1$) to be $f_{2r} = 4$ Hz, not smaller, to avoid severe noise and vibration at start (when $f_1 = f_{2r}$) and to reduce secondary winding losses.

From (5.12)—with $G_{ei} = 1$ at $\omega_1 = \omega_{2r} = 2\pi f_{2r}$—the secondary slot useful height h_{s2} is obtained:

$$h_{s2} = \frac{1 \times 2\pi \times 1.0 \times 10^{-3} \times 1.25 \times 1.4 \times 1.1 \times 1.5}{1.256 \times 10^{-6} \times 2\pi \times 4 \times 0.069^2 \times 3.2 \times 10^7 \times 0.45} = 13.16 \times 10^{-3} \text{m} \quad (5.18)$$

The primary length is as follows:

$$(2p+1)\tau = 7 \times 0.069 = 0.483 \text{m} \quad (5.19)$$

The active primary slot area is, thus, A_{ps}:

$$A_{ps} = \frac{w_1 I_1}{pqj_{cor} \cdot K_{fill}} = \frac{3.28 \times 10^3}{3 \times 2 \times 4 \times 10^6 \times 0.60} = 227 \text{ mm}^2 \quad (5.20)$$

The slot filling factor is $K_{fill} = 0.6$, because slots are open and thus preformed coils are inserted.

Now, with $b_{s1} = 6.32 \times 10^{-3}$ m, the active primary slot height h_{s1} is

$$h_{s1} = \frac{A_{ps}}{b_{s1}} = \frac{226 \text{ mm}^2}{6{,}32 \text{ mm}} = 35.8 \text{ mm} \tag{5.21}$$

The slot aspect ratio is rather high ($b_{s1}/h_{s1} = 5.67$) but, eventually, still acceptable for a reasonable power factor of SLIM.

The peak normal force F_{nK}, of attractive character, as the secondary slots are semi-closed, is approximately

$$F_{nK} = \frac{B^2{}_{g1K}}{2\mu_0} \cdot 2p\tau l_{stack} = \frac{0.7^2 \times 0.483 \times 0.138}{2 \times 1.256 \times 10^{-6}} = 13 \text{ kN} \tag{5.22}$$

The thrust is only 500 N and thus the ratio $F_{nK}/F_{xn} = 26/1$. If the total mover (table) weight is lower than F_{nK}, it is feasible to reduce B_{g1} by a larger secondary frequency at F_{xn}, at the price of higher peak current and total winding losses:

$$F_x \approx \frac{3I_1^2 R_2' sG_e}{2\tau f_1(1+(sG_e)^{-2})} \tag{5.23}$$

For $sG_e = 1$ and, with $G_e = \omega_1 L_m / R_2' K_{l2}$, (5.23) becomes

$$F_{xn} = \frac{3\pi}{2\tau} I_{1n}^2 \frac{L_m}{K_{l2}} \tag{5.24}$$

From (5.6) L_m is

$$L_m = \frac{6 \times 1.256 \times 10^{-6} \times 0.069 \times 0.138 \times 0.933^2}{\pi^2 \times 1.25 \times 1.0 \times 10^{-3} \times 3 \times 1.4} \cdot w_1^2$$

$$= 0.804 \times 10^{-6} \times 1.5 \times w_1^2$$

$$= 1.2067 \cdot 10^{-6} w_1^2 \tag{5.25}$$

The rated thrust (at $s_n G_e = 1$) is now verified:

$$F_{xn} = \frac{3\pi}{2\tau}(w_1 I_{1n})^2 \times 0.804 \times 10^{-6} \times 1.5 \times 11.5$$

$$= \frac{3\pi}{20.069}(3.28 \times 10^3)^2 \times 0.804 \times 10^{-6}$$

$$= 589.7 \text{ N} > 500 \text{ N} \tag{5.26}$$

4. The number of turns and equivalent circuit parameters

Let us remember that $f_{2r} = 4$ Hz, the required speed is $u_r = 3$ m/s, and the pole pitch $\tau = 0.069$ m. Consequently, the primary required frequency f_{1r} is

$$f_{1r} = f_2 + \frac{u_r}{2\tau} = 4 + \frac{3}{2 \times 0.069} = 25.74 \text{ Hz} \tag{5.27}$$

The available voltage (RMS value) per phase is approximately

$$V_{10} \approx 0.95 \times V_{1line}/\sqrt{3} = 0.95 \times 220/\sqrt{3} = 120.81 \text{ V} \tag{5.28}$$

The way to calculate w_1 is to first prepare the circuit parameters by factorizing them to w_1^2. From (5.1),

$$R_1 = \frac{2\rho_{co}(l_{stack} + l_{ec})l_{cor}w_1^2}{w_1 I_{1r}}$$

$$= \frac{2\times 2.3\times 10^{-8}\left(0.138 + 1.5\times 0.069\times 4\times 10^6\right)}{3.475\times 10^3}\times w_1^2$$

$$= 1.2570\times 10^{-5}\times w_1^2 \quad (5.29)$$

To calculate L_{1l} (5.2), λ_{diff1} and λ_{ec1} are (5.3) through (5.5)

$$\lambda_{s1} = \frac{h_{s1}}{3b_{s1}} + \frac{h_{sp}}{b_{sp}} = \frac{35.8}{3\times 6.32} + \frac{1}{6.32} = 2.126 \quad (5.30)$$

(open primary slots as $b_{s1}/g = 6.32/1.5 = 4.2$ is acceptable)

$$\lambda_{diff1} = \frac{5K_c g/b_{sp}}{5 + 4K_l g/b_{sp}} = \frac{5\times 1.25\times 1.5/6.32}{5 + 4\times 1.5\times 1.5/6.32} = 0.2315 \quad (5.31)$$

$$\lambda_{ec1} \approx 0.3\times(3\times 516 - 1)\times 2 = 0.9 \quad (5.32)$$

$$L_{1l} = \frac{2\mu_0}{pq}\left[\left(\lambda_{s1} + \lambda_{diff1}\right)\cdot l_{stack} + \lambda_{ec1}\cdot l_{ec}\right]\cdot w_1^2$$

$$= \frac{2\times 1.256\times 10^{-6}}{3\times 2}\left[\left(2.126 + 0.2315\right)\times 0.138 + 0.9\times 0.08625\right]w_1^2$$

$$= 0.1687\times 10^{-6}\times w_1^2 \quad (5.33)$$

With semiclosed slots in the secondary, the slot pitch $\tau_{s2} = \tau_{s1}\times 0.9 = \tau/6\times 0.9 = 0.0069/6\times 0.9 =$ 0.01035 m, secondary slot width $b_{s2} \approx 5.5$ mm, slot opening $b_{ss} = 2g = 3$ mm, and $h_{ss} = 1.5$ mm, we may proceed to calculate L'_{2l} (5.8) with $K_{ladder} \approx 0.1$:

$$\lambda_{s2} = \frac{13.11}{3\times 5.5} + 1.5/3 = 1.2975;\quad \lambda_{diff2} \approx 0.15 \quad (5.34)$$

$$L'_{2l} = 24\times 1.256\times 10^{-6}(1.2975 + 0.15)\times 0.138\times\frac{0.933^2}{40}(1 + 0.1)\times w_1^2$$

$$= 0.144\times 10^{-6}\times w_1^2 \quad (5.35)$$

The secondary resistance ((5.7) and (5.9)) is gradually

$$A_{s2} = h_{s2}\cdot b_{s2} = 13.16\times 5.5 = 72.38\,\text{mm}^2;\quad \alpha_{es} = \frac{2\pi\cdot p}{N_{s2}} = \frac{\pi\times 6}{40} \quad (5.36)$$

$$A_{lad} = \frac{A_{s2}}{2\sin\alpha_{es}/2} = \frac{72.38}{2\times\sin(\pi\cdot 6/80)} \approx 155\,\text{mm}^2 \quad (5.37)$$

$l_{lad} = 2p\ \tau/N_{s2} = 2 \times 3 \times 69/40 = 10.35$ mm, so

$$R_2' = 12\rho_{Al}\frac{(K_{w1}w_1)^2}{N_{s2}}\left(\frac{l_{stack}}{A_{s2}} + \frac{2l_{lad}}{A_{lad}}\right)$$

$$= \frac{12\cdot 0.933^2}{3.20\times 10^7\times 40}\left(\frac{0.138}{72.38\times 10^{-6}} + \frac{2\times 0.01035}{316\times 10^{-6}}\right)\times w_1^2$$

$$= 1.609\times 10^{-5}\times w_1^2 > R_1 \quad (5.38)$$

To calculate w_1 (turns/phase), we have to introduce the rated $w_1 I_1 = 3280$ A turns/pole

$$S = f_2/f_1 = 4/25.74 = 0.155$$

$$V_{1r} = I_{1r}\left[R_1 + j\omega_{1r}L_{1l} + \frac{j\omega_1 L_m(R_2'/s + j\omega_1 L_{2l}')}{R_2'/s + j\omega_1(L_m + L_{2l}')}\right]$$

$$= w_1\cdot(w_1 I_{1n})\left[1.275\times 10^{-5} + j2\pi\times 25.74\times 0.1687\times 10^{-6}\right.$$

$$\left. + \frac{j2\pi\times 25.74\times 1.206\times 10^{-6}\times(16.09/0.155 + j2\pi\times 25.74\times 0.144)}{1.609\times 10^{-5}/0.155 + j2\pi\times 25.74\times(1.206 + 0.144)}\right]$$

$$= w_1\times(w_1 I_{1n})\times(79.37 + j72.3)\times 10^{-6} \quad (5.39)$$

with $V_{1r} = 120$ V, finally $w_1 \approx 327$ turns/phase.

The RMS phase current for rated thrust

$$I_{1n} = (w_1 I_{1n})/w_1 = 3.28\times 10^3/327 = 10.02\ \text{A} \quad (5.40)$$

The input power P_{1n} is

$$P_{1n} = 3V_{1n}I_{1n}\cos\varphi_{1n} = 3\times 120\times 10.02\times 0.707 = 2596\ \text{W} \quad (5.41)$$

From (5.46),

$$\cos\varphi_{1n} = \frac{R_e(Z_e)}{|Z_e|} = \frac{79.3}{79.3\sqrt{2}} = 0.707 \quad (5.42)$$

With core losses neglected ($f_1 = 25$ Hz), the electromagnetic power P_{elm} is

$$P_{elm} = F_x\cdot U_s = P_1 - p_{cu1} = P_1\times\frac{66.8}{79.3} = 2186\ \text{W} \quad (5.43)$$

The synchronous speed

$$u_s = 2\tau f_1 = 2\times 0.069\times 25.74 = 3.552\ \text{m/s} \quad (5.44)$$

So the thrust may be recalculated from

$$F_x = \frac{P_{elm}}{u_s} = \frac{2186}{3.552} = 615 \text{ N} > 500 \text{ N} \tag{5.45}$$

Note: The difference between the required rated thrust of 500N and the obtained thrust at given current (voltage) and slip frequency is due to the fact that the deteriorating influence of secondary leakage inductance on goodness factor (by $K_{l2}=1.5$) was exaggerated, in order to be "on the safe side" in the design.

Now the efficiency η_n is

$$\eta_n = \frac{F_x u_r}{P_{1n}} = \frac{615 \times 3}{2596} = 0.711 \tag{5.46}$$

Discussion

- By choosing a small airgap ($g=1$ mm) for a small-speed (3 m/s) flat SLIM with cage secondary, acceptable power factor and efficiency were obtained ($\eta \approx \cos\varphi_{1n} \approx 0.7$).
- To obtain acceptable performance, the thrust density was reduced to 0.88 N/cm^2.
- Also, the airgap flux density is rather high (0.7 T) for rated conditions $\left(f_{2_n} = 4\,\text{Hz}\right)$, and thus the normal force is very large (26 times larger than the thrust); if the total mover mass is larger than the normal force by partial compensation of weight, a smooth ride can be obtained by placing the primary (mover) below the long (fix) secondary.
- The rest of the SLIM design is very similar to rotary IM design; for a primary core depth of 15 mm, the total active (copper + iron core) primary weight is about 36 kg. Supposing that the machine-tool table weighs 1500 kg (normal force is 13 kN), with a thrust of 615 N, an acceleration of 0.4 m/s^2 can be secured.
- Though rather high, the performances here are notably inferior to those of linear PMSMs for similar applications, but the ruggedness of the SLIM and the absence of PMs along the track length make the SLIM less costly and, in a harsh environment, desirable.
- To prepare the SLIM for design of the control system, the equivalent circuit parameters are calculated:

$$R_1 = 12.75 \times 10^{-6} \cdot w_1^2 = 1.363\,\Omega; \quad R_2' = 16.09 \times 10^{-6} \times w_1^2 = 1.72\,\Omega$$

$$L_m = 1.206 \times 10^{-6} \cdot w_1^2 = 0.129\,\text{H}; \quad L_{1l} = 0.1687 \times 10^{-6} \times w_1^2 = 0.01839\,\text{H}$$

$$L_{2l}' = 0.144 \times 10^{-6} \cdot w_1^2 = 0.0154\,\text{H}; \quad w_1 \approx 327\,\text{turns/phase}$$

- Once the primary and secondary winding losses are known (easy to calculate via the equivalent circuit) and the complete geometry of both primary and secondary is known already from the previous equation, the thermal design—strongly dependent on the operation cycle (speed and thrust versus time)—may be approached; thermal design is beyond our scope here; it is, however, somewhat similar to that for rotary IMs.
- Again, we left out the Al-on-iron secondary SLIM because the performance (efficiency, power factor, and thrust density) is, in our view, unpractical, despite the small normal force/thrust ratio (2.5 ÷ 3.5/1), which may be advantageous in some light mover weight applications. There are some LPMSMs with passive laminated iron track (such as transverse flux type), which also enjoy small normal force/thrust ratio but at much higher efficiencies, as shown later in the book.

5.3 TUBULAR SLIM WITH CAGE SECONDARY

For limited travel (a few meters), and zero normal force applications, with a single-track-long secondary tubular structure to hold the inner cage secondary of SLIM, the tubular configuration may be a practical solution for tables, pumps, or vibrators at a low frequency.

The small airgap (1–1.5 mm) cage (copper/aluminum)-ring-secondary with disk-shape laminations that are provided with radial slits in the back cores, to reduce core losses in the primary (with circular coils in slits) and secondary (Chapter 3) cores, completes a practical topology.

For convenience, a generic configuration is presented in Figure 5.3.

A small part of circumferential length may be sacrificed and used for linear bearings to secure central running of primary through the secondary at the small airgap and speed up to 3–6 m/s.

We will proceed as for the flat SLIMs with ladder secondary for similar specifications:

1. *Specifications*
   a. Rated thrust F_{xr} = 2.5 kN.
   b. Max travel length: 2 m.
   c. Max (and rated) speed: 2 m/s.

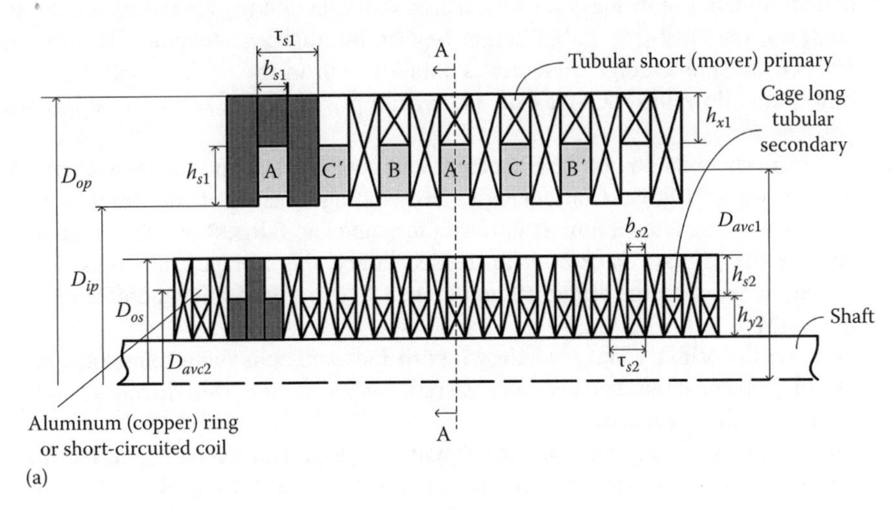

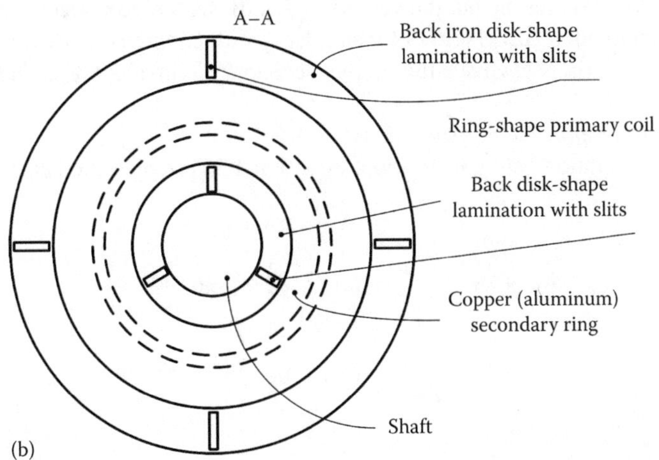

FIGURE 5.3 Tubular SLIM with cage long secondary: (a) longitudinal cross section and (b) transverse cross section.

d. Power source: three-phase ac 220 V (60 Hz) line voltage with diode rectifier and PWM voltage source inverter.
e. Let us add that the maximum outer diameter of the tubular SLIM: $D_{os} \approx$ 160mm (say, for an in-pipe pump application, etc.).

2. *Additional experience-based data*
    a. The mechanical (hereby, equal to magnetic) airgap is taken as 1.5 mm, feasible as total radial force is zero or small (when the short primary is eccentrically placed within the fix secondary).
    b. The primary and secondary slots have to be open as the two magnetic cores are made of disk-shape silicon steel laminations, to reduce the costs of core material and fabrication.
    c. Since the slots are open, to keep the Carter coefficient reasonable (and reduce surface core losses due to mmf harmonics and keep the power factor acceptable), the slot openings should not be larger than four airgaps in the primary and three airgaps in the secondary (this is one reason that the airgap is not too small).
    d. As the primary coils are circular in shape, a single full pitch layer three-phase winding with ($q_1 = 1$ as in Figure 5.3 or $q_1' = 2$) may be easily inserted in slots.
    e. As the airgap flux/pole distributes in the primary back iron at a notably larger diameter area, the primary back-iron area radial height (h_{y1}) is rather small; on the contrary, the back-iron radial height (h_{y2}) in the interior secondary is larger than τ/π because the average diameter is smaller than in the airgap zone; in fact, the minimum allowable back-iron secondary interior diameter seems to be the design bottleneck.
    f. The primary coils do not have apparent end connections, and the contact area with the slot walls is large. Consequently, the winding thermal energy transfer is good and the winding temperature is uniform; for complete fairness, we should notice that because the primary is exterior to the secondary, the average coil (turns) length is already larger than the airgap circumference so there is additional coil length (as in flat SLIMs).
    g. Stacking the primary and secondary laminations and coils (with some local holes to exit the coil ends) thus becomes an easy task, very similar to electric transformer standard fabrication practice.
    h. The equivalent circuit, in the absence of dynamic end and transverse edge effect (the secondary cooper rings are flowed by circular currents), is still that of Figure 5.2, and only the skin effect is present, but if inverter fed, $f_2 < 10$ Hz, the skin effect may be neglected. In fact, the tubular SLIM allows for smaller f_2 with copper rings, without losing much thrust density (in N/cm²) due the smaller secondary equivalent resistance for the given primary.

3. *Equivalent circuit parameter expressions*

    For further exploration before utilization, Equation 4.36 are presented again here:

$$R_{1t} = \rho_{Co}\frac{\pi D_{avc1}}{I_{1r}\sigma_{Al}w_1}\cdot j_{cor}\cdot 2\cdot w_1^2 = K_{R1t}\cdot w_1^2$$
$$L_{1lt} = 2\mu_0\frac{\pi D_{avc1}}{pq_1}\cdot 2\left(\lambda_{s1} + \lambda_{diff1}\right)w_1^2 = K_{1l}\cdot w_1^2 \tag{5.47}$$

Factor 2 accounts for the fact that two circular-shape coils mean a "pole pitch equivalent" coil.

$$L_{mt} = \frac{6\mu_0}{\pi^2}\left(K_{w1}\cdot w_1\right)^2 \frac{\tau\pi D_{ip}}{pgK_c(1+K_s)} w_1^2 = K_{lm}\cdot w_1^2$$

$$L'_{2lt} = 2\mu_0\pi D_{avc2}\cdot 12\frac{\left(K_{w1}w_1\right)^2}{N_{s2}}\left(\lambda_{s2}+\lambda_{diff2}\right) = K_{2l}\cdot w_1^2 \quad (5.47)'$$

$$R'_{2t} = \rho_{Al(Co)}\cdot\frac{\pi D_{avc2}}{A_{ring}}\cdot 12\frac{\left(K_{w1}w_1\right)^2}{N_{s2}} = K_{R_2}\cdot w_1^2$$

$A_{ring}=h_{s2}\cdot b_{s2}\cdot K_{fill2}$; $K_{fill2}=0.9$ for solid bars and 0.6 for short-circuited copper coils.
The diameters D_{avc}, D_{ip}, D_{avc2} and the secondary slot area $b_{s2}\cdot h_{s2}$ are all visible in Figure 5.3.

4. *Pole pitch τ_{max} and its design consequences*
The equivalent stack length is here πD_{ip}, which for $D_{ip}=80$ mm would mean 0.2512 m. The airgap was chosen as 1.5 mm (not 1 mm) to allow large enough slot openings to easily insert the coils and the secondary conductor rings in slots, though it increases the magnetization current and thus decreases the power factor. As a minimum of 40 mm solid-iron shaft is needed to form with the secondary a rigid body, it means that the secondary slot and back-iron height available would be

$$h_{s2}+h_{y2} = \frac{D_{ip}-2g-40}{2} = \frac{70-2\times1.5-40}{2} = 18.5 \text{ mm} \quad (5.48)$$

With $h_{s2}=8$ mm for the secondary slot depth, only 10.5 mm remains for the back-iron height of the secondary. Even neglecting the redistribution of flux density due to the limited length of primary (Chapter 3), and choosing a back core flux density $By_{2max}=1.8$ T, the maximum pole pitch τ_{max} would be

$$\pi D_{ip} = \frac{\tau_{max}}{\pi}B_{g1max} = B_{y2\,max}\pi\frac{\left(\left(D_{os}-2h_{s2}\right)^2-\left(D_{os}-2h_{s2}-2h_{y2}\right)\right)}{4} \quad (5.49)$$

with $D_{ip}=80$ mm, $B_{g1max}=0.6$ T (not higher here as D_{ip} is rather small) we may calculate the maximum pole pitch τ_{max}:

$$\tau_{max} = \frac{1.8\times\pi\times\left(61^2-30^2\right)}{0.6\times80\times4} = 83 \text{ mm} \quad (5.50)$$

This is quite a reasonable value for the scope; smaller values would further decrease power factor (via lower goodness factor).

At this pole pitch, the number of slots per pole/phase $q_1=3$ to yield a slot pitch in the primary $\tau_{s1}=\tau_{max}/9=83/9=9.22$. The primary slot width $b_{s1}=5.5$ mm$<4g=6$ mm and the primary tooth $b_{t1}=3.72$ mm to secure a maximum flux density in the tooth $B_{t1max}\approx B_{g1max}\cdot\tau_{s1}/b_{t1}=0.6\times9.22/3.72=1.48$ T (thinner tooth than 3.72 mm may not be mechanically stiff enough, even after proper primary impregnation). With 8 mm deep secondary slots, we have to adopt the secondary slot pitch, which should not be for away from the primary slot pitch (to avoid important torque pulsations, noise, and vibration, as in rotary IMs).

With the secondary slot opening limited to three airgaps (4.5 mm), we may adopt a secondary tooth of $b_{t1}=4.0$ mm.

Now the problem is to avoid any iteration in our preliminary design. To accomplish that, we have to choose the number of poles. Adapting a reasonably larger f_{tr} (for a larger airgap), $f_{tr}=2$ N/cm^2, the number of poles yields

$$2p=\frac{F_{xr}}{f_{tr}\cdot\pi\cdot D\cdot\tau_{\max}}=\frac{2500}{2\times\pi\times 8\times 8.3}\approx 6 \tag{5.51}$$

From now on, the design methodology follows the same path as for the flat SLIMs, but using the dedicated expressions of parameters.

5. *The rated slip frequency* f_{2r}

   The condition of maximum thrust per primary current is maintained. This means an about 0.707 secondary (airgap) power factor and in general smaller primary power factor, which, due to the large airgap/pole pitch ratio, is natural:

$$S_r\cdot G_e=1 \tag{5.52}$$

$$G_e\approx\frac{\omega_1 L_m}{K_{l2}R_2^{'}};\quad SK_{l2}\approx\sqrt{1+\left(\frac{\omega_1 L_{2l}^{'}}{R_2^{'}}\right)} \tag{5.53}$$

At this moment, we may calculate $L_{2l}^{'}/R_2^{'}$ as we know all the data on secondary conducting ring, but ω_1 is still missing.

From (5.53), the term $\omega_1 L_m/R_2^{'}$ is calculated:

$$\frac{\omega_1 L_m}{R_2^{'}}\approx\frac{\mu_0\omega_{1r}\tau^2\sigma_{AL}\cdot h_{s2}\left(1-b_{t2}/\tau_{s2}\right)}{\pi^2 gK_c(1+K_s)}$$

$$=\frac{1.256\times 10^{-6}\times 0.083^2\times 3.25\times 10^7\times 8\times(1-3.75/9.25)}{\pi^2\times 1.5\times 1.4\times 1.15}\times\omega_{1r}$$

$$=0.056\times\omega_{1r} \tag{5.54}$$

The saturation factor is taken as $K_s=0.15$ as the airgap is rather large and the pole pitch is not large; the Carter coefficient $K_c\approx 1.4$ (it may be calculated "exactly" as in Chapter 3).

To avoid any iteration, the factor K_{l2}, which accounts for $L_{2l}^{'}, K_{l2}\approx 1.1$, from (5.52) to (5.54),

$$S_r\frac{0.056\times\omega_1}{1.1}=1$$

Consequently,

$$S_r\omega_1=\frac{1.1}{0.056}=19.64 \tag{5.55}$$

Finally

$$f_{2r}=S_r f_1=\frac{19.64}{2\pi}=3.128\text{ Hz}$$

This is apparently a rather small value for LIMs, but in tubular SLIMs we may accept it because the net radial force is ideally zero so the noise and vibration at start (when $f_{2r}=f_1=3.1278$ Hz) will be acceptably low.

As the speed is $u_r=2$ m/s, the primary frequency at rated speed f_{1r} is

$$f_{1r}=f_{2r}+\frac{u_r}{2\tau}=3.1278+\frac{2}{2\times 0.083}=15.176\text{ Hz} \tag{5.56}$$

It becomes evident that the primary rated frequency is only about five times the secondary (slip) frequency, so the efficiency will be below 80%, despite the fact that the core losses may be neglected by comparison with winding losses.

6. *The primary phase mmf: w_1I_{1r} and primary slot depth h_{s1}*
With $sG_e=1$ and airgap flux density $B_{g1\max}=0.6$ T from (5.23)

$$B_{g1\max} = \frac{\mu_0 3\sqrt{2}K_{w1}\cdot w_1I_{1r}}{\pi pgK_c(1+K_s)2}$$

$$= \frac{1.256\times3\times\sqrt{2}\times0.925\times10^{-6}}{\pi\times3\times1.5\times10^{-3}\times1.4\times1.15\times2}\times w_1I_{1r}$$

$$= 0.1080\times10^{-3}w_1I_{1r} \quad (5.57)$$

It follows that $w_1I_{1r}=5555$ A turns/phase.

Let us verify quickly if we can obtain the required thrust (5.24):

$$F_{xn} = \frac{3\pi}{2\tau}I_{1n}^2\frac{L_m}{K_{l2}} \quad (5.58)$$

with L_m from (5.47)

$$L_m = \frac{6\mu_0K_{w1}^2}{\pi^2}\frac{\tau\pi D_{ip}}{pgK_c(1+K_s)}w_1^2$$

$$= \frac{6\times1.256\times10^{-6}\times0.925^2\times0.083\times\pi\times0.08}{\pi^2\times3\times1.5\times10^{-3}\times1.4\times1.15}\times w_1^2$$

$$= 1.882\times10^{-6}\times w_1^2 \quad (5.59)$$

with (5.58) in (5.57) the thrust is

$$F_{xn} = \frac{3\times3.14}{2\times0.083}\times(5555)^2\times\frac{1.882\times10^{-6}}{1.1} = 2925 > 2500\ \text{N} \quad (5.60)$$

It is fortunate that Equation 5.60 leads to a thrust larger than required and thus the design stands (on the safe side). In case the thrust had been smaller, the whole design should have been repeated with a larger outer diameter (and pole pitch, etc.) and a larger number of poles ($2p=8$, for example). The rather large specific force (2 N/cm^2) "is paid for" by the large phase ampere turns w_1I_{1r}, and thus larger primary copper losses, for limited stator slot depth h_{s1} (to at most 5–6 times the slot width $b_{s1}=5.5$ mm, $h_{s1}=35$ mm).

The phase mmf (in RMS terms) is divided into $2pq$ slots/phase. So the mmf per slot $n_{coil}I_{1r}$ is

$$n_{coil}I_{1r} = \frac{w_1I_{1r}\times2}{2pq} = \frac{5555\times2}{2\times3\times3} = 617.2\ \text{A turns/slot} \quad (5.61)$$

(Two circular coils "qualify" for a regular coil with pole pitch τ.)

So the rated current density J_{cor} is obtained from

$$h_{s1} = \frac{n_{coil} I_{1r}}{J_{cor} \cdot K_{fill} \cdot b_{s1}} = \frac{617.2}{J_{cor} \times 0.5 \times 5.5} = 35 \text{ mm} \quad (5.62)$$

From (5.62),

$$J_{cor} = 7.833 \text{ A/mm}^2$$

This current density value is rather high for continuous duty cycle unless forced air cooling is applied, with a primary back-iron depth of 7 mm (it is 10.5 mm in the secondary at a much smaller diameter), mainly because of mechanical rigidity and room to assemble longitudinal bars; the outer diameter is $D_{op} = D_{ip} + 2h_{s1} + 2h_{y1} = 80 + 70 + 14 + 2 = 166$ mm.

We may now proceed to calculate all equivalent circuit parameters, then the number of turns/phase, wire gauge, and performance (thrust, efficiency, and power factor).

7. *Number of turns per phase w_1 and equivalent circuit parameters*
   To calculate the number of turns/phase w_1, the equivalent parameter coefficients (5.47) should be computed first:

$$R_{1t} = \rho_{Co} \frac{\pi D_{avc1}}{I_{1r} w_1} J_{cor} \cdot 2w_1^2$$

$$= 2.3 \times 10^{-8} \times \frac{\pi \times 0.116}{5555} \times 7.833 \times 10^6 \times 2w_1^2$$

$$= 2.3636 \times 10^{-5} \times w_1^2 = K_{R1t} \cdot w_1^2 \quad (5.63)$$

$$L_{1lt} = \frac{2\mu_0 \pi D_{avc1}}{pq_1} \left(\lambda_{s1} + \lambda_{diff1}\right) w_1^2 \times 2$$

$$= \frac{2 \times 1.256 \cdot 10^{-6} \times \pi \times 0.116}{3 \times 3} (2.12 + 0.112) \times 2 \times w_1^2$$

$$= 2 \times 0.2269 \times 10^{-6} \times w_1^2 = K_{1lt} \cdot w_1^2 \quad (5.64)$$

$$\lambda_{s1} \approx \frac{h_{s2}}{3b_{s2}} = \frac{35}{3 \times 5.5} \approx 2.12; \quad \lambda_{diff} \approx \frac{5 \times 1.4 \times 1.5/5.5}{5 + 4 \times 1.4 \times 1.5/5.5} = 0.112 \quad (5.65)$$

From (5.47)′,

$$L_{mt} = 1.882 \times 10^{-6} \times w_1^2 = K_{mt} \cdot w_1^2 \quad (5.66)$$

$$R_{2t}' = \rho_{Al} \cdot \frac{\pi D_{avc2}}{A_{ring}} \times \frac{12 \times K_{w1}^2}{N_{s2}} \times w_1^2$$

$$= 3.25 \times 10^{-8} \times \frac{\pi \times 0.069}{36 \times 10^{-6}} \times \frac{12 \times 0.925^2}{58} \times w_1^2$$

$$= 3.462 \times 10^{-5} \times w_1^2 = K_{R2t} \cdot w_1^2 \quad (5.67)$$

$$A_{ring} = h_{s2} \cdot b_{s2} = 8 \times 4.5 = 36 \tag{5.68}$$

$$N_{s2} = N_{s1} \cdot \frac{\tau_{s1}}{\tau_{s2}} = 6 \times 3 \times 3 \times \frac{9.22}{8.5} \approx 58 \text{ secondary slots/primary length} \tag{5.69}$$

$$L'_{2lt} = 2\mu_0 \pi D_{avc2} \times \frac{12 \times K_{w1}^2 \cdot w_1^2}{N_{s2}} \cdot (\lambda_{s2} + \lambda_{diff2})$$

$$= 2 \times 1.256 \times 10^{-6} \times \pi \times 0.069 \times 12 \times 0.925^2 \left(\frac{0.5926 + 0.1}{58} \right) \times w_1^2$$

$$= 0.066 \times 10^{-6} \times w_1^2 = K_{2lt} w_1^2 \tag{5.70}$$

$$\lambda_{s2} \approx \frac{h_{s2}}{3b_{s2}} = \frac{8}{3 \times 4.5} = 0.5936; \quad \lambda_{diff2} = 0.1$$

We are now using again the equivalent circuit (as in (5.39)) to calculate the number of turns/phase for the max voltage (120 V/phase (RMS)) and $f_{1r} = 15.176$ Hz, $f_{2r} = s_r \cdot f_{1r} = 3.1278$ Hz:

$$V_{1r} = w_1 \cdot (w_1 I_{1r}) \cdot \left| K_{R1t} + j\omega_{1r} K_{1lt} + j \frac{K_{mt}(K_{R2t}/s_r + j\omega_{1r} K_{2lt})}{K_{2Rt}/(s_r \omega_{1r}) + j(K_{mt} + K_{2lt})} \right| \tag{5.71}$$

$$120 = w_1 \times 5555 \times 10^{-6} \left| \begin{array}{l} 18.82 + j95.3 \times 0.2269 \times 2 \\ + j1.882 \dfrac{(34.62/0.2067 + j \times 95.3 \times 0.06)}{34.62/19.64 + j(1.882 + 0.06)} \end{array} \right|$$

$$120 = w_1 \times 5555 \times 10^{-6} \left| 23.626 + 85.97 + j127 \right| \tag{5.72}$$

So, $w_1 \approx 126$ turns/phase. As two coils make a pole-span equivalent coil, the number of turns/slot is

$$n_{c1} = \frac{w_1 \times 2}{2pq} = \frac{252}{2 \times 3 \times 3} \approx 15 \text{ turns/coil} \tag{5.73}$$

So $w_{1f} = 135$ turns/phase.

The current in the coils (all 18 coils in series) is I_{1r}:

$$I_{1r} = \frac{w_1 I_{1r}}{w_1} = \frac{5555}{126} \approx 44.1 \text{ A/phase (RMS)} \tag{5.74}$$

Finally, the parameters of the equivalent circuit are

$$\begin{aligned}
R_{1t} &= K_{R1t} \cdot w_1^2 = 2.3626 \times 10^{-5} \times 126^2 = 0.375\,\Omega \\
L_{1lt} &= K_{1lt} \cdot w_1^2 = 2 \times 0.2269 \times 10^{-6} \times 126^2 = 0.007204\,\text{H} \\
L_{mt} &= K_{mt} \cdot w_1^2 = 1.882 \times 10^{-6} \times 126^2 = 0.02987\,\text{H} \\
R_{2t}' &= K_{R2t} \cdot w_1^2 = 3.462 \times 10^{-5} \times 126^2 = 0.5496\,\Omega \\
L_{2lt}' &= K_{2lt} \cdot w_1^2 = 0.066 \times 10^{-6} \times 126^2 = 0.0010478\,\text{H}
\end{aligned} \tag{5.75}$$

Now that the number of turns was reduced from 129 to 126, to maintain the rated current (for given $w_1 I_{1r} = 5555$ A turns/phase (RMS)), we need to further reduce the voltage per phase to $V_{1rf} = V_{1r} \cdot 126/129 = 120 \times 126/129 = 117.2$ V/phase (RMS); so there is a further (useful) voltage reserve that will probably eliminate the necessity for overmodulation in the PWM inverter for more sinusoidal phase currents.

8. *Power factor and efficiency*

The power factor is evident from (5.72):

$$\cos\varphi_1 = \frac{\text{Re}(\underline{Ze})}{|\underline{Ze}|} = \frac{109.596}{167.00} \approx 0.6 \tag{5.76}$$

To calculate the efficiency, we first have to determine the electromagnetic power P_{elm}.

$$P_{elm} = F_{xr} \cdot U_s; \quad U_s = 2\tau f_{1r} = 2 \times 0.083 \times 15.176 = 2.5192 \text{ m/s} \tag{5.77}$$

But

$$\begin{aligned}
P_{elm} &= P_1 - 3R_1 I_{1r}^2 = 3V_{1r} I_{1r} \cos\varphi_1 - 3R_1 I_{1r}^2 \\
&= 3 \times 117.2 \times 44.1 \times 0.6 - 3 \times 0.375 \times 44.1^2 = 7115\,\text{W}
\end{aligned} \tag{5.78}$$

So the thrust is in fact

$$F_x = 7115/2.5192 = 2824 \text{ N} \tag{5.79}$$

The efficiency is in fact

$$\eta_1 = \frac{F_x U_r}{P_1} = \frac{2824 \times 2}{9303} = 0.607 \tag{5.80}$$

Discussion

- The required 2.5 kN thrust was surpassed in the preliminary design to 2.824 kN at a specific force of almost 2.4 N/cm^2 of primary; this is a rather large value for an LIM.
- The power factor is rather good for LIMs ($\cos\varphi_1 = 0.60$); the price to pay for high force density is the lower efficiency, which implies forced air cooling, at least for the primary.
- The forecasted 0.37 of $\eta_1 \cos\varphi_1$ product is a fairly good LIM performance for 2 m/s

- For given travel length and duty cycle, the temperature of the fix secondary may be assessed by fairly simple equivalent thermal circuit models [5].
- For better efficiency at the price of a heavier primary and secondary, the overall D_{op} and primary bore D_{ip} diameters have to be increased and the design routine has to be repeated (a simple MATLAB®-code would take only a few seconds to do it).

9. *Note on optimization design*
For low-speed (negligible dynamic end effect) SLIMs with ladder secondary, the optimization design is very similar to that of rotary IMs [6]. Only parameter and performance expressions and the objective function should be adapted for the scope. For medium- and high-speed (transportation) LIMs, the optimization design will be approached in some detail, as there are more notable differences with respect to the case of rotary IMs (Chapter 6).

5.4 SUMMARY

- By low-speed LIMs it is meant low (negligible end effect) LIMs ($(\gamma_{2r}\tau)^{-1} < 2p/10$, Chapter 3); speeds up to 5–6 m/s in adequate design LIMs may fall into this category.
- To raise the efficiency × power factor product from 0.2–0.25 to 0.36, only cage-secondary SLIMs with 1–1.5 mm airgaps are considered.
- Only for medium- and high-speed LIMs (Chapter 6), the aluminum-iron secondary will be investigated, in long travel of hundreds of meters to tens and hundreds of kilometers.
- The design airgap flux density in ladder secondary, for 1–1.5 mm airgap, low speed (up to 5–6 m/s) should be in the range of 0.6–0.7 T to yield reasonable thrust density (1–2 N/cm^2).
- In flat SLIMs with cage secondary, the normal (attraction) force is even more than 20 times the thrust and is recommended to be used for the "almost" entire mover (plus load) suspension to relieve the linear bearing of too high mechanical stresses.
- In tubular SLIMs, the net normal force is ideally zero, and in up to 1–2 m long travels (even longer) it may be the preferred solution.
- For the design of flat (or tubular) low-speed SLIMs, the equivalent circuit of IMs is used with dedicated expressions for the parameters: $R_1, L_{1l}, L_m, R_2', L_{2l}'$; as both cores are laminated, and the secondary frequency is lower than 10 Hz, the skin effect in the secondary conductor rings in slots is moderate, so that only magnetic saturation has some but limited influence (the airgap/pole pitch ratio is not very small, as in rotary IMs).
- By design we mean here dimensioning (sizing), that is, to find a suitable geometry for a SLIM of given specifications.
- The key design concept used here is the maximum thrust per primary mmf (current), which leads to $S_rG_e = 1$; this represents a practical compromise between acceptable efficiency and primary weight/thrust.
- The slip (secondary) frequency for rated thrust is chosen here (for the case studies in this chapter $f_{2r} = 4$ Hz, respectively, 3 Hz for the flat, respectively tubular, rather medium thrust 500 N (2500 N) and 3(2) m/s).
- With a low (0.88 N/cm²) specific thrust, 0.707 power factor and 0.711 efficiency were obtained for the flat ladder-secondary SLIM for 615N at 3 m/s, with a short (mover) primary of about 36 kg; this may be considered competitive performance in some applications where ruggedness and low initial costs are prevalent.
- On the contrary, for the tubular ladder-secondary SLIM at 2.4 N/cm², cos $\varphi = 0.6$, at 2824 N at 3 m/s for an 0.166 m outer primary diameter (0.08 m inner primary diameter), 0.5 m long primary (about 66 kg short primary weight).
- The tubular SLIMs with cage secondary and disk-shape laminations in both primary and secondary cores lead to negligible core losses in most low-speed designs and represent an easy manufacturable topology.

- Flat SLIMs with ladder long secondary and short primary are already commercial in some machine-tool table, etc., applications.
- Even for low-speed (say, at 5–6 m/s) SLIM designs, dynamic end effect may be accepted by choosing higher goodness factor topologies (higher frequency and pole pitch) when both conventional performance and the deteriorating dynamic end effect rise; but the compromise at optimum goodness factor (Chapter 4)—zero end effect force at zero slip—may end up in better performance; this case is implicit, however, in medium- or high-speed SLIMs design, to be treated in Chapter 6.
- Optimization analytical design of low-speed (negligible dynamic end effect) SLIMs is too similar to that of rotary IMs treated here [see Ref. 6].
- 2(3)D FEM-based direct geometrical design after optimization analytical design should follow, but this is beyond our scope here.

REFERENCES

1. I. Boldea and S.A. Nasar, *Simulation of High Speed LIM End Effects is Low Speed Tests*, *Proc. IEEE*, 121(9), 1974, 961–964.
2. I. Gieras, *Linear Induction Drives*, Clarendon Press, Oxford, U.K., 1994.
3. I. Boldea and S.A. Nasar, *Linear Electric Actuators and Generators*, Cambridge University Press, Cambridge, U.K., 1997.
4. I. Boldea and S.A. Nasar, *Linear Motion Electromagnetic Devices*, Taylor & Francis, New York, 2001.
5. I. Boldea and S.A. Nasar, *Induction Machine Design Handbook*, 2nd edn., CRC Press, Taylor & Francis Group, New York, 2010.
6. I. Boldea and L. Tutelea, *Electric Machines: Steady State Transients and Design with Matlab*, CRC Press, Taylor & Francis, New York, 2009.

6 Transportation (Medium- and High-Speed) SLIM Design

6.1 INTRODUCTION

We define transportation as urban and interurban public people or freight movers with vehicle suspension on wheels or via MAGLEVs.

The main difference between magnetic and wheel suspension of people movers from the SLIM design point of view is the treatment of net normal force F_n.

As shown in Chapter 3, F_n is reasonably low (three times rated propulsion force in aluminum-on-iron SLIMs) if the efficiency is the main target of the design. F_n switches from repulsion to attraction force as the slip (secondary) frequency decreases; the switching point condition, however, corresponds to a rather large slip frequency (Chapter 3), which may not be acceptable as it produces high losses (lower efficiency) in the SLIM.

It is thus advisable for SLIM-MAGLEVs to arrange the secondary track above the SLIM, on the two sides of the vehicle, and to use the attraction normal force of SLIMs at lower slip frequency, that is, better efficiency.

To complement the main (SLIM) magnetic suspension system, actively controlled dc electromagnets with (or without) PMs are placed also on board (Figure 6.1a). For vehicles on wheels, the net attraction normal force may be limited to a certain value that does not produce excessive noise and vibration at vehicle start while it enhances a little the mechanical adhesion of wheels (Figure 6.1b). Also for some SLIM-MAGLEVs, the SLIM is still placed above the track, and thus, the net normal force of LIM has to be counteracted by the dc-controlled electromagnets that provide vehicle levitation (Figure 6.1c); the SLIM slip frequency may also be controlled to have the net normal force close to zero, but that may compromise efficiency too much.

The previous chapter has dealt with the design of tubular-secondary SLIMs, while in this chapter, only the SLIM with aluminum-on-iron secondary will be treated, as it seems the practical way to follow for a reasonable passive track cost for long travels, as in transportation (even for industrial transport, over a few tens of meters).

Again, the electromagnetic design aspects will be the main target, but at least for urban SLIM vehicles, some issues of secondary losses and temperature around stop stations will be approached in some detail.

Case studies will accompany the design methodologies, for urban SLIM vehicles (max speed >34 m/s) and for interurban vehicles (100 m/s).

6.2 URBAN SLIM VEHICLES (MEDIUM SPEEDS)

The design methodology by a numerical example includes (1) the specifications; (2) additional data; (3) main dimensions, parameters, and performance; (4) secondary thermal design issues and performance numerical results.

1. Specifications
   a. Peak thrust up to base speed ($u_b = 12$ m/s): 12 kN
   b. Thrust at max speed ($u_{bmax} = 34$ m/s): 3 kN
   c. Starting current 500–600 A

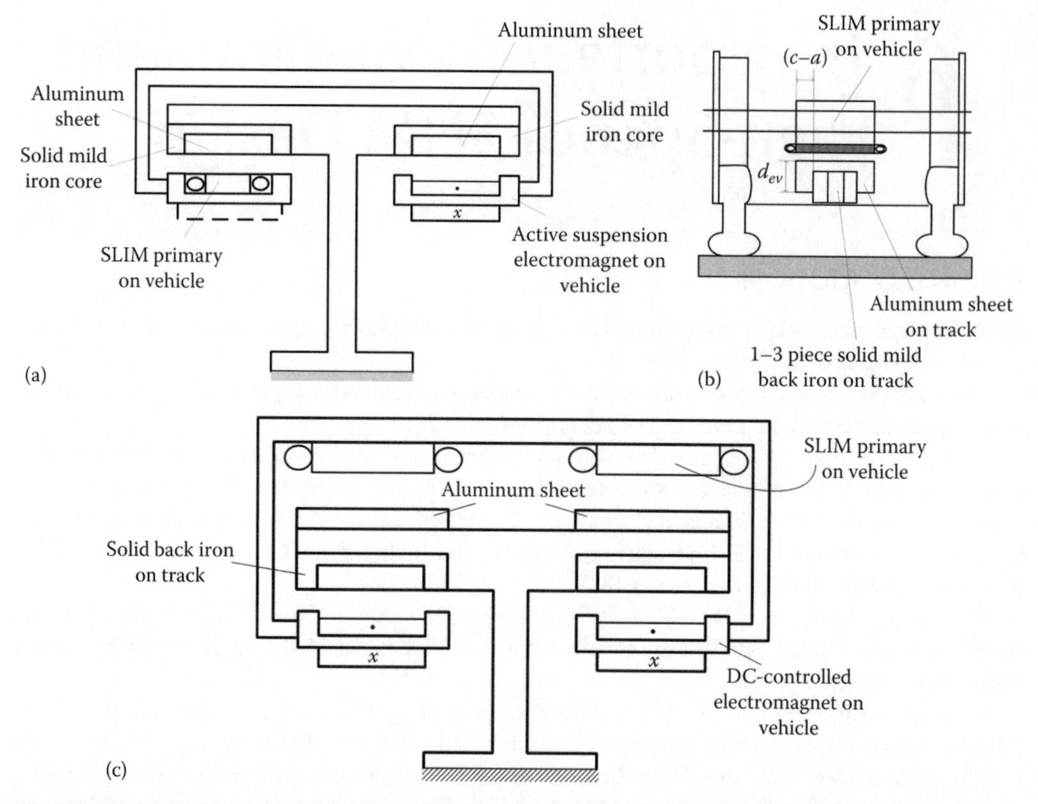

FIGURE 6.1 SLIM arrangements for transportation: (a) for MAGLEVs, (b) for vehicles on wheels, (c) alternative MAGLEV.

d. Supply: DC supply at 700 V_{dc} and PWM voltage-source IGBT inverter on vehicle (max voltage/phase (RMS): 300 V)
e. Aluminum-on-iron secondary track

$$(\eta_1 \cos \varphi_1)_{u_{\max}} \geq 0.35$$

2. Additional data from prior experience
   a. As even with three-piece secondary solid mild back iron, the field penetration depth, under heavy saturation, is not going to reach more than 15–20 mm, the pole pitch has to be limited to perhaps 0.22–0.27 m; a three-piece solid back-iron secondary is chosen here.
   b. The tight curves to be handled in urban transportation restrict the primary on-board length to 2–3 m.
   c. To reduce end-coil length, the primary winding should have two layers and chorded coils, which leads to two-layer windings with $2p+1$ poles and half-filled end-pole slots.
   d. The number of poles for urban SLIM vehicles is thus limited to 7, 9, and 11; the lower limit is imposed by the necessity to limit the dynamic (and static) end effect.
   e. The airgap flux density at maximum (peak) thrust has to be limited due to limited secondary back-iron skin effect, but also to limit the net normal force, with its consequences, as discussed earlier. Values of $B_{1g\max} = 0.20 - 0.35$ T are considered feasible.
   f. The optimum equivalent goodness factor G_{eo} (Chapter 3) is considered here the key design variable; with these restrictions, for $2p+1 = 11$ and $\tau = 0.255$ m, from Chapter 3

$(G_{eo})_{2p=10} = 24$; the equivalent goodness factor varies with respect to slip frequency, current (due to magnetic saturation), and with the airgap variation due to track irregularities, so G_{eo} value has to be observed at rated slip (and peak speed).

g. The dynamic end effect may not be neglected even for this urban application, though the optimum goodness factor strikes a good compromise between thrust/weight and $\eta_n \cos \varphi_n$ (which, in turn, is related to energy usage rate and inverter initial costs).
h. In presence of dynamic end effects, it seems more practical to design (size) the SLIM rather in the 1D technical field theory (Chapter 3) than in equivalent circuit terms; the end effect correction coefficient $K_e(G_e, p, s)$, inserted in the circuit model, may be added at the end to the other circuit parameters; their dependence on current and slip frequency makes them less useful in assessing quickly the performance, unless convenient curve-fitting approximations of their variations are not produced.
i. The mechanical airgap $g = 8$–10 mm for urban SLIM vehicles (with safety devices to secure it), for handling realistic track irregularities.
j. Aluminum and back-iron pieces are added together to make the track; they produce some speed-tied frequency pulsations in the thrust and normal force, with only 2%–3% average thrust reduction for secondary sections in the 1–2 m length range; bolt joints may be added, but they are too costly for a small beneficial effect [1].
k. The thickness of aluminum sheet is $d_{Al} = 4$–6 mm, as a compromise between track costs and SLIM performance and costs; a large thickness leads to better goodness factor for given mechanical gap. But at the price of higher magnetization current (lower power factor), d_{Al} may become a variable to optimize, for a given application, when a multiple-target objective function is used.
l. The primary stack width $2a = 0.2$–0.3 mm, again to limit the secondary track costs; as $2a/\tau \approx \sim 0.8$–1.2, the edge effect is also limited this way; finally, the aluminum sheet overhang cross section should be $\tau d_{Al}/\pi$. The depth of the aluminum overhang when placed adjacent to secondary back iron $d_{ov} < 3d_{Al}$, to limit the skin effect in this zone also, so $(c-a)_e = \tau d_{Al}/d_{ov} > \tau/12$, while for "pure" aluminum sheet, $(c-a) \geq \tau/\pi$.
m. This additional data have to be corroborated with the specifications, by adjusting the number of poles $2p+1$, pole pitch, and stack width, to secure a specific thrust in the range of 2 N/cm^2 for the case in point.
n. The secondary back-iron width $w_i \approx 2a + 0.02$ m.
o. The rated slip (once the optimum goodness factor is chosen for rated thrust as a compromise between conventional and with-end-effect-considered performance and primary weight) is chosen not very far away from the condition $S_r G_{eo} = 1$, which corresponds to max conventional thrust/mmf. For urban SLIM vehicles, we may choose $S_r G_{eo} \approx 2$ for max speed $U_{max} = 34$ m/s corresponding thrust $F_{xn} = 3$ kN.
p. For the peak thrust at standstill and then up to base speed ($U_b = 12$ m/s), the same conditions may be kept; that is, constant slip (secondary) frequency $f_2 = f_{2r} = S \cdot f_{1max}$; the primary frequency f_{1max} corresponds to max speed (here, $U_{max} = 34$ m/s).
q. If, as the required thrust decreases, the slip frequency is increased, the conventional losses increase, but the average end effect also diminishes (slip increases); a slight increase in f_2 with speed becomes thus practical; care must be exercised, as f_2 influences notably the total normal force also.

3. Main dimensions

First, let us choose from paragraph (2) a few notable SLIM sizing data

- Aluminum sheet thickness $d_{Al} = 6$ mm (initial value)
- Airgap $g = 8$ mm

- Pole pitch $\tau=0.25$ m (initial value)
- Slip frequency (constant) $f_{2r}=6$ Hz, at base speed
- Stack width $2a=0.27$ m
- The aluminum equivalent overhang $(c-a)=\tau/\pi$ $c-a=\tau/\pi$
- $2p+1=11$ poles two-layer primary winding with half-filled end-pole slots

The design start is related to the optimum goodness factor

$$G_{eo}=24=\frac{\mu_0\omega_{1\max}\tau^2\sigma_e d_{Al}}{\pi^2 g_e\left(1+K_{ss}\right)} \tag{6.1}$$

where σ_e is the equivalent aluminum conductivity including skin effect at rated slip frequency, edge effect, and solid (3 laminations) back-iron thrust and loss contribution: $g_e \approx (g+d_{Al})^*K_cK_{Fg}$

Assuming $f_{2r}=8$ Hz at max speed, with $\tau=0.25$ m and $U_{\max}=34$ m/s, the maximum frequency $f_{1\max}$ becomes

$$f_{1\max}=f_{2r}+\frac{u}{2\tau}=8+\frac{34}{2\times 0.25}=76\,\text{Hz} \tag{6.2}$$

From (6.1) to (6.2), the ratio $\sigma_e/(1+K_{ss})$ may be found:

$$\frac{\sigma_e}{\left(1+K_{ss}\right)}=\frac{24\pi^2\left(g+d_{Al}\right)\times 1.15\times 1.1}{\mu_0\omega_{1\max}\tau^2 d_{Al}}$$

$$=\frac{24\pi^2\left(8+6\right)10^{-3}\times 1.15\times 1.1}{1.256\times 10^{-6}\times 2\pi\times 76\times 0.25^2\times 6\times 10^{-3}}$$

$$=2.131\times 10^7\left(\Omega\ \text{m}\right)^{-1} \tag{6.3}$$

Let us remember that the aluminum electrical conductivity is $\sigma_{Al}\approx 3.25\times 10^7$ $(\Omega\ \text{m})^{-1}$. This shows that the aluminum edge effect and the solid back-iron contribution to σ_e have to be substantial (or the pole pitch may simply be decreased, even below $G_{eo}=24$). Let us consider first an approximate value of aluminum edge effect K_{tr} on its electrical conductivity (though the more involved expression derived in Chapter 3 may be used in a design computer code):

$$K_{tr}=\frac{1}{1-\left(\dfrac{\tanh\dfrac{\pi a}{\tau}}{\dfrac{\pi a}{\tau}}\right)\Bigg/\left[1+\tanh\dfrac{\pi a}{\tau}\tanh\dfrac{\pi\left(c-a\right)}{\tau}\right]}$$

$$=\frac{1}{1-\left(\dfrac{\tanh\dfrac{\pi\times 0.135}{0.25}}{\dfrac{\pi\times 0.135}{0.25}}\right)\Bigg/\left[1+\tanh\dfrac{\pi\times 0.135}{0.25}\tanh\dfrac{\pi}{\tau}\dfrac{\tau}{\pi}\right]}=1.475 \tag{6.4}$$

So the iron contribution to secondary conductivity $K_{i\sigma}$ (3.46) and the airgap through K_{ss} (3.44) should amount to

$$\frac{1+K_{i\sigma}}{1+K_{ss}} \approx \frac{\sigma_e K_{tral}}{\sigma_{Al}} = \frac{1.864\times10^7\times1.475}{3.25\times10^7} = 0.845 \tag{6.5}$$

But

$$K_{ss} \approx \frac{2\tau^2\mu_0}{\pi^2 2(g+h_{Al})K_e\delta_{iron}\mu_{iron}}, \quad K_{i\sigma} = \frac{\sigma_{iron}}{\sigma_{Al}}\cdot\frac{\delta_{iron}K_{tral}}{d_{Al}K_{tri}} \tag{6.6}$$

From (3.42) for iron field penetration depth δ_{iron} (3.15), K_{tri} is

$$K_{tri} \approx \frac{1}{1-\left(\tanh\frac{\pi}{\tau}\frac{a_e}{i}\right)\Big/\left(\frac{\pi}{\tau}\frac{a_e}{i}\right)} = \frac{1}{1-\left(\tanh\frac{\pi}{0.25}\frac{0.14}{3}\right)\Big/\left(\frac{\pi}{0.25}\frac{0.14}{3}\right)} \approx 9.93 \tag{6.7}$$

($i=3$ "laminations" in the secondary back iron)

$$\delta_{iron} \approx \sqrt{\frac{1}{\pi f_2\mu_{iron}\cdot(\sigma_{iron}/K_{tri})}} \tag{6.8}$$

The secondary back-iron permeability may only approximately be calculated as in (3.17):

$$B_{t\max}\cdot\delta_{iron}\left(1-e^{-1}\right) = B_{g1}\frac{1}{\pi}\tau\left(1+K_{ss}\right) \tag{6.9}$$

with B_{g1} the airgap flux density at max speed and corresponding thrust (3 kN). Considering $B_{g1\max}=0.3$ T at max thrust (12 kN, to limit the normal force), it follows that for thrust proportional to current squared and flux density proportional to current $B_{g1} \approx B_{g1\max}/2=0.15$ T; considering end effect, we may use $B_{g1}=0.22$ T.

For $B_{t\max}=2.2$ T and $H_m=50{,}000$ A/m for mild solid steel of secondary, μ_{iron} (on its surface) is

$$\frac{\mu_{iron}(2.2\,\text{T})}{\mu_0} \approx \frac{2.2}{1.256\times10^{-6}\times50{,}000} = 35 \tag{6.10}$$

so

$$\delta_{iron} = \sqrt{\frac{1}{\pi\times6\times35\times1.256\cdot10^{-6}\times3.5\times10^{-6}\times19.93}} = 5.852\times10^{-2}\text{ m} \tag{6.11}$$

The depth of penetration looks large, but let us remember that at same $f_{2r}=8$ Hz, the maximum flux density has to be flown at start and up to base speed (for $F_{x\max}=12$ kN). So condition (6.9) has to be satisfied rather for $B_{g1\max}=0.3$ T:

$$2.2\times5.8\times10^{-2}\left(1-e^{-1}\right) > 0.3\times\frac{1}{\pi}\times0.25\times1.5 \tag{6.12}$$

$$\text{or} \quad 0.0806 > 0.03582 \tag{6.13}$$

The factor $K_{ss}=1.5$ in (6.9) takes care of back-iron field redistribution with doubled flux density, due to primary limited length (Chapter 3).

However, it seems clear that even an $\delta_i \approx 25$ mm would suffice, which may be obtained with an iron surface flux density of around 2.0 T and $H_m=30{,}000$ A/m.

Additionally, two-lamination back iron may be used, for which $K_{iron} \approx 4$, $\mu_{iron} \approx 53\mu_0$, and $\delta_1 \approx 25$ mm. Consequently from (6.6),

$$K_{ss} = \frac{2\times 0.25^2 \times \mu_0}{\pi^2 \times 2(10+6)\times 10^{-3} \times 1.15 \times 2.5\times 10^{-2} \times 53 \times \mu_0} = 0.26 \tag{6.14}$$

$$K_{i\sigma} = \frac{3.5\times 10^6}{3.25\times 10^7} \times \frac{25}{6} \times \frac{1.475}{4.0} = 0.1655 \tag{6.15}$$

So, returning to (6.5),

$$\frac{1+K_{i\sigma}}{1+K_{ss}} = \frac{1+0.1655}{1+0.26} = 0.925 > 0.845 \tag{6.16}$$

The difference between the required and obtained ratio is acceptably small, and thus, the 25 mm thick solid back iron of secondary made of two-lamination (pieces) stands. But it has to be noticed that the iron contribution (to thrust, losses, and magnetization current) depends heavily on primary current (via magnetic saturation) for given slip frequency; a robust thrust control loop is required as also the airgap may vary due to track irregularities, etc.

For simplicity, the optimum goodness factor at $B_{g1\max}=0.22$ T and $f_{2r}=8$ Hz, $G_{eo}=24$ will be considered further on in the design, though above base speed ($u_b=12$ m/s) up to $u_{b\max}=34$ m/s, the thrust decreases and so does the current, but less than proportional, because of smaller penetration depth of iron and thus a larger demand on magnetization current. Also, the goodness factor decreases somewhat below $G_{eo}=24$. In a rather complete design, such conflicting influences should be accounted for.

One more aspect is related to the occurrence of a leakage inductance effect of iron eddy currents in the secondary back iron, because the skin effect inflicts eddy current density phase shifting, besides magnitude reduction, along the back-iron depth. At 50 Hz, it has been proved that $\omega_1 L'_{2li} \approx R'_{2i}$ of iron; at 8 Hz, the influence is notably smaller, and so is the iron influence on losses (and thrust), so a first approximation $L'_{2li} \approx 0$. Also the airgap fringing coefficient K_{fg} accounts approximately for the skin effect leakage inductance in aluminum, which, again, is rather small.

4. Peak thrust capability

   To continue the design, the peak thrust capability required below base speed is checked in terms of primary slotting and winding loss suitability for $2p+1=11$ poles, $\tau=0.25$ m, and $2a=0.27$ m for 12 kN thrust.

   First, let us calculate the peak specific thrust f_{xspeak}:

$$f_{xspeak} \approx \frac{F_{xpeak}}{(2p-1/2)\tau\cdot 2a} = \frac{12{,}000}{(10-1/2)\times 0.25\times 0.27} = 1.8713\ \text{N/cm}^2 \tag{6.17}$$

This is a reasonable value, to expect good efficiency, especially at lower values of thrust required above base speed.

It is not sure that the LIM will provide at, $f_1=f_{2r}=6$ Hz at start, the peak thrust at reasonable losses with the main geometry already in place. For 6 Hz, the goodness factor at start will be

$$\left(G_{es}\right)_{S=1} = \frac{f_{2r}}{f_{1\max}} \cdot G_{eo} = \frac{6}{75} \times 24 = 1.92 \tag{6.18}$$

This value is far away from the maximum thrust/current at start where $G_{eso}=1$. Though reducing further $f_2 < 6$ Hz at start would lead to vibration and noise in wheeled vehicles, the higher normal force in MAGLEVs may be advantageously used at start to assist the dc-controlled electromagnets in magnetic suspension.

The flux density $B_{g1\max}=0.3$ T is (5.24)

$$B_{g1\max} = \frac{\mu_0 3\sqrt{2}\left(w_1 K_{w1}\right) I_{1peak}}{\left(g + d_{Al}\right) K_c K_{fg} (1 + K_{ss}) \sqrt{1 + G_{es}^2} \times p \times \pi} = 0.3 \text{ T} \tag{6.19}$$

With $\tau=0.25$ m, slot openings as wide as $b_{s1}=18$ mm are allowed ($b_{s1}/(g+d_{Al})=18/14<2/1!$), without heavy influence of Carter coefficient on airgap flux-density space harmonics augmentation (Figure 6.2).

So $q_1=3$ slots/pole/phase, $\gamma/\tau=7/9$ (chorded coils) with slot pitching, $\tau_{s1}=\tau/q_1 m_1=0.25/9=0.027$ m $=27.77$ mm, and $b_{s1}=18$ mm; the primary tooth width $b_{t1}=\tau_{s1}-b_{s1}=27.77-18=9.77$ mm (with up to 0.3 T in the airgap, no stator teeth heavy saturation may occur).

From (6.19), the peak phase Ampere turns, $w_1 I_{1peak}$ (RMS value), is

$$w_1 I_{1peak} \approx \frac{\pi \times 0.3 \times 16 \cdot 10^{-3} \times 1.15 \times 1.1 \times 1.26 \times \sqrt{1 + 1.96^2 \times 5}}{3\sqrt{2} \times 1.256 \times 10^{-6} \times 0.925} = 68.503 \times 10^3 \text{ A turns/phase} \tag{6.20}$$

The number of coils/phase corresponds to 10 (not 11) poles, but fully filled (two layers) and thus the peak Ampere turn per slot $2n_c I_{1peak}$ is

$$2n_c I_{1peak} = \frac{w_1 I_{1peak}}{p q_1} = \frac{68.05 \times 10^3}{5 \times 2} = 4.567 \times 10^3 \text{ A turns/slot} \tag{6.21}$$

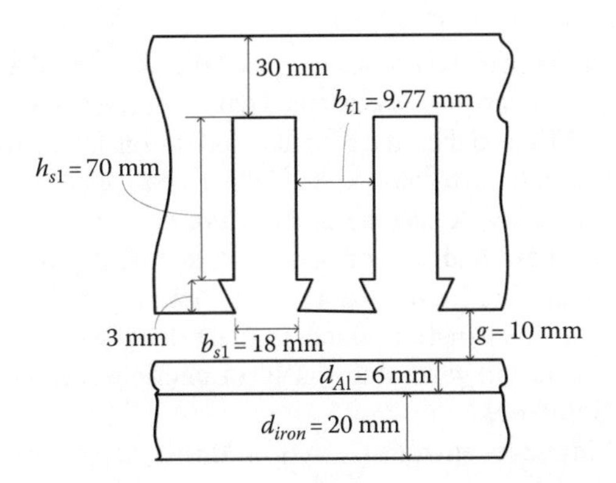

FIGURE 6.2 Longitudinal geometry of designed SLIM for urban transportation.

This mmf has to be sustained for $B_{g\max}=0.3$ T and $f_{20}=6$ Hz up to base speed. It is not yet sure that it will provide the required thrust. But before that, let us see if reasonable slot geometry is feasible.

For a $b_{s1}=18$ mm wide and $h_{s1}=70$ mm deep slot, with a slot filling factor $K_{fill}=0.55$, the current density j_{cop} for peak thrust would be (Figure 6.2)

$$j_{cop}=\frac{2n_cI_{1peak}}{h_{s1}\cdot b_{s1}\cdot K_{fill}}=\frac{4.567\times 10^3}{70\times 18\times 0.55}=6.59\ \text{A/mm}^2 \tag{6.22}$$

The peak thrust check for this peak Ampere turns/phase value is now done, at standstill (4.11):

$$F_x=\left(\frac{3\pi L_m\cdot I_{1peak}{}^2\left(SG_e\right)}{\tau\left(1+S^2G_{es}{}^2\right)}\right)_{S=1} \tag{6.23}$$

where L_m is the magnetization inductance (5.6):

$$L_m=\frac{6\mu_0K_{w1}^2\cdot\tau\cdot 2a}{\pi^2\cdot p(g+d_{Al})K_c(1+K_{ss})K_{fg}}\times w_1^2$$

$$=\frac{6\times 1.256\times 10^{-6}\times 0.925^2\times 0.25\times 0.27}{3.14^2\times 5\times 16\times 10^{-3}\times 1.15\times 1.26\times 1.1}w_1^2$$

$$=0.3277\times 10^{-6}w_1^2 \tag{6.24}$$

From (6.23),

$$\left(F_{xpeak}\right)_{s=1}=\frac{3\pi}{0.25}\times\frac{0.3277\times 10^{-6}\left(68.05\times 10^3\right)^2\times 1.92}{\left(1+1.92^2\right)}=23{,}429\ \text{N}\rangle\rangle 12{,}000\ \text{N} \tag{6.25}$$

So a smaller flux density than 0.3 T in the airgap is necessary to produce the peak thrust. Consequently w_1I_{1peak} has to be reduced approximately by the square root of thrust overestimation:

$$\left(w_1I_{1peak}\right)_{trial}=68.05\times 10^3\sqrt{\frac{12{,}000}{23{,}429}}=48{,}665\ \text{A turns/phase RMS} \tag{6.26}$$

So the peak current density in (6.22) becomes $j_{copt}=6.59\times 48{,}665/68{,}050=4.712$ A/mm^2. This is quite a reasonable value, which may be sustained during acceleration cycles from zero to base speed. Also $B_{g1\max}\approx 0.215$ T, and thus, the secondary back-iron depth may be reduced to 20 mm.

Let us consider $w_1=90$ turns/phase, to yield a peak phase current $I_{1peak}=(w_1I_{1peak})/w_1=48{,}665/90=540\ \text{A}<600\ \text{A}$ (see the specifications).

With $w_1=90$ turns/phase and 30 coils/phase in series, it follows that a coil has only three turns, which means that twisted-cable turns should be used in preformed coils. The nonuniform flux-density distribution along primary length due to dynamic end effects above base speed precludes in general the usage of multiple current paths in parallel.

5. Dynamic end effect influence

Dynamic end effect influence manifests itself on thrust, secondary and primary winding losses, and, to a smaller degree, on power factor.

The key influence, as described in Chapters 3 and 4, is the ratio $f_e = F_{xe}/F_{xc}$, with F_{xe}—end effect force and F_{xc}—conventional thrust (without end effect).

From (3.70),

$$f_e = \frac{F_{xe}}{F_{xc}} \approx \left(1 + S^2 G_e^{\,2}\right)\mathrm{Re}\left[\frac{j\left(\dfrac{\gamma_1 \tau}{\pi} + SG_e\right)\left(\exp(2p\pi\left(\gamma_2^* \dfrac{\tau}{\pi} - j\right)\right) - 1\right)}{SG_e \dfrac{\tau}{\pi}\left(\gamma_2^* - \gamma_1^*\right)\left(1 - jSG_e\right)2p\pi\left(\gamma_2^* \dfrac{\tau}{\pi} - j\right)}\right] \tag{6.27}$$

Alternatively Ref. [2] has introduced the emf coefficient $K_e = E_{1e}/E_{1c}$, where E_{1e} is the end effect emf and E_{1c} is the conventional emf (without end effect), as followed in Chapter 4 (4.37).

So the total thrust F_x is

$$F_x = F_{xe}\left(1 - f_e\right) \tag{6.28}$$

as the total emf per phase E_1 is

$$E_1 = E_{1c}\left(1 - K_e\right) \tag{6.29}$$

As evident, both f_e and K_e depend at least on speed and slip (or on goodness factor, which includes primary frequency and slip) for given SLIM geometry.

Only in the low slip region (as described in Chapter 3) coefficient f_e varies monotonously with slip and increases rather monotonously with speed.

The slip used, in the present case study, at maximum speed (34 m/s) corresponds to 6–8 Hz in the secondary and 74(76) Hz in the primary.

$$S = \frac{f_2}{f_1} = \frac{8}{76} = 0.105 \tag{6.30}$$

Considering constant influences of edge and skin effects f_e, (6.27) may be straightforwardly calculated as a function of speed (frequency f_1, that is G_e) for given slip S, and then its dependence on speed is evident $u = 2\tau(f_1 - f_2)$.

For the case in point, and similar ones [3], a rather linear dependence of f_e with f_1 (speed) is obtained.

When designing the SLIM at optimum goodness factor for a slip $S \approx 0.1$, the end effect influence on thrust (F_e), which depends on the goodness factor G_e and on the number of poles (the speed is implicit in G_e), for $2p = 10$, calculated from (6.27), amounts in general to maximum 20%.

Similarly from (3.74) to (3.75), the secondary aluminum iron eddy current losses p_2 and the airgap reactive power Q_2 are calculated.

By the balance of powers, input active power P_1 and reactive power Q_2, efficiency η_n, and power factor $\cos\varphi_1$ are as follows:

$$\begin{aligned} &P_1 \approx 3R_1I_1^2 + F_x \times U + p_2 = 3V_1I_1\cos\varphi_1 \\ &Q_1 = Q_2 + 3\omega_1 L_{1l} I_1^2 \\ &\cos\varphi_1 = \frac{P_1}{\sqrt{P_1^2 + Q_1^2}};\quad \eta_n = \frac{F_x \cdot U}{P_1} \end{aligned} \tag{6.31}$$

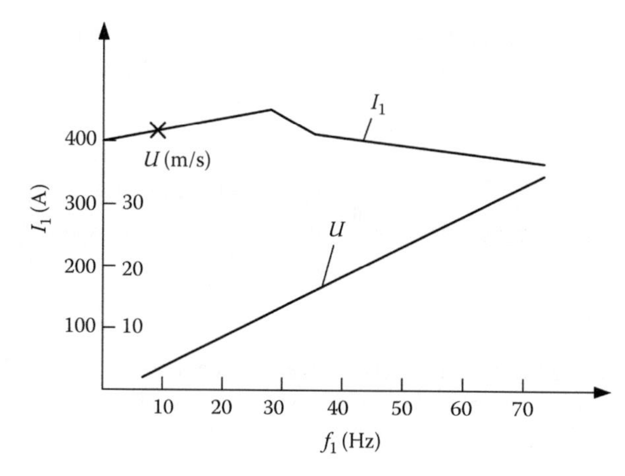

FIGURE 6.3 Imposed current and speed versus primary frequency.

This way, with a limited effort, the main SLIM performance for given I_1, s, $G_e(f_1)$ is obtained, including all main influences of dynamic end effect (on thrust, losses, and power factor).

Finally, the modified equivalent circuit to account for dynamic end effects (Figure 4.14) may be explicit and curve fitted, if so needed for further simplification.

For control purposes, the factor K_e (dependent on speed, slip, current) alone may be used in the space phasor model (4.74) to account for dynamic end effects.

Note: As the secondary back-iron saturation has a heavy influence on equivalent goodness factor, at high speed, the end effect influence changes notably, and thus, a robust thrust (or at least speed) control is required.

To save space with tedious, but, as seen earlier, rather straightforward mathematical developments, Figures 6.3 through 6.5 present final results on performance of SLIM with data very close to ones in our case study, but with an aluminum sheet of only 4 mm thickness, where dynamic end effect influence was considered by a coefficient f_e (or K_e), which is only 0.15 at max speed and $S=0.1$ and varies linearly with speed, for $f_2=8$ Hz [3]. The results in Figures 6.4 through 6.5 warrant remarks such as the following:

- A rather good efficiency is obtained (around 0.8) but the power factor is only around 0.5, which imposes notable oversizing of the inverter.
- The normal force is in general about three times larger than thrust, which should be acceptable, especially in a wheeled vehicle.
- A sizeable amount of power is retrieved during regenerative breaking, and the dc power bus should be able to handle it (to transfer it back to the ac power system or to the other vehicles).

6. Secondary thermal design

The thermal design of SLIM primary is similar to that of rotary IMs, with only a few peculiarities, such as the fact that it is placed in the way of the ambient air (below the vehicle) at vehicle speed and the heat transfer from secondary to primary (and vice versa) may be neglected due to the large mechanical airgap. As the thrust at low speed is high (peak thrust), this situation has to be considered when designing the cooling system for SLIM primary.

The secondary cooling raises special problems, especially around stop stations when the frequency of vehicle passage at low speed is large.

It may be practical to calculate the minimum head (lead) time between two successive vehicles that start from the same secondary section such that the maximum temperature of the latter does not surpass 120°–150° in general.

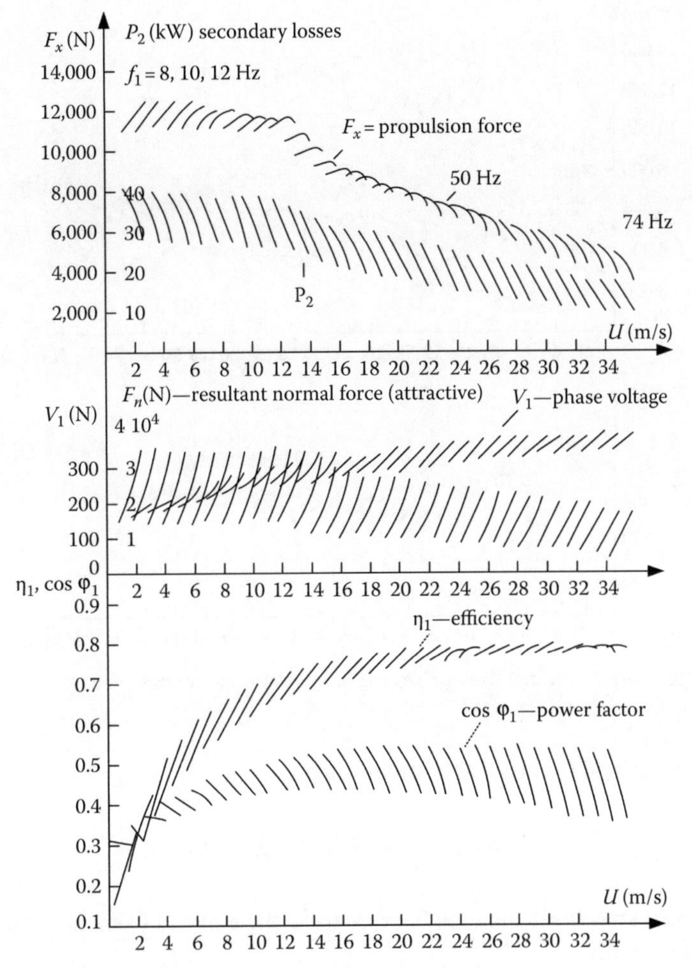

FIGURE 6.4 Thrust F_x, secondary losses p_2, normal force F_n, phase voltage V_1, efficiency η_1, and power factor $\cos\varphi_1$ versus speed for urban SLIM.

The secondary temperature influences the goodness factor, and thus, the thrust and secondary losses produced at standstill vary and should be accounted for by a verification of the worst scenario startup conditions (maximum temperature of secondary).

Let us consider peak thrust F_{xp} with vehicle mass M; the acceleration a_{ms} is

$$a_{ms} \approx \frac{F_{xp}}{M} \tag{6.32}$$

The thermal energy, W_p, in 1 m length of secondary, while a vehicle passes over it, is

$$W_p = \frac{n}{L}\int_0^{t_1} P_2(t)dt; \quad t_1 \approx \sqrt{\frac{2L}{a_{ms}}} \tag{6.33}$$

where

n is the number of successive motors on a vehicle

L is the length of the primary ($L \approx (2p+1)\tau$)

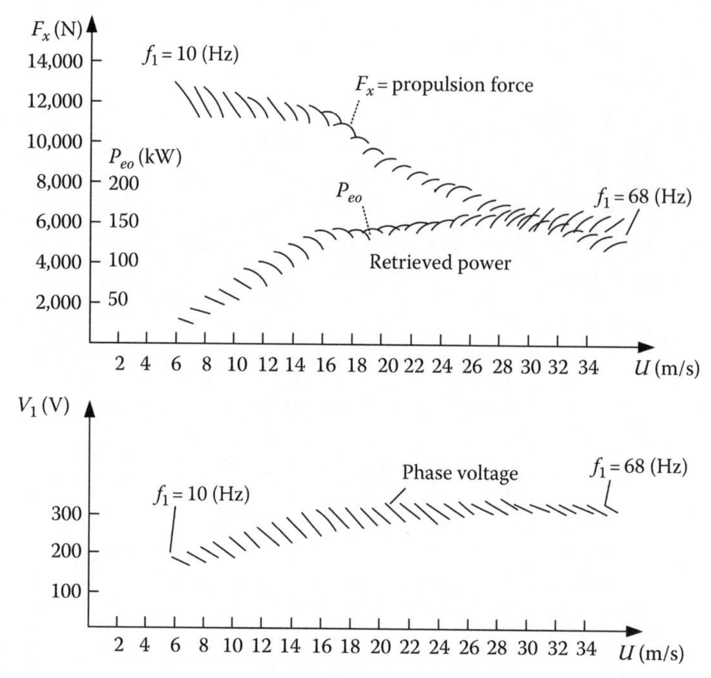

FIGURE 6.5 Regenerative braking: thrust F_x, retrieved power p_2, and phase voltage.

But the secondary losses at start are simply

$$\left(p_2\right)_{S=1} = F_{xp} \cdot 2 f_{1s} \cdot \tau; \quad f_{1s} = f_{2r} \tag{6.34}$$

$(p_2)_{S=1}$ is in fact the worst situation because the vehicle moves from the secondary section of start quite rapidly.

With T as the minimum interval between two successive vehicles and m vehicles present over the same section since their start in the morning (when secondary temperature was equal to the ambient temperature), the secondary temperature $\theta(t)$ becomes

$$\theta(t) = \theta(m'T + \varepsilon)\exp\left[-\frac{\alpha L_p}{c_s \gamma S_t}(t - m'T)\right]$$
$$\text{for} \quad m'T + \varepsilon < t < (m'+1)T - \varepsilon \tag{6.35}$$

When the temperature stabilizes,

$$\theta'\left[(m'+1)T - \varepsilon\right] = \theta(m'T - \varepsilon) \tag{6.36}$$

And thus

$$\theta'(m'T - \varepsilon) = -\frac{W_p}{C_s \gamma S_t} + \theta'(m'T + \varepsilon) \tag{6.37}$$

From (6.34) through (6.37), it follows that

$$T = \frac{C_s \gamma S_t}{L_p} \ln \frac{\theta(m'T+\varepsilon)}{\theta(m'T-\varepsilon)} \tag{6.38}$$

with

C_s as the equivalent specific heat of secondary
α as the specific heat transfer (W/m^2 C)
γ as the specific weight of secondary (kg/m^3)
L_p as the periphery along heat transmission area
S_t as the cross section of the secondary

The final temperature $\theta'(m'T+\varepsilon)$ is selected, and from (6.34) through (6.37), the minimum head time T is calculated. In order to start with lower secondary losses, a ladder-type secondary may be used in the vicinity of stop stations. However, the control has to be very robust to handle quickly the change in equivalent parameter circuits L_m, R_2', and L_{2l}'. It may be also feasible to use forced cooling of secondary within stop stations.

6.3 HIGH-SPEED (INTERURBAN) SLIM VEHICLES DESIGN

High-speed SLIM vehicles on wheels are characterized by a more demanding thrust/speed envelope (Figure 6.6).

The braking force due to magnetic suspension by dc electromagnets with solid iron track is not considered here, but in general it is within 10% of cruising speed thrust, even for speeds of 60–100 m/s.

A rather constant acceleration of 1 m/s^2 up to 80 m/s was considered in Figure 6.6.

As the starts are not so frequent, the hot spot in the design rests on the performance at cruising (maximum) speed with thrust capability verification for start and acceleration.

Again, the design is based on the optimum goodness factor concept, but the length of the SLIM primary has to be allowed to be longer (up to 3.5 m), and thus, the number of pole pitches for $\tau=(0.22 \div 0.28)$ m becomes higher. With $2p=14$ and $\tau=0.25$, $G_{eo}=35$, and with $d_{Al}=4\text{–}6$ mm, $g=10$ mm, the frequency is large, and with $S=0.07$,

$$f_{1\max} = \frac{U_{\max}}{(1-S)2\tau} = \frac{100}{(1-0.07)\times 2\times 0.25} = 215\,\text{Hz}; \quad f_2 = Sf_1 = 15\,\text{Hz} \tag{6.39}$$

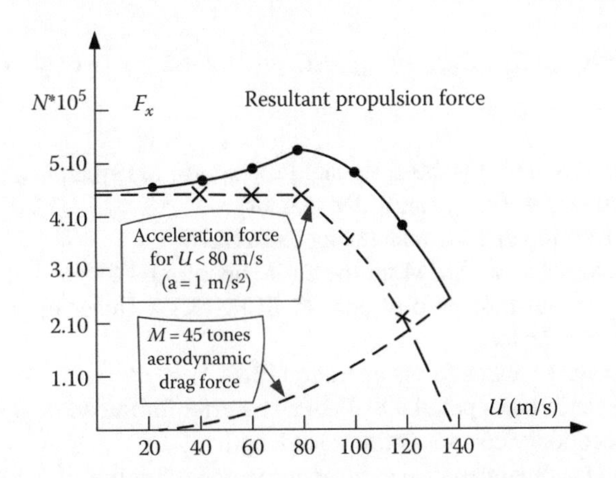

FIGURE 6.6 Typical propulsion force profile for a 45 ton high-speed MAGLEV.

To comply with $G_{eo}=35$ at 215 Hz, $g=10$ mm, and $d_{Al}=4$–6 mm, the equivalent secondary conductivity is (6.3)

$$\frac{\sigma_e}{1+K_{ss}}=\frac{G_{eo}\cdot\pi^2(g+d_{Al})K_eK_{fg}}{\mu_0\cdot 2\pi\cdot f_{1\max}\cdot\tau^2\cdot d_{Al}}=\frac{35\times\pi^2\times(10+6)\times1.15\times1.1}{1.256\times10^{-6}\times2\pi\times215\times0.25^2\times6}=1.098\times10^7(\Omega\text{ m})^{-1} \quad (6.40)$$

As the aluminum electrical conductivity $\sigma_{Al}\approx 3.25\times10^7\ (\Omega\text{ m})^{-1}$, the concerted influence of edge effect and of solid back-iron saturation together with the counteracting effect of solid back-iron conductivity can corner the 2.96 electrical equivalent conductivity reduction ratio. A reduction of aluminum plate overhangs will contribute to bring the electrical conductivity into range; also a reduction of aluminum plate thickness d_{Al} will produce similar effects but will also reduce the magnetization mmf, as added benefit.

The rather large slip $S\approx0.07$ at maximum speed ($f_2=15$ Hz) is justified by the necessity to keep the dynamic end effect adverse influences on performance within practical limits. The design, from this point, may follow the path used for urban vehicles with SLIM propulsion. For typical results, see the examples in Chapters 3 and 4, for high-speed SLIMs (Figure 3.23).

6.4 OPTIMIZATION DESIGN OF SLIM: URBAN VEHICLES

The preliminary electromagnetic design methodologies developed in Chapter 5 should constitute a good ground for a closer to target initialization of the optimization design and the SLIM model required in this enterprise.

For optimization design, a few steps are appropriate:

- Specifications and constraints
- Variables vector
- Cost (fitting) objective function
- Development of an SLIM model (realistic, but still reasonable in terms of computer effort for its repeated usage in optimization process)
- Choosing a mathematical optimization method (deterministic or evolutionary as for rotary machines [4])
- Developing a computer code to easily handle the path from specifications to final design with ready-to-use documentation for manufacturing

A possible cost function may be

$$F(|x|)=n_{SLIM}\left(C_a+C_{f+m}+C_{losses}+C_{inv}+C_{loss}^{inv}+C_{maint}\right)+C_{seci}+C_{secmaint} \quad (6.41)$$

with

n_{SLIM} as the number of SLIMs for the designated transportation system
C_a as the cost of active materials in an SLIM primary
C_{f+m} as the cost of framing and manufacturing of an SLIM
C_{losses} as the cost of losses in an SLIM for the entire operating life
C_{inv} as the inverter initial cost proportional to SLIM KVA (inversely proportional to SLIM efficiency and power factor)
C_{loss}^{inv} as the cost of inverter losses for its operation life
C_{maint} as the maintenance costs per one SLIM + its inverter for the entire operation life
C_{seci} as the initial secondary cost for entire track length
$C_{secmaint}$ as the secondary maintenance cost for entire operation life

Typical constraints are related to costs or losses or both, but translated in terms of analytical SLIM model; they are efficiency x power factor of SLIM at base speed, primary winding maximum temperature or peak current for given dc power bus voltage and peak starting thrust and given peak thrust versus speed envelope or maximum normal force of SLIM, etc.

In Ref. [5], a thorough optimization design process for an SLIM-MAGLEV system with 33.3 kN peak thrust for vehicles that contain 6, 8, and 10 SLIM modules is introduced. The cost function in [5] has a multi-objective character but still less demanding than that of (6.41).

In short, the two main cost function options in [5] are related to efficiency and SLIM primary weight and, respectively, secondary material cost and SLIM primary weight. As expected, a better efficiency is accompanied by larger SLIM primary weight and lower secondary materials cost, while lower SLIM primary weight leads to notably lower efficiency; the power factor is smaller when efficiency is larger and vice versa. Efficiency below 64% and power factor below 0.647 are typical but not for the same optimal design, directed to SLIM-MAGLEVs at 200 km/h.

The results in previous paragraphs, based on the optimum goodness factor concept, suggest better performance than in [5] even at higher speed (up to 100 m/s), with track costs low but comparable to competitive solutions for wheeled vehicles or MAGLEVs. The reader is advised to proceed with care as the SLIM optimal design is far from an exhausted subject, even in the presence of a few commercial, urban, wheeled, successful SLIM vehicles.

6.5 SUMMARY

- By design, dimensioning (sizing) for given specifications is meant.
- Transportation LIMs refer to urban and interurban people (or freight) movers on wheels or in MAGLEVs when dynamic end effect has to be considered [3,6].
- In MAGLEV applications, the SLIMs are placed on both sides of the vehicles below (or even above) the track structure; in one case, the net normal force helps the magnetic suspension while in the other impedes on it; zero normal force SLIM slip frequency control leads to efficiency below 0.65, even for urban applications.
- Only aluminum-on-iron long track secondary SLIMs are considered here practical, due to overall cost limitations.
- 1–3 pieces (laminations) may constitute the mild steel secondary back iron, to increase the field penetration depth and thus allow sufficient airgap flux density; also, for the same reason, the length of the primary pole pitch is limited to 0.22–0.27 m.
- To allow tight curves in the track, the total length of the short primary on board should be below 2–2.75 m for urban applications and beyond 3.5 m/per unit in interurban applications.
- Thus, the number of primary poles $2p+1=7 \div 11(15)$ per unit.
- $2p+1$ poles with double-layer windings and half-filled end poles with chorded coils are recommended to reduce end-coil length and primary copper losses.
- The mechanical airgap varies from 8 to 12 mm, and the aluminum sheet should be 4–6 mm thick.
- The slip frequency should be in the range of 6–8 Hz in urban SLIM vehicles and 10–15 Hz in high-speed (interurban) vehicles, to provide enough thrust but also to reduce noise and vibration at start and dynamic end effect at maximum speed.
- The key design concept used here is the optimum goodness factor $G_{eo}(f_{1\max})$ (zero end effect thrust, at zero slip and maximum speed). For urban transportation, even $G_e < G_{eo}$ may be adopted for lower cost reasons.
- $(G_{eo})_{f_{1\max}}$ imposes 2(3) back-iron laminations in the secondary and a thickness $h_{y2} \leq$ 25 mm.
- At G_{eo}, the secondary back iron is heavily saturated [3,5].

- Peak thrust, F_{xpeak} (from zero to base speed), has to be obtained, eventually for $(G_{eo})_{fr2} = 1 - 2$; this is a key design verification.
- F_{xpeak} leads to the peak phase current mmf $w_1 I_{1peak}$ and the peak airgap flux density B_{g1max}, both essential in designing the primary slots and back iron.
- Then, the second verification is related to peak thrust at base speed and full voltage, which leads to the number of turns w_1 per phase.
- The third verification is related to thrust at peak speed at max. inverter voltage including (also at base speed) the end effect thrust, f_e, or emf correction coefficient, K_e.
- In the presence of dynamic end effect, it seems to be more practical to use the field theory and power balances to calculate performance for the entire speed range; only at the end, the circuit parameters are worth calculating to serve for the control system design.
- Designing for optimum goodness factor brings the end effect deteriorating influence within limits, but new end effect compensation schemes (as discussed in Chapter 4) should not be left out, as addition for performance boosting.
- Despite of the SLIM low power factor (below 0.6), which leads to inverter overratings, its ruggedness and around 75%–80% efficiency may be acceptable in the long run, for medium- and high-speed applications where initial cost is critical.

REFERENCES

1. J.F. Gieras, *Linear Induction Drives*, Oxford University Press, Oxford, U.K., 1994.
2. I. Boldea, and S.A. Nasar, *Linear Motion Electromagnetic Systems*, Chapter 6, Wiley Interscience, New York, 1985.
3. T. Higuchi, T. Nishimoto, S. Nonaka, and H. Muramoto, Design optimization of single sided linear induction motors for MAGLEV vehicles, *Record of LDIA-2001*, Nagasaki, Japan, pp. 25–29.
4. I. Boldea and L. Tutelea, *Electric Machines: Steady State Transients and Design with Matlab*, CRC Press, Taylor & Francis, New York, 2009.
5. C. Cabral, L.J. Concalves, E. Pappalardo, and C. Cabrita, A new model and performance of SLIM taking into account the back iron saturation, *Record of LDIA-1998*, Tokyo, Japan, pp. 359–362.
6. S. Nonaka and T. Higuchi, Design strategy of SLIMs for propulsion of vehicles, *Record of International Conference on MAGLEV & Linear Drives, IEEE*, Las Vegas, NV, 1987, pp. 1–6.

7 DC-Excited Linear Synchronous Motors (DCE-LSM)

Steady State, Design, Transients, and Control

7.1 INTRODUCTION AND TOPOLOGIES

DC-excited linear synchronous motors (DCE-LSM) in flat active ac stator (Figure 7.1) configuration [1–3] are favored for MAGLEVs in low, medium, and high-speed active guideway applications because

- They provide, by multiple on board dc exciters, suspension control.
- They produce, by interaction of dc exciters on board field and ac-fed long primary (stator) active guideways, MAGLEV propulsion.
- The propulsion and suspension functions may be decoupled essentially by pure i_q vector control in the ac stators fed by inverters placed on ground, along the track, and synchronized with the vehicle motion.
- The propulsion efficiency is reasonable, above 0.75 for 1–1.5 km long active stator sections; power factor is 0.55–0.6 (lagging) for the same conditions.
- The suspension power (consumed in the dc exciters on board) is around 3 kVA/ton for an airgap of about 10 mm, which is quite acceptable.
- The stator has a laminated core with open slots and $q = 1$, single-layer, aluminum cable three-phase windings.
- There is no power collection system because the active guideway is supplied on ground and because the needed power on board is produced there through dedicated coils on dc exciter poles, which collect the mmf space harmonics field energy of the stator, representing a linear electric generator on board.
- Single-sided configurations are favored for MAGLEVs, but double-sided ones may be used when zero normal force is needed in the application.
- The PMs (in Figure 7.1b) are used to produce the bulk of the excitation field (say to cover the magnetic suspension of empty vehicle weight), while the dc coils are used only for controlling the static and dynamic stability of magnetic suspension. If all the dc control coils are connected through dc–dc converters from a common dc power bus, the average control power is very low (ideally zero) because of the multiple dc coil energy exchange.
- Both the dc exciter mover and the primary (guideway) iron cores are made of laminated silicon steel, while the dc coils are made of copper (or aluminum). Multiphase ac windings of the stator are made of aluminum cables to reduce the initial system costs. Considered too expensive decades ago, the inclusion of PMs in the dc exciter mover seems today much more practical, especially because of the lower total average power in the control coils.

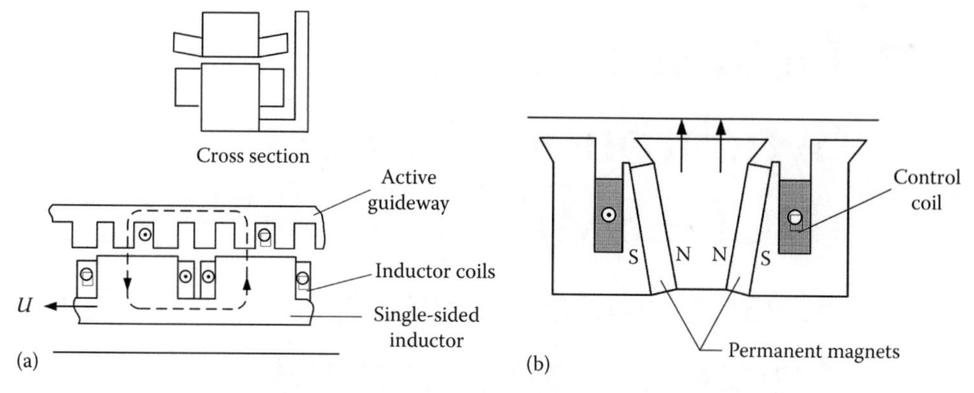

FIGURE 7.1 DCE-LSM with active guideway (stator): (a) with dc excitation on mover; (b) hybrid (dc + PM) mover. (After Boldea, I. and Nasar, S.A., *Linear Motion Electromagnetic Systems*, John Wiley, New York, 1985.)

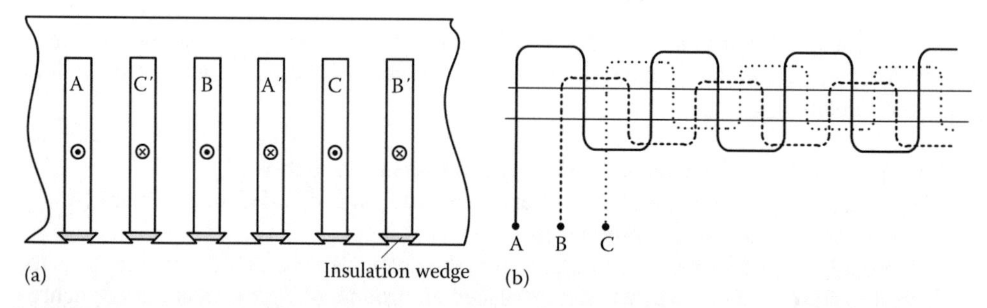

FIGURE 7.2 Three-phase armature winding (a) phase slot allocation; (b) cable winding. (After Boldea, I. and Nasar, S.A., *Linear Motion Electromagnetic Systems*, John Wiley, New York, 1985.)

- The ac wound primary guideway is placed on both sides of the vehicle (Figure 7.1a) with the dc exciters below the former, to provide laterally balanced suspension.
- The active guideway comprises a silicon-laminated magnetic core with uniform open slots (Figure 7.2). The core is made of sections, 1–1.2 km long, fixed to the track structure.
- The single-turn-cable single-layer winding has three slots/pole ($q = 1$). The premade (preinsulated) cable, one per winding section (1000–1200 m or more), is easy to insert and leads to the reduction of one of two ends of a standard winding coil (per pole pitch). Thus, the aluminum weight, cost, and Joule losses are reduced by about 20%.

7.2 DC EXCITER (INDUCTOR) DESIGN GUIDELINES

A typical dc exciter mover module for MAGLEVs is shown in Figure 7.3.

It contains four full-poles and two half-poles at the two ends and is about 1.3 m long, to allow the manufacture of the magnetic core of thin laminations from one sheet piece. To avoid notable lateral forces, when lateral displacement of dc exciter versus stator occurs, the width of the stator stack is slightly shorter than the width of the former.

Also, to harvest energy on board of mover (vehicle), a two-phase ac winding is placed in dedicated slots on the dc exciter poles. Two-phase tooth-wound coils on the dc exciter poles "collect" by electromagnetic induction the mmf open-slot stator harmonics energy. So, a linear generator on dc exciter pole pitch (slot pitch) $\tau_{s2} = \tau_{s1}/2$ is obtained, τ_{s1}—stator slot pitch.

The double slotting on stator and on mover—for an airgap of 10–12 mm in MAGLEVs—produces magnetic flux density harmonics that cause thrust and normal pulsations, which, by

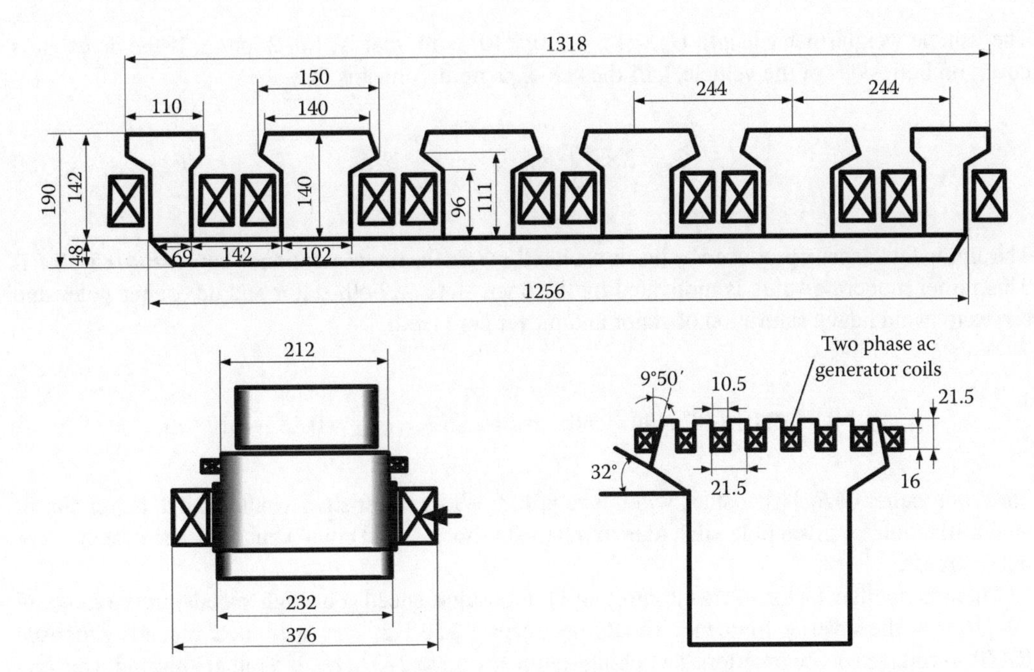

FIGURE 7.3 Typical dc exciter (inductor) module of DCE-LSM (tentative dimensions in millimeters). (After Boldea, I. and Nasar, S.A., *Linear Motion Electromagnetic Systems*, John Wiley, New York, 1985.)

careful design, may be kept to a reasonable p.u. level, in order to prevent strong vibrations and noise of DCE-LSM.

The coordination of stator and dc exciter slot opening and the dc exciter pole shoe per pole pitch ratio $\tau_{p2}/\tau \approx 2/3$ produces pulsations below 3% in thrust and 1% in the normal force [5].

For preliminary design of the dc exciter, we may suppose that the contribution of the stator mmf field to the normal force is neglected. Such an approximation becomes acceptable if the propulsion control is built around pure I_q control (q axis current in the dq orthogonal model of LSM); the dc excitation field is aligned with axis d. Also, the armature reaction field is notably smaller than the dc excitation airgap field.

The normal force F_n is the attraction force between the dc exciter and the stator core.

The average airgap flux density, B_{fg}, produced by the dc exciter, is

$$B_{fg} = \frac{\mu_o w_f I_f}{g K_c \left(1 + K_{sat}\right)} \tag{7.1}$$

where

w_f is the number of turns per pole (coil)
I_f is the field current
K_c is the Carter coefficient (see Chapter 2)
K_{sat} is the saturation factor (includes the influence of the magnetic core to the equivalent magnetic airgap)

The normal force for $\tau_{p2}/\tau \approx 2/3$ is

$$F_n = \frac{B_{fg}^2}{2\mu_0} \times l_{stack} \times \frac{\tau_{p2}}{\tau} \times 2p\tau \tag{7.2}$$

The vehicle weight/meter length $G_e \approx (1.5 - 2.0)\times 10^4$ N/m, that is, 1.5–2 ton/m. If the dc exciters cover, on both sides of the vehicle, half the vehicle length, it means that G_e is

$$G_e = 2\times\frac{B_f^2}{2\mu_0}\times l_{stack}\times\frac{2}{3}\times\frac{1}{2}\ \text{N/m} \tag{7.3}$$

The airgap flux density produced by the dc exciter for G_e in the range mentioned here is $B_f \approx 0.5 - 0.7$ T. This rather moderate value is motivated by the open slots on both stator and dc exciter poles and serves to avoid heavy saturation of stator and mover core teeth.

With

$$G_e \approx (1.5\text{–}2.0)\times 10^4\ \text{N/m} \quad \text{and} \quad B_f = 0.65\,\text{T},\quad l_{stack} = 0.133\text{–}0.177\ \text{m}$$

For lower values of B_f, larger stack width is required. Wider stator stack would lead to better aluminum utilization for given pole pitch (due to relatively shorter "end coils") but, still, the cost of stator will increase.

There is another factor in the design that is important: speed. For high speeds, in the range of 90–110 m/s, the inverter frequency should not surpass 250 Hz, even with most recently proposed IGCTs—voltage source inverters (switching frequency up to 2–3 kHz), if a rather sinusoidal current (field-oriented) control is to be applied for propulsion.

For $f_1 = 250$ Hz and 110 m/s, the stator and the mover pole pitch τ is

$$\tau = \frac{U_s}{2f_1} = \frac{110}{2\times 250} = 0.22\ \text{m} \tag{7.4}$$

Though l_{stack}/τ ratio may be smaller than unity, the solution is acceptable as the cable winding eliminates one of two end connections (Figure 7.2b), while the cost of the active stator core is reduced with shorter stack width ($l_{stack}/\tau \approx 0.6\text{–}0.8$).

For smaller speeds, the pole pitch will be decreased, but not proportionally, as still considerable room is needed per dc exciter pole to produce the 0.5–0.75 T airgap flux density, unless the airgap is decreased notably for low speeds.

The dc exciter mmf per pole $w_f I_f$, for $g = 10$ mm, $B_{fg} = 0.65$ T, and $K_c = 1.2$, $K_{sat} = 0.2$, is (7.1)

$$w_f I_f = \frac{0.65\times 0.01\times 1.2\times 1.2}{1.256\times 10^{-6}} = 7452\ \text{A turns/pole} \tag{7.5}$$

For a fill factor $K_{fill} = 0.45$ and $j_{co2r} = 3.5\times 10^6$ A/m^2, the required window area per pole (coil) A_{win} is

$$A_{win} = \frac{w_f I_f}{K_{fill}\cdot j_{co2r}} = \frac{7452}{0.45\times 3\times 10^6} = 5520\ \text{mm}^2 \tag{7.6}$$

The space available for the dc excitation coils in the typical dc exciter is quite acceptable for ac coil width of 50 mm and a height of 112 m.

For dynamic control of magnetic suspension, higher $w_{f\max} I_{f\max}$ is required, but not above 4.4 A/mm^2 and thus the cooling of the dc exciter may be provided (most probably) by the air flow around it (resulted from the motion of the vehicle).

The dc exciter back-iron depth h_{y2} is

$$h_{y2} \approx \frac{B_F \frac{1}{3}\tau}{B_{y2}} = \frac{0.65 \times \frac{0.22}{3}}{1.3} = 0.0366 \text{ m} \tag{7.7}$$

So the 0.048 m available in Figure 7.3 will allow even 1.0 T in the back iron, besides space left for eventual bolts for framing.

The dc exciter module is supposed to be supplied separately from a voltage source of $V_{dc} = 220$ V. To allow room for fast field current variation in order to stabilize the magnetic suspension, a 5/1 voltage boost is allowed here. Consequently, the rated dc voltage of the dc exciter V_{dc_r} is

$$V_{dc_r} = \frac{V_{dc}}{5} = \frac{220}{5} = 44V = R_{Fr} i_{Fr} \tag{7.8}$$

With R_F, the dc exciter module resistance is

$$R_F = 5\rho_{co} \frac{l_{coF} w_F}{I_{Fr}} j_{co2r} \tag{7.9}$$

so

$$R_F I_{Fr} = 5\rho_{co} l_{coF} w_F j_{co2r} = 44 V_{dc} \tag{7.10}$$

Consequently

$$w_F = \frac{44}{5 \times 2.3 \times 10^{-8} \times 3 \times 10^6 \times 0.88} = 144.9 \approx 145 \text{ turns/coil} \tag{7.11}$$

with $l_{stack} \approx 0.2\ m$; l_{coF} dc exciter turn length is

$$l_{coF} \approx 2\left(l_{stack} + 0.02 + \tau\right) \tag{7.12}$$

or

$$l_{coF} \approx 2(0.2 + 0.01 \times 2 + 0.22) = 0.88 \text{ m} \tag{7.13}$$

The field rated current I_{Fr} yields

$$I_{Fr} = \frac{w_F I_{Fr}}{w_F} = \frac{5520}{144} = 38.33 \text{ A} \tag{7.14}$$

The copper total cross section area per turn A_{Co} is thus

$$A_{Co} = \frac{I_{Fr}}{j_{Co2r}} = \frac{38.33}{3} = 12.77 \text{ mm}^2 \tag{7.15}$$

As the ac field current components will add to the steady-state dc component, to provide dynamic suspension stability, the dc exciter coils may be made of copper foils, after careful assessment of the skin effect on total copper losses.

The steady-state power for suspension is (in W/kg)

$$P_{sus} = \frac{V_{Fr} I_{Fr}}{G_e / a_{grav}} \times \frac{1}{6\tau} = \frac{44 \times 38.33}{2000 \times 6 \times 0.22} = 0.62 \text{ (W/kg)} \tag{7.16}$$

For the 5/1 voltage boost, the peak kVA of the dc exciter pole dc–dc converter will be about $P_{sus} \times 5 \approx 3$ kVA. This is quite a small value in comparison with the true propulsion requirements, as seen later.

7.3 STATOR (ARMATURE) CORE DESIGN

With the pole τ and the stack length (average value for stator and dc exciter) already in place, and $q = 1$ slot/pole/phase, the main question is related to the mmf per phase required to produce the peak thrust $w_1 I_{1peak}$ and the number of turns w_1 per phase, for given dc supply power station on ground and inverter voltage, at cruising/maximum speed.

Let us consider the peak acceleration $a_{peak} = 1$ m/s² (to observe adequate passenger comfort) available up to 40 m/s, for a maximum speed $u_{s\,max} = 110$ m/s; the aerodynamic drag adds at 40 m/s only 5%–6% to the accelerating thrust, and the thrust at maximum speed is 40% of the peak thrust.

Let us also consider 2 ton weight per vehicle meter length.

So the peak thrust occurs (Figure 7.4) at 40 m/s; this may be considered the base speed of the drive.

With pure I_q control (Figure 7.4), the phase current is in phase with emf E_1 (to decouple suspension from propulsion function):

$$E_{1b} = \pi\sqrt{2} w_1 K_{w1} \Phi_{p1} f_{1b}; \quad \Phi_{p1} \approx \frac{4\sqrt{3}}{\pi^2} B_F \tau l_{stack} \tag{7.17}$$

$$f_{1b} = \frac{u_b}{2\tau} = \frac{40}{2 \times 0.22} = 90.90 \text{ Hz} \tag{7.18}$$

$$F_{xpeak} = \frac{3 E_{1b} I_{1peak}}{u_b} = \frac{3\pi\sqrt{2} K_{w1} f_{1b}}{2\tau f_{1b}} \times \frac{4\sqrt{3}}{\pi^2} B_F \tau l_{stack} w_1 I_{1peak} \tag{7.19}$$

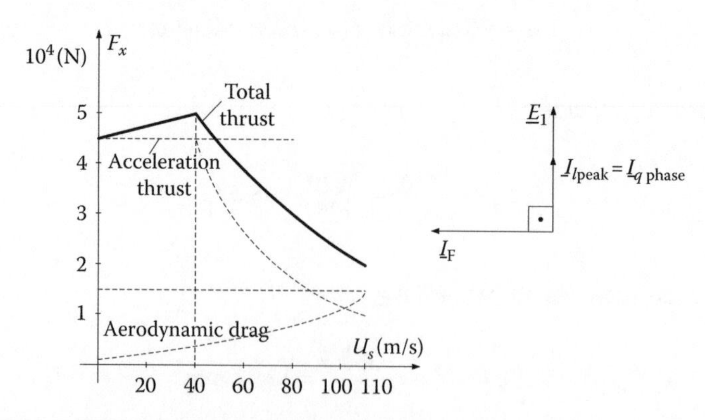

FIGURE 7.4 Thrust/speed for a 45 ton MAGLEV at 110 m/s.

With 2 ton per vehicle meter length, there is 22.5 m for the length of a 45 ton vehicle. The peak thrust per 1 m of vehicle length is thus

$$f_{xpeak} = \frac{F_{xpeak}}{l_{vehicle}} = \frac{48{,}000\ \text{N}}{22.5\ \text{m}} = 2133\ \text{N/m} \tag{7.20}$$

As the dc exciters occupy half of the vehicle length on the left and right side, the situation is equivalent to one side with a continuous dc exciter module row. But for 1 m length of vehicle, there are about five stator poles ($\tau = 0.22$ m), and thus, the number of equivalent turns per phase w_1 is (single-layer windings)

$$w_{1meter} = 2\ \text{turns in series/m} \tag{7.21}$$

The number 2 (not 5/2) was chosen for design safety.

Let us note that there is one bar (cable) per coil and one coil per pole per phase.

From (7.19) to (7.21), we find $w_{1meter} I_{1peak}$:

$$w_{1meter} I_{1peak} = 2I_{1peak} = \frac{\pi f_{xpeak}}{6\sqrt{6} l_{stack} B_F} = \frac{\pi \times 2133}{6\sqrt{6} \times 0.22 \times 0.65} = 3185\ \text{A turns/m/side} \tag{7.22}$$

So the number of peak A turns per slot is I_{peak}

$$I_{peak} = 3185/2 = 1592.5\ \text{A}\ (\textit{RMS value}) \tag{7.23}$$

Again, with a fill factor of $K_{fill} = 0.27$ (cable winding) and $j_{Co1p} = 4$ A/mm^2 (aluminum cable, short duty cycle), the slot area A_{slot1} is

$$A_{slot1} = \frac{I_{peak}}{K_{fill} j_{Co1p}} = \frac{1592.5}{0.27 \times 4 \times 10^6} = 1474\ \text{mm}^2 \tag{7.24}$$

With a slot pitch $\tau_{s1} = \tau/3 = 0.22/3 = 0.0733$ m and a tooth $w_{t1} = 0.03$ m, the slot width is $w_{s1} = \tau_{s1} - w_{t1} = 0.0433$ m. So the slot height h_{s1} is

$$h_{s1} = \frac{A_{slot1}}{w_{s1}} = \frac{1474}{43.3} = 34\ \text{mm} \tag{7.25}$$

This leads to a rather small slot aspect ratio $h_{s1}/w_{s1} = 34.00/43.3 = 0.786$, which, in turn, means a low slot leakage inductance of stator winding; thus, an acceptable power factor (and converter kVA) can be expected, as 1–1.5 km sections of primary are connected on both sides of the vehicle at any time.

But if the aluminum Joule losses are too large, a smaller current density is adopted, and thus, the primary slot becomes deeper, and consequently, the slot leakage inductance becomes larger. So the power factor decreases and the converter kVA rating increases.

An optimum problem is implicit here, but this is beyond our scope to follow it up further.

7.4 DCE-LSM PARAMETERS AND PERFORMANCE

Let us now consider the case of a DCE-LSM with $2p$ poles per vehicle length for $2p' > 2p$ poles per stator active section length.

Also, on each side of the vehicle, only half of the poles are active (because only half of the length is occupied by dc exciter movers).

Furthermore, we may assume that the active track (stator sections) on the two sides of the vehicle is supplied by separate inverters on ground, to damp eventual transverse vertical axis motions of the vehicle.

The parameters to consider here are the emf per phase E_1, the stator active section resistance R_1, leakage inductance L_{a1}, and synchronous magnetization inductances L_{dm}, L_{qm}, all needed for performance evaluation through the phasor diagram in steady state and then for dynamics modeling and control.

The emf E_1 (per phase section length, per vehicle side) is (single layer, cable winding, $q=1$)

$$E_1 = \pi\sqrt{2} f_1 \left(\frac{p}{2}\right) \Phi_{p1f} \text{ (RMS)}; \quad \Phi_{p1f} \approx B_F \frac{4\sqrt{3}}{\pi^2} \tau \times l_{stack} \tag{7.26}$$

The stator active section resistance (per vehicle side) R_1 is

$$R_1 \approx \rho_{Al} \frac{(\tau + l_{stack} + 0.06) \cdot 2p' j_{Co1p}}{I_{1peak}} \tag{7.27}$$

The magnetization inductances are straightforward in approximated expressions [6]

$$X_{dm} = \frac{6\mu_0}{\pi^2} \frac{l_{stack} \times \tau \times (p/2)^2}{gK_c(1+K_{sat})(p/2)} K_{dm} \tag{7.28}$$

$$X_{qm} = X_{dm} \cdot \frac{K_{qm}}{K_{dm}} \tag{7.29}$$

with

$$K_{dm} \approx \frac{\tau_p}{\tau} + \frac{1}{\pi} \sin\left(\frac{\tau_p}{\tau}\pi\right) \tag{7.30}$$

$$K_{qm} \approx \frac{\tau_p}{\tau} - \frac{1}{\pi} \sin\left(\frac{\tau_p}{\tau}\pi\right) + \frac{2}{3\pi} \cos\left(\frac{\tau_p}{\tau}\frac{\pi}{2}\right) \tag{7.31}$$

With

$$\frac{\tau_p}{\tau} \approx \frac{2}{3}, \; K_{dm} \approx 0.978, \quad K_{qm} \approx 0.385$$

As the airgap is rather large, more precise values of L_{dm}, L_{qm} that include slot opening and magnetic saturation influences, may be derived from FEM investigations.

The leakage inductance per phase per side per active section L_{a1} contains three terms: one corresponding to the part occupied by the dc exciters L_a, one slot and end-connection leakage inductance of primary part free of the dc exciters L_{as}, and one corresponding to the field below the "empty" stator section (unoccupied by the dc exciters), L_{es}:

$$L_{a1} = L_a + L_{as} + L_{es} \tag{7.32}$$

L_a is straightforward [6]:

$$L_a \approx \mu_0 p\left[(\lambda_{s1} + \lambda_{d1}) l_{stack} + \lambda_{e1}(\tau + 0.06)\right] \tag{7.33}$$

$$\lambda_s \approx \frac{h_{s1}}{3w_{s1}} + \frac{0.004}{w_{s1}}; \quad \lambda_d \approx \frac{5g/w_{s1}}{5+4g/w_{s1}}; \quad \lambda_{el} \approx 0.6 \tag{7.34}$$

$$L_{as} = \mu_0(2p' - p/2)\left[\lambda_{s1} l_{stack} + \lambda_{el}(\tau + 0.06)\right] \tag{7.35}$$

$$L_{es} = \frac{6\mu_0}{\pi^2}\frac{\tau l_{stack}}{(\tau/\pi)}(p' - p/2) \tag{7.36}$$

In (7.36), the distribution of the field below the "empty" stator was considered to correspond to a large airgap (τ/π), which, for a 2D field analysis, is acceptable (3D FEM analysis may be used for even better precision).

The efficiency η_1 and the power factor $\cos\varphi_1$ may now be defined for an active section (and one vehicle side):

$$\eta_1 = 1 - \frac{\left(3R_1 I_1^2 + p_{irona} + p_{ironp}\right)}{F_x \cdot U} \tag{7.37}$$

where p_{irona} represents the core losses in the active part of the stator that faces the dc exciter (where the flux density is large in iron) and p_{ironp} represents the core losses in the "empty" part of slots, $(p'-p/2)\tau$ in length, where the flux density in iron is small but the iron weight is very large. Again, the propulsion force F_x is

$$F_x = \frac{3E_1 I_1}{U}, \textit{ for pure } I_q \textit{ control} \tag{7.38}$$

$$\cos\varphi_1 = \frac{F_x U}{3V_1 I_1 \eta_1} \tag{7.39}$$

For the control, the relationship between I_f and emf E_1 is to be expressed in terms of mutual (motion) inductance M_{aF}:

$$M_{aF} = \frac{E_1}{\omega_1 I_F} = \frac{\dfrac{\pi\sqrt{2}}{2\pi f_1} f_1\left(\dfrac{p}{2}\right)\cdot\dfrac{4\sqrt{3}}{\pi^2}\tau l_{stack}\cdot\dfrac{\mu_0 w_F I_F}{gK_c(1+K_{sat})}}{I_F} = \mu_0\frac{\sqrt{6}}{\pi^2}\cdot\frac{\tau l_{stack} w_F p}{gK_c(1+K_{sat})} \tag{7.40}$$

In terms of the dq orthogonal model, the thrust may be expressed, in general, by the formula

$$F_x = \frac{3\pi}{2\tau}\left[M_{aF} i_F i_q + (L_d - L_q) i_d i_q\right] \tag{7.41}$$

Park transformation writes (in mover coordinates)

$$\begin{vmatrix} i_d \\ i_q \\ i_0 \end{vmatrix} = \frac{2}{3}\begin{vmatrix} \cos(-\theta_{er}) & \cos\left(-\theta_{er} + \dfrac{2\pi}{3}\right) & \cos\left(-\theta_{er} - \dfrac{2\pi}{3}\right) \\ \sin(-\theta_{er}) & \sin\left(-\theta_{er} + \dfrac{2\pi}{3}\right) & \sin\left(-\theta_{er} - \dfrac{2\pi}{3}\right) \\ \dfrac{1}{2} & \dfrac{1}{2} & \dfrac{1}{2} \end{vmatrix} \tag{7.42}$$

where

$$\theta_{er} = \frac{\pi}{\tau}\int U dt + \theta_0; \quad \frac{d\theta_{er}}{dt} = \frac{\pi}{\tau}U$$

The presence of the homopolar components is related to the possible presence of the Δ connection of phases, to allow section feeding from both ends, with simplified power switching devices.

Also, the synchronous inductances L_d, L_q are

$$L_d = L_{dm} + L_{al}; \quad L_q = L_{qm} + L_{al} \tag{7.43}$$

It should be noticed that $L_{al} \gg L_{dm}$ or L_{qm} because the “empty” part of the active primary section is much longer than the one “covered” by the dc exciters on board of vehicle (mover).

To complete the steady-state general analysis motoring and generating, phasor diagrams are given in Figure 7.5, for pure I_q control or zero phase lag between emf and stator current in each phase.

It might be argued that this way the reluctance thrust is not used (I_d=0); true, but, $L_d/L_q \approx 1$, and, instead, notable decoupling of propulsion and suspension function is obtained, except for the cross-coupling saturation effect, which is small because the airgap field-current-produced flux density is reasonably low (0.65 T) and the stator peak mmf reaction airgap flux density is less than 0.15 T in general.

Generating mode is required for vehicle (mover) regenerative braking, provided the converter that feeds the active stator section can handle (retrieve) the recuperated energy.

It is evident that, with pure I_q control, the DCE-LSM is always underexcited, which, however, allows easily regenerative braking through a bidirectional dual-voltage source converter.

Typical efficiency and power factor performance, obtained for the case study in the design performed earlier in this chapter, are shown in Figure 7.6.

Though the results are orientative, it is clear that, to secure good performance, the active section length to vehicle length ratio (p'/p)—where only half of the vehicle length is covered by the dc exciters on the mover—should not be above 25–100. For a 50 m long vehicle of 100 ton, this would mean 1.25 km of active track sections; the latter is traveled (at 110 m/s) in roughly 11 s.

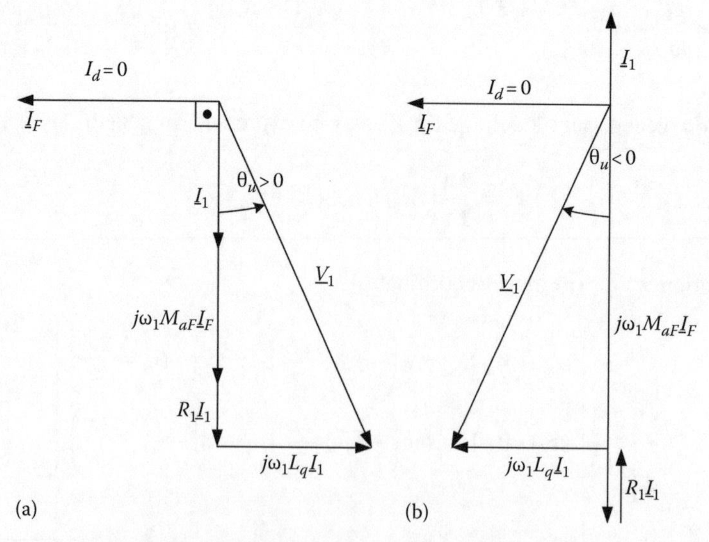

FIGURE 7.5 Phasor diagrams of DCE-LSM: (a) motoring, (b) generating.

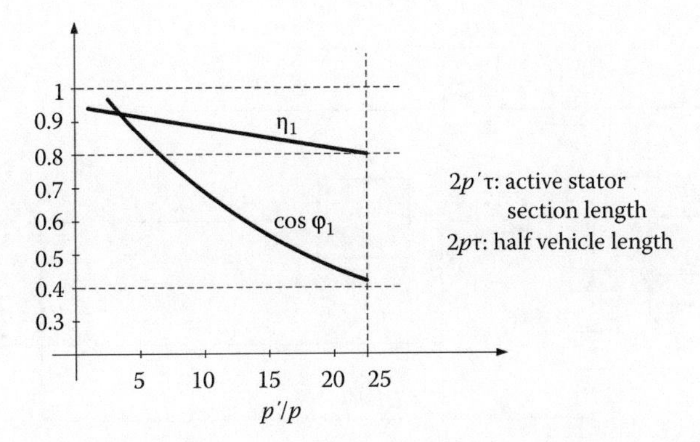

FIGURE 7.6 Typical efficiency η_1 and power factor $\cos\varphi_1$ of a DCE-LSM 45 ton MAGLEV at 110 m/s.

As some sections supply overlapping is necessary at active section switching, there is enough time to achieve seamless transition of vehicle from one active section to another.

7.5 CIRCUIT MODEL FOR TRANSIENTS AND CONTROL

The circuit model for transients and control stems from the rotary synchronous machine orthogonal model with pertinent adaptations (per vehicle side):

$$\begin{aligned}
&i_d R_1 - V_d = -\frac{\partial \Psi_d}{dt} + \omega_r \Psi_q; \quad \Psi_d = L_{a1} i_d + L_{dm}(i_F' + i_d) \\
&i_q R_1 - V_q = -\frac{\partial \Psi_q}{dt} - \omega_r \Psi_d; \quad \Psi_q = L_q i_q \\
&i_F' R_F' - V_F' = -\frac{\partial \Psi_F'}{dt}; \quad \Psi_F' = L_{Fl} i_F' + L_{dm}(i_F' + i_d) \\
&\frac{i_F'}{i_F} = \frac{M_{aF}}{L_{dm}} = K_F; \quad V_F' = \frac{1}{K_F} V_F; \quad \omega_r = \frac{\pi}{\tau} U \\
&F_x = \frac{2}{3}\frac{\pi}{\tau}(L_{dm} i_F' + (L_{dm} - L_{qm}) i_d) i_q; \quad 2\pi\tau\text{—}\textit{half vehicle length} \\
&\frac{M}{2}\frac{dU}{dt} = F_x - F_{load}
\end{aligned} \tag{7.44}$$

In the absence of a secondary suspension system,

$$\frac{M}{2}\frac{dg}{dt} = M \cdot a_g - F_n; \quad F_n \approx \frac{B_F^2}{2\mu_0} l_{stack} \frac{2p\tau}{2}\frac{2}{3} \tag{7.45}$$

M is the vehicle mass ($M/2$ corresponds to one vehicle side served by one inverter for one active stator section). If magnetic suspension is not used, Equation 7.45 is not needed (or controlled).

Note: When a secondary (mechanical) active or passive suspension is used, besides the primary magnetic suspension, more equations are added. Also separate equations for each group of dc exciters that are controlled from same source, if this is the case, are to be used with track irregularities accounted for, to allow for ride comfort assessment (as shown in later chapters, dedicated to magnetic suspension and MAGLEVs).

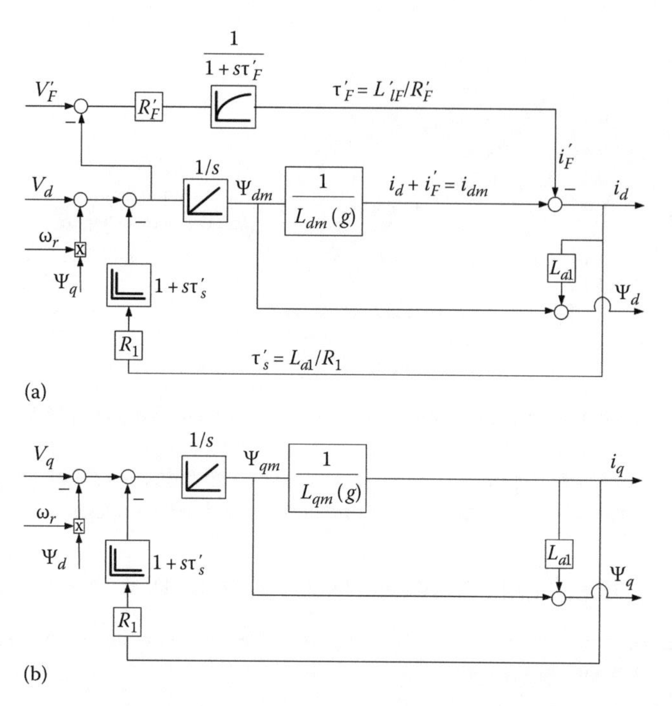

FIGURE 7.7 Structural diagram of DCE-LSM at given speed ω_r (a) axis d; (b) axis q.

Model linearization may be used to investigate stability under various controls but, due to strong nonlinearities in the system and large variation of variables, robust control systems are needed for DCE-LSM.

A structural diagram of DCE-LSM equations is shown in Figure 7.7 to serve in the control system analysis and design. As shown, for $i_d=0$, the structural diagram is notably simplified because no damper cage is present and the influence of the linear generator placed on the pole shoes of the dc exciter is not considered.

Only the stator leakage τ_l' and dc exciter leakage τ_F' time constants are present. By extracting power from stator mmf harmonics field, the linear generator placed on the dc exciter pole shoes does not interfere directly with the structural diagram that depicts space fundamental properties of the machine.

The equations of motion, horizontal (propulsion) and vertical (suspension, in (7.44) and (7.45)), may be added in a separate structural diagram (Figure 7.8).

If a secondary (and tertiary) suspension was used, the vertical motion structural diagram (Figure 7.8b) would include their pertinent equations. Only for zero i_d, approximately, the normal force F_n is

$$F_n \approx \frac{3}{2}\frac{\partial L_{dm}}{\partial g}\frac{\left(i_F'\right)^2}{2} \approx -\frac{3}{4}\frac{L_{dm}}{g}\left(i_F'\right)^2 \tag{7.46}$$

Magnetic saturation is neglected due to the large airgap and for simplicity, but the L_{dm}, L_{qm} dependence on airgap, which varies notably in MAGLEVs (say, from 20 to 10 mm), is visible all over in the structural diagrams; track irregularities may also be included as they further modify the airgap, but this is beyond our scope here. Though nonzero i_d is considered in the structural diagrams

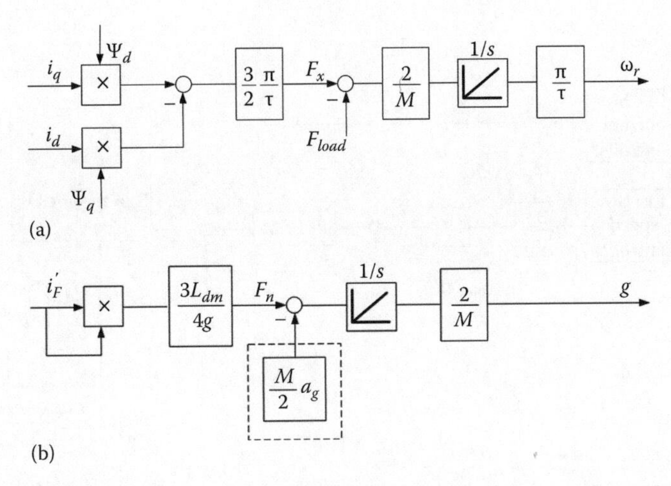

FIGURE 7.8 Motion structural diagrams of DCE-LSM/vehicle side: (a) propulsion and (b) vertical (suspension) motion ($i_d=0$ in normal force F_n).

in Figure 7.7a and b, zero i_d current control is targeted to decouple the propulsion and suspension functions in the control system.

7.6 FIELD-ORIENTED CONTROL OF DCE-LSM

The zero i_d field-oriented control (FOC) of DCE-LSM is similar to that of rotary SMs, but the field current may be handled differently:

- With given i_F' versus speed dependence when suspension control is absent (eventually to provide maximum thrust).
- With the i_F' controlled to keep the airgap versus time $g^*(t)$ as commanded by the suspension system central control. Each dc exciter module (or group of them) is controlled to satisfy magnetic suspension performance $\left(\Delta g_{\max}^*\right)$ and ride comfort in MAGLEVs.

A typical FOC system is shown in Figure 7.9.

While Figure 7.9 treats the case of both propulsion and suspension decoupled control, in the absence of suspension control, the reference current i_F' (in Figure 7.9b) is given based on dedicated objectives such as i_F' versus speed envelope or operation at unity power factor, etc.

The emf decoupling (in Figure 7.9a) is introduced to improve the control at high speeds and allow open-loop space vector PWM in the converter, to exploit best the limited switching frequency to maximum fundamental frequency rather small ratio at high speeds, even when multilevel inverters are used on ground to supply the vehicle.

The measured variables are, in general, mover position θ_{er} (used to calculate speed $U=(d\theta_{er}/dt)(\tau/\pi)$), stator currents i_a, i_b, and dc field current i_F (the field current reduced to stator i_F' is used in the control); moreover, for suspension control, the airgap g is measured (perhaps also the vertical acceleration should be measured, by inertial accelerometers) to provide a robust vertical speed observer used for a second-order sliding mode (SM) functional. SM stands for sliding mode control, a very robust and practical control strategy [7].

Alternatively, direct stator flux and thrust control (DTFC) can be applied instead of FOC, but the zeroing of i_d current, to reduce propulsion-suspension interaction, may render the later as more practical.

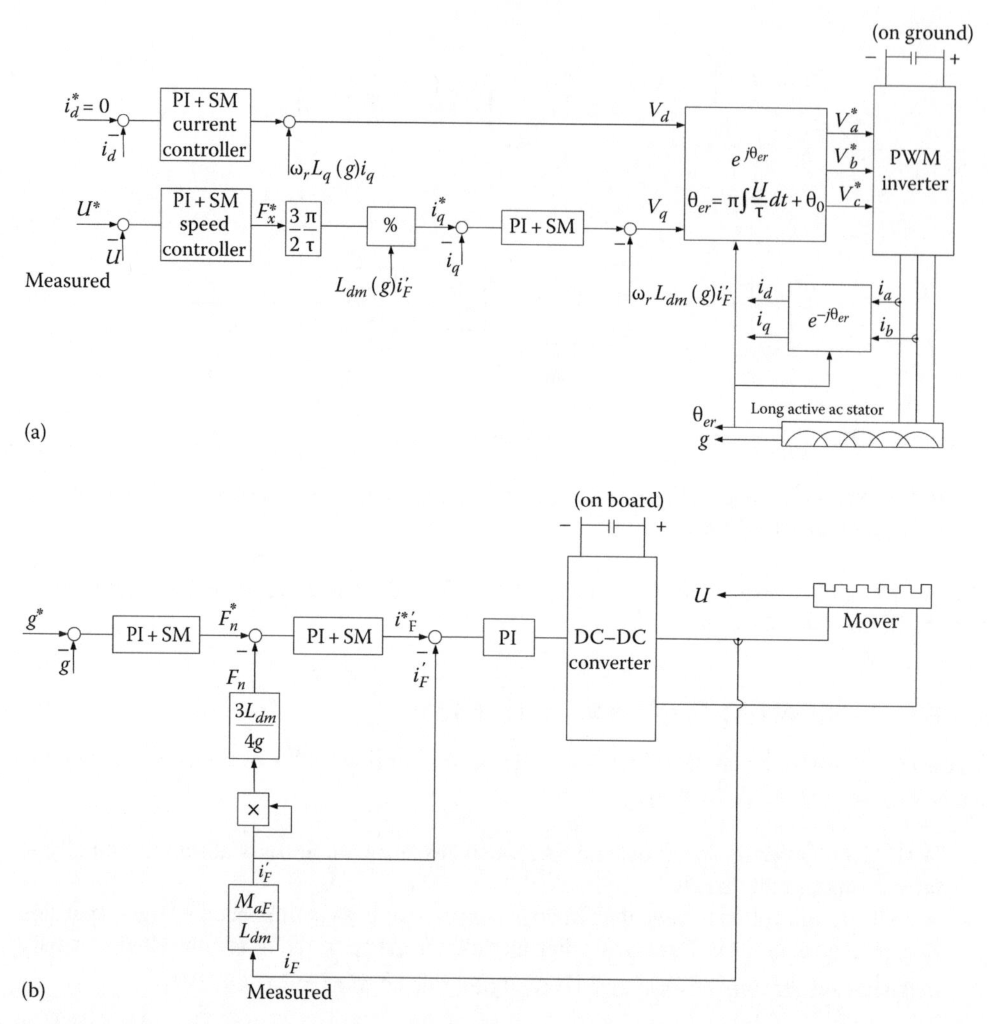

FIGURE 7.9 General FOC of DCE-LSM: (a) propulsion FOC control for zero i_d and (b) decoupled airgap (suspension) control.

7.7 NOTE ON PM + DCE-LSM

For the PM + dc coils exciters, the design [8] model for transients and control is very similar but

- The electromagnetic airgap in axis d is increased by the presence of PMs, so L_{dm} becomes smaller than L_{qm} and M_{aF} also becomes smaller.
- The flux in axis d gets an additional component

$$\Psi_d = \Psi_{PM} + L_{dm}\left(i_F' + i_d\right) + L_{al} i_d \tag{7.47}$$

- The normal force F_n (per vehicle side or 50% mover occupancy) formula gets a new expression to include the PM contributions:

$$F_n \approx \frac{\left(B_{PM} + B_F\left(i_F\right)\right)^2}{2\mu_0} \frac{2}{3} l_{stack} \frac{2p\tau}{2} \tag{7.48}$$

- The main advantages of hybrid (PM + dc exciters) for LSM are excitation power reduction and low total suspension control power for MAGLEVs.
- At the limit, if the field coils are eliminated from the dc exciter (mover) and replaced by PMs, still the developments in this chapter can be adapted to the situation, but, for integral suspension–propulsion control, DTFC may be used (zero i_d current control is not good anymore) with stator flux control for propulsion control and thrust control for suspension control.

7.8 SUMMARY

- DCE-LSM means dc-excited linear synchronous motor with short dc exciter mover and long ac-fed stator along the track (excursion length).
- DCE-LSM may be applied for short travel (a few meters) transport on linear bearings or on magnetic suspension (MAGLEVs) or to medium or high-speed MAGLEVs.
- The dc exciter (mover) contains a heteropolar concentrated dc coil magnetic laminated core, which produces by motion a traveling magnetic field in the airgap at the linear speed u_s and pole pitch τ; a four full-pole and two end-pole configuration produces a well-behaved heteropolar airgap magnetic field distribution that allows minimum back-iron thickness (weight).
- The dc exciter full poles contain a few slots (say, 6) to host a two-phase winding with tooth-wound coils and a pole pitch $\tau_{s2}=\tau_{s1}/2$; τ_{s1}—slot pole pitch in the three-phase ac-fed primary stator. A linear electric generator (LEG) that collects the power by electromagnetic induction, from the stator mmf and slot harmonics, is thus obtained; the LEG produces enough energy, above a certain speed, to recharge the backup battery on board of mover and supply the dc excitation circuit to provide LSM dc heteropolar excitation and (if applied) controlled magnetic suspension (in MAGLEVs).
- A basic design methodology for dc exciter is put into place for an initial 110 m/s, 2 ton/m length MAGLEV, propulsion and suspension system.
- The airgap flux density by dc exciter is chosen in the interval of 0.5–0.7 T, the maximum frequency in the stator $f_{1max}=250$ Hz, and thus, the pole pitch τ for $u_{s\,\max}=110$ m/s is $\tau=0.22$ m; the stack length is chosen as $l_{stack}=0.2$ m and thus occupying half of vehicle length on each side, for a MAGLEV of 2 ton/m and an average airgap of 10 mm, only 0.62 W/kg is needed for suspension of the vehicle. For dynamic airgap control, via field current control, a 5/1 over voltage is provided in the dc supply on board (from $44V_{dc}$ to $220V_{dc}$); we end up with 3 kVA/ton peak rating of the dc–dc converter that feeds the field circuit.
- Though the dc exciter design methodology was exercised on a high-speed MAGLEV, it may be directly applied to a low-speed application as well.
- The ac long stator contains a three-phase aluminum cable winding with $q=1$ slot/pole/phase and a single layer; this way, a low-cost, easy to install in open slots, solution is obtained; the laminated core along the track length has uniform open slots that house the three-phase ac winding (one on each side of the vehicle in a MAGLEV, for balanced integrated suspension propulsion). Small slot height/slot width ratio is obtained to secure small leakage slot inductance. As an active stator section is more than 1 km long (for a high-speed MAGLEV), this is an asset, in limiting total machine inductance, and provides not so bad a power factor, thus limiting the on-ground converter kVA rating (costs).
- 2(3)D-FEM or magnetic circuit or multilayer [9] analytical field theory may be applied to calculate rather precisely thrust and normal force.
- To decouple the propulsion from suspension, the pure i_q vector control of propulsion is adopted for the active stator section. The stator mmf for maximum thrust per slot is thus calculated easily (emf in phase with ac current).

- The circuit model parameters of DCE-LSM are calculated observing the "busy" and the "empty" or idle part of active stator section inductances and resistances. Finally R_1, L_d, L_q, E_1, L_{dm}. L_{qm}, M_{aF} are given design expressions, starting from rotary SM theory.
- Efficiency, power factor, thrust, and normal force are calculated based on these parameters, for our case study; efficiency in the range of 80% and a power factor of 0.5–0.6 (lagging) is obtained $2p'/2p=25$, with 2p active stator poles and $2p'$ active stator section poles. At 110 m/s, 2 ton/m vehicle, 11 s is required for a 50 m long vehicle to travel over a 1.25 km active stator section; this is enough time to switch on and off successive stator sectors with the vehicle in synchronism.
- The *dq* model of SMs is adopted here for the study of transients and control, and *dq* structural diagrams become rather simplified in the absence of damper windings on mover and at zero i_d ($i_d=0$).
- Separate structural diagrams are added for horizontal (propulsion) motion and for vertical (suspension) motion equations. Airgap variation during suspension control in the presence of track (stator) irregularities is paramount in leading to a heavily nonlinear system. The system may be linearized, but if g varies widely (in 2/1 ratio for MAGLEVs), a robust suspension and propulsion coordinated control is required for stable and high-quality performance of both propulsion and suspension.
- For the DCE-LSM control, FOC is proposed; with $i_d^*=0$ for decoupled propulsion and levitation, i_q^* is given by the output of the speed regulator; emf compensation is added to (i_d, i_q) close-loop controllers, and open-loop PWM is performed in the limited switching frequency inverters located in on-ground stations.
- The airgap (suspension) control is operated through field current i_F^* and normal force control; as i_F^* variation leads also to propulsion force variation, the speed and airgap (suspension) control have to be harmonized for safe and comfortable rides.
- If active magnetic suspension is not provided, the field reference current i_F^* function (versus speed or thrust) is given by an alternative energy conversion criterion.
- To reduce the excitation power and the dynamic control power of magnetic suspension, PMs may be added to the dc coils; the so-called "zero control power" principle may be used when all dc exciters with rather average airgap control, in groups, are fed from the same dc power bus on board; the motions of vehicle are damped reciprocally to reach almost zero net average suspension control power, if the PMs provide compensation for all vehicle weight.
- At the limit, the dc coils on mover may be eliminated when PMs remain alone; this case will be treated in a separate chapter on LPMSM as it has many new peculiarities for short travel (low speed), medium- or high-speed applications.
- As the only commercial high-speed MAGLEV so far uses DCE-LSMs for integrated propulsion- suspension (lateral guidance is provided by dedicated controlled dc electromagnets), the DCE-LSM technologies are expected to develop further in the near future with more application sites in view. More information on active magnetic suspension control and MAGLEVs will be available in later dedicated chapters of this book.

REFERENCES

1. H. Weh, The integration of functions of magnetic levitation and propulsion (in German), *ETZ. A.*, 96(9), 1975, 131–135.
2. H. Weh, Synchronous long stator motor with controlled normal forces (in German), *ETZ. A.*, 96(9), 1975, 409–413.
3. H. Weh and M. Shalaby, Magnetic levitation with controlled permanentic excitation, *IEEE Trans.*, MAG-13(5), 1977, 1409–1411.

4. I. Boldea and S.A. Nasar, *Linear Motion Electromagnetic Systems*, Chapter 8, John Wiley, New York, 1985.
5. H. May, N. Mosebach, and H. Weh, Pole-force oscillations caused by armature slots in the active guideway synchronous motor, *Arch. Elektroteh.*, 59, 1977, 291–296 (in German).
6. I. Boldea and L. Tutelea, *Electric Machines: Steady State, Transients and Design with MATLAB*, Chapter 6, CRC Press, Taylor & Francis, New York, 2009.
7. V. Utkin, J. Guldner, and J. Shi, *Sliding Mode Control in Electromechanical Systems*, 2nd edn., CRC Press, Taylor & Francis, New York, 2009.
8. H.W. Cho, H.S. Han, J.M Lee, B.S. Kim, and S.-Y. Sung, Design considerations of EM-PM hybrid levitation and propulsion device for magnetically levitated vehicle, *IEEE Trans.*, MAG-45(10), 2009, 4632–4635.
9. M.S. Hosseini and S. Vaez-Jadeh, Modelling and analysis of linear synchronous motors in high speed Maglev vehicles, *IEEE Trans.*, MAG-46(7), 2010, 2656–2664.

8 Superconducting Magnet Linear Synchronous Motors

Superconducting magnet linear synchronous motors (SM-LSM) with active guideway have been proposed for MAGLEVs, at speeds up to 550 km/h so far. This chapter treats the following aspects of SM-LSMs:

- Topologies of practical interest
- Superconducting magnet (SM)
- A technical field and circuit theory of SM-LSM
- Normal and lateral forces
- SM-LSM with figure eight-shape-coil stator
- Direct thrust and suspension (airgap) via flux control (DTFC) of SM-LSM
- Summary

8.1 INTRODUCTION

The SM-LSM consists of dc superconducting coils (magnets) placed on the mover (vehicle) and an active guideway with three-phase air-core cable windings placed horizontally (Figure 8.1) or vertically (laterally on the two sides of the track) (Figure 8.2).

While both configurations (with rectangular and, respectively, figure 8-shape coils) of stator windings develop in interaction with the SM field, propulsion, suspension, and lateral forces, the latter is better in force density and dynamics and was proposed to integrate the three functions in a MAGLEV.

In this chapter, to elucidate the fundamentals, we will treat first the configuration in Figure 8.1 and only later the configuration in Figure 8.2, as the latter is more complicated to analyze.

Other early configurations such as the Magneplane [2] are not considered here as they have not been proved practical though they are very intuitive.

The air-core cable three-phase ac windings on ground (on both sides of the vehicle for MAGLEVs) allow for longer energized sections because the total inductance per section is lower in air. The SM-LSM is preferred for even higher speeds as the airgap is in the order of 5–10 cm or more, while it is 10 mm for dc-excited LSMs (Chapter 7). The larger airgap means rougher guideways, at lower cost. The alternate polarity SMs placed along the vehicle length produce a traveling magnetic field by motion, at speed $u_s = u$. The pole pitch of SMs and of stator windings is the same (τ), and the frequency f_1 of stator currents speed u_s is

$$f_1 = \frac{u_s}{2\tau} \tag{8.1}$$

As for any synchronous motor, to vary speed u_s, the frequency f_1 has to be varied from zero. The key element of SM-LSM is the SM (coil).

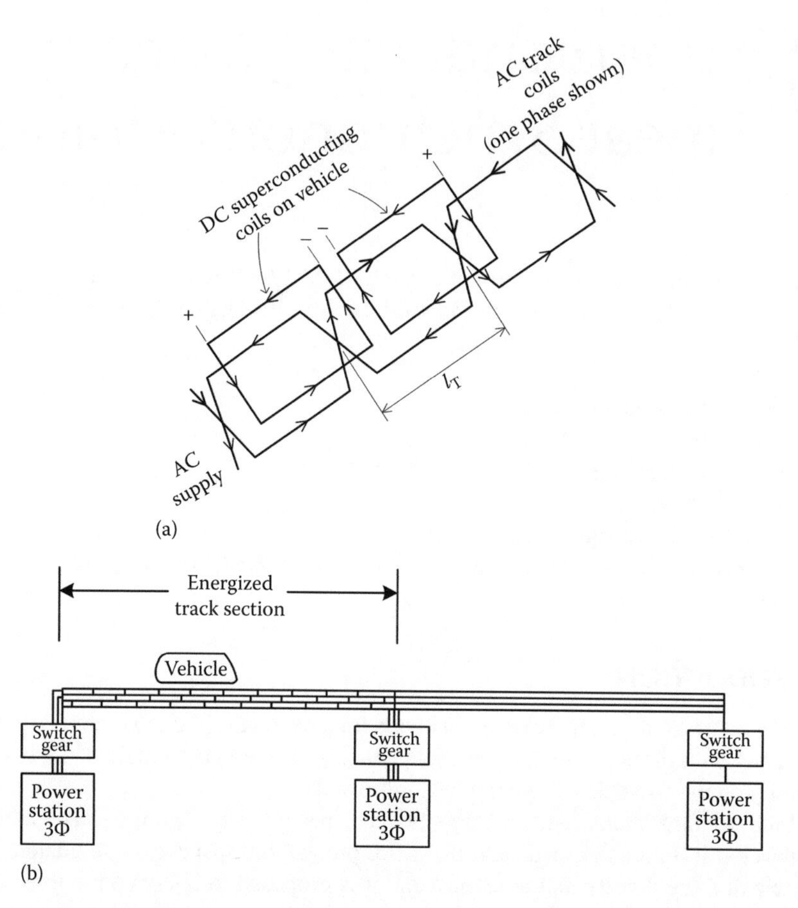

FIGURE 8.1 SM-LSM: (a) SMs on vehicle and air-core three-phase cable windings on ground and (b) three-phase track energized sections. (After Nasar, S.A. and Boldea, I., *Linear Motion Electric Machines*, John Wiley & Sons, New York, 1976.)

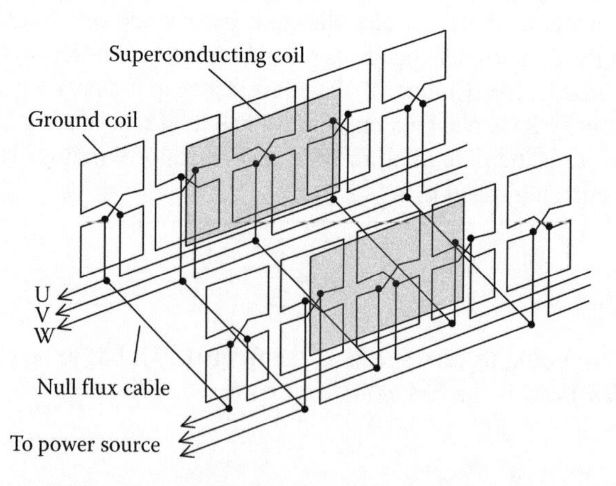

FIGURE 8.2 Figure eight-shape-coil three-phase ac winding with propulsion, suspension, and guidance force production. (After Toshiaki, M. and Shunsuke, F., Design of coil specifications in EDS Maglev using an optimization program, *Record of LDIA-1998*, Tokyo, Japan, pp. 343–346, 1998.)

8.2 SUPERCONDUCTING MAGNET

Kamerlingh Onnes has observed around 1911 that some metals lose most of their electrical resistance below 10 K; the transition from natural to superconducting behavior is called critical temperature T_c (T_c=9.1 K for niobium); aluminum, indium, tin, lead, niobium, and their alloys are superconductors at very low temperatures. A regular supply of liquid helium and liquid nitrogen on board of vehicle (or a dedicated refrigerator) are required to maintain this range of temperatures (bellow 10 K). More recently, high-temperature (70 K) superconductors that need only liquid nitrogen cooling have been discovered (for more on applied superconductivity, see [3,4]).

A typical SM of low temperature (Figure 8.3) contains

- Superconducting coil
- Liquid helium dewar
- Liquid nitrogen dewar
- Evacuated fiberglass insulation

The liquid nitrogen dewar is made of aluminum, and the liquid helium dewar is fabricated from stainless steel (to reduce eddy current losses).

The evacuated fiberglass insulation has very low thermal conductivity, can transmit high pressure, and makes the force-density distribution uniform.

The force components between SM and the three-phase ac conductors in an "air core" are exerted directly on the superconducting coil, at gas pressures of 100 μmHg, maintained by a vacuum pump. The SM coil (currents) leads have to be well cooled by additional tubes and posts in the dewars, to streamline the supply of cooling agents properly stored on board of vehicle.

The design, the charging, and discharging of SM such that to avoid quenching (transition to normal conductor from superconducting mode) is an art in itself, which is not pursued here further (either for low- and high-temperature SMs [3]).

In fact, the SM is a superconducting short-circuited coil that maintains the dc current in it constant (if periodical replenishing with current by a special static converter is done properly (slowly), as required, without quenching the superconductor).

The SM may also be compared with a very strong and large permanent magnet; the latter behaves like an equivalent room temperature superconducting coil, but its energy is stored in a large hysteresis cycle strong magnetic material.

It is argued that for large geometries (0.5–1 m in one direction) the SM may produce more thrust (or suspension, guidance) force per given weight or per one USD than a permanent magnet.

With the advent of high-temperature SMs and of ever stronger PMs (B_r > 1.5 T), the competition becomes stronger than ever.

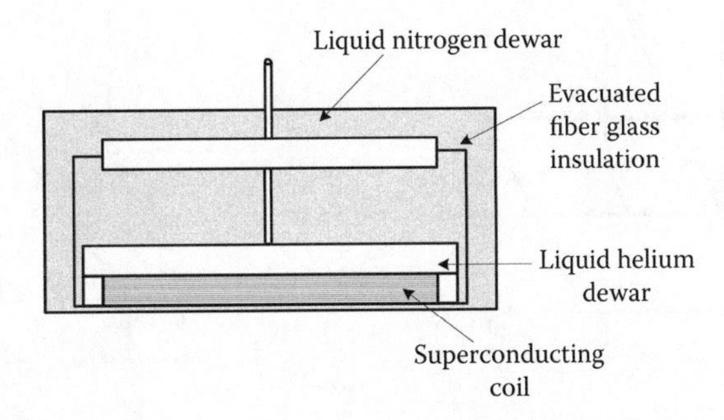

FIGURE 8.3 Low-temperature SM. (After Nasar, S.A. and Boldea, I., *Linear Motion Electric Machines*, John Wiley & Sons, New York, 1976.)

8.3 TECHNICAL FIELD AND CIRCUIT THEORY OF SM-LSM

Typical SM-LSMs for MAGLEVs have power in the range of 5 MW or more at speeds of 500–550 km/h with energized sections of 3–5 km in length. The dc mmf of LSM is in the order $(3 \div 7)10^5$ A turns to produce a magnetic flux density of 0.2–0.45 T that interacts with the stator currents to produce propulsion, suspension, (levitation), and lateral forces for an effective gap of 5–10 cm at least (the superconducting coil to winding center to center distance may be as large as 30 cm).

In essence, we have to start with the magnetic field produced by a rectangular SM in air, and then we add the contribution of neighboring SMs of alternate polarity, to obtain the needed magnetic field distribution in the stator three-phase air-core winding zone.

Further on, we will calculate the emfs in the three-phase ac windings; there are three-phase ac winding inductances and resistances to form a phasor diagram that, for the fundamental, describes the steady-state operation. A third space harmonic, which is typical, may be added to reveal essential thrust pulsations.

After steady-state characteristics are obtained, the thrust, normal force, and lateral force versus power angle are calculated, to illustrate the complex behavior of SM-LSM. This is called here the technical circuit theory of SM-LSM.

8.3.1 Magnetic Field of a Rectangular SM in Air

After approximating the conductors by infinitely thin filaments, the flux density produced in point (P) (Figure 8.4a) by a finite length conductor [1] is obtained via Biot-Savart law (Figure 8.4b):

$$\overline{B_1} = \frac{\mu_0 I_0}{4\pi} \int_{-l}^{l} \frac{\overline{dx_1} \times \overline{dr}}{r^2} \tag{8.2}$$

with

$$r_1 = \sqrt{(y+b)^2 + z^2}\ ; \quad r = \sqrt{(x-x_1)^2 + r_1^2} \tag{8.3}$$

and

$$d\overline{x}_1 \times d\overline{r} = \frac{r_1 d\overline{a}_1}{\sqrt{(x-x_1)^2 + r_1^2}} \tag{8.4}$$

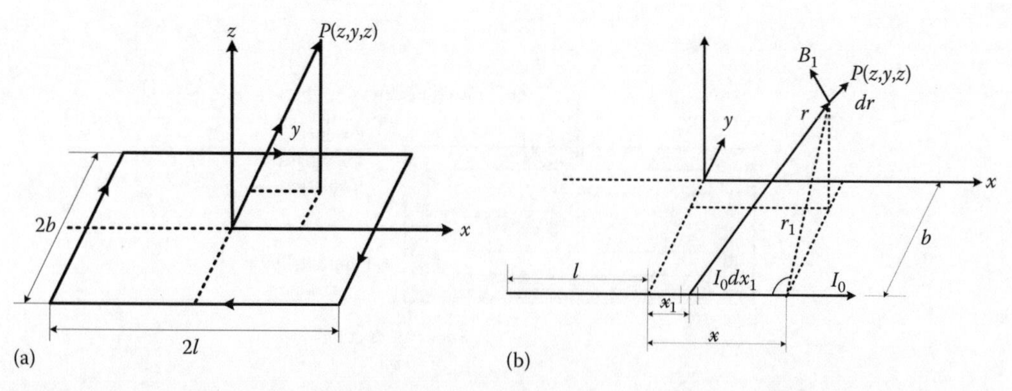

FIGURE 8.4 Rectangular SM (a) and the magnetic field produced by a finite length conductor (b). (After Nasar, S.A. and Boldea, I., *Linear Motion Electric Machines*, John Wiley & Sons, New York, 1976.)

$\overline{a_1}$ is the unit vector in the direction of the plane formed by point P and the conductor finite length. As visible in Figure 8.4, $\overline{a_1}$ has two components $\overline{a_y}$ and $\overline{a_z}$:

$$\overline{a_1} = \frac{z}{\sqrt{(y+b)^2+z^2}}\overline{a_y} + \frac{(b+y)}{\sqrt{(y+b)^2+z^2}}\overline{a_z} \tag{8.5}$$

Finally, from (8.2) to (8.5),

$$B_{1y} = \frac{\mu_0 I_0}{4\pi}\frac{z}{\sqrt{(y+b)^2+z^2}}\int_{-l}^{l}\frac{\sqrt{(y+b)^2+z^2}\,dx_1}{\left[(x-x_1)^2+(y+b)^2+z^2\right]^{3/2}} \tag{8.6}$$

$$B_{1z} = \frac{\mu_0 I_0}{4\pi}\frac{b+y}{\sqrt{(y+b)^2+z^2}}\int_{-l}^{l}\frac{\sqrt{(y+b)^2+z^2}\,dx_1}{\left[(x-x_1)^2+(y+b)^2+z^2\right]^{3/2}} \tag{8.7}$$

A rectangular SM has four sides. Thus, the flux density produced at point P adds up the components produced by the four SM sides and exhibits all three components B_x, B_y, B_z:

$$B_x = \frac{\mu_0 I_0}{4\pi}z\left\{\frac{1}{(x-l)^2+z^2}\left[\frac{y+b}{\sqrt{(x-l)^2+(y+b)^2+z^2}} - \frac{y-b}{\sqrt{(x-l)^2+(y-b)^2+z^2}}\right] - \frac{1}{(x+l)^2+z^2}\left[\frac{y+b}{\sqrt{(x+l)^2+(y+b)^2+z^2}} - \frac{y-b}{\sqrt{(x+l)^2+(y-b)^2+z^2}}\right]\right\} \tag{8.8}$$

$$B_y = \frac{\mu_0 I_0}{4\pi}z\left\{\frac{1}{(y-b)^2+z^2}\left[\frac{x+l}{\sqrt{(x+l)^2+(y-b)^2+z^2}} - \frac{x-l}{\sqrt{(x-l)^2+(y-b)^2+z^2}}\right] - \frac{1}{(y+b)^2+z^2}\left[\frac{x+l}{\sqrt{(x+l)^2+(y+b)^2+z^2}} - \frac{x-l}{\sqrt{(x-l)^2+(y+b)^2+z^2}}\right]\right\} \tag{8.9}$$

$$B_z = \frac{\mu_0 I_0}{4\pi}\left\{\frac{(y+b)}{(y+b)^2+z^2}\left[\frac{x+l}{\sqrt{(x+l)^2+(y+b)^2+z^2}} - \frac{(x-l)}{\sqrt{(x-l)^2+(y+b)^2+z^2}}\right] - \frac{(y-b)}{(x+l)^2+z^2}\left[\frac{x+l}{\sqrt{(x+l)^2+(y-b)^2+z^2}} - \frac{(x-l)}{\sqrt{(x-l)^2+(y-b)^2+z^2}}\right]\right\} \tag{8.10}$$

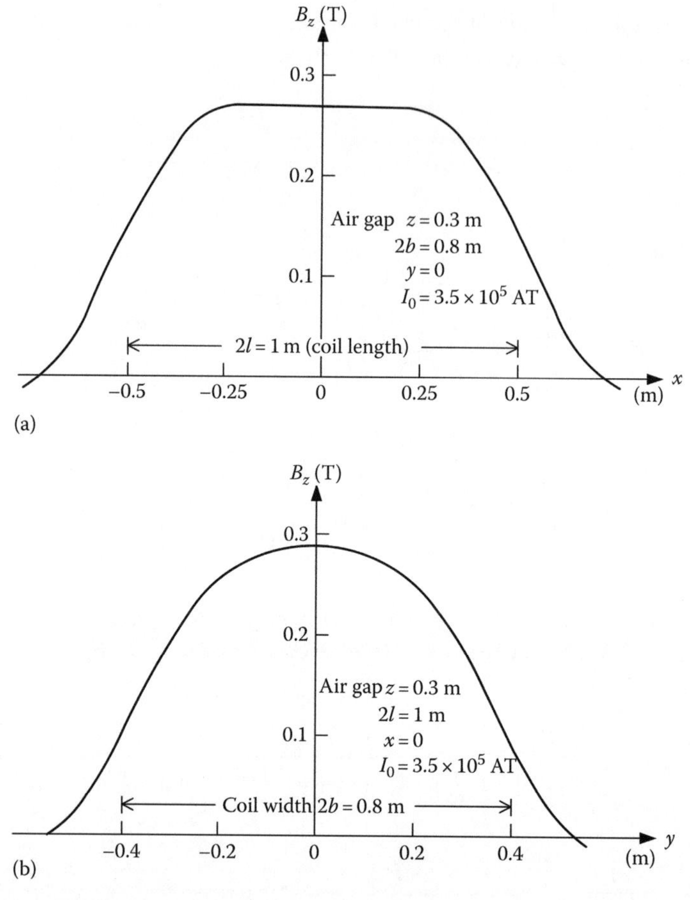

FIGURE 8.5 Rectangular SM coil: Normal flux density B_z distribution (a) versus motion direction Ox, (b) versus transverse direction Oy. (After Nasar, S.A. and Boldea, I., *Linear Motion Electric Machines*, John Wiley & Sons, New York, 1976.)

For a numerical example with the data speed $u_s=130$ m/s; $f_1=50$ Hz; $2b=0.8$ m; $2l=1$ m, $z=0.3$ m, $l_T=1.3$ m (Figure 8.4). $I_0=3.5\times10^5$ A turns, the B_z distribution along motion direction x and transverse direction y are shown in Figure 8.5a,b.

There is a notable flux density outside the SM ($|x|>l$), which means that if the distance between neighboring SMs is about the same as the airgap length (z), the magnetic field sensed by the armature winding at a point comprises contributions of a few (say three) SMs.

8.3.2 EMF E_1, Inductance and Resistance L_s, R_s per Phase

The flux variation in a stator coil, placed at distance x with respect to the SM center, is thus

$$\frac{d\Phi}{dx}=\int_{-b}^{b}B_z(x,y,z)\,dy \tag{8.11}$$

The voltage induced in a track coil by a single SM, E_{SM}, is

$$E_{SM}=u_s\frac{d\Phi}{dx} \tag{8.12}$$

For three-SM contribution, the total emf in a stator coil E_0 is

$$E_0 = u_s \sum_{n=1}^{3} \frac{d\Phi}{dx} \tag{8.13}$$

If the vehicle has N SMs per side, the emf in the stator phase E_1 is

$$E_1 = NE_0 \tag{8.14}$$

The shape of $d\Phi/dx$ is essential for current shape control. For sinusoidal (vector) current control, the emf has to be almost sinusoidal, in order to produce low thrust pulsations.

For $\tau=1$ m and SM length/pole pitch $L/\tau=0.7$, $z_0=0.2$ m (airgap), $I_0=5\times10^5$ A turns, $2b=0.9$ m, $d\Phi/dx$ is close to a sinusoidal (Figure 8.6); not so for $\tau=4$ m. Investigating for optimal pole pitch and L/τ and "stack" width/pole pitch ratio, $2b/\tau$, is a must if the emf shape is to be the desired one (sinusoidal in general).

The armature (stator) mmf produced magnetic field is notably smaller than that of the SMs. Consequently, the armature inductance per phase, L_{ph}, for an energized section of $2p'$ poles may be calculated as for long electric power lines:

$$L_{ph} \approx \frac{\mu_0}{\pi}\left\{4p'l\times\ln\left[\frac{\tau}{d}+\left[\left(\frac{\tau}{d}\right)^2-1\right]^{1/2}\right]+2p'\tau\ln\left[\frac{l}{d}+\left[\left(\frac{l}{d}\right)^2-1\right]\right]^{1/2}\right\} \tag{8.15}$$

where

d is the diameter of armature coil conductor

l is the armature coil width

The resultant inductance L_s per phase, including the influence of the other two phases, is

$$L_s \approx L_{ph}\left(1+\frac{1}{3}+\frac{1}{6}\right) \tag{8.16}$$

The armature resistance/phase R_s writes

$$R_s \approx \frac{\rho_{Al}}{2q_{Al}}(l+\tau)2p'\times K_l K_{skin} \tag{8.17}$$

where

ρ_{Al} is the aluminum electric resistivity

q_{Al} is the aluminum cable cross section

$K_l=1.05$–1.2 accounts for the departure of stator coils from rectangular shape

$K_{skin}>1$ accounts for skin effect in the stator conductors

The factor 2 accounts for the situation when two cables in parallel make the phase winding (Figure 8.1).

Note on skin effect: As the stator current with cable winding is rather large, in the 2000 A or more range, for a more than 5 MW propulsion vehicle, the skin effect should be reduced and, anyway, considered in the cable design (multiple-conductor twisted cable is to be used).

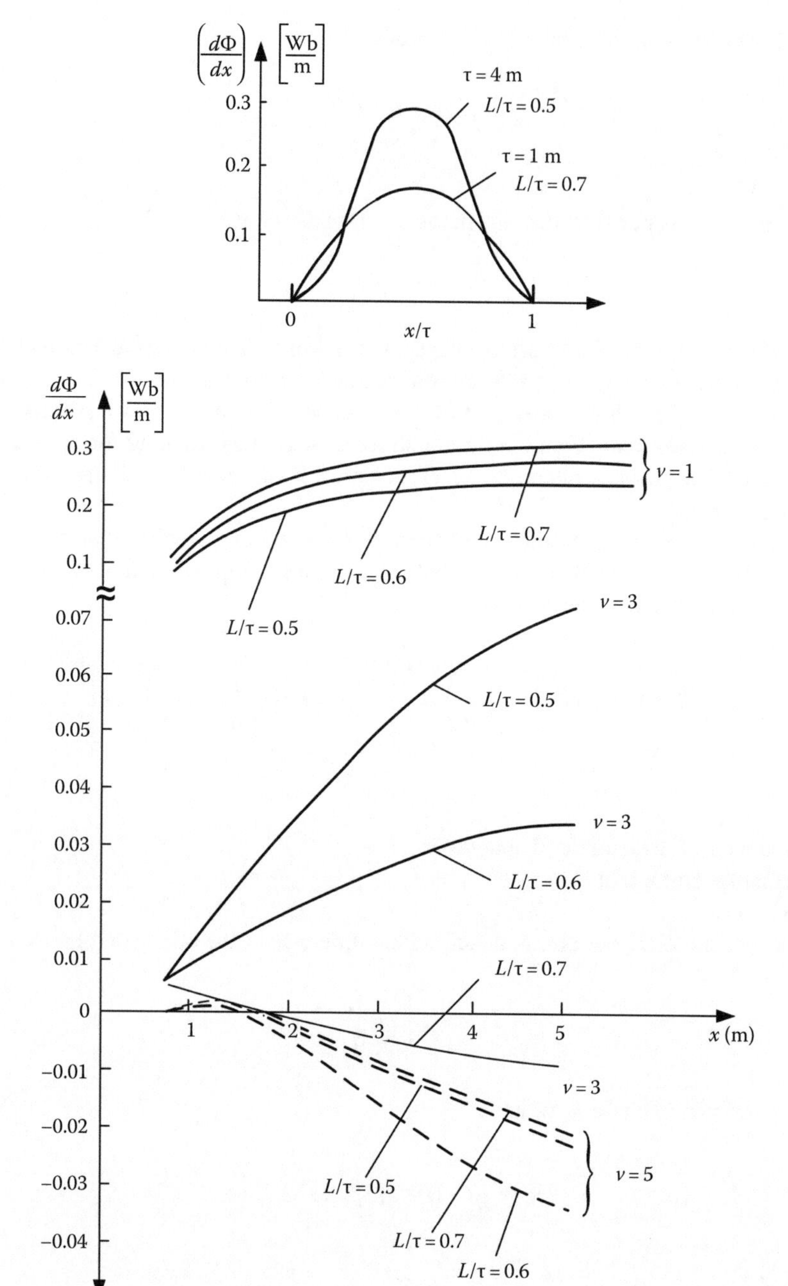

FIGURE 8.6 Rate of SM flux change with position. (After Boldea, I. and Nasar, S.A., *Linear Motion Electromagnetic Systems*, Wiley & Sons, New York, 1985.)

8.3.3 Phasor Diagram, Power Factor, and Efficiency

The phasor diagram is valid only for sinusoidal emf; a practical waveform is shown in Figure 8.7; it contains some third and fifth time harmonics, which may be generating thrust pulsations if the third harmonic occurs in current (delta connections of phases).

Such delta connection of phases is feasible as it allows connection of an energized section from both ends (Figure 8.8).

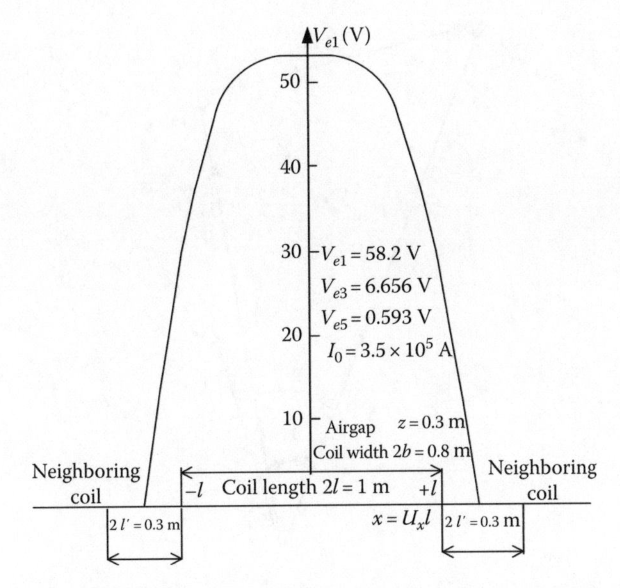

FIGURE 8.7 Emf in a track coil. (After Nasar, S.A. and Boldea, I., *Linear Motion Electric Machines*, John Wiley & Sons, New York, 1976.)

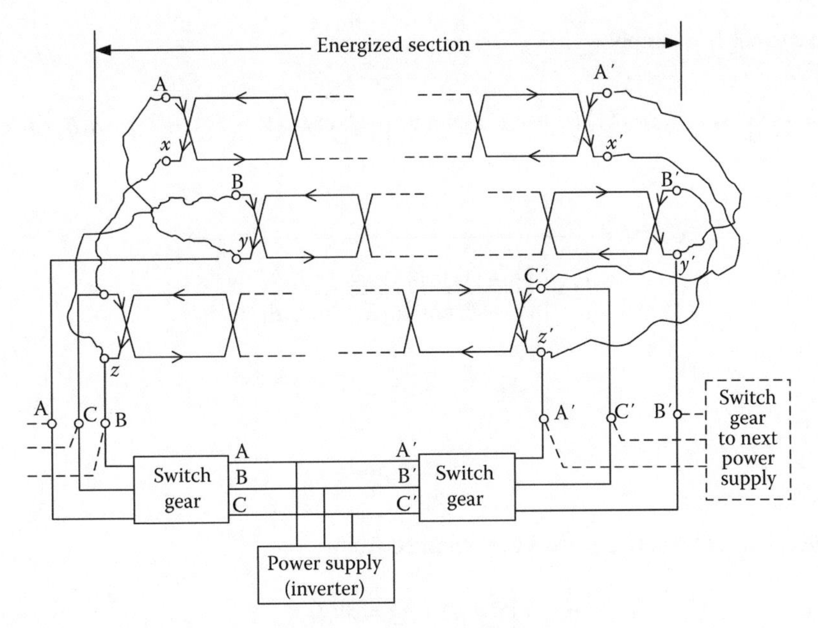

FIGURE 8.8 Delta connection of stator-phase sections. (After Nasar, S.A. and Boldea, I., *Linear Motion Electric Machines*, John Wiley & Sons, New York, 1976.)

For the fundamental, the voltage equation per phase is

$$\underline{V_1} = R_s \underline{I_1} + j\omega_1 L_s \underline{I_1} - E_1 e^{-j\delta_v}; \quad \omega_1 = \frac{\pi u_s}{\tau}, \tag{8.18}$$

where

V_1 is the applied phase voltage fundamental
I_1 is the phase current fundamental
δ_v is the power angle (Figure 8.9)

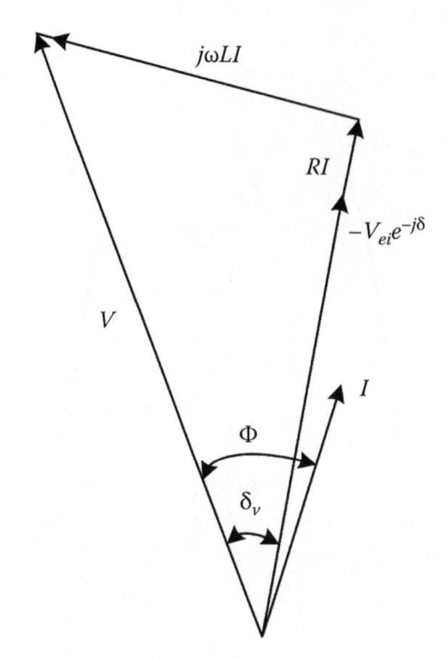

FIGURE 8.9 Phasor diagram. (After Nasar, S.A. and Boldea, I., *Linear Motion Electric Machines*, John Wiley & Sons, New York, 1976.)

The phase current $\underline{I_1}$ is simply

$$\underline{I_1} = \frac{1}{Z_1^2}\left\{\left[(V_1 - E_1\cos\delta_v)R_s + \omega_1 L_s E_1 \sin\delta_v\right] - j\left[\omega_1 L_s (V_1 - E_1\cos\delta_v) - R_s E_1 \sin\delta_v\right]\right\} \quad (8.19)$$

$$\cos\varphi_1 = \frac{1}{\left\{1 + \left[\frac{\omega_1 L_s (V_1 - E_1\cos\delta_v) - R_s E_1 \sin\delta_v}{(V_1 - E_1\cos\delta_v)R_s + \omega_1 L_s E_1 \sin\delta_v}\right]^2\right\}^{1/2}} \quad (8.20)$$

with

$$Z = R_s + j\omega_1 L_s \quad (8.21)$$

Finally, the active and reactive powers are obtained from

$$|S| = |P - jQ| = 3V_1 I_1 \quad (8.22)$$

$$P_1 = \frac{1}{Z_s^2}\left[V_1 R_s (V_1 - E_1\cos\delta_v) + \omega_1 L_s V_1 E_1 \sin\delta_v\right]$$
$$Q = \frac{1}{Z_s^2}\left[\omega_1 L_s V_1 (V_1 - E_1\cos\delta_v) - R_s V_1 E_1 \sin\delta_v\right] \quad (8.23)$$

In a dc-excited LSM, the power factor could be adjusted by controlling the dc field current. This is not the case with SM-LSM, where equivalent dc mmf in SM is rather constant.

The thrust F_{x1} is

$$F_{x1} = \frac{P_1 - 3R_s\left|I_1^2\right|}{U_s} = \frac{3E_1\left|I_1\right|\cos(\delta_v - \phi)}{U_s} \tag{8.24}$$

For a Δ connection of phases, the third harmonic current $\underline{I_3}$ in the stator, due to emf third harmonic emf, is obtained from the equation

$$R_{s_3}\underline{I_3} + j3\omega_1 L_s \underline{I_3} - E_3 e^{-j3\delta_v} = 0 \tag{8.25}$$

The apparent third harmonic power $\underline{S_3}$ is

$$\underline{S_3} = P_3 - jQ_3 = \frac{\left|E_3\right|^2}{R_{s_3} + j3\omega_1 L_s} \tag{8.26}$$

Skin effect is more important at $3\omega_1$, and thus, we have introduced R_{s_3} instead of R_s.

In a three-phase circuit, the third harmonic currents are in phase, and thus, the instantaneous power pulsates at frequency $6\omega_1$, leading to additional losses $\left(3R_{s_3}\left|\underline{I_3}\right|^2\right)$ and to thrust F_{x_3}:

$$F_{x_3} = \frac{3P_3}{U_s}\left(1 - \cos 6\omega_1 t\right) \tag{8.27}$$

The total thrust F_x is

$$F_x = F_{x_1} + F_{x_3} \tag{8.28}$$

The efficiency is thus

$$\eta \approx \frac{F_{xav} U_s}{3V_1 I_1 \cos\varphi_1} \tag{8.29}$$

8.3.4 Numerical Example 8.1

Let us illustrate the SM-LSM performance, as extracted from the so far developed technical theory, versus voltage power angle δ_v, for the following initial specifications:

Input voltage $V_1=4200$ V (peak phase value), $f_1=50$ Hz, SM mmf $I_0=3.5\times10^5$ A turns, $u_s=130$ m/s, SM dimensions $2l=1$ m, $2b=0.8$ m, number of $SMs=32$, energized track length=2 km, delta connection $R_s=0.18$ Ω, $L_s=4.22$ mH, and airgap $z=0.3$ m.

From Figure 8.7 (which uses the same data), we may calculate E_1, E_3(E_5, I_5 are neglected):

$$E_1 = 32\times 58.21\sqrt{2} = 1316 \text{ V}$$
$$E_3 = 32\times 6.65\sqrt{2} = 150 \text{ V} \tag{8.30}$$

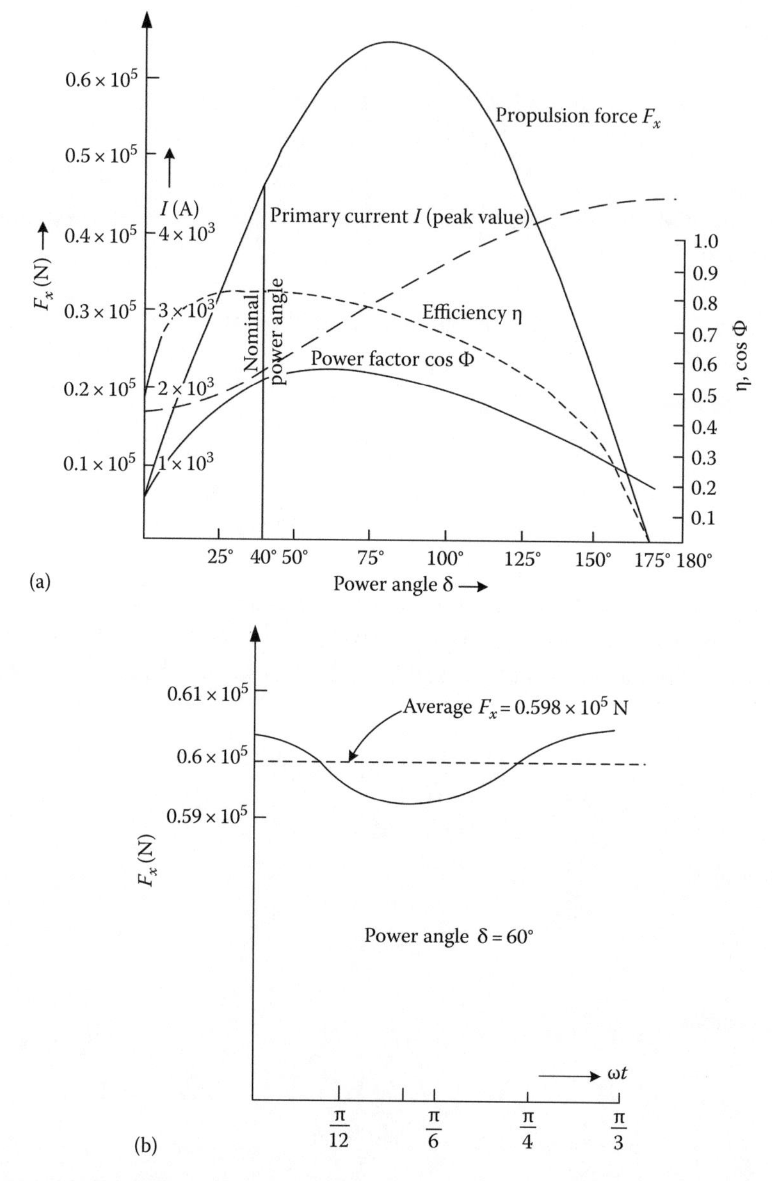

FIGURE 8.10 SM-LSM performance (a) and thrust pulsations (b). (After Nasar, S.A. and Boldea, I., *Linear Motion Electric Machines*, John Wiley & Sons, New York, 1976.)

From (8.19) to (8.29), the results in Figure 8.10 are obtained.

For the rated power angle $\delta_v = 40°$, $F_{xav} = 0.469 \times 10^5$ N, $\eta_n = 0.82$, $\cos \varphi_1 = 0.53$, and $I_n = 2241$ A (peak value).

The thrust pulsations with $\delta_v = 60°$ are shown in Figure 8.10b.

8.4 NORMAL AND LATERAL FORCES

As already stated, besides propulsion force (thrust) the SM-LSM develops also normal and lateral force. The normal force F_z acts perpendicularly to the motion direction and, in our case (of horizontal stator), it is vertical; F_z may be an attraction or a repulsive force, which is nonzero for symmetric or asymmetric placement of stator and SMs (Figure 8.11a).

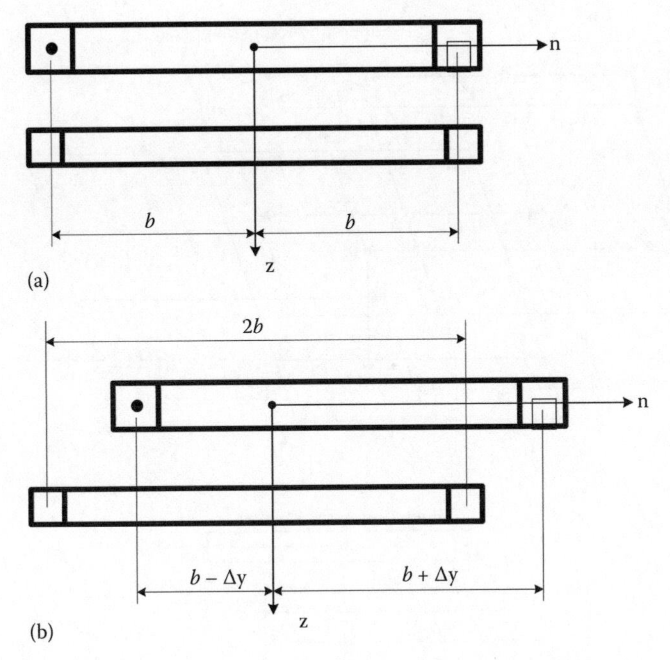

FIGURE 8.11 Symmetric (a) and asymmetric placement of SMs and track coils (b). (After Nasar, S.A. and Boldea, I., *Linear Motion Electric Machines*, John Wiley & Sons, New York, 1976.)

The lateral force F_y occurs only in case of laterally asymmetric placement of stator and SMs (Figure 8.11b); this asymmetry may be produced by wind or other perturbation forces.

In general an attraction (unstable) normal force is accompanied by a restoring (stabilizing) lateral force, and a repulsion-type normal force coexists with a decentralizing lateral force, to fulfill the Earnshaw's theorem.

It would be ideal to stabilize by the converter control both propulsion and suspension, from zero (or small) speed upward and, eventually, stabilize separately the guidance (produced by lateral forces). The figure eight-shape-coil vertical SM-LSM attempts just that. But, to reveal the modeling of normal and lateral forces, we keep the simple rectangular coil stator winding placed asymmetrically (Figure 8.11b). For one phase of the three-phase stator, the normal and lateral forces (F_z, F_y) (with N SMs on vehicle) are [6]

$$\left(F_z\right)_{phase} = NI_0 i_a \frac{\partial M_{st}}{\partial z}$$
$$\left(F_y\right)_{phase} = NI_0 i_a \frac{\partial M_{st}}{\partial y} \tag{8.31}$$

where

I = SM mmf
i_a is the phase current
M_{st} is the mutual inductance between the SM and phase a track coil

The total forces add up the contributions of all three stator phases.

As the instantaneous phase currents i_a, i_b, i_c may be calculated from (8.19) for given V_1, speed (frequency), and power angle δ_v, only the mutual inductances M_{st_i} ($i = a, b, c$) have to be calculated.

A generic SM and one-phase track coil placed asymmetrically are shown in Figure 8.12.

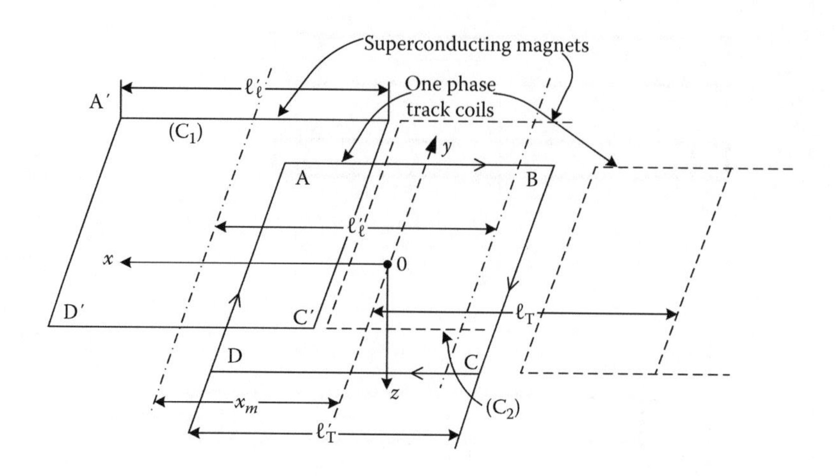

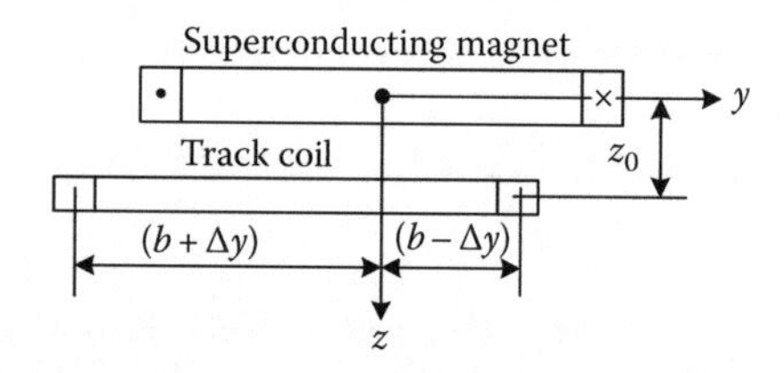

FIGURE 8.12 SM and stator coil, and asymmetric SM/stator placement. (After Nasar, S.A. and Boldea, I., *Linear Motion Electric Machines*, John Wiley & Sons, New York, 1976.)

The SM and track coil lengths are different from each other $\left(l_T' \neq l_t'\right)$, but the pole pitch is the same, and the lateral asymmetry is portrayed by $b_1(b + \Delta y) \neq b_2(b - \Delta y)$. The Neumann inductance formula writes

$$M_{st0} = \frac{\mu_0}{4\pi} \int_{c_1} \int_{c_2} \frac{dxdx' + dydy' + dzdz'}{\left[\left(x - x'\right)^2 + \left(y - y'\right)^2 + \left(z - z'\right)^2\right]^{1/2}} \tag{8.32}$$

where xyz and $x'y'z'$ are coordinates of one point on the stator coil and, respectively, on SM, with c_1, c_2 as the contours of the two coils. The contours c_1 and c_2 are defined by $A', B', C', D'\left(x_{oi}', y_{oi}', z_{oi}', i = 1,2,3,4\right)$ and by A, B, C, D (x_{oj}, y_{oj}, z_{oj}, $j = 1, 2, 3, 4$).

The integral in (8.32) leads to

$$M_{st0} = \frac{\mu_0}{4\pi} \sum_{j=1}^{4} \sum_{i=1}^{4} (-1)^{i+j} \left[\psi\left(x_{0j}, y_{oj}, 0; x_{0i}', y_{0i}', z_0\right) + \psi\left(y_{oj}, x_{0j}, 0; y_{0i}', x_{0i}', z_0\right)\right] \tag{8.33}$$

with

$$\psi\left(x, y, z, x', y', z'\right) = -\left(x - x'\right) \ln\left[\left(x - x'\right) + \sqrt{\left(x - x'\right)^2 + \left(y - y'\right)^2 + \left(z - z'\right)^2}\right] \tag{8.34}$$

There are many (n) stator coils of alternate polarity that interact with one SM, and thus, the total mutual inductance adds up all these components:

$$M_{st} = (-1)^{n+1}\frac{\mu_0}{4\pi}\sum_{n=0}^{\infty}\sum_{i=1}^{4}\sum_{j=1}^{4}(-1)^{i+j}\left[\psi\left(x_{nj}, y_{nj}, 0; x'_{0i}, y'_{0i}, z_0\right) + \psi\left(y_{nj}, x_{nj}, 0; y'_{0i}, x'_{0i}, z_0\right)\right]$$

$$+(-1)^{n}\frac{\mu_0}{4\pi}\sum_{n=0}^{\infty}\sum_{i=1}^{4}\sum_{j=1}^{4}(-1)^{i+j}\left[\psi\left(x_{-nj}, y_{-nj}, 0; x'_{0i}, y'_{0i}, z_0\right) + \psi\left(y_{-nj}, x_{-nj}, 0; y'_{0i}, x'_{0i}, z_0\right)\right] \quad (8.35)$$

Both left- and right-side stator coils around each SM have been considered in (8.35).

The coordinates in (8.35)—Figure 8.12—are

$$x'_{01} = x'_{04} = u_s t + x_m + \frac{1}{2}l'_T$$

$$x'_{02} = x'_{03} = u_s t + x_m - \frac{1}{2}l'_T$$

$$y'_{01} = y'_{02} = b + \Delta y$$

$$y'_{03} = y'_{04} = -b + \Delta y$$

$$x_{n1} = x_{n4} = \frac{1}{2}l'_t + (n-1)l_t$$

$$x_{n2} = x_{n3} = -\frac{1}{2}l'_t + (n-1)l_t$$

$$x_{-n1} = -x_{-n4} = \frac{1}{2}l'_t - nl_t$$

$$x_{-n2} = x_{-n3} = -\frac{1}{2}l'_t - nl_t$$

$$y_{n1} = y_{n2} = b = y_{-n1} = y_{-n2}$$

$$y_{n3} = y_{n4} = -b = y_{-n3} = y_{-n4} \quad (8.36)$$

where x_m is the position of the center of SM at t (time) = 0. As expected, M_{st} is a periodic function with respect to position x (or time):

$$M_{st} = \sum_{k=1}^{\infty} M_{stk}\cos\frac{k}{l_t}\left(\pi u_s t + \pi x_m\right) \quad (8.37)$$

The position x_m is related to the power angle δ_v (Figure 8.9):

$$\frac{\pi x_m}{l_t} = \frac{\pi}{2} - \delta_v; \quad \omega_1 = \frac{\pi u_s}{l_t} \quad (8.38)$$

M_{st} may be written as

$$M_{st} = \sum_{k=1}^{\infty} M_{stk}\cos k\left(\omega_1 t + \frac{\pi}{2} - \delta_v\right) \quad (8.39)$$

Let us now suppose that the phase stator current has two components:

$$i_a = i_1 + i_3 = I\sqrt{2}\cos(\omega_1 t - \varphi) - I_3\sqrt{2}\cos(3\omega_1 t - \delta_v - \varphi_3) \tag{8.40}$$

with

$$\varphi_3 = \tan^{-1}(3\omega_1 L_s/R_s). \tag{8.41}$$

From (8.31) to (8.41) and considering all three phases, the normal and lateral forces F_z and F_y are

$$F_{z,y} = 3NI_0 \left\{ \begin{aligned} & I\frac{\partial(M_{st1})}{\partial z,y}\sin(\delta_v - \varphi_1) - I_3\frac{\partial(M_{st3})}{\partial z,y}\left[\sin\varphi_3 - \sin(6\omega_1 t - 6\delta_v - \varphi_3)\right] \\ & + I\frac{\partial(M_{st5})}{\partial z,y}\sin(6\omega_1 t - 5\delta_v - \varphi_1) \end{aligned} \right\} \tag{8.42}$$

where N is the number of SMs per vehicle.

8.4.1 Numerical Example 8.2

For the same data as in Example 8.1, and no lateral asymmetry, applying the earlier formulae, the normal force F_n is calculated and shown in Figure 8.13a,b.

From Figure 8.13a, it is evident that the normal force changes from attraction to repulsion mode. The lateral force is zero as $\Delta y = 0$ (symmetric SM stator location).

Next, we consider a lateral displacement $\Delta y = 3.5$ cm. Only the lateral force is shown in Figure 8.14a,b.

At $\delta_v = 55°$, again, the lateral force switches from restoring to decentralizing mode.

Discussion

- Propulsion force is produced at reasonable efficiency and power factor for $\delta_v = 40° - 100°$; for vector control, the power angle may be controlled as needed.
- Unfortunately, in its rectangular stator-coil configuration, the normal and lateral forces are not very large for $\delta_v = 40° - 100°$, but they both pass through zero at $\delta_v = 55°$.
- As the inverter on ground may control only the current amplitude I_1 and the power angle δ_v, it follows that only two forces may be controlled at any time.
- Also, the propulsion and normal forces are of the same order of magnitudes, and thus, the latter is very far from the level needed for full suspension of the vehicle through SM-LSM. For 1 m/s^2 acceleration, the normal force has to be 10 times larger than the propulsion force, to fully suspend magnetically the vehicle.
- These remarks suggest that for fully integrated propulsion, suspension, and some guidance, the eight-shape-stator-coil lateral stators on ground configuration are required.

8.5 SM-LSM WITH EIGHT-SHAPE-STATOR COILS

The eight-shape-stator-coil configuration (Figure 8.1) was conceived to produce thrust and normal and guidance (lateral) forces in a controlled manner to be sufficient for a MAGLEV. This is a departure from the standard system, which has special short-circuited coils on track for suspension and guidance.

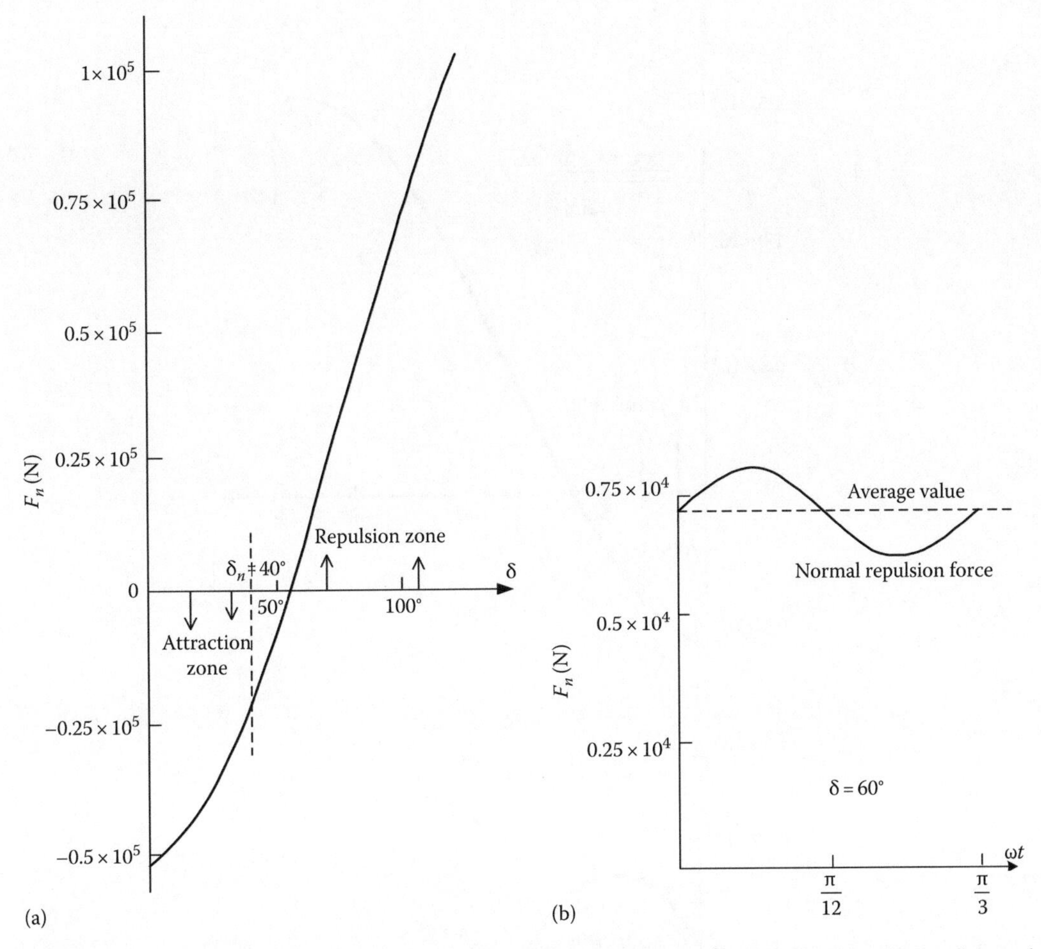

FIGURE 8.13 Average normal force, (a) and normal force pulsations at $\delta_v = 60°$ (b). (After Nasar, S.A. and Boldea, I., *Linear Motion Electric Machines*, John Wiley & Sons, New York, 1976.)

Figure 8.15a adds the generator coils onboard the vehicle, placed between the SMs and the stator coils, and the eight-shape-stator coils. SM-LSM main geometry parameters are given in Figure 8.15b.

The linear generator "collects" the power through electromagnetic induction from the stator mmf space harmonics (especially the 5th harmonic). The generator tooth-wound coils pole pitch $\tau_2 = (4/15)\tau$; τ-main pole pitch.

The eight-shape-stator coils (Figure 8.3) are connected in series, and they are concentrated coils (with $q = 0.5$): three of them side by side make a pole pitch (Figure 8.15c), $3\tau_1 = \tau$. This leads to a lower winding factor (0.867), but it is easy to build and install.

There are also connections of eight-shape-stator coils of the two sides of the vehicle (Figure 8.3) to create opposite lateral forces to "centralize" the vehicle, for damping of the eventual oscillations. These connections are called null-flux cables on Figure 8.2.

To elevate the stability between vehicle guidance and rolling motion, the eight-shape figure is made asymmetric, with the lower part taller than the upper part (Figure 8.15a), though this merit is accompanied by an increase in the take-off velocity [7].

The reverse current direction in the upper and lower parts of eight-shape coils enhances levitation (propulsion) force, while the reverse current between left- and right-side coils increases the

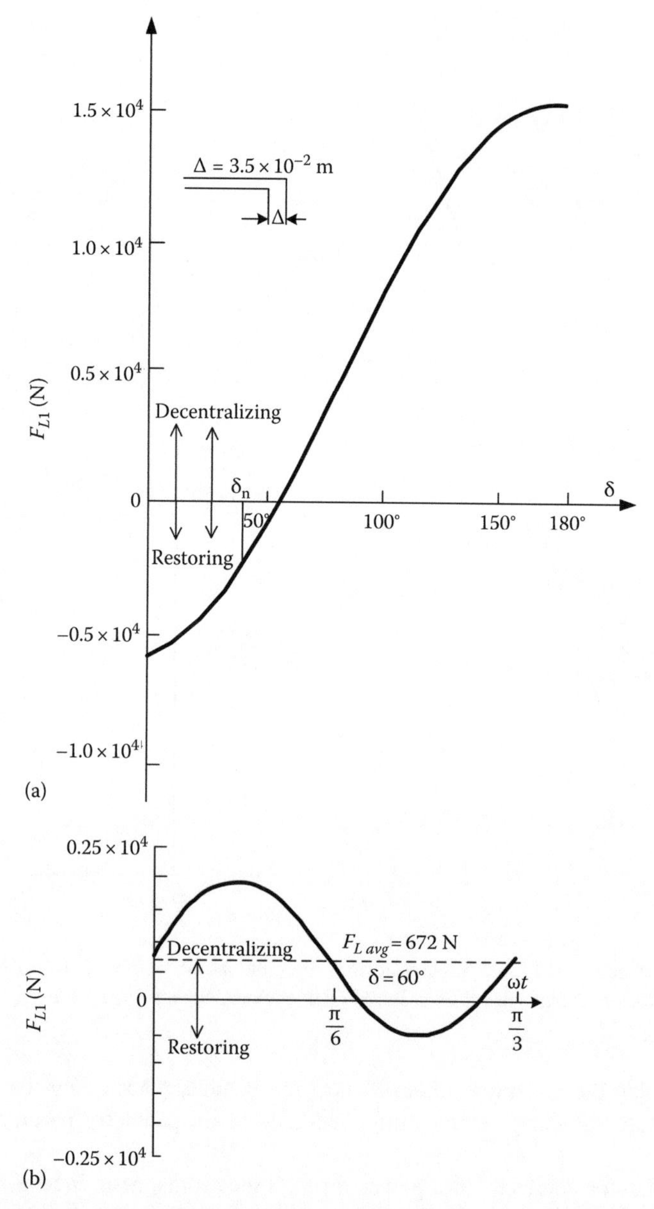

FIGURE 8.14 Average lateral force ($\Delta y = 3.5$ cm) (a) and its pulsations at $\delta_v = 60°$ (b). (After Nasar, S.A. and Boldea, I., *Linear Motion Electric Machines*, John Wiley & Sons, New York, 1976.)

lateral force. The circulating currents induced by SMs through motion are producing the bulk of levitation and lateral forces.

A design optimization attempt should observe quite a few constraints such as

- Levitation (vertical) force F_z (in this case) balances the required force (vehicle weight) at both take-off (u_1) and cruising (u_2) speed, when the thrust levels are different from each other.
- Equivalent guidance and rolling stiffness, F^*_{yy} and $M^*_{\phi\phi}$, should exceed the requirements at take-off speed u_1, where they are minimum by nature.

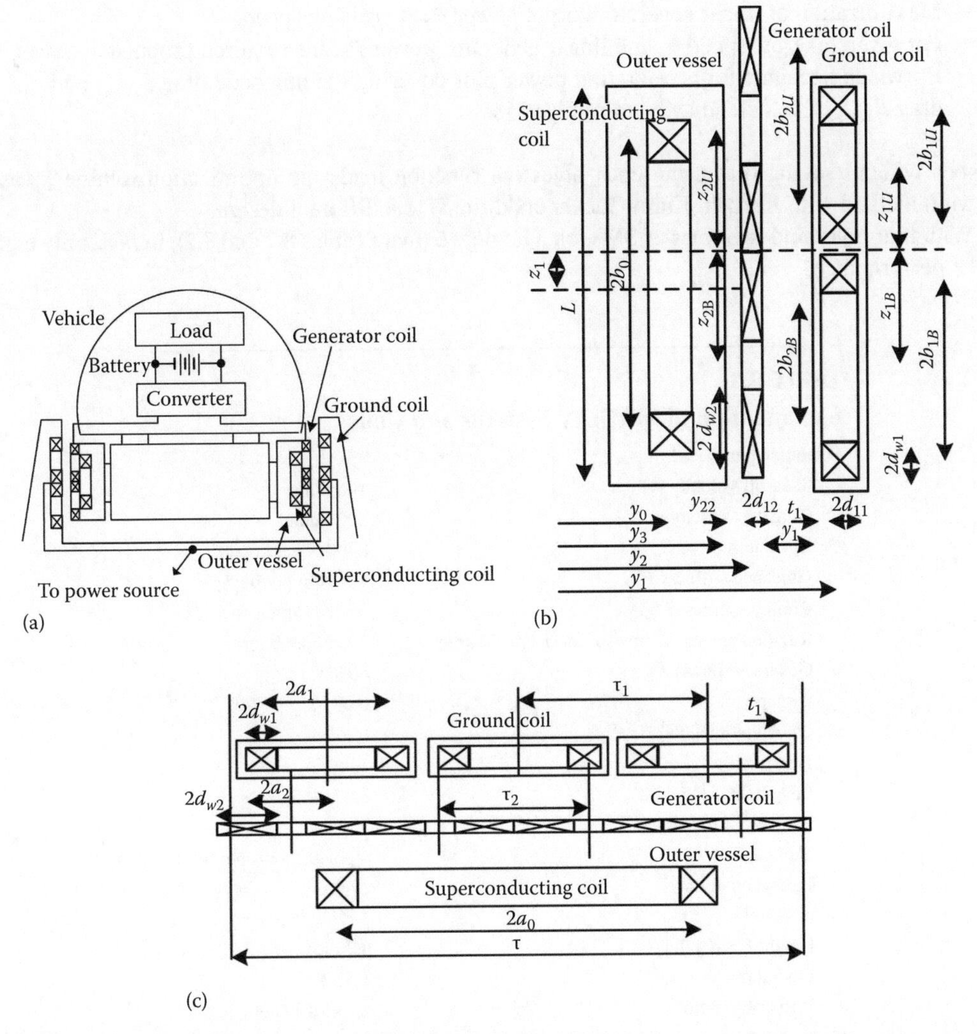

FIGURE 8.15 Integrated propulsion-levitation guidance system. (a) coil composition, (b) transverse cross section, (c) cross and longitudinal geometries. (After Toshiaki, M. and Shunsuke, F., Design of coil specifications in EDS Maglev using an optimization program, *Record of LDIA-1998*, Tokyo, Japan, pp. 343–346, 1998.)

- Propulsion force F_x balances the total drag force load F_{xload} at cruising speed u_2, where, in general, propulsion requirements are maximum (acceleration force during starting is not larger).
- Linear generator power P_g exceeds the requirements above a certain speed (it could be cruising speed u_2, but, better, a smaller value).

Besides constraints such as earlier, quite a few objective functions (or combination of them) can be put into place:

- Minimization of magnetic drag F_{mdrag} produced by the eight-shape-stator-coil circulating currents by which most levitation (suspension) is produced, at take-off and (or) at cruising speed u_2).
- Maximization of guidance stiffness F_{yy}^* at take-off speed u_1.
- Optimization of propulsion performance in terms of efficiency or power factor.

- Maximization of linear generator output power P_2 at cruising speed.
- For given take-off speed u_1 and linear generator power P_2, the required propulsion power P_1 (which accounts for accelerating power plus covering the magnetic drag F_{mdrag} and air drag F_{adrag} of vehicle) may be minimized [8].

It goes without saying that using each objective function leads the optimization routine (quasi-Newton method with Karush–Kuhn–Tucker conditions) to a different design.

With four poles and two rows of SMs, on a bogie we treat (Tables 8.1 and 8.2), in fact, only eight SMs' performance.

TABLE 8.1
Specification of MAGLEV Systems and Coils

Operation parameter	
Take-off velocity v_1	100 km/h
Cruising velocity v_2	500 km/h
Levitation force F_{zs}	230 kN per bogie
Guidance stiffness F_{yys}	1 MN/m per bogie
Rolling stiffness $M_{\Phi\Phi s}$	3 MN/m/rad per bogie
Running resistance (except for magnetic drag) F_{xa}	20 kN per bogie
Generated power P_{2s}	50 kW
Airgap y_a	0.08 m
Superconductiong coil	
Pole pitch τ	1.35 m
Length $2a_0 \times$ height $2b_0$	1.07×0.5 m
Magnetomotive force I_0	700 kA
Numbers	4 poles 2 rows per bogie
Lateral position y_0	1.49 m
Outer vessel	
Height $L \times$ depth t_3	0.7 m $\times$ 0.01 m
Lateral position y_3	1.55 m
Resistance ratio	3.1×10^{-7} Ω cm
Ground coil	
Coil pitch τ_1	0.45 m
Insulator thickness t_1	0.01 m
Cross-sectional area S_1	0.003 m^2
Resistance ratio	3.1×10^{-7} Ω cm
Space factor	0.9
Section length	12 poles per SC
Propulsion current I_1	1000 A
Conductor divided number N_p	4
Generator coil	
Pitch τ_2	0.36 m
Resistance ratio	1.9×10^{-7} Ω cm
Space factor	0.75
Numbers	30 per bogie
Efficiency	0.9

Source: Toshiaki, M. and Shunsuke, F., Design of coil specifications in EDS Maglev using an optimization program, *Record of LDIA-1998*, Tokyo, Japan, pp. 343–346, 1998.

TABLE 8.2
Maximized Specifications for Each System

Maximized Function	Levitation	Guidance	Propulsion	Generator
Ground coil				
Coil length $2a_1$ (m)	0.356	0.370	0.397	0.349
Upper coil height $2b_{1U}$ (m)	0.330	0.221	0.256	0.227
Lower coil height $2b_{1B}$ (m)	0.415	0.284	0.333	0.356
Cross-sectional ratio $2d_{w1}/2d_{d1}$ (m)	1.71	1.13	0.33	2.09
Lateral position from SC $y_1 - y_0$ (m)	0.191	0.196	0.218	0.194
Generator coil				
Coil length $2a_2$ (m)	—	—	—	0.229
Upper coil height $2b_{2U}$ (m)	—	—	—	0.352
Lower coil height $2b_{2B}$ (m)	—	—	—	0.262
Cross-sectional width $2d_{w2}$ (m)	—	—	—	0.121
Cross-sectional depth $2d_{d2}$ (m)	—	—	—	0.01
Lateral position from outer vessel y_{23} (m)	—	—	—	0.014
Performance				
Magnetic drag by loop loss F_{x1} (kN)	2.1	3.7	3.7	2.6
Guidance stiffness F^*_{yy} (MN/m)	0.4	2.0	1.4	1.0
Rolling stiffness $M^*_{\Phi\Phi}$ (MN/m)	2.0	1.8	1.3	3.0
Generated power P_2 (kW)	—	—	—	60.5

System	PLG System		PLG and Linear Generator System
Magnetic Drag by Eddy Current	**Not Considered**	**Considered**	**Considered**
Ground coil			
Coil length $2a_1$ (m)	0.380	0.395	0.366
Upper coil height $2b_{1U}$ (m)	0.294	0.278	0.265
Lower coil height $2b_{1B}$ (m)	0.365	0.421	0.372
Cross-sectional ratio $2d_{w1}/2d_{d1}$ (m)	0.78	0.36	1.26
Conductor section ratio width/depth	—	0.55	0.56
Lateral position from SC $y_1 - y_0$ (m)	0.201	0.216	0.199
Generator coil			
Coil length $2a_2$ (m)	—	—	0.223
Upper coil height $2b_{2U}$ (m)	—	—	0.365
Lower coil height $2b_{2B}$ (m)	—	—	0.260
Cross-sectional width $2d_{w2}$ (m)	—	—	0.127
Cross-sectional depth $2d_{d2}$ (m)	—	—	0.01
Lateral position from outer vessel y_{23} (m)	—	—	0.013
Performance			
Magnetic drag by loop loss F_{x1} (kN)	2.4	2.7	2.4
Magnetic drag by eddy current F_{xe} (kN)	—	4.1	4.5
Guidance stiffness F^*_{yy} (MN/m)	1.2	1.1	1.0
Rolling stiffness $M^*_{\Phi\Phi}$ (MN/m)	3.0	3.0	3.0
Power capacity P_s (kW)	—	4.7	4.9
Generated power P_2 (kW)	—	—	50

Source: Toshiaki, M. and Shunsuke, F., Design of coil specifications in EDS Maglev using an optimization program, *Record of LDIA-1998*, Tokyo, Japan, pp. 343–346, 1998.

Table 8.1 [7] shows typical specifications for one 230 kN bogie, 500 km/h SM-LSM MAGLEV. Table 8.2 shows typical objective function specification design results [7].

Table 8.2 shows clearly that four different designs are obtained based on the four main objective functions, and thus, a global cost function has to be added to discriminate between them. Combining 1–2 objective functions would lead to yet other designs.

8.6 CONTROL OF SM-LSM

Combined propulsion and levitation control has to be attempted from start (especially for eight-shape-coil stators). Both flux-oriented control (FOC) and direct thrust and flux control (DTFC) may be adopted for this air-core machine with constant dc mmf excitation but variable average airgap z (Figure 8.16).

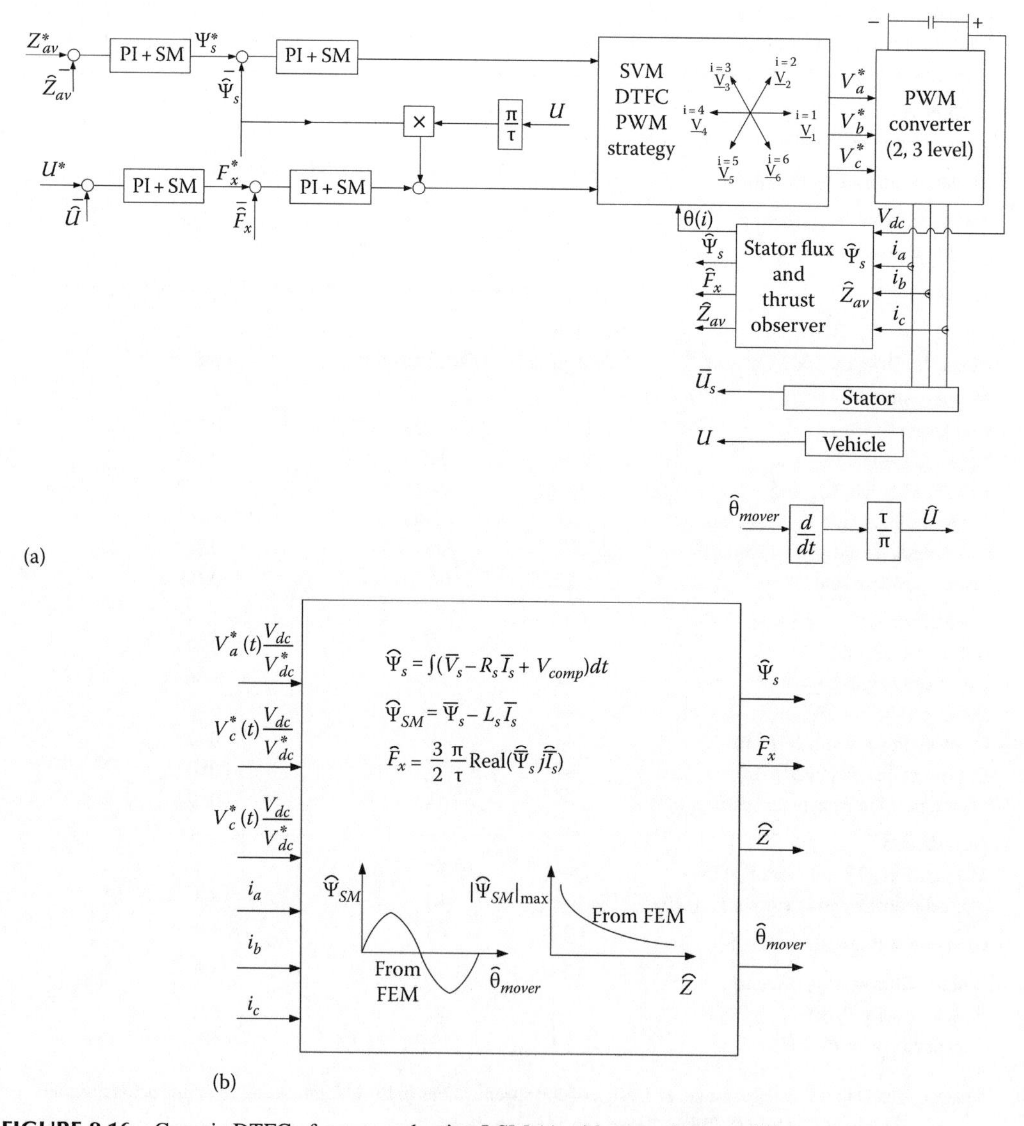

FIGURE 8.16 Generic DTFC of superconducting LSM (a) with state observer (b).

Today's power electronics is better prepared for FOC or DTFC, even for powers of 5–10 MW, at 6 kV and $f \approx U_{max}/2\tau = 130/(2 \cdot 1) = 65$ Hz. Even pole pitches τ smaller than 1 m are feasible for a lateral distance of 0.2–0.3 m between SMs and stator (the mechanical gap is 5–10 cm).

Note: We should mention again that the bulk of the repulsion-type (statically stable) levitation and guidance is produced by the circulating currents induced by the SMs (by motion) in the stator eight-shape coils at "the cost" of a magnetic drag force F_{mdrag}. These aspects will be treated in a later chapter dedicated to magnetic suspension systems.

A generic DTFC for SM-LSM is shown in Figure 8.16, where the thrust is controlled directly for propulsion and the SM flux channel is used to dynamically control levitation. This means implicit control of stator currents I_1 and power angle δ_v.

DTFC does not need SM position feedback for control directly, other than for the flux observer at low speeds; the speed u is anyway measured and used to compensate emf and provide for cruising control.

A peculiarity of SM-LSM is that the stator inductance L_s and resistance are rather constant (rather independent of airgap z): the temperature influence on R_s may be needed during vehicle starting, for more robust control.

A special control method for starting may be needed for a motion-sensorless control.

The compensator voltage V_{comp} in Figure 8.16b may be based on a current model of the machine:

$$V_{comp} = P_i\left[\overline{\Psi}_s - \overline{\Psi}_{si}\right] \tag{8.43}$$

The current model based observer $\overline{\Psi}_{si}$ used back and forth in stator to mover to stator coordinates is shown in Figure 8.17. Both $\hat{\theta}_{mover}$ and $\overline{Z}$ estimation are obtained, as needed for motion-sensorless control.

Safety precautions impose multiple airgap sensors on board vehicle in key points; their output has to be transmitted wirelessly to the ground station for control. Especially for protection, as all controls are provided on ground, the DTFC system is basically sensorless and it works as such but, in practice, it may be used merely for redundancy.

Quenching of SMs (eight in all on a bogie in our case study) would severely disturb the vehicle stable ride, and thus, a symmetrically placed additional SM is quenched intentionally [9]. But the faulty bogie has still to be capable to handle the entire weight for suspension although at a smaller (but not dangerous) airgap z_{min}. The SMs experience induced eddy current losses due to both the ac fields of stator mmf and of linear generator mmf [10]. Other specific phenomena pertaining to SM-LSM such as active track inverter supply section by section in a synchronized manner will be dealt with in a later chapter dedicated to active guideway MAGLEVs.

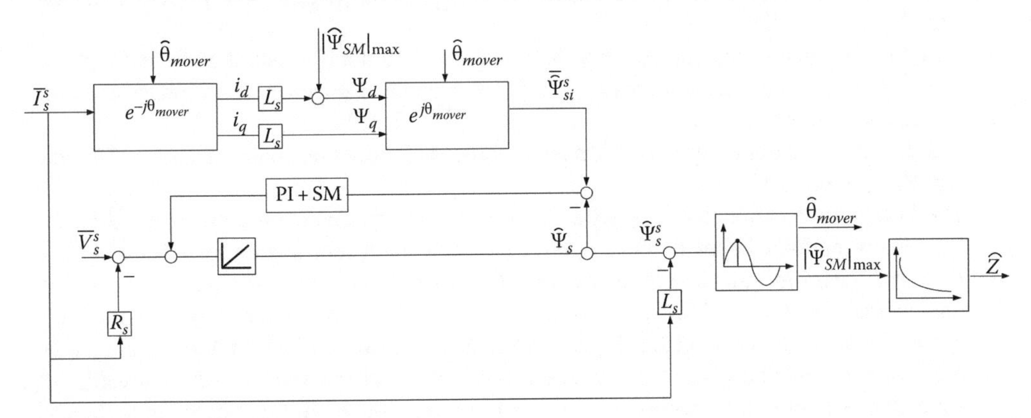

FIGURE 8.17 Flux $\hat{\Psi}_s^s$ and airgap $\hat{Z}$ observer.

8.7 SUMMARY

- SMs are air-core coils enclosed in dewars to keep them at low temperatures (10 K or less respectively 77 K or less) such that the conductor resistivity drops 10^5 times with respect to normal temperatures; so once the SM is charged with a dc current the latter stays there for hours and days and may be slowly refilled, such that to avoid notable eddy current losses that may quench (turn from superconductor to normal conductor mode) the SM [11].
- It is argued that for large flat coil sizes the SM is economically feasible and is even better than the large permanent magnet (for same weight or cost), in terms of electromagnetic propulsion, levitation, and guidance at very high speeds and large effective gaps between vehicle and guideway (5–10 cm).
- A row of alternate polarity SMs on the vehicle behaves like a constant field current excitation (or PM) for the air-core stator three-phase winding placed along the track (travel) length and ac fed through inverters on ground in sections of a few kilometers.
- Horizontal placement of SMs below the floor of the vehicle with flat rectangular three-phase cable windings with stator (active track) is natural, and this configuration was first used to investigate thoroughly the SM-LSM modeling and performance via a technical field and circuit theory.
- The technical theory starts with the calculation of the magnetic flux-density components B_x, B_y, B_z in a point in air by an air-core rectangular SM of given dc mmf I_0 (Ampere turns); fortunately there is an analytical approach that saves time, when a computer code is used.
- For adequate combinations of airgap z, pole pitch of SM spreading $\tau(l_T)$ per coil length and coil width/pole pitch, the emf E_1, produced by motion by the SM in a track (stator) coil, shows close to a sinusoidal waveform in time for given vehicle speed.
- The active section air-core stator windings cyclic inductance per phase (L_s) is rather independent of SM-LSM airgap and is calculated as for long power transmission lines; the active section (phase) resistance R_s expression is also straightforward.
- With E_1, R_s, L_s of SM-LSM, steady-state (or transient) equations, as a nonsalient-pole synchronous machine, are straightforward; the accompanying phasor diagram serves to calculate the phase current $\underline{I}_1$ (amplitude and phase angle) for given speed u, stator phase voltage V_1, and power angle δ_v.
- Through a numerical example with $V_1 = 4200/\sqrt{2}$ RMS value phase, $u=130$ m/s, $I_0=3$, 5×10^5 A turns, $2l=1$ m, $2b=0.8$ m, 32 SMs per vehicle, $z=0.3$ m (theoretical airgap), $f_1=50$ at $\delta_v=40°$, $F_{xav}=0$, 47×10^5 N, $\eta=0.82$, cos $\varphi_1=0.53$, which could be considered acceptably good performance.
- To investigate normal force F_z, and lateral force F_y of an SM-LSM, the vehicle is placed laterally asymmetric.
- Based on Neumann inductance formula applied for the double row of SMs and respectively, stator coils, all three forces F_x, F_y, F_z are calculated for given stator-phase current, speed, and power angle δ_v.
- It is shown that in our case study, the normal force is of attraction character for $\delta_v<55°$ and repulsive for $\delta_v > 55°$.
- The lateral force is nonzero only when lateral asymmetry (Δy) is present; it also changes from restoring force below $\delta_v=55°$ (in our case study) to decentralizing force for $\delta_v > 55°$.
- However, the normal and lateral forces are of the order of magnitude of propulsion force for the rectangular stator-coil SM-LSM investigated so far; so they are far away from the high values (10 times higher) needed for full suspension. Separate short-circuited coils along the guideway may be placed to produce full levitation of the vehicle. Alternatively, the eight-shape-stator-coil (null-flux) winding is used to produce all three forces in sufficient amplitudes and fully controlled.

- The eight-shape-stator coils of stator interact three ways with the SMs row in vehicle:
  - The motion-induced circulating currents in the eight-shape coils interact with the SMs and produce a levitation force F_n (besides a magnetic drag force F_{mdrag}).
  - The same motion-induced circulating currents, which also circulate laterally from left- to right-side coils produce the lateral force F_y that restores the lateral motion equilibrium.
  - Finally, the ac stator currents interact with the SM row field to produce three forces F_x, F_y, F_z; the first one F_x propulses the vehicle; the other two F_y, F_z related to one another and of similar magnitude serve to control (damp) various vehicle oscillatory parasitic motions.
- Magnetic levitation performance will be discussed in a dedicated later chapter.
- The SM-LSM control system may control only the thrust and, say, the stator flux to control suspension/levitation via average airgap z control.
- A generic DTFC system is introduced and detailed in the situation of any motion-sensor absence; an actual vehicle has quite a few airgap sensors (at least for safety), and thus, sensorless control may be used only for redundancy.
- Design optimization methods [7] for eight-shape-stator-coil SM-LSM have been proven capable to lead to good performance, and the full-scale vehicle at Yamanashi track in Japan seems to validate them satisfactorily.
- Very similar in behavior to SM-LSM, the air-core PM active guideway LSM has been proposed for urban MAGLEVs first in Germany and then recently in USA; so the air-core SM-LSM or PM-LSM with active guideway is open for new developments.

REFERENCES

1. S.A. Nasar and I. Boldea, *Linear Motion Electric Machines*, Chapter 5, John Wiley & Sons, New York, 1976.
2. M.M. Kolm, R.D. Thornton, Y. Kwasa, and W.S. Brown, The magneplane system, *Cryogenics*, 15(7), July 1975, 377–383.
3. I. Sakamoto, Vertical electromagnetic force of a superconducting LSM vehicle based on the formulation in dq axes, *Record of LDIA-2001*, Nagano, Japan, pp. 149–153.
4. J.R. Schreiffer and J.S. Brooks (eds.), *Handbook of High Temperature Superconductivity: Theory end Experiment*, Springer Verlag, New York, 2007.
5. I. Boldea and S.A. Nasar, *Linear Motion Electromagnetic Systems*, Chapter 5, Wiley & Sons, New York, 1985.
6. E. Ohno, M. Iwamoto, and T. Yamada, Characteristics of superconductive suspension and propulsion for high speed trains, *Proc. IEEE*, 61(5), 1973, pp. 579–586.
7. M. Toshiaki and F. Shunsuke, Design of coil specifications in EDS Maglev using an optimization program, *Record of LDIA-1998*, Tokyo, Japan, pp. 343–346.
8. T. Murai and S. Fujiwara Characteristics of combined propulsion levitation and guidance system with asymmetric figure between upper and lower coils in EDS, *Record of LDIA-1995*, Nagasaki, Japan, pp. 37–46.
9. O. Shunsuke, O. Hiroyuki, and E. Masada, Stabilizing the motion of the superconducting magnetically levitated train against coil quenching, *Record of LDIA-1998*, Tokyo, Japan, pp. 339–342.
10. H. Hasegawa, T. Murai, and T. Sasakawa, Electromagnetic analysis of eddy current loss in superconducting magnet for MAGLEV, *Record of LDIA-2001*, Nagano, Japan, pp. 315–322.
11. F.C. Moon, *Superconducting Levitation*, John Wiley & Sons, New York, 1997.

9 Homopolar Linear Synchronous Motors (H-LSM)

Modeling, Design, and Control

Homopolar linear synchronous motors (H-LSM) are characterized, in general, by a passive long (fixed) variable reluctance (segmented) ferromagnetic secondary and a short primary on board of the vehicle (mover) that contains an iron core, which hosts both dc coils as long as the core and a three-phase ac winding [1–15].

The dc coil(s) produces a homopolar, pulsating, airgap magnetic field, which is used for controlled electromagnetic suspension of the vehicle; its $2p$ pole fundamental ac component interacts with the $2p$ pole three-phase ac winding to produce propulsion (braking) force. Thus, H-LSM potentially provides integrated suspension and propulsion functions for the vehicle (MAGLEV) with a passive iron low-cost, variable-reluctance-solid-ferromagnetic-segmented track. The 10/1 or more ratio between levitation (attraction) and propulsive force can "cover" entirely the vehicle suspension for an acceleration of about 1 m/s^2, if the H-LSM primary weighs up to 10% of vehicle weight.

The recent progress in forced cooling and the IGBT (IGCT) inverter technologies lead to a low enough propulsion (levitation) equipment weight that allows enough room (weight) for a competitive payload. Besides transportation, H-LSM has been proposed for levitated transport in clean industrial rooms. The combined controlled propulsion-levitation properties of H-LSM are obtained at rather good overall efficiency (about 80% in low-speed industrial and urban transportation and about 85% for interurban (100 m/s or more) MAGLEVs). The power factor, above 0.8 in general, is very favorable for keeping the PWM converter on-board size (weight) and cost within reasonable bounds.

Though thoroughly investigated in the 1970s and 1980s up to a 4 ton experimental prototype (MAGNIBUS 01 [16]), the H-LSM has not been applied commercially yet for urban and interurban transport in spite of its high potential performance, as only two active guideway MAGLEV projects (in Germany and Japan) survived through the last decades among global economics ups and downs.

9.1 H-LSM: CONSTRUCTION AND PRINCIPLE ISSUES

A generic layout of H-LSM (Figure 9.1) emphasizes the following:

- The short primary core may be made of longitudinal laminations separated by a solid body core (Figure 9.1a) or of transverse (U shape) lamination packs (Figure 9.1b), separated by insulation distancers to (from) the primary slots and teeth.
- The dc winding may be built into one or two long coils that embrace the entire core length longitudinally.
- The H-LSM is essentially a transverse field machine.
- The passive, notched, ferromagnetic secondary, made of solid iron, exhibits magnetic anisotropy along the longitudinal (motion) direction; alternatively, it may be made of solid segments separated by air with 2τ periodicity (Figure 9.1c).
- The τ –pole pitch ac winding may be made of two parts, each embracing one leg of the U core (Figure 9.2a) or may be made in one winding with eight-shape coils or straight coils (Figure 9.2b,c).

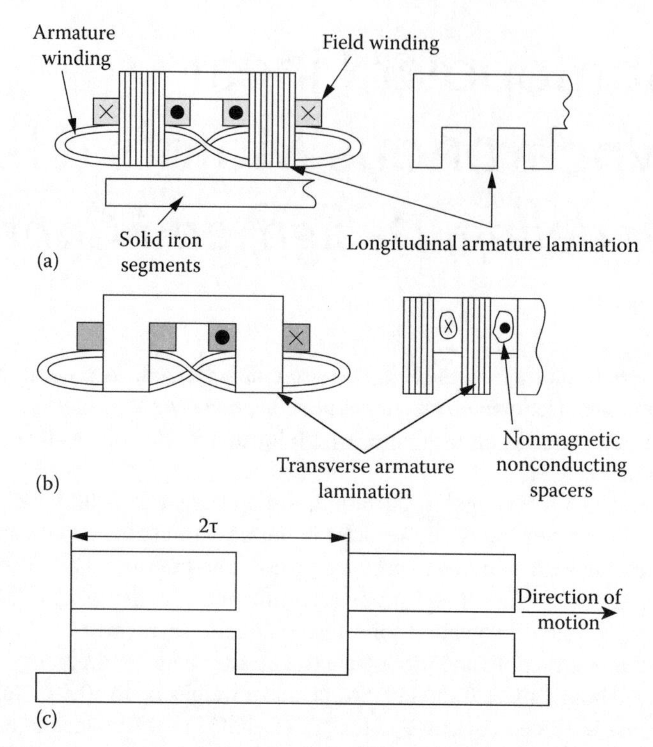

FIGURE 9.1 H-LSM layout (a) with longitudinal laminations primary core, (b) with transverse laminations, and (c) segmented and standard secondary solid care. (After Boldea, I. and Nasar, S.A., *Linear Motion Electromagnetic Systems*, John Wiley & Sons, New York, 1985.)

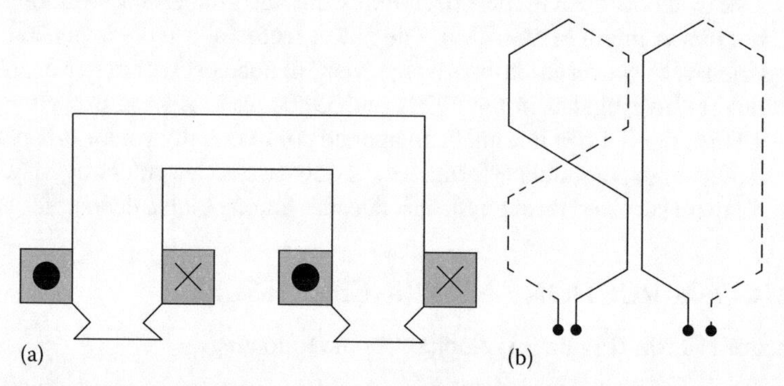

FIGURE 9.2 Armature coils: (a) segmented windings, (b) 8-shape (twisted) coils (c) straight coils. (After Boldea, I. and Nasar, S.A., *Linear Motion Electromagnetic Systems*, John Wiley & Sons, New York, 1985.)

- The eight-shape coils or the separate ac winding (Figure 9.2a) corresponds either to continuous solid, mild-steel-shifted (by τ) notched secondary or to straight secondary segments (Figure 9.1c). A typical winding is shown in Figure 9.3.
- The separate windings may be executed with standard 1(2) layer chorded coils (to reduce end connections) or with three layers (only two of them are filled out) and $y/\tau = 2/3$, when a further reduction of end connections is realized (Figure 9.4), if $2a/\tau > 0.6–0.7$.
- As the three phases occupy different positions in the slots with respect to stator core, and their leakage inductances differ from each other, to balance the currents, the two windings have phases A and C in different positions in slots.
- Only two out of three layers (in Figure 9.4) are occupied, but the free one may be used to blow air through, or a liquid, for better cooling.

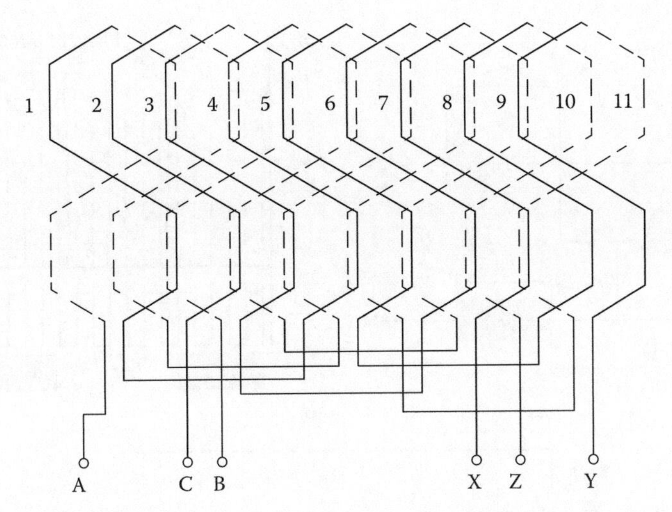

FIGURE 9.3 Typical $q=1$ two-layer ac winding with 2/3 τ coil span with 8-shape coils ($2p+1=5$-half-filled end poles as in LIMs). (After Boldea, I. and Nasar, S.A., *Linear Motion Electromagnetic Systems*, John Wiley & Sons, New York, 1985.)

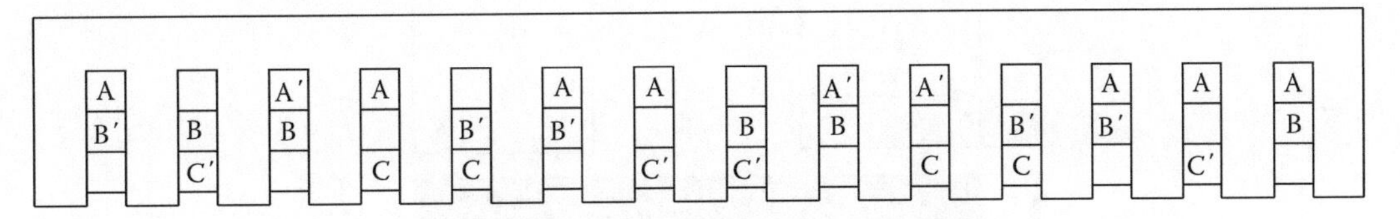

FIGURE 9.4 Three-layer ac winding (with short end connections) $y/\tau=2/3$, $q=1$, $2p+1=5$).

9.2 DC HOMOPOLAR EXCITATION AIRGAP FLUX DENSITY AND AC EMF E_1

As the dc coils embrace the total H-LSM length, their airgap flux density pulsates with a period of 2τ but keeps the same polarity: opposite though below the two U core legs (Figure 9.5).

The secondary segment span (b_{ps}) varies around the pole pitch τ value; if larger levitation force is needed, $\tau_p=(1\div 1.2)\cdot\tau$, while when larger thrust is needed, $\tau_p=(0.7\div 0.8)\cdot\tau$; in the latter case, the normal (levitation) force is smaller. In general, to compromise between levitation and propulsion, $\tau_{ps}\approx 1.0\times\tau - g$, where $g=10\div 12$ mm for urban (interurban) transportation and may go down to $1.0\div 2.0$ mm for industrial applications with a few meter long travels.

The calculation of the maximum airgap flux density, B_{0g}, is rather straightforward:

$$B_{0g}=\frac{\mu_0\cdot W_F\cdot I_F}{2g_1\cdot k_C\left(1+k_S\right)} \tag{9.1}$$

where

$W_F I_F$ is the field current mmf
g_1 is the airgap
k_C is the Carter coefficient:

$$k_C\approx\frac{1}{1-\gamma\cdot\dfrac{b_{slot}}{\tau_{slot}}};\quad \gamma\approx\frac{\left(\dfrac{b_{slot}}{g_1}\right)^2}{5+\dfrac{b_{slot}}{g_1}} \tag{9.2}$$

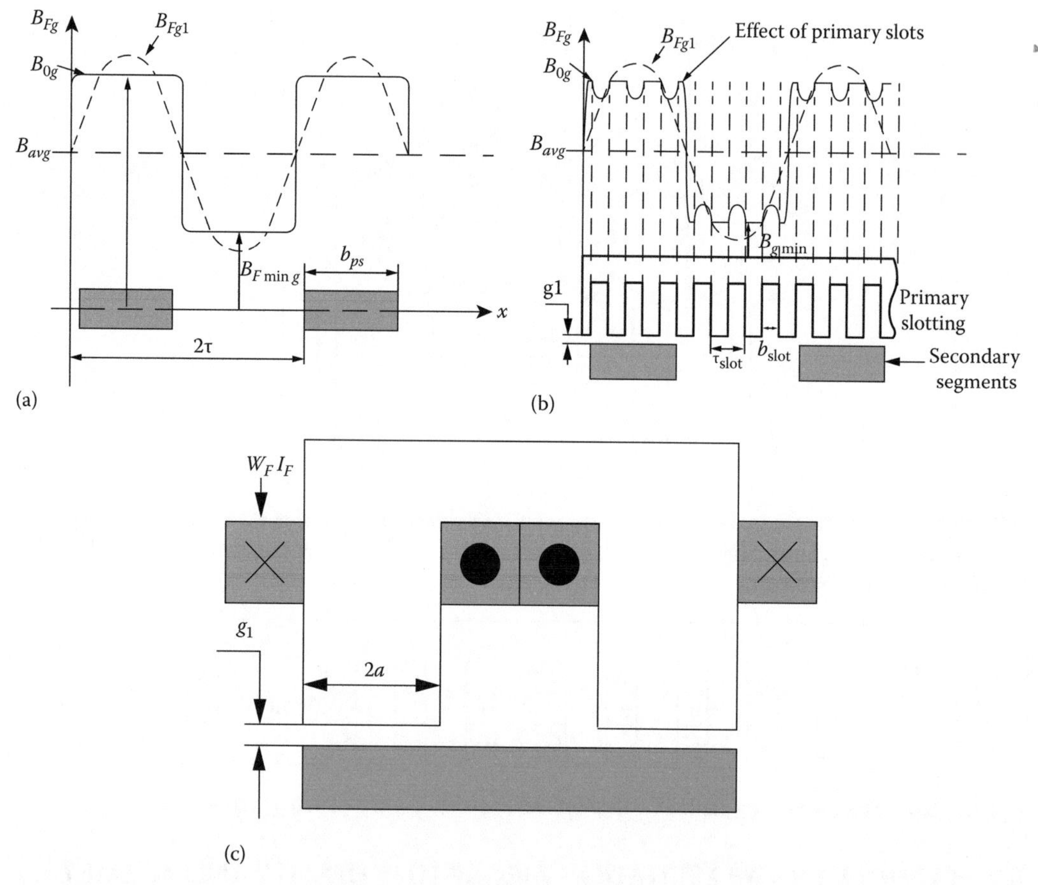

FIGURE 9.5 DC homopolar excitation airgap flux density: (a) ideal, (b) with primary slotting accounted for, and (c) transverse view.

k_S is the magnetic saturation coefficient; due to large airgap, even for heavy magnetic saturation, the value of k_S is expected to remain below 0.25. To obtain high normal and propulsion force densities, the value of $B_{0g} \approx 1.3 \div 1.5$ T. With an armature reaction flux density in the airgap of maximum 0.35 T, standard silicon *U*-shape transformer lamination cores may be used.

The minimum flux density in the airgap may be calculated using magnetic circuit method or FEM, but in general for $g_1 = 10$ mm, $B_{gmin} \approx (0.2 \div 0.24) B_{0g}$.

In a simpler calculation, a larger equivalent fictitious airgap g_2 may be used in between the track poles: $g_2 \approx 4 \cdot g$, if $\tau/g > 10$.

So the minimum airgap flux density $B_{Fmin\,g}$ is

$$B_{F\,min\,g} \approx \frac{g_1}{g_2} \cdot B_{0g} \cdot \frac{1}{1 + k_{Fringe}} \tag{9.3}$$

Using FEM, the more precise values of B_{0g} and $B_{Fmin\,g}$ may be found, and k_{Fringe} in (9.3) may be determined as a function of dc excitation mmf $W_F \cdot I_F$ for given machine geometry and airgap.

The airgap flux density fundamental B_{Fg1} is

$$B_{Fg1} \approx \frac{2}{\pi} \cdot \left(B_{ag} - B_{F\,min\,g}\right) \cdot \sin\left(\frac{\pi}{2} \cdot \frac{\tau_p}{\tau}\right) \tag{9.4}$$

In general, $\tau_p \approx \tau$, as discussed before, for MAGLEVs; again, if only propulsion is needed, to minimize normal force, $\tau_p/\tau \approx 0.67 \div 0.7$.

The emf per both ac winding legs is

$$E_1 = \pi \cdot \sqrt{2} \cdot f_{1r} \cdot \phi_{p1} \cdot k_{W_1} \cdot (2W_1); \quad \phi_{p1} = \frac{2}{\pi} B_{g1} \cdot \tau \cdot 2a \tag{9.5}$$

where

E_1 is the phase RMS value of emf
W_1 is the turns per phase/primary leg
ϕ_{p1} is the ac flux/pole (peak value)

In (9.5), the two sides are connected in series (as for eight-shape-coil system); they may be connected in parallel also, to balance the normal force against track irregularities above the two U core legs alternatively; the two ac windings may be controlled by two 50% rating PWM inverters for more redundancy.

9.3 ARMATURE REACTION AND MAGNETIZATION SYNCHRONOUS INDUCTANCES L_{dm} AND L_{qm}

The three-phase ac winding produces, in turn, an armature reaction magnetic field in the airgap. Let us consider the ac traveling mmf $J_1(x, t)$ as

$$J_1(x,t) = J_{1m} \cdot \sin\left(\frac{\pi}{\tau} x - \omega_1 t + \delta_i\right) \tag{9.6}$$

where

$$J_{1m} = \frac{3\sqrt{2} \cdot k_{W_1} \cdot W_1 \cdot I_1}{\pi \cdot p} \tag{9.7}$$

and δ_i is the current power angle.

When in axis d, the armature reaction field "sees" the small airgap g_1 for the pole span and the large airgap g_2 in the interpole airgap zone; with transverse laminations, the magnetic field paths flow in the transverse plane and $g_2 \approx \tau/\pi$. Pure q axis current control is adequate for integrated suspension and propulsion control for decoupling the two functions; this situation is shown in Figure 9.6.

Based on the magnetic energy in the airgap W_{md} concept,

$$\left(W_{md}\right)_{\delta_i=0} = 3X_{dm} \cdot I_d^2; \quad \left(W_{mq}\right)_{\delta_i=\frac{\pi}{2}} = 3X_{qm} \cdot I_q^2 \tag{9.8}$$

with $\alpha = \pi \cdot \tau_p/\tau$, $k = g_2/g_1$:

$$X_{dm} \approx \frac{2}{\pi}\left[\left(\frac{\alpha}{2} + \frac{1}{4}\sin 2\alpha\right) \cdot \left(1 - \frac{1}{k}\right) + \frac{\pi}{2k}\right] X_m \tag{9.9}$$

$$X_{qm} \approx \frac{2}{\pi}\left[\left(\frac{\alpha}{2} - \frac{1}{4}\sin 2\alpha\right) \cdot \left(1 - \frac{1}{k}\right) + \frac{\pi}{2k}\right] X_m \tag{9.10}$$

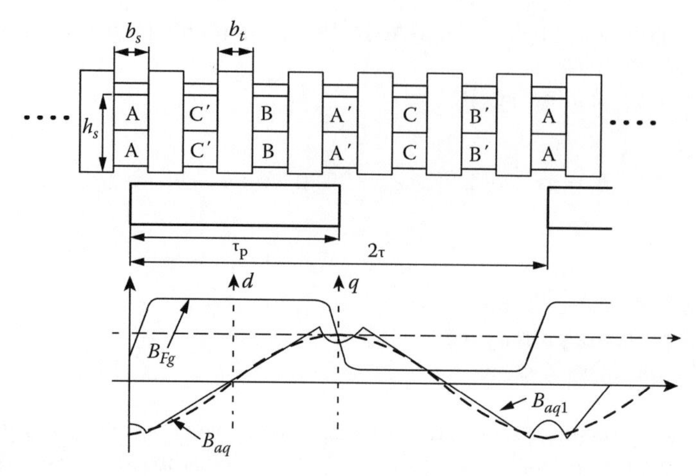

FIGURE 9.6 q axis reaction flux density.

where X_m is the magnetization inductance of the ac winding for smooth airgap ($2g_1 \cdot k_C$) and including both U-shape leg ac coils in series ($2W_1$):

$$X_m = \frac{\omega_1 \cdot \mu_0 6\,(2W_1 \cdot k_{W_1})^2 \cdot \tau \cdot L}{2g_1 \cdot k_C \cdot (1+k_S) \cdot p} \tag{9.11}$$

Again

- $2p$ is the poles of ac winding
- k_C is the Carter coefficient
- $L=2a$ is the width of one U core leg

The ac winding leakage reactance L_{ls} includes mainly the slot leakage and the end-connection leakage components. For "open slots" at both ends, the double-slot width ($2b_S$) is considered in the slot leakage inductance:

$$L_{ls} \approx 2\mu_0 \frac{(2W_1)^2}{p_1 \cdot q_1}(\lambda_s \cdot L + \lambda_c \cdot l_{lc}); \quad X_{ls} = \omega_1 \cdot L_{ls} \tag{9.12}$$

$$\lambda_s \approx \frac{h_s}{6 \cdot b_s}; \quad \lambda_c \approx 0.66 \cdot q_1; \quad l_{lc} \approx 2 \cdot \tau \left(\frac{2}{3}\right) \tag{9.13}$$

q_1 is the slots per pole per phase.

To complete the main parameter expressions, we add the ac winding resistance:

$$R_s = \rho_{co} \cdot \frac{(2W_1) \cdot j_{co} \cdot W_1 \cdot l_c}{(I_{1n} \cdot W_1)}; \quad l_c = 2L + 0.02 + l_{lc} \tag{9.14}$$

For the eight-shape coil in (9.14), $2W_1$ becomes W_1, and l_c becomes $2l_c$, where j_{co} is the rated (design) current density. The no-load steady-state voltage equation/phase is thus, as in a rotary synchronous machine,

$$\underline{V}_S = R_S \underline{I}_S + jX_d \underline{I}_d + jX_q \underline{I}_q - \underline{E}_1 \tag{9.15}$$

$$\begin{aligned} \underline{I}_S &= \underline{I}_d + \underline{I}_q; \quad \underline{E}_1 = -jX_{ma}\underline{I}_F \\ X_d &= X_{ls} + X_{dm}; \quad X_q = X_{ls} + X_{qm} \end{aligned} \tag{9.16}$$

From (9.5), (9.1), and (9.3) and (9.4), the field armature coupling reactance X_{ma} is

$$X_{ma} \approx \mu_0 W_1 \sqrt{2} k_{W_1} (2W_1) W_F \cdot \frac{\tau L}{k_C \cdot \pi^2} \left(\frac{1}{g_1} - \frac{1}{g_2} \right) \frac{1}{(1 + k_S)} \sin \frac{\pi}{2} \frac{\tau_p}{\tau} \tag{9.17}$$

Let us consider the general case of $\tau_p/\tau \neq 1$ ($X_{dm} \neq X_{qm}$ or $X_d \neq X_q$) and $I_d \neq 0$ in the phasor diagram of Figure 9.7.

The electromagnetic power P_{elm} is, as in any SM,

$$P_{elm} = 3 \cdot E_1 \cdot I_q + 3 \cdot \omega_1 \cdot (L_d - L_q) \cdot I_d \cdot I_q = F_x \cdot 2\tau \cdot f_1 \tag{9.18}$$

Similarly the reactive power

$$Q_1 = 3 \cdot E_1 \cdot I_d + 3 \cdot \left(X_d \cdot I_d^2 + X_q \cdot I_q^2 \right) \tag{9.19}$$

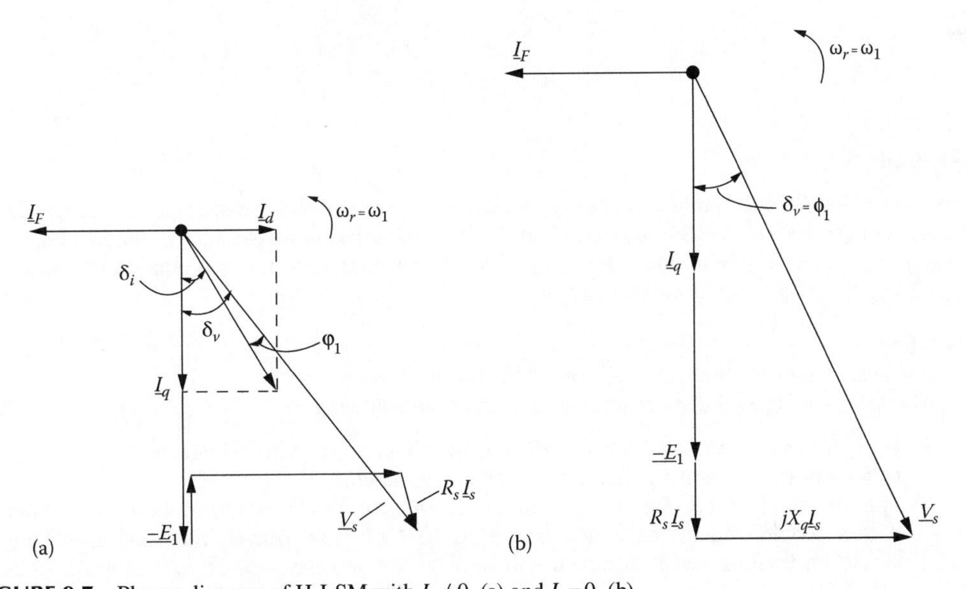

FIGURE 9.7 Phasor diagram of H-LSM with $I_d \neq 0$, (a) and $I_d = 0$, (b).

So

$$F_x = \left(3L_{ma} \cdot I_F \cdot I_q + 3\left(L_d - L_q\right) \cdot I_d \cdot I_q\right) \cdot \frac{\pi}{\tau} \tag{9.20}$$

Note: I_d, I_q are phase variables (RMS values of them).

The total magnetic energy in the machine W_{mt} when neglecting magnetic saturation is

$$W_{mt} \approx \frac{Q_1}{\omega_1} \tag{9.21}$$

and the normal (attraction) force F_n is

$$F_n = \frac{\partial W_{mt}}{\partial g_1} \approx -\frac{W_{mt}}{g_1} \tag{9.22}$$

Thus, from the phasor diagram, I_d and I_q are

$$I_d = \frac{\left(E_1 - V_S \cdot \cos\delta_V\right) \cdot X_q + R_S \cdot V_S \sin\delta_V}{R_S^2 + X_d \cdot X_q}$$
$$I_q = \frac{V_S \cdot \left(R_S \cdot \cos\delta_V + X_d \cdot \sin\delta_V\right) - R_S \cdot E_1}{R_S^2 + X_d \cdot X_q} \tag{9.23}$$

Finally, the power factor cos φ and efficiency η are

$$\cos\varphi = \frac{F_X \cdot 2\tau f_1 + 3 \cdot R_S \cdot I_S^2}{3 \cdot V_S I_S}; \quad \eta \approx \frac{1}{1 + \dfrac{3 \cdot R_S \cdot I_S^2}{F_X \cdot 2\tau \cdot f_1}}; \quad I_S^2 = I_d^2 + I_q^2 \tag{9.24}$$

Example 9.1

For the so far developed preliminary theory, let us consider the following data: $U_s = 90$ m/s, $2p = 12$ poles, $\tau = 0.45$ m, $L = 2a = 0.25$ m, $g_1 = 0.02$ m (!), $W_1 = 305$ turns per phase, $k_{W_1} = 1$ (winding factor, say, $q_1 = 1$, $y/\tau = 1$), $W_F I_F = 30{,}000$ A turns, $V_S = 3$ kV(RMS value); $\tau_p/\tau \approx 1$. It is required to calculate F_x, F_y, (F_n), η, cos φ versus power angle δ (δ_V).

Solution

Applying Equations 9.1 through 9.24, we get the results in Figure 9.8:

The results in Figure 9.8 warrant remarks such as the following:

- For $I_d = 0$, which corresponds to $\delta_v \approx 45°$, sizeable rated thrust, normal force, and good power factor (cos φ > 0.8) and efficiency (η > 0.95) are obtained.
- The dc winding copper (or aluminum) losses are not considered here but, even if they would, the losses would be up to 10% of rated power; the total efficiency would be then above 85%; but it will provide for both suspension and propulsion in MAGLEV applications.

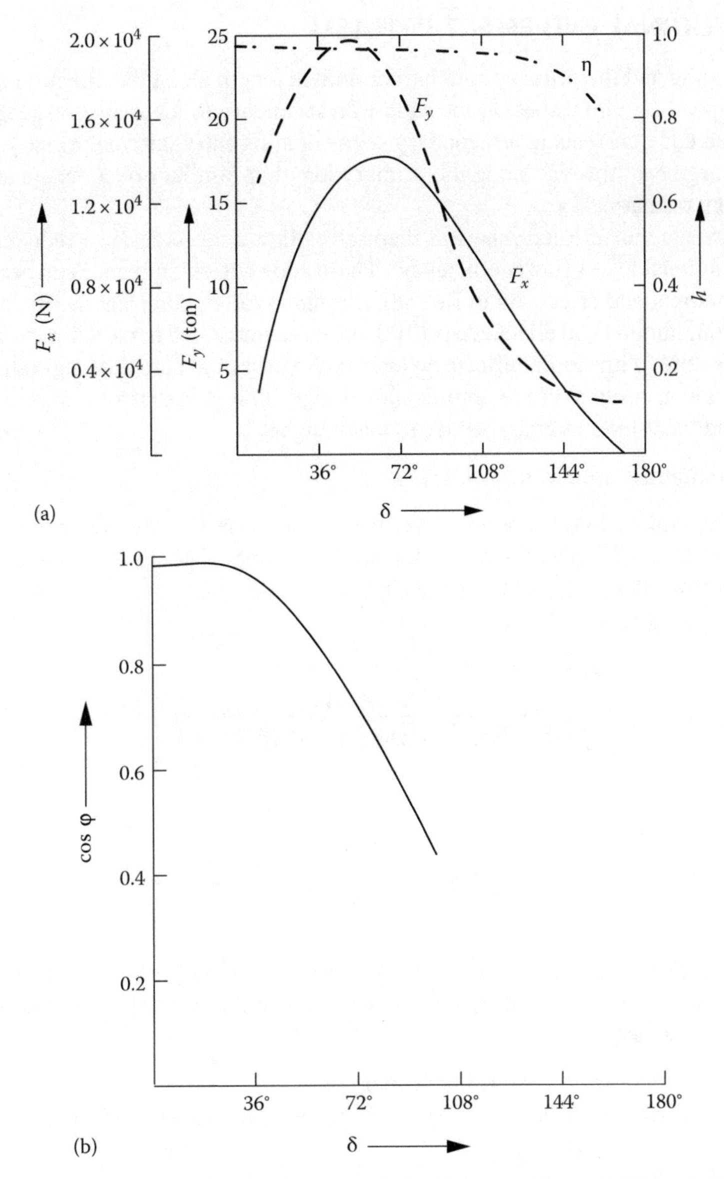

FIGURE 9.8 H-LSM steady-state characteristics: (a) thrust F_x, normal force F_y (F_n) and efficiency, (b) power factor. (After Boldea, I. and Nasar, S.A., *Linear Motion Electric Machine*, John Wiley & Sons, New York, 1976.)

- The normal force to thrust ratio for $\delta_v = 45°$ is more than 12/1, which means that more thrust is available, if the maximum vehicle acceleration is limited to 1 m/s^2 with the H-LSM making 10% of vehicle total weight. When fully loaded with people or freight, the normal force to thrust will be around 10/1.
- The earlier example is rather general, and no attempt at optimal design was made; the pole pitch $\tau = 0.45$ m is rather high; $0.2 \div 0.25$ m would be more adequate, as the U-shape core legs width $L = 2a \approx 0.2 \div 0.25$ m, to reduce the track segments weight (and cost).
- The maximum dc produced airgap flux density B_{0g} could go as high as 1.6 T, for high saturation flux density, and, with an airgap $g_1 = 10$ mm, outstanding performance is to be expected, for integrated propulsion and suspension functions on MAGLEVs.

9.4 LONGITUDINAL END EFFECT IN H-LSM

The dc field winding and the primary core have a limited length along the direction of motion; the latter moves at speed U_s with respect to the solid-iron secondary (track) segments (Figure 9.9).

These induced eddy currents in a secondary segment at its entry into the primary will decay in time, while the segment "travels" along the primary length; a similar effect occurs at a secondary segment exit from primary.

The eddy currents thus created manifest themselves in a drag force F_d, a reduction of normal force, and by additional losses (lower efficiency). These eddy current influences on performance are called the longitudinal end effect. As in Ref. [8], through a rather complete theory, it was demonstrated that the longitudinal end effect even at 100 m/s has a thrust and normal force reduction effect of less than 10%–15% (with similar effects on losses). We may treat here the longitudinal end effect by a simplified model, easier to use in optimization design. This is in sharp contrast to LIMs, where dynamic longitudinal effects at high speeds are much higher.

The main assumptions are the following:

- The current induced in the secondary segments exhibits only a transverse (J_y) component (Figure 9.9); the inevitable presence of longitudinal component (J_x) is accounted for by the Russel-Northworthy coefficient (see the chapters on LIMs) k_t:

$$k_t \approx \frac{1}{\left[1-\frac{\tanh \beta a_e}{\beta a_e \cdot \left(1+\tanh \beta a_e \cdot \tanh \beta (c-a_e)\right)}\right]} \tag{9.25}$$

with

$$a_e \approx \frac{L}{2}+g_1; \quad c \approx \frac{L_W}{4}+a; \quad \beta = \frac{\pi}{\tau_p} \tag{9.26}$$

where L_w is the window area width between the primary U core legs; the approximation for c takes into account the single secondary segment with two primary U core leg interaction in defining the eddy current density lines:

- The magnetic field is zero outside the primary core.
- The induced current distribution along O_x (direction of motion) is uniform inside and outside the core.

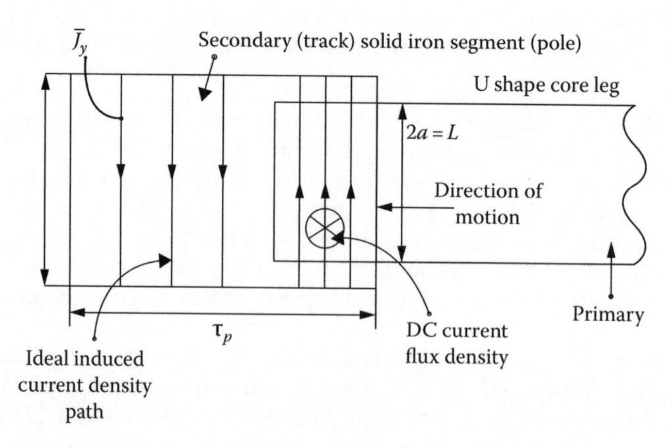

FIGURE 9.9 Ideal secondary induced currents at H-LSM primary entry.

- The secondary segment magnetic permeability is constant but it is iteratively calculated.
- The depth of field penetration in the secondary segment is also iteratively calculated [9].
- The reaction field of induced currents is neglected.
- The eddy currents in the entry and exit secondary segments die out quickly and thus only their influence on longitudinal end effect is considered here (see Ref. [8] for a more detailed analysis).

Consequently, with ψ_F as dc field current flux in the entry segment, speed U, and segment resistance R, we have

$$-\frac{d\psi_F}{dt} = \frac{d\psi_F}{dt} \cdot U = R \cdot I \tag{9.27}$$

$$R = \frac{2k_t a_e}{\sigma_i \cdot di}\left(\frac{1}{x} + \frac{1}{\tau_p - x}\right) \tag{9.28}$$

Approximately

$$\frac{d\psi_F}{dx} = 2B_{0g}a_e \tag{9.29}$$

with B_{0g} from (9.1).

The effect of armature traveling field on longitudinal end effect is neglected here. From the previous equations, the current I in the secondary segment is (Figure 9.9)

$$I = \frac{U \cdot B_{0g}}{\frac{\mu_0 U}{2g_0} + \frac{k_t \cdot \tau_p}{d_i \cdot \sigma_i x(\tau_p - x)}} \tag{9.30}$$

where σ_i is the iron electrical conductivity.

The drag force, for both U core legs and entry segment F_d, is

$$F_d = I \cdot B_{0g} 4a_e = \frac{4U \cdot B_{0g}^2 a_e}{\frac{\mu_0 U}{2g_0} + \frac{k_t \cdot \tau_p}{d_i \cdot \sigma_i x(\tau_p - x)}} \tag{9.31}$$

The maximum drag force $F_{d\max}$ occurs at $x = \tau_p/2$. The average drag force (considering both entry and exit segments, with $2p$ poles per primary) is $F_d \approx F_{d\max}$:

$$F_{dav} = F_{d\max} \approx \frac{4U \cdot B_{0g}^2 a_e}{\frac{\mu_0 U}{2g} + \frac{k_t \cdot U}{di \cdot \sigma_i \tau_p}} \tag{9.32}$$

Note: As the reaction of the eddy currents I has been neglected, their influence on normal force is also neglected. Ref. [8] includes the reduction of normal force due to longitudinal end effect.

Experimental results (Ref. [8]) confirm the theory, for a 30 m/s max speed test platform (Figure 9.10) in relation to normal force reduction and drag force increase with speed ($2p\tau = 0.62$ m, $2a = 0.03$ m, $\tau_p = 0.069$ m $= \tau$, $g_1 k_C = 6.7$ mm).

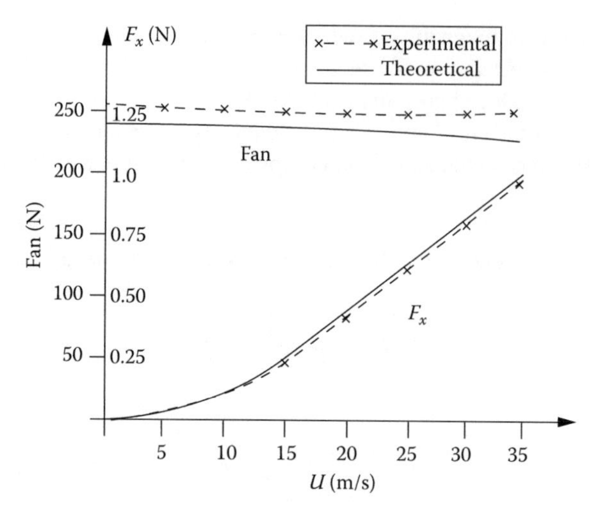

FIGURE 9.10 Average drag (F_x) and total average normal force F_n versus speed for an H-LSM (longitudinal end effect). (After Boldea, I. and Nasar, S.A., *EME J. (now EPCS)*, Taylor & Francis, 2(2), 254, 1978.)

It is estimated that for a 1 MW H-LSM with a total stack width ($2\times 2a=0.15$ m), 2.5 m long at 100 m/s, the drag force power ($F_{x\,\text{av}}\,U$)-longitudinal effect losses are less than 65 kW. This is quite acceptable and proves the suitability of the H-LSM for high-speed MAGLEVs with solid mild steel track segments. The normal force decrease with speed due to longitudinal end effect at 100 m/s is less than 15%, to prove once more this claim.

9.5 PRELIMINARY DESIGN METHODOLOGY BY EXAMPLE

The design methodology starts with preliminary specifications such as

- Rated (or peak) thrust: $F_{xu}=6$ kN (8 H-LSMs for a 36 ton MAGLEV)
- Rated normal force $F_n=4.5$ ton
- Rated (peak) speed $U_n=100$ m/s
- Rated ac phase voltage (maximum RMS value): $V_S=1080$ V

Pure I_q control

- Rated (and boost) dc voltage for field winding supply: not given here.

Besides this data, quite a few parameters are given from prior experience, such as the following:

- Airgap $g_1=1.5\times 10^{-2}$ m (from mechanical reasons, it could be lowered to $g_1=1.0\times 10^{-2}$ m).
- Secondary track segments total width $L_s<0.35$ m ($L_s=2\times 2a+L_W$); this limitation should keep the track cost within reasonable bounds.
- Maximum H-LSM length: $2p\tau<3.7$ m (to accept tight curves).

With this data in mind, we return to a few important items that provide the mathematics for the preliminary design. Then we use the case study data to produce sample design results.

9.5.1 Armature AC Winding Specifics and Phasor Diagram

The two practical ac windings of interest have been already introduced in Section 9.1. The eight-shape-coil ac winding (Figure 9.11) seems the first choice, especially with three slots/pole ($q=1$ slot/pole/phase) and diametrical coils ($y/\tau=1$) to yield unity fundamental winding factor k_{W_1}.

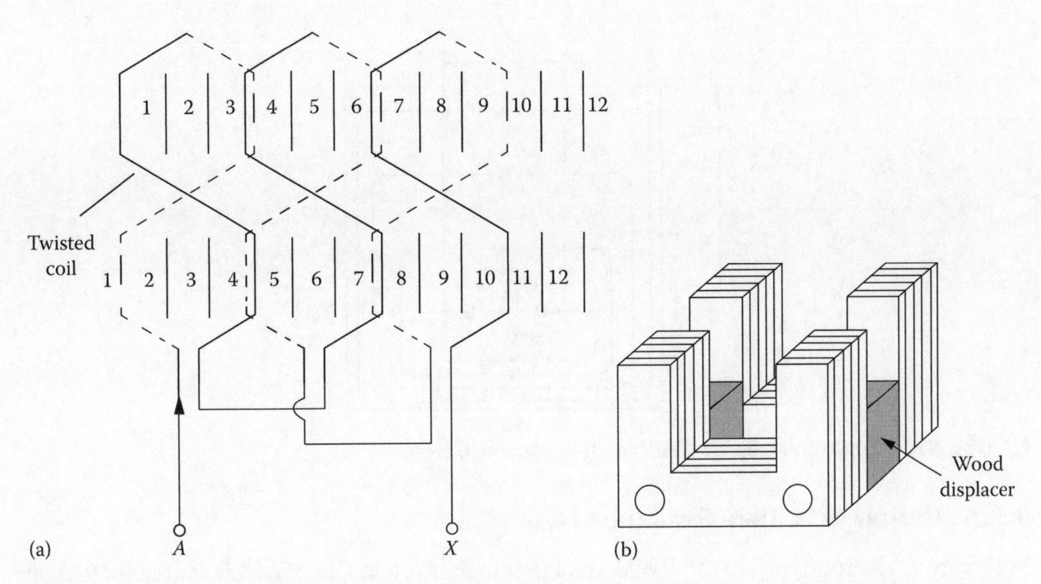

FIGURE 9.11 Eight-shape ac coil winding for H-LSM (a), transverse (U shape) primary cores, (b). (After Boldea, I. and Nasar, S.A., *EME J.* (now *EPCS*), Taylor & Francis, 4(2–3), 125, 1979.)

- The three-layer winding (Figure 9.4) has its own merits in reducing the end connections of coils (mainly for $2a \le \tau$) and in providing space for airflow between layers, but its fundamental winding factor is only 0.867.

For transverse (transformer) lamination primary cores, the armature reaction field is also transverse (ac coil slots are open at both ends), and thus, the interaction between propulsion and suspension functions for pure I_q control (Figure 9.11) is notably reduced.

The phasor diagram in Figure 9.7b stands for MAGLEVs, $L_d = L_q = L_s$ ($\tau_p/\tau \approx 1$), to provide enough normal force.

Note: When small normal force design is required, either the ratio τ_p/τ is reduced to 0.65–0.7 or a double-sided structure is used. At the limit when short travel of primary on linear bearings is adopted, the dc coils may be replaced by primary long PMs [14] (Figure 9.12).

Again, to reduce the radial force on each U core, $\tau_p/\tau = 0.65–0.70$, after optimization attempts.

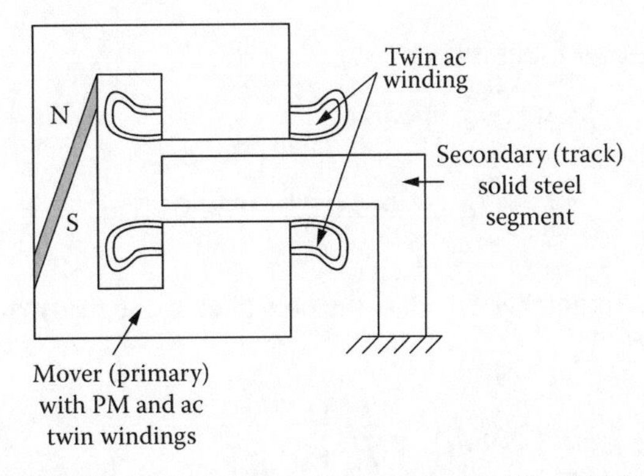

FIGURE 9.12 Ideally zero normal force H-LSM with concentrated flux PM homopolar excitation. (From Evers, W. et al., *Record of LDIA-1998*, Tokyo, Japan, pp. 46–49, 1998.)

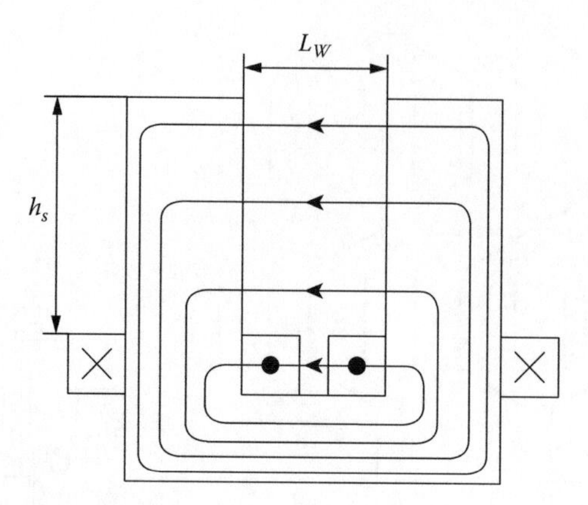

FIGURE 9.13 Transverse leakage flux lines of dc excitation coils.

9.5.2 Primary Core Teeth Saturation Limit

With pure I_q control (typical with PWM voltage-source inverters), in MAGLEVs (Figure 9.6), the teeth on the second half of the pole pitch experience both the dc excitation and full ac magnetic flux density B_{0g} (9.1).

The armature reaction airgap flux density translates into a larger teeth flux density. So its influence should be enlarged by the slot pitch/tooth length ratio: on top of that, the leakage flux density B_{W_1} in the U core window has to be added as it is notable (Figure 9.13).

We may not add flux densities unless magnetic saturation is ignored (or kept constant (known)). So the total maximum primary tooth flux density $B_{t\,\max}$ is approximately

$$B_{t\max} \approx \left[\frac{\mu_0 W_F I_F}{k_C}\left(\frac{1}{2g_1} + \frac{h_s}{\tau \cdot L_W}\right) + \frac{\tau_s}{b_t} \cdot \frac{3\sqrt{2}(2W_1)I_1 k_{w1}}{p(2g_1)k_C}\right]\frac{1}{1+k_s(B_{t\max})} \tag{9.33}$$

where

τ_s is the tooth pitch
b_t is the tooth span

Equation 9.33 unveils an iterative process, as heavy magnetic saturation with $B_{t\,\max} = 1.85 \div 1.9$ T (even 2.2 T for special, high-saturation flux density, silicon lamination) is practical. $B_{t\,\max}$ is a design parameter as it limits how low L_W could go, together with the constraint of enough room for ac end coils.

9.5.3 Preliminary Design Expressions

With L_S limited to 0.3–0.35 m, we may infer that

$$L_S \approx 4a + \frac{\tau \cdot c_1}{\sqrt{3}} + 0.03 < 0.3 \div 0.35 \text{ m} \tag{9.34}$$

Also $2p \cdot \tau < 3.7$ m; $c_1 \geq 1$. To quickly calculate the pole pitch τ, the thrust and normal force expressions are recalculated here:

$$F_{xn} = \frac{3E_1 I_1\left(1 - \dfrac{F_d}{F_{xn}}\right)}{2\tau f}; \quad \frac{F_d}{F_{xn}} \approx 0.05 \div 0.1 \tag{9.35}$$

$$F_n \approx \frac{B_{0g}^2 \cdot 2a \cdot 2p\tau}{2\mu_0}; \quad (I_q \text{ control}) \tag{9.36}$$

where F_d is the longitudinal effect (braking) force (9.31).

With a large airgap ($g_1 = 1.5 \times 10^{-2}$ m), the airgap flux density of dc coils $B_{0g} \approx 0.85$ T (much larger values (1.5–1.6 T) could be adopted for superior laminations with $B_{sat} = 2.35$ T!), to limit the U core window weight. The U core leg width ($2a$) can now be calculated, by fixing $2p\tau$ (primary length), from (9.35), with given normal force F_n.

Then, from (9.34), the pole pitch τ is obtained. The number of Ampere turns per winding ($2W_1I_1$) follows then from (9.35) for given thrust F_{xa}. From here on the slot height h_S, calculation for given τ and $q_1 = 1$ is straightforward.

And so all parameters and performance based on technical theory developed in previous paragraphs may be calculated.

Typical results for our numerical example are as follows:

$j_F = 15$ A/mm²—rated current density in the field coils (forced cooling)
$j_{Co} = 20$ A/mm²—rated current density in ac windings (forced cooling)
$\tau = 0.18$ m—pole pitch
$2p = 20$—number of poles
$L_W = 0.158$ m—window width
$2a = 7.08 \times 10^{-2}$ m—U core one leg width
$f_n = 274$ Hz—rated frequency
$I_n \approx 250$ A
$W_1 \approx 114$ turns/phase
$V_S \approx 1080$ V
$\cos \varphi_1 = 0.8368$
$W_F I_F \approx 22{,}332$ A turns—field coils mmf
$F_d \approx 280$ N (longitudinal effect drag force at $d_i = 0.6 \times 10{-}3$ m)
$\eta_1 \approx 0.95$—efficiency without dc winding losses
$\eta_{net} = 0.836$—total efficiency (with dc winding losses)
$F_{xn} = 6$ kN—total thrust
$F_n = 4.5$ ton—normal force
$T_F = 0.04$ s—dc field circuit time constant
$p_{exc} = 65$ kW—dc field coil losses
$h_s = 0.09$ m—ac coil slot height

The preliminary design may be discussed as follows:

- As the dc current produced airgap flux density B_{0g} is only $B_{0g} = 0.85$ T, the ratio of 7.5/1 of normal force/thrust is obtained at an airgap $g_1 = 1.5 \times 10^{-2}$ m. A smaller airgap ($g_1 = 1 \times 10^{-2}$ m) with $B_{0g} = 1.3$–1.5 T (for high saturation flux density laminations) would allow notably better performance; still the preliminary design yields acceptable performance for integrated propulsion suspension of a MAGLEV at 100 m/s, such as 0.886 kg/kW for the H-LSM at $\eta_{net} = 0.836$ and $\cos \varphi = 0.8368$.
- The previous preliminary design should be used only as a starting point; FEM analysis is to be performed to calculate $B_{g\,\min}$ and thus more precisely B_{g1} (fundamental airgap flux density in the airgap produced only by the dc winding); B_{g1} is crucial for thrust force computation F_{xn}.
- A more detailed analytical model with 2 (3) D FEM correction factors may be used for an optimization design methodology; finally, again, key 2 (3) D FEM should be used to validate the optimal analytical design.

- An example of an H- LSM as applied to an urban MAGLEV 4 ton research prototype (Magnibus-01) will be presented in more detail in a later chapter on "Passive Guideway MAGLEVs" [16].
- The application of H-LSM with PM excitation [14] for short travel (a few meters) on linear bearings or on wheels, though interesting in nature, is not pursued here.
- The H- LSM with 1.3–1.5 T dc airgap flux density leads to overall propulsion and suspension performance, which makes the latter very performance/cost competitive for MAGLEVs, where its large normal force is fully exploited.

9.6 H-LSM MODEL FOR TRANSIENTS AND CONTROL

H-LSM behaves like a rotary synchronous machine with dc excitation, $L_d = L_q = L_s$ (only in MAGLEVs), and with a very weak damper cage represented by the solid-iron secondary segments. For simplicity, we neglect this cage effect.

It should also be noticed that there are basically two sets of equations of motion: vertical (suspension) and longitudinal (propulsion) motion. The drag force, $F_d = k_d \cdot U$, acts as a damper for the propulsion motion longitudinal oscillations.

We may now simply write the dq model of H-LSM in stator coordinates:

$$\bar{V}_S = R_S\bar{I}_S + \frac{d\bar{\psi}_S}{dt} - j\omega_r\bar{\Psi}_S; \quad \omega_r = \frac{\pi}{\tau}U; \quad \bar{V}_S = V_d + jV_q; \quad \bar{I}_S = I_d + jI_q$$

$$\bar{\psi}_S = L_{Sl}I_d + L_{dm}(g_1)(I_d + I_F^p) + j(L_{Sl}I_q + L_{qm}(g_1)I_q)$$

$$I_F^p = k_{I_F} \cdot I_F; \quad k_{I_F} = \left(\frac{L_{dm}}{M_{aF}}\right)^{-1}; \quad I_F^p \text{ is the field current reduced to primary}$$

$$\mathrm{I}_F R_F - V_F = -L_{Fl}\frac{dI_F}{dt} - \frac{dL_{dm}(g_1)}{dt}(k_{I_F} \cdot I_F + I_d) \cdot k_{I_F} \tag{9.37}$$

$$m\frac{du}{dt} = F_x - F_{Load} - F_d; \quad F_x = \frac{3}{2}\frac{\pi}{\tau}\mathrm{Real}(j\bar{\psi}_S \cdot \bar{I}_S^*); \quad F_d \approx k_d' \cdot \left(\frac{I_F}{g_1}\right)^2 \cdot U$$

$$\text{or} \quad F_x = \frac{3}{2}\frac{\pi}{\tau}\left[L_{dm}(g_1)I_F^p + (L_{dm}(g_1) - L_{qm}(g_1))I_d\right] \cdot I_q$$

$$m\frac{dU_V}{dt} = m \cdot g - F_n; \quad F_n \approx \frac{2a \cdot 2p\tau}{2\square_0} \cdot \left(\frac{\square_0 W_F I_F}{2g_1 k_C(1+k_S)}\right)^2; \quad \frac{dg_1}{dt} = U_V$$

No track irregularities are yet considered previously. We should add the Park transformation:

$$\begin{vmatrix} I_d \\ I_q \\ I_0 \end{vmatrix} = \frac{2}{3}\begin{vmatrix} \cos(-\theta_{er}) & \cos\left(-\theta_{er} + \frac{2\pi}{3}\right) & \cos\left(-\theta_{er} - \frac{2\pi}{3}\right) \\ \sin(-\theta_{er}) & \sin\left(-\theta_{er} + \frac{2\pi}{3}\right) & \sin\left(-\theta_{er} - \frac{2\pi}{3}\right) \\ \frac{1}{2} & \frac{1}{2} & \frac{1}{2} \end{vmatrix}\begin{vmatrix} I_a \\ I_b \\ I_c \end{vmatrix} \tag{9.38}$$

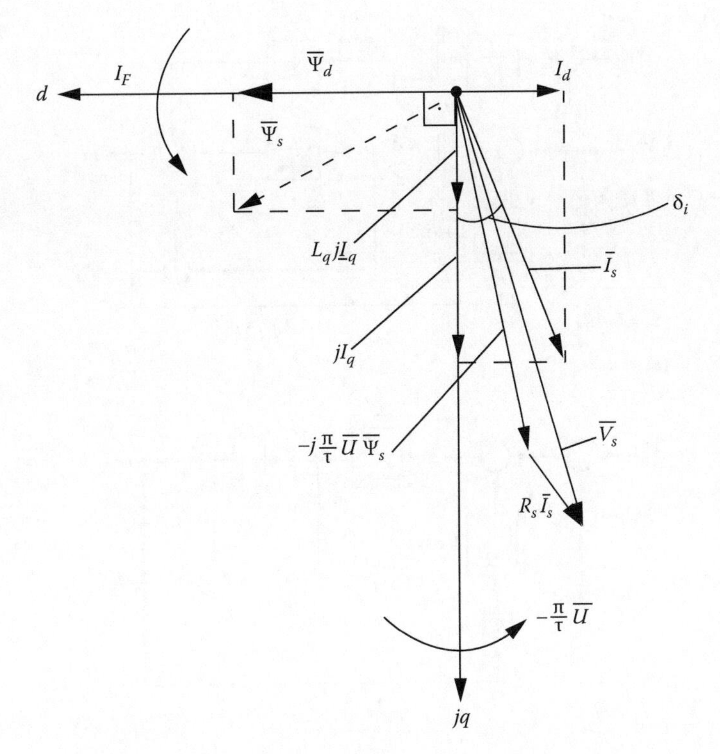

FIGURE 9.14 The vector (space phasor) diagram of H-LSM (steady state, $I_d \neq 0$).

The sign (−) for θ_{er} in (9.39) stems from the fact that the primary is moving against its traveling field (Figure 9.14):

$$\frac{d\theta_{er}}{dt} = -\frac{\pi}{\tau}u + \frac{\pi}{2} + \delta_i \tag{9.39}$$

As expected, Figures 9.7 and 9.14 are very similar in terms of angles, but it is time angles in Figure 9.7 and space angles in Figure 9.14 as the latter deals with space vectors and not with phase variables.

The same Park transformation is valid for the voltage vectors. Once the voltages $V_{a,b,c}(t)$, F_{load}, k_d', $V_F(t)$, and H-LSM parameters are given, numerical methods allow investigating any transients by (9.36) through (9.38).

For given airgap, the vertical equations of motion are eliminated and the order of the system degenerates from 7th (I_d, I_q, I_F, U, θ_{er}, U_v, g_1), to 5th (I_d, I_q, I_F, U, θ_{er}). The propulsion and the suspension close-loop control add more equations (and orders) to this system of equations.

A generic structural diagram for Equation 9.37 for constant airgap, pure propulsion control, is given in Figure 9.15.

Note: In Figure 9.15, the V_F and I_F are actual field winding input voltage and current.

As noticed in Figure 9.15, there are only two leakage electric time constants to consider, in the absence of the secondary damping cage (see Ref. [13] for an SM with damper cage).

Magnetic saturation occurs in ψ_{dm} and ψ_{qm}. The airgap variation also marks its presence in ψ_{dm} and ψ_{qm}.

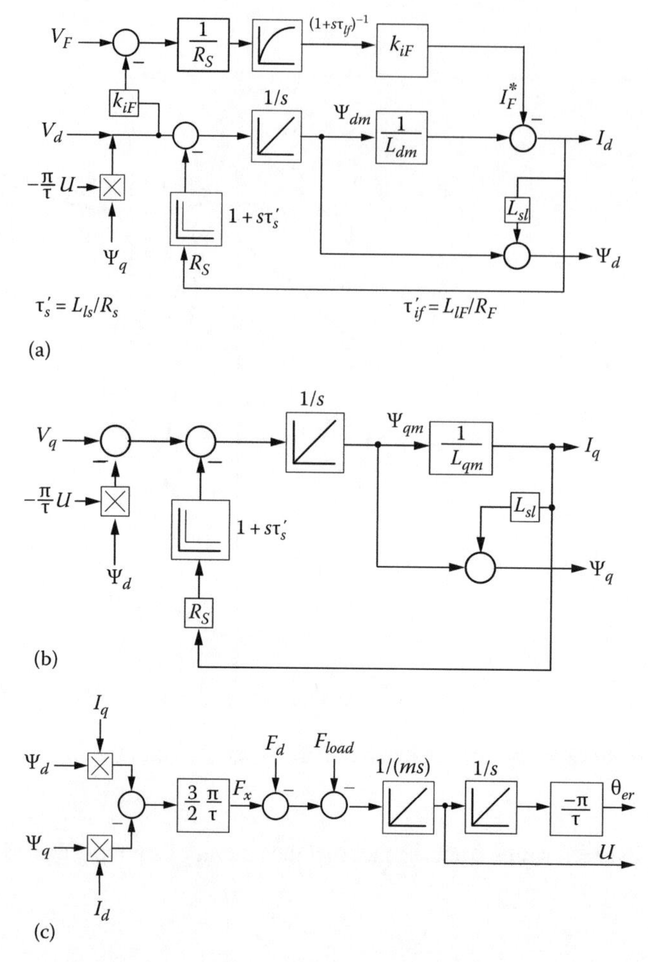

FIGURE 9.15 Generic structural diagrams of H-LSM: (a) axis d, (b) axis q, and (c) motion equations (constant airgap, g_1).

9.7 VECTOR THRUST (PROPULSION) AND FLUX (SUSPENSION) CONTROL

Based on the circuit model, a recently developed generic vector control system of H-LSM is shown in Figure 9.16.

The rather complex control in Figure 9.16 may be somewhat clarified by remarks such as the following:

- zero I_d^* vector control is chosen to reduce propulsion-suspension interaction (in practice, it is not zero but it is small).
- PI + SM (sliding mode) control for airgap, speed, and current controllers is chosen to provide robust control in the presence of airgap and magnetic saturation (inductance) variations.
- In the presence of a secondary mechanical suspension system for each H-LSM on a MAGLEV, rather independent but highly robust, partly decentralized, airgap control is feasible.

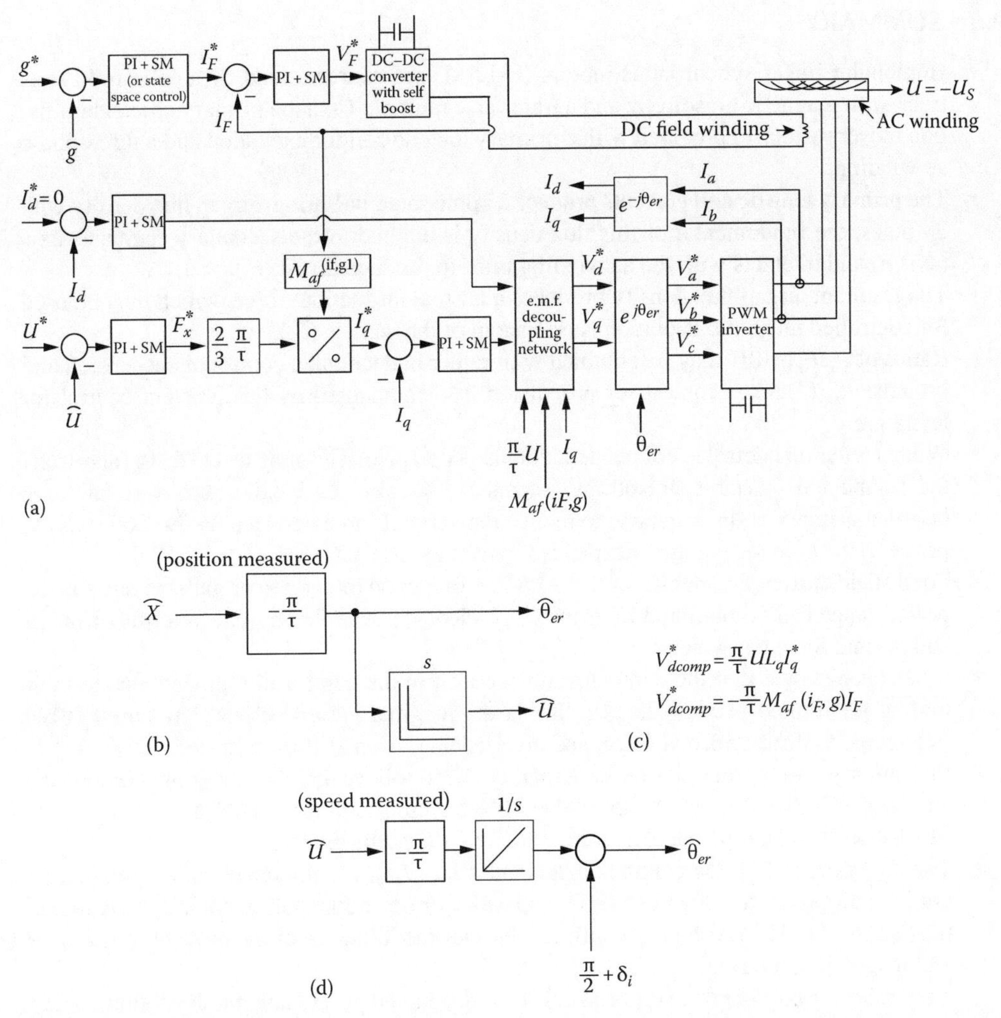

FIGURE 9.16 Vector control of H-LSM (integrated propulsion and suspension control): (a) vector control, (b) position angle and speed from measured position, (c) emf decoupling network, and (d) position angle θ_{er} from measured speed.

- For propulsion-only control, the reference field current may be made dependent either on reference thrust F_d^* (to minimize total copper losses, for example) or it is kept constant up to base speed and then decreased with speed for constant power wide-speed range.
- Either mover position (with respect to secondary (track) segments) or mover speed may be measured or estimated, for motion sensorless control; for nonhesitant start, initial mover position has to be known (or estimated).
- The emf compensation network improves performance at high speeds by relaxing somewhat the contribution of PI + SM I_d, I_q current controllers.

Typical vector control results for a four H-LSM (in parallel) MAGLEV (MAGNIBUS-01) are given in Ref. [16].

While, in the 1980s, current source inverters were used for the H- LSM control [6,17], today PWM voltage-source inverters, one for each H-LSM control, provide higher efficiency and lower converter weight in integrated vector MAGLEV propulsion and suspension control.

9.8 SUMMARY

- Homopolar linear synchronous motors (H-LSM) are characterized by a solid mild steel track with magnetic anisotropy and a transverse multiple *U*-shape primary silicon lamination mover (primary) provided with a primary long dc coil for excitation and a three-phase ac winding.
- The primary long dc coil currents produce a homopolar, pulsating, airgap flux density with $2p$ poles; the fundamental of this flux density is aligned to the secondary segments axis (axis d) and interacts with the ac winding mmf to produce thrust (F_x).
- The resultant airgap flux density produces a normal (attraction) force, which may be used for controlled magnetic levitation (suspension) of the vehicle (F_n).
- Ratios of $F_n/F_x \geq 10/1$ may be obtained with rather undecoupled control of suspension and propulsion, if pure I_q control for propulsion is accompanied by field current control for levitation.
- When levitation control is not needed, double -sided primaries may be used "to sandwich" the segmented (variable anisotropy) secondary (track), for ideally zero vertical force between primary and secondary; to reduce the vertical force between the two sides of the primary, the secondary segment span is reduced to $\tau_p/\tau = 0.6 \div 0.7$.
- For dc field current flux density of $1.3 \div 1.5$ T, rather good integrated propulsion-suspension performance can be obtained in terms of efficiency, power factor, and in terms of thrust and normal force per weight.
- It has been shown that the eddy currents induced in the solid iron segments due to their motion for limited primary length, that is the longitudinal end effect, have mild (10%) influences on thrust, normal force, and on efficiency, even at 100 m/s speed.
- Preliminary design examples indicate total efficiency above 85% (for integrated propulsion and suspension) and power factor above 80% for up to 100 m/s MAGLEVs, but 2 (3) D FEM should be used to calculate more precisely the performance.
- The *dq* model of H-LSM contains inductances L_{dm}, L_{qm}, which depend on airgap; in general, the airgap varies due to track irregularities or during levitation control. This makes the *dq* model of H-LSM highly nonlinear, besides the influence of the product of variables and magnetic saturation.
- Very robust vector control [18] of H-LSM for decoupled propulsion and levitation control is required to account for these nonlinearities.
- Passive guideway H-LSM MAGLEVs, for urban and interurban transportation, should be investigated aggressively now that high-performance PWM inverters and mechanical electric power collection (through controlled pantographs) in the MW power range are available up to 120 m/s and more.

REFERENCES

1. E. Rummich, Linear synchronous machines—Theory and construction, *Bull. ASE*, 23, 1972, 1338–1344 (in French).
2. M. Guarino, Jr., Integrated linear electric motor propulsion system for high speed transportation, *Record of International Symposium on Linear Electric Motors*, Lyon, France, 1974.
3. E. Levi, Preliminary design studies of iron core synchronous operating linear motor, Polytechnical & Institute of Brooklyn, Report no. 76/005, 1976, DOT ORD, Washington, DC.
4. I. Boldea and S.A. Nasar, *Linear Motion Electric Machine*, Chapter 9, John Wiley & Sons, New York, 1976.
5. H. Lorenzen and W. Wild, The Synchronous linear motor, Intern Report, Technical University of Munich, 1976 (in German).
6. B.T. Ooi, Homopolar linear synchronous motor, *Record of ICEM-1979*, Brussels, Belgium.

7. T.R. Haden and W.R. Mischler, A comparison of linear induction and synchronous homopolar motors, *IEEE Trans.*, Mag-14(5), 1978, 924–926.
8. I. Boldea and S.A. Nasar, Field winding drag and normal forces in linear synchronous homopolar motors, *EME J.* (now *EPCS*), Taylor & Francis, 2(2), 1978, 254–268.
9. I. Boldea and S.A. Nasar, Linear synchronous homopolar motor—Design procedure for propulsion and levitation, *EME J. (now EPCS)*, Taylor & Francis, 4(2–3), 1979, 125–136.
10. I. Boldea and S.A. Nasar, Thrust and normal force pulsations of current inverter—Fed linear induction motors, *EME J. (IBID)*, 7(2), 1982.
11. G.R. Slemon, A homopolar linear synchronous motor, *Record of ICEM-1979*, Brussels, Belgium.
12. I. Boldea and S.A. Nasar, *Linear Motion Electromagnetic Systems*, Chapter 7, John Wiley & Sons, New York, 1985.
13. I. Boldea and L. Tutelea, *Electric Machines: Steady State, Transients and Design with Matlab*, Chapter 9, CRC Press, Taylor & Francis, New York, 2009.
14. W. Evers, G. Henneberger, H. Wunderlich, and A. Seelig, *Record of LDIA-1998*, Tokyo, Japan, 46–49.
15. S.-M. Jang and S.-B. Jeong, Design and analysis of the linear homopolar synchronous motor for integrated magnetic propulsion and suspension, *Record of LDIA-1998*, Tokyo, Japan, 74–77.
16. I. Boldea, I. Trica, G. Papusoiu, and S.A. Nasar, Field tests on a Maglev with passive guideway linear induction motor transportation system, *IEEE Trans.*, VT-37(5), 1988, 213–219.
17. T.A. Nondahl, Design studies for single-sided linear electric motors: Homopolar synchronous and induction, *EME Journal (now EPCS)*, Taylor & Francis, ISSN: 0361 -6967, Hemisphere Publish Company, 5(1), 1980, 1–14.
18. C.-T. Liu and J.-L. Kuo, Improvements on transients of transverse flux homopolar linear machines using artificial knowledge based strategy, *IEEE Trans.*, EC-10(2), 1995, 275–286.

10 Linear Reluctance Synchronous Motors *Modeling, Performance Design, and Control*

Linear reluctance synchronous motors (L-RSM) are characterized by a short primary magnetic core with uniform slots that host a distributed three-phase ac winding and a long variable reluctance (magnetically anisotropic) secondary (Figure 10.1a and b). The distributed three-phase ac windings produce a traveling magnetic field in the airgap; its inductances (self and mutual) vary rather sinusoidally with $2\theta_{er}$ (θ_{er} rotor electrical angle between phase *a* axis and the secondary *d* (high inductance) axis). The key to high performance is a high ratio (and difference) between *d* and *q* axis synchronous inductances: an $L_d/L_q \geq 3$ is a minimum requirement, but $L_d/L_q > 7$ (8) would be desirable.

Essentially, there are two main applications, which refer to

- Medium- and high-speed transportation
- Low-speed short travel industrial applications

The configuration in Figure 10.1a, especially with segmented–laminated secondary, may be considered for medium- and high-speed MAGLEVs ($g = 6 \div 10$ mm) even if only moderate saliency is obtained ($L_d/L_q = 2.5 \div 3.5$), because its large normal force serves to levitate the vehicle, "saving" thus the rather low power factor at an acceptably good efficiency, for a reasonably low-cost passive guideway, but for a more expensive PWM inverter on board (higher kVA). In contrast, the multiple flux barrier (MFB) secondary or axially laminated secondary (Figure 10.1b) produces larger saliency for a low airgap, $g = 0.5 \div 1$ mm for short travel (a few meters), heavy weight load transporters (primaries), fed from a mechanically flexible ac power cable.

Note: The MFB (higher saliency) secondary may also be used for large airgaps ($g = 6 \div 10$ mm), but for long travel (tens or hundreds of kms), the track cost is increased significantly, though a better power factor means a lower cost and weight of the on-board PWM inverter.

The long primary/short secondary mover topology has to be ruled out for L-RSM due to the "vanishing" of the magnetic saliency ($L_d/L_q = 1.1 \div 1.15$) and thus operation at very low power factors ($0.1 \div 0.2$) and, consequently, at huge kVA rating of on-ground PWM inverter sections.

10.1 L_{dm}, L_{qm} MAGNETIZATION INDUCTANCES OF CONTINUOUS SECONDARY (STANDARD) L-RSM

Intuitively the derivation of L_d and L_q should start with the fundamental airgap flux density produced by the primary mmf in the airgap in axes *d* and *q*, B_{ad1}, B_{aq1}, and then compare them with the same but produced for a uniform airgap g_1, B_{a1}. The magnetization inductances would then be

$$L_{dm} = \frac{B_{ad1}}{B_{a1}} \cdot L_m = k_{dm1} \cdot L_m; \quad L_{qm} = \frac{B_{aq1}}{B_{a1}} \cdot L_m = k_{qm1} \cdot L_m \tag{10.1}$$

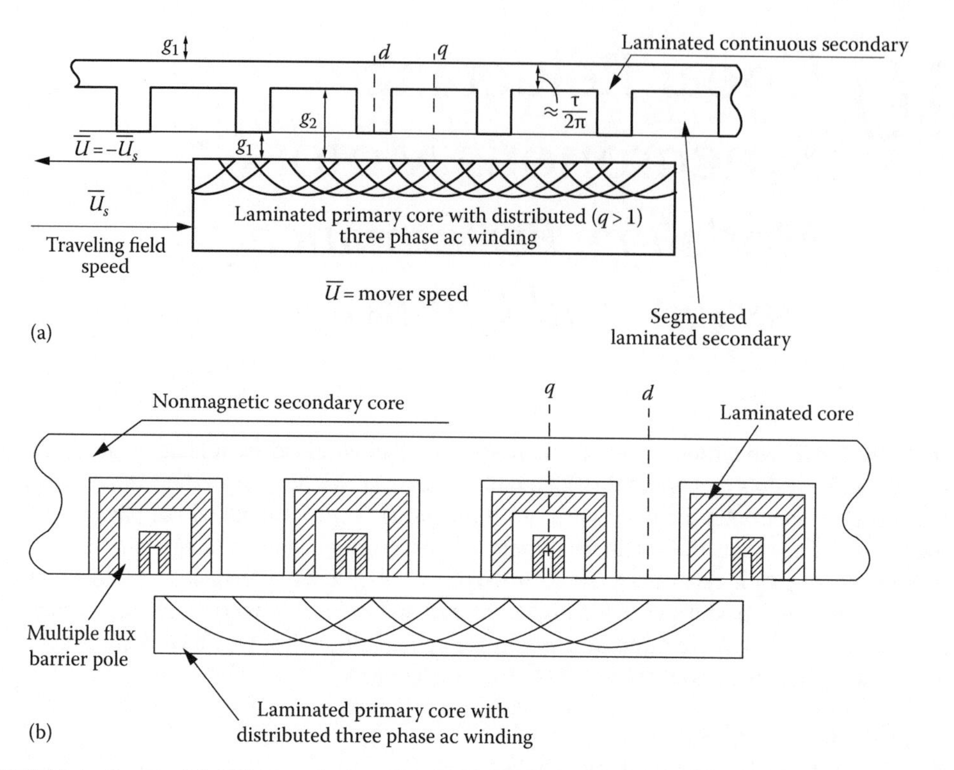

FIGURE 10.1 Practical L-RSM topologies: (a) with low saliency (for large airgap: $g > 1$ mm) and (b) with higher saliency (for smaller airgap: $g < 1$ mm).

$$L_m = \frac{6\mu_0 \left(W_1 k_{W_1}\right)^2 \tau \cdot L}{\pi^2 p \cdot k_C \cdot g_1 \left(1 + k_S\right)} \tag{10.2}$$

where

W_1 is the turns/phase in series

k_{W_1} is the fundamental winding factor ($k_{W_1} = k_{q1} \cdot k_{y1} \cdot k_{skew}$)

τ is the pole pitch of ac winding and of secondary saliency

L is the stack length

g_1 is the mechanical gap

k_S is the magnetic saturation factor $k_S = 0.1 \div 0.2$ for large airgap machine

The coefficients k_{dm1}, k_{dm1} for the conventional salient-pole continuous (laminated) secondary (Figure 10.1a) are approximately [4]

$$k_{dm1} \approx \frac{g_1}{g_2} + \left(1 - \frac{g_1}{g_2}\right)\left(\frac{\tau_p}{\tau} + \frac{1}{\pi}\sin\frac{\pi}{\tau}\tau_p\right) \tag{10.3}$$

$$k_{qm1} \approx \frac{g_1}{g_2} + \left(1 - \frac{g_1}{g_2}\right)\left(\frac{\tau_p}{\tau} - \frac{1}{\pi}\sin\frac{\pi}{\tau}\tau_p\right) \tag{10.4}$$

For large airgap ($g_1 = 10$ mm), $\tau_p/\tau = 0.66$; $g_2 = 5g_1$ (to reduce secondary weight), $k_{dm1} = 0.9467$ $k_{qm1} = 0.484$. Only a saliency ratio $L_{dm}/L_{qm} = 1.956$ is obtained. This will eventually lead to a power factor below 0.4, and the secondary height $h_{\text{sec}} = \tau/2\pi + 6 \cdot g_1$ is large and thus costly (and heavy).

Note: For a small airgap ($g = 0.4 \div 0.5$ mm) and a pole pitch $\tau \geq 0.06$ m, the low power factor may be acceptable (as for LIMs) in short travel, small-speed (even at 5 Hz) direct drives due to the simplicity of the secondary and its small cost.

10.2 L_{dm}, L_{qm} MAGNETIZATION INDUCTANCES FOR SEGMENTED SECONDARY L-RSM

In an effort to increase magnetic saliency ratio (L_{dm}/L_{qm}), the segmented secondary [1] may be adopted (Figure 10.2a); as a bonus, the secondary weight is reduced, though to reduce eddy current losses (due to airgap d and q axis field space harmonics), a laminated–segmented structure is needed. The qualitative distribution of the airgap flux density in axes d and q for the segmented secondary is shown in Figure 10.2a,b.

A few remarks are in order:

- Even the airgap flux density in axis d has a notable fifth harmonic and a smaller third harmonic.
- The airgap flux density in axis q crosses zero at points A_1, A_2, as the magnetic potential of the secondary segment $P \neq 0$.
- Consequently, the fundamental airgap flux density B_{aq} may be reduced (to create saliency) but at the "expense" of large space harmonics (especially the third one).
- The presence of stator slots openings introduces further field space harmonics, which create some additional core losses in the secondary.
- It becomes thus evident that laminated secondary segments are needed to limit these space harmonics core losses in the secondary.

The magnetic scalar potential of the secondary segment (neglecting magnetic saturation) may be obtained by equalizing the flux along AB and BC:

$$\frac{\mu_0 \tau}{\pi g_1} \int_{\left(1-\frac{\tau_p}{\tau}\right)\frac{\pi}{2}}^{\frac{\pi}{2}} (F_{1qm} \sin\theta - P)d\theta = \mu_0 P \cdot \frac{h_{SS}}{g_2} \tag{10.5}$$

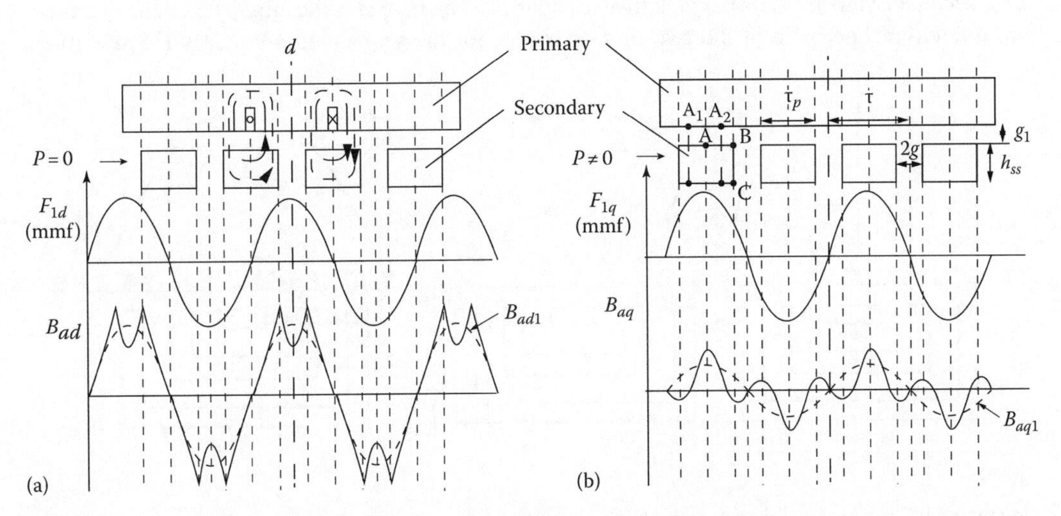

FIGURE 10.2 L-RSM with segmented secondary (a) in axis d and (b) in axis q.

Solving for P, we get

$$P = \frac{\dfrac{\tau}{\pi \cdot g_1} \cdot F_{1qm} \cdot \sin\dfrac{\pi}{2} \cdot \dfrac{\tau_p}{\tau}}{\dfrac{h_{SS}}{g_2} + \dfrac{\tau_p}{2\pi \cdot g_1}} \tag{10.6}$$

Finally, we get approximately [2]

$$k_{dm1} \approx \frac{\tau_p}{\tau} - \frac{1}{\pi}\sin\pi \cdot \frac{\tau_p}{\tau} \tag{10.7}$$

$$k_{qm1} \approx \frac{\tau_p}{\tau} + \frac{1}{\pi}\sin\pi\frac{\tau_p}{\tau} - \frac{4P}{\pi F_{1qm}}\sin\frac{\pi}{2}\frac{\tau_p}{\tau} \tag{10.8}$$

with $\tau = 0.24$ m, $g_1 = 10$ mm, $2g_2 = 40$ mm, $\tau_p/\tau = (\tau - 2g_2)/\tau = 200/240 = 0.833$, we obtain $k_{dm1} = 0.673$ and $k_{qm1} \approx 0.1$ ($h_{ss} = 40$ mm).

Now the ratio $L_{dm}/L_{qm} = 6.73$ but the difference $L_{dm} - L_{qm} = 0.573 \cdot L_m$, 24% above the $0.46 \cdot L_m$ for the standard (continuous) secondary. So the thrust for given stator mmf and power angle of the segmented secondary L-RSM will be 20% larger and the power factor notably larger in comparison with the standard (continuous) salient-pole-laminated secondary L-RSM, and again, the weight of the secondary laminations will be notably smaller (less than 50%) in comparison with the continuous variable reluctance secondary.

10.3 L_{dm}, L_{qm} (MAGNETIZATION) INDUCTANCES IN MULTIPLE FLUX BARRIER SECONDARY L-RSM

Figure 10.1b presents a MFB secondary, mainly feasible in low-speed short travel industrial applications of L-RSM. An analytical approach to calculate the magnetization inductance is given in [3–5] for rotary RSM. In principle, the airgap flux density produced by the primary mmf looks like that in Figure 10.3, if we neglect the primary slot openings.

It should be noticed that the airgap flux density in axis d has space harmonics related to the number of secondary flux barriers per pole (five in Figure 10.3a). In axis q, the airgap flux density crosses zero a few times per pole (it did two times per pole for the segmented secondary (Figure 10.2)).

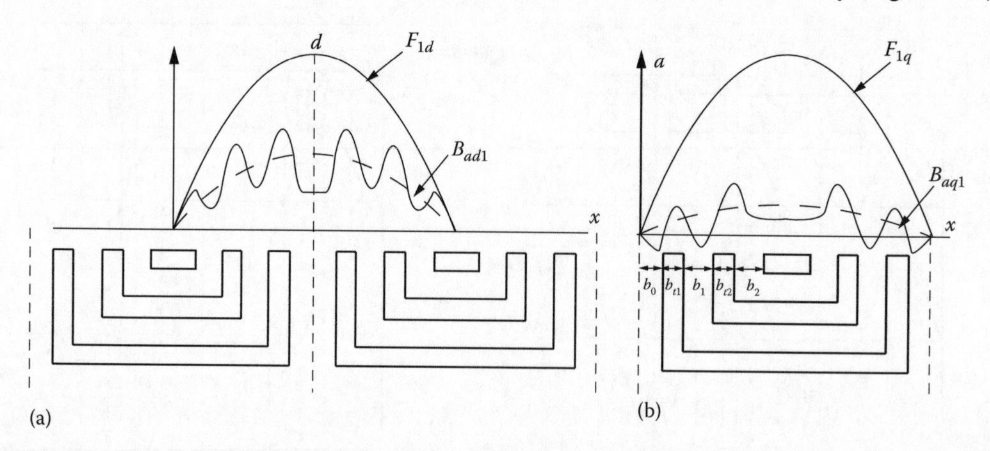

FIGURE 10.3 Qualitative airgap flux-density variation for MFB secondary L-RSM (a) in axis d and (b) in axis q.

This multiple zero crossing produces increased space harmonics, but also it reduces the fundamental airgap flux density B_{aq1}. In Figure 10.3b, zero slot openings in the stator have been considered. In reality, the open slot openings restrict the number of zero crossings per pole of B_{aq} unless $(b_1, b_2) > b_{S1}$. Mainly for $q_1 > 2$, 3 slots per pole per phase in the primary and five flux barriers (open slots) per pole in the secondary, the previous condition may be approached.

But for low airgap, low pole pitch, semiclosed primary slots, multiple zero crossing of q axis airgap flux density is feasible without increasing too much the slot leakage inductance of primary. This is why for large airgap L-RSM the MFB secondary does not seem practical, besides the higher cost of its manufacturing. For practical designs, 2D FEM (at least) or multiple magnetic circuit model [3] should be used to calculate L_{dm} and L_{qm} in MFB secondary. The axially laminated anisotropic (ALA) secondary is just a high number MFB case.

To a first approximation, L_{dm} may be calculated with (10.7) for $\tau = 2(b_{t1} + b_{t2} + b_{t3})$ and L_{qm} with the same formula but

$$\frac{L_{qm}}{L_{dm}} \approx \frac{4}{\pi} \frac{k_C g_1}{k_C g_1 + b_2 + b_1 + b_0/2} \cdot \left(1 + k_{fringe}\right) \tag{10.9}$$

k_{fringe} depends on the number of flux barriers per pole, airgap to primary slot opening, q_1–slots/pole/phase, etc. [5]. 2(3)D FEM produces more realistic results [6].

10.4 REDUCTION OF THRUST PULSATIONS

L-RSM with all these secondary topologies presented in previous paragraphs shows notable thrust pulsations. The τ_p/τ ratio may be optimized to reduce thrust pulsations without severe average thrust reduction. For the MFB secondary, further reduction of thrust pulsations may be obtained by the so-called reluctance equalization design (Figure 10.4) [7].

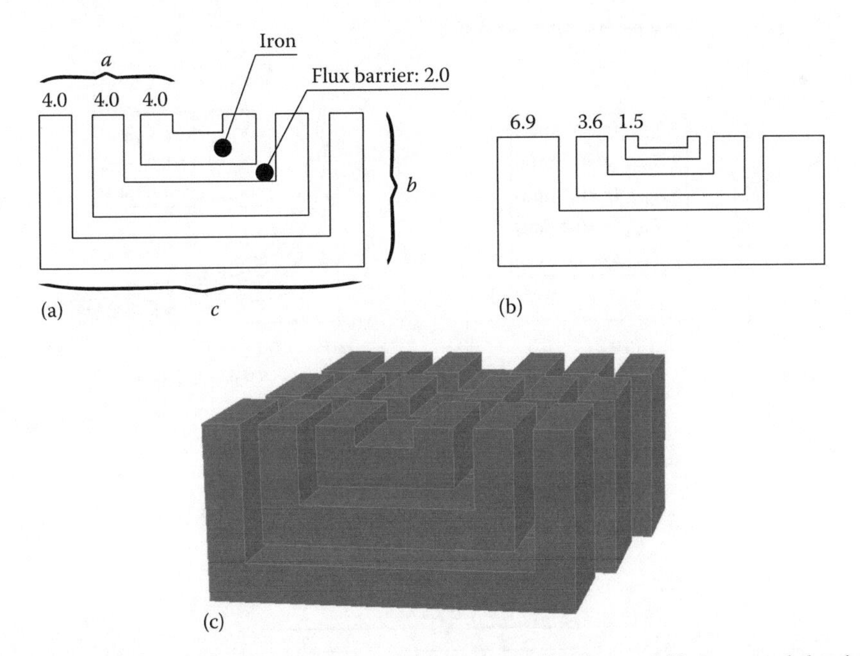

FIGURE 10.4 MFB design (a) conventional, (b) reluctance equalization, and (c) segmented skewing. (After Masayuki. S. et al., Thrust ripple improvement of linear synchronous motor with segmented mover construction, *Record of LDIA-2001*, Nagano, Japan, pp. 451–455, 2001.)

But still, the main means to reduce thrust pulsations is skewing, segmented skewing in particular (Figure 10.4c). A thrust ripple of only 10.9% is obtained with 5/4 stator slots long three segments skewing (6 slots per primary pole), with reluctance equalization [7] (Figure 10.5).

Typical k_{dm1} and k_{qm1} approximate values for an axially laminated secondary (with many laminations and insulation layers per pole) are given in Figure 10.6 versus pole pitch τ and for a few small airgap values, which are typical for short travel, low-speed, industrial applications.

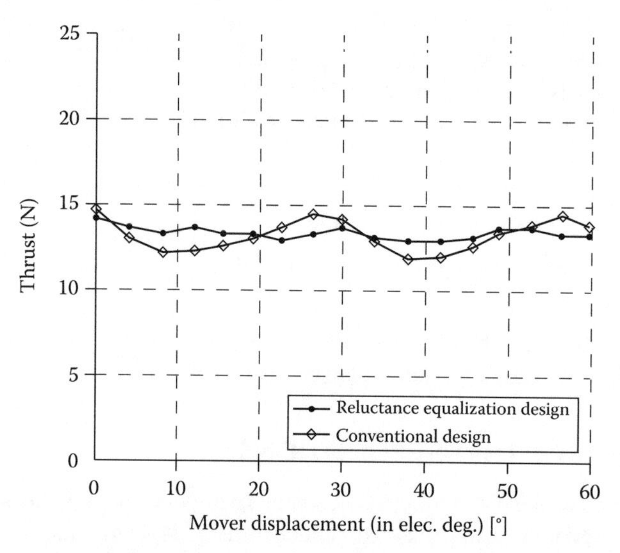

FIGURE 10.5 Thrust ripple with segmented (3/4) slot pitch skewing. (After Boldea, I., *Reluctance Synchronous Machines and Drives*, Clarendon Press, Oxford, U.K., 1996.)

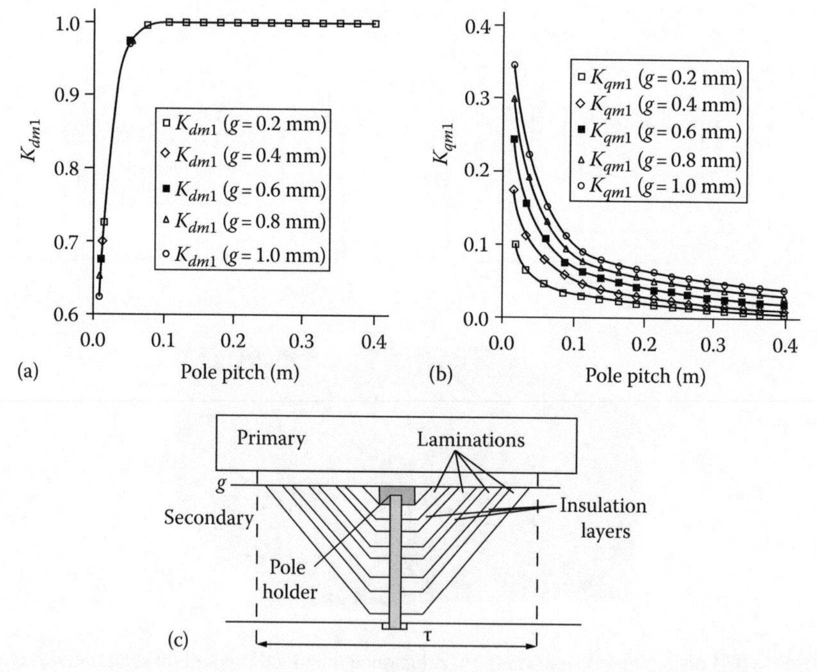

FIGURE 10.6 $k_{dm1} = L_{dm}/L_m$, (a), $k_{qm1} = L_{qm}/L_m$, (b) or a multiple lamination insulation layer ALA secondary, (c). (After Boldea, I., *Reluctance Synchronous Machines and Drives*, Clarendon Press, Oxford, U.K., 1996.)

It may be seen that even at $\tau = 50$ mm and $g = 1$ mm, $k_{dm1} = 0.95$ and $k_{qm1} = 0.2$; so a good saliency is obtained: $L_{dm}/L_{qm} = k_{dm1}/k_{qm1} = 0.95/0.2 = 4.75$ (magnetic saturation was ignored). As the expressions of primary leakage inductance L_{sl} and resistance R_S per phase are similar to those for LIMs, we skip their derivation here to investigate the *dq* model of L-RSM.

10.5 *dq* (SPACE PHASOR) MODEL OF L-RSM

As no dynamic or static longitudinal end effects have been considered—due to large enough number of primary poles ($2p \geq 8$)—the *dq* (space phasor) model of rotary RSM may be adopted here. For mini L-RSM, with $2p = 2,4$, 2(3)D FEM should be used to capture the total thrust, normal force, self and mutual phase inductances, etc. The asymmetry of inductances may be alleviated by using a two-phase (*dq*) primary winding.

The space phasor *dq* model of L-RSM in secondary (stationary) coordinates is thus

$$\bar{I}_S R_S - \bar{V}_S = -\frac{d\bar{\psi}_S}{dt} - j\omega_r \bar{\psi}_S; \quad \omega_r = -\frac{\pi}{\tau} U \text{ (primary as mover)} \tag{10.10}$$

$$\bar{\psi}_S = L_d I_d + jL_q I_q; \quad \bar{I}_S = I_d + jI_q; \quad \bar{V}_S = V_d + jV_q \tag{10.11}$$

$$L_d = L_{sl} + L_{dm}; \quad L_q = L_{sl} + L_{qm} \tag{10.12}$$

$$F_x = \frac{3}{2} \cdot \frac{\pi}{\tau} \text{Real}\,(j\bar{\psi}_S \cdot \bar{I}_S^*) = \frac{3}{2}\frac{\pi}{\tau} \cdot (L_d - L_q) \cdot I_d \cdot I_q \tag{10.13}$$

$$F_n = \frac{3}{2} \frac{\partial (L_{dm} I_d^2 + L_{qm} I_q^2)}{\partial g_1} \tag{10.14}$$

For steady state $d/dt = 0$, in secondary coordinates (the primary is the mover), and thus the space phasor (vector) diagram of L-RSM is as shown in Figure 10.7. All variables are basically dc in Figure 10.7.

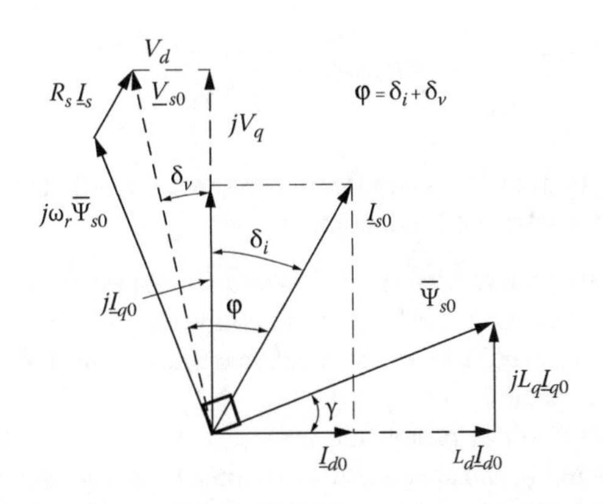

FIGURE 10.7 The space-phasor (*dq*) model of L-RSM in steady-state and stationary coordinates.

From Figure 10.7,

$$\begin{aligned} I_S \cos \varphi_1 &= I_{q0} \cos \delta_v - I_{d0} \sin \delta_v; \quad V_{S0} = V\sqrt{2} \\ I_{d0} \omega_r L_d &= V_{S0} \cos \delta_v - I_{q0} R_S \\ I_{q0} \omega_r L_q &= V_{S0} \sin \delta_v + I_{d0} R_S \end{aligned} \tag{10.15}$$

V is the RMS value of phase voltage.

Consequently I_{d0} and I_{q0} are

$$\begin{aligned} I_{d0} &= \frac{V_S \left(\omega_r L_q \cos \delta_v - R_S \sin \delta_v\right)}{R_S^2 + \omega_r^2 L_d L_q} \\ I_{q0} &= \frac{V_S \left(\omega_r L_d \sin \delta_v + R_S \cos \delta_v\right)}{R_S^2 + \omega_r^2 L_d L_q} \end{aligned} \tag{10.16}$$

The electromagnetic (mechanical) power P_m is

$$P_m = F_x U = \frac{3}{4} \frac{V_S^2 \left(\omega_r L_d - \omega_r L_q\right)}{\left(R_S^2 + \omega_r^2 L_d L_q\right)^2} \cdot \left[\left(\omega_r^2 L_d L_q - R_S^2\right) \sin 2\delta_v + R_S\left(\omega_r L_d + \omega_r L_q\right) \cos \delta_v - R_S(\omega_r L_d - \omega_r L_q)\right] \tag{10.17}$$

Considering the winding and core (iron) losses, the efficiency and power factor η and $\cos \varphi$ are

$$\eta = \frac{P_m}{P_m + 3/2\left(I_{d0}^2 + I_{q0}^2\right) R_S + p_{iron}}; \quad I_{S0} = \sqrt{I_{d0}^2 + I_{q0}^2} \tag{10.18}$$

$$\cos \varphi = \frac{P_m}{\eta (3/2) V_{S0} I_{S0}} \tag{10.19}$$

Neglecting all losses, the maximum ideal power factor $\cos \varphi_{i\max}$ is

$$\cos \varphi_{i\max} \approx \frac{L_d - L_q}{L_d + L_q} \tag{10.20}$$

Note: For LIMs (with no longitudinal end effect), the same formula is valid but $L_q \rightarrow L_S$ (no load inductance) and $L_q \rightarrow L_{Sc}$ (short-circuit inductance).

As it will be demonstrated later, low-speed (low-power) applications of L-RSM may show better power factor and efficiency than LIMs, and thus, lower kVA in the PWM inverter is needed. Including the machine losses would raise the ideal maximum power factor in motoring and will reduce it in generating, as expected.

A typical qualitative variation of propulsion force F_x, attraction force F_n, η, $\cos \varphi$ is shown in Figure 10.8, for given voltage V_{S0} and speed $U(\omega_r) = 90$ m/s, for a segmented secondary L-RSM with $g_1 = 10$ mm airgap.

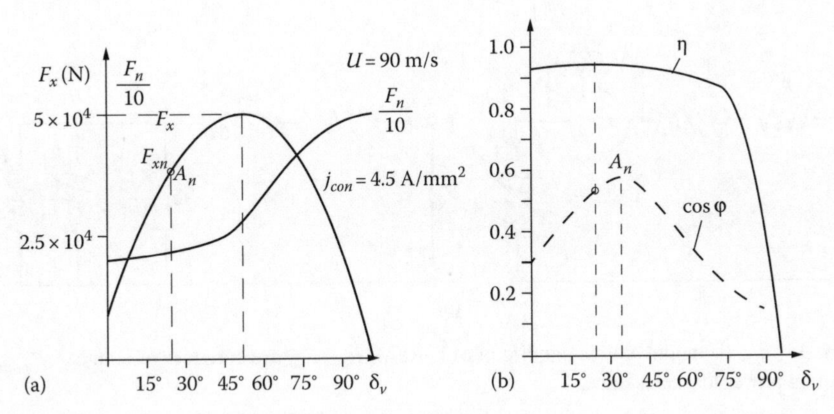

FIGURE 10.8 Typical steady-state performance on a high-speed L-RSM with segmented (laminated) secondary (track): (a) thrust and normal force and (b) η and cos φ.

A few remarks may be made at this point:

- The efficiency is larger (above 90%) over a wide range of power angle (loads).
- The power factor for rated thrust may be around 0.5, which is low, but L-RSM provides at least an 8/1 larger normal force than thrust and thus may serve for integrated propulsion–levitation in MAGLEVs ($a_{max} \approx 1.25$ m/s^2), if the L-RSM primary represents 12.5% of vehicle weight: this seems feasible as L-RSM is simpler and lighter than H-LSM. However, the low power factor implies a heavier and costlier PWM converter; today's forced cooled inverters may show less than 1 kg/1 kVA specific weight, which may render the system overall competitive.
- The vector control of L-RSM may provide both levitation and propulsion control for a multiple module primary on each vehicle. Propulsion control may be coordinated between units, while vehicle levitation (via flux) control may be robust and decentralized when a secondary and tertiary mechanical suspension are added to the bogie and, respectively, to the cabin. We should remember that even in active guideway dc-excited LSMs for propulsion and suspension systems, power factor per activated section is around 0.6; only H-LSM may show a power factor around (above) 0.8, thus saving weight (and cost) in the PWM converters on board, at the expense of heavier primary of H-LSM on board of vehicle.

10.6 STEADY-STATE CHARACTERISTICS FOR VECTOR CONTROL STRATEGIES

L-RSMs are always supplied from PWM voltage source inverters. Consequently, vector (or direct thrust and flux [DTFC]) control may be applied by using the space phasor (*dq*) model.

Three main vector control strategies are investigated here:

- Constant I_d control
- Constant $\alpha = I_d/I_q$ control
- Primary flux and thrust control

The L-RSM equivalent circuits (based on (10.10) through (10.14)), for transients in *d* and *q* axes, are shown in Figure 10.9.

From (10.10) to (10.13), after I_q elimination,

$$V_S^2 = \left(R_S^2 + \omega_r^2 L_d^2\right) I_d^2 + \left(R_S^2 + \omega_r^2 L_q^2\right)\left[\frac{2F_x\tau}{3\pi\left(L_d - L_q\right)I_d}\right]^2 - \frac{4R_S\omega_r\tau F_x}{3\pi} \quad (10.21)$$

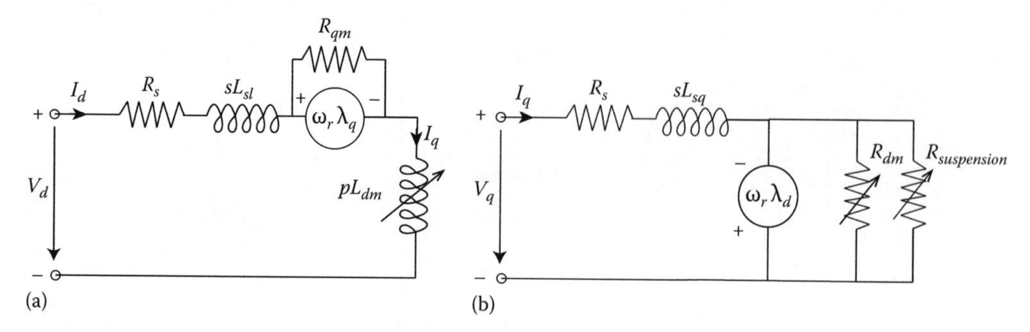

FIGURE 10.9 *dq* axis model equivalent circuit of L-RSM: R_{dm}, R_{qm} are core loss resistances, $R_{suspension}$ is the vertical motion power loss resistance.

with I_d = const., a mild (dc series motor)-type thrust/speed curve (F_x/ω_r), $\omega_r = \pi \cdot U/\tau$, is obtained; the no load ($F_x = 0$) speed U_0 is

$$U_0 = \tau\frac{\omega_r}{\pi} = \tau\frac{\sqrt{V_S^2 - R_S I_d^2}}{\pi L_d I_d}; \quad I_q = \frac{2}{3\pi}\frac{\tau F_x}{(L_d - L_q) I_d} \tag{10.22}$$

with I_d = const. and constant airgap g_1, the normal force varies with thrust. The total airgap flux is directly related to normal force:

$$F_n \approx -\frac{1}{g_1}\left(L_{dm}^2 I_d^2 + L_{qm}^2 I_q^2\right)\frac{3}{2} \tag{10.23}$$

So even with I_d = const. and constant g_1 (airgap), the normal force increases with thrust (Figure 10.10a). Also F_n is not related directly to speed (10.22) if I_d, I_q control is performed.

For $I_d/I_q = \alpha$, the stator flux λ_S is

$$\lambda_S = \sqrt{(L_d I_d)^2 + (L_q I_q)^2} = I_d\sqrt{L_d^2 + \left(L_q^2/\alpha^2\right)} \tag{10.24}$$

$$F_x = \frac{3}{2}\frac{\pi}{\tau}\left(L_d - L_q\right)\frac{\lambda_S^2\alpha}{L_d^2 + \left(L_q^2/\alpha^2\right)} \tag{10.25}$$

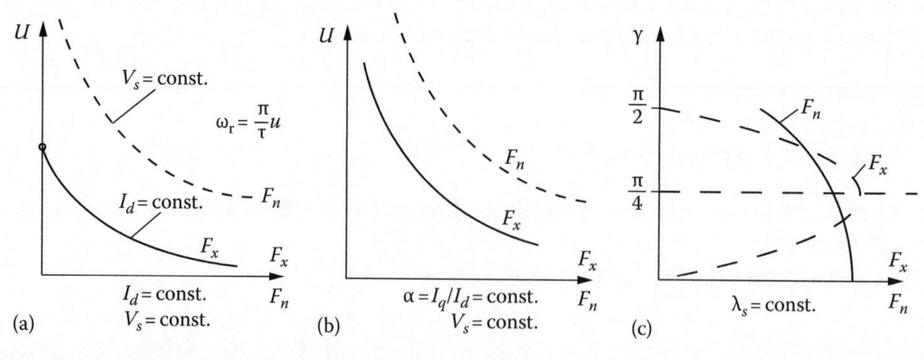

FIGURE 10.10 Three typical ways for vector control: steady-state characteristics; (a) I_d = const., V_S = const., (b) $\alpha = I_d/I_q$ = const., V_S = const., and (c) λ_S = const.

For given voltage $V_S \approx \omega_r \lambda_S$, λ_S is given and thus the maximum thrust $F_{x\max}$/stator flux λ_S is obtained for $\alpha = L_d/L_q$:

$$F_{x\max} = \frac{3\pi}{4\tau}\frac{(L_d - L_q)}{L_d L_q}\lambda_S^2 \tag{10.26}$$

The normal force is

$$F_n = -\frac{3}{2}\frac{1}{g_1}\frac{\lambda_S^2 (L_{dm} + L_{qm}\alpha^2)}{L_d^2 + (L_d^2/\alpha^2)} \tag{10.27}$$

Again, where α varies for constant V_S, when speed decreases, λ_S decreases, and thus both F_x and F_n decrease. Care must be exercised for noise and vibration variations (due to normal force) with thrust control. It is also feasible to directly control the DTFC.

In general, primary flux λ_S control will take care of airgap (levitation) control (or will maintain $\lambda_S \approx$ constant to keep the normal force within bounds), and the flux angle γ control will do thrust control:

$$\cos\gamma = \frac{L_d \cdot I_d}{\lambda_S}; \quad \sin\gamma = \frac{L_q \cdot I_q}{\lambda_S} \tag{10.28}$$

So

$$\begin{aligned} F_x &= \frac{3}{4}\frac{\pi}{\tau}(L_d - L_q)\frac{\lambda_S^2}{L_d \cdot L_q}\sin 2\gamma; \quad -\frac{\pi}{4} \le \gamma \le \frac{\pi}{4} \\ F_n &\approx -\frac{3}{2}\frac{1}{g_1}\lambda_S^2\left(\frac{L_{dm}}{L_d^2}\cos^2\gamma + \frac{L_{qm}}{L_q^2}\sin^2\gamma\right) \end{aligned} \tag{10.29}$$

The maximum thrust is obtained again for $\gamma = \pi/4$ and thus $L_d \cdot I_d = L_q \cdot I_q$, but now the thrust may be controlled through γ and normal force by λ_S; though the two functions are not fully decoupled, as seen in (10.29), F_n is less dependent on γ than F_x. So, if F_x and F_n are regulated in different frequency bands, their simultaneous control seems feasible.

10.7 DESIGN METHODOLOGY FOR LOW SPEED BY EXAMPLE

Let us consider the design of a short travel ($l_t \approx 2$ m) L-RSM that is able to produce a peak thrust $F_{x\max} = 1.6$ kN from zero to the maximum speed $U_{\max} = 2$ m/s for a short duty cycle (20%). The continuous duty thrust is $F_x = 0.8$ kN. An airgap $g = 0.5$ mm is mechanically feasible; the ALA secondary (track) is placed above the primary (mover); the latter is fed by a three-phase ac power flexible cable.

Solution

(a) Primary geometry and maximum mmf per slot

First, a reasonable saliency ratio L_{dm}/L_{qm} has to be secured. Consequently, a high ratio between the pole pitch τ and airgap is chosen: $\tau = 200$; $g_1 = 200 \cdot 5 \cdot 10^{-3} = 0.1$ m. For $U_{\max} = 2$ m/s, it would mean a frequency $f_{1max}=U_{max}/2\tau$, the unsaturated $k_{dm1}=0.97$; with the leakage inductance $L_{ls}=0.17\ L_m$, $L_{dm}/L_{qm}=15/1$, and a magnetic saturation coefficient $k_{Sd}=0.3$, the ratio L_d/L_q becomes

$$\frac{L_d}{L_q}=\frac{0.97(L_m/(1+k_{Sd}))+L_{lS}}{0.97(L_m/20)+L_{lS}}=\frac{(0.97/1.3)+0.07}{(0.97/15)+0.07}=6.06 \tag{10.30}$$

The maximum thrust is considered to be obtained in conditions of maximum thrust per flux: for $L_dI_d=L_qI_q$. If the d axis airgap flux density produced by I_d is B_{d1p}, the resultant airgap flux density B_{g1k} for maximum thrust

$$B_{d1p}=\frac{B_{1gk}}{\sqrt{1+(L_d/L_q)^2(L_{qm}(1+k_{Sd})/L_{dm})^2}} \tag{10.31}$$

Making sure that B_{g1k} is within reasonable limits—$B_{g1k}=0.75$ T—B_{d1p} is from (10.31):

$$B_{d1peak}=\frac{0.75}{\sqrt{1+6.06^2\cdot(1.3/15)^2}}=0.664\ \text{T} \tag{10.32}$$

On the other hand, this airgap flux-density component B_{d1peak} is related to the I_d component of phase current by

$$B_{d1p}=\frac{\mu_0\cdot 3\sqrt{2}\cdot k_{W_1}W_1\cdot I_{dpeak}\cdot k_{dm1}}{\pi\cdot k_C(1+k_{Sd})g_1\cdot p_1} \tag{10.33}$$

where $k_C\approx 1.3$ (Carter coefficient), p_1 – pole pairs, W_1 – turns per phase, $k_{W_1}=0.925$ ($q_1=2$, $y/\tau=5/6$).

From (10.33), we may calculate the phase d axis current component mmf, for peak thrust, W_1I_{dpeak}, as

$$W_1I_{d\,peak}=\frac{0.664\times\pi\times1.3\times1.3\times0.5\times10^{-3}\times p}{3\sqrt{2}\times1.256\times10^{-6}\times0.925}=358.5\times p_1 \tag{10.34}$$

Note: I_{df} and I_{qf} refer to RMS phase current components.

The q axis mmf peak value W_1I_{qpeak} is

$$W_1I_{q\,peak}=W_1I_{d\,peak}\cdot\frac{L_d}{L_q}=358.5\times6.06=2172\times p_1\ \text{A}\cdot\text{turns} \tag{10.35}$$

The peak thrust F_{xpeak} is

$$F_{x\,peak}=\frac{3}{2}\frac{\pi}{\tau}L_m\left(\frac{k_{dm1}}{1+k_{Sd}}-\frac{k_{qm1}}{k_{qm1}}\right)\left(I_{df\,p}\sqrt{2}\right)\left(I_{qf\,p}\sqrt{2}\right) \tag{10.36}$$

with $k_{dm1}/k_{qm1}=15$, $k_{dm1}=0.97$, $1+k_{Sd}=1.3$, $\tau=0.1$ m and

$$L_m = \frac{6\mu_0 \cdot \left(k_{W_1} \cdot W_1\right)^2 \cdot \tau \cdot L}{\pi^2 \cdot k_C \cdot g_1 \cdot p_1} \tag{10.37}$$

where again, $\tau=0.1$ m, $k_{W_1}=0.925$, $k_C=1.3$, $g_1=0.5$ mm, $F_{x\,\text{peak}}=1.6$ kN and we find

$$1600 = \frac{16\times1.256\times10^{-6}\times0.925^2\times358.5\times2172\times(p_1\cdot L)((0.97/1.3)-(1/15))}{\pi\cdot1.3\times0.5\times10^{-3}} \tag{10.38}$$

so $p_1 \cdot L=0.359$ m. It is now time to choose the number of poles, say $2p=6$, and thus, the stack width $L=0.359/3=0.12$ m. The peak thrust density $f_{x\,\text{p}}$ is

$$f_{x\,\text{p}} = \frac{F_{x\max}}{2p\tau L} = \frac{1600}{2\times3\times0.1\times0.12} = 2.22\times10^4\ \text{N/m}^2 = 2.22\ \text{N/cm}^2 \tag{10.39}$$

This may be termed as a good value (even for a linear PM synchronous motor of similar size and output).

(b) Primary slot design (Figure 10.11)
With $q_1=2$ slots/pole/phase, the peak value of slot mmf (RMS value), $n_1 I_{1\text{p}}$, is

$$n_1 \cdot I_{1\,\text{peak}} = \frac{W_1 \cdot I_{1\text{p}}}{p \cdot q} = \frac{358.5 \cdot \not{p_1} \cdot \sqrt{1+6.06^2}}{\not{p_1} \cdot 2} = 1100.945\ \text{A} \cdot \text{turns/slot} \tag{10.40}$$

n_1 is the turns per slot (two coils per slot)

For a duty cycle of 20%, the average peak slot mmf (RMS) $n_1 I_{avp}$ is

$$n_1 I_{avp} = n_1 I_{peak}\sqrt{0.2} = 1100.745\times\sqrt{0.2} = 492\ \text{A} \cdot \text{turns} \tag{10.41}$$

The slot pitch τ_s is

$$\tau_s = \frac{\tau}{2\times3} = \frac{100}{6} = 16.66\ \text{mm} \tag{10.42}$$

With teeth width $b_{t1}=\tau_s/2 \approx 8$ mm. So the slot depth h_s is

$$h_s = \frac{A_{slot}}{b_{bs1}} = \frac{307.5}{116.66-8} = 35.5\ \text{mm} \tag{10.43}$$

The primary stack thickness h_{c1} is

$$h_{c1} = \frac{B_{g1p}}{B_{cp}} \cdot \frac{\tau}{\pi} = \frac{0.75}{1.5} \times \frac{0.1}{\pi} = 0.016\ \text{m} \tag{10.44}$$

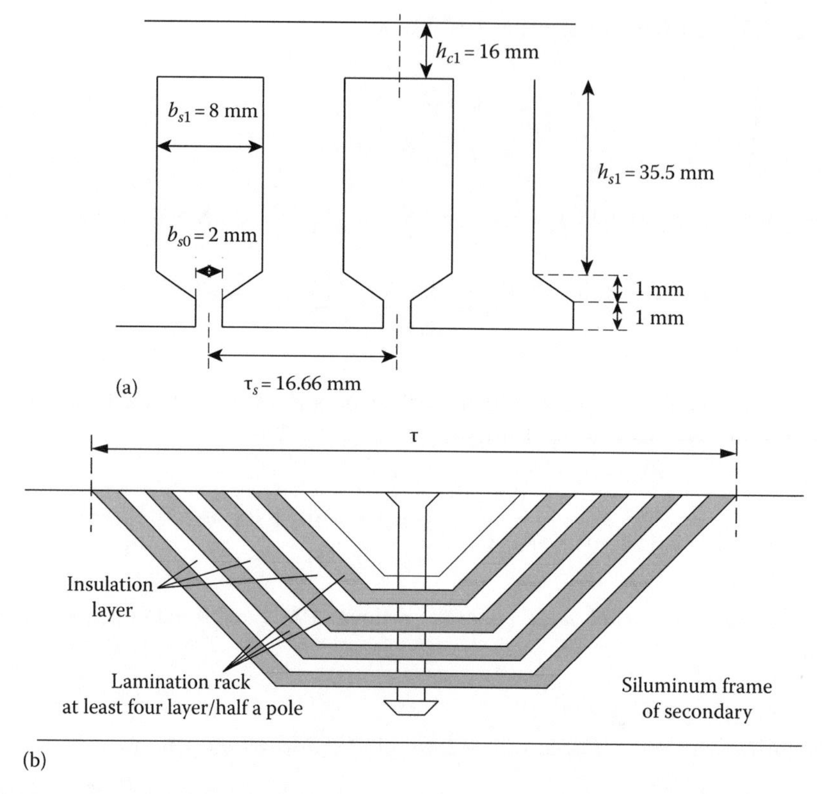

FIGURE 10.11 Primary slotting geometry (a) and ALA secondary topology (b) with $2p = 6$, but for a double-layer chorded ($y/\tau = 5/6$) coils.

The end slots are half field, and thus, the machine has $(2p+1)\cdot q\cdot m-1=41$ slots and 42 teeth. The total length of primary L_p is

$$L_p \approx 41\times\tau_s+\tau_s=42\times\tau_s\approx 700\text{ mm}=0.7\text{ m} \tag{10.45}$$

We may now calculate the circuit model parameters (Figure 10.9).

(c) Circuit parameter expressions

Let us notice that the number of turns/phase W_1 is not yet calculated, for a given voltage.

The phase resistance R_S is

$$R_S = 2\cdot\delta_{Co}\cdot\frac{(L+0.01+1.5y)\cdot W_1^2}{W_1 I_{avp}}J_{avco}$$

$$=2\cdot 2.3\cdot 10^{-8}\frac{(0.12+0.01+1.5\cdot(5/6)\cdot 0.1)}{6\times 492}\times 4\times 10^6\cdot W_1^2=1.589\times 10^{-5}\cdot W_1^2 \tag{10.46}$$

$$n_1\cdot p_1\cdot q=W_1;\quad W_1\cdot I_{avp}=p_1\cdot q\cdot n_1 I_{avp}=3\times 2\times 492 \tag{10.47}$$

$$l_C=2\cdot(L+0.01+1.5\cdot y)=2\times 0.255=0.51\text{ m/turn} \tag{10.48}$$

The unsaturated uniform airgap inductance L_m (10.2) is

$$L_m = \frac{6\mu_0 \left(k_{W_1} \cdot W_1\right)^2 \cdot \tau \cdot L}{\pi^2 p_1 \cdot k_C g_1} = \frac{6\times 1.256\times 10^{-6} \cdot 0.925^2 \times 0.12\times 0.1\times W_1^2}{\pi^2 \cdot 3\times 1.3\times 0.5\times 10^{-3}} = 4.239\times 10^{-6} \cdot W_1^2 \tag{10.49}$$

Consequently

$$L_{dm} = \frac{k_{dm1}}{1+k_{Sd}} \cdot L_m = \frac{0.97}{1.3} \times 4.239\times 10^{-6} \cdot W_1^2 = 3.1637\times 10^{-6} \cdot W_1^2 \tag{10.50}$$

$$L_{qm} = \frac{k_{qm1}}{k_{dm1}} \cdot L_m = \frac{1}{15} \times 4.239\times 10^{-6} \cdot W_1^2 = 0.2826\times 10^{-6} \cdot W_1^2 \tag{10.51}$$

The leakage inductance is

$$L_{ls} = \frac{2\mu_0}{p_1 \cdot q} W_1^2 \left(\lambda_{ss} \cdot L + \lambda_l \cdot l_{lc}\right) \tag{10.52}$$

The slot permeance λ_{ss} is

$$\lambda_{ss} = \frac{h_{s1}}{3b_{s1}} + \frac{h_{sa}}{b_{sa}} = \frac{35.5}{3\times 8} + \frac{1}{2} + \frac{2}{2+8} = 2.18 \tag{10.53}$$

$$\lambda_l = 0.3 \cdot q \cdot (3 \cdot \beta_1 - 1) = 0.3\times 2\times \left(\frac{3\times 5}{6} - 1\right) = 0.9 \tag{10.54}$$

The end connection (side) length

$$l_{lC} = 0.01 + 1.5y = 0.01 + 1.5\times \frac{5}{6} 0.1 = 0.135 \text{ m} \tag{10.55}$$

Finally, from (10.52) to (10.55),

$$L_{ls} = \frac{2\times 1.256\times 10^{-6}}{3\times 2} \cdot \left(2.18\times 0.1 + 0.9\times 0.135\right) \cdot W_1^2 = 0.142\times 10^{-6} \cdot W_1^2 \tag{10.56}$$

As can be seen, the leakage inductance ratio $L_{ls}/L_m = 0.142/3.1637 = 0.0449$; it is smaller than presumed ($L_{ls}/L_m = 0.078$), and thus, the design is safe.

The L_d/L_q ratio becomes now

$$\frac{L_d}{L_q} = \frac{\left(0.142 + 3.1637\right)\times 10^{-6} \cdot W_1^2}{\left(0.142 + 0.2826\right)\times 10^{-6} \cdot W_1^2} = 7.77 > 6.06 \tag{10.57}$$

(d) Turns per phase W_1

Let us consider that we have a $V_{dc} = 500$ Vdc power source available for the L-RSM drive. This voltage should be sufficient to deliver rated thrust, at good efficiency, and the peak (maximum) thrust at maximum speed $U_{max} = 2$ m/s.

The RMS phase voltage $V_{ph\max}$ is

$$V_{ph\max} = \frac{4k \cdot \text{Vdc}}{\pi\sqrt{6}} = \frac{4 \times 0.85 \times 500}{\pi\sqrt{6}} \approx 221 \text{ V (RMS/phase)} \tag{10.58}$$

To yield good performance at rated thrust, we may choose to produce it at maximum power factor, which corresponds to

$$\frac{I_q}{I_d} = \sqrt{\frac{L_d}{L_q}} = \sqrt{7.77} = 2.787 \tag{10.59}$$

With the same saturation level for maximum and rated thrust, the ratio of d axis mmf for the two forces is simply

$$\frac{F_x}{F_{x\max}} = \left(\frac{W_1 I_{d\max}}{W_1 I_{d\text{phase}}}\right)^{-2} \times \frac{\sqrt{L_d/L_q}}{L_d/L_q} \tag{10.60}$$

So

$$W_1 I_{d\text{phase}} = 3 \times 358.5 \times \sqrt{\frac{800}{1600} \times \frac{6.606}{2.781}} = 3 \times 423.26 \text{ A} \cdot \text{turns} \tag{10.61}$$

Consequently

$$W_1 I_{q\text{phase}} = W_1 I_{d\text{phase}} \cdot 2.787 = 1121.38 \times 2.787 = 3125.3 \text{ A} \cdot \text{turns} \tag{10.62}$$

These are phase mmf RMS values.

Let us now use the space phasor diagram where $V_{ph} \to V_S = V_{ph}\sqrt{2}$, $I_{d\text{ phase}} \to I_d = I_{d\text{ phase}}\sqrt{2}$, $I_{q\text{ phase}} \to I_q = I_{q\text{ phase}}\sqrt{2}$ to calculate the number of turns per phase W_1:

$$\begin{aligned} V_d &= R_S \cdot I_d - \omega_r \cdot L_q \cdot I_q \\ V_q &= R_S \cdot I_q + \omega_r \cdot L_d \cdot I_d \end{aligned} \tag{10.63}$$

$$V_{d\text{ phase}} = \frac{V_d}{\sqrt{2}} = 1.589 \times 10^{-5} \cdot \frac{W_1^2 I_d}{\sqrt{2}} - 62.8 \times 0.4246 \times 10^{-6} \cdot \frac{W_1^2 I_q}{\sqrt{2}} = -0.0652 \cdot W_1 \tag{10.64}$$

$$V_{q\text{ phase}} = \frac{V_q}{\sqrt{2}} = 1.589 \times 10^{-5} \cdot W_1^2 I_q + 62.8 \times 3.305 \times 10^{-6} \cdot W_1^2 I_d = 0.281 \cdot W_1 \tag{10.65}$$

$$V_{ph} = \sqrt{V_{d\,\text{phase}}^2 + V_{q\,\text{phase}}^2} = W_1\sqrt{0.0652^2 + 0.281^2} = 220 \tag{10.66}$$

$$W_1 = \frac{220}{0.2886} = 762 \text{ turns/phase} \tag{10.67}$$

The number of turns per coil n_c is

$$n_c = \frac{n_1}{2} = \frac{W_1}{2pq} = \frac{762}{6\times 2} \approx 63 \text{ turns/coil} \tag{10.68}$$

Now,

$$I_{d\,\text{phase}} = \frac{W_1 I_{d\,\text{phase}}}{W_1} = \frac{1121.38}{756} = 1.4833 \text{ A} \tag{10.69}$$

$$I_{q\,\text{phase}} = \frac{W_1 I_{q\,\text{phase}}}{W_1} = \frac{3125.3}{756} = 4.134 \text{ A} \tag{10.70}$$

The phase current (RMS) for rated thrust ($F_{xn} = 800$ N) is

$$I_1 = \sqrt{I_{d\,\text{phase}}^2 + I_{q\,\text{phase}}^2} = \sqrt{1.484^2 + 4.134^2} = 4.392 \text{ A} \tag{10.71}$$

So the product $\eta_n \cos\varphi_n$ is

$$\eta_n \cos\varphi_n = \frac{F_x U_{\max}}{3V_{\text{phase}} I_1} = \frac{800.2}{3.22\times 4.392} = 0.552 \tag{10.72}$$

The copper losses p_{con} are

$$p_{con} = 3R_S \cdot I_1^2 = 3\times 4.392^2 \times 1.589\times 10^{-5} \times 756^2 = 525 \text{ W} \tag{10.73}$$

With $f_{1n} = 10$ Hz, core losses may be neglected, and thus, efficiency and power factor at $F_{xn} = 800$ N and $U_{\max} = 2$ m/s are

$$\eta_n = \frac{1}{1 + p_{con}/F_{xn}U_{\max}} = 0.7529; \quad \cos\varphi_n = \frac{\eta_n \cos\varphi_n}{\eta_n} = \frac{0.552}{0.7529} = 0.733 \tag{10.74}$$

This is quite good performance at 2 m/s and 1.6 kW of mechanical power.

(e) Verification of peak thrust at maximum speed

As we designed the L-RSM at rated thrust and maximum voltage, a verification of machine capability to produce the maximum thrust at the same (maximum) speed is required. For the purpose, we use again the voltage equations, but now with

$$W_1 I_{d\,\text{peak}} = 1075.5 \text{ A}\cdot\text{turns} \tag{10.75}$$

$$W_1 I_{q\,\text{peak}} = 1075.5 \times 6.06 = 6453 \ \text{A} \cdot \text{turns} \tag{10.76}$$

From (10.63),

$$V_{d\,\text{phase}} = (1.589 \times 10^{-5} \times 1075.5 - 62.8 \times 0.4246 \times 10^{-6} \times 6453) \times 756 = -117.44 \ \text{V} \tag{10.77}$$

$$V_{q\,\text{phase}} = \left(1.589 \times 10^{-5} \times 6.453 + 62.8 \times 3.305 \times 10^{-6} \times 1075.5\right) \times 756 = 246.2 \ \text{V} \tag{10.78}$$

$$V_{\text{phase}} = \sqrt{117.4^2 + 246.2^2} = 272.84 \ \text{V} \tag{10.79}$$

Note: As seen from (10.77), the L-RSM is not capable to produce the maximum thrust $F_{x\max} = 2F_{xn} = 1.6$ kN at 2 m/s at 220 V. This simply means that the voltage for rated thrust should be chosen smaller; this would lead to a smaller number of turns per coil and larger current, but same efficiency and power factor. The main drawback of this choice is that the peak kVA of the inverter will end up larger, and so will its cost.

(f) Primary active weight

The primary active weight comprises iron core (G_{iron}) and copper weight, G_{Co}:

$$G_{Co} = 3W_1 \frac{I_{av}}{J_{av\,Co}} \times 2\left(L + 0.01 + 1.5 \times \frac{5}{6} \times 0.1\right) \times \gamma_{Co}$$

$$= 3.756 \times \frac{492}{63 \times 2 \times 4 \times 10^6} \times 2 \times 0.255 \times 8900 = 10.05 \ \text{kg} \tag{10.80}$$

$$G_{iron} = G_{teeth} + G_{core} \approx (43 \times 37.5 \times 10^{-3} \times 8 \times 10^{-3} \times 0.12 + 0.7 \times 0.12 \times 0.016) \times 7600$$

$$= 21.98 \ \text{kg} \tag{10.81}$$

So the active weight of primary G_{a1} is

$$G_{a1} = G_{Co} + G_{iron} = 10.05 + 21.98 \approx 32 \ \text{kg} \tag{10.82}$$

The peak thrust/primary weight = 1600/32 ≈ 50 N/kg, which would mean an ideal acceleration of more than 5 g_{grav}.

10.8 CONTROL OF L-RSM

So far we have dealt with the steady-state theory and performance of L-RSM with notable attention to its electromagnetic design. Some main vector control possibilities have been also introduced through their potential characteristics, to assist in the design. Here, we will introduce a rather new vector control scheme for propulsion and the direct thrust and normal force control for integrated propulsion and suspension control.

10.8.1 "Active Flux" Vector Control of L-RSM

The active flux [8] ψ_d^a is defined as

$$\psi_d^a = \overline{\psi}_S - L_q \cdot \overline{I}_S = \left(L_d - L_q\right) I_d \tag{10.83}$$

The thrust F_x is thus

$$F_x = \frac{3}{2}\frac{\pi}{\tau}\psi_d^a \cdot I_q \tag{10.84}$$

The *dq* model may be written in terms of active flux by replacing $\overline{\psi}_S$ in (10.10) by using (10.81):

$$\overline{I}_S \cdot R_S - \overline{V}_S = -\frac{d\psi_d^a}{dt} - j\omega_r \cdot \psi_d^a - L_q \frac{d\overline{I}_S}{dt} - j\omega_r \cdot L_q \frac{d\overline{I}_S}{dt} \tag{10.85}$$

So in fact by the active flux concept, we turn the anisotropic (salient pole) machine into an isotropic machine with inductance L_q along both *d* and *q* axes.

For a direct vector control scheme, the active flux has to be estimated. It suffices to estimate properly the stator flux $\overline{\psi}_S$ (by a combined voltage and current model) and then use (10.83) to obtain ψ_d^a (Figure 10.12). As evident from (10.83), the active flux is always aligned to axis *d*, irrespective of thrust, which means simpler mover position estimation in a sensorless control system.

For very low speeds, including standstill, the same active flux observer may be used but with an additional high-frequency voltage signal injection, which, after filtering, yields the initial mover position (for nonhesitant heavy starting) and the position at better precision for very low speeds. The signal injection position observer part corrects the fundamental one, and then it is made to fade away with increasing speed. For absolute positioning of the mover, either a linear encoder is provided or, at least, certain calibration position information is available.

A typical vector control system is shown in Figure 10.13.

The control scheme in Figure 10.13 is characterized by

- Combined PI and sliding mode (robust) active flux and I_q close-loop regulation and a robust PI + SM speed regulator.
- The ac voltages are not measured, but a combination of reference voltages V_a^*, V_b^*, V_c^* and the dc voltage info is used, instead. The active flux observer is provided with PLL position and speed estimators (Figure 10.12).
- An emf compensation may be added to yield V_d^* and V_q^*, but since active flux is regulated and the regulators are robust, such an addition may not be necessary.

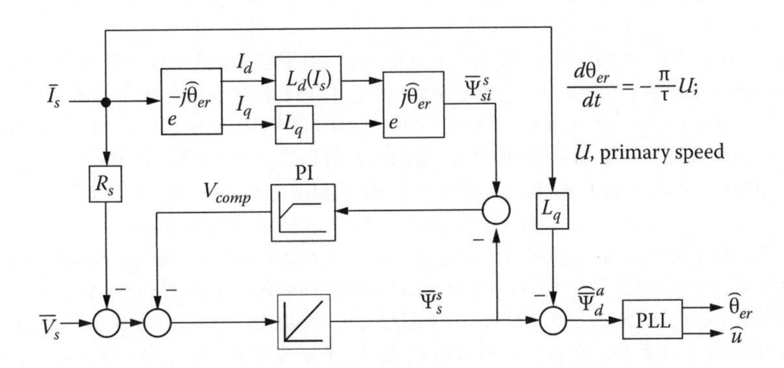

FIGURE 10.12 Generic active flux observer with PLL mover position and speed estimator.

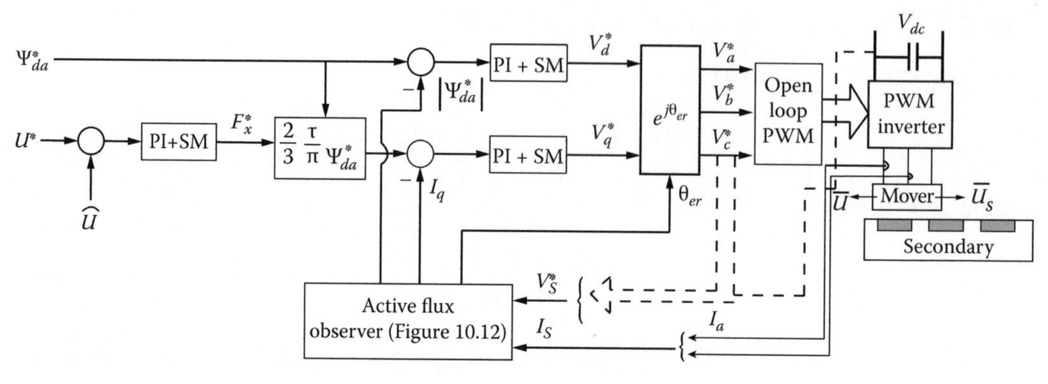

FIGURE 10.13 Generic active flux vector speed sensorless control of L-RSM.

10.8.2 Direct Thrust and Normal Force (Levitation) Control

When integrated propulsion–levitation control is targeted (MAGLEVs), the control is more elaborated but the mover position and speed estimation are again necessary.

In this case, however, the thrust as well as the normal force have to be estimated:

$$\widehat{F}_x = \frac{3}{2}\frac{\pi}{\tau}\,\text{Real}\left(j\overline{\psi}_S \cdot \widehat{I}_S^*\right) = \frac{3}{2}\frac{\pi}{\tau}\left(\psi_\alpha \cdot I_\beta - \psi_\beta \cdot I_\alpha\right) \tag{10.86}$$

$$F_n \approx -\frac{3}{2}\frac{1}{g_1}\left(\widehat{L}_{dm} \cdot I_d^2 + \widehat{L}_{qm} \cdot I_q^2\right) \tag{10.87}$$

From the active flux observer (Figure 10.12), we may further estimate

$$\left(\widehat{L}_{dm} - \widehat{L}_{qm}\right) \cdot I_d = \left|\overline{\psi}_d^a\right| \tag{10.88}$$

In a first approximation, $\widehat{L}_{dm}/\widehat{L}_{qm} \approx$ constant and known from design.

Consequently, both $\widehat{L}_{dm}$ and $\widehat{L}_{qm}$ (magnetization inductances) may be estimated. The airgap may be measured by a dedicated sensor or (at least for redundancy) it may be estimated, as $\widehat{L}_{dm}$ is (for constant saturation) inversely proportional to airgap.

The currents I_d and I_q are available in the active flux observer and so are the mover position and speed estimations. It is inferred here that a robust (PI+SM) airgap control is feasible for a multiunit vehicle.

A potential thrust and airgap via flux control (DTFC) for L-RSM for integrated propulsion–levitation are introduced in Figure 10.14.

- There are three regulators in row, for airgap normal force and active flux control; as they are all robust, the numerous nonlinearities of L-RSM and the track irregularities may be handled even in presence of normal force perturbations from other units on board of vehicle.
- The transformation of coordinates is skipped for DTFC, but a table of switching sequences in the inverter is defined as the angles of both stator flux $\overline{\psi}_S - \widehat{\theta}_{\overline{\psi}_S}$ and of mover position (axis d) $\overline{\psi}_d^a - \widehat{\theta}_{\overline{\psi}_d^a}$ are available from the active flux extended observer.
- The active flux observer implies that L_d and L_q are known from design or from a commissioning sequence at high speeds, which may simply be invented for the purpose.
- In addition to the active flux observer estimates, both the propulsion and normal forces and finally also the airgap are estimated, if $L_{dm}(g_1)$ is known from the design stage or measured at standstill.

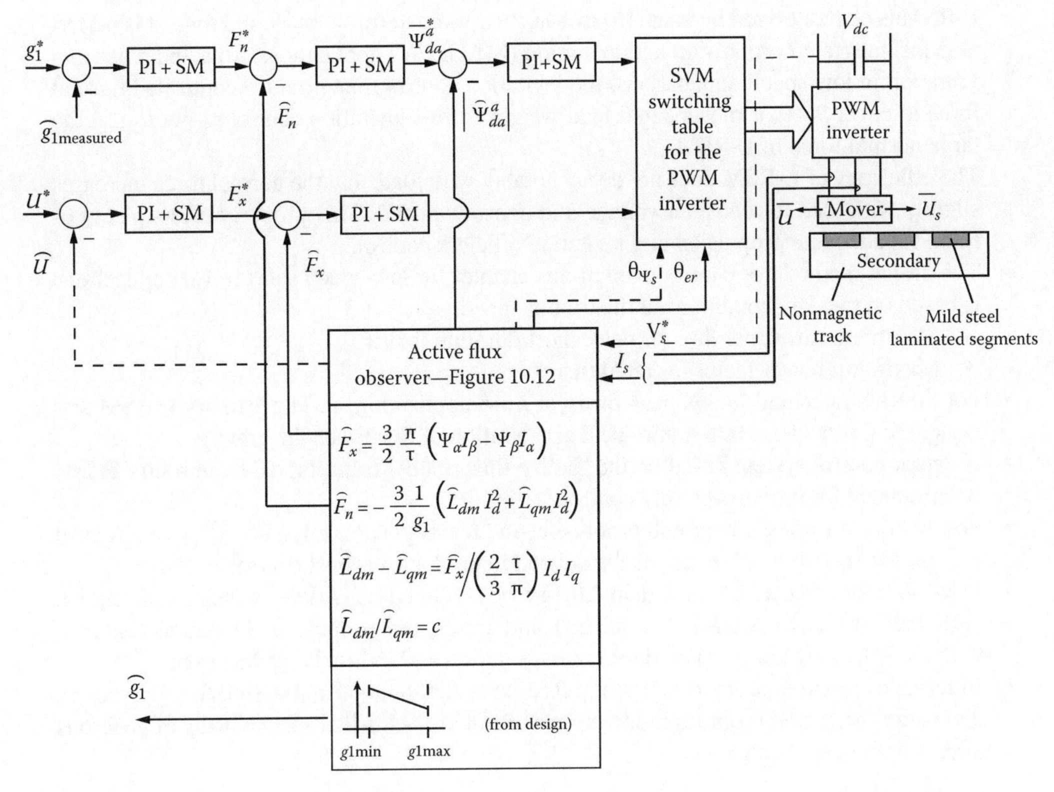

FIGURE 10.14 DTFC of L-RSM for integrated propulsion and suspension control in MAGLEVs.

10.9 SUMMARY

- Linear reluctance synchronous motors have a uniformly slotted primary core that hosts a three-phase ac winding and a variable reluctance laminated passive long secondary; the absence of PMs and of a secondary winding is the main merit of L-RSM in terms of costs/performance.
- Medium- and high-speed transportation and low-speed short travel industrial propulsion (and controlled magnetic suspension) in MAGLEVs or wheel (or linear bearing) suspension are the main applications of L-RSMs.
- The rather large airgap ($6 \div 10$ mm) in medium- to high-speed MAGLEVs leads to a laminated–segmented pole secondary with medium saliency ratio (for costs reason) that provides good efficiency, but at $0.5 \div 0.6$ power factor, which means high PWM inverter kVA on board of vehicle; as the L-RSM is capable to provide integrated propulsion and levitation control with standard PWM inverter vector DTFC, this drawback may be an acceptable compromise.
- For low-speed ($1 \div 3$ m/s) short travel transport applications in industry at an airgap of $0.5 \div 1$ mm), good saliency ($L_{dm}/L_{qm} = 6 \div 8$) may be obtained with MFB (axially laminated anisotropic equivalent) secondary; thus, a power factor above 0.7 may be obtained for reasonable efficiency ($75\% \div 80\%$): this compares reasonably even with linear PM synchronous motors of similar thrust and speed, as seen in subsequent chapters.
- With $2p \geq 6$ poles, the dynamic (and static) longitudinal effect is negligible, and thus, the theory of rotary RSMs may be applied to L-RSM; for small L-RSMs (with $2p \geq 2, 4$), direct 2(3)D FEM analysis is to be used to precisely assess the performance, as current asymmetry due to static longitudinal end effect is notable.

- L-RSM is characterized by an 8 (15) to 1 normal force to thrust ratio, and thus, it is preferable for integrated propulsion and levitation (MAGLEV) applications. To avoid noise and vibration in low-speed applications, the levitation control may produce controlled normal force to cover 90% of mover rated load weight or full levitation control to put to use the large normal force of L-RSM.
- The efficiency of L-RSM does not decay notably with load, and the normal force increases slightly with load for constant voltage and constant speed. This property offers plenty of room for adequate propulsion and levitation efficient control.
- The design methodology introduced in this chapter for low-speed short travel applications is based on two key conditions at maximum speed:
  - Maximum thrust per flux for peak short duration thrust
  - Maximum power factor for rated thrust
- For a 0.8 kN rated and 1.6 kN peak thrust at 2 m/s application, a 32 kg primary L-RSM was designed; it provides a maximum ideal acceleration of five times the gravity.
- A vector control system based on the "active flux" model transplanted from rotary RSMs is introduced for propulsion-only control.
- For motion sensorless integrated propulsion and levitation control, a DTFC system (based also on the "active flux" model) is introduced to produce a robust control system.
- L-RSM, better in performance than LIM in well-defined conditions, simple, and rugged, with low-cost passive track (secondary) and integrated propulsion–levitation features, without PMs, may not be ruled out in transportation and industrial applications.
- In terms of research, core loss, thrust pulsation reduction, optimal design for application, and better integrated propulsion–levitation control are fields that seem worthy of generous talents in the near future.

REFERENCES

1. P. Lawrenson and S.K. Gupta, Developments in the performance and theory of segmental rotor reluctance motors, *Proc. IEEE*, 114(5), 1967, 645–653.
2. B.J. Chalmers and AC Williams, *Ac Machines: Electromagnetics & Design*, John Wiley & Sons, New York, 1991, 75–85.
3. A.I.O. Cruickshank and R.W. Menzies, Axially laminated anisotropic rotors for reluctance motors, *Proc. IEEE*, 113, 1966, 2058–2060.
4. I. Boldea, *Reluctance Synchronous Machines and Drives*, Clarendon Press, Oxford, U.K., 1996.
5. A. Vagati, G. Franceskini, I. Marongiu, and G.P. Troglia, Design criteria for high performance synchronous reluctance motors, *Record of IEEE—IAS-1992*, Houston, TX, vol. 1, pp. 66–73.
6. I. Boldea, Z.H. Fu, and S.A. Nasar, Performance evaluation of ALA rotor reluctance synchronous machine, *Record of IEEE—IAS-1992*, Houston, TX, vol. 1, pp. 212–218.
7. S. Masayuki, S. Morimoto, and Y. Takeda, Thrust ripple improvement of linear synchronous motor with segmented mover construction, *Record of LDIA 2001*, Nagano, Japan, pp. 451–455.
8. I. Boldea, M.C. Paicu, and G.D. Andreescu, Active flux concept for motion-sensorless unified ac drives, *IEEE Trans.*, PE-23(5), 2008, 2612–2618.

11 Linear Switched Reluctance Motors (L-SRM)

Modeling, Design, and Control

Linear switched reluctance motors (L-SRMs) are counterparts of rotary switched reluctance motors [1–3]. The finite length of their magnetic circuit along the direction of motion introduces, for a small number of poles, some longitudinal static effects (asymmetries between phases). Also, the normal force has its peculiarities with L-RSM, but other than that, the theory developed for rotary SRMs may be adapted easily for linear SRMs.

So, as for rotary SRMs, L-SRM produces a thrust and motion by the tendency of a ferromagnetic secondary to hold a position where inductance of the active (excited) primary phase is maximum. Though it is in general a multiphase machine, L-SRM operates with 1(2) phases as active at any time. The turning on and off of each phase is triggered by a linear position sensor or by an observer so as to fully exploit the thrust capability of that phase. So the turning on and off of each phase is "in tact" (synchronous) with rotor position; however, each phase is fully energized and deenergized for each thrust cycle, and thus L-SRMs are not traveling field synchronous machines (like L-RSMs in the previous chapter). This brings L-SRM closer to linear stepper motors, but with position control of each phase current. Finally, the mutual inductances between phases in L-SRMs are small, which is good for fault tolerance but it is the reason why all the energization and deenergization cycling process takes place through the special PWM inverter (multiphase dc–dc converter) that supplies it.

The rather low thrust density of L-SRM (unless very heavily saturated) in comparison with linear PM synchronous motors) is another secondary effect of its simplicity and low cost. The L-SRM's ruggedness and rather low losses recommend it for linear motion application in various industries.

11.1 PRACTICAL TOPOLOGIES

L-SRMs have a longitudinal or transverse lamination uniformly slotted primary core with tooth-wound (concentrated) 1, 2, 3, 4, or more phase coils and a variable reluctance longitudinal or a transverse-lamination secondary core. The pole pitches τ_s, τ_p of the slottings of the primary and secondary are in general

- $\tau_p = \tau_s$ ($N_s = N_r$) one-phase cores for single-phase L-SRMs with "stepped" airgap for self-starting (Figure 11.1)
- $\tau_p/\tau_s = 4/6$ or 8/6 ($N_s = 6$, $N_r = 4$, 8 or multiples of these values) for three phases (Figure 11.2)
- $\tau_p/\tau_s = 6/8$ ($N_s = 8$, $N_r = 6$, 10 etc.) for four phases (Figure 11.3)

The principle is that each secondary pole is attracted to the active phase poles when the latter is supplied with a PWM voltage to produce a mover-position-triggered phase unipolar current in that phase.

While single-sided, flat, three-phase configurations with longitudinal and transverse flux are shown in Figure 11.2, a double-sided flat four-phase configuration is shown in Figures 11.3. In Figures 11.1 through 11.3, the mover is active (primary), and thus it must be supplied by a flexible ac power cable in short travel applications or by a mechanical (or inductive) power transfer on board in transportation (or special clean room) applications.

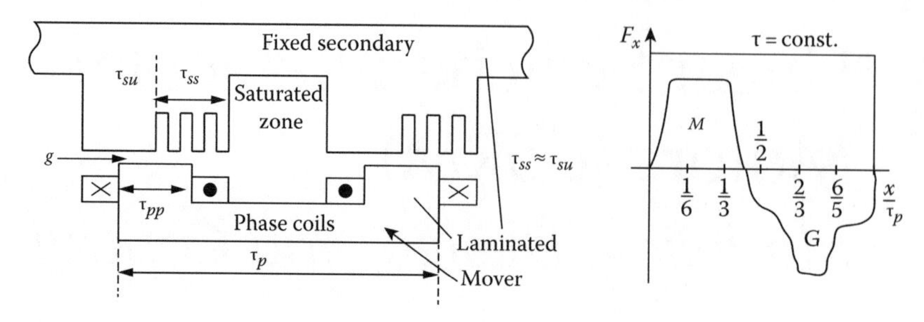

FIGURE 11.1 Single-phase, saturated, half-secondary pole, flat, single-sided L-RSM and its thrust versus position for constant current fixed secondary.

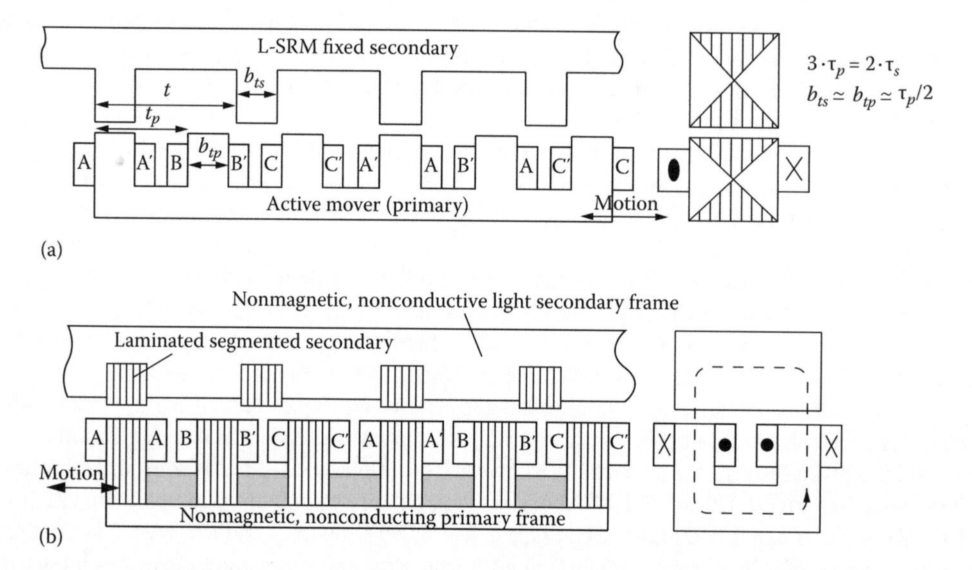

FIGURE 11.2 Three-phase, flat, single-sided L-SRM: (a) longitudinal and (b) transverse.

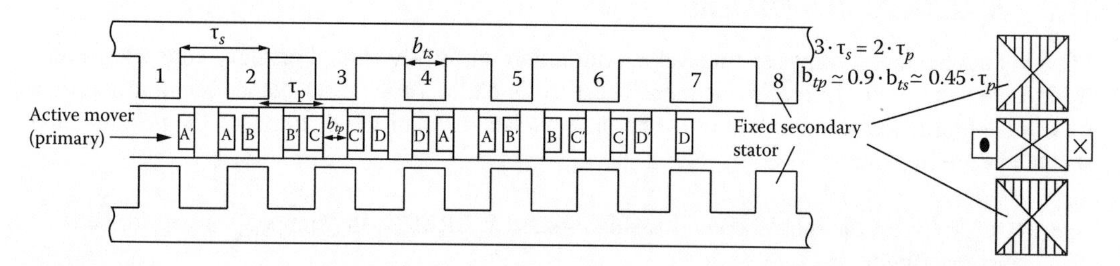

FIGURE 11.3 Four-phase, double-sided, flat L-RSM with active mover (primary) and longitudinal flux.

Tubular configurations may also be adopted for short travel taking advantage of the "blessing" of circularity (Figure 11.4). The tubular L-SRM cores may be made of soft magnetic composites (tooth by tooth in the primary) or from two types of disk-shaped laminations with radial slits, to reduce eddy current losses in the back cores (in not so high fundamental frequency (below 100 Hz) applications).

The ratio between normal force and thrust in L-RSM is rather large ($F_n/F_x > 10$), and thus, unless F_n is required for levitation, it is just a source of noise, vibration, and, consequently of stress in the iron cores and in the frame. Double-sided flat and tubular topologies implicitly provide zero-ideal normal force on the interior part of L-SRM [2–5].

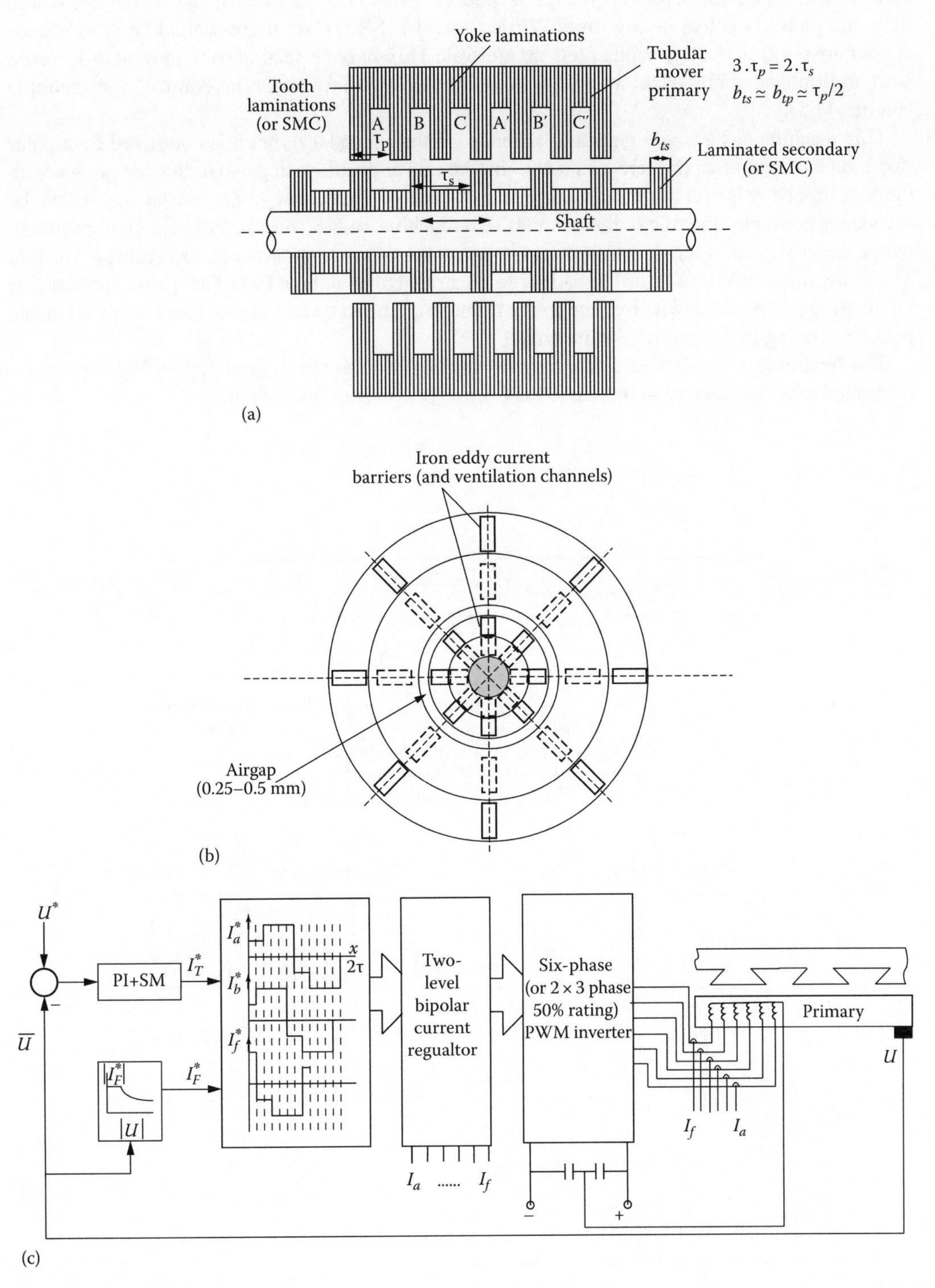

FIGURE 11.4 Three-phase tubular L-RSM with disk-shaped laminations, (a) radial slits (as barriers for eddy currents) (b), and generic control system (c).

The transverse-flux TF-L-SRM is characterized by shorter flux lines and possibly lower core losses, while the iron core usage is poorer, especially in three-phase L-RSMs, where only one phase is active at any time. While typical L-SRMs use tooth-wound coils to reduce copper losses drastically, segmented or multiple flux-barrier secondaries may also be used with multiphase diametrical windings and bipolar two-level current control, for example (Figure 11.5).

This machine has a single magnetic saliency (in the secondary), but it is controlled by bipolar two-level current pulses triggered "in tact" with the mover position. In general, the two phases with the coils around axis q (of lower inductance) play the role of "excitation" coils, and the ones below the secondary poles play the role of thrust coils. This machine, in fact (here in its linear configuration), rather directly mimics the dc brush machine with separate excitation but in its brushless version. Therefore, this would be a "true" brushless dc linear machine without PMs. One phase commutates all the time but the rest are active, and thus a rather high thrust density is provided, with reasonable peak kVA rating in the multiphase inverter.

The field current control level I_p^* may be performed to control the airgap and thus provides controlled magnetic levitation in a MAGLEV with good overall performance.

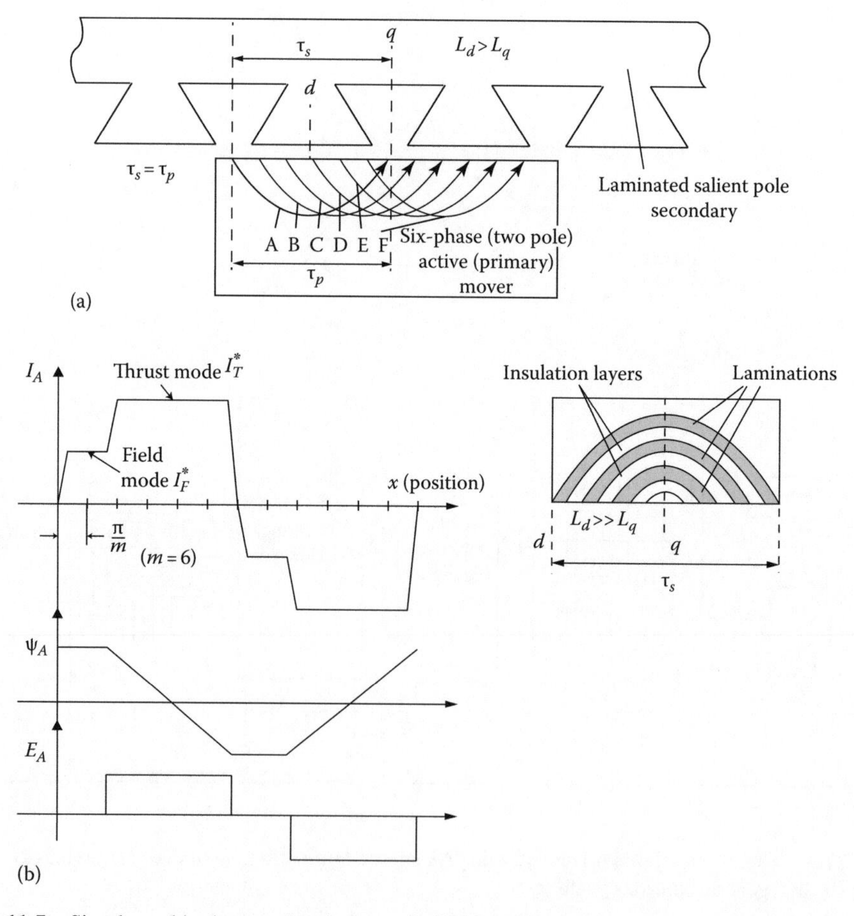

FIGURE 11.5 Six-phase, bipolar, two-level current L-SRM (a), its phase current waveform (b)

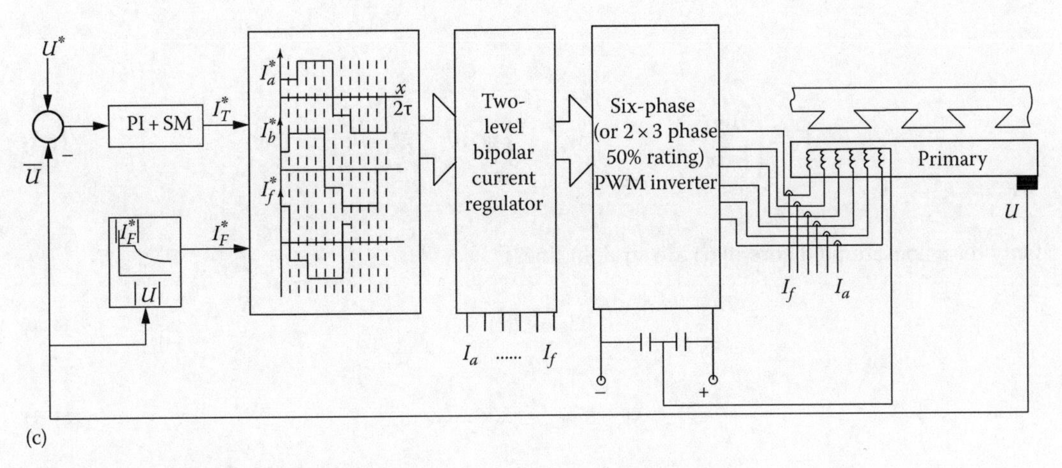

FIGURE 11.5 (continued) An ALA secondary generic control system (c).

11.2 PRINCIPLE OF OPERATION

Let us consider an elementary single-phase L-SRM (Figure 11.6).

To assess the thrust correctly, the energy conservation principle has to be applied:

$$\underset{\substack{\text{electric}\\\text{input}\\\text{energy}}}{dW_e} = \underset{\substack{\text{mechanical}\\\text{output}\\\text{energy}}}{dW_{mec}} + \underset{\substack{\text{stored}\\\text{magnetic}\\\text{energy}}}{dW_m}; \quad dW_{mec} = F_x dx \tag{11.1}$$

As demonstrated in Chapter 1 the thrust is

$$F_x = -\left(\frac{\partial W_m}{\partial x}\right)_{\psi=const.}; \quad W_m = \int_0^{\psi} i \cdot d\psi \tag{11.2}$$

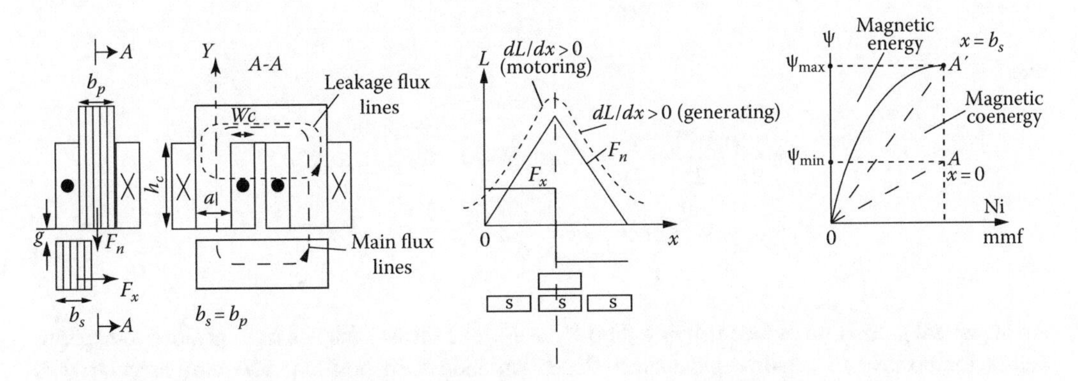

FIGURE 11.6 Elementary single-phase L-SRM.

or

$$F_x = \left(\frac{\partial W_m'}{\partial x}\right)_{i=const.}; \quad W_m' = \int_0^i \psi \cdot di; \quad W_m + W_m' = i \cdot \psi \tag{11.3}$$

Only for a nonsaturated core ($\psi(i)$ are straight lines)

$$L = L(x) \tag{11.4}$$

$$W_m = W_m' = \frac{L(x) \cdot i^2}{2} \tag{11.5}$$

And thus the thrust F_x and the normal attraction force F_n are

$$F_{xi} = \frac{i^2}{2}\frac{\partial L}{\partial x}; \quad F_{ni} = \frac{i^2}{2}\frac{\partial L}{\partial g} \tag{11.6}$$

For a simplified case, when all fringing flux lines (at the airgap) are neglected and the iron cores permeability is infinite, the inductance L in Figure 11.6 may be written as

$$L = L_{leakage} + \mu_0 W_1^2 \frac{a \cdot x}{2g}; \quad 0 \le x \le b_p = b_s \tag{11.7}$$

$$L = L_{leakage} + \mu_0 W_1^2 \frac{a \cdot (2b_p - x)}{2g}; \quad b_p \le x \le 2b_p \tag{11.8}$$

In this case,

$$F_{xi} \approx \frac{\mu_0 W_1^2 i^2}{2} \cdot \frac{a}{2g} = \text{constant} > 0 \quad \text{for } 0 \le x \le b_p = b_s$$

$$F_{ni} \approx -\frac{\mu_0 W_1^2 i^2}{2} \cdot \frac{a \cdot x}{2g^2} < 0 \tag{11.9}$$

and

$$F_x = -\frac{\mu_0 W_1^2 i^2}{2} \cdot \frac{a}{2g} = \text{constant} < 0 \quad \text{for } b_p \le x \le 2b_p$$

$$F_n = -\frac{\mu_0 W_1^2 i^2 a(2b_p - x)}{g^2} < 0 \tag{11.10}$$

As in general $g<b_p/4$ (to reduce fringing flux) $F_y \gg F_x$. The thrust switches from positive to negative values for positive to negative inductance slopes variation with position. The current polarity is irrelevant for thrust and normal force and thus, in general, the L-SRM is supplied from a multiphase dc–dc converter.

Example 11.1

Let us consider an elementary single-phase L-SRM with the data: airgap $g=0.3$ m, U-shape core leg width $a=10$ mm, the primary and secondary pole length $b_p=b_s=10$ mm. Calculate

1. Thrust and maximum normal force for $\mu_{iron}=\infty$, $Ni=500$ A turns
2. The normal airgap flux density B_{g1}
3. Copper losses for $j_{Co}=3$ A/mm² and coil equal width and height ($w_c=h_c$)

Solution

1. We may use directly Equation 11.9 for $x=b_p$:

$$F_x = \frac{1.1256\times10^{-6}\times500^2}{2}\times\frac{10^{-2}}{2\times0.3\times10^{-3}} = 2.616\text{ N} \tag{11.11}$$

$$F_{nm\,ax} = -\frac{1.1256\times10^{-6}\times500^2\times10^{-2}\times10^{-2}}{4\times0.3^2\times10^{-6}} = -87.22\text{ N} \tag{11.12}$$

With a total active area of primary $A=2a\cdot b_p=2$ cm², it means a specific thrust $f_x=$ 1.308 N/cm² and a specific maximum normal thrust $f_{nmax}=-43.61$ N/cm².

2. The normal airgap flux density B_n is

$$B_n = \frac{\mu_0 WI}{2g} = \frac{1.256\times10^{-6}\times500}{2\times0.3\times10^{-3}} = 1.0466\text{ T} \tag{11.13}$$

3. There are two semicoils. The cross section of each of them, A_{Co}, is

$$A_{Co} = \frac{\left(\frac{WI}{2}\right)}{j_{Co}k_{fill}} = \frac{\frac{500}{2}}{3\times0.50} = 166.66\text{ mm}^2 \tag{11.14}$$

But $A_{Co} = h_c\cdot w_c = h_c^2 = 166.66\text{ mm}^2$, $h_c=w_c=12.91$ mm.

The turn average length l_c is

$$l_c \approx 2a+2b_p+4W_c = 2\times10^{-2}+2\times10^{-2}+4\times1.291\times10^{-2} = 9.164\times10^{-2}\text{m} \tag{11.15}$$

So the copper losses are

$$p_{Co} = \frac{2\delta_{Co}l_cWj_{Co}i^2}{i} = 2\delta_{Co}l_cWj_{Co}i = 2.21\times10^{-8}\times9.164\times10^{-2}\times250\times3\times10^6 = 2.886\text{ W} \tag{11.16}$$

Note: Though the numerical example is elementary, the force densities f_x, f_n, and the thrust/losses (N/W) are typical for well-designed practical L-SRMs.

11.3 INSTANTANEOUS THRUST

The thrust in (11.9) and (11.10) is constant with mover position but, in reality, due to magnetic saturation variation and due to airgap fringing flux, it is far away from constancy and varies notably with position x and current as seen in what follows. To yield instantaneous thrust, the flux linkage/current/position curves have to be calculated, either analytically (with a multiple magnetic circuit model) or by 2(3) D-FEM; alternatively, it may be measured at standstill from current-decay tests with the primary to secondary position fixed in numerous different situations.

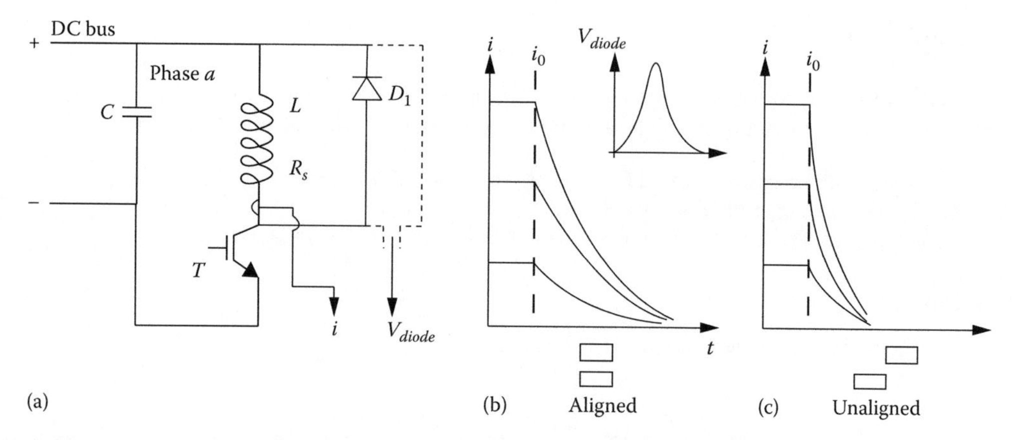

FIGURE 11.7 Standstill test setup to yield flux/current/position curve family decay in aligned and unaligned position: (a) electrical circuit, current decay for aligned, (b) unaligned, and (c) positions.

The dc–dc converter typical for the test is shown in Figure 11.7.

By PWM a certain initial dc current i_0 is installed, and then the switch S is turned off and the current and freewheeling diode voltage decay versus time are recorded. The initial flux linkage in the phase ψ_0, corresponding to initial current i_0, for given mover position x, is obtained by

$$\psi_0(i_0,x) = L(i_0,x)\cdot i_0 = \int i\cdot R_S \cdot dt + \int V_{diode}\cdot dt \tag{11.17}$$

The time integrals may be done numerically down to 1%–2% of initial current, to avoid notable errors. Results as in Figure 11.8 are obtained for the family of $\psi(i,x)$ curves. In practice, this family of curves is to be curve-fitted and then used to calculate the thrust, as seen in Figure 11.8:

$$F_x = \frac{\partial W_m'}{\partial x} = \left(\frac{\Delta W_m'}{\Delta x}\right)_{i=const.} ; \quad \Delta x = x_A - x_B \tag{11.18}$$

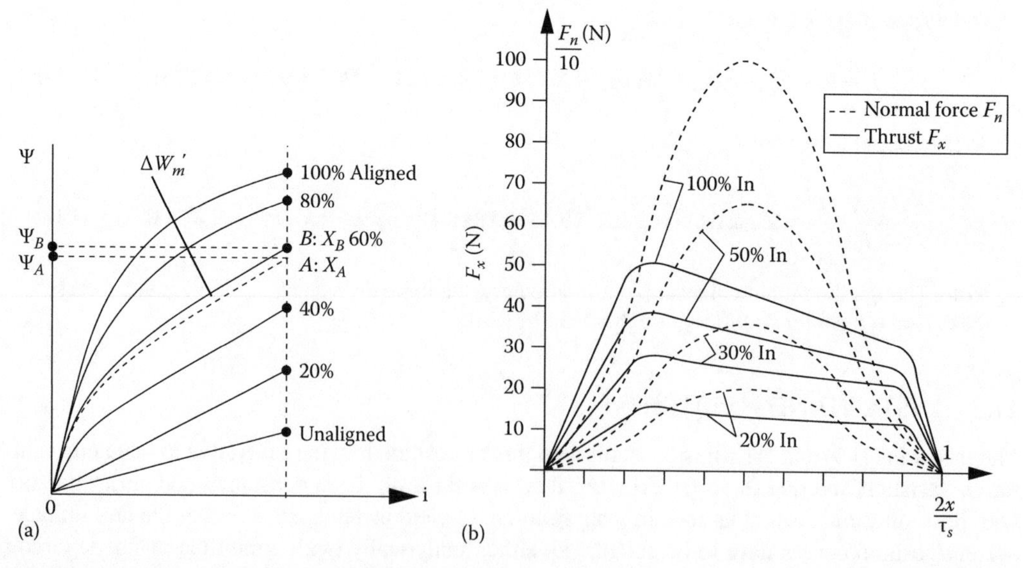

FIGURE 11.8 Flux linkage/current/position curves (a) and thrust (F_x) and normal force F_n per phase versus current and position (typical results) (b).

In addition, the normal force may be calculated if the test results are obtained at two close-to-each-other airgaps:

$$F_n = \frac{\partial W_m'}{\partial g} = \frac{(\Delta W_m')_{\Delta g}}{\Delta g}; \quad W_m' = \int_0^i \psi \cdot di \tag{11.19}$$

The thrust nonuniformity with position is due to both magnetic saturation and fringing airgap flux that vary with mover position. In real operation, the current is flat-top controlled at low speeds and is the result of one voltage pulse per cycle at high speeds (Figure 11.9).

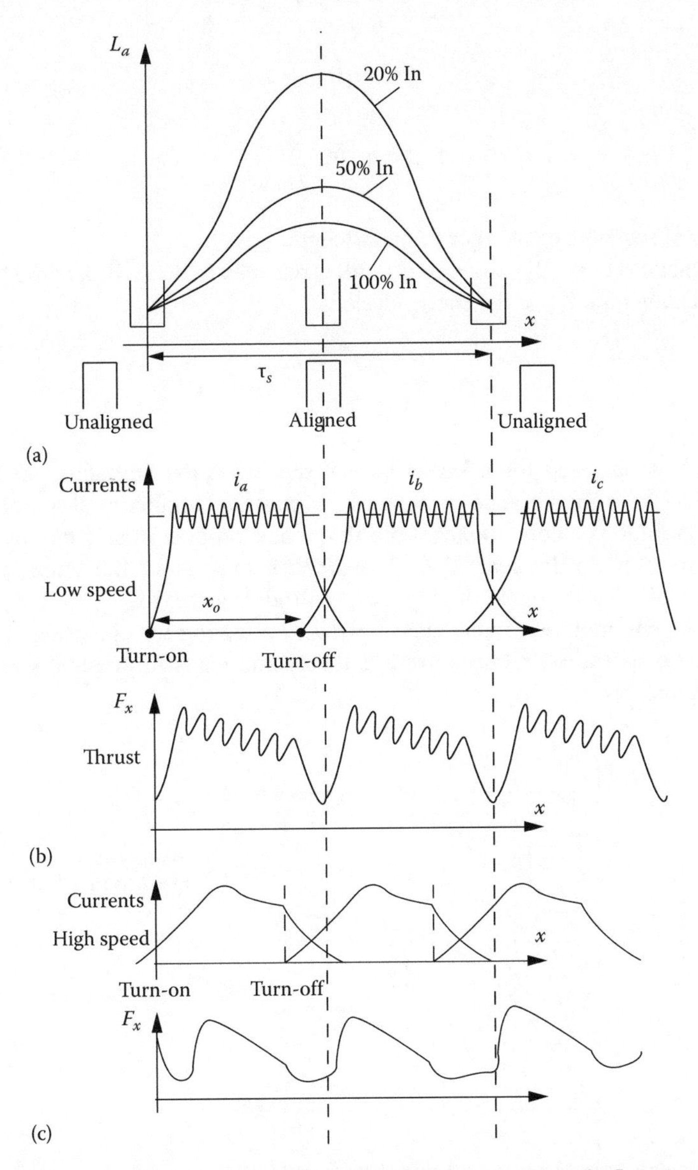

FIGURE 11.9 Three-phase L-SRM: (a) phase inductance (with saturation), (b) flat-top currents and thrust at low speeds, and (c) single voltage pulse current and thrust at high speeds. (After Boldea, I., *Variable Speed Generators*, CRC Press, Taylor & Francis, Boca Raton, FL, 2006.)

A few remarks are as follows:

- The lack of enough voltage at high speeds leads to one pulse voltage/phase/cycle, and anticipated turn on and turn off angles (position) are adopted for higher thrust.
- The thrust pulsations inevitably increase with speed and, for three-phase L-SRM, large thrust pulsations remain and thus four-phase L-SRMs with two active phases at any time are needed for low (servo drive class) thrust pulsations (10% or less).

11.4 AVERAGE THRUST AND ENERGY CONVERSION RATIO

$$\left(F_x\right)_{phase}^{av} = \frac{W_{mec}}{l_{stroke}}; \quad l_{stroke} = \frac{\tau_s}{2} \tag{11.20}$$

For m phases

$$F_{xav} = m \cdot N_r \cdot \frac{W_{mec}}{l_{stroke}} \tag{11.21}$$

N_r is the number of secondary poles per primary length.

As some magnetic energy (W_{mag} in Figure 11.10) is returned to the dc link of the PWM converter, the energy conversion ratio η_{ec} of the energy cycle is

$$\eta_{ec} = \frac{W_{mec}}{W_{mec} + W_{mag}} \geq 0.5 \tag{11.22}$$

The value of 0.5 is obtained for a linear flux/current variation with flux versus position for all situations; that is, in the absence of magnetic saturation (implicitly at a larger airgap). The energy conversion ratio is a correspondent of the $\eta \cdot \cos\varphi$ product in ac winding, which, in general, is larger than 0.5. So the peak kVA of the PWM inverter for the same speed and thrust is larger for L-SRM. The presence of magnetic saturation increases η_{ec} to 0.65 ÷ 0.67 while it also decreases the commutation inductance $L_t = d\psi_t/di$ allowing for operation at higher speeds. However, magnetic saturation reduces average thrust and leads to larger thrust pulsations for given machine geometry.

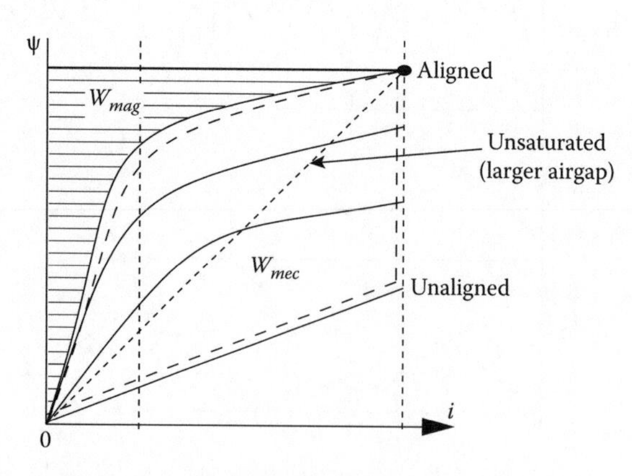

FIGURE 11.10 Energy cycle and average thrust/phase.

11.5 CONVERTER RATING

As already stated, the PWM converter has to handle the entire energy per cycle of all phases: $m \cdot (W_{mec} + W_{mag})$. To calculate the converter kVA rating, we start with the flux linkage per phase ψ_C:

$$\psi_C \approx V_{dc} t_C = \frac{V_{dc}}{U}(x_{OH} - x_n) \tag{11.23}$$

U is the L-SRM speed.

In terms of energy,

$$W_{mec} + W_{mag} = \frac{W_{mec}}{\eta_{ec}} = k \cdot \psi_C \cdot I_C \tag{11.24}$$

where

I_C is the peak current per phase

k is the so-called converter utilization factor [1]

So the peak kVA S_{peak} of the m phase L-SRM is (from (11.21) to (11.24))

$$S_{peak} = mV_0 I_C = \frac{F_{xm} U}{k \cdot \eta_{ec}} \cdot \frac{l_{stroke}}{x_{OH} - x_n} = \frac{P_{elm}}{k \cdot \eta_{ec}} \cdot \frac{l_{stroke}}{x_{OH} - x_n} \tag{11.25}$$

For $\eta_{ec} \le 0.65$, $k \approx 0.7$–0.8, $l_{stroke}/(x_{OH} - x_n) = 1.25$, $S_{peak}/P_{elm} = 2.747$. Such values are larger than for typical L-PMSMs (to be treated in a future chapter); it is the price to pay for L-SRM simplicity, fault tolerance, and smaller initial cost.

11.6 STATE SPACE EQUATIONS AND EQUIVALENT CIRCUIT

Let us neglect the mutual flux between L-SRM phases as it is small (less than 3%–4% in general). It means that each phase may be treated separately. The voltage equation for one phase is thus

$$V = R_S I + \frac{d\psi(x,i)}{dt}; \quad \psi = L(x,i) \cdot i \tag{11.26}$$

Using the derivative of the flux linkage as product of inductance and current, we obtain first

$$V = R_S I + L_t \frac{di}{dt} + \frac{\partial L}{\partial x} \cdot \frac{i \cdot dx}{dt}; \quad L_t = L(x,i) + \frac{i \cdot \partial L(x,i)}{\partial i} \tag{11.27}$$

$L_t(x, i)$ is the so-called transient (or commutation) inductance, which, under magnetic saturation, is even lower than the saturated $L(x, i)$. The third term in (11.27) is speed dependent and may be called the pseudo emf:

$$E_p = \frac{\partial L}{\partial x} \cdot i \cdot U; \quad U = \frac{dx}{dt} \tag{11.28}$$

Multiplying by i, Equation 11.27 may be written as

$$v \cdot i = R_S I^2 + \frac{d}{dt}\left(\frac{1}{2}L(x,i)i^2\right) + \frac{1}{2}i^2\frac{dL(x,i)}{dx}U \tag{11.29}$$

The second term coincides with stored magnetic variation in time only when magnetic saturation is constant ($\partial L/\partial i = 0$); consequently, only in this case the last term is the mechanical power from which thrust F_x varies:

$$\left(F_x\right)_{\partial L/\partial i=0} = \frac{1}{2}i^2\frac{dL(x)}{dx} \tag{11.30}$$

When magnetic saturation is considered, Equations 11.2 and 11.3 for thrust are to be used. For the normal force, Equations 11.29 and 11.30 are to be considered. But in this case, typical when both propulsion and levitation control are performed, the voltage equations (11.27) and (11.28) may be extended as

$$V = R_S I + \frac{d\psi(x,i)}{dt} + E_p + E_S; \quad E_S = i\frac{\partial L(x,i,g)}{\partial g}\cdot\frac{dg}{dt} \tag{11.31}$$

And again,

$$E_p = i\frac{\partial L(x,i,g)}{\partial x}\cdot U \tag{11.32}$$

Similarly, the instantaneous power/phase equation gains an additional term due to vertical (levitation) motion speed:

$$v \cdot i = R_S I^2 + \frac{d}{dt}\left(\frac{1}{2}L(x,i)i^2\right) + \frac{1}{2}i^2\frac{\partial L(x,i,g)}{\partial x}U + \frac{1}{2}i^2\frac{\partial L(x,i,g)}{\partial g}\cdot\frac{dg}{dt} \tag{11.33}$$

Based on (11.32), though, with E_p and E_S as pseudo emfs an equivalent circuit may be drawn (Figure 11.11).

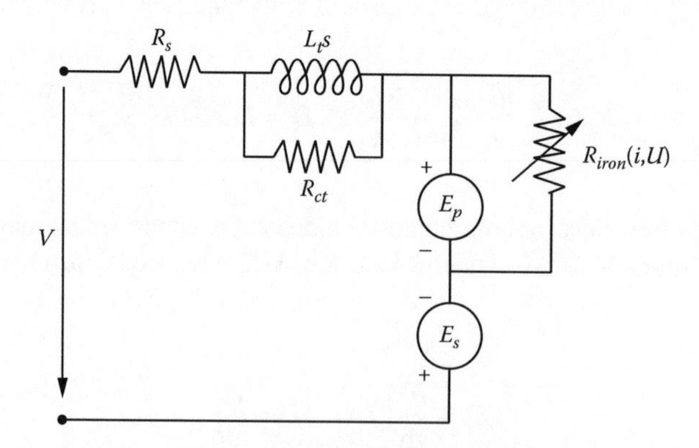

FIGURE 11.11 L-SRM single-phase circuit with propulsion (E_p) and suspension (E_S) pseudo-emfs.

The core losses occur in L-SRM both in the primary and in the secondary because the machine does not have a traveling field. A rather complete study of iron losses in rotary SRMs is offered in Ref. [3]. The iron core resistance R_{iron} in the equivalent circuit may thus be calculated or may be estimated from measurements during motion at constant current while R_{ct} refers to core losses due to current time variations (*di/dt*) during current control, mainly; R_{ct} may act even at standstill.

The multiphase equation of L-SRM may be simply written as

$$V_{a,b,c,d,..} = R_S I_{a,b,c,d,..} + \frac{d\psi_{a,b,c,d,..}\left(x, i_{a,b,c,d,..}\right)}{dt} \tag{11.34}$$

$$F_{x\,a,b,c,d,..} = \frac{\partial}{\partial x}\int_0^{i_{a,b,c,d,..}} \psi_{a,b,c,d,..}\left(x, i_{a,b,c,d,..}\right) di_{a,b,c,d,..} \tag{11.35}$$

As already mentioned, the flux linkage/current/position curve family is essential to calculating thrust F_x and normal force F_n. Its approximation is crucial for control system design, for example,

$$\psi(i,x) = a_1(x)(1-e^{-a_2(x)i}) + a_3(x)i \tag{11.36}$$

with

$$a_{1,2,3}(x) = \sum_{k=0}^{\infty} A_{1,2,3}^{k}\cos\left(\frac{2\pi}{\tau_s}x\right) \tag{11.37}$$

or

$$\psi_j = \psi_{sat}\left(1-e^{-i_j f_j(x)}\right); \quad i_j \geq 0$$

$$f_j(x) = a_0 + \sum_{k=1}^{3} a_n\cos\left(\frac{2\pi}{\tau_s}x - (i-1)\frac{2\pi}{m}\right); \quad j = 1,2,3,4\left(\text{for } m = 4 \text{ phases}\right) \tag{11.38}$$

Approximations such as these are useful for rather high-fidelity curve fitting, allowing also analytical thrust and normal force expressions. However, for control design purposes two-piece linear approximations may be acceptable (Figure 11.12).

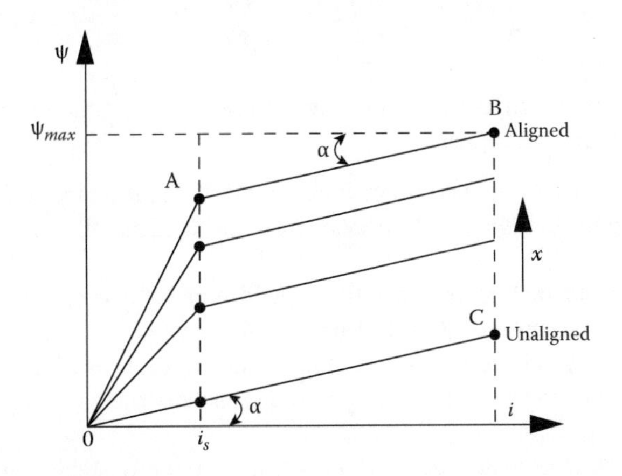

FIGURE 11.12 Simplified $\psi(i)x$ curves.

$$\psi_j = i_j\left(L_U + \frac{k_S(x - x_{on})}{i_S\frac{\tau_p}{2}}\right); \quad i \le i_S$$

$$\psi_i = i_jL_U + \frac{k_S(x - x_{on})}{\frac{\tau_p}{2}}; \quad i \ge i_S \tag{11.39}$$

For this simplification, it suffices to perform 3D-FEM calculations, corresponding to points A, B, and C [9].

11.7 SMALL SIGNAL MODEL OF L-SRM

Let us revisit the voltage Equation 11.26 and add the motion equation (for propulsion only):

$$G_m\frac{dU}{dt} = F_x - F_{Load} - BU; \quad \frac{dx}{dt} = U; \quad G_m, \text{mover mass} \tag{11.40}$$

and neglect magnetic saturation to yield

$$F_x = \frac{1}{2}\sum_{j=1}^{3(4)..} i_j^2\frac{\partial L_j}{\partial x} \tag{11.41}$$

Linearizing Equations 11.26 and 11.40 starts with

$$i = i_0 + \Delta i; \quad U = U_0 + \Delta U; \quad V = V_0 + \Delta V; \quad F_{load} = F_{Lo} + \Delta F_{Load} \tag{11.42}$$

and yields

$$\left(s + \frac{1}{\tau_e}\right)\Delta i + \frac{k_b}{L_{av}}\Delta U = \frac{\Delta V}{L_{av}}; \quad R_e = R_S + \frac{\partial L}{\partial x}U_0; \quad L_{av} \approx \frac{(L_{\max} + L_{\min})}{2}$$

$$-\frac{k_b}{G_m}\Delta i + \left(s + \frac{1}{\tau_m}\right)\Delta U = -\frac{\Delta F_l}{G_m}; \quad \tau_e = \frac{L_{av}}{R_e}; \quad k_b = \frac{\partial L}{\partial x}i_0; \quad \tau_m = \frac{G_m}{B} \tag{11.43}$$

Note: The machine transient inductance is considered here as L_{av} (instead of L_t), while the phase inductance L varies linearly with position (x).

With $\Delta E = k_b\Delta U$, the structural diagram of Equation 11.42 is presented in Figure 11.13.
The structural diagram in Figure 11.13 may be characterized as follows:

- The equivalent resistance R_e includes the pseudo-emf influence and thus increases with speed; so the electric time constant τ_e decreases with speed.
- The structural diagram is very similar to that of the dc series brush motor.
- The equivalent time constants (τ_1 and τ_2) in the reduced structural diagram are dependent on i_0, U_0, $\partial L/\partial x$, L_{av}, R_S.
- The response of this second-order system is thus inherently stable; the response may be aperiodic or periodic but always attenuated.
- The L-SRM operates as a motor for $\partial L/\partial x > 0$ and as generator for $\partial L/\partial x < 0$.

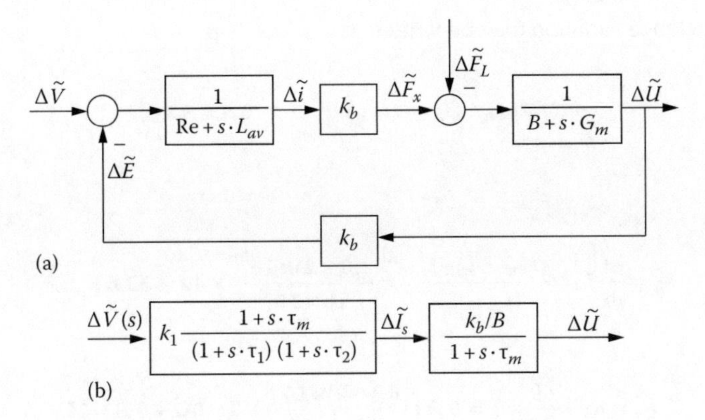

FIGURE 11.13 L-SRM, per phase structural linear diagram (a) and its reduced form (b).

Example 11.2

Let us consider a three-phase L-SRM with $L_{min} = 2$ mH, $L_{max} = 10$ mH, and $R_S = 0.2\ \Omega$, which is operated at $U_0 = 1$ m/s and controlled at a constant current value I_0 over the entire $x_{d\,wel} = \tau_p/2 = 10$ mm, which corresponds to equal pole and interpole primary lengths, for voltage $V_0 = 42$ Vdc.

Calculate

1. Current I_0, peak flux linkage ψ_{max}, average thrust F_{xav}, and load force F_{load} at $U_0 = 1$ m/s.
2. For constant load and a 10% increase in average voltage V_0 ($\Delta V = +0.1\ V_0$) indicate the way for calculating the eigen values and current and speed transients using the small signal approach, with $\tau_1 = G_m/B = 3$ s, $B = 5.6$ Ns.

Solution

1. The conducting time/phase t_{on} is

$$t_{on} = \frac{x_{d\,wel}}{U_0} = \frac{10 \times 10^{-3}}{1} = 0.01\text{s} \tag{11.44}$$

Using voltage equation yields (at constant speed)

$$V_0 t_{on} = R_S I_0 t_{on} + (L_{max} - L_{min}) I_0 \tag{11.45}$$

$$I_0 = \frac{V_0 t_{on}}{R_S t_{on} + (L_{max} - L_{min})} = \frac{42 \times 0.01}{0.2 \times 0.01 + (10 - 2) \times 10^{-3}} = 42\text{ A} \tag{11.46}$$

The average thrust F_{xav}/phase is in fact total average thrust as only one phase is active at any time:

$$F_{xav} = \frac{1}{2} i_0^2 \frac{\partial L}{\partial x} = \frac{i_0^2}{2} \cdot \frac{(L_{max} - L_{min})}{x_{dwell}} = \frac{42^2}{2} \cdot \frac{(10 - 2)10^{-3}}{10 \times 10^{-3}} = 705.6\text{ N} \tag{11.47}$$

Now the load force F_{load} is

$$F_{load} = F_{xav} - BU = 705.6 - 5.6 \times 1 = 700\text{ N} \tag{11.48}$$

2. The inductance variation may be written as

$$L(x) = L_{min} + (L_{max} - L_{min})\frac{x}{x_{dwell}}; \quad 0 \le x \le x_{dwell} \tag{11.49}$$

So,

$$k_b = \frac{\partial L}{\partial x} I_0 = \frac{(L_{max} - L_{min})}{x_{dwell}} \cdot I_0 = \frac{(10-2)\times 10^{-3}}{10\times 10^{-3}} \times 42 = 33.6 \text{ Wb/m} \tag{11.50}$$

$$R_e = R_S + \frac{\partial L}{\partial x} \cdot U_0 = 0.2 + \frac{(10-2)\times 10^{-3}}{10\times 10^{-3}} \times 1 = 0.2 + 0.8 = 1\Omega \tag{11.51}$$

The mover mass $M_G = \tau_m \cdot B = 3 \times 5.6 = 16.8$ kg, and the equivalent electrical time constant

$$\tau_e = \frac{L_{av}}{R_e} = \frac{6\times 10^{-3}}{1} = 6\times 10^{-3} \text{ s} \tag{11.52}$$

The two small signal equations (11.43) write

$$\begin{aligned} &\left(s + \frac{1}{6\times 10^{-3}}\right)\Delta i + \frac{33.6}{6\times 10^{-3}}\Delta U = \frac{\Delta V}{6\times 10^{-3}} \\ &-\frac{33.6}{16.8}\Delta i + \left(s + \frac{1}{3}\right)\Delta U = \frac{-\Delta F_L}{16.8} \end{aligned} \tag{11.53}$$

Now solving the characteristic equation (11.53) for $\Delta V = 0.1V_0 = 4.2$ V and $\Delta F_L = 0$, the two eigen values $s_{1,2}$ are obtained. Once the eigen values are calculated (most probably in our case both are real and negative, say, $-\tau_1$, $-\tau_2$), the speed U is

$$U(t) = U_0 + \Delta U(t) = U_0 + A_1 e^{-\frac{t}{\tau_1}} + A_2 e^{-\frac{t}{\tau_2}} + (U_{final} - U_0) \tag{11.54}$$

The new conducting time t_{on}' (at the new steady-state speed) is

$$(V_0 + \Delta V)t_{on}' = R_S I_0 t_{on}' + (L_{max} - L_{min})I_0 \tag{11.55}$$

$$U_{final} = U_0 \cdot \frac{t_{on}}{t_{on}'} \tag{11.56}$$

Now also at $t=0$, $U=U_0=1$ m/s, and $(dU/dt)_{t=0}=0$ and thus from (11.54), both A_1 and A_2 constants are obtained. The current variation Δi may be obtained from (11.42)

$$\Delta i = \left(\frac{\Delta U(t)}{\tau_m} + \frac{d(\Delta U(t))}{dt}\right)\frac{G_m}{k_b} \tag{11.57}$$

As expected, the final speed increase is +10% as is voltage increase from constant load; and the current increases and then decreases to its initial value I_0, because the load is unchanged. The structural diagram with its reduced form may be used directly in the design of control system (Figure 11.14).

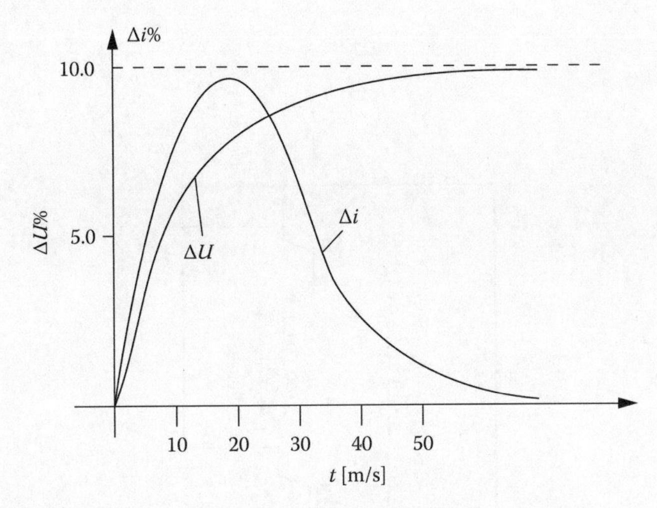

FIGURE 11.14 +10% voltage step response at constant load (qualitative response). (After Krishnan, R., *Switched Reluctance Motor Drives*, CRC Press, Boca Raton, FL, 2001.)

11.8 PWM CONVERTERS FOR L-SRMs

Typically, L-SRMs are supplied by unipolar currents per phase, and thus a two-switch per phase multiphase dc–dc converter with a front-end diode rectifier and capacitor filter constitutes the first option (Figure 11.15).

For typical hysteresis hard (fast) current control ± Vdc (with both switches on and off) is performed (Figure 11.4b); for soft (slower) current control +Vdc and zero voltage control is applied (Figure 11.14c). The minimum voltage rating of power devices is the dc link voltage. To provide equal RMS current in the power devices and the same average current in the freewheeling diodes, a distinct PWM strategy (different from those in Figure 11.14) is needed.

In essence, only one power switch is turned off to reduce the phase current, but the two power switches play this role one after the other. In an effort to reduce the number of power switches, quite a few alternative converter topologies have been proposed [3]. Among them, the C-dump converter stands out in terms of performance/cost Figure 11.15.

The total number of power switches is m + 1, and energy recovery when one (two) phase(s) is (are) active is secured from C-dump through T_r and L_r. There are five operation modes visible in Figure 11.16. The C-dump capacitor is calculated from its voltage variation ΔV_0 during T_1 turn off (from switch current ripple):

$$C_d = \frac{(1-d_1)I}{f_c \Delta V_0} \tag{11.58}$$

Where

d_1 is the minimum value T_1 of PWM index
I is average phase current
f_c is the switching frequency of current

The current ripple Δi, for the interval, which corresponds to phase inductance variation ΔL (rotor position changes), is

$$\Delta i = \frac{V_{dc} \cdot d_1}{f_c \cdot L_{av}} - \frac{I \cdot \Delta L}{L_{av}} \tag{11.59}$$

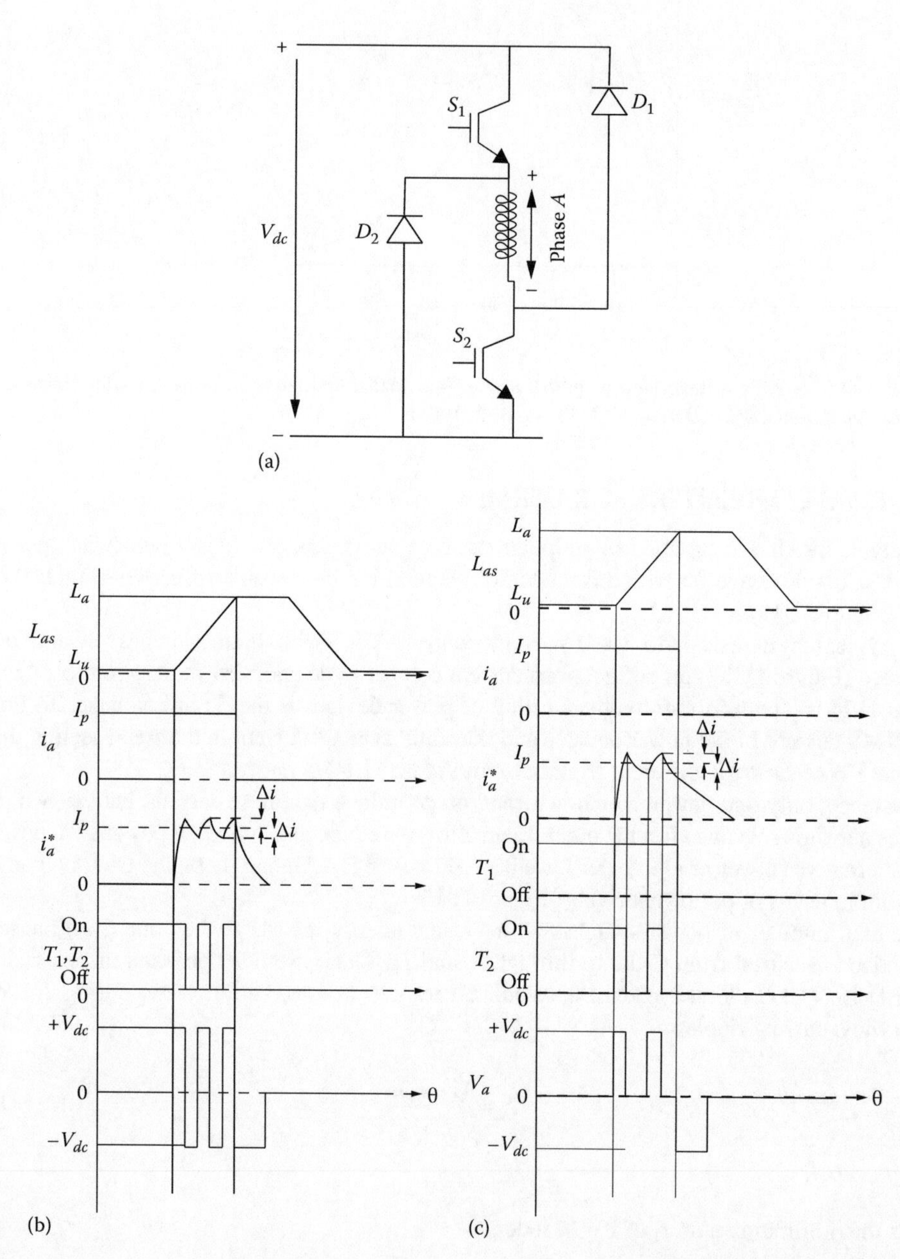

FIGURE 11.15 Two switches/phase dc–dc converter (a), hysteresis "hard" current control (b), and hysteresis "soft" current control (c). (After Krishnan, R., *Switched Reluctance Motor Drives*, CRC Press, Boca Raton, FL, 2001.)

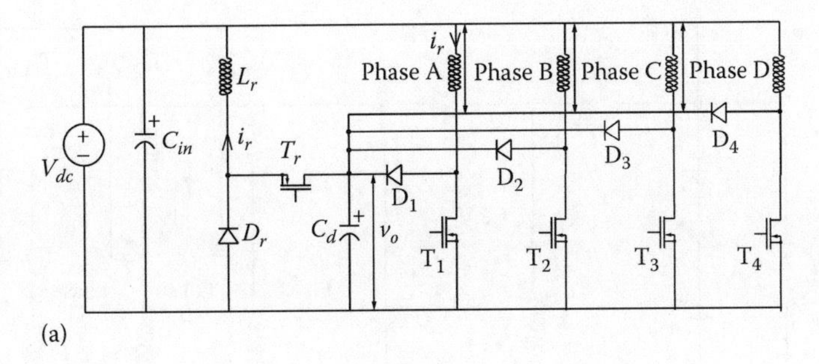

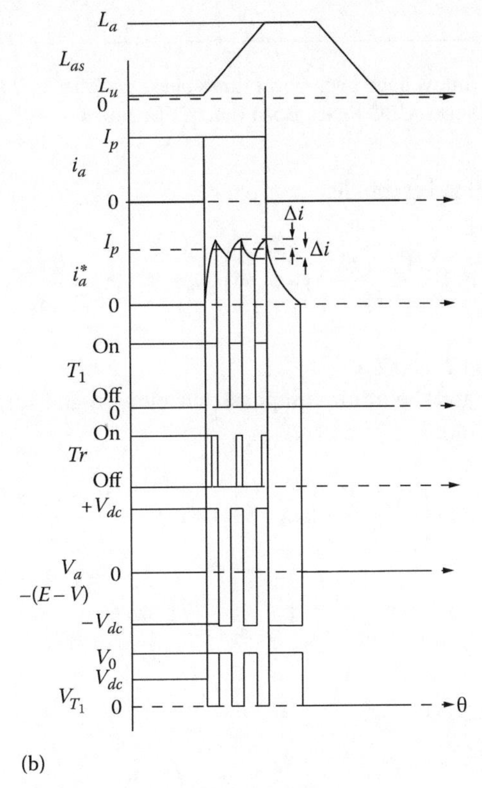

FIGURE 11.16 C-dump converter with energy recovery (a) and its main variable waveforms, (b). (After Krishnan, R., *Switched Reluctance Motor Drives*, CRC Press, Boca Raton, FL, 2001.)

Current ripple leads to torque pulsations and additional copper and iron losses. The minimum switching frequency is calculated by equating the duty cycle d_l in (11.58) and in (11.59):

$$f_c = \frac{V_{dc}}{I \cdot \Delta L + L \cdot \Delta i + V_{dc} \cdot C_d \cdot S \cdot \Delta V_0} \tag{11.60}$$

Δi and ΔV_0 are imposed in relative values for minimum I and L. We may continue with the design of the recovery inductance L_r (Figure 11.17), based on imposed ripple on the recovery current Δi_r:

$$L_r \Delta i_r = \frac{d_2}{f_c} \frac{(V_0 - V_{dc})}{V_0} E_p; \quad E_p = U \cdot I \frac{\partial L}{\partial x}; \quad U\text{—speed} \tag{11.61}$$

where

I is the average phase current
E_p is the average pseudo emf per phase

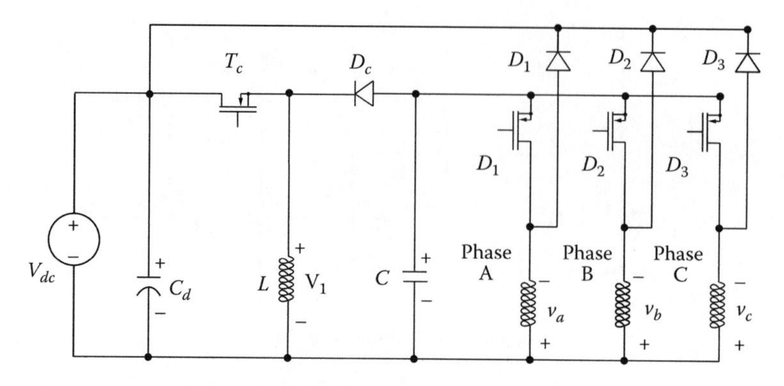

FIGURE 11.17 Variable dc link voltage buck-boost three-phase converter for L-SRM. (After Krishnan, R., *Switched Reluctance Motor Drives*, CRC Press, Boca Raton, FL, 2001.)

The phase voltage equation is such that

$$i(t) = \frac{V_{dc} - E_p}{R_S}\left(1 - e^{-\frac{t}{\tau_e}}\right) + i_0 e^{-\frac{t}{\tau_e}}; \quad \tau_e = \frac{L(x)}{R_S} \tag{11.62}$$

during t_{on}, when T_1 is on and T_r is off.

If $R_S = 0$, the maximum and the minimum phase currents $i(t)$ and $i_0(t)$ are obtained from linear variations (hard commutation):

$$i(t) = i_0(t) + \left(\frac{V_{dc} - E_p}{L}\right)t_{on}; \quad i_0(t) = i_1(t) - \frac{\left(V_0 - V_{dc} + E_p\right)}{L}t_{off} \tag{11.63}$$

$$t_{on} = d_1\frac{1}{f_c}; \quad t_{off} = \frac{\left(1 - d_1\right)}{f_c} \tag{11.64}$$

So the average PWM index d_1 is

$$d_1 = \frac{V_0 - V_{dc}}{V_0} + \frac{E_p}{V_0} \tag{11.65}$$

Usually, the pseudo emf E_p is defined as equal to base voltage V_b. So at base speed U_b:

$$V_0 = V_b = \frac{\partial L}{\partial x}\cdot U_b \cdot I_b \tag{11.66}$$

Substituting in (11.63) yields

$$d_1 = 1 - \frac{V_{dc}}{V_b} + \frac{U}{U_b}\cdot\frac{I}{I_b}; \quad \frac{V_{dc}}{V_b} = 0.7\text{–}0.8 \tag{11.67}$$

As evident from (11.67), if V_{dc}/V_b = const., d_1 has to increase with speed U/U_b. Increasing $V_0(V_b)$ for limited current limits the machine maximum speed. Adequate current (i_r) and voltage (V_0) control in the secondary circuit (L_r, C_d) leads to reasonable recovery current ripple Δi_r [3]. Yet another variable dc link voltage buck-boost converter with $m+1$ power switches is shown in Figure 11.17.

The switch T_c, diode D_c, inductor L, and capacitor C form the buck-boost part of the converter.

The machine phase voltage V_i may reach up to 2 V_{dc} (say, for wider constant power-speed range), but then the power switches have to be designed to three times V_{dc} for the scope. In the buck mode $V_i < V_{dc}$, and thus better low-speed performance may be obtained.

For more in L-SRM converters, see Ref. [3], Chapter 4.

11.9 DESIGN METHODOLOGY BY EXAMPLE

Let us develop an electromagnetic design methodology for an L-SRM with specifications as given in the following:

- Rated thrust to base speed: $F_{xn}=400$ N

Base speed $U_b=2.4$ m/s

- DC link voltage $V_{dc}=V_0=300$ Vdc
- Number of phases $m=3$ (6 coils in all)
- Topology: tubular
- Travel length: 1 m
- Free acceleration to U_b length ≤ 0.2 m
- Flux/current/position curves: piece-wise linear

First, a thrust density by $f_{xn}=1.6$ N/cm^2 for the stroke length $l_{stroke}=\tau_p/2=15$ mm and the airgap $g=0.3$ mm is adopted. To retain enough time to turn off the active phase before producing drag force, we choose

$$\frac{x_0 - x_u}{l_{stroke}} = 0.8 \tag{11.68}$$

From (11.23), we my calculate (with $R_S=0$) the maximum flux linkage in the phase ψ_b:

$$\psi_b = \frac{V_{dc}}{U} \cdot 0.8 \cdot l_{stroke} = \frac{300}{2.4} \times 0.8 \times 15 \times 10^{-3} = 1.5 \text{ Wb} \tag{11.69}$$

With equal primary slot and pole span, the primary (pole) length is $b_p=l_{stroke}=15$ mm. The secondary slot pitch τ_S is thus

$$\tau_S = 3 l_{stroke} = 45 \text{ mm} \tag{11.70}$$

The maximum thrust (current) is produced when (Figure 11.18) $L_S=L_u$ (unaligned).

$$\frac{W_{mec}}{k_{safe}} \approx \psi_0' \left(\frac{x_0 - x_u}{l_{stroke}} \right) I_b - \frac{\psi_0'^2}{2L_{au}}; \quad k_{safe} < 1; \quad \psi_0' = \psi_b - L_S \cdot I_b \tag{11.71}$$

k_{safe} is a safety factor.

Equation 11.71 may be used as verification after the design is done. At this stage, we can calculate the L-SRM diameter at airgap D_{pi}, as there are only two coils/phase and only one active phase at a time:

$$F_{xn} = \pi \cdot D_{pi} \cdot 2(2 l_{stroke}) \cdot f_{xn}$$
$$D_{pi} = \frac{400}{\pi \times 4 \times 2 \times 15 \times 10^{-3} \times 1.6 \times 104} = 0.0664 \text{ m} \tag{11.72}$$

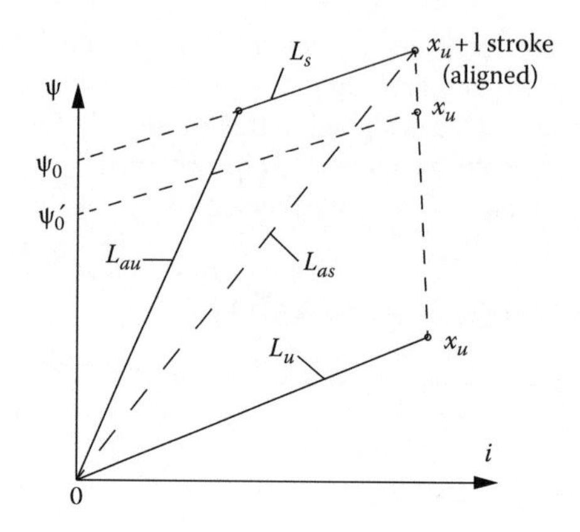

FIGURE 11.18 Simplified ψ(*i*)*x* curves.

Let us allow for maximum of airgap flux density in the aligned position of a phase to be B_b = 1.4 T. The maximum flux linkage in the two coils per phase in series ψ_b is

$$\psi_b = 2 \cdot B_b \cdot b_b \cdot \pi \cdot D_{pi} \cdot W_C; \quad k_d = \frac{x_0 - x_u}{l_{stroke}} = 0.8 \tag{11.73}$$

W_C is the turn/coil

$$B_b \cdot W_C = \frac{1.5}{2 \times 0.015 \times \pi \times 0.0664} = 239 \text{ T} \cdot \text{turns} \tag{11.74}$$

The unaligned L_u and aligned L_{au} inductances per phase are

$$L_u = 2P_{mu} \cdot W_C^2; \quad L_{au} = 2P_{ma} \cdot W_C^2 \tag{11.75}$$

P_{mu}, P_{ma} magnetic permeances per coil in unaligned and aligned position (see Figure 11.19 per phase A)—unsaturated:

$$P_{ma} \approx \frac{\mu_0 \cdot \pi \cdot D_{Si} \cdot b_p}{2g} \tag{11.76}$$

For saturated conditions in aligned position,

$$L_{aS} = 2 \cdot P_{mas} \cdot W_C^2; \quad P_{mas} = \frac{P_{ma}}{(1 + k_S)}; \quad k_S\text{—saturation factor.} \tag{11.77}$$

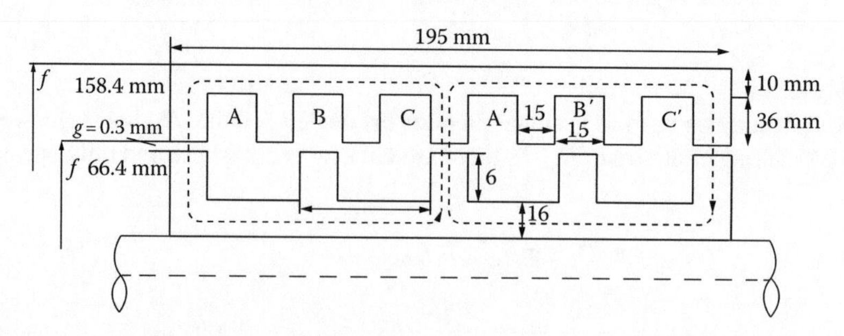

FIGURE 11.19 Tubular three-phase L-SRM.

Now with $B_b = 1.4$ T from (11.74), the number of turns per coil W_C is

$$W_C = \frac{W_C B_b}{B_b} = \frac{239}{1.4} \approx 170 \text{ turns/coil} \tag{11.78}$$

So the unsaturated aligned position inductance yields

$$L_{au} = \frac{2\mu_0 \pi D_{pi} b_p W_C^2}{2g} = \frac{2\times 1.1256\times 10^{-6}\times \pi \times 0.0664\times 0.015\times 170^2}{2\times 0.3\times 10^{-3}} = 0.37884 \text{ H} \tag{11.79}$$

From field calculations (analytical [3] or 2D-FEM), the unaligned inductance L_u is $L_u = L_{au}/10 = 0.03784$ H. With $k_{safe} = 0.75$, $k_d = 0.8$, from (11.71) we may directly calculate the base (flat-top) phase current I_b as

$$W_{mec} = F_{xav} \cdot l_{stroke} = 400\times 15\times 10^{-3} = 6 \text{ J} \tag{11.80}$$

$$\frac{6}{0.75} \approx (1.5 - 0.03784 \cdot I_b) 0.8 \cdot I_b - \frac{(1.5 - 0.03784 \cdot I_b)^2 0.8^2}{2\times 0.03784} \tag{11.81}$$

$$I_b \approx 8.35 \text{ A}$$

The saturated inductance L_{as} (Figure 11.18) is thus obtained from

$$\psi_b = L_{as} \cdot I_b \cdot k_d; \quad L_{as} = \frac{1.5}{8.25\times 0.8} = 0.22 \text{ H} \tag{11.82}$$

The RMS current/phase $(I_b)_{RMS}$ is approximately

$$(I_b)_{RMS} = \frac{I_b}{\sqrt{3}} = \frac{8.25}{\sqrt{3}} = 4.768 \text{ A} \tag{11.83}$$

With $J_{Co\,RMS} = 3$ A/mm^2 and slot filling factor $k_{sfill} = 0.5$, the slot area of primary A_{slotp} is

$$A_{slotp} = \frac{I_{b\,RMS} \cdot W_C}{J_{Co\,RMS} \cdot k_{sfill}} = \frac{4.768\times 170}{3\times 0.5} = 540 \text{ mm}^2 \tag{11.84}$$

But the slot $b_{sp} = b_p = 15$ mm and thus the slot height is

$$h_{sp} = \frac{A_{slotp}}{b_{sp}} = \frac{540}{15} = 36 \text{ mm} \tag{11.85}$$

The secondary tooth height (to reduce flux fringing) $h_{ss} = 20$ g $= 20 \times 0.3 = 6$ mm.

The saturation factor k_S is

$$k_S = \frac{L_{au}}{L_{as}} - 1 = \frac{0.3784}{0.22} - 1 = 0.72 \tag{11.86}$$

There is enough space for reducing the primary and secondary back iron to reach this value of k_S. A routine utilization of Ampere's law along the field line as in Figure 11.19 will yield k_S for $B_b = 1.4$ T. It is imperative to get $B_b = 1.4$ T for aligned position for $W_C \cdot I_b = 170 \times 8.4768 = 1441.056$; otherwise, the rated thrust will not be obtained.

The copper losses are

$$P_{Con} = 3 I_{bRMS}^{2} \cdot R_S \tag{11.87}$$

with

$$R_S = 2\rho_{Co}\pi(D_{pi} + h_{sp})\frac{W_C}{I_{bRMS}} J_{CoRMS}$$

$$= 2.21 \times 10^{-8} \times \pi \times (0.064 + 0.036) \times \frac{170}{4.788} \times 3 \times 10^{6} = 0.8866\ \Omega \tag{11.88}$$

So,

$$P_{Con} \approx 3 \times 0.8866 \times 4.768^2 = 60.46\ \text{W} \tag{11.89}$$

The frequency f_n is

$$f_n = \frac{U_b}{\tau_s} = \frac{2.4}{45 \times 10^{-3}} = 53.33\ \text{Hz} \tag{11.90}$$

The approximate weight of primary is

$$G_p \approx \frac{\pi\left(\left(D_{pi} + 2h_{sy} + 2h_{cp}\right)^2 - D_{pi}^2\right)}{4} \cdot 13 \cdot b_p \cdot \gamma_{i+c}$$

$$= \frac{\pi((66.4 + 2 \times 36 + 2 \times 10)^2 - 66.4^2)}{4} \times 10^{-6} \times 13 \times 15 \times 10^{-3} \times 8200 = 25.96\ \text{kg} \tag{11.91}$$

With a thrust of 400 N, the ideal acceleration of primary $a_{i\max}$ is

$$a_{i\max} = \frac{F_{xa}}{G_p} = \frac{400}{25.96} = 15.408\ \text{m/s}^2 \tag{11.92}$$

The ideal acceleration travel l_{travel} to $U_b = 2.4$ m/s from standstill is

$$l_{travel} = \frac{U_b^2}{2a} = \frac{2.4^2}{2 \times 15.408} = 0.187\ \text{m} < 0.2\ \text{m} \tag{11.93}$$

So the ideal acceleration to full speed may be accomplished under 0.2 m of travel out of the total travel of 1 m.

The efficiency may be calculated as

$$\eta_n = \frac{P_{mec}}{P_{mec} + p_{Com} + p_{iron} + p_{mec}} \tag{11.94}$$

With about 60% iron in the primary (15 kg) and 5 kg in the secondary and 2.5 W/kg iron losses and 10 W mechanical losses, the total efficiency is

$$\eta_n = \frac{400 \times 2.4}{960 + 60.46 + 20 \times 2.5 + 10} = 0.888 \tag{11.95}$$

Note: At this speed (2.4 m/s), the designed L-SRM is better than the LIM for same size and output in terms of efficiency; in terms of converter kVA, it is about the same (cos φ of an LIM is about 0.4 ÷ 0.5). A distinct design for a ship elevator of 55,000 N with multimodular, flat, double-sided, four-phase L-SRMs with active movers is described in Ref. [10] and could be a valuable guide of L-SRM potential.

Efforts to reduce undesired high normal force in single-sided flat L-SRM by pole shaping design are presented in Ref. [11], while separated-phase tubular topologies of L-SRM are designed in Ref. [12].

11.9.1 L-SRM Control

By L-SRM control, we mean position or speed or thrust close-loop regulation.

Instantaneous current close-loop control is, in general, implicit. In L-SRM control, we may suppose that

- Position sensing (by linear encoders or resolvers) is available
- Motion-sensorless control is performed (in general) for speed and (or) thrust control

In view of the complexity of the flux/current/position and thrust/current/position curve families, the position, speed, or thrust control is rather nonlinear.

The linearized second-order structural diagram (11.13b) may be used for a preliminary design of, first, the current regulator and then the speed regulator (Figure 11.20; this design may be considered as preliminary. Thorough verifications of stability and performance of current and speed control in extreme conditions of thrust perturbations and machine parameter detuning are then required.

The generic control system in Figure 11.20 includes the following:

- A fast phase current regulator with limiter (to avoid trespassing of maximum voltage available in the converter and for current fast protection).
- 1–2 kHz critical frequency with a damping factor ξ=0.6–0.7.

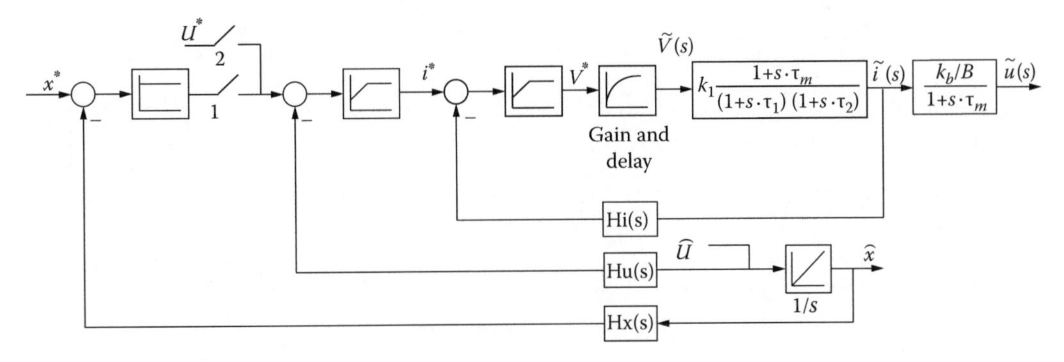

FIGURE 11.20 Generic cascaded position, speed, and current control; one phase in terms of current control is shown. (After Lim, H.S. et al., *IEEE Trans.*, IE-55(2), 534, 2008.)

- A mildly fast speed regulator with a critical frequency around 100 Hz or more and, again, a damping factor $\xi = 0.6–0.75$.
- A proportional (if any) position controller for a positioning application.
- Other more robust controllers may be applied.

For details of current and speed regulator design, see Ref. [3]. Once the preliminary design of current speed and position regulators is done, a more complete view of the current control may be obtained, by establishing the waveforms of phase reference currents for each phase so as to produce the required dynamic response and to exhibit low thrust pulsations and thus avoid noise and vibrations. Reference current waveforms for a given average torque and speed (to fulfill, in the presence of magnetic saturation) are not easy to find.

However, if magnetic saturation is not considered, the situation simplifies to some extent as the thrust of phase k is

$$F_{xk} = \frac{1}{2} g_k(x) \cdot i_k^2; \quad g_k(x) = \frac{\partial L_k}{\partial x} \tag{11.96}$$

For a four-phase L-SRM, the phase inductances $L_{a,b,c,d}$ vary with position (magnetic saturation is absent or constant) as in Figure 11.21 [4].

As seen in Figure 11.20, the waveforms of the current should be obtained by reversing (11.96):

$$i_k^*(x) = \sqrt{\frac{2F_{xk}^*}{g_k(x)}} \tag{11.97}$$

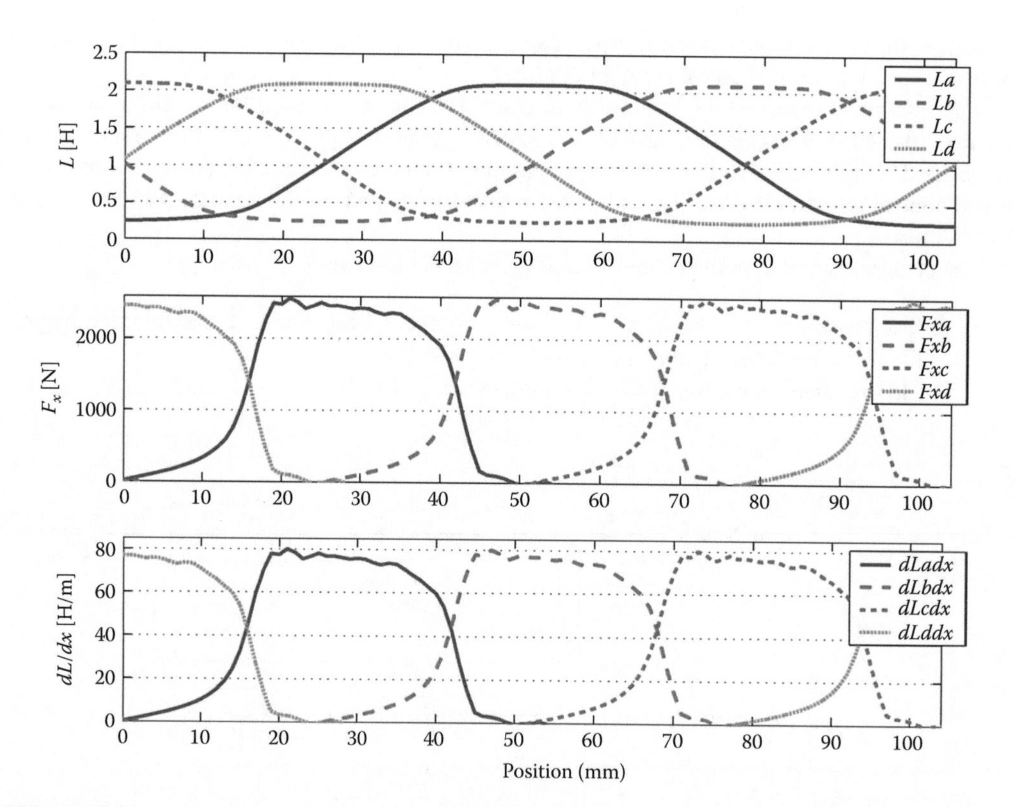

FIGURE 11.21 Inductances (a, b, c, d) (a), thurst (Fx) (b), and dL/dx (c) versus position for a four-phase L-SRM. (After Lim, H.S. et al., *IEEE Trans.*, IE-55(2), 534, 2008.)

But, more than that, the reference thrust F_{xk}^* of each phase should be modulated with a function $f_k(x)$ such that the total reference thrust is obtained always with most efficient (in thrust) phases:

$$F_x^* = \sum F_x^* \cdot f_x(k); \quad \sum_{k=a,b,c,d} f_k = 1 \tag{11.98}$$

f_x is called force distribution factor (FDF) [4], and its function choice is a matter of experience or trial and error, to minimize, again, the thrust pulsations with mover position in a feed forward manner.

A typical result with such an FDF system is shown in Figure 11.21 [4].

The control system in such a case may be defined as in Figure 11.22.

Though $f_k(x)$ and $g(k)$ are a priori functions of mover position, the design of position, speed, and current regulator for a single phase with constant equivalent current could be a good start for more involved investigations. Digital simulations and experiments for critical dynamic operation modes will then validate the simplified regulator designs. Such dynamic simulation results [4] show good performance (Figures 11.23 and 11.24).

Controlled power-speed region extension control may be obtained by just changing FDF, that is, $f_k(x)$ functions with speed (and, evenly, with reference total thrust F_x^*). In an effort to reduce the off- and on-line computation effort, very visible in Figure 11.22, direct average thrust control may be attempted [17]. The average thrust may be estimated by starting with phase flux estimation:

$$\frac{d}{dt}\psi_k = V_k - R_S I_k; \quad k = a,b,c,d\ldots \tag{11.99}$$

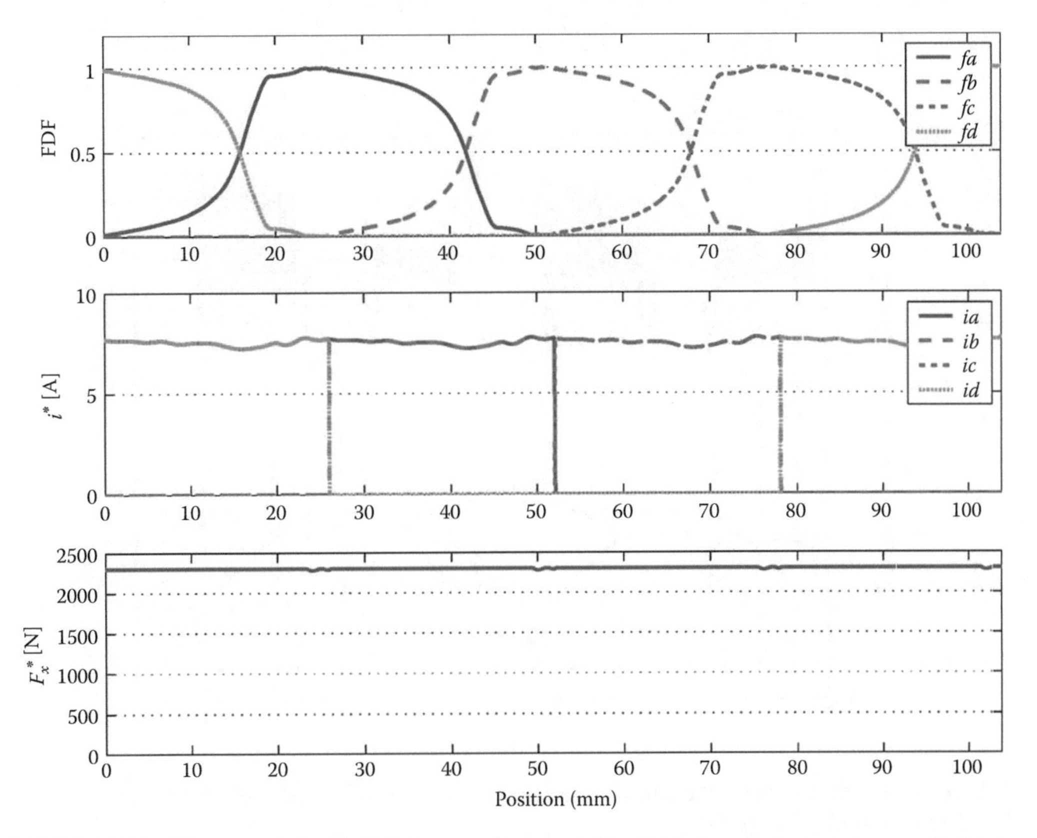

FIGURE 11.22 Thrust control with FDF thrust reference at $U=-0.5$ m/s and advance phase excitation by 2 mm. (After Lim, H.S. et al., *IEEE Trans.*, IE-55(2), 534, 2008.)

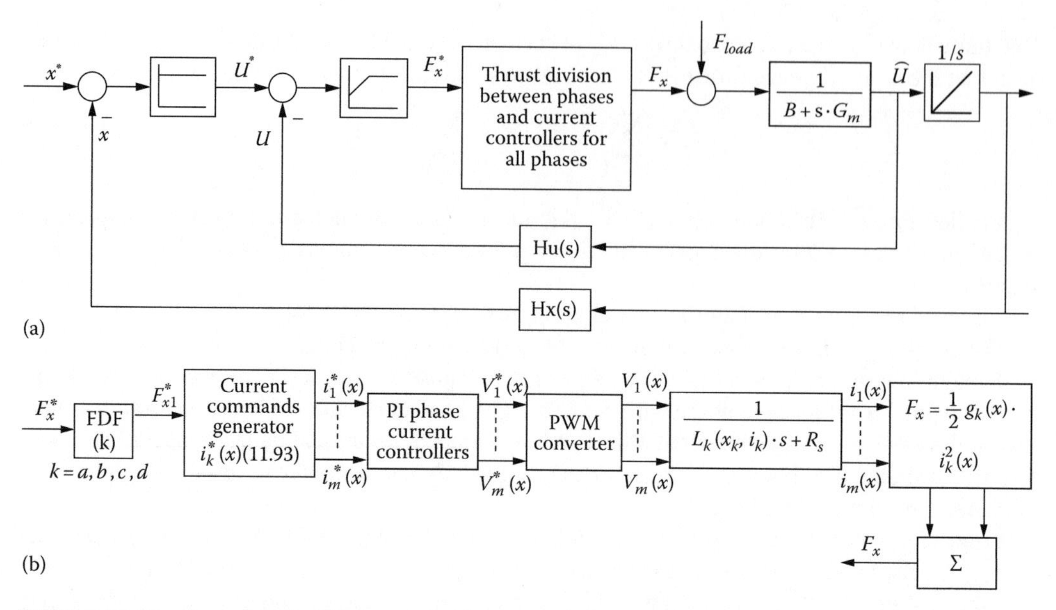

FIGURE 11.23 Four-phase L-SRM control: (a) structural diagram, (b) thrust division between phases and current control. (After Lim, H.S. et al., *IEEE Trans.*, IE-55(2), 534, 2008.)

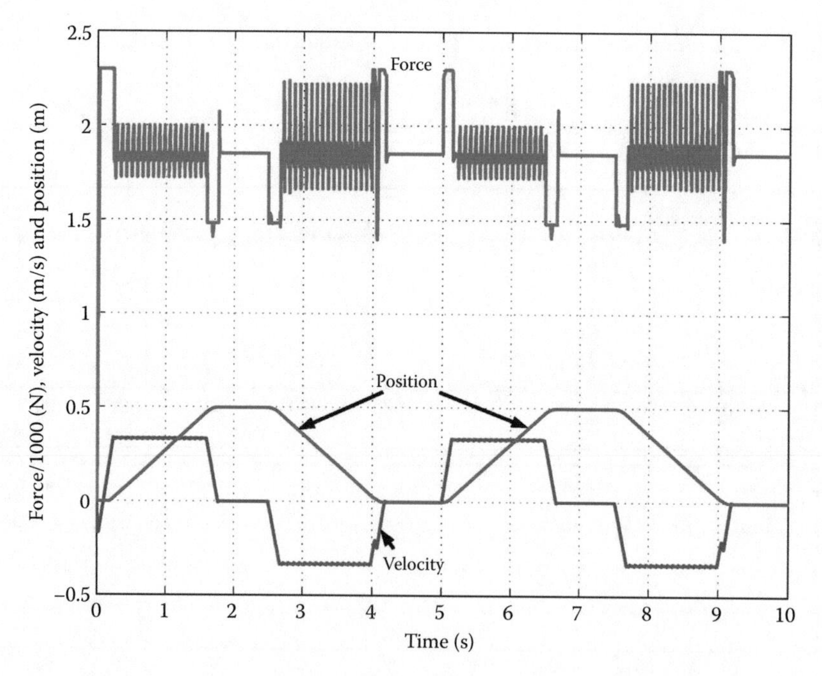

FIGURE 11.24 Position and speed repetitive response (flat-top speed is + 0.33 m/s). (After Lim, H.S. et al., *IEEE Trans.*, IE-55(2), 534, 2008.)

The mechanical energy W_{mec} per cycle (from zero to zero current per phase) is

$$W_{mec} = \int \frac{d\psi}{dt} i \cdot dt = \int_{x\,on}^{x\,off} \frac{\partial W_m{}'}{\partial x} \cdot dx = \int_{x\,on}^{x\,off} \int_{0}^{i} \frac{\partial \psi_k(x,i)}{\partial x} \cdot di \cdot dx \tag{11.100}$$

This integral should be reset to zero after the current reaches zero and thus the magnetization—demagnetization energy portions per phase cancel each other, and the average thrust of the machine per stroke is

$$F_{xav} = m \frac{W_{mec}}{l_{stroke}} \tag{11.101}$$

m is the number of phases

The average torque estimator is illustrated in Figure 11.25a.

The generic direct average thrust control in Figure 11.25 may be characterized as follows:

- It uses an average thrust/phase estimator based on a flux calculator from the voltage model: integral offset, converter nonlinearities, and phase resistance correction are all needed for good operation down to low speed.
- The average thrust may be estimated for each phase and added for faster average thrust control, as anyway current regulators are in general needed for all phases.

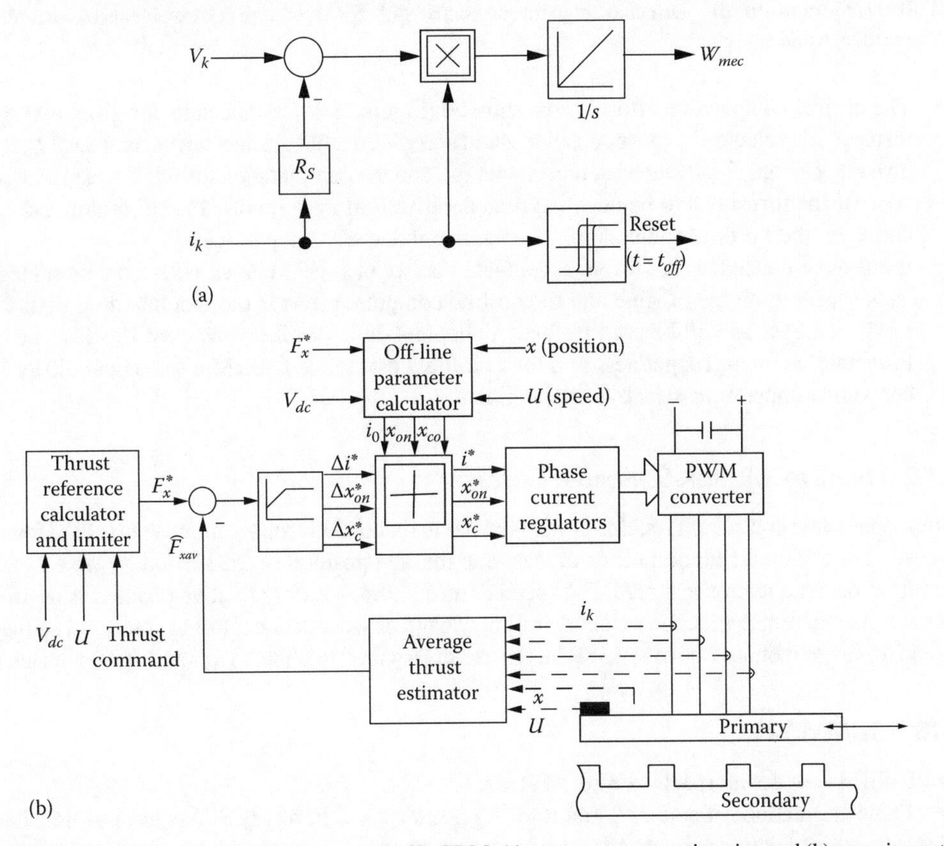

FIGURE 11.25 Direct average torque control of L-SRM: (a) average torque estimation and (b) generic control. *(continued)*

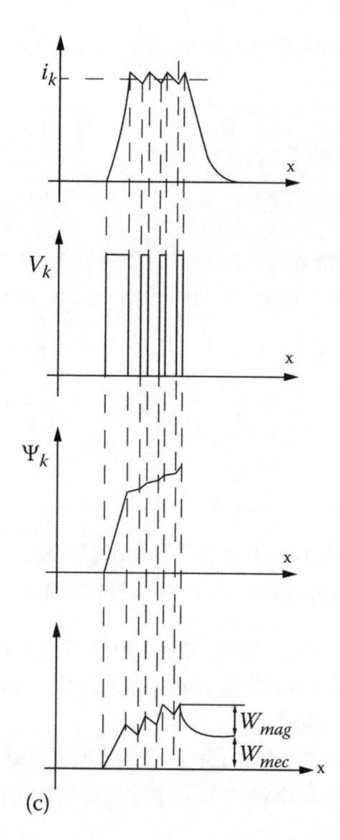

FIGURE 11.25 (continued) Direct average torque control of L-SRM: (c) typical waveforms of current, voltage, flux linkage and energy.

- The off-line computation effort is concentrated (Figure 11.25) to calculate, for given average thrust F_x, dc voltage V_{dc}, mover position x and speed U, the flat average top current level I_0, the turn on position x_{on} (advanced at higher speeds), and the commutation (turn off) position x_{co}.
- Also, if the thrust has to be negative (regenerative braking) the off-line calibrator sets x_{on} and x_{co} at their proper values (into the negative inductance slope region).
- It has to be conceded that for an acceptable control of L-SRM, even with a position feedback sensor, both the off-line and the on-line computer effort is more demanding than for LIMs or LSMs as will be seen in further chapters on L-PMSMs; however, the absence of PMs and the motor ruggedness and lower initial costs make L-SRM a strong candidate in cost- and temperature-sensitive applications.

11.9.2 Note on Motion-Sensorless Control

Motion-sensorless control of L-SRMs is very similar to that of their rotary counterparts [3]. However, the absolute position estimation needs at least one marked position in the secondary (fixed part, in general) to detect a reference signal. For speed control, however, only relative position is required.

Other than these remarks, the rich body of knowledge of sensorless control of rotary SRM may be adapted for linear motion control by L-SRM. An exemplary investigation, to follow, is given in Ref. [13].

11.10 SUMMARY

- L-SRMs are counterparts of rotary SRMs.
- The combinations of stator N_S and rotor N_R poles typical to rotary SRMs have to translate into ratios of active primary to passive secondary saliency period (τ_p/τ_s).
- L-SRMs may be flat, single sided, double sided, and tubular.

- As the normal force F_n to thrust F_x ratio in flat single-sided L-SRMs tends to be larger than 10/1, either the normal force is controlled for magnetic levitation or double-sided flat or tubular topologies are to be used. Recent new dual primary topologies reduce the F_n/F_x ratio considerably [18].
- As for rotary machines, longitudinal and transverse flux topologies are feasible; transverse topologies are in general free of longitudinal end static effects, and thus with 3(6) or 4(8) primary pole small L-SRMs such configurations for 3(4) phase operation are recommended.
- A typical, single (secondary) saliency topology with 5, 6, 7 phases and two-level bipolar current control, where m_F phases produce the "excitation" and the other $m - m_F$ phases are thrust phases, is also introduced as potentially it may yield good thrust and normal force density for good efficiency and reasonable peak kVA ratings in the multiphase inverter topology.
- To illustrate the operation principle, a simple U core primary single coil with I-secondary is investigated for thrust, normal force, and losses/thrust performance. The attraction of the laminated secondary to the primary, when the latter is supplied, and the typical coenergy variation with position (x and g-airgap) are the basis for thrust and normal force production in L-SRM. In a numerical example, 2.6 N/cm^2 of thrust, 87 N/cm^2 of normal force, and 0.9 N/W of copper losses are typical performance indexes for L-SRM.
- The instantaneous thrust calculation from the coenergy $\left(F_x = -\left(\partial W_m{}'/\partial x\right)_{i=const.}\right)$ requires the flux linkage/current/position family of curves, which may be calculated via complex magnetic circuit or FEM models or measured (from standstill current-decay tests).
- Due to magnetic saturation, the thrust/position curve for constant current is not a horizontal line and thus commutation phases are added—there may be notable thrust pulsations with flat-top current control in three- or even in four-phase L-SRMs.
- But magnetic saturation reduces the transient (commutation) inductance ($L' = L + i\partial L/\partial i$) and thus allows higher speeds.
- From the flux/current/position, $\psi/i/x$, family of curves, the average mechanical energy per energy cycle/phase may be determined $W_{mec} = F_{xav} \cdot l_{stroke}(lstroke \approx \tau_p/2)$. A part of energy W_{mag} is returned to the dc link through the PWM converter that supplies each phase.
- The ratio $\eta_{ec} = W_{mec}/(W_{mec} + W_{mag})$ is called the energy cycle conversion ratio and it is 0.5 for nonsaturated L-SRM and 0.6–0.67 for saturated conditions. A low η_{ec} is an indication of high kVA rating requirements from the PWM converter supply. In general, the peak kVA of the converter is larger than for most ac linear motors.
- As the thrust sign (for motoring or generating) does not depend on current polarity, homopolar current is in general supplied to L-SRM phases.
- The phase magnetic coupling is rather small in L-SRMs, which is good for fault tolerance but not so good for thrust density.
- The state space equations contain phase voltage equations and the equations of motions.
- The phase voltage equations contain a pseudo-emf

$$\left(E = i\frac{\partial L(x,i,g)}{\partial x}\frac{dx}{dt} + i\frac{\partial L(x,i,g)}{\partial g}\frac{dg}{dt}\right),$$

  which is related to linear (along x) and vertical (along g) motions: however, the thrust may be calculated from $F_x = (i^2/2)(\partial L/\partial x)$ only in the absence of magnetic saturation (all $\psi/i/x$ curves are linear); otherwise, the coenergy formula has to be used $\left(F_x = -(\partial W_m'/\partial x)_{i=const.}\right)$.
- As the L-SRM is not a traveling field machine, there are core losses both in the primary and in the secondary; they have to be carefully assessed in any realistic efficiency calculation attempt.
- For linear piece-wise $\psi/i/x$ curves (Figure 11.12), with state space equations linearization, a second-order structural diagram for transients, typical to series dc excited brush motor, are obtained. This result helps notably in the design of the current and speed close-loop regulators for L-SRM.

- From a myriad [3] of unipolar current multiphase dc–dc converters suitable for L-SRMs, only the one with two power switches per phase (2 m switches in general) and the C-dump converter with $m+1$ power switches are introduced and characterized here as very practical for the scope. An analytical design methodology based on linear-wise $\psi/i/x$ curve (magnetic saturation not considered) has been developed in this chapter via a numerical example: the design yields also all circuit parameters tuned for close-loop current and speed regulators. For a 400 N, 2.4 m/s, tubular L-SRM case study, a total efficiency of 0.888 is obtained for an active primary of 26 kg; an ideal acceleration of active primary mover of 15 m/s^2 is obtained, which allows for acceleration to 2.4 m/s within 0.2 m, which may be enough for many short travel transportation jobs in industrial applications.
- The control of L-SRM is approached first through the cascaded—current, speed, position—strategy that makes use of the linearized second-order model of L-SRM for control design.
- The "force distribution factor" (FDF) f_k dependent on $\partial L/\partial x$, when supplied for a nonsaturated four-phase L-SRM to produce the required reference current waveforms that reduce the thrust pulsation to less than 10%, is good enough in many servo applications.
- Positioning steady-state error of 3.5 μm has been claimed with a rather involved computer intensive control of three-phase L-SRM [14,15].
- In an effort to reduce the off- and on-line computation effort, a direct average thrust control system is presented in some detail.
- A note on motion-sensorless control sends to the sensorless control of rotary SRM [3] wherefrom a lot can be used for L-SRMs.
- As the use of PMs in electromagnetic devices increases (especially in automotive and wind and small hydro sectors), the SRMs and especially L-SRMs may become more competitive due to their simplicity, ruggedness, and low enough initial system costs.

REFERENCES

1. T.J.E. Miller, *Switched Reluctance Motors and Their Control*, Magna Physics Publishing, Oxford, U.K., 1993.
2. I. Boldea and S.A. Nasar, *Linear Electric Actuators and Generators*, Cambridge University Press, Cambridge, U.K., 1997.
3. R. Krishnan, *Switched Reluctance Motor Drives*, CRC Press, Boca Raton, FL, 2001.
4. H.S. Lim, R. Krishnan, and N.S. Lobo, Design and control of linear propulsion system for an elevator using linear switched reluctance motor drives, *IEEE Trans.*, IE-55(2), 2008, 534–542.
5. K.B. Saad, M. Benrejeb, and P. Brochet, One two control methods for smoothing the linear tubular switched reluctance stepping motor position, *Record of LDIA-2007*.
6. I.D. Law, A. Chertok, and T.A. Lipo, Design and performance of the field regulated reluctance machine, *Record of IEEE-IAS*, 1992, vol. 1, Houston, TX, pp. 234–241.
7. I. Boldea, *Reluctance Synchronous Machines and Drives*, Chapter 2, Oxford University Press, U.K., 1996.
8. I. Boldea, *Variable Speed Generators*, Chapter 9, CRC Press, Taylor & Francis, Boca Raton, FL, 2006.
9. A. Radun, Design considerations on SRMs, *Record of IEEE-IAS*, 1994.
10. H.S. Lim, Design and control of a linear propulsion system for an elevator using L-SRM drives, *IEEE Trans.*, IE-55(2), 2008, 534–542.
11. E. Santo, M.R. Calado, and C.M. Cabrita, Influence of pole shape in linear switched reluctance actuator performance, *Record of LDIA*, 2007.
12. W. Missaoui, L. El Ambraoni Ouni, F. Gillon, M. Benrejeb, and P. Brochet, Comparisons of magnetic behavior of SR and PM tubular linear three-phase synchronous linear motors, *Record of LDIA*, 2007.
13. G. Pasquesoone and I. Husain, Position estimation at starting and lower speed in 3 phase SRMs using pulse injection and two thresholds, *Record of IEEE-ECCE*, 2010, pp. 2660–2666.

14. W.-C. Gan, N.C. Cheung, and L. Qiu, Position control of linear SRMs for high precision applications, *IEEE Trans.*, IA-39(5), 2003, 1350–1362.
15. S.H. Lee and Y.S. Baek, Contact free switched reluctance linear actuator for precision stage, *IEEE Trans.*, Mag-40(4), 2004, 3075–3077.
16. G. Baoming, A.T. de Almeida, and F. Ferreira, Design of transverse flux linear switched reluctance motor, *IEEE Trans.*, Mag-45(1), 2009, 113–119.
17. R.B. Inderka and R.W. de Doncker, Simple average torque estimation for control of SRMs, *Record of EPE-PEMC*, 2000, vol. 5, Kosice, Slovakia, pp. 176–181.
18. W. Wang, C.H. Lin, and B. Fahimi, Comparative analysis of double stator switched reluctance machine and PSMS, *Record of IEEE-ISIE*, 2012.

12 Flat Linear Permanent Magnet Synchronous Motors

By flat linear permanent magnet synchronous motors (F-LPMSM), we mean here single- or double-sided, short mover or long stator three-phase ac-fed primary and long (stator) or short (mover) secondary configurations, with sinusoidal or trapezoidal emf or current waveform control. Both distributed ($q \geq 1$) and tooth-wound three-phase windings in iron or air-core primaries are considered. The tubular linear PM motors (T-LPMSMs) are treated in a separate chapter as their design and control is related to the rather short travel length (below 2–3 m). Linear PM brushless motors with variable reluctance short (mover) or long (stator) secondary and PM primaries that contain both the PMs and three-phase ac windings will also be treated separately under the name of pulse linear PM brushless motors (P-LPMBM).

The F-LPMSM may be applied, in the long iron-core primary (stator) and short PM mover configuration, for industrial urban and interurban transportation. Short (mover) primary and long PM-secondary (stator) F-LPMSM configurations are used in short travel (a few meters) both in single-sided versions (as MAGLEV carriers, especially) and in double-sided PM-secondary (tube) versions, when the normal force on the short primary (mover) is to be zero, in order to avoid noise, vibration, and large stray PM magnetic fields.

In what follows, we will first discuss practical topologies, analytical (technical) and finite element method (FEM) field theories for F-LPMSMs, cogging force computation and reduction, circuit parameters, and circuit models for dynamic and field oriented control (FOC) and direct thrust and flux control (DTFC); then a design methodology is introduced by numerical example.

12.1 A FEW PRACTICAL TOPOLOGIES

Though there are many "novel" topologies for F-LPSM, only a few are representative and practical:

- Single-sided type (Figure 12.1a,b, and d)
- Double-sided type (Figure 12.1c and f)
- Short PM secondary (mover) type (Figure 12.1a and d)
- Short ac primary (mover) type (Figure 12.1c,e,f, and g)
- Distributed ($q=1$ slot/pole/phase) three-phase ac winding (Figure 12.1a through d)
- Concentrated (tooth-wound) three-phase winding (Figure 12.1e,f, and g)
- Iron-tooth primary core (Figure 12.1a through f)
- Air-core primary (three-phase) (Figure 12.1c)
- With surface PM secondary (Figure 12.1a,c through g)
- With Halbach array PMs secondary (Figure 12.1b)
- With four sides (Figure 12.1g)

An investigation of these configurations reveals the following:

- The cable-coil $q=1$, single layer, three-phase winding (Figure 12.1a and d) is easy to install along track, with a fixed primary (stator), but it requires open slots, and thus the airgap should be about the size of the slot openings (or less) to allow small emf harmonics and low enough cogging force. The configuration would be typical for urban or interurban MAGLEVs with an active guideway (airgap from 6 to 10 mm).

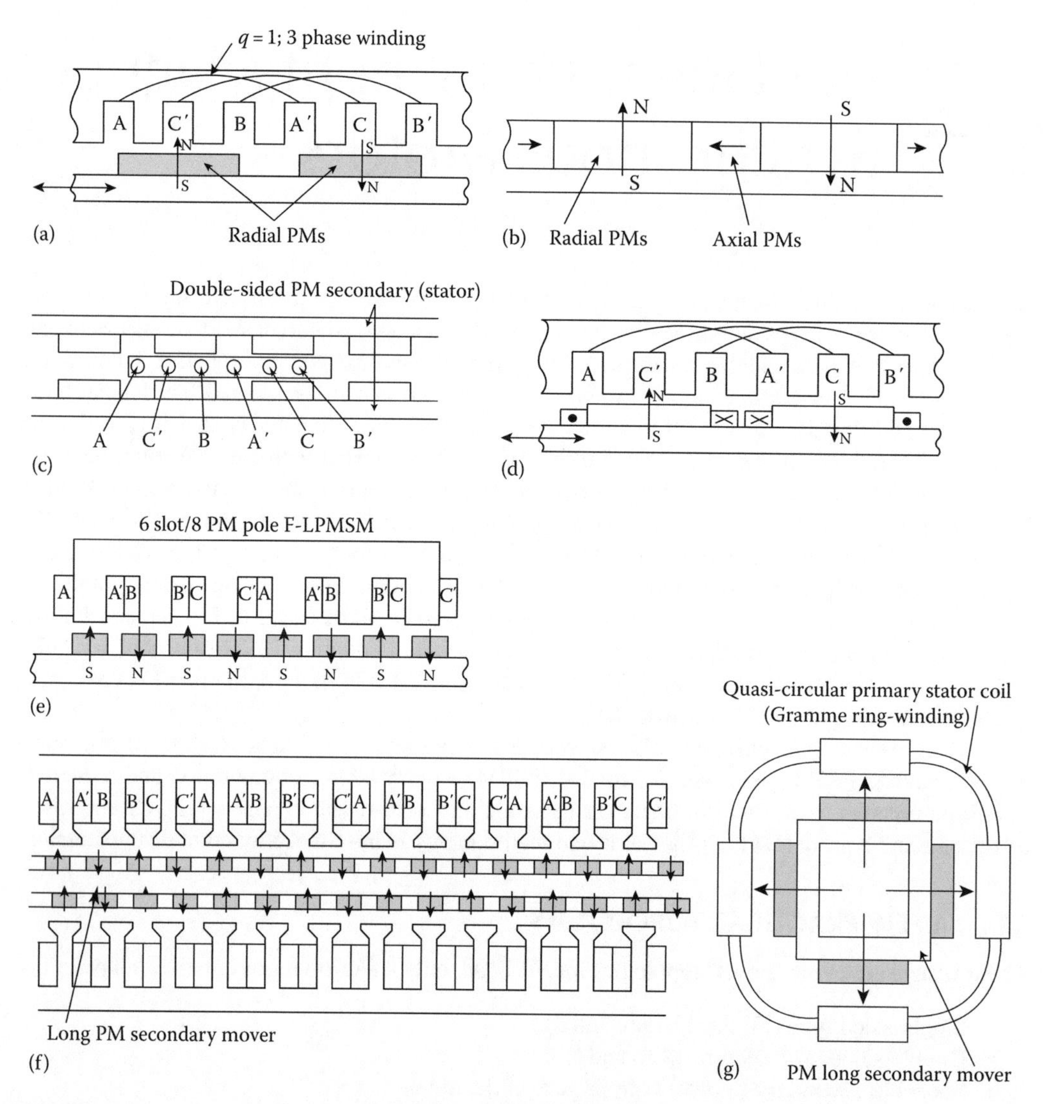

FIGURE 12.1 A few practical F-LPMSMs topologies (a) with $q=1$ slot/pole/phase winding primary and radially surface PMs, (b) radial+ axial (Halbach) PM secondary, (c) double-sided F-LPMSM with hybrid PM+ dc coil secondary for flux weakening or levitation control, (e) single-sided F-LPMSM with tooth-wound three phase winding (6 slot/8 pole), (f) double-sided PM secondary with shifted PMs F-LPMSM (12 slot/16 pole), (g) four sided configuration with quasi-circular coils Gramme winding.

- The Halbach PM array secondary (Figure 12.1b) leads to less emf ripple, higher thrust and lower cogging (zero current) force, and thinner secondary back core; it may be used with both PM mover or mild length (travel) PM stator.
- The double-sided PM-secondary stator with air-core primary (mover) is typical for industrial linear motor automation when the cogging force has to be small (less than 1%), and the stray PM fields still have to be limited.
- The hybrid PM + dc coil secondary mover configuration (Figure 12.1d) is typical for active guideway (primary) MAGLEVs.
- The single-sided 6 slot/8 PM pole configuration (Figure 12.1e) is just an example of tooth-wound three-phase winding; to reduce cogging force, the PM span per pole pitch is around 0.6 (very close to teeth shoe span).

- The double-sided primary and secondary configuration in Figure 12.1f provides for ideally zero normal force on dual PM-secondary mover; the two rows of PMs are shifted to reduce cogging force; the 12/16 pole module combination is used to mitigate between reasonably low cogging force and low iron losses, with reasonably low inductance for generator operation mode. 12/14 (in general, $N_S - 2p = \pm 2k$) configurations reduce further cogging force but, due to larger mmf harmonics, they lead to larger core losses in the primary and secondary under load and show a larger synchronous inductance.
- The four-sided configuration in Figure 12.1g is close to a tubular structure (due to quasi-circular ac winding coils), but it retains the regular structure of the laminated stack of primary; the small displacement of the four PM rows leads to a small cogging force. For wave energy generators, tooth-wound single-layer windings with quasi-circular coils may lead to small pole pitches (the speed is around 1 m/s and the power in hundreds of kW), which allow for hopefully acceptable thrust density (in N/kg) and efficiency; this is yet to be proved.

In what follows, we will deal with the field theory of F-LPMSMs first by an analytical multilayer model, then by the magnetic circuit model, and finally with FEM (through results, mainly).

12.2 MULTILAYER FIELD MODEL OF IRON-CORE F-LPMSMs WITH SINUSOIDAL EMFs AND CURRENTS

Let us consider a single-sided F-LPMSM with iron-core primary and investigate the PM and armature-reaction airgap fields and forces (Figure 12.2). To simplify the calculations, we replace the slots/teeth area by a nonslotted area, but the magnetic permeability is different along the two axes x (longitudinal) and y (vertical) [1,2]:

$$\mu_x = \frac{\mu_0 \cdot \mu_r}{1 + (b_s/\tau_s)(\mu_r - 1)} \tag{12.1}$$

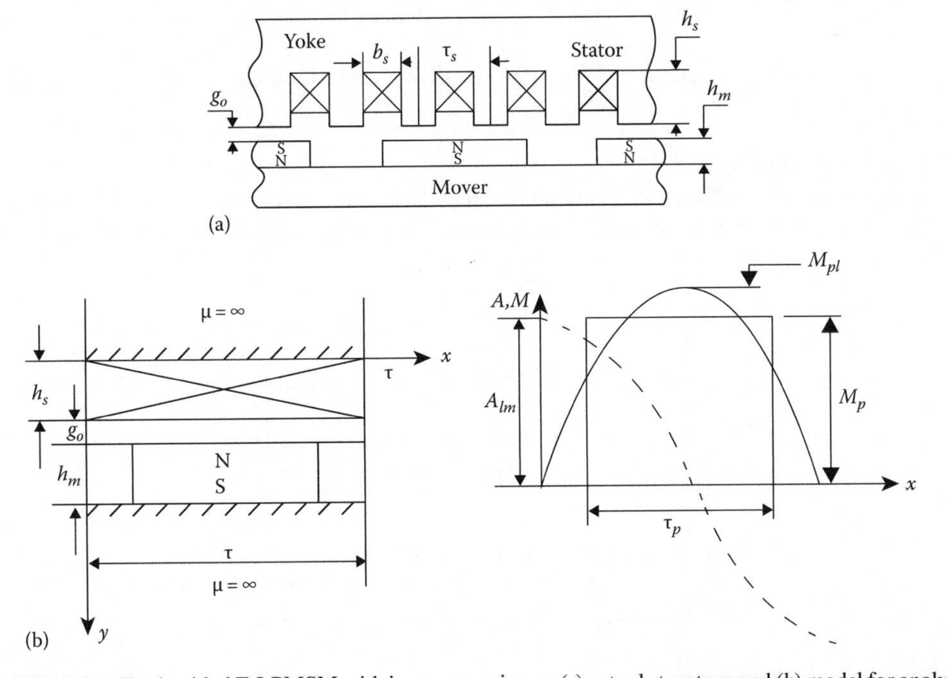

FIGURE 12.2 Single-sided F-LPMSM with iron-core primary (a) actual structure and (b) model for analysis.

$$\mu_y = \mu_0 \left[\frac{b_s}{\tau_s} + \mu_r \left(1 - \frac{b_s}{\tau_s} \right) \right] \tag{12.2}$$

where
μ_r is the average magnetic permeability in the primary teeth
b_s is the slot width
τ_s is the slot pitch

A multilayer structure (Figure 12.3) is thus obtained. There are three main layers and the PM layer. The armature structure equivalent mmfs are represented by their space fundamentals distributed ac windings ($q=1, 2$), placed at random heights in their layers. The regions (II + III′), containing the PMs, are isotropic ($\mu_x = \mu_y \approx \mu_0$); iron regions are infinitely permeable. The fundamentals of PM magnetization $M_{\nu PM}$ harmonics are (in PM-secondary coordinates) as follows:

$$M_{\nu PM} = \frac{4}{\pi \nu_2} M_{PM} \sin \frac{\nu_2 \pi \cdot \tau_p}{\tau} \sin \nu_2 \frac{\pi}{\tau} x; \quad M_{PM} = H_C \cdot h_m; \quad \nu = 1,2... \tag{12.3}$$

where
H_C is the coercive field of PMs
h_m is the PM depth

Its current sheet $A_{\nu PM} = \partial M_{\nu PM} / \partial x$, where τ_p is the PM span/pole and τ is the pole pitch of both primary and secondary. The armature-reaction mmf fundamental and harmonics in PM-secondary coordinates would be

$$A_{\nu s} = A_{\nu sm} \cos \left(\frac{\nu \pi x}{\tau} + \gamma_0 \right) \tag{12.4}$$

$$A_{\nu sm} = \frac{3\sqrt{2}}{\pi \tau} k_{W\nu} W_1 I_\nu \text{-for distributed windings} \tag{12.5}$$

where $k_{W\nu}$ is the winding factor.

For tooth-wound windings, $A_{\nu sm}$ has many large (sub and super) space harmonics around the fundamental. A full derivation of field equations with boundary conditions is presented in [1,2]

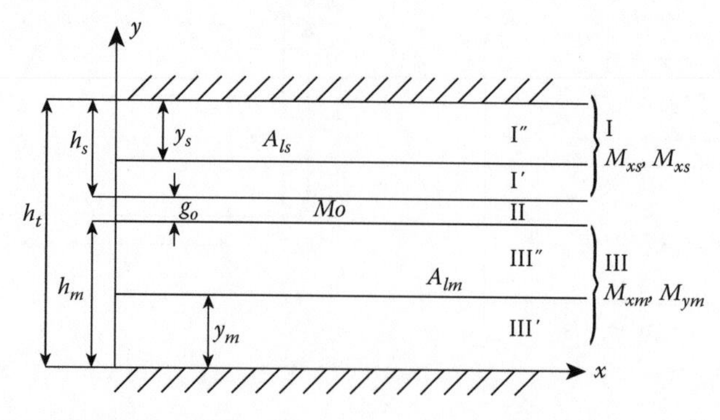

FIGURE 12.3 Multilayer model.

for the layers I′ + I′, II, and III + III′. In Ref. [1], it was shown that the heights y_m and y_s, where the equivalent PM and armature mmfs, within the layer depth, are placed, are not important.

Finally, for the fundamental PM flux density B_{yPM_1} in the airgap

$$E_\nu = \frac{\sqrt{2}}{\pi}\omega_1 \cdot k_{W_\nu} \cdot W_1 \cdot \tau \cdot l_{stack} \cdot \left(B_{yav}\right)_{av}; \quad \nu = 1 \text{ for the fundamental} \tag{12.6}$$

(For the air-core machine, this flux density has to be calculated as an average along the air-core coil height.) The airgap inductance L_m is thus

$$L_{m\nu} = \frac{\sqrt{2}}{\pi I} k_{W_\nu} \cdot \frac{\tau}{\nu} \cdot l_{stack} \cdot \left(B_{yav}\right)_{av}; \quad \nu = 1 \text{ for the fundamental} \tag{12.7}$$

The airgap space harmonics $(L_m)_{\nu>1}$ are considered as leakage inductances and added to give the total synchronous cyclic inductance. For $b_s/\tau_s = 0.53$, $h_s = 35$ mm, $\tau = 114.3$ mm, $\tau_p/\tau = 0.67$, $\mu_{rel} = 400$, $g = 2$ mm, $h_m = 6.35$ mm, $B_r = 0.86$ T, $H_C = 0.7$ MA/m, and $l_{stack} = 101.6$ mm, the normal airgap flux density was calculated as mentioned earlier, measured and verified by 2D FEM for zero current and for pure I_q current ($\gamma = \pi/2$) as in Figure 12.4a and b [1].

Satisfactory fundamental value agreement between the analytical model and FEM is obtained while FEM fits well the Hall-probe measurements of airgap flux density. A full analytical field theory may be developed by also considering the slot openings. In addition to our isotropic slot-teeth

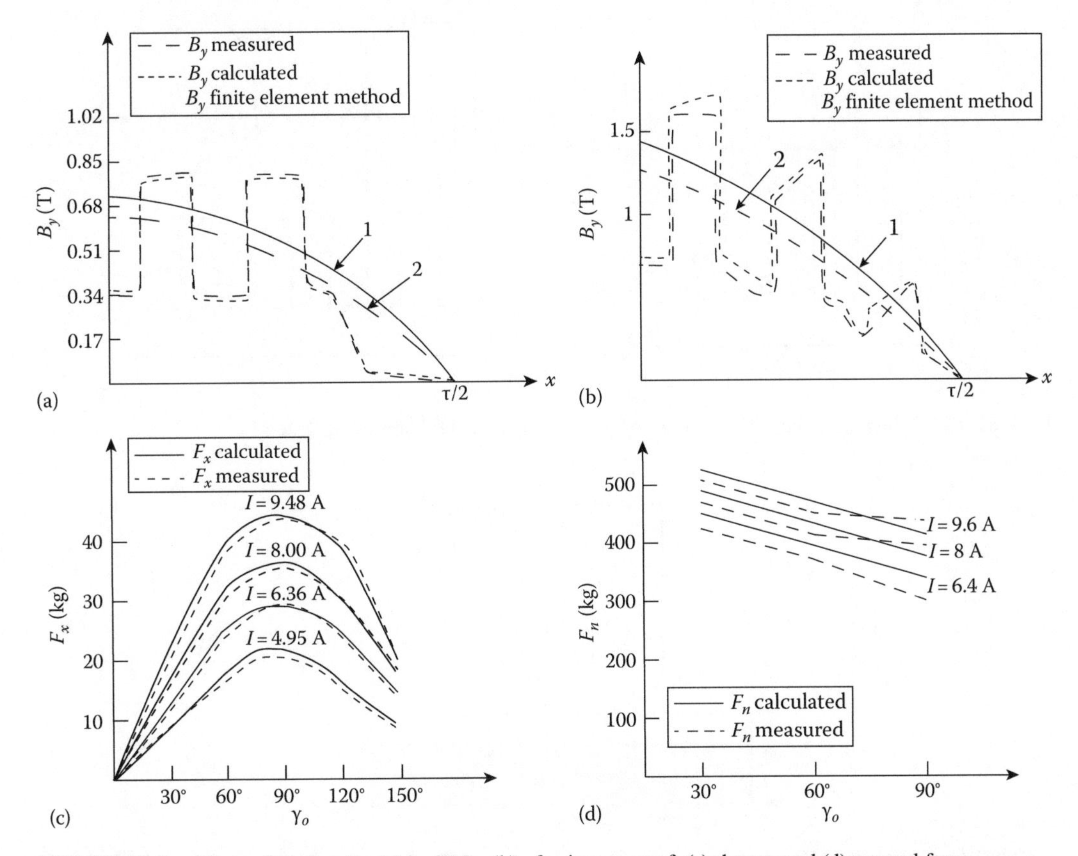

FIGURE 12.4 Airgap flux density (a) by PMs, (b) of primary mmf, (c) thrust, and (d) normal force.

area with $\mu_x \neq \mu_y$, which, in fact, accounts for mmf step-like distribution due to slots and allows to use the concept of current sheet by [3]

$$B_{yPM\nu}(x) = (B_{yPM\nu})_{slotless} \cdot P_g(x); \quad \frac{\partial B_{yPM\nu}(x,y)}{\partial y} + \frac{\partial B_{xPM\nu}(x,y)}{\partial x} = 0 \tag{12.8}$$

$$B_{av}(x) = (B_{y\nu})_{slotless} \cdot P_g(x) \tag{12.9}$$

where P_g is the inverse per-unit airgap (or airgap) permeance function (Figure 12.5), which may be described rather precisely by conformal mapping; other simplified curve fitting is feasible.

To this scope the theory for rotary brushless dc motors [4,5] may be adopted for F-LPMSMs. In a simplified method, force at zero current F_{cog} may be calculated analytically by adding the forces on the primary open slot walls (Figure 12.6):

$$F_{cog} \approx \frac{l_{stack}}{2\mu_0} \sum_{n=1}^{N_S} \sum_{\nu=1}^{\infty} B_{xPM\nu}{}^2 \left(\frac{\nu \cdot 2\pi n}{N_S} + \theta_1 \right) (g_\alpha \cdot ssg) \tag{12.10}$$

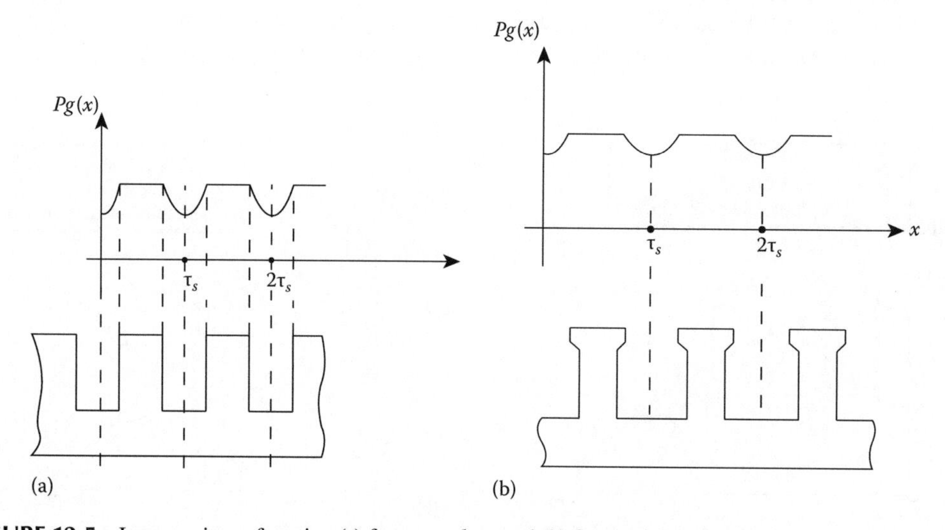

FIGURE 12.5 Inverse airgap function (a) for open slots and (b) for semiclosed slots.

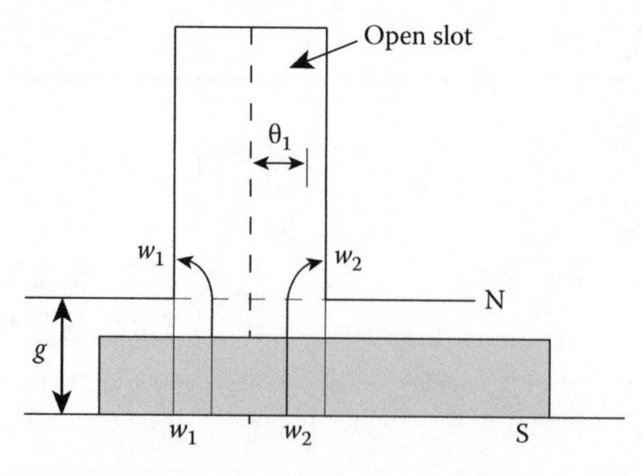

FIGURE 12.6 Slot flux ideal path.

where

$g_\alpha = 0$ and $ssg = 0$ outside slots openings
$g_\alpha = w_1 + g$ and $ssg = 1$ for left side of slots
$g_\alpha = w_2 + g$ and $ssg = -1$ for the right side of slots
N_S is the number of primary slots
l_{stack} is the primary stack length
θ_1 is the angle of slot center with respect to PM axis
$B_{xPM\nu}$ is obtained from (12.8)

Now the thrust may be calculated directly from the analytical field theory:

$$F_x = l_{stack} \sum_{\nu=1}^{\infty} \int_0^{2p\tau} B_{yPm\nu}(x) A_{s\nu}(x) dx \tag{12.11}$$

$$F_y \approx \frac{l_{stack}}{2\mu_0} \sum_{\nu=1}^{\infty} \left[\int_0^{2p\tau} \left[B_{yPM\nu}(x) + B_{a\nu}(x)\right]^2 dx + 2\int_0^{2p\tau} \left[\left(B_{yPM\nu}(x) + B_{a\nu}(x)\right) \cdot B_{a\nu}(x)\right] \cdot dx \right] \tag{12.12}$$

A comparison between calculated and measured forces (Figure 12.4c and d) shows satisfactory agreement [2]. An analytical multilayer theory may be used even for the Halbach PM array secondary (Figure 12.7) air-core F-LPMSM [6].

In this case, the PM magnetization has two components M_{xn} and M_{zn} (Figure 12.7). The PM flux linkage ψ_a per phase is [6]

$$\psi_{PM} = wn_0 p_m \sum_{n=-\infty,odd}^{\infty} \frac{1}{jk_m \gamma_m} \left(\frac{j\mu_0 M_{xn}}{k_m} + \frac{\mu_0}{\gamma_n} M_{zn} \right)$$
$$\times \left(1 - e^{-\gamma_n \Delta}\right) \left(e^{\frac{j\pi n}{3}} - 1 \right) \left(1 - e^{-\gamma_n \Gamma}\right) \times e^{-\gamma_n x_0} e^{jk_n x_0} \tag{12.13}$$

$$\psi_a = 4\mu_0 w p_s n_0 \tau^3 J_0 \sum_{n=-\infty,odd}^{\infty} \frac{1}{(n\pi)^4} \left(1 - \cos\frac{\pi}{3} n\right) \left(\Gamma + \frac{e^{-\gamma_n \Gamma - 1}}{\gamma_n}\right) \tag{12.14}$$

$$E = \frac{d\psi_{PM}}{dz} \cdot U; \quad U = \frac{\pi}{\tau}\omega_1; \quad L_m = \frac{\psi_a}{I} \tag{12.15}$$

where, say, $\mu_0 M_0 = 1.1$ T, $n_0 = 1.47 \times 10^6$ turns/m^2 ($W_C = 150$ turns/coil), J_0 is the current density of the coils for phase current I, $k_n = 2\pi n/2\tau$, where n is the mmf space harmonics order, $\gamma_n = |k_n|$, $x_0 = 5$ mm, $\tau = 51$ mm, $\Gamma = 6$ mm, $\Delta = 25.5$ mm, "stack" length $w = 25.5$ mm; the Halbach array contains $p_m = 4 + 1/2$ pitches; the pole pairs of one-phase coil $p_s = 7$. In Figure 12.7b, a good agreement between theory and experiments is proven [6].

Therefore, apparently, the analytical multilayer theory seems self-sufficient and easy to use for optimization design attempts, after preliminary design with key 2(3) D-FEM corrections.

Accounting for slot openings (and slotting in general) with magnetic saturation consideration requires the magnetic circuit method [4.7], besides 2(3) D-FEM.

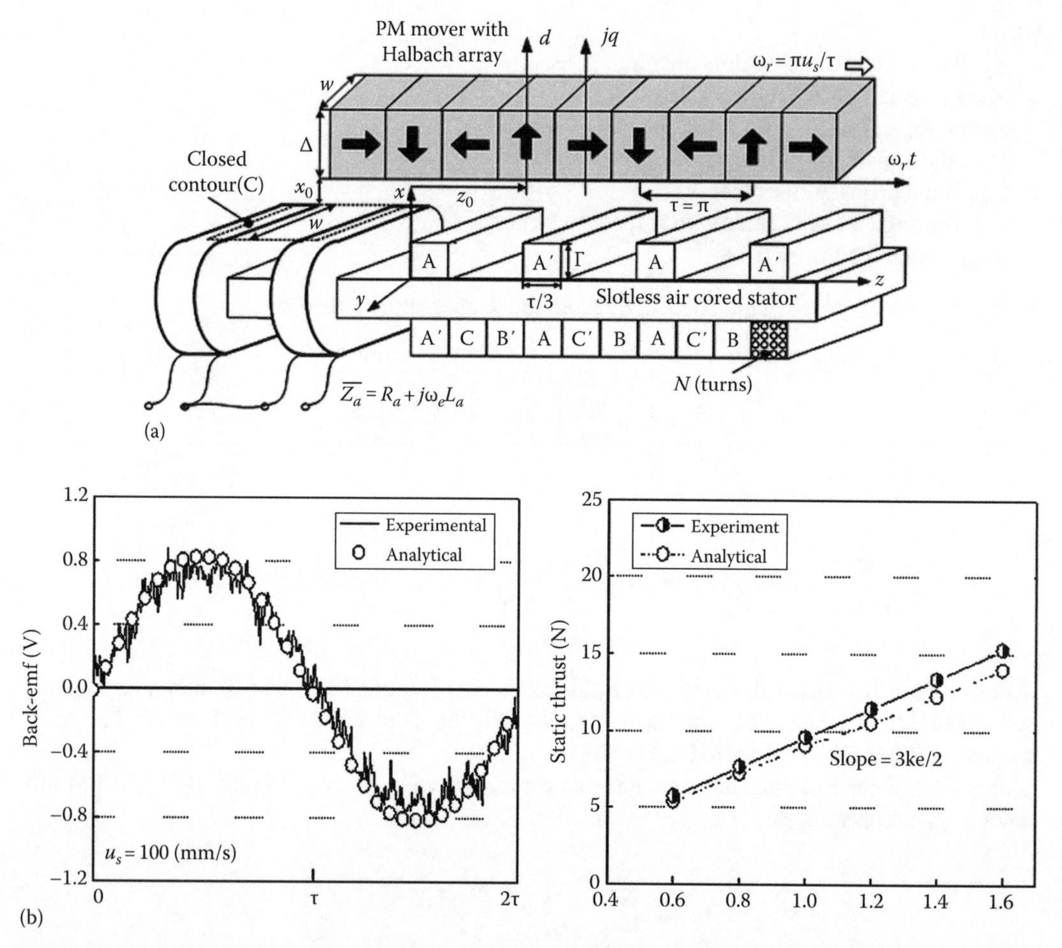

FIGURE 12.7 Air-core F-LPMSM with Halbach PM array secondary (a) emf and thrust versus *iq* current (b). (After Trumper, D.L. et al., *IEEE Trans.*, IA-32(2), 371, 1996.)

12.3 MAGNETIC EQUIVALENT CIRCUIT (MEC) THEORY OF IRON-CORE F-LPMSM

Calculating the airgap flux density distribution accounting for slot openings (in open slot magnetic cores), the actual distribution of windings in slots and magnetic saturation, within a limited computation effort means, according to the MEC method, to use realistic flux tubes and define their reluctance dependence of F-LPMSM geometry, mover position, and magnetic saturation, slot by slot and tooth by tooth (Figure 12.8), all assembled in a complete MEC (magnetostatic) model, to be solved by algebraic methods. In a flux tube, the flux density is considered uniform; this idea leads to the level of complexity (number of flux tubes—reluctances of MEC).

The structure of the MEC has to be fixed during motion, to secure limited computation effort. A two PM pole structure with its primary section of MEC is shown in Figure 12.8 [7].

A bi-section for airgap and yet another one for the PM secondary are added [7]. Also visible in Figure 12.8b are the longitudinal leakage reluctances R_{gz}, "attached" to primary (to the middle of the airgap); they are constant; only the tooth reluctance R_t varies due to magnetic saturation. The other side of airgap MEC (Figure 12.9) contains the reluctances R_{mg1}, R_{mg2}, and

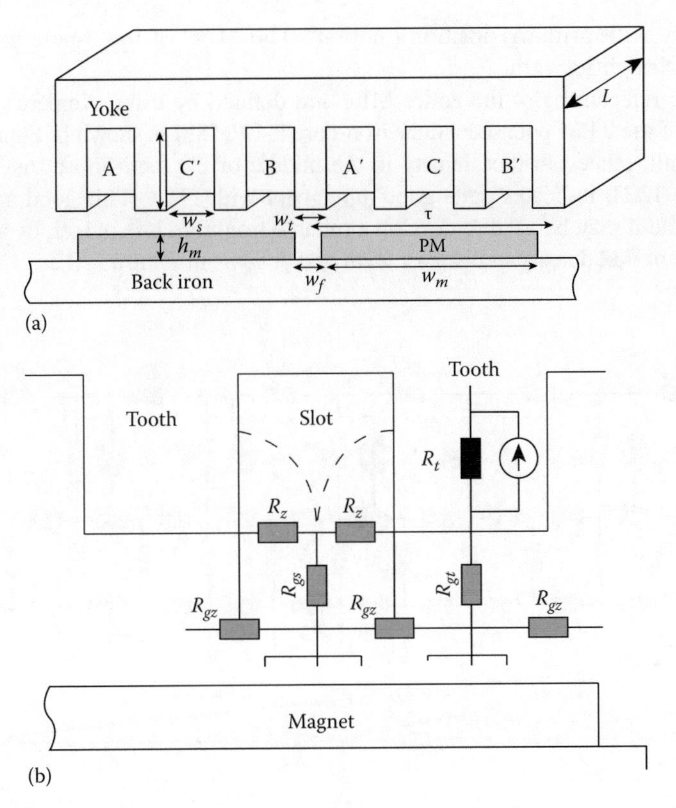

FIGURE 12.8 Iron-core F-LPMSM geometry (a) and elementary MEC, (b). (After Sheikh-Ghalavand, B. et al., *IEEE Trans.*, MAG-46(1), 112, 2010.)

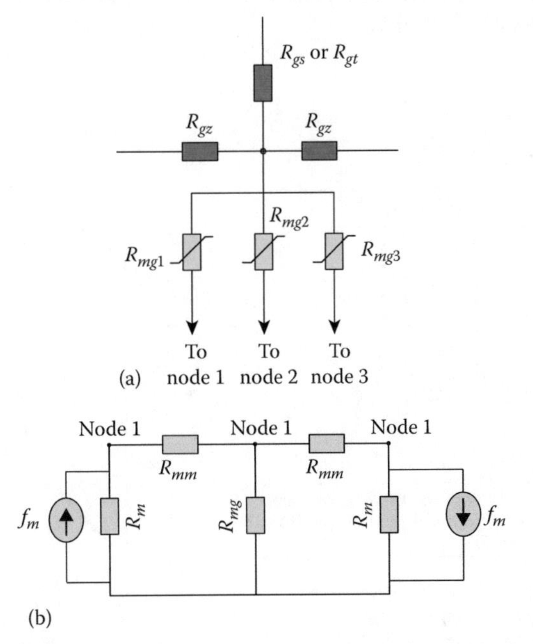

FIGURE 12.9 Airgap secondary MEC (a) and PM-secondary MEC (b). (After Sheikh-Ghalavand, B. et al., *IEEE Trans.*, MAG-46(1), 112, 2010.)

R_{mg3}, which vary with primary position (motion). The MEC of the 2-pole secondary (Figure 12.9b) is rather straightforward.

The magnetic reluctances of the entire MEC are defined by using Ampere's law. Finally, the complete MEC of the 2 PM pole secondary iron-core F-LPMSM is shown in Figure 12.10 [7].

A typical result, related to flux density in the middle of the teeth in saturated conditions and shown in Figure 12.11, indicates quite good agreement with FEM. This good agreement is valid also for thrust, and it may be used to develop a reliable iron-core loss model, by assuming a linear variation in time of flux density in primary teeth and yoke as in Figure 12.12.

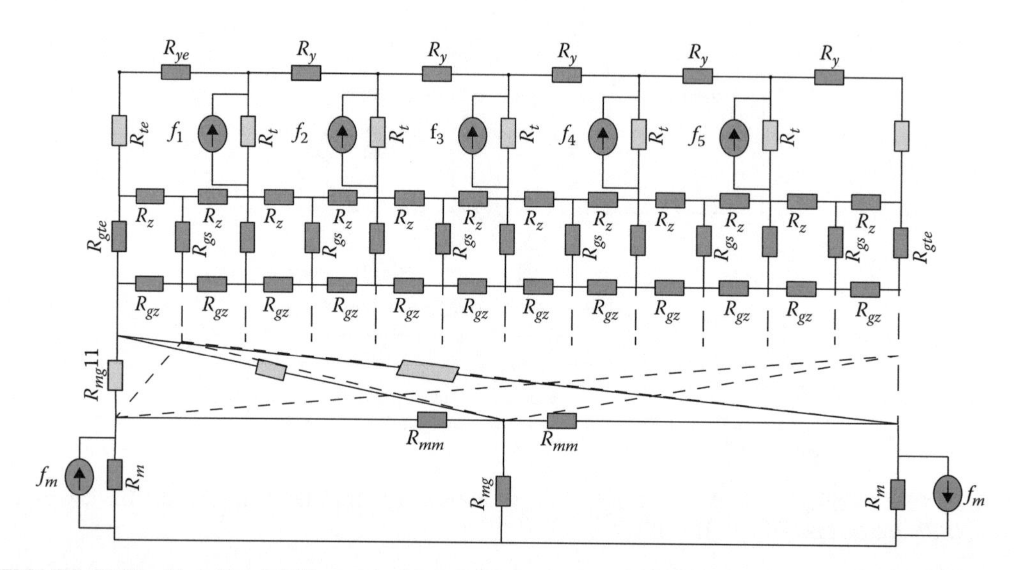

FIGURE 12.10 Complete MEC model of 2-pole PM-secondary F-LPMSM. (After Sheikh-Ghalavand, B. et al., *IEEE Trans.*, MAG-46(1), 112, 2010.)

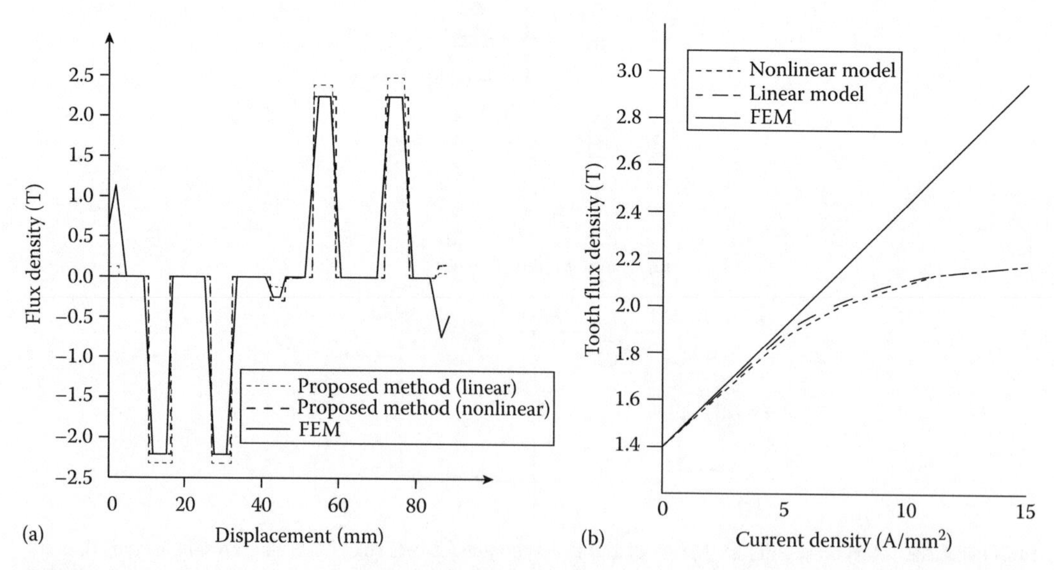

FIGURE 12.11 Flux density in the middle of the teeth (a) and flux density in tooth 2 versus primary current density, (b). (After Sheikh-Ghalavand, B. et al., *IEEE Trans.*, MAG-46(1), 112, 2010.)

Traditionally, the primary core losses are separated into two terms: the hysteresis losses p_h and eddy current losses p_{eddy}:

$$p_{iron} = p_h + p_{eddy} = k_h \cdot B_{\max}^{\alpha}\left(\frac{f_1}{50}\right) + k_e \cdot B_{\max}^2\left(\frac{f_1}{50}\right)^2 \ \mathrm{W/m^3} \tag{12.16}$$

where

$\alpha = 1.8–2.2$

k_h and k_e are core losses per m³ of laminated core at 1 T and 50 Hz

As (12.16) is valid for sinusoidal flux density-time variation, it gives large errors for waveforms as in Figure 12.12 and thus

$$p_{eddy} = \frac{2k_e}{T}\int_0^T \left(\frac{dB(t)}{dt}\right)^2 dt \ (\mathrm{W/m^3}) \tag{12.17}$$

$B_{t\max}$ and $B_{y\max}$ are calculated by the MEC method or by FEM. The total eddy current losses for the waveforms in Figure 12.12 are

$$p_e \approx 12q_1k_ek_{cl}\left(\frac{U}{\tau}B_{t\max}\right)^2 V_{teeth} + 8q_1k_ek_{cl}\frac{\tau}{W_m}\left(\frac{U}{\tau}B_{y\max}\right)^2 V_{yoke} \ (\mathrm{W}) \tag{12.18}$$

with

W_m is the PM span per pole

$k_{cl} = 1.1–1.2$ accounts for longitudinal eddy currents in the core

q_1 is slots/pole/phase

The PM-secondary losses, due to slot openings and primary mmf space harmonics fields, have not been considered here. Results from FEM and MEC indicate a total relative error below 5% in total primary core losses [7].

Note: The tooth-wound primary exhibits a rich content of mmf space harmonics, but the MEC method can also be used for it. Finally, the current-time particular variations end up in flux density—time variations, and sufficiently small time periods may handle the situation. Ultimately, perhaps, only FEM can produce reliable enough results for total core losses during steady state and transients, accounting for both space harmonics and time harmonics.

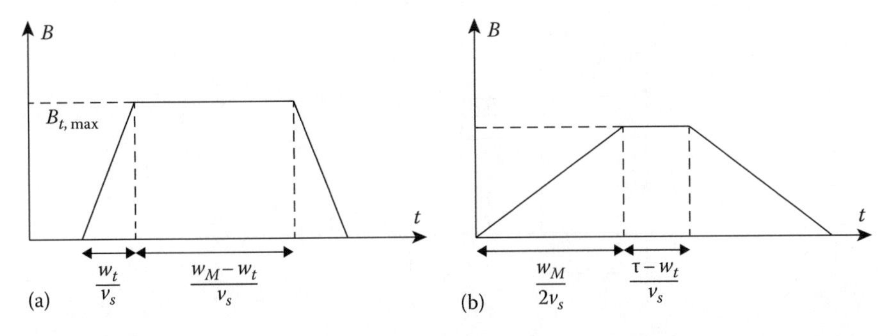

FIGURE 12.12 Ideal (linear) time variation of flux density in primary teeth (a) and yoke (b). (After Sheikh-Ghalavand, B. et al., *IEEE Trans.*, MAG-46(1), 112, 2010.)

12.4 ANALYTICAL MULTILAYER FIELD THEORY OF AIR-CORE F-LPMSM

The multilayer field theory of air-core F-LPMSM (Figure 12.13a) may be approached with the same method as in Section 12.2, with the simplification that the slot zone is the ac coils zone and $\mu=\mu_0$. To add more generality to the model, the case of rectangular current control is considered (Figure 12.13b) [8].

The PM magnetization $M(x)$ is decomposed in Fourier series:

$$M(x)=\frac{4}{\pi}\frac{B_r}{\mu_0}\sum_{\nu=1}^{\infty}\frac{1}{2\nu-1}\sin(2\nu-1)\frac{\pi}{2}\frac{\tau_p}{\tau}\sin(2\nu-1)\frac{\pi}{\tau}h_m\times$$
$$\times\exp\left[-(2\nu-1)\frac{\pi}{\tau}y\right]\sin(2\nu-1)\frac{\pi}{\tau}x \tag{12.19}$$

The final B_{xI}, B_{yI}, PM flux density components in region I [8], are

$$B_{xI}(x,y)=\frac{4B_r}{\pi}\sum_{\nu=1}^{\infty}\frac{1}{2\nu-1}\sin(2\nu-1)\frac{\pi}{2}\frac{\tau_p}{\tau}\sinh(2\nu-1)\frac{\pi}{\tau}h_m\times\exp\left(-(2\nu-1)\frac{\pi}{\tau}y\right)\cdot\sin(2\nu-1)\frac{\pi}{\tau}x$$

$$B_{yI}(x,y)=\frac{4B_r}{\pi}\sum_{\nu=1}^{\infty}\frac{1}{2\nu-1}\sin(2\nu-1)\frac{\pi}{2}\frac{\tau_p}{\tau}\sinh(2\nu-1)\frac{\pi}{\tau}h_m\times\exp\left(-(2\nu-1)\frac{\pi}{\tau}y\right)\cdot\cos(2\nu-1)\frac{\pi}{\tau}x \tag{12.20}$$

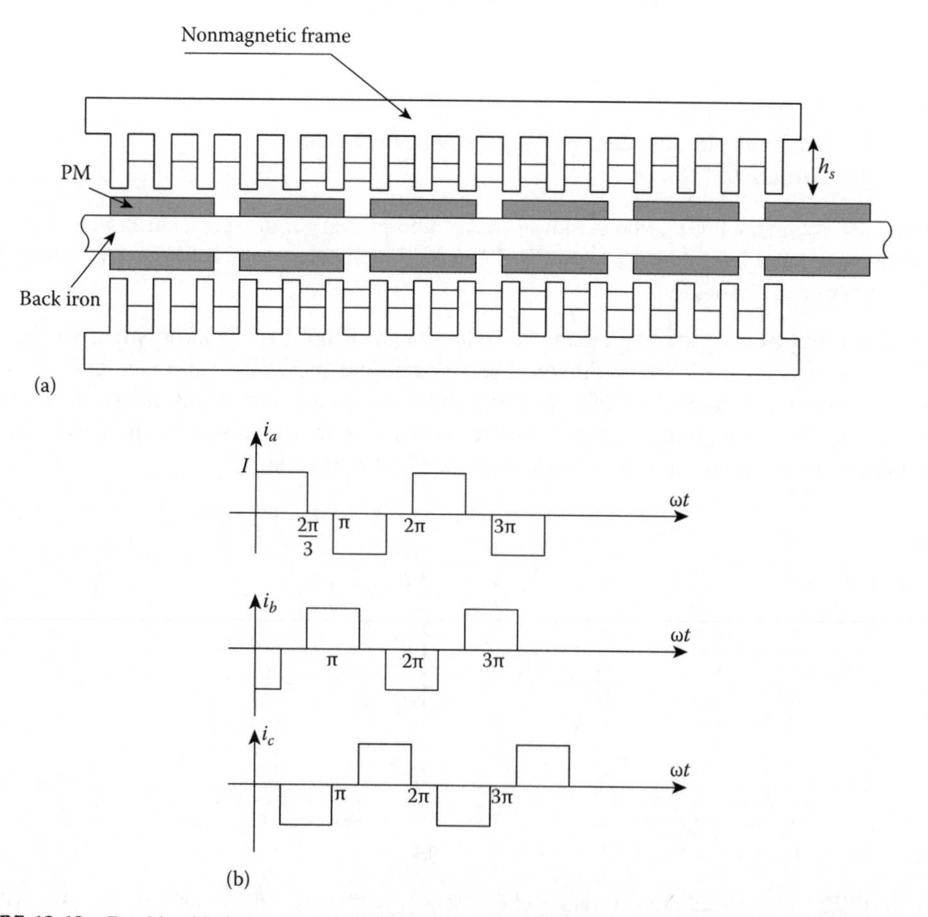

FIGURE 12.13 Double-sided air-core F-LPMSM, (a) and its ideal (rectangular) phase currents, (b).

B_{xI} and B_{yI} are dependent on B_r, τ_p/τ, h_m/τ, and they decay along Oy (airgap and winding depth). The fundamental components decay e times along τ/π airgap depth. To secure a good interaction between PM field and the primary air-core winding, the latter's depth h_s should be smaller than τ/π; so larger pole pitch allows deeper windings. In a rectangular, current-controlled F-LPMSM, the two active phase mmf angle shifts with respect to PM flux density axis from 60° to 120° electrical degrees (Figure 12.14).

Using the Lorenz force formula, the thrust F_x and normal force F_y force (on primary side) are

$$F_x(t) = \frac{64 \cdot n_c I \cdot l_{stack}}{w_s} \sum_{\nu=1}^{\infty} C_\nu \cos(2\nu-1)\frac{\pi}{6}\left[\cos(2\nu-1)\left(\frac{\pi}{\tau}U \cdot t - \frac{\pi}{6}\right)\right]$$

$$F_y(t) = \frac{64 \cdot n_c I \cdot l_{stack}}{w_s} \sum_{\nu=1}^{\infty} C_\nu \sin(2\nu-1)\frac{\pi}{6}\left[\cos(2\nu-1)\left(\frac{\pi}{\tau}U \cdot t + \frac{\pi}{6}\right)\right] \tag{12.21}$$

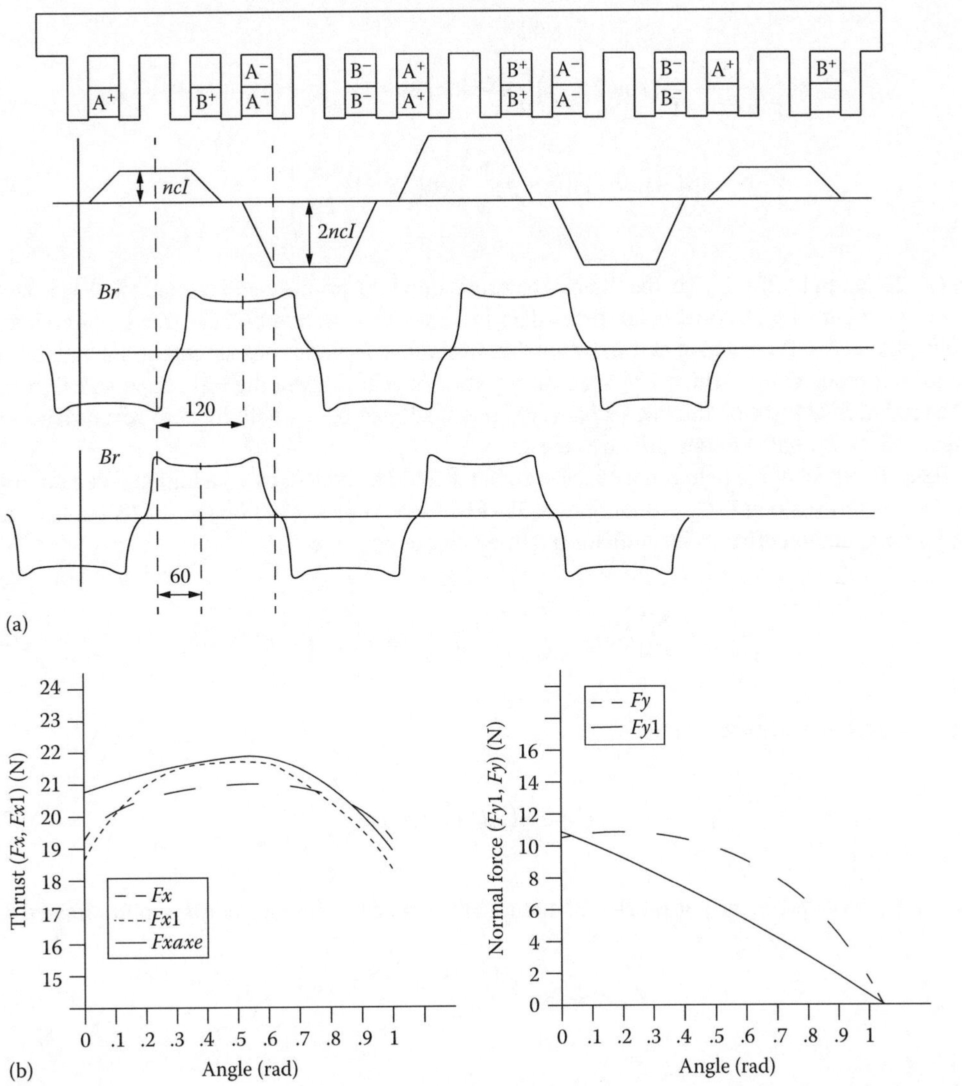

FIGURE 12.14 Primary mmf angle PM axis; from 60° to 120° for rectangular current control (a) and F_x, F_y forces (b) for the data in Table 12.1.

TABLE 12.1
Air-Core F-LPMSM Data

	B_r	0.9 T
	H_C	0.87 MA/m
PM height	h_m	12.5 mm
PM span	τ_p	25 mm
Pole pitch	τ	29 mm
Airgap	g	1 mm
"Slot" height	h_s	9 mm
"Slot" pitch	τ_s	$\tau/3(q=1)$
"Slot" width	w_s	$0.7\ \tau_s$
"Plastic tooth" width	w_t	$\tau_s - w_s$

with

$$C_\upsilon = \frac{4B_r\tau^2}{\pi^3 h_s(2\nu-1)^3}\sin\left(2\nu-1\right)\frac{\pi}{2}\frac{\tau_p}{\tau}\sinh\left(2\nu-1\right)\frac{\pi}{2}h_m\exp\left[-(2\nu-1)\frac{\pi}{\tau}h_m\right]$$
$$\times\left[1-\exp\left(-(2\nu-1)\left(h_s+g\right)\frac{\pi}{\tau}\right)\sin\left(2\nu-1\right)\frac{\pi h_s}{2g}\right] \tag{12.22}$$

In (12.21) and (12.22), l_{stack} is the "stack" or active depth of primary and secondary (PM). Plots of forces in Figure 12.14 correspond to the data in Table 12.1, over a 60° electrical degree interval, when phases A + B are active and show notable variations. F_x variations are important and they may be further reduced by varying PM span ratio τ_p/τ, h_s/τ, g/τ, or even the PM shape and W_s/τ_s ratio. The total normal force on the PM secondary is practically zero, or small, for zero eccentricity. (F_y in Figure 12.14 is related to one primary side.)

The 3D aspect of the field is not considered but it could be added, by an additional decomposition in Fourier series along z (l_{stack}) direction. A 3D-FEM would yield even better results but for a much larger computation effort. The emfs for the three phases $e_{a,b,c}$ are

$$e_{a,b,c} = \sum C_n U\cos\left(\frac{\pi}{\tau}Ut+\gamma_0-(i-1)\frac{2\pi}{3}\right);\quad i=1,2,3 \tag{12.23}$$

So the thrust may also be written as

$$F_x = \frac{1}{U}\left(e_a i_a + e_b i_b + e_c i_c\right) \tag{12.24}$$

For $q=1$ (3 slots/pole), the phase self and mutual inductances L_{aa} and L_{ab} may be approximated as [9]

$$L_{aa} = L_{sl} + \frac{\mu_0 W_1^2\tau\cdot l_{stack}\left(1+k_{fringe}\right)}{p\left(g+h_m\right)} \tag{12.25}$$

$$L_{ab} = -\frac{L_{aa}-L_{sl}}{3};\quad L_{sl}\text{—leakage inductance} \tag{12.26}$$

The fringing coefficient $k_{fringe}>0$ accounts for longitudinal flux density and transverse (3D) field components; k_{fringe} may be calculated from the multilayer theory (in 3D form), from 3D-FEM or from current decay measurements at standstill.

Before approaching the circuit model of F-LPMSMs, required for performance evaluation and for dynamics and control design, let us first focus a little on the cogging force and its particular end effect in F-LPMSMs.

12.5 COGGING FORCE AND LONGITUDINAL END EFFECTS

The cogging (zero current), or detent, force is the result of the variation of the coenergy stored in the PMs due to the iron-core slot openings. It is a periodic function that, for rotary PMSMs, has the period of LCM (N_s, $2p$), where N_s is the number of rotor poles:

$$\mathrm{LCM}(N_s,2p)=\frac{N_s}{\mathrm{GCD}(N_s,2p)} \tag{12.27}$$

GCD is the largest common divisor.

For L-PMSM, $2p\tau$ in the secondary should be considered to correspond to $N_s\tau_s$ in the primary (τ_s is the slot pitch). For tooth-wound windings, LCM becomes larger than for distributed windings (even with $q_1=1$ slot/pole/phase) and, consequently, the cogging force gets smaller.

Two tooth-wound F-LPMSMs situations, which correspond to 9 slots/12 poles and 10 slots/9 poles, are shown in Figure 12.15.

The configuration with 10 slots and 9 poles is somewhat unusual as it still hosts 3×3 phase coils, but it is used purposely to drastically reduce the cogging force. As LCM (9, 12) is 36 and LCM (9, 10) is 90, a high number of cogging force periods is expected. With the latter solution and in the form of 10 slots and 9 PM poles per module length, the periodicity of cogging force is even higher (in the ratio of 10/3 and not 10/4) [10]:

$$\begin{aligned} F_{9/12}(x) &= 9\sum A_{3n}\sin\left(3n\frac{\pi}{\tau}x+\alpha_{3n}\right) \\ F_{10/9}(x) &= 10\sum A_{10n}\sin\left(10n\frac{\pi}{\tau}x+\alpha_{10n}\right) \end{aligned} \tag{12.28}$$

The results in Figure 12.16 show that even if the peak cogging force of each slot is about the same in both cases, their phase shift is such that the total cogging force for the 10 slot/9 pole F-LPMSM is

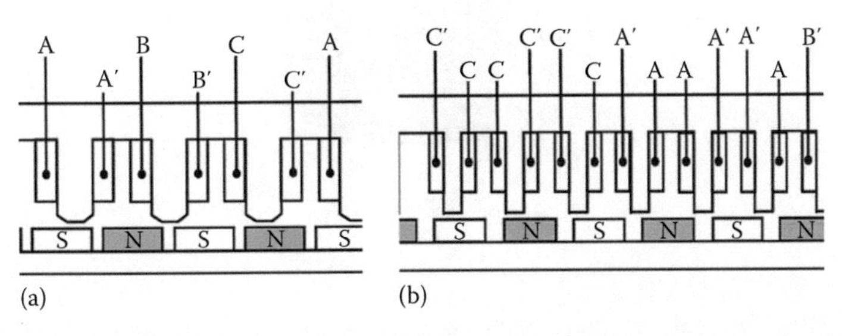

FIGURE 12.15 Two tooth-wound F-LPMSMs: (a) 9/12. (b) 10/9. (After Baatar, N. et al., *IEEE Trans.*, MAG-45(10), 4562, 2009.)

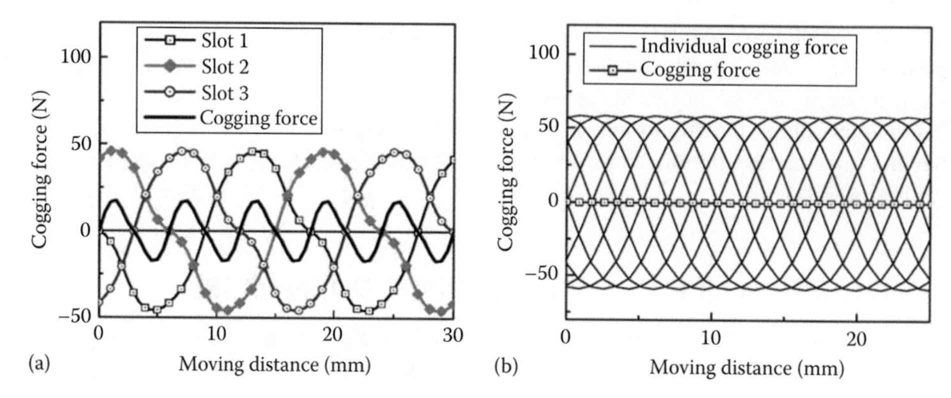

FIGURE 12.16 Cogging force: (a) for 9 slots/12 poles, (b) for 10 slots/9 poles. (After Baatar, N. et al., *IEEE Trans.*, MAG-45(10), 4562, 2009.)

almost zero [10]. Even a single slot cogging force, if uncompensated, can lead to a large cogging force. This may be called the end effect in cogging force. A typical 9 slot/8 pole combination would show it.

Apart from slot pitch PM to pole pitch ratio, primary shape optimization may be used to reduce the drag force due to primary (or PM secondary) finite length, especially in short-length modules. Chamfering and widening the end tooth is a way to do it (Figure 12.17)

The question is if these methods that reduce the cogging (and drag) force will not impede on average total thrust; not so for the aforementioned configuration (10 slots/9 poles) and chamfered

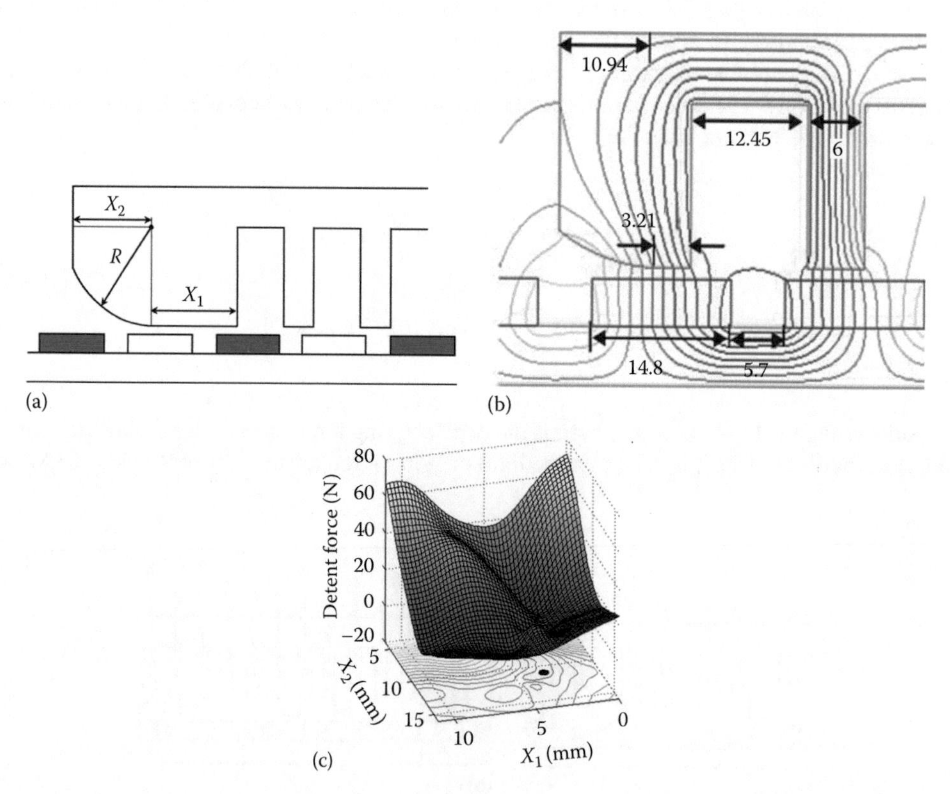

FIGURE 12.17 Drag force due to finite primary length at zero current: (a) variables *X*1, *X*2, (b) final geometry, and (c) drag force. (After Baatar, N. et al., *IEEE Trans.*, MAG-45(10), 4562, 2009.)

end teeth of primary [10]. Now that the principle of cogging force occurrence is elucidated, let us remember that all methods to reduce cogging torque in rotary PMSMs may be applied for LPMSMs, to reduce the cogging force. Among them we enumerate here

- Optimization of PM span τ_p per pole pitch τ [11]
- Pole displacement (in pairs), slot opening optimization
- Making a PM pole of a few sectors as in PWM signals [12] with notable average thrust reduction
- Simple (or double) skewing [13], with notable average thrust reduction
- PM shaping (bread loaf shape) for sinusoidal emf
- Double sided topology for zero normal force [14]
- Current waveform shaping

12.5.1 End Effects in 2(4) Pole PM-Secondary F-LPMSMs

In some applications where, say, short length PM movers are required, a 2-pole short PM secondary may be adopted while the long primary is the stator, supplied in sections to keep the copper losses low. In such a configuration (Figure 12.18), with air-core windings, two phase (rather than three-phase) ac windings are used (to avoid nonbalanced phase current due to non-equal self and mutual inductances) [15]. The PM flux per one stator coil versus mover position (Figure 12.18b) shows that the emf in the coil has one full-scale positive and two half-scale negative peak values. To smooth the total thrust, two coils in series have to be connected for the negative polarity of emf, according to mover position [15]. This change in coil connection needs a special power electronics system, to feed the long stator sequentially (Figure 12.18).

The usage of 2(4) pole PM mover is to be avoided whenever possible as it creates problems in the switching of the active stator windings section after section. In any case, 3D-FEM is to be used to analyze such topologies with air core (or with iron core).

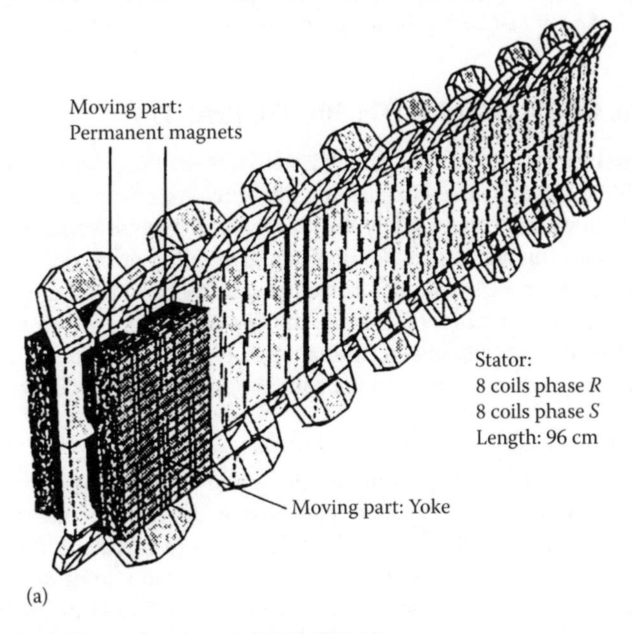

FIGURE 12.18 Two-pole PM mover air-core F-LPMSM (a).

(continued)

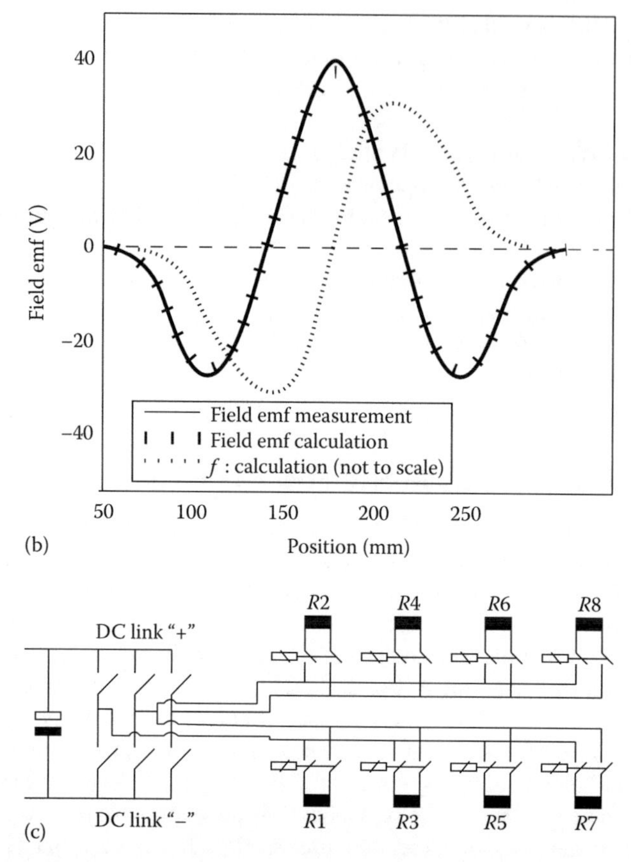

FIGURE 12.18 (continued) The emf coil vs. mover position, (b) converter plus coil switches for one stator phase (there are two phases in all) (c). (After Henneberger, G. and Reuber, C., A linear synchronous motor for a clean room system, *Record of LDIA-1995*, Nagasaki, Japan, pp. 227–230, 1995.)

12.6 *dq* MODEL OF F-LPMSM WITH SINUSOIDAL EMF

As seen earlier in this chapter, proper design may lead to rather sinusoidal emfs in the 3(2) phases of primary of F-LPMSMs. If we further treat the cogging force (the force at zero current) as a disturbance (load) force, then we may apply safely the *dq* model; for tooth-wound primaries, this model reflects the action of mmf fundamental as for distributed-winding primaries.

Thrust pulsations due to space harmonics consequences in emf or mmf are not considered in the standard *dq* model of electric machinery. Despite the approximation, the *dq* model is widely used to assess basic steady-state and dynamic performance of F-LPMSMs. The *dq* model also provides the ideal tool for the design of position, speed, or thrust (and normal force, eventually) control via FOC or DTFC.

So far we inferred that only surface PM secondaries are practical for F-LPMSM, but, for the sake of generality, different *d*–*q* inductances L_d and L_q are considered in the *dq* model. This is the case in mixed (PM + dc controlled coils) excitation secondary with long active stator MAGLEVs with buried PMs on mover, to leave room for the addition of slots to host an ac two-phase generator on board the mover, based on stator mmf space harmonics field ($q_1 = 1$), Figure 12.19 ($L_d < L_q$). The role of PMs in Figure 12.19b is quite different [16]: they are placed along axis *q* in multiple flux barriers and the dc coils in axis *d* ($L_d > L_q$). The PMs serve for wide constant power speed range and for operation at unity power factor (minimum converter kVA).

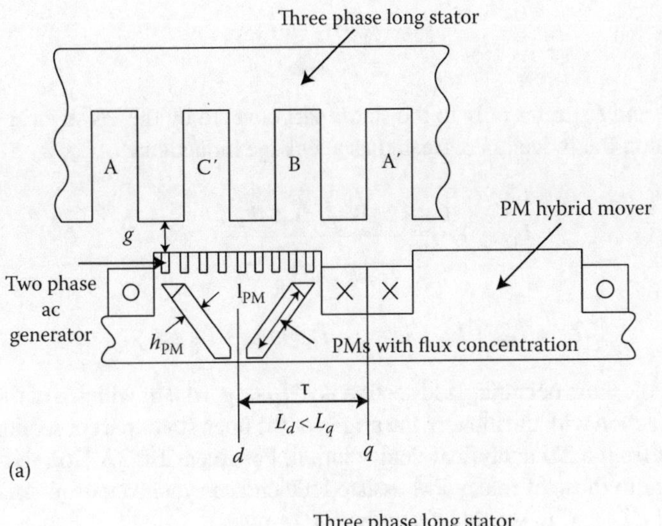

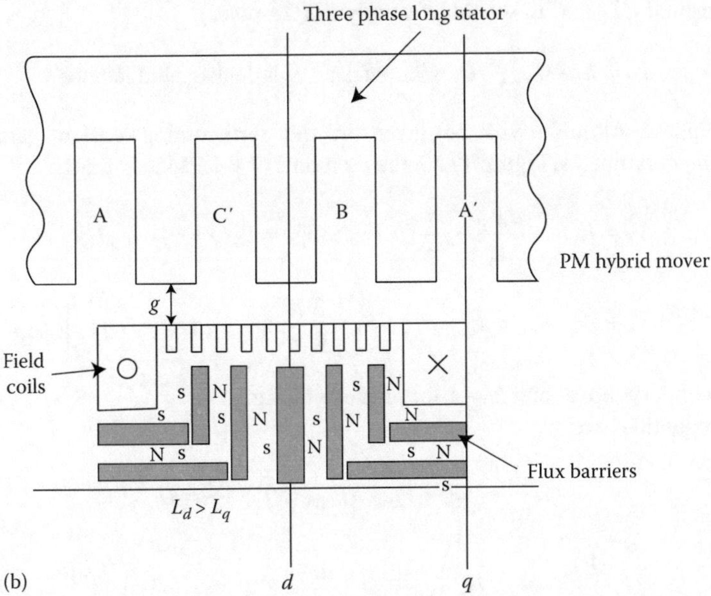

FIGURE 12.19 PM-hybrid secondary (mover) and active stator F-LPMSM (a) with PM+dc coils in axis *d* and (b) dual-axis-field linear BEGA. (After Boldea, I. and Tutelea, L., *Electric Machines: Steady State, Transients and Design with Matlab*, CRC Press, Taylor & Francis, New York, 2009.)

To a first approximation, the magnetization inductances L_{md} and L_{mq} in Figure 12.19 are (as for its rotary machine counterparts) [17] as follows:

$$L_{dm} = L_m \cdot \frac{k_{dm}}{1+k_{sd}}; \quad L_{qm} = L_m \cdot \frac{k_{qm}}{1+k_{sq}} \tag{12.29}$$

$$k_{dm} = \left(\frac{\tau_p}{\tau} + \frac{1}{\pi}\sin\frac{\tau_p}{\tau}\pi\right) \cdot \frac{g}{g + h_{PM}\dfrac{\tau}{2l_{PM}}} \tag{12.30}$$

$$k_{qm} = \frac{\tau_p}{\tau} - \frac{1}{\pi}\sin\frac{\tau_p}{\tau}\pi + \frac{2}{3\pi}\cos\frac{\tau_p}{\tau}\frac{\pi}{2} \tag{12.31}$$

$$L_m = \frac{6\mu_0 \left(k_{W_1} W_{1p}\right) \tau \cdot l_{stack}}{\pi^2 p \cdot g \cdot k_C} \tag{12.32}$$

For a long stator, L_{md} and L_{mq} refer only to the stator part covered by the PM-hybrid secondary. For the rest of "open" stator, the inductance, treated as a leakage inductance $L_{p'-p}$, is

$$L_{p'-p} \approx L_{lp'-p} + \frac{6\mu_0 \left(k_{W_1} W_{1p'-p}\right)^2 \tau \cdot l_{stack}}{\pi^2 (p'-p)\frac{\tau}{\pi}} \tag{12.33}$$

$$L_d = L_{dm} + L_{lp} + L_{p'-p}; \quad L_q = L_{qm} + L_{lp} + L_{p'-p}$$

The total number of pole pairs per energized section is p' ($p \ll p'$) out of which p of them correspond to active part (in interaction with the mover); the airgap of the open stator part of section is considered τ/π as it would result from a 2D analytical field solution. For linear BEGA [16], the expressions of parameters are similar to those of rotary PM assisted reluctance synchronous motors.

For a short primary, L_d and L_q would be simply (for $2p$-poles)

$$L_d = L_{ls} + L_{dm}; \quad L_q = L_{ls} + L_{qm}; \quad L_{ls}\text{—leakage inductance} \tag{12.34}$$

The resistance/phase formula will be given in the forthcoming design paragraph. So, in PM-secondary dq coordinates, Figure 12.19, the dq model of F-LPMSMs writes

$$\bar{V}_S = R_S \bar{I}_S + \frac{d\bar{\psi}_S}{dt} + j\frac{\pi}{\tau} Uk \cdot \bar{\psi}_S; \quad \bar{\psi}_S = L_d I_d + \psi_{PM} + jL_q I_q \tag{12.35}$$

$$\bar{V}_S = V_d + jV_q; \quad \bar{I}_S = I_d + jI_q; \quad F_x = \frac{3}{2}\frac{\pi}{\tau}\left[\psi_{PMd} + \left(L_d - L_q\right) I_d\right]\left(I_{qk}\right) \tag{12.36}$$

$k=1$ for PM-secondary mover and $k=-1$ for primary mover.

The motion equations are

$$G_v \frac{dU}{dt} = F_x - F_{load} - F_{cogg}(x); \quad \frac{dx}{dt} = U \tag{12.37}$$

$$G_m \frac{dU_g}{dt} = F_n - G_m \cdot g - F_{dis}; \quad \frac{dg}{dt} = U_g; \quad \gamma = \tan^{-1}\left(\frac{I_d}{I_q}\right) \tag{12.38}$$

$$F_n \approx \frac{\left(B_{PMg}{}^2 + B_{ag}{}^2 + 2B_{PMg} \cdot B_{ag} \cdot \sin\gamma\right)}{2\mu_0} \frac{2p \cdot \tau \cdot l_{stack}}{2} \tag{12.39}$$

B_{PM}, B_{ag} are PM and armature airgap flux density fundamentals.

U_g is the vertical speed in case the airgap varies (due to track irregularities or during levitation control). The Park transformation completes the system:

$$\begin{vmatrix} V_d \\ V_q \\ V_0 \end{vmatrix} = \frac{2}{3} \begin{vmatrix} \cos\left(-\frac{\pi}{\tau}x\right) & \cos\left(-\frac{\pi}{\tau}x + \frac{2\pi}{3}\right) & \cos\left(-\frac{\pi}{\tau}x - \frac{2\pi}{3}\right) \\ \sin\left(-\frac{\pi}{\tau}x\right) & \sin\left(-\frac{\pi}{\tau}x + \frac{2\pi}{3}\right) & \sin\left(-\frac{\pi}{\tau}x - \frac{2\pi}{3}\right) \\ \frac{1}{2} & \frac{1}{2} & \frac{1}{2} \end{vmatrix} \begin{vmatrix} V_a \\ V_b \\ V_c \end{vmatrix} \tag{12.40}$$

Given the phase voltages $V_a(t)$, $V_b(t)$, $V_c(t)$, the initial values of variables (I_{d0}, I_{q0}, U_0, X_0, U_{g0}, g_0) and force perturbations ($F_{cogg}(x)$ and $F_{dis}(x)$), and the load force F_{load}, together with machine parameters values, the *dq* model equations may be solved numerically for any steady-state or transient operation mode, with constant or variable airgap *g*. However, for propulsion control purposes, a few typical control law steady-state characteristics are worthy of discussion. Before that, the equivalent circuits for transients and the structural diagram are given for completeness. Adding the primary and secondary core loss via equivalent resistances (R_{pcore}, R_{score}), the *dq* model equations at constant speed, U=const., and constant airgap, g=const., yield the equivalent circuits in Figure 12.20. They are similar to those of rotary IPMSMs. The structural diagram in Figure 12.21 is valid for propulsion only (constant airgap).

The space phasor diagrams for motoring and generating are similar to those of rotary PMSMs [17].

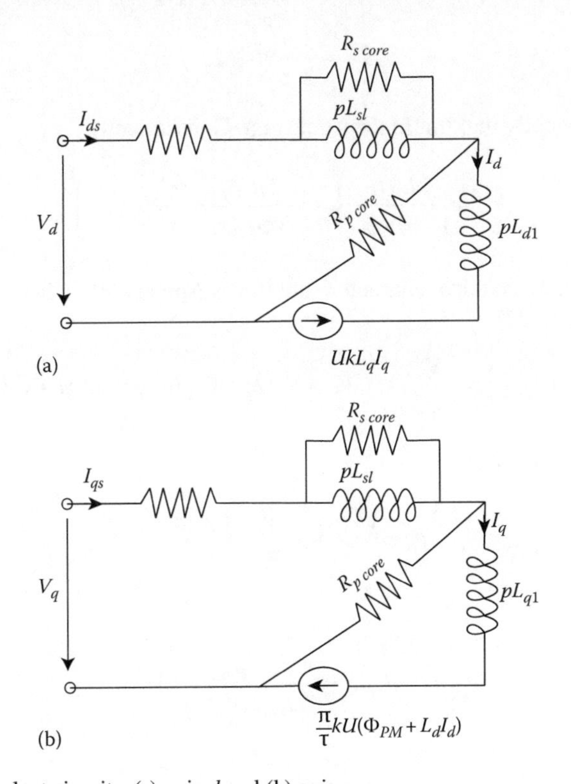

FIGURE 12.20 Equivalent circuits: (a) axis *d* and (b) axis *q*.

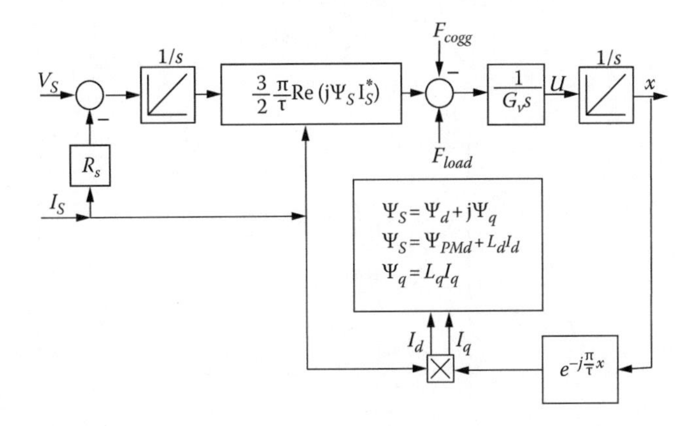

FIGURE 12.21 F-LPMSM structural diagram for propulsion only.

12.7 STEADY-STATE CHARACTERISTICS FOR TYPICAL CONTROL STRATEGIES

There are quite a few control strategies of practical interest for F-LPMSMs. We select here two of them:

1. Maximum thrust/current control
2. Flux weakening control

12.7.1 Maximum Thrust per Current Characteristics ($L_d = L_q = L_s$)

Core losses are neglected for simplicity:

So, from (12.33) and (12.34), for surface PM-secondary ($L_d = L_q = L_s$),

$$F_x = \frac{3}{2}\frac{\pi}{\tau}\psi_{PMd} I_q \tag{12.41}$$

Zero I_d is thus automatically maximum thrust/current. Consequently,

$$V_S^2 = \left(\frac{2L_s U F_x}{3\psi_{PMd}}\right)^2 + \left(\frac{2R_s F_x}{3\pi\psi_{PMd}} + \frac{\pi}{\tau} U \psi_{PMd}\right)^2 \tag{12.42}$$

This is a typically mild mechanical characteristic: (thrust/speed curve), but with a definite ideal no load speed U_{i0} (Figure 12.22).

$$\left(U_{i0}\right)_{I_q = I_S = 0} = \frac{V_S \tau}{\pi \psi_{PMd}}; \quad V_S = V\sqrt{2}; \quad V\text{: phase voltage (RMS)} \tag{12.43}$$

The normal force ($\gamma = 0$) is (12.39)

$$F_n = \frac{\left(B_{PMq_1}{}^2 + B_{ag}{}^2\right) p \tau l_{stack}}{2\mu_0} \tag{12.44}$$

but

$$B_{PMq_1} = \frac{\pi}{2}\frac{\psi_{PMd}}{l_{stack}\tau W_1 k_{W_1}} \tag{12.45}$$

and

$$\left(B_{ag1}\right)_{I_d = 0} = \frac{L_{qm} I_S}{\frac{\pi}{2} l_{stack} \tau W_1 k_{W_1}} \tag{12.46}$$

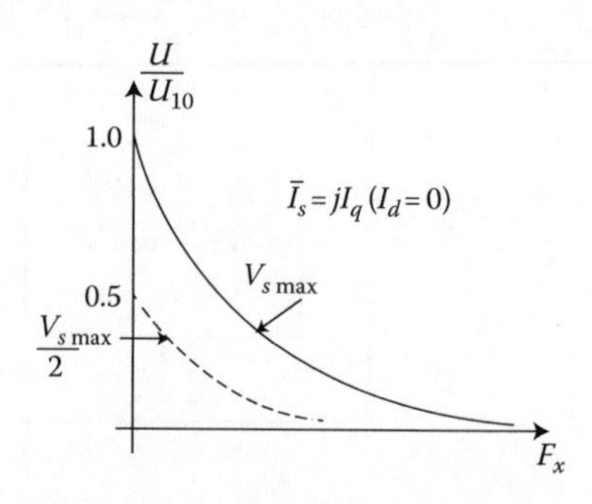

FIGURE 12.22 Constant ($I_d = 0$) speed/thrust curves ($f = U/2\tau$, variable).

So

$$F_n = \left(\frac{\pi}{2}\frac{1}{l_{stack}\tau W_1 k_{W_1}}\right)^2 \psi_{PMd}{}^2\left[1+\left(\frac{L_{qm}I_S}{\psi_{PMd}}\right)^2\right]\frac{p\tau l_{stack}}{2\mu_0} \tag{12.47}$$

Consequently, the normal force increases with load ($I_S = jI_q$, that is, with F_x). This may lead to increased noise and vibration in the mover.

12.7.2 Maximum Thrust/Flux ($L_d = L_q = L_s$)

Flux weakening is today attempted with surface PM secondary for tooth-wound primary whose total inductance L_s is rather large (due to added large differential leakage inductance of primary mmf space harmonics). A large L_s means a not so large demagnetizing current $I_d<0$. For constant ψ_S

$$\psi_S{}^2 = (\psi_{PMd} + L_s I_d)^2 + L_s{}^2 I_q{}^2; \quad F_x = \frac{3}{2}\frac{\pi}{\tau}\psi_{PMd} I_q \tag{12.48}$$

with I_q from F_x in the flux ψ_S:

$$\left(\frac{V_S}{\frac{\pi}{\tau}U}\right)^2 = \psi_S{}^2 = \left(\psi_{PMd} + L_s I_d\right)^2 + L_s{}^2\left(\frac{3}{2}\frac{\pi}{\tau}\psi_{PMd}\right)^{-2} F_x{}^2 \tag{12.49}$$

The maximum thrust F_x per given ψ_S (or given speed U and voltage V_S) in (12.48):

$$\psi_{PMd} + L_s I_{dk} = 0 \tag{12.50}$$

So

$$I_{dk} = -\frac{\psi_{PMd}}{L_s} \tag{12.51}$$

However, in such conditions for $F_x = 0$ (ideal no load) in (12.49), $U_0 = \infty$ (infinite ideal no load speed). The force/speed curve for different values of $I_d \neq 0$ are shown in Figure 12.23 for given voltage $V_S = V_{S\,max}$.

This time the thrust and normal force from (12.39) and (12.44) through (12.48) are

$$F_x = \frac{3}{2}\frac{\pi}{\tau}\psi_{PMd} I_S \cos\gamma; \quad \text{for } F_x < \gamma > 90° \tag{12.52}$$

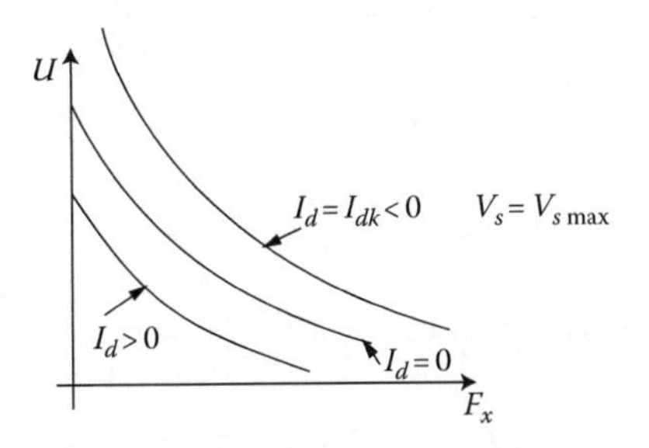

FIGURE 12.23 Speed/thrust curve for nonzero I_d.

$$F_n = \left(\frac{\pi}{2}\frac{1}{l_{stack}\tau W_1 k_{W_1}}\right)^2 \psi_{PMd}{}^2\left[1+\left(\frac{L_{qm}I_s}{\psi_{PMd}}\right)^2 + 2\frac{L_{qm}I_s}{\psi_{PMd}}\sin\gamma\right]\cdot\frac{p\tau l_{stack}}{2\mu_0} \tag{12.53}$$

$$\gamma = \tan^{-1}\left(\frac{I_d}{I_q}\right);\quad \gamma<0 \quad \text{for } I_d<0$$

$$I_s = \sqrt{I_d{}^2 + I_q{}^2};\quad L_{qm} = L_m \quad L_d = L_q \tag{12.54}$$

The normal force may be controlled by the angle $\gamma \neq 0$, and the thrust may be varied by the current amplitude, I_s. F_x and F_n both depend on γ and I_s, so their control is not decoupled. This has consequences when integrated propulsion and levitation are targeted.

Note: For $L_d \neq L_q$, the maximum thrust/flux conditions yield a different $I'_{dk} \neq I_{dk} = -\psi_{PMs}/L_s$ (see Ref. [18], for more details).

Example 12.1

F-LPMSM with $L_d = L_q = L_S$ characteristics

Let us consider a flat F-LPMSM with a surface PM secondary and the following data: $V_{s\max} = 120$ V (RMS) per phase, $\tau = 0.06$ m, $B_{gM1} = 0.7$ T, $L_d = L_q = L_s = 30$ mH, $R_S = 3\ \Omega$, stack length $l_{stack} = 0.1$ m, $2p = 4$ poles, $q = 1$ slot/pole/phase single layer winding, and $W_1 = 400$ turns/phase.

Calculate:

(a) The PM flux linkage per phase ψ_{PMd}.
(b) For $I_d = 0$ the ideal no load speed U_{i0} and the corresponding (maximum) frequency $\omega_{1\max}$.
(c) For $U_S = U_{i0}/2$ and $I_{q0} = -10$ A ($I_d = 0$), determine the thrust F_x and the voltage, frequency, thrust/watt losses (N/W), and specific thrust (N/cm²).
(d) For same $U_S = U_{i0}/2$ and $I_d = I_{dk}/3 = -\psi_{PMd}/3L_S$ calculate I_q, thrust, and voltage, and discuss the results; calculate the efficiency and power factor.
(e) What current density is needed in the winding design; are the data realistic?

Solution

(a) The PM flux linkage per phase ψ_{PMd} is

$$\psi_{PMd} \approx \frac{2}{\pi}B_{PMg1}\tau\cdot l_{stack}\cdot W_1 k_{W_1} = \frac{2}{\pi}\times 0.7\times 0.06\times 0.1\times 800\times 1 = 1.07\ \text{Wb} \tag{12.55}$$

(b) The ideal no load speed U_{i0} is

$$U_{i0} = \frac{V_S\cdot\tau}{\pi\cdot\psi_{PMd}} = \frac{120\sqrt{2}\times 0.06}{\pi\times 1.07} = 3.02\times\frac{\pi}{0.06} = 158.0\ \text{rad/s} \tag{12.56}$$

So

$$f_{1\max} = \frac{\omega_{1\max}}{2\pi} = 25\ \text{Hz}$$

(c) For $I_d = 0$ and $I_q = -10$ A (as $kU = -U < 0$, $I_q < 0$ for motoring) the thrust (12.41) is

$$F_x = \frac{3}{2}\frac{\pi}{\tau}\psi_{PMd}I_{q0} = \frac{3}{2}\frac{\pi}{0.06}\times 1.07\times(-10) = -840\ \text{N} \tag{12.57}$$

The voltage components V_d and V_q at $U=U_{i0}/2=1.51$ m/s are

$$V_d = R_S \cdot I_{d0} - k\frac{\pi}{\tau}U \cdot L_s \cdot I_{q0}; \quad k=-1 \text{ (PM secondary long stator)}$$
$$V_q = R_S \cdot I_{q0} + k\frac{\pi}{\tau}U \cdot \Psi_{PMd} \tag{12.58}$$

So

$$V_d = 2\times 0 + \frac{\pi \times 1.5}{0.06} = 23.707 \text{ V}$$
$$V_q = 3\times(-10) - \frac{\pi\times 1.51\times 0.003\times(-10)}{0.06}1.07 = 114.55 \text{ V} \tag{12.59}$$

$$V_S = \sqrt{V_d^2 + V_q^2} = 116.98 \text{ V} \tag{12.60}$$

The specific thrust f_x is

$$f_x = \frac{|F_x|}{2p\tau l_{stack}} = \frac{840}{4\times 0.06\times 0.1} = 3.5\times 10^4 \text{ N/m}^2 \tag{12.61}$$

The copper losses are

$$P_{Co} = \frac{3}{2}R_S I_{q0}^2 = \frac{3}{2}\times 3\times 10^2 = 450 \text{ W} \tag{12.62}$$

$$\frac{F_x}{P_{Co}} = \frac{840 \text{ N}}{450 \text{ W}} \approx 1.866 \text{ N/W} \tag{12.63}$$

(d) For

$$I_{d0} = \frac{-\psi_{PMd}}{3L_S} = \frac{-1.07}{(3\times 0.03)} = -11.88 \text{ A} \tag{12.64}$$

This time the thrust is the same as it depends only on I_{q0} but the voltage required is

$$V_d' = 3\times(-11.88) - \frac{\pi}{0.06}\times 1.51\times 0.03\times(-10) = -59.347 \text{ V}$$
$$V_q' = 3\times(-10) - \frac{\pi}{0.06}\times 1.51\times(1.07 - 0.03\times 11.88) = -56.39 \text{ V} \tag{12.65}$$

$$V_S' = \sqrt{V_d' + V_q'} = 81.865 \text{ V} << V_S = 116.98 \text{ at the same speed and thrust but with } I_d = 0 \tag{12.66}$$

This simply means that the maximum voltage allows larger thrust at the same voltage or higher thrust at speeds above $U_{i0}/2$ to U_{i0}.

The efficiency η at $I_d=0$, $I_{q0}=-10$ A, $F_x=-840$ N, $U=U_{i0}/2=1.51$ m/s is (zero core losses)

$$\eta = \frac{F_x \cdot k \cdot U}{k\cdot F_x \cdot U + P_{Co}} = \frac{(-840)\times(-1.51)}{(-840)\times(-1.51)+450} = 0.738 \tag{12.67}$$

The power factor for this case is

$$(\cos\varphi)_{I_{d0}} = \frac{|V_q|}{V_S} = \frac{114.55}{116.98} \approx 0.98 \quad (12.68)$$

This performance looks good in comparison with previous linear electric motor examples (LIMs and L-RSMs or L-SRMs) but is it realistic?

(e) Let us first check what current density would be needed for the 4 pole $q=1$, single layer winding for, say, $I_{q0}=-10$ A, $I_{d0}=0$ ($I_n=10/\sqrt{2}$ A per phase, RMS value). We do have for 4 poles only 2 coils per phase, so the number of turns per coil $n_C=W_1/2=400/2=200$ turns/coil. The turn length l_c is

$$l_C \approx 2\cdot l_{stack} + 0.02 + 2\times 1.4\cdot\tau = 2\times 0.1 + 0.02 + 2.8\times 0.06 = 0.388\,\text{m} \quad (12.69)$$

So the phase resistance R_S yields

$$R_S = 2\cdot\rho_{Co}\cdot\frac{l_C W_C J_{CO}}{I_n} = 3\,\Omega \quad (12.70)$$

$$J_{Co} = \frac{3\times 10/\sqrt{2}}{2.1\times 10^{-8}\times 400\times 0.388} = 6.583\times 10^{6} = 6.583\,\text{A/mm}^2 \quad (12.71)$$

This looks OK but let us check the slot depth as the slot pitch $\tau_s=\tau/3=20$ mm and the slot width may be $W_S=0.55\cdot\tau_s=11$ mm.

With a slot filling factor $k_{fill}=0.5$, and one coil per slot, the slot height h_S is

$$h_S = \frac{n_C I_{q0}/\sqrt{2}}{J_{Co}\cdot k_{fill}\cdot W_S} = \frac{200\times 10/\sqrt{2}}{6.528\times 0.5\times 11} = 39.50\,\text{mm} \quad (12.72)$$

So the slot aspect ratio is OK: $h_S/W_S=39.5/11=3.59<4$–5, required to limit the slot leakage inductance. Still there is one more crucial step in checking: the practicality of the data.

An ultimate verification refers to the machine inductance, though the electric time constant $\tau_e=L_s/R_s=30\cdot 10^{-3}/3=10$ ms seems feasible.

For $B_{gPM}=0.7$ T, with bonded NeFeB PMs; $B_r=0.8$ T, $\mu_{rec}=1.05\cdot\mu_0$, for an airgap $g=1$ mm, the PM thickness is obtained from (PMs cover full pole span in our example):

$$\frac{\pi}{4}B_{gPM1} = B_r\frac{h_m}{h_m+g}\cdot\frac{1}{1+k_{fringe}}; \quad k_{fringe}\approx 0.20 \quad (12.73)$$

So

$$\frac{h_m}{h_m+g} = \frac{0.7\times 4\times 1.2}{\pi 0.8} = 0.8245 \quad (12.74)$$

$$\text{So } h_m \approx 5.7\,\text{mm}$$

So the airgap inductance L_{sg} (12.24) is

$$L_{sg} = \frac{4}{3}(L_{aa}-L_{sl}) = \frac{4}{3}\frac{\mu_0 W_1^2\cdot\tau\cdot l_{stack}(1+k_{fringe})}{2p\cdot(g+h_m)}$$

$$= \frac{4}{3}\frac{1.256\times 10^{-6}\times 400^2\times 0.06\times 0.1(1+0.1)}{4\times(1+5.7)\times 10^{-3}} = 0.0671\,\text{H} = 67.1\,\text{mH} \quad (12.75)$$

If we add the leakage inductance ($L_s = L_{sl} + L_{sg}$), we could end up with 90 mH instead of 30 mH assumed in our numerical example. So finally, the example data are not practical but they serve to demonstrate the multitude of aspects encountered in F-LPMSM design. The forthcoming paragraph on design methodology by example will deal with practical data and results.

12.8 F-LPMSM CONTROL

As MAGLEV detailed control will be dealt with in a separate later chapter, here the FOC and the DTFC for propulsion only will be treated in what follows. Most applications, in such conditions, refer to industrial cases (low speed and a few meter long travels), where primarily position control is the target, and thus we will concentrate on this aspect from the start.

12.8.1 Field Oriented Control (FOC)

Making use of the *dq* model of F-LPMSM makes the FOC rather straightforward. There are two main types of FOC: indirect and direct, and then again with dc current controllers or with ac current controllers [19], and, finally, with position sensor [13] and without position sensor:

- Indirect FOC
- Direct FOC
- With ac current controllers
- With dc current controllers
- With position sensor
- Without position sensor

For constant I_d, the *dq* equations (12.32 through 12.34) of F-LPMSM may be written as

$$V_d = R_S I_{d0} - \frac{\pi}{\tau} kU \cdot L_q I_q; \quad F_x = \frac{3}{2}\frac{\pi}{\tau}\left(\psi_{PM} + \left(L_d - L_q\right) I_{d0}\right) I_q k$$

$$V_q = R_S I_q + L_q \frac{dI_q}{dt} + \frac{\pi}{\tau} Uk\left(\psi_{PM} + L_d I_{d0}\right) \tag{12.76}$$

$$G_m \frac{dU}{dt} = F_x - F_{load} - BU; \quad \frac{dx}{dt} = U; \quad T_m = \frac{G_m}{B}$$

With $d/dt = s$ and $L_q/R_s = T_{eq}$, they lead to the structural diagram in Figure 12.24a.

The structural diagram resembles those of dc PM brush (when $I_d = 0$) or of rotor PM brushless ac motors, as expected.

We choose here the indirect FOC with ac current controllers due to its simplicity and because the fundamental frequency is small enough, to avoid notably phase error lag in current at highest speed for industrial applications. Also, position control is the target (Figure 12.24) where PM secondary is the mover ($k = 1$). Speed control is optional, and the FOC in Figure 12.24 is very robust if the current controllers' response has reasonable dynamic and steady-state errors, up to maximum speed and current. As the airgap may vary, ψ_{PMd} varies; also the dc link voltage may vary and then the current controllers cannot respond properly for large thrust commands; for stability, I_d^* and I_q^* have to be suppressed until the current controller errors become reasonable for extreme conditions.

If the F-LPMSMs are used for industrial (or transportation) MAGLEV carriers, say with four F-LPMSM, one in each corner, additional control coils may be used to control the levitation of the vehicle (Figure 12.25) [20]. To a first approximation, the resultant thrust per vehicle side and the

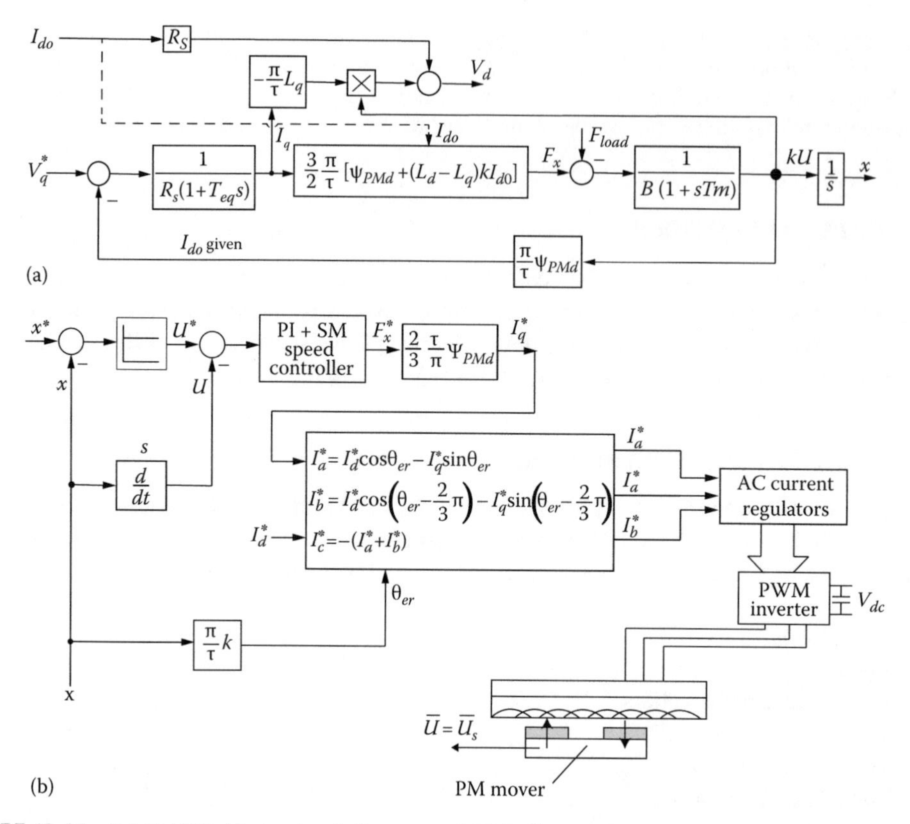

FIGURE 12.24 F-LPMSM: (a) structural diagram and (b) indirect ac current vector control.

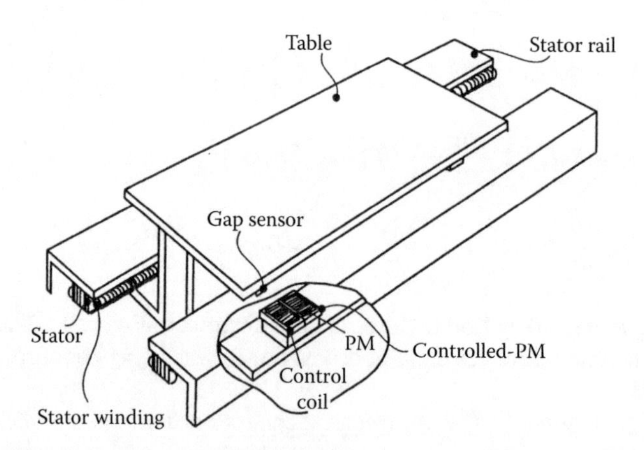

FIGURE 12.25 Industrial MAGLEV with four F-LPMSMs.

total thrust may not be influenced too much by levitation control (due to compensation of vertical motions between vehicle corners) and thus still pure I_q^* control may be used for propulsion, with the dc control coils used for dynamic stabilization of suspension. As a more sophisticated system, both I_d^* and I_F^* (in the dc coil) of each of the four F-LPMSMs may be controlled concurrently, for optimum propulsion and levitation control (for levitation) and I_q^* control (for propulsion), if the machine inductance is large enough to allow reasonable $I_{d\max}^*$ values (negative and positive!). In this case, however, the inverter utilization will be poorer.

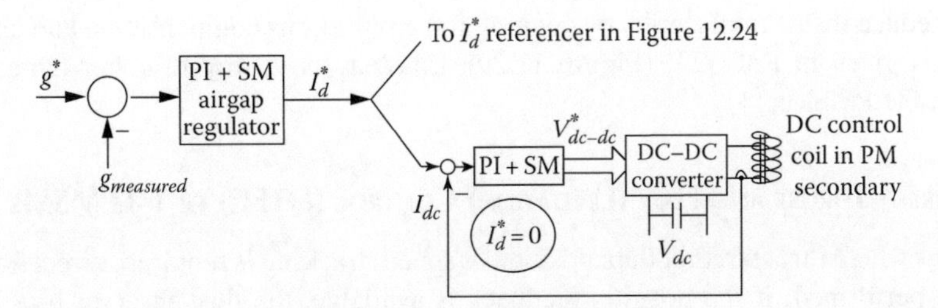

FIGURE 12.26 Alternative for integrated levitation (and propulsion) control of F-LPMSM MAGLEV carrier.

The two control alternatives identified here are depicted in Figure 12.26.

It goes without saying that the addition of dc control coils may lead to probably better levitation control, but levitation control by I_d^* control (by adding I_d dc regulators) may suffice in some applications.

Besides the here-suggested PI + SM (sliding mode) controllers for speed control plus P controller for position control and interior dc current regulators (Figure 12.24), recently fuzzy-neural-network controllers proposed to replace the position and speed cascaded controllers have been introduced [21,22]; Figure 12.27 [21].

Sample results in Figure 12.28 show that the dynamic errors are still up to 0.2 mm for a trapezoidal periodic positioning of ± 5 mm with a period of 2 s [21]; and all these with position sensor resolution of a few μm.

Thrust pulsations due to nonsinusoidal emf, airgap variations, cogging force, and current ripple and vibration due to large normal force might be pertinent reasons why the dynamic positioning errors are still large.

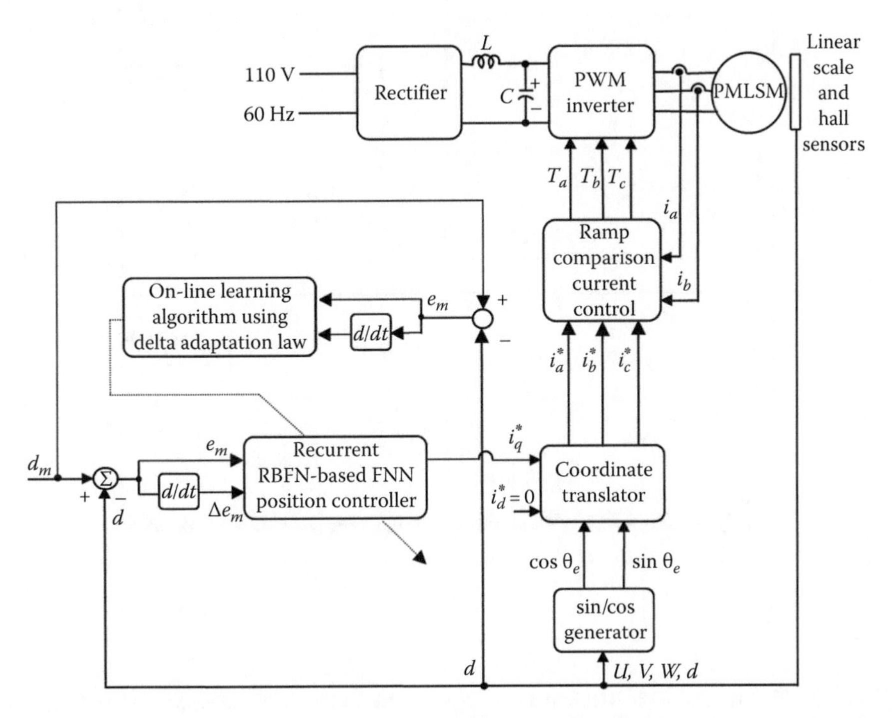

FIGURE 12.27 FOC with fuzzy neural network position control of F-LPMSMs. (After Lin, F.-J. et al., *IEEE Trans.*, MAG-42(11), 3694, 2006.)

A way to reduce thrust total ripple, by cogging force pulsations compensation and using primary coordinates, is given in Ref. [23] (Figure 12.29); this may be a way to better dynamic position tracking with F-LPMSMs.

12.8.2 Direct Thrust and Flux (Levitation) Control (DTFC) of F-LPMSMs

In applications where thrust (rather than position or speed) tracking is required, direct flux and thrust estimation is performed; if the position feedback is available, the flux observer may be designed rather easily to work satisfactorily above 1–2 Hz; if position-sensorless control is required, then all the "arsenal" of motion sensorless control of rotary PMSMs may be "summoned to duty" for F-LPMSM [19,24]. Let us present such a generic position-sensorless DTFC system for F-LPMSM (Figure 12.30).

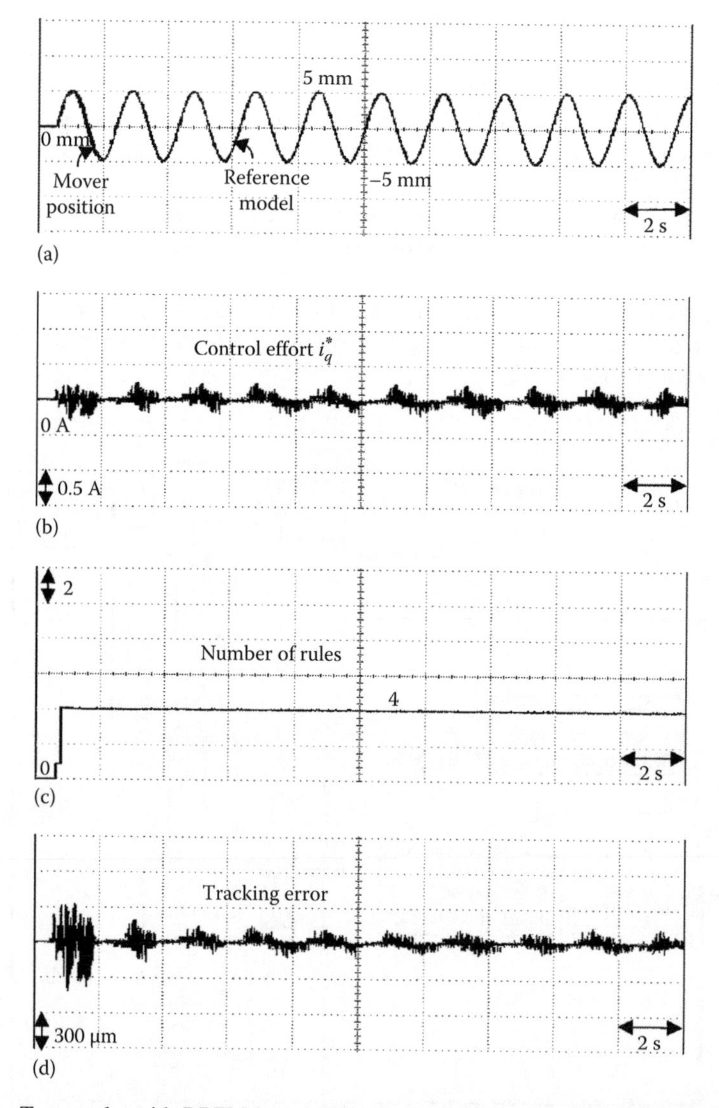

FIGURE 12.28 Test results with RBFM-based FNN position FOC of F-LPMSM. (After Lin, F.-J. et al., *IEEE Trans.*, MAG-42(11), 3694, 2006.) (a) tracking response at nominal condition, (b) control effort at nominal condition, (c) number of rules at nominal condition, (d) tracking error at nominal condition.

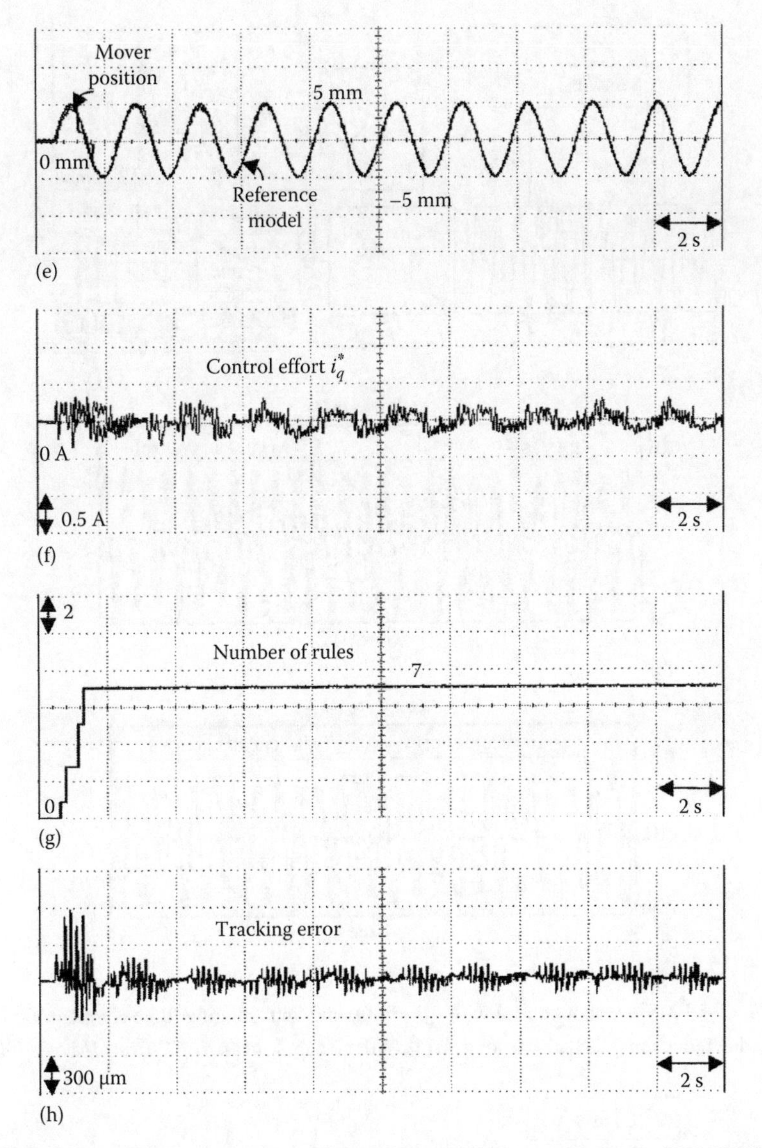

FIGURE 12.28 (continued) Test results with RBFM-based FNN position FOC of F-LPMSM. (After Lin, F.-J. et al., *IEEE Trans.*, MAG-42(11), 3694, 2006.) (e) tracking response at parameter variation condition, (f) control effort at parameter variation condition, (g) number of rules at parameter variation condition, and (h) tracking error at parameter variation condition.

For the primary flux $\overline{\psi}_S$ observer (Figure 12.31), we may use here a combined current and voltage model [18]:

$$\overline{\Psi}_S^s = \int\left(\overline{V}_S - R_S\overline{I}_S + V_{comp}\right)dt \tag{12.77}$$

V_{comp} compensates the inverter nonlinearities and the primary resistor variations while it uses a PI loop on the error between the current and the voltage model (12.68), in stator coordinates, so as to produce both low-speed and higher-speed (frequency) primary flux good estimation (Figure 12.31).

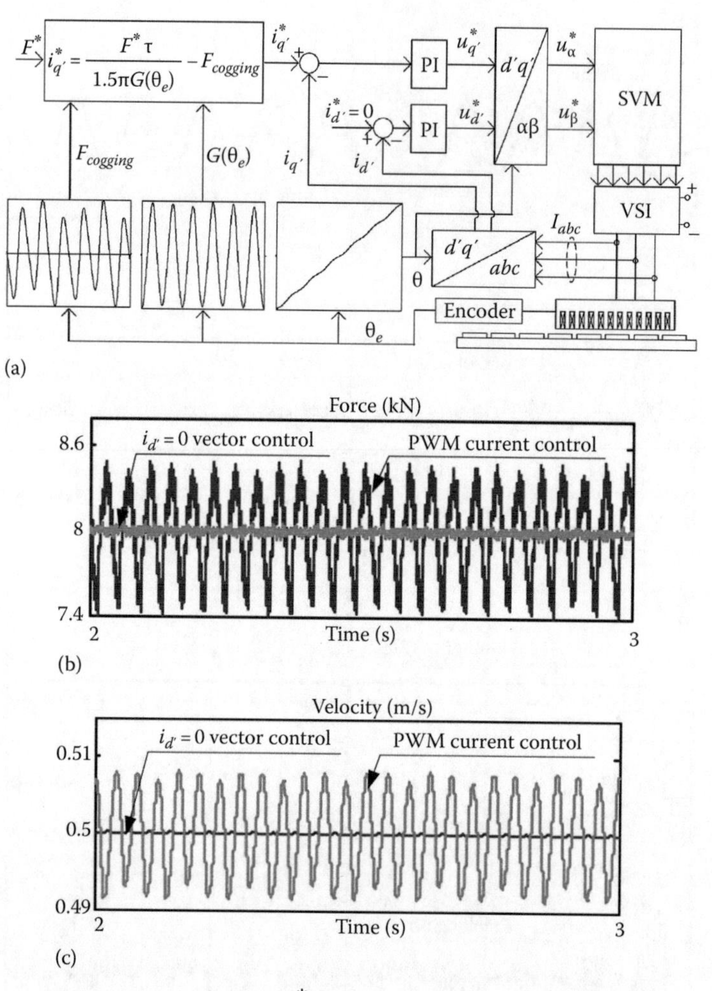

FIGURE 12.29 F-LPMSM vector control with $i_d^{*\prime} = 0$ (a) and with cogging force compensation and (b) thrust and speed pulsation reduction (with $i_d^{*\prime} = 0$ control) (c). (After Zeng, L. et al., *IEEE Trans.*, MAG-46(3), 954, 2010.)

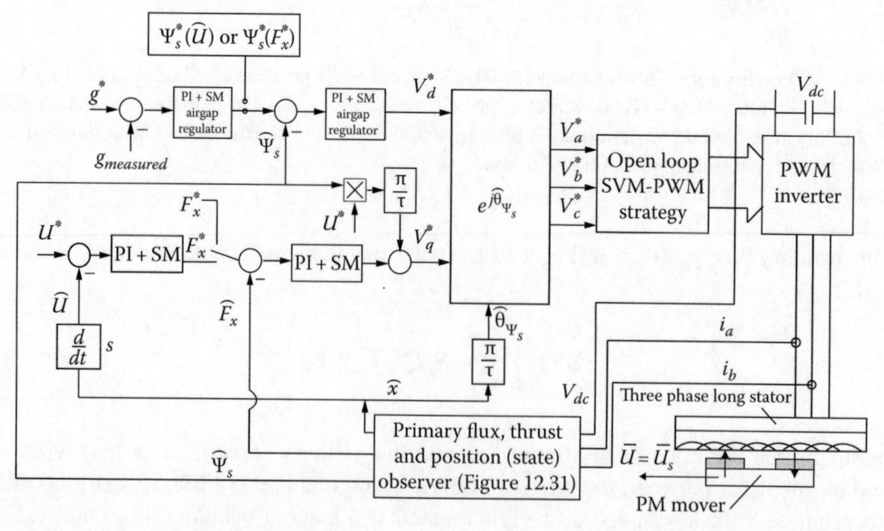

FIGURE 12.30 Generic DTFC of LPMSM.

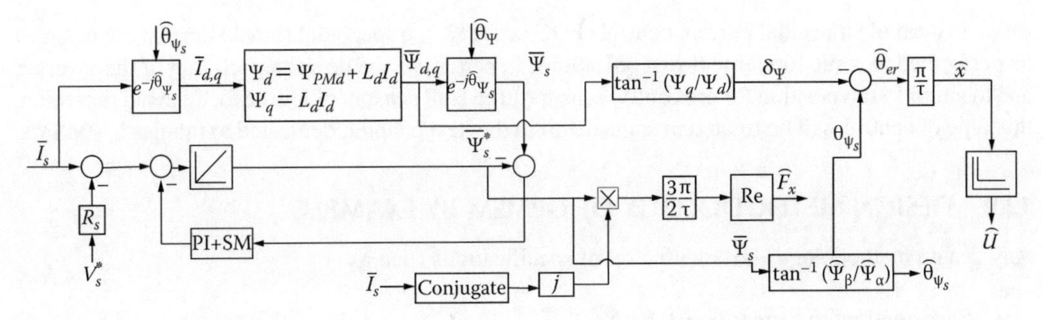

FIGURE 12.31 F-LPMSM state observer for sensorless DTFC: the primary flux $\bar{\psi}_S$, thrust F_x, primary flux angle $\widehat{\theta}_{\psi S}$, speed $\widehat{u}$, and position $\widehat{x}$ (relative).

A few comments on DTFC are as follows:

- The current to flux model (Figure 12.31) requires the stator flux angle $\widehat{\theta}_{\psi S}$ and is strictly valid in steady state; otherwise, the mover position angle $\widehat{\theta}_{er}$ may be used instead.
- The speed $\widehat{u}$ is estimated from relative position $\widehat{x}$ by a time derivative; in reality, a PLL or other digital filter may be used to improve $\widehat{x}$ and obtain a better $\widehat{u}$ (speed estimation).
- The relative position estimation $\widehat{x}$ may be used for drive positioning only if some proximity sensors exist along the track for periodic position calibration.
- For initial position, or a nonhesitant start, a signal injection addition to $\widehat{x}$ is needed [19]. The primary flux amplitude control, Figure 12.30, is used to control the airgap (via normal force); a normal force estimator may also be added; then an additional normal force regulator is required before the stator flux ψ_S regulator.
- The presence of PM flux makes the stator flux (and normal force) regulations (by I_d essentially) rather stiff but still worthy of consideration as no addition of power hardware is required for levitation control.
- If cogging force cancelation is required, it is possible to add $(-F_{cog}(x))$ to F_x^* and thus, with enough bandwidth in the thrust regulator, the F-LPMSM may compensate cogging force through the control system.
- For industrial (low speed) travel (a few meters) with short primary mover MAGLEV carriers, it may be feasible to use interior PM fixed secondary with weaker PMs and higher magnetic saliency $L_d/L_q \geq 3$ (airgap: 0.5 mm) (Figure 12.32).

Note on trapezoidal current control: As already discussed, both with air-core and iron-core primary, the emf may be trapezoidal (with $q_1 = 1$ and with $q_1 < 0.5$ (tooth-wound windings); in such

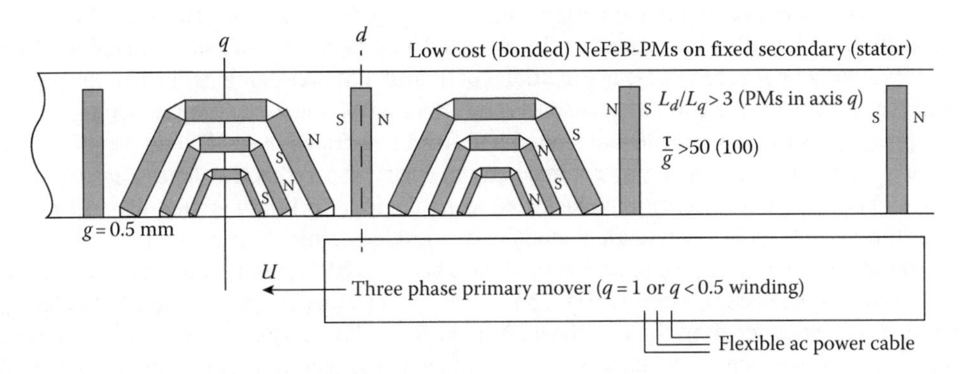

FIGURE 12.32 PM-assisted linear reluctance synchronous motor.

cases, instead of sinusoidal current control (FOC or DTFC), trapezoidal (block) current control may be performed to avoid too large thrust pulsations, to better use the dc voltage coiling of the inverter, and to simplify the position (to proximity) sensor (three Hall sensors of low cost). To avoid repetition, this type of control will be treated in some detail in the next chapter, dedicated to tubular L-PMSMs.

12.9 DESIGN METHODOLOGY OF L-PMSM BY EXAMPLE

Any design methodology starts with a set of specifications such as

- Base continuum thrust: $F_{xn}=0.8$ kN
- Base speed: $U_b=1.5$ m/s
- DC voltage: $V_{dc}=300$ V
- Primary: as mover (short)
- Normal force: zero (double-sided PM long secondary (stator))
- Primary core: iron core
- Airgap: $g=1$ mm
- Travel length: 3.5 m
- Maximum speed: $U_{\max}=U_b=1.5$ m/s
- Peak thrust at max. speed: $F_{xpeak}=1.2$ kN
- Efficiency at base power (1.2 kW) at base speed>0.85
- Power factor at base power (1.2 kW) at base speed>0.8
- Maximum ideal acceleration of primary $a_{\max}>40$ m/s^2
- Primary mover supply: by ac power flexible cable

What is required may be summarized as follows:

- Use a realistic analytical model, size the PM long secondary and the short primary core.
- Calculate, by the same model, the machine circuit parameters (inductances and resistances).
- Use the circuit model (and vector diagrams) to calculate the performance at base (maximum) speed and base and peak thrust for $V_{dc}=300$ V.
- Redo the process until all specifications are met.
- Eventually, use a numerical code optimization design by choosing a variable vector, its domain of allowed values, and the aforementioned analytical model (or a more complex one) to evaluate a global objective function with constraints; then use an optimization algorithm (Hooke–Jeeves and genetic algorithms in [17]) to find an optimal design.

12.9.1 PM-Secondary Sizing

As there is no flux weakening in the design data ($U_b=U_{\max}$), to yield good efficiency (high thrust/copper losses), surface PMs are planted on the two fixed long secondaries (stators); to reduce further thrust pulsations (cogging force), we use multiples of 6 slot/8 PM pole combinations for the primary/secondary and displace the two PM secondaries by 1/6 of a PM pole pitch (step skewing).

The primary teeth are premade and wound and fixed to a frame at stack ends; the bolts in the trapezoidal open holes are made of nonmagnetic material; the respective holes cause a further reduction of cogging force. Other manufacturing technologies of primary are feasible. As a 6 teeth (slot) primary/8 PM pole combination module is considered, the fundamental winding factor $k_{W_1}=0.866$. A 12 slot/10 PM pole module would have had $k_{W_1}=0.933$, both for a double and a single layer winding (as the one in Figure 12.33). The $2p>N_S$ (8/6) means less maximum PM flux per coil for larger thrust per given mmf than for the 4 PM pole, 6 slot case. Also, the yoke flux is smaller, so the stator yoke of fixed long secondary (stator) is thinner; consequently, cost is reduced notably. The machine has two airgaps, but this is "the price" for ideally zero normal force on primary movers.

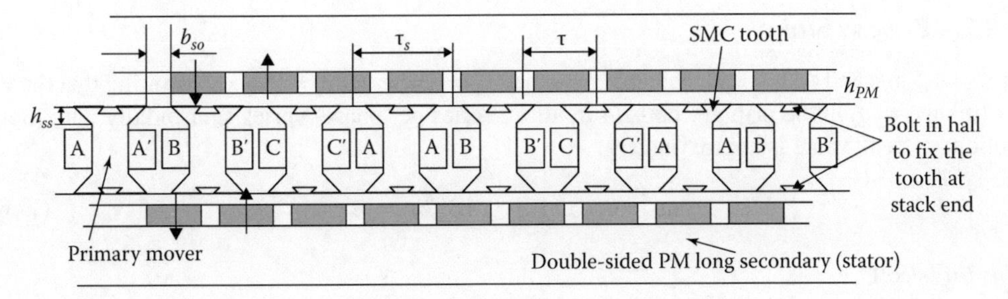

FIGURE 12.33 Double-sided F-LPMSM with primary mover (6 slots/8 PMs poles).

An FEM study considering variations for τ_{PM}/τ, τ_{PM}/τ_s, b_{s0}/τ_s, g/h_m would lead to a practical design that mitigates between more average thrust density (N/cm²) at a good loss factor (N/W) with small enough cogging force. However, $\tau_{PM} \approx (\tau_s - b_{s0})0.8$, $\tau = 30$ mm, $\tau_s = \tau \cdot 2p/N_S = 30 \times 8/6 = 40$ mm would be a good start ($g = 1$ mm, $b_{s0} = (3\text{–}4)g = 3\text{–}4$ mm). After choosing the airgap PM flux density $B_{PMg} = 0.7$ T for a bonded NeFeB PMs with $B_r = 0.8$ T, $\mu_{rec} = 1.07\ \mu_0$, the PM height h_m is calculated from (12.65)

$$B_{PMg}\frac{\pi}{4} = B_{PMgu} = B_r \cdot \frac{h_m}{h_m + g} \cdot \frac{1}{1 + k_{fringe}}; \quad \frac{h_m}{h_m + g} = \frac{0.7 \times 1.2 \times 4}{\pi 0.8} = 0.825 \tag{12.78}$$

The fringing factor k_{fringe} (due to loss of PM flux between teeth necks), for surface PMs, if $b_{s0}/g = 4\text{–}5$, $k_{fringe} \leq 0.2$ (it may be checked by FEM). B_{PMgu} is in fact the "useful" flux density that produces PM flux in the primary coils. For $g = 1$ mm, it follows that $h_m = 5.7$ mm. This rather thick PM leads also to smaller main inductance and thus to a better power factor (lower inverter kVA); since bonded NeFeB PMs are considered, their total cost should be reasonable. The PM span $\tau_{PM} = (\tau_S - b_{S0}) \times 0.8 = (40 - 5)0.8 = 28$ mm $< \tau = 30$ mm; the inter-PM space may be filled with adhesive to better fix the magnets.

The yoke of PM-secondary h_{yS} is

$$h_{yS} \approx \frac{\tau_{PM} \cdot B_{gPMu}}{B_{yS}} = 28 \times \frac{0.7}{1.7} = 11.5 \text{ mm} \tag{12.79}$$

To allow for armature-reaction contribution at peak thrust, $h_{yS} = 15$ mm; solid mild steel may be used for PM-secondary yoke; the magnetic airgap $(g + h_m)$ is large and thus the subharmonics flux of the primary mmf does no "reach" the solid yoke of stator to produce large eddy currents. Still to find is the primary stack length l_{stack}, almost equal to PM (and stack) length. Considering a base thrust density $f_{xb} = 2$ N/cm² (3 N/cm² for peak thrust) and an integer number of primary modulus, the thrust is

$$F_{xn} = f_{xb} \cdot l_{stack} \cdot 6\tau_s \cdot n \tag{12.80}$$

Hence, the choice between l_{stack} and number of primary-mover modules, n, depends on the secondary (travel) length, which, then, determines its initial costs. For a 4 m length track, the primary length could not be above 0.5 m (travel length 3 m); for $n = 2$, the primary length is $6\tau_s \cdot n = 6 \times 40 \times 2 = 480$ mm $= 0.48$ m.

Consequently, l_{stack} is

$$l_{stack} = \frac{800}{2 \times 10^4 \times 0.48} = 0.0833 \text{ m} \tag{12.81}$$

12.9.2 Primary Sizing

For $U_b=1.5$ m/s and $\tau=0.03$ m, the base frequency $f_{1b}=1.5/(2\times0.03)=25$ Hz. Assuming that the PM flux linkage ψ_{PMd} in the primary coils (4 in all, in series per phase) varies sinusoidally with mover position, its peak value is approximately

$$\psi_{PMd} = B_{PMgu}\tau_{PM}l_{stack}4W_C k_{W_1};\quad p=4 \tag{12.82}$$

W_C is turns/coil

$$\psi_{PMd} = 0.7\times0.028\times0.0833\times4\times W_C\times0.867 = 5.66\times10^{-3}\cdot W_C = k_{PM}W_C \tag{12.83}$$

The rated (base) thrust F_{xn} is

$$F_{xn} = 800 = \frac{3}{2}\frac{\pi}{\tau}\psi_{PMd}\cdot I_{qn} = \frac{3}{2}\frac{\pi}{0.03}\times5.66\times10^{-3}\times W_C\cdot I_{sn} = 0.0888\cdot W_C\cdot I_{sn}$$
$$I_{qn} = I_{sn}\quad (I_d=0) \tag{12.84}$$

So $W_C\cdot I_{sn}=900$ A turns/coil (peak value). As there are two coils in a primary slot, the active area of the slot, A_{slot} is

$$A_{slot} = \frac{2W_C\left(I_{sn}/\sqrt{2}\right)}{j_{Co}k_{fill}} = \frac{2\times900/\sqrt{2}}{45\times0.4} = 792\ \text{mm}^2 \tag{12.85}$$

The slot width $b_s\approx0.5\tau_s=0.5\times40=20$ mm, so the active slot height h_{su} is

$$h_{su} = \frac{A_{slot}}{b_s} = \frac{792}{b_s} \approx 40\ \text{mm} \tag{12.86}$$

Note: We have to add on each side of the doubly opened slot about 5 mm ($h_{ss}=4$ mm) of height, to leave room for the rods that support the primary cores.

The rated current density $j_{Con}=4$ A/mm² and the slot fill factor $k_{fill}=0.4$, as there is space between the two coils in slot.

There is no yoke in the primary mover, in order to reduce its weight.

12.9.3 Circuit Parameters and Vector Diagram

The synchronous inductance comprises the "magnetization" one $L_m=L_{dm}=L_{qm}$ and the leakage one L_l. The magnetization inductance is straightforward (and apparently includes space harmonics contribution):

$$L_m \approx \frac{3}{2}\frac{\mu_0\left(\tau_s-b_{s0}\right)l_{stack}W_C{}^2 4}{4\left(g+h_m\right)k_C};\quad p=4=\text{number of coils/phase} \tag{12.87}$$

or

$$L_m = \frac{3}{2}\cdot\frac{1.256\times10^{-6}\times\left(0.04-0.005\right)\times0.0833\times4\times W_C{}^2}{4\times\left(1+5.7\right)10^{-3}\times1.1} = 0.7453\times10^{-6}\cdot W_C{}^2 \tag{12.88}$$

The leakage inductance of the four coils per phase includes basically slot leakage and end coil leakage components:

$$L_l = 4\times 2\mu_0 \cdot l_{stack}\left(\lambda_{slot} + \frac{l_e}{l_{stack}}\cdot\lambda_{end}\right)\cdot W_C^{\,2} \tag{12.89}$$

The slot is doubly opened and the variation of leakage field H along slot height is as in Figure 12.34.

The situation is equivalent to two semiclosed slots of half-height, but with doubled slot width:

$$\lambda_{slot} \approx 2\left(\frac{h_{su}}{3\cdot(2\cdot b_s)} + \frac{h_{ss}}{b_s + b_{s0}}\right) = 2\left(\frac{\frac{40}{2}}{3\times 2\times 20} + \frac{4}{20+5}\right) = 0.652! \tag{12.90}$$

The end coil geometrical permeance $\lambda_e \approx 0.63$ (as in $q=1$ windings) [19].

$$L_l = 4\times 2\times 1.256\times 10^{-6}\times 0.0833\times\left(0.652 + \frac{1.2\times 0.04}{0.0833}\times 0.33\right)\cdot W_C^{\,2} = 0.7049\times 10^{-6}\cdot W_C^{\,2} \tag{12.91}$$

As seen $L_l \approx L_m$, since the machine has four large equivalent airgaps per flux closed path. So $L_S = L_m + L_l = 1.45\times 10^{-6}\, W_C^{\,2}$.

The phase resistance R_S is simply

$$R_S = 4\rho_{Co} l_C \frac{W_C^{\,2} j_{Con}}{W_C\, I_{sn}/\sqrt{2}} = 4\times 2.1\times 10^{-8}\times 0.2826\frac{4\times 10^6}{900/\sqrt{2}}\cdot W_C^{\,2} = 1.4876\times 10^{-4}\cdot W_C^{\,2} = k_R\cdot W_C^{\,2} \tag{12.92}$$

$$l_C \approx 2\cdot(l_{stack} + 1.2\cdot\tau_s + 0.01) = 2\cdot(0.0833 + 1.2\times 0.04 + 0.01) = 0.2826\ \text{m} \tag{12.93}$$

As the $f_{1b} = 25$ Hz, the core losses are to be notably smaller than copper losses (less than 1 W/kg at 1.5 T and 25 Hz). Standard transformer (oriented grain) laminations (0.35 mm thick) may be used.

The vector diagram "spells" the voltage equation (Figure 12.35):

$$\overline{V}_{S0} = R_S \overline{I}_{S0} + j\frac{\pi}{\tau}(Uk)\overline{\psi}_S;\quad \overline{\psi}_S = \psi_{PMd} + jL_q I_q;\quad I_d = 0 \tag{12.94}$$

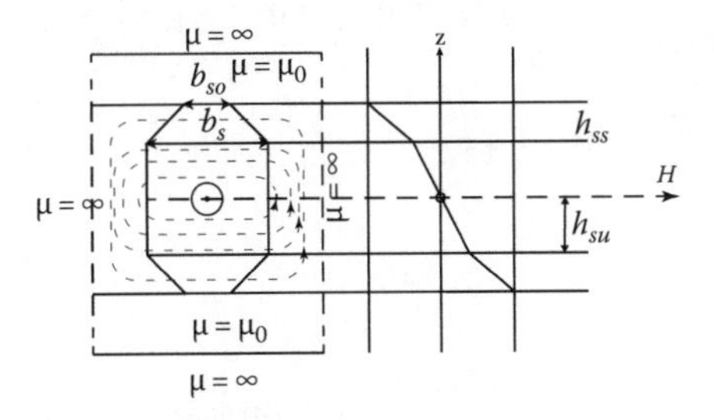

FIGURE 12.34 Slot leakage field.

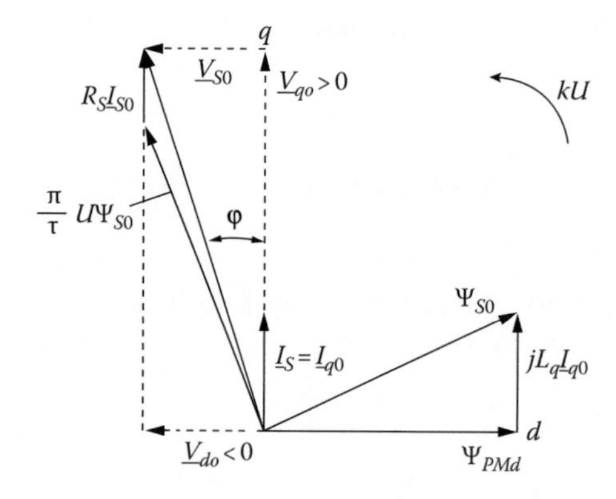

FIGURE 12.35 F-PLMSM vector diagram at steady state with $I_d=0$.

$$V_{d0} = R_S I_{d0} - \frac{\pi}{\tau}(Uk)L_q I_{q0}$$

$$V_{q0} = R_S I_{q0} + \frac{\pi}{\tau}(Uk)\psi_{PMd} \tag{12.95}$$

$$V_{S0} = \sqrt{{V_{d0}}^2 + {V_{q0}}^2}$$

$k=1$ for PM mover
$k=-1$ for primary mover

12.9.4 Number of Turns per Coil W_t and Wire Gauge D_{Co}

To calculate the number of turns W_t, we have to "test" the vector diagram at peak thrust (rather than base thrust) at maximum voltage $V_{s\max}$:

$$V_{s\max} \approx \frac{2V_{dc}}{\pi}\cdot 0.95 = \frac{2\times 300}{\pi}\times 0.95 = 181.52\,\text{V (peak phase value)} \tag{12.96}$$

For 1200 N of thrust $W_C \cdot I_{Speak} = 1.5 \cdot W_C \cdot I_{Sn} = 900\times 1.5 = 1350$ A-turns (peak value per coil); there are four coils in series per phase. Knowing the parameters' expressions (but W_C), we may now use the vector diagram:

$$V_{s\max} = W_C\sqrt{\left(k_{PM}\frac{\pi}{\tau}\cdot U + k_R\left(W_C\cdot I_{Speak}\right)\right)^2 + \left(\frac{\pi}{\tau}\cdot U\cdot k_L W_C\cdot I_{Speak}\right)^2}$$

$$V_{s\max} = W_C\sqrt{\left(5.66\times 10^{-3}\times\frac{\pi}{0.03}\times 1.5 + 10^{-4}1.4876\times 1350\right)^2 + \left(\frac{\pi}{0.03}\times 1.5\times 1.45\times 10^{-6}\times 1350\right)^2}$$

$$= W_C\sqrt{1.08944^2 + 0.3073^2} = W_C\cdot 1.132 \tag{12.97}$$

So

$$W_C = \frac{181.52}{1.132} = 160 \text{ turns/coil}$$

As

$$W_C I_{sn} = 900; \quad I_{sn} = \frac{900}{160} = 5.666\text{ A}; \quad I_n = \frac{I_{sn}}{\sqrt{2}} = 4.014\text{ A (RMS per phase)} \tag{12.98}$$

With $j_{Con}=4$ A/mm², the wire diameter is

$$d_{Co} = \sqrt{\frac{4}{\pi}\cdot\frac{I_n}{j_{Con}}} = \sqrt{\frac{4}{\pi}\times\frac{4.014}{4}} = 1.13\text{ mm (bare wire)} \tag{12.99}$$

12.9.5 Efficiency, Power Factor, and Voltage at Base Thrust

At base thrust the phase current $I_n=4.014$ A, and thus the copper losses are

$$p_{Con} = 3R_S I_n^2 = 3\times 1.4876\times 10^{-4}\times\left(\frac{900}{\sqrt{2}}\right)^2 = 180.7\text{ W} \tag{12.100}$$

So the efficiency at rated thrust is

$$\eta_n = \frac{F_{xn}U_b}{F_{xn}U_b + p_{Con}} = \frac{800\times 1.5}{1200+180.74} = 0.869 > 0.85 \tag{12.101}$$

Given the low speed (1.5 m/s), the efficiency, for 2 N/cm² thrust density, is considered really good.

The power factor can be calculated from (12.87) with $W_C I_{sb} = W_C I_{peak}\times 2/3$; with W_C known, the voltage V_S can be obtained from (12.97):

$$V_S = 160\sqrt{\left(5.66\times 10^{-3}\times\frac{\pi}{0.03}\times 1.5 + 10^{-4}\times 1.486\times 900\right)^2 + \left(\frac{\pi}{0.03}\times 1.5\times 1.45\times 10^{-6}\times 900\right)^2}$$

$$= 160\sqrt{1.02248^2 + 0.20488^2} = 160\times 1.0428 = 166.848\text{ V (peak phase value)} \tag{12.102}$$

The power factor may be calculated simply as

$$\cos\varphi_n = \frac{1.02248}{1.0428} = 0.98 \tag{12.103}$$

Even for peak thrust the copper losses are

$$p_{Cop} = p_{Con}\left(\frac{F_{xpeak}}{F_{xn}}\right)^2 = 180\times\left(\frac{3}{2}\right)^2 = 405\text{ W} \tag{12.104}$$

$$\eta_{peak\,thrust} = \frac{1200\times 1.5}{1800+405} = 0.816 \tag{12.105}$$

And from (12.87)

$$\cos\varphi_{peak} = \frac{1.0814}{1.132} = 0.955 \tag{12.106}$$

12.9.6 Primary Active Weight

The primary active weight is made of core weight W_{core} and copper weight W_{Copper}:

$$W_{core} = 12 l_{tack}\left[\tau_s \cdot (h_{su} + 2h_{ss}) - \left(h_{su} \cdot b_s + (b_s + b_{s0}) h_{ss}\right)\right]\gamma_{iron}$$
$$= 12 \times 0.0833 \times \left[0.04 \times (0.04 + 2 \times 0.005) - \left(0.04 \times 0.02 + (0.02 + 0.005) \times 0.005\right)\right]$$
$$\times 7600 = 8.138 \text{ kg} \quad (12.107)$$

$$W_{Copper} = \frac{3 \times l_c W_c \times 4 \times I_n \gamma_{Co}}{j_{Con}} = \frac{3 \times 0.286 \times 160 \times 4 \times 4.014 \times 8900}{(4 \times 10^6)}$$
$$= 4.846 \text{ kg} \quad (12.108)$$

$$W_{ap} = W_{core} + W_{Copper} = 8.138 + 4.846 = 12.98 \text{ kg} \quad (12.109)$$

Let us suppose that the primary framing and rolling facility of mover leads to a total mover weight $W_{mover} = 20$ kg. Consequently, the ideal maximum acceleration (available up to base (maximum) speed of 1.5 m/s for $F_{xpeak} = 1200$ N) is

$$a_{i\max} = \frac{1200}{20} = 60 \text{ m/s}^2 \approx 6 \cdot a_{gravity} >> 40 \text{ m/s}^2 \quad (12.110)$$

12.9.7 Design Summary

- Primary (mover) active and total weight: 12.98 kg/20 kg.
- Maximum ideal acceleration (up to 1.5 m/s) is 60 m/s^2.
- The efficiency and power factor at rated thrust 800 N and U_b are 0.869 and 0.98, respectively, $f_n = 25$ Hz, $V_{sn} = 161.848$ V (peak phase value).
- For the peak thrust (1200 N) and $U_b = 1.5$ m/s at $V_{s\max} = 181.52$ V ($V_{dc} \approx 300$ V), efficiency is still 0.816 and the power factor is still very good: 0.955.
- The rather very good power factor is due to the small fundamental frequency $f_n = 25$ Hz, and due to the small machine inductance $L_S = 1.45 \times 10^{-6} \times 160^2 = 37.12$ mH; the stator resistance $R_S = 1.4876 \times 10^{-4} \times 160^2 = 3.808\ \Omega$.
- In turn L_S is small due to the large total magnetic airgap $4(g + h_m)$ and small leakage inductance due to doubly opened slots and short end-coils (small pole pitch).
- As power factor is high, the design at higher thrust for higher current density, smaller number of turns, but lower efficiency is also possible.
- It is quite disputable that such performance, at such a low speed, can be surpassed in aggregated performance by many other configurations.

12.9.8 Note on F-LPMSM as Three-Phase Generators

For wave generators, speeds in the range of 0.7–1.5 m/s are required for very large thrust (tens of kN for tens of kW and hundreds of kN for hundreds of kW) for short (a few meters) travel length. For such designs, square-shape (four-sided) F-LPMSM has been proposed for 16 kW generators with PM-mover, operating at 10 Hz for 0.9 m/s [25]. As the thrust density for rather small pole pitch ($\tau = 45$ mm) is not very high—as seen earlier, the weight and cost of the device are still large. A 250 kW (speed around 1 m/s) F-LPMSM wave generator has been calculated in [26]. It uses a tooth-wound winding in the double-sided short stator and double-sided PM long mover (translator) with interior PMs. By proper skewing with semiclosed stator slots, the cogging force was reduced to well below 1% of rated thrust. However, at moderate pole pitch (in latter case [26] $\tau = 60$ mm in the PM mover) and slot pitch $\tau_s = 165$ mm!) and such a small speed, it is still hard to say that $q = 1$ or $q < 0.5$ winding stator configurations will result in acceptable power/weight and efficiency for large wave generators. Shifted tooth wound $\tau = \tau_s$ primaries (by half slot pitch) pitch phase double-sided modules for F-LPMSM have been proven to allow larger thrust density with low armature reaction [27] and may constitute a way to lower weight for such large thrust for reasonable generator weight cost, efficiency, and power factor (which influences heavily the PWM inverter kVA ratings and cost).

12.10 SUMMARY

- Linear PM synchronous motors are the counterparts of the rotary PM synchronous machine, which are characterized essentially by sinusoidal emf and thus by sinusoidal current control.
- The primary has a three-phase ac winding area built with distributed coils ($q \geq 1$) or tooth-wound (nonoverlapping) coils ($q \leq 0.5$), placed in the uniform slots of a laminated iron core or in an air (plastic) core.
- The primary may be the stator, placed along the track, or the mover; in the latter case, the electric power is transferred to the mover by sliding power collector (pantographs) or by flexible power cable for short travel.
- The PM secondary or primary may be the mover.
- To cancel the normal force, double-sided primaries or secondaries are used.
- To calculate the emf thrust F_x and normal force F_n, typical to flat linear PM synchronous motors (F-LPMSMs), a 2(3) D multilayer analytical field model is first applied; even slot openings can be considered by defining a proper inverse-airgap function.
- Also 2(3) D magnetic equivalent circuit (MEC) methods are used to fully characterize F-LPMSMs, including magnetic saturation, besides all geometry parameters (and slot openings).
- These two methods allow to calculate emfs, force, and inductances —circuit parameters— to prepare all data for phase variable and then for the *dq* model of F-LPMSMs.
- In addition, the moderate computation time of multilayer methods recommends them for optimization design methodologies.
- For "ultimate" precision, 2(3) D-FEM is used (in its magnetostatic, time stepping, and field-circuit modes) to describe F-LPMSM; the only trouble is the computation time; to check (and correct) or to perform direct geometric optimization design around the analytical optimum design, FEM is the way to go.
- Field theories yield the data (dB/dt in essence) for calculation of core losses.
- F-LPMSM with iron-core primary experience a longitudinal force at zero current, called cogging force (the counterpart of cogging torque in rotary PM machines); it is caused by the variation of coenergy stored in the PMs with slot openings; as expected its average is practically zero; the cogging force pulsates at the number of periods equal to LCM (N_{slots}, $2p$ PM poles); the larger the LCM, the smaller the cogging force peak.
- Methods to reduce cogging force [28] are similar to those used in rotary machines: PM span/pole/pitch ratio, PM poles displacement, skewing, end-teeth shaping etc; besides, thrust ripple disturbance rejection by control is also feasible [29].
- The *dq* model of F-LPMSM can include not only the propulsion motion (horizontal, in general) but also the levitation motion (vertical, in general along airgap).

- There are two FOC methods for F-LPMS control: zero I_q propulsion control and, respectively, flux weakening ($I_d^* < 0$ control) and thrust (I_q control); alternatively, I_d^* can be used to control the levitation.
- The structural diagrams for constant I_d (zero or not) and FOC of F-LPMSMs are very similar to that of PM brush motor, ideal for motion control.
- On the other hand, primary thrust and flux control (DTFC) allow direct thrust control and it may be used to control levitation by flux control.
- Advanced fuzzy neural network and other robust linear position tracking with F-LPMSM have shown very good steady-state precision but with still large dynamic tracking errors. Using a disturbance force, the observer may substantially enhance relative position tracking by F-LPMSM [30].
- A DTFC for F-LPMSM without a position sensor reveals the possibility of even position sensorless tracking if a few calibration proximity position sensors are placed at known points along the track.
- A design methodology with a case study for 1200 N (peak value up to 1.5 m/s) shows efficiency above 80%, a very good power factor for a 13 kg active weight (20 kg total) primary mover.
- As recent research efforts have shown, the application of F-LPMSM as wave generators is thoroughly investigated (at about 1 m/s) and for 250 kW at 250 kN); the power/weight and efficiency and power factor product need further drastic improvements, before becoming economical.
- New configurations with ever better performance keep surfacing [31–33].

REFERENCES

1. Z. Deng, I. Boldea, and S.A. Nasar, Fields in PM-LSMs, *IEEE Trans.*, MAG-22, 1986, 107–112.
2. Z. Deng, I. Boldea, and S.A. Nasar, Forces and parameters of PM-LSMs, *IEEE Trans.*, MAG-23, 1987, 305–309.
3. Z.Q. Zhu, D. Howe, E. Bolte, and B. Ackerman, Instantaneous magnetic field distribution in brushless PM dc motors, Parts I–IV, *IEEE Trans.*, MAG-29(1), 1993, 124–158.
4. V. Ostovic, *Dynamics of Saturated Electric Machines*, John Wiley & Sons, New York, 1989.
5. W.B. Jang, S.-M. Jang, and D.-Y. Yon, Dynamic drive analysis through base speed determination for optimum control boundary in PMLSM with self-load, *IEEE Trans.*, MAG-41(10), 2005, 4027–4029.
6. D.L. Trumper, W.-J. Kim, and M.E. Williams, Design and analysis framework for linear PM motors, *IEEE Trans.*, IA-32(2), 1996, 371–379.
7. B. Sheikh-Ghalavand, S. Vaez-Zadeh, and A.H. Isfahani, An improved magnetic equivalent circuit model for iron-core linear permanent magnet synchronous motors, *IEEE Trans.*, MAG-46(1), 2010, 112–120.
8. I. Boldea, S.A. Nasar, and Z. Fu, Fields forces and performance of air-core linear self-synchronous motor with rectangular current control, *IEEE Trans.*, MAG-24, 1988, 2194–2203.
9. T.J.E. Miller, *Brushless PM and Reluctance Motor Drives*, Oxford University Press, Oxford, U.K., 1989.
10. N. Baatar, H.S. Yoon, M.T. Pham, P.S. Shin, and C.S. Koh, Shape optimal design of a 9 slot 10 pole PM LSM for detent force reduction using adaptive response surface method, *IEEE Trans.*, MAG-45(10), 2009, 4562–4565.
11. D.-Y. Lee, Ch.-G. Jung, K.-I. Yoon, and G.T. Kim, A study on the efficiency optimum design of a PM type LSM, *IEEE Trans.*, MAG-41(5), 2005, 1860–1863.
12. S. Chaithongsuk, N. Takorabet, and F. Meybody-Tabar, On the use of PWM method for the elimination of flux density harmonics in the airgap of surface PM motors, *IEEE Trans.*, MAG-45(3), 2009, 1736–1739.
13. J. Gieras, *Linear Synchronous Motors*, Chapter 5, CRC Press, Boca Raton, FL, 2000 (second edition, 2010).
14. B.-Q. Kou, W.H.-Xing, L.-L. Yi, Z.-L. Liang, Z. Zhe, and C.-H. Chuan, The thrust characteristics investigation of double sided PMLSM for EML, *IEEE Trans.*, MAG-45(1), 2009, 501–505.
15. G. Henneberger and C. Reuber, A linear synchronous motor for a clean room system, *Record of LDIA-1995*, Nagasaki, Japan, 1995, pp. 227–230.

16. I. Boldea and V. Coroban, BEGA—Vector control for wide constant power speed range at unity power factor, *Record of OPTIM-2006*, Brasov, Romania, 2006 (IEEExplore).
17. I. Boldea and L. Tutelea, *Electric Machines: Steady State, Transients and Design with Matlab*, Chapters 6 and 9, CRC Press, Taylor & Francis, New York, 2009.
18. I. Boldea and S.A. Nasar, *Linear Motion Electromagnetic Devices*, Chapter 4, Taylor & Francis Group, New York, 2001.
19. I. Boldea and S.A. Nasar, *Electric Drives*, 2nd edn., Chapters 10 and 11, CRC Press, Taylor & Francis Group, New York, 2009.
20. K. Yoshida, Y. Tsubone, and T. Yamashita, Propulsion control of controlled PMLSM Maglev carrier, *Record of ICEM-1992*, Manchester, U.K., 1992, pp. 726–740.
21. F.-J. Lin, P.-H. Shen, S.-L. Yang, and P.-H. Chou, Recurrent radial basis function network-based fuzzy neural network control for PMLSM servodrive, *IEEE Trans.*, MAG-42(11), 2006, 3694–3705.
22. F.-J. Lin, P.-H. Shen, S.-L. Yang, and P.-H. Chou, Recurrent functional-link-based fuzzy neural network controller with improved particle swarm optimization for a linear synchronous motor drive, *IEEE Trans.*, MAG-45(8), 2009, 3151–3165.
23. L. Zeng, X. Chen, X. Luo, and X. Li, A vector control method of LPMBDCM considering effects of PM flux linkage and cogging force, *IEEE Trans.*, MAG-46(3), 2010, 954–959.
24. R. Krishnan, *Permanent Magnet Synchronous and Brushless Motor Drives*, CRC Press, Taylor & Francis, New York, 2010.
25. N.M. Kionoulakis, A.G. Kladas, and J.A. Tegoupoulos, Power generation optimization from sea waves by using PM linear generator drive, *IEEE Trans.*, MAG-44(6), 2008, 1530–1533.
26. J. Faiz, M.E. Salari, and G. Shahgolian, Reduction of cogging force in linear PM generators, *IEEE Trans.*, MAG-46(1), 2010, 135–140.
27. W.R. Canders, F. Laube, and H. Mosebach, High thrust double sided permanent magnet excited linear synchronous drive with shifted stators, *Record of LDIA-2001*, Nagano, Japan, 2001, pp. 435–440.
28. Y.-J. Kim, M. Wataba, and H. Dohmeki, Reduction of cogging force at the outlet edge of a stationary discontinuous primary linear synchronous motor, *IEEE Trans.*, MAG-43(1), 2007, 40–45.
29. Y.-W. Zhu and Y.-H. Cho, Thrust ripples suppression of permanent magnet linear synchronous motor, *IEEE Trans.*, MAG-43(6), 2007, 2537–2539.
30. Y.-D. Yoon, E. Jung, and S.-K. Sul, Application of a disturbance observer for a relative position control system, *IEEE Trans.*, IA-46(2), 2010, 849–856.
31. Y. Du, K.T. Chau et al., Design and analysis of linear stator PM vernier machines, *IEEE Trans.*, MAG-47(10), 2011, 4219–4222.
32. G. Zhou, X. Huang, H. Jiang, L. Tan, and J. Dong, Analysis method to a Halbach PM ironless linear motor with trapezoidal windings, *IEEE Trans.*, MAG-47(10), 2011, 4167–4170.
33. C.-C. Hwang, P. Li, and C.-T. Liu, Optimal design of a PM-LSM with low cogging force, *IEEE Trans.*, MAG-48(2), 2012, 1039–1042.

13 Tubular Linear Permanent Magnet Synchronous Motors

By tubular linear permanent magnet synchronous motors (T-LPMSMs), we mean the three-phase distributed ($q \geq 1$) or tooth-wound ($q \leq 0.5$) winding primary and PM-secondary tubular configurations with sinusoidal or trapezoidal emfs, supplied by sinusoidal or trapezoidal bipolar currents, to produce a low ripple electromagnetic thrust in industrial applications for travel lengths lower than $1.5 \div 3$ m. The single-phase (or coil) PM actuators, used for short stroke (up to ±15 mm) resonant oscillator linear motion production, will be investigated in a chapter dedicated to linear oscillatory PM machines.

All linear motors mentioned so far in this book have ac coils with a span around one pole pitch (half of flux period either in integer q (slots/pole/phase) or fractionary q ($q \leq 0.5$) windings). Multiple-pole coils three-phase linear PM reluctance brushless motors, which include the so-called transverse flux, flux reversal, flux switch, hybrid stepper configurations, which are essentially three single-phase models added together (as in their rotary counterparts), will also be dealt with in a separate chapter.

With regard to T-LPMSMs, the following issues will be detailed in the sections that follow:

- Characterization of a few practical topologies
- Analytical simplified theory of three-phase windings
- An analytical multilayer field theory
- A magnetic equivalent circuit theory
- FEM T-LPMSM analysis
- Iron losses in T-LPMSM
- Circuit space phasor d–q equivalent circuit for transients and the structural diagram model
- Field oriented control (FOC) and direct thrust and flux control (DTFC) of T-LPMSMs
- Design methodology by example
- Generator mode design and control issues

13.1 A FEW PRACTICAL TOPOLOGIES

As in flat LPMSMs, T-LPMSMs are built with a three-phase winding primary and a PM-secondary. Each of them—primary and secondary—may be the mover or the stator; also each of them may be the shorter or the longer part; as in T-LPMSMs, travel length is limited to (perhaps) $3 \div 3.5$ m (in high thrust applications). More than that, they may have outer or inner PM-secondary.

In summary, T-LPMSMs may be, in general,

- With short-mover-primary and long-PM-secondary stator (Figure 13.1a)
- With short-stator-primary and long-PM-secondary mover (Figure 13.1a and d)

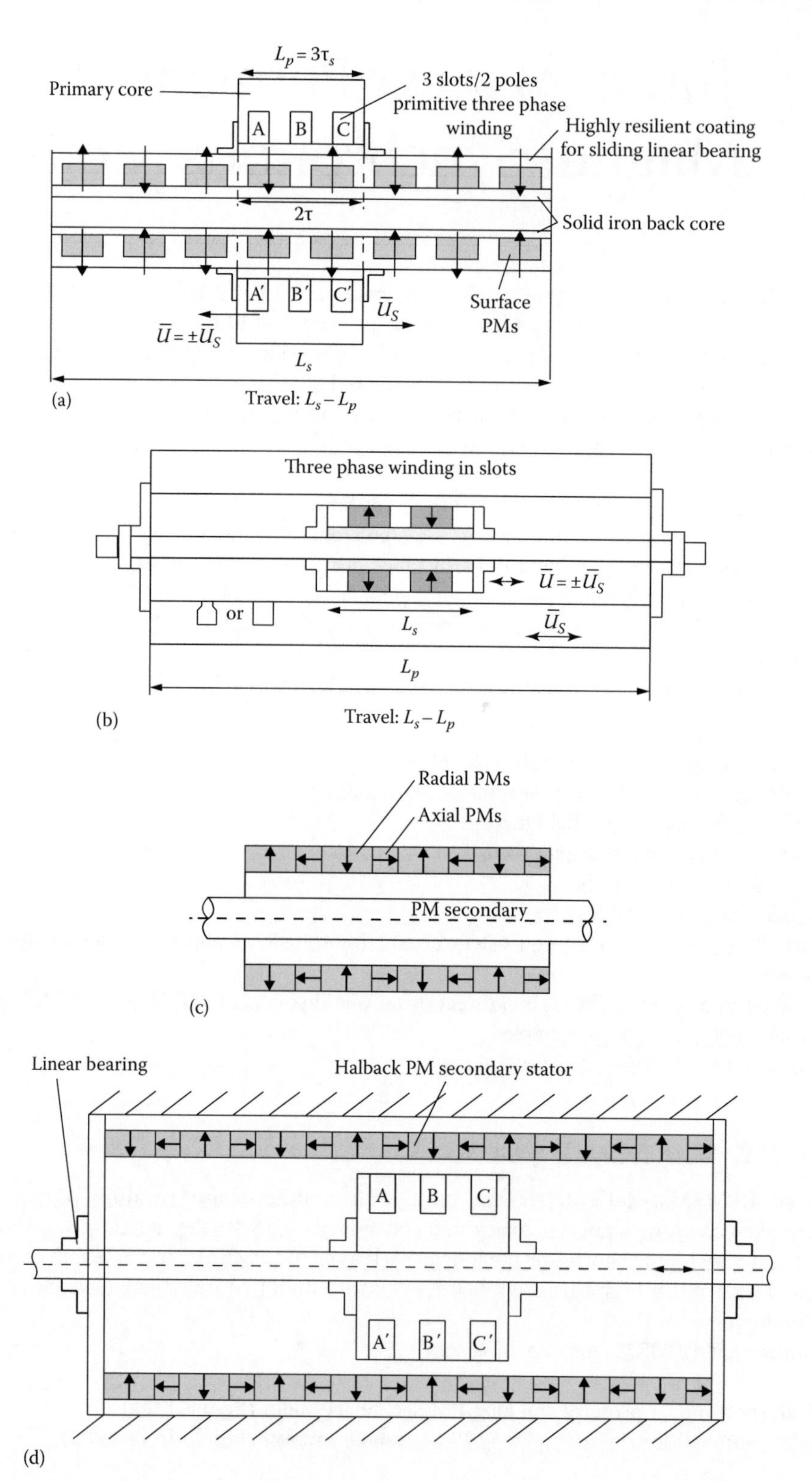

FIGURE 13.1 Basic T-LMSM topologies, (a) exterior short primary and surface PM-secondary, (b) exterior long primary stator and short PM-secondary mover, (c) Halbach PM interior secondary, (d) long Halbach PM-secondary (stator).

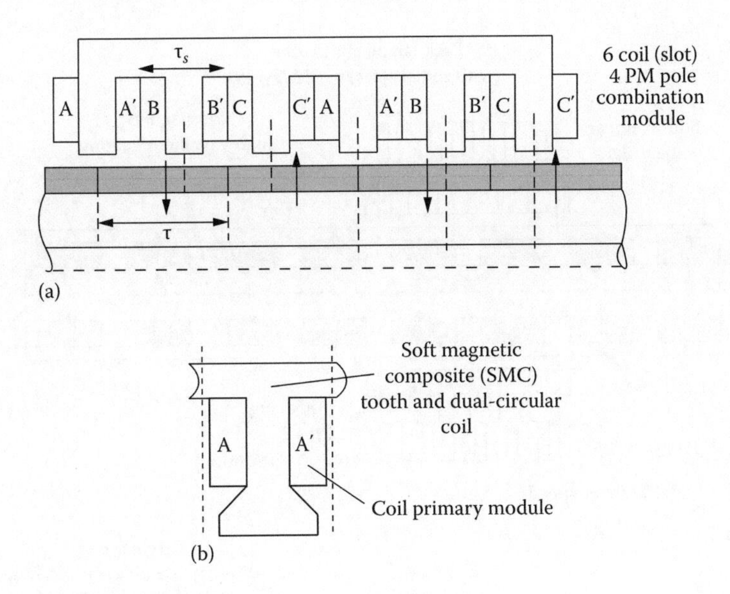

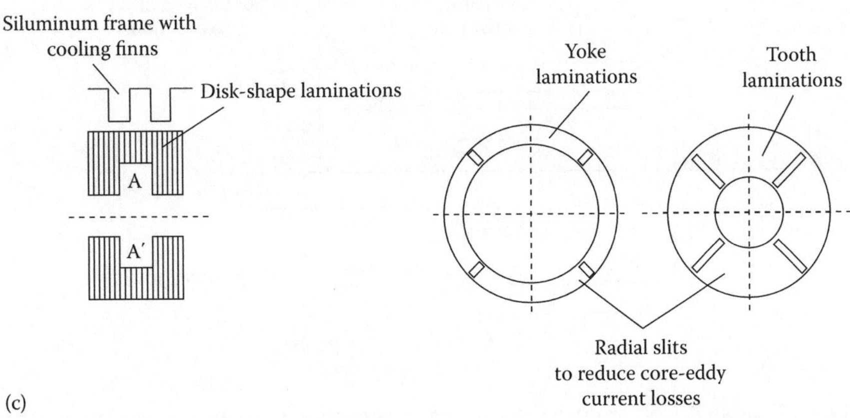

FIGURE 13.2 Six-coil (slot)/4 PM pole combination T-LPMSM (a) SMC preformed wound tooth, (b) disk-shaped lamination primary core (c).

- With long-stator-primary and short-PM-secondary mover (Figure 13.1b)
- With outer primary (Figure 13.1a through c)
- With outer PM-secondary (Figure 13.1d)
- With $q=1$ (3 slots/pole) three-phase windings primary
- With $q<0.5$ (tooth-wound) three-phase windings primary (Figures 13.1a and d, 13.2a)
- With iron core (slotted) primary (Figures 13.1 and 13.2)
- With air-core (slotless) primary (Figure 13.3)
- For driving (motor/generator) (Figures 13.1 through 13.3)
- For generator mode only (Figure 13.4)
- With surface PM-secondary (Figure 13.1a and b)
- With Halbach PM-secondary (Figure 13.1c and d)

The configurations in Figures 13.1 through 13.4 may be characterized as follows.

The long PM-secondary-stator short T-LPMSM (Figure 13.1a) is preferable when the energy consumption should be minimum while tolerance to stray PM fields is large; global initial cost tends to be larger when the travel approaches $2 \div 3$ m; a mechanically resilient nonmagnetic coating on PM-secondary is needed for sliding-contact linear bearings fixed to the primary mover.

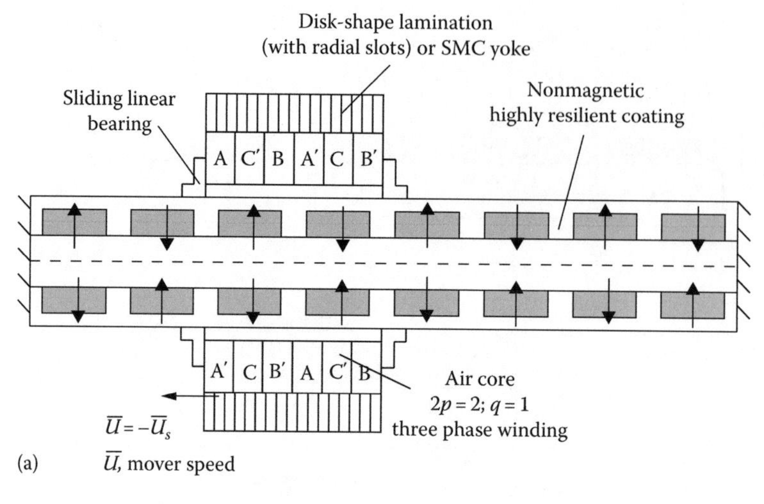

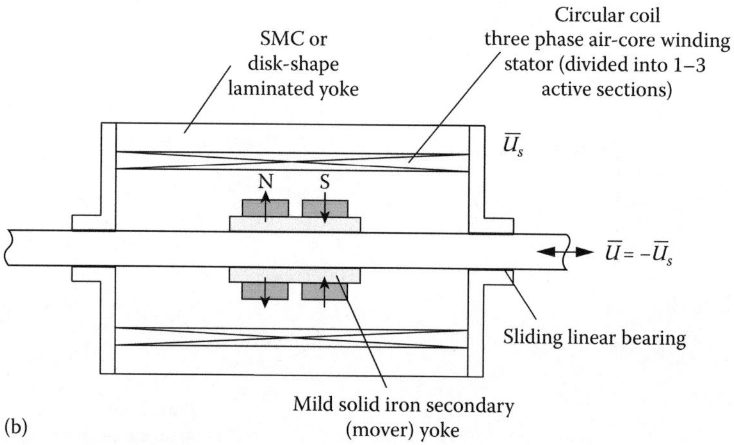

FIGURE 13.3 T-LPMSM short primary with long PM-secondary, (a) long air-core winding primary (stator) and short PM-secondary (mover) (b).

The outer-long-primary stator with interior-short-PM-secondary mover (Figure 13.1b) leads to the lowest mover weight (and highest mover acceleration) and to zero stray fields, but, unless the stator primary is supplied in a few sections, the energy consumption tends to be large. A compromise would be a 2–3 interspaced sections primary stator and PM-secondary, which is as long as one stator section plus the interspace between two stator sections. To increase the airgap flux density, thrust (by +25% or so) efficiency, axially magnetized PMs may be placed between radially magnetized ones (in a ½ ratio in general) in a kind of Halbach array (Figure 13.1c).

To cancel the stray PM fields but still reduce the mover weight and energy consumption, the long outer Halbach array PM-secondary stator with short primary mover may be adopted (Figure 13.1d); in not so cost-sensitive applications, this solution is worth trying; additional, insulated from shaft power aluminum bars with brushes are needed for power collection.

As the travel in T-LPMSMs is rather short (up to $3 \div 3.5$ m at most), the pole pitch is to be small, to limit the yokes depth, for limited speed; for pitches lower than 60 mm, in general, at worst $q = 1$ (3 slot/pole) ac windings can be used; but, more probably, fractionary q ($q < 0.5$) three-phase ac

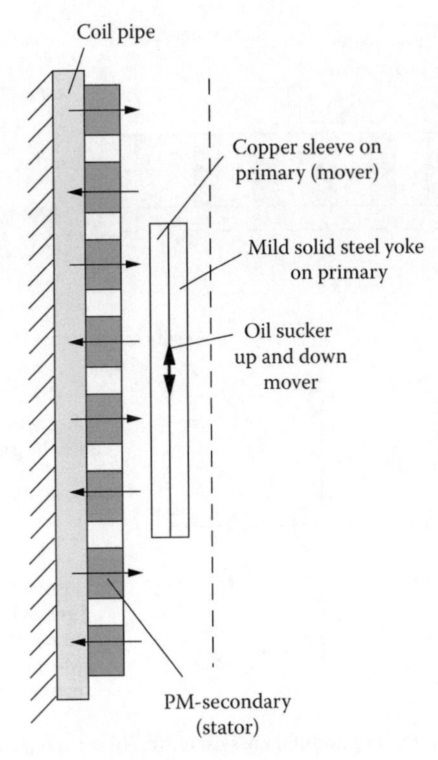

FIGURE 13.4 T-LPMSM: generator (copper sleeve heater in an oil pumping unit to avoid paraffin formation).

windings may be applied (a 3 slot/2 PM pole combination is shown in Figure 13.1a and a 6 slot/4 PM pole module is visible in Figure 13.2a).

Semiclosed slots lead to reduced cogging force but may be built easily only from "prefabricated" SMC teeth with the two circular phase coils already attached (Figure 13.2b). Alternatively, disk-shape laminations may be used, both in teeth and in the yokes; to reduce eddy current iron loss, when ac field transverses the yoke zones perpendicularly to laminations, and radial slits are stamped into the disk-shape laminations; four per circle is enough; the "saliency" at the laminations outer rims is to be "embedded" in the siluminum frame of primary (Figure 13.2c).

Short or long primary T-LPMSMs may be built with air-core three-phase ac winding (Figure 13.3a and b); each phase zone per pole defines the equivalent $q=1$ or $q<0.5$ winding type, to produce a trapezoidal or a sinusoidal emf; the air-core primary T-LPMSMs may lead to low thrust pulsations; cogging force is manifest only at primary iron yoke ends.

The T-LPMSM may be used purposely only as a generator, driven by a free-piston engine ($70\div150$ mm travel length), a wave energy converter, or by an oil-sucker up and down mover in an oil pumping fixture (Figure 13.4); in the latter case, the ac winding is replaced by a copper slab with solid-iron yoke mover primary.

Interpole (interior) PM-secondary for T-LPMSM (Figure 13.5) has been also investigated [1,2] but the limited yoke area at lower diameters makes it hardly adequate in terms of performance/costs; except, perhaps, for very low PM pole pitches ($\tau<10\div12$ mm) [3] or for larger diameters; the Halbach array surface PM-secondaries (Figure 13.1c and d) are considered in more detail in what follows as they have been proven the best solution for T-LPMSMs, where thrust density, efficiency, and mover weight are critical issues.

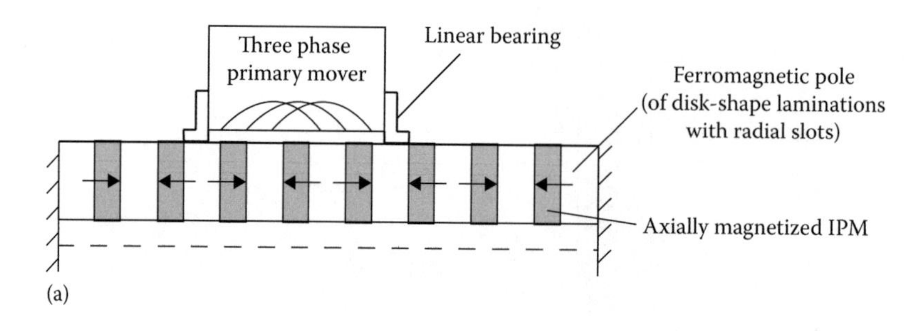

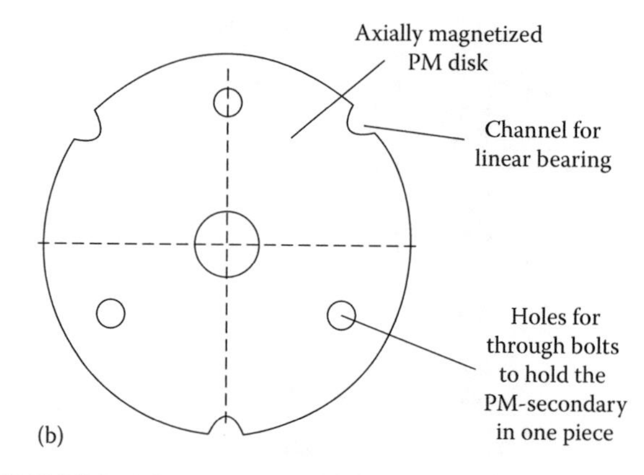

FIGURE 13.5 IPM- T-LPMSM; (a) longitudinal cross section, (b) transverse cross section.

13.2 FRACTIONARY ($q \leq 1$) THREE-PHASE AC WINDING

As already mentioned, due to limited length, even limited maximum speed (6 m/s for, say, 70 mm travel in 40 Hz oscillatory motion) the pole pitch is limited and thus, except for special cases, maximum 3 slots per pole are practical ($q=1$); this case is good especially for generator applications as it provides for notably smaller synchronous inductance (L_S) than the fractionary q ($q \leq 0.5$) windings, because the latter exhibits a very large additional differential leakage inductance due to each armature mmf space harmonic. A smaller L_S leads to smaller voltage regulation, which is good for autonomous generators.

However, for motor-generator (drive) applications used for positioning tracking, the smaller the pole pitch (to same lower limit) the smaller the weight of the system. The rationale for choosing the primary slots count N_p and the secondary PM-poles $2p$ combinations developed for rotary tooth-wound PMSMs applies here as well. To avoid an "inflation" of (N_p, $2p$) combinations, we will leave out most of those that have been proved to produce notable primary and secondary eddy current losses and thus have too many large sub and super speed mmf harmonics. They are governed by the rule [4]

$$N_p = 2p \pm 1 \tag{13.1}$$

We will choose mainly those that correspond to

$$N_p = 2p \pm 2 \quad \text{or} \quad \pm 4 \tag{13.2}$$

Modular structures are encouraged for periodicity in linear machines that have limited longitudinal core length and thus exhibit a nonzero end effect drag force due to end teeth, if not attenuated by proper shaping of end teeth. Consequently, except for micro T-LPMSMs, which may have a short primary with only three slots and thus host three coils when a single layer winding with one coils per slot and $2p=2$ or 4 is used, the following combinations are considered here, which avoid multiple coils per phase in row (as in $N_p=9$, 15 …): $N_p/2p=6/4(8)$ (not exclusively) and multiples;

$N_p/2p=12/10$ and multiples of it when very small cogging force is imperative. The number of poles $2p$ corresponds in length to the N_p slots (teeth surrounded by coils): $2p\tau=N_p\tau_S$.

13.2.1 Cogging Force

Cogging force manifests itself as in F-LPMSMs, and the methods to eliminate it, including the drag force, are the same for T-LPMSMs. Ultimately, when very low cogging force is required, the module may be made of $N_p/2p=6/5(7)$, 12/11 (13), but care must be exercised in observing the very large total synchronous inductance L_S (or L_d, L_q in IPM secondaries). The asymmetry at primary ends due to the odd number of PM poles serves to reduce the end effect drag force at zero current. Double layer windings (Figure 13.2a) are recommended to reduce the armature mmf reactive field and thus allow for larger thrust density for the given power factor. Typical winding factors for the emf (mmf fundamentals) calculated as for rotary machines, using the emf arrows diagrams [5], are given in Table 13.1. Table 13.2 shows the lowest common multiplier (LCM) of N_p and $2p$, which is the number of periods of cogging force per short member length (either primary or secondary) [6]. The larger the LCM, the lower the cogging force amplitude. The almost zero mutual inductance between phases in some (N_p, $2p$) combinations may be used for fault tolerant-sensitive applications.

13.3 TECHNICAL FIELD THEORY OF T-LPMSM

By the technical theory of T-LPMSM, we mean here a simplified 1D field theory that accounts, however approximately, for PM flux fringing in the airgap in the primary slot openings and, for IPM secondary, at its IPM's smaller diameter.

To a first approximation, the slotting may be considered by the Carter coefficient (especially for semiclosed slots), k_C; then the radial PM airgap flux density B_g may be considered flat-topped but reduced by IPM flux fringing k_{fringe} and by magnetic saturation k_S coefficients[3]:

$$B_{gsPM}=B_r\left(1-\frac{(2\cdot g\cdot k_C+h_{PM})}{D_{ip}}\right)\frac{1}{1+\mu_{rec}\left(g\cdot k_C/\tau_m\right)\left(1-(2\cdot g\cdot k_C+h_{PM})/D_{ip}\right)}\cdot\frac{1}{(1+k_{fringe})(1+k_S)} \tag{13.3}$$

for the surface PM-secondary, and

$$B_{gIPM}=\frac{B_r}{2}\frac{\tau_{PM}\left[1-\left(D_{sh}/D_{ip}\right)^2\right]}{\mu_{rec}\left[gk_C\left[1-\left(D_{sh}/D_{ip}\right)^2\right]+(\tau_{PM}/\mu_{rec})(\tau-\tau_{PM})/D_{ip}\right]}\cdot\frac{1}{(1+k_{fringe})(1+k_S)} \tag{13.4}$$

for IPM secondary.

Note: The theory in [3] was completed here by adding influence of k_C for slot opening, k_{fringe} for PM flux fringing, and k_s for magnetic saturation.

The slot openings' influence is more important in IPM secondaries (as g replaces $g + h_{PM}$ in SPM secondaries); the magnetic saturation influence is larger for IPMs while PM flux fringing is comparable; k_{fringe} can be calculated with good precision by FEM; 2D FEM suffices, as the T-LPMSM mover is placed symmetrically in the airgap. More importantly, in the presence of primary current (load), the magnetic saturation conditions change notably, especially for IPM secondaries.

In general, for well-designed T-LPMSM, $k_C=1.05\div1.25$; $k_{fringe}=1.1\div1.2$; $k_S=0.10\div0.25$ (for IPMs); ignoring k_{fringe} and k_S leads to $30\div50\%$ errors in thrust if $k_S=k_{fringe}=1.22$.

TABLE 13.1
Winding Factors of Concentrated Windings

	2p—Poles															
	2		4		6		8		10		12		14		16	
N_s slots	*	**	*	**	*	**	*	**	*	**	*	**	*	**	*	**
3	*	0.86	*	*	*	*	*	*	*	*	*	*	*	*	*	*
6	*	*	0.866	0.866	*	*	0.866	0.866	0.5	0.5	*	*	*	*	*	*
9	*	*	0.736	0.617	0.667	0.866	0.96	0.945	0.96	0.945	0.667	0.764	0.218	0.473	0.177	0.175
12	*	*	*	*	*	*	0.866	0.866	0.966	0.933	*	*	0.966	0.933	0.866	0.866
15	*	*	*	*	0.247	0.481	0.383	0.621	0.866	0.866	0.808	0.906	0.957	0.951	0.957	0.951
18	*	*	*	*	*	*	0.473	0.543	0.676	0.647	0.866	0.866	0.844	0.902	0.96	0.931
21	*	*	*	*	*	*	0.248	0.468	0.397	0.565	0.622	0.521	0.866	0.866	0.793	0.851
24	*	*	*	*	*	*	*	*	0.93	0.463	*	*	0.561	0.76	0.866	0.866

Note: *, one layer; **, two layers.

TABLE 13.2
Lowest Common Multiplier LCM of N_p and $2p$

N_s	2	4	6	8	10	12	14	16
3	6	12	*	*	*	*	*	*
6	*	12	*	24	30	*	*	*
9	*	36	18	72	90	36	126	144
12	*	*	*	24	60	*	84	48
15	*	*	30	120	30	60	210	240
18	*	*	*	72	90	36	126	144
21	*	*	*	168	210	84	42	336
24	*	*	*	*	120	*	168	48

Let us suppose that we extract the fundamental airgap flux density amplitude $B_{g1}PM$ and operate from now on with it, only

$$B_{g1PM} \approx \frac{4}{\pi} B_{gPM} \cdot \sin\frac{\pi}{2}\left(\frac{\tau_{PM}}{\tau}\right); \quad B_{g1}(x,t) = B_{g1PM} \cdot \sin\left(\frac{\pi}{\tau}x - \omega_1 t\right); \quad \omega_1 = \frac{\pi}{\tau}U \tag{13.5}$$

where U is the linear speed.

As Equations 13.3 and 13.4 account for flux conservation in the presence of secondary curvature, they have to be used as they are, unless $\tau/D_{ip}<0.1$, in general (for large diameters and small pole pitches).

The emf in a primary phase E_{1a} is

$$E_{1a} = -\pi \cdot D_{ip} \cdot n_c \cdot k_k \cdot k_{W1} \cdot \frac{d}{dt}\int_0^\tau B_{g1}(x,t)dx = 2\omega_1 D_{ip}\tau\left(k_{W_1} k_k n_c\right) B_{g1PM} \sin \omega_1 t \tag{13.6}$$

$$E_{1a} = \omega_1 \cdot \psi_{PM} \cdot \sin(\omega_1 t); \quad \psi_{PM1} = 2D_{ip} \cdot \tau \cdot k_{W_1} \cdot k \cdot n_c \cdot B_{g1PM} \tag{13.7}$$

Now k_k is the number of equivalent coils per phase

$k_k = N_p/3$ for $q \le 0.5$ modules (double layer windings)

$k_k = N_p/6$ for $q \le 0.5$ modules (one layer windings)

n_c is turns per one f coil

As in double layer windings (Figure 13.1), there are two coils in one slot and in one layer winding there is only one coil/slot (Figure 13.6b); the two situations are equivalent.

So the three-phase emfs are

$$E_{1a,b,c} = E_1\sqrt{2}\sin\left(\omega_1 t - (i-1)\frac{2\pi}{3}\right); \quad E_1\sqrt{2} = \psi_{PM} \cdot U\frac{\pi}{\tau} \tag{13.8}$$

The phase inductances may be derived rather simply for $q=1$ and $q \le 0.5$ windings (Chapter 12, with πD instead of stack length etc):

$$L_d = L_q = L_S = L_l + L_m; \quad L_m = \frac{\mu_0\left(k_k \cdot n_c \cdot k_{w_1}\right)^2 \tau \cdot \pi D_{ip}(1 + k_{fringe})}{p \cdot k_c(g + h_{PM})} \tag{13.9}$$

for smaller PM secondaries and

$$L_q = L_l + L_{qm}; \quad L_{qm} \approx \frac{\mu_0 \left(k_k \cdot n_c \cdot k_{w_1}\right)^2 \tau \cdot \pi D_{ip}}{p \cdot k_c (g + h_{PM})} \cdot k_{qm} \tag{13.10}$$

$$k_{qm} \approx \left(1 - \frac{\tau_{PM}}{\tau}\right) + \frac{1}{\pi} \sin \pi \left(1 - \frac{\tau_{PM}}{\tau}\right) \tag{13.11}$$

$$L_d = L_l + L_{dm}; \quad L_{dm} = \frac{\mu_0 \left(k_k n_c k_{w_1}\right)^2 \tau \cdot \pi D_{ip} (1 + k_{fringe})}{p \cdot k_c \left(1 + \frac{\tau_{PM}}{2g} \frac{\tau}{2 l_{PM}}\right)(1 + k_s)} \tag{13.12}$$

for IPM—secondaries.

k_q=4/3 for q=1; it may go to 1.5 for some $q \geq 0.5$ windings. It may go to 1 when the mutual inductance between phases is zero (for some selected ($N_p/2p$) combinations).

Note: For more exact expressions of L_{qm} and L_{dm} for IPMs in Figure 13.6b, we may adopt the respective expressions obtained for segmented-secondary linear reluctance synchronous motors (Chapter 10, Equations 10.5 through 10.8, with πD instead of l_{stack} and axis d for q axis).

The leakage inductance comprises essentially the slot leakage component only:

$$L_l = \mu_0 \pi (D + h_S) k_k \cdot n_C^2 \cdot \lambda_{ss}; \quad \lambda_{ss} = \frac{h_{s0}}{b_{s0}} + \frac{2h_w}{b_{s0} + b_s} + \frac{h_s}{3b_s} \tag{13.13}$$

All geometrical variables in (13.13) are visible in Figure 13.6.

Note: For double layer windings, there is some leakage-flux coupling between the adjacent coils in the same slot pertaining to different phases, but as ($-i_a + i_b$) varies in time its average value might be neglected. Apparently, there are no end coils in the circular shape coils of T-LPMSM; in reality, the turn average length is larger than the πD by $(D + h_s)/D$ and thus there is additional leakage inductance in comparison with F-LPMSMs; however, its effect is notably less. We have lumped up all main flux lines in calculating L_S, L_{dm}, and L_{qm} in (13.7) through (13.9) and thus approximately, the whole inductance is considered; 2D FEM verifications are required to document this assertion thoroughly (i.e., the inclusion of different leakage inductance components). We still need the phase resistance R_S:

$$R_S = \rho_{Co} l_C \frac{k_k n_C j_{Co}}{I_n}; \quad l_c = \pi (D + h_s) \tag{13.14}$$

where

ρ_{Co} is the copper resistivity
j_{Con} is the rated current density for rated RMS phase current I_n
L_C is the turn average length
k_k is the coils/phase

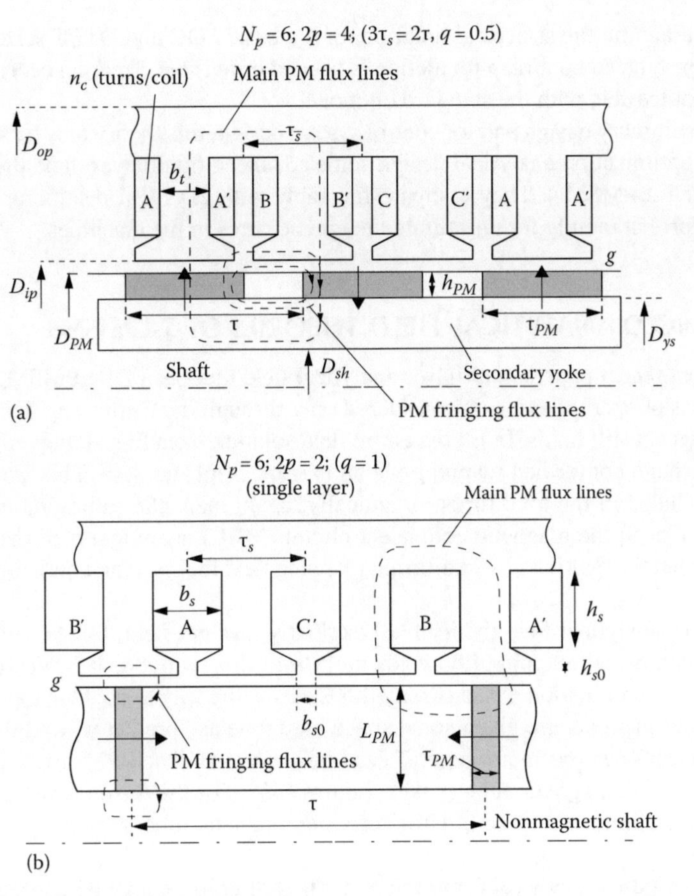

FIGURE 13.6 T-LPMSM: (a) with surface PM-secondary (N_p, $2p=6/4$, $q=0.5$), (b) with IPM secondary (N_p, $2p=6/2$, $q=1$).

13.4 CIRCUIT *dq* MODEL OF T-LPMSM

We use here the same *dq* model and the vector diagram as for F-LPMSMs. Let us summarize here for completeness:

$$\bar{V}_S = I_S R_S + \frac{d\bar{\psi}_S}{dt} + j\omega_r \bar{\psi}_S; \quad \omega_r = \frac{\pi}{\tau} kU \tag{13.15}$$

$k=1$ for PM mover and $k=-1$ for primary mover.

$$\bar{\psi}_S = \psi_{PMd} + L_d I_d + jL_q I_q \tag{13.16}$$

$$F_x = \frac{3}{2}\frac{\pi}{\tau}\left[\psi_{PMd} + (L_d - L_q)I_d\right]kI_q \tag{13.17}$$

$$G_m \frac{dU}{dt} = F_x - F_{load}k - BUk; \quad \frac{dx}{dt} = U \tag{13.18}$$

As the equations are identical and only the expressions of ψ_{PMd}, L_d, L_q are different for T-LPMSMs (in contrast to F-LPMSM), there is no need to repeat here the equivalent circuits along axis d and q

and, the vector diagram, the structural diagram, or the basic FOC and DTFC schemes. The airgap is considered constant, so no airgap regulation is needed or feasible. Treating eccentricity, radial or (and) axial, is not feasible with the standard *dq* model.

While for preliminary design and for control system design, this theory may be satisfactory, even in industry, for optimization analytical design some advanced (multilayer) field theories have been proposed (as for F-LPMSMs). They compare favorably with 2D FEM results for a notably lower computation effort but mainly for unsaturated magnetic cores in the machines.

13.5 ADVANCED ANALYTICAL FIELD THEORIES OF T-LPMSMs

In a few pertinent recent papers, multilayer analytical field theories of T-LPMSMs have been proposed [7,8]. The slot opening effect is considered only through the Carter coefficient, and thus the current sheet concept still holds. In [9] the entire field solution including slot openings, one by one, is considered through conformed mapping via an existing Tool Box [10]. This way in [9] the cogging force is included in the total thrust organically; again magnetic saturation is either constant or ignored. Also in [9] the magnetic equivalent circuit (MEC) analytical field theory is shown to perform worse than the multilayer + conformal mapping field theory, when both are compared with 2D FEM results.

The multilayer analytical field theory [7,8] explicitly lays out field, forces, and circuit parameters while the conformal mapping [9] avoids multilayer theory, but as it is, yields only force via Maxwell tensor. A combination of the two methods seems the way to go. Due to their high degree of generality, both methods are given some space here (one as "predictor" and the other as "corrector"). Let us consider the more general case of a quasi-Halbach PM array inner secondary with magnetic or a nonmagnetic tube (yoke) (Figure 13.7). For the ferromagnetic secondary tube (yoke), region 3 is eliminated ($\mu = \infty$), while for a nonmagnetic tube, region (3) holds from D_{ys} to longitudinal axis.

The field is periodic along axis *x*, but the end effect drag force may be considered by a fudge factor a posteriori placed in the thrust expression [7,8]. Due to the cylindrical symmetry, the magnetic potential holds along polar axis $A(x, r, t)$ and must be calculated first on no load and then on load.

13.5.1 PM Field Distribution

The vector potential equations in cylindrical coordinates (Chapter 1) in the two regions ($\mu_{rec} \approx 1$ for PMs), winding + airgap and PM zones, are

$$\frac{\partial}{\partial x}\left(\frac{1}{r}\frac{\partial}{\partial x}(rA_{1\theta})\right)+\frac{\partial}{\partial r}\left(\frac{1}{r}\frac{\partial}{\partial r}(rA_{1\theta})\right)=0;\quad \text{region 1 (zero currents)} \tag{13.19}$$

$$\frac{\partial}{\partial x}\left(\frac{1}{r}\frac{\partial}{\partial x}(rA_{2\theta})\right)+\frac{\partial}{\partial r}\left(\frac{1}{r}\frac{\partial}{\partial r}(rA_{2\theta})\right)=-\mu_0\nabla\times\bar{\mathrm{M}} \tag{13.20}$$

In cylindrical coordinates, the PM magnetization vector $\bar{\mathrm{M}}$ of the quasi-Halbach array—has two components M_r and M_x:

$$\bar{\mathrm{M}} = \mathrm{M}_r\bar{e}_r + \mathrm{M}_x\bar{e}_x \tag{13.21}$$

The distributions of M_r and M_x are rectangular—periodic along *x* (Figure 13.7b)

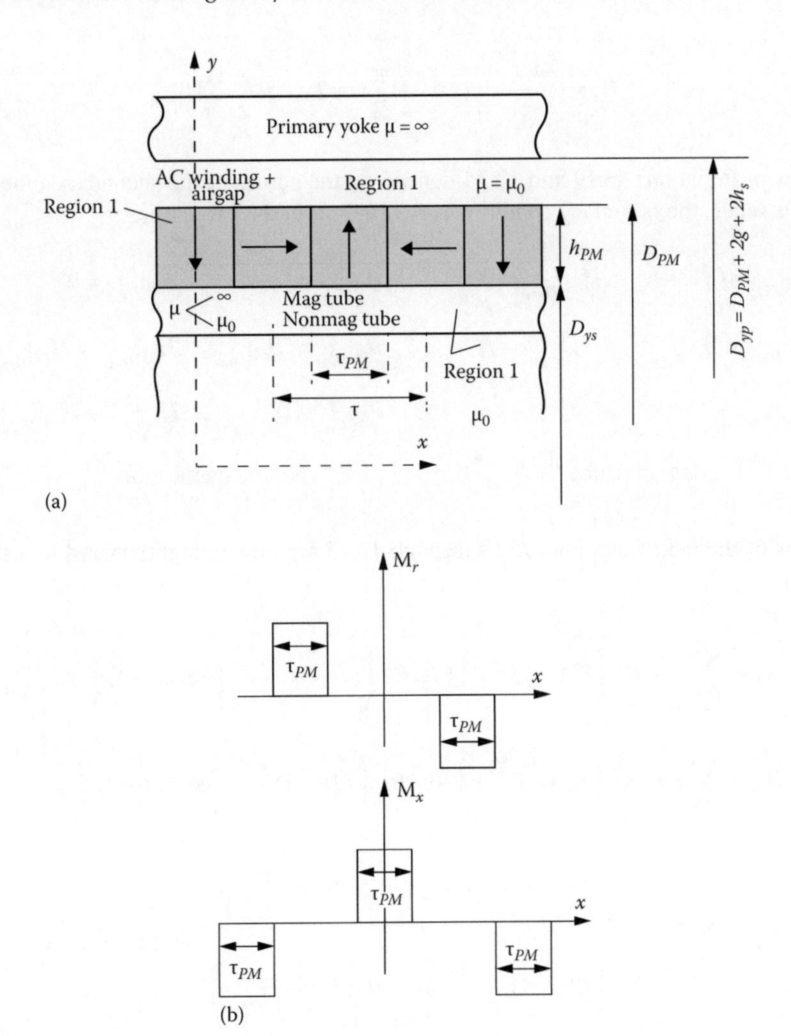

FIGURE 13.7 (a) Field layers and (b) PM magnetization vector components.

After expansion into a Fourier series

$$\mathrm{M}_r = \sum_{\upsilon=1,2}^{\infty} \mathrm{M}_{rn} \sin\left(\frac{2\upsilon-1}{\tau}\right)\pi x; \quad \mathrm{M}_x = \sum_{\upsilon=1,2}^{\infty} \mathrm{M}_{xn} \sin\left(\frac{2\upsilon-1}{\tau}\right)\pi x \tag{13.22}$$

$$\begin{aligned} \mathrm{M}_{rn} &= \frac{4}{\pi}\frac{B_r}{\mu_0(2\upsilon-1)}\sin\left(\frac{(2\upsilon-1)}{2}\pi\right)\sin(2\upsilon-1)\frac{\pi}{2}\frac{\tau_{PM}}{\tau} \\ \mathrm{M}_{xn} &= \frac{4}{\pi}\frac{B_r}{\mu_0(2\upsilon-1)}\cos(2\upsilon-1)\frac{\pi}{2}\frac{(\tau-\tau_{PM})}{\tau} \end{aligned} \tag{13.23}$$

With (13.22 and 13.23) in (13.20) the latter yields:

$$\frac{\partial}{\partial x}\left(\frac{1}{r}\frac{\partial}{\partial x}(rA_{2\theta})\right) + \frac{\partial}{\partial r}\left(\frac{1}{r}\frac{\partial}{\partial r}(rA_{2\theta})\right) = \sum_{\upsilon=1}^{\infty} P_{\upsilon}\cos(2\upsilon-1)\frac{\pi}{\tau}x \tag{13.24}$$

$$P_\upsilon = -\frac{4B_r}{\tau}\sin(2\upsilon-1)\frac{\pi}{2}\sin(2\upsilon-1)\frac{\pi}{2}\frac{\tau_{PM}}{\tau} \quad (13.25)$$

The solutions of Equations 13.19 and 13.24 depend on the nature of the secondary tube that holds the PMs; in essence, the boundary conditions are different in the two cases:

$$\begin{array}{ll|ll}
B_{x1}\big|_{D_{yp}} = 0; & H_{x2}\big|_{D_{PM}} = 0 & B_{x1}\big|_{D_{yp}} = 0; & A_{\theta 3}\big|_{D=0} = 0 \\
B_{r1}\big|_{D_{PM}} = B_{r2}\big|_{D_{PM}}; & H_{x1}\big|_{D_{PM}} = H_{x2}\big|_{D_{PM}} & B_{r1}\big|_{D_{PM}} = B_{r2}\big|_{D_{PM}}; & H_{x1}\big|_{D_{PM}} = H_{x2}\big|_{D_{PM}} \\
 & & B_{r2}\big|_{D_{ys}} = B_{r3}\big|_{D_{ys}}; & H_{x2}\big|_{D_{ys}} = H_{x3}\big|_{D_{ys}}
\end{array} \quad (13.26)$$

Magnetic tube Nonmagnetic tube

The solutions of the field Equations 13.19 through 13.25 are now straightforward by using Bessel functions [7]:

$$B_{1r} = \sum_{\upsilon=1,2}^{\infty}\left[a_{1n}BI_1\left((2\upsilon-1)\frac{\pi}{\tau}r\right) + b_{1n}Bk_1\left((2\upsilon-1)\frac{\pi}{\tau}r\right)\right]\sin(2\upsilon-1)\frac{\pi}{\tau}x$$

$$B_{1x} = \sum_{\upsilon=1,2}^{\infty}\left[a_{1n}BI_0\left((2\upsilon-1)\frac{\pi}{\tau}r\right) - b_{1n}Bk_0\left((2\upsilon-1)\frac{\pi}{\tau}r\right)\right]\cos(2\upsilon-1)\frac{\pi}{\tau}x$$

$$B_{2r} = \sum_{\upsilon=1,2}^{\infty}\left\{\begin{array}{l}\left[F_{An}\left((2\upsilon-1)\frac{\pi}{\tau}r\right) + a_{2n}\right]BI_1\left((2\upsilon-1)\frac{\pi}{\tau}r\right) + \\ \left[-F_{Bn}\left((2\upsilon-1)\frac{\pi}{\tau}r\right) + b_{2n}\right]Bk_1\left((2\upsilon-1)\frac{\pi}{\tau}r\right)\end{array}\right\}\sin(2\upsilon-1)\frac{\pi}{\tau}x \quad (13.27)$$

$$B_{2x} = \sum_{\upsilon=1,2}^{\infty}\left\{\begin{array}{l}\left[F_{An}\left((2\upsilon-1)\frac{\pi}{\tau}r\right) + a_{2n}\right]BI_0\left((2\upsilon-1)\frac{\pi}{\tau}r\right) - \\ \left[-F_{Bn}\left((2\upsilon-1)\frac{\pi}{\tau}r\right) + b_{2n}\right]Bk_0\left((2\upsilon-1)\frac{\pi}{\tau}r\right)\end{array}\right\}\cos(2\upsilon-1)\frac{\pi}{\tau}x$$

Ferromagnetic tube

$$B_{3r} = \sum_{\upsilon=1,2}^{\infty} a'_{3n}BI_1\left((2\upsilon-1)\frac{\pi}{\tau}r\right)\sin(2\upsilon-1)\frac{\pi}{\tau}x$$

$$B_{3r} = \sum_{\upsilon=1,2}^{\infty} a'_{3n}BI_0\left((2\upsilon-1)\frac{\pi}{\tau}r\right)\cos(2\upsilon-1)\frac{\pi}{\tau}x \quad (13.28)$$

Nonmagnetic tube

B_{1r}, B_{1x}, B_{2r}, B_{2x}, have similar expressions as for ferromagnetic tube but with a'_{1n}, b'_{1n}, a'_{2n}, b'_{2n}, instead of a_{1n}, b_{1n}, a_{2n}.

The constants F_{An}, F_{Bn}, a_{1n}, b_{1n}, a_{2n}, b_{2n}, a'_{1n}, b'_{1n}, a'_{2n}, b'_{2n}, a'_{3n} are found from the aforementioned boundary conditions; BI_0, BI_1 are modified Bessel functions of first kind, Bk_0, Bk_1 are modified Bessel functions of the second kind of order 0 and 1 [7];

$$F_{An} = \frac{P_\upsilon}{(2\upsilon-1)\frac{\pi}{\tau}} \int_{(2\upsilon-1)\frac{\pi}{\tau}\frac{D_{ys}}{2}}^{(2\upsilon-1)\frac{\pi}{\tau}r} \frac{Bk_1(x)}{BI_1(x)Bk_0(x)+Bk_1(x)\cdot BI_0(x)}dx$$

$$F_{Bn} = \frac{-P_\upsilon}{(2\upsilon-1)\frac{\pi}{\tau}} \int_{(2\upsilon-1)\frac{\pi}{\tau}\frac{D_{ys}}{2}}^{(2\upsilon-1)\frac{\pi}{\tau}r} \frac{BI_1(x)}{BI_1(x)Bk_0(x)+Bk_1(x)\cdot BI_0(x)}dx \tag{13.29}$$

$$a_{1n} = \frac{A_{1n}}{c_{1n}};\quad b_{1n} = \frac{B_{1n}}{c_{1n}};\quad a_{2n} = -\mu_{rec}\frac{A_{2n}}{c_{5n}};\quad b_{2n} = \frac{-B_{2n}}{c_{10n}}$$

$$B_n = \frac{4B_r}{\mu_{rec}(2\upsilon-1)\pi}\sin(2\upsilon-1)\frac{\pi}{\tau}\left(1-\frac{\tau_{PM}}{\tau}\right)$$

$$c_{1n} = BI_0\left((2\upsilon-1)\frac{\pi}{\tau}\frac{D_{yp}}{2}\right);\quad c_{2n} = Bk_0\left((2\upsilon-1)\frac{\pi}{\tau}\frac{D_{yp}}{2}\right)$$

$$c_{3n} = BI_1\left((2\upsilon-1)\frac{\pi}{\tau}\frac{D_{PM}}{2}\right);\quad c_{4n} = Bk_1\left((2\upsilon-1)\frac{\pi}{\tau}\frac{D_{PM}}{2}\right)$$

$$c_{5n} = BI_0\left((2\upsilon-1)\frac{\pi}{\tau}\frac{D_{PM}}{2}\right);\quad c_{6n} = Bk_0\left((2\upsilon-1)\frac{\pi}{\tau}\frac{D_{PM}}{2}\right) \tag{13.30}$$

$$c_{7n} = BI_1\left((2\upsilon-1)\frac{\pi}{\tau}\frac{D_{ys}}{2}\right);\quad c_{8n} = Bk_1\left((2\upsilon-1)\frac{\pi}{\tau}\frac{D_{ys}}{2}\right)$$

$$c_{9n} = BI_0\left((2\upsilon-1)\frac{\pi}{\tau}\frac{D_{ys}}{2}\right);\quad c_{10n} = Bk_0\left((2\upsilon-1)\frac{\pi}{\tau}\frac{D_{ys}}{2}\right)$$

A_{1n}, B_{1n}, A_{2n}, B_{2n} are found from the matrix:

$$\begin{bmatrix} 1 & -\frac{c_{2n}}{c_{4n}} & 0 & 0 \\ \frac{c_{3n}}{c_{1n}} & 1 & \frac{c_{3n}}{c_{5n}} & \frac{c_{4n}}{c_{10n}} \\ \frac{c_{5n}}{c_{1n}} & -\frac{c_{6n}}{c_{4n}} & 1 & -\frac{c_{6n}}{\mu_{rec}c_{10n}} \\ 0 & 0 & -\frac{\mu_{rec}c_{9n}}{c_{5n}} & 1 \end{bmatrix} \cdot \begin{bmatrix} A_{1n} \\ B_{1n} \\ A_{2n} \\ B_{2n} \end{bmatrix}$$

$$= \begin{bmatrix} c_{3n}F_{An}\left((2\upsilon-1)\frac{\pi}{\tau}\frac{D_{ys}}{2}\right) - c_{4n}F_{Bn}\left((2\upsilon-1)\frac{\pi}{\tau}\frac{D_{ys}}{2}\right) \\ \frac{1}{\mu_{rec}}\left[c_{5n}F_{An}\left((2\upsilon-1)\frac{\pi}{\tau}\frac{D_{ys}}{2}\right) \square\, c_{6n}F_{Bn}\left((2\upsilon-1)\frac{\pi}{\tau}\frac{D_{ys}}{2}\right)\right] \\ \mu_{rec}B_n \\ \end{bmatrix} \tag{13.31}$$

Similarly [7]

$$a_{1n}' = \frac{A_{1n}'}{c_{1n}}; \quad b_{1n}' = \frac{B_{1n}'}{c_{4n}}; \quad a_{2n}' = -\mu_{rec}\frac{A_{2n}'}{c_{5n}}; \quad b_{2n}' = \frac{B_{2n}'}{c_{8n}}; \quad a_{3n}' = -\frac{A_{3n}'}{c_{9n}} \tag{13.32}$$

with A_{1n}', B_{1n}', A_{2n}', B_{2n}', and A_{3n}' from

$$\begin{bmatrix} 1 & -\frac{c_{2n}}{c_{4n}} & 0 & 0 & 0 \\ \frac{c_{3n}}{c_{1n}} & 1 & \mu_{rec}\frac{c_{3n}}{c_{5n}} & -\frac{c_{4n}}{c_{8n}} & 0 \\ c_{1n} & -\frac{c_{6n}}{c_{4n}} & 1 & \frac{c_{6n}}{\mu_{rec}c_{8n}} & 0 \\ 0 & 0 & \mu_{rec}\frac{c_{7n}}{c_{5n}} & 1 & \frac{c_{3n}}{c_{9n}} \\ 0 & 0 & -\frac{c_{9n}}{c_{5n}} & -\frac{c_{10n}}{\mu_{rec}c_{8n}} & 1 \end{bmatrix} \cdot \begin{bmatrix} A_{1n}' \\ B_{1n}' \\ A_{2n}' \\ B_{2n}' \\ A_{3n}' \end{bmatrix}$$

$$= \begin{bmatrix} 0 \\ c_{3n}F_{An}\left((2\upsilon-1)\frac{\pi}{\tau}\frac{D_{ys}}{2}\right) - c_{4n}F_{Bn}\left((2\upsilon-1)\frac{\pi}{\tau}\frac{D_{ys}}{2}\right) \\ \frac{1}{\mu_{rec}}\left[c_{5n}F_{An}\left((2\upsilon-1)\frac{\pi}{\tau}\frac{D_{ys}}{2}\right) \square\, c_{6n}F_{Bn}\left((2\upsilon-1)\frac{\pi}{\tau}\frac{D_{ys}}{2}\right)\right] \\ 0 \\ B_n \end{bmatrix} \tag{13.33}$$

The radial flux density B_r (r=0.025 m—in the airgap) for D_{yp}=0.051 m, D_{pm}=0.049 m, D_{ys}=0.039 m, τ=0.01 m, τ_{PM}=0.006 m, B_r=1.15 T, μ_{rec}=1.05 [7] compares favorably with 2D FEM results (Figure 13.8a), and the beneficial effect of the Halback PM array is very visible in Figure 13.8b.

Similar to the aforementioned, the armature reaction and design optimization may be proceeded [7]; armature field calculation is followed by the computation of self and mutual inductances. The total force pulsations [7] shown in Figure 13.8c prove the power of the multilayer analytical method.

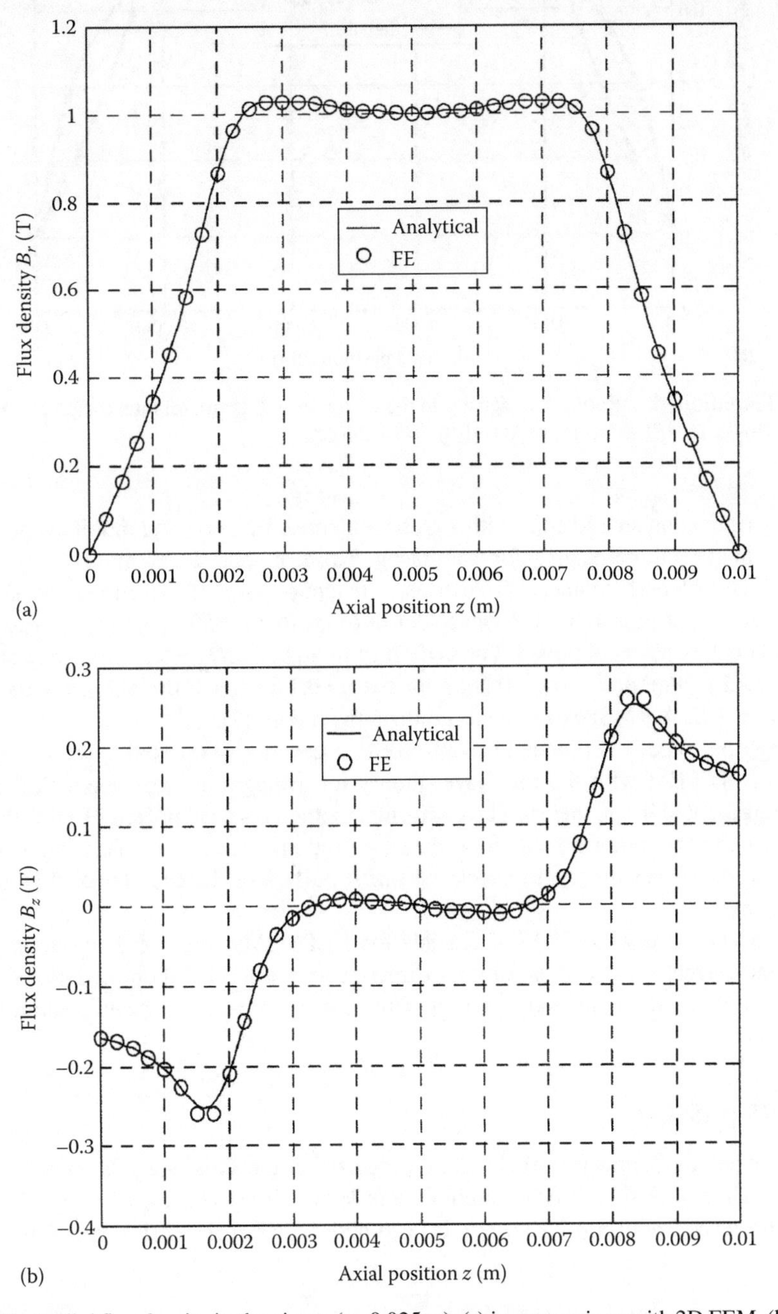

FIGURE 13.8 Radial flux density in the airgap (r=0.025 m): (a) in comparison with 2D FEM, (b) radial PMs versus Halbach array.

(continued)

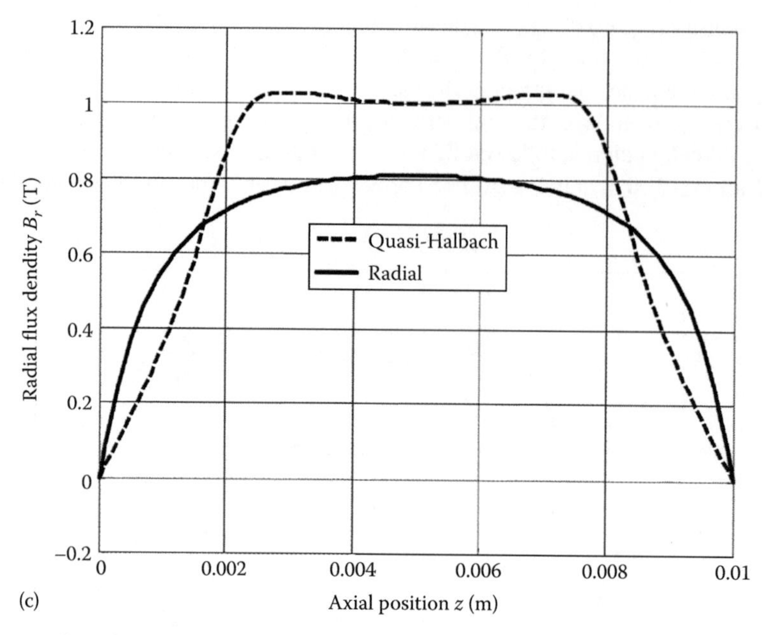

FIGURE 13.8 (continued) Radial flux density in the airgap ($r=0.025$ m): (c) total thrust versus time. (After Wang, J. and Howe, D., *IEEE Trans.*, MAG-41(9), 2470, 2005.)

The design optimization should start with a global objective function and multiple constraints. The variables are, in general, geometric, for example: g, h_{PM}, τ, τ_{PM}, b_{s0}, b_s, τ_S, and D_{ys}, D_{PM}, D_{yp}.

The sensitivity of performance [7] such as, efficiency, power factor, force density (N/m^3 or N/kg of mover, thrust pulsations for sinusoidal current, (for a 250 N average thrust, $U=6$ m/s, for $\tau_{PM}/\tau=0.7$) below 5% is obtained. For $\tau=0.0125$ m and $D_{PM}/D_{yp}=0.6$, efficiency above 94% is obtained for 3×10^5 N/m^3 and above 60 N/kg for cos $\varphi>0.9$ for linear thrust/current dependence up to 250 N at $I_n = 7\sqrt{2}$ A RMS; experiments confirm the theory [7].

As the cogging force was reduced by optimization design (by optimal $\tau_{PM}/\tau=0.7$ mainly), the compliance of 2D FEM with the multilayer theory for average thrust is good. Still the cogging force is not represented in the theory. The conformal method (as used in Ref. [9]) for thrust calculations shows clearly the rather notable total thrust pulsations and may be used, after optimization design through multilayer theory, to check the thrust pulsations, before 2D FEM is used for trial key verifications.

It may be argued that with 2D FEM feasible for T-LPMSMs, and, for given main geometrical parameters and current density as variables, a direct-geometric 2D FEM optimization design is now possible; true, but still to be done and proven. This is very probable for heavily saturated or small $2p$ T-LPMSMs.

13.6 CORE LOSSES

As already noticed, the primary mmf is rich in space harmonics. This aspect is visible in Figure 13.9 for a 9 teeth primary (9 slot/10 pole combination). When the $i_a=I_{\max}$, $i_b=i_c=-I_{max}/2$, the current sheet $J_a(x)$ (the derivative of mmf: $dF(x)/dt$) is decomposed in Fourier series. For phase a only

$$J_a(x)=\sum_{\upsilon'=1}^{\infty} J_{\upsilon'}\sin\frac{2\pi\upsilon'}{N_p\tau_s} \tag{13.34}$$

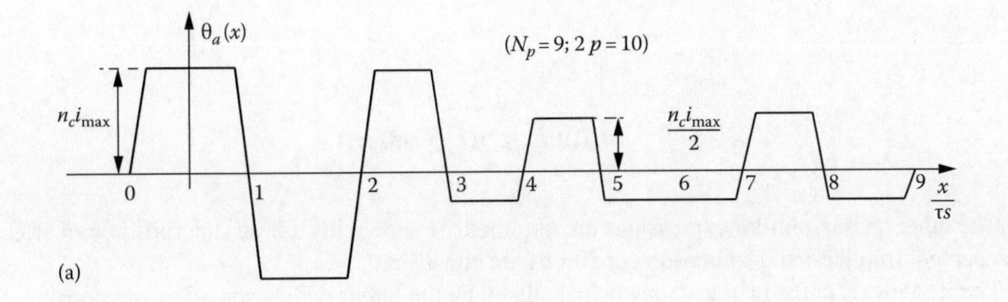

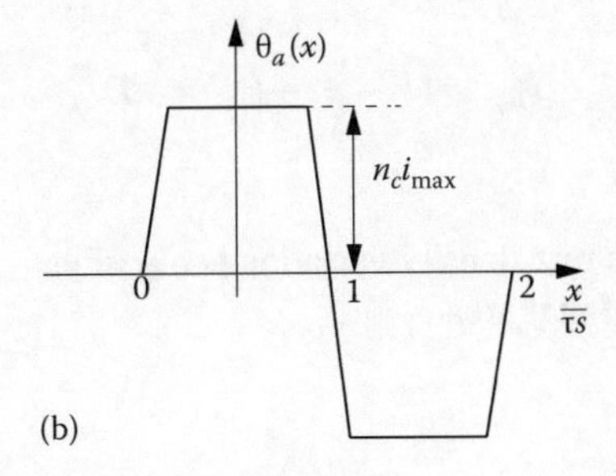

FIGURE 13.9 Primary three-phase mmf versus x when $i_a = i_{max}$, $i_b = i_c = -i_{max}/2a$, phase (a) only (b).

$$J_{\upsilon'} = \frac{2n_c i}{\tau} k_{d\upsilon'} k_{p\upsilon'}; \quad \alpha_{\upsilon'} = \left(\frac{2\pi\upsilon'}{N_p \tau_s} \times \frac{b_{sp}}{2}\right); \quad k_{d\upsilon'} = \frac{\sin \alpha_{\upsilon'}}{\alpha_{\upsilon'}} \tag{13.35}$$

$$k_{p\upsilon'} = \frac{2\tau}{N_p \tau_s}\left[2\sin\left(\frac{\upsilon'\pi}{N_p}\right) - \sin\left(\frac{3\upsilon'\pi}{N_p}\right)\right] \tag{13.36}$$

Following up the calculation of the armature flux density through the multilayer theory (as in the previous paragraph), we can use Equation 13.24 of magnetic vector potential in the airgap (and PM zone) with the following boundary conditions:

$$\begin{aligned} &B_x\big|_{D_{ys}} = 0 \quad \text{for nonmagnetic tube;} \quad B_r\big|_{D_{ys}} = 0 \quad \text{for magnetic support tube;} \\ &H_x\big|_{D_{yp}} = J_{y'(x)} \end{aligned} \tag{13.37}$$

So

$$B_r^p = -\sum_{\upsilon=1}^{\infty} A_n B I_1\left(\frac{2\pi\upsilon'}{\tau_s} r\right)\frac{2\pi\upsilon'}{\tau_s}\cos\frac{2\pi\upsilon'}{\tau_s}x \tag{13.38}$$

$$B_x^p = \sum_{\upsilon=1}^{\infty} A_n B I_0\left(\frac{2\pi\upsilon'}{\tau_s} r\right)\frac{2\pi\upsilon'}{\tau_s}\sin\frac{2\pi\upsilon'}{\tau_s}x \tag{13.39}$$

with

$$A_n = \frac{\mu_0 J_{\upsilon'}}{BI_0((2\pi\upsilon'/\tau_s)D_{yp})(2\pi\upsilon'/\tau_s)}$$

For the other phases, similar expressions are obtained. So, implicitly, phase superposition of effects is accepted. Implicit-fixed saturation conditions are considered.

The iron losses in the primary may be calculated by the rather widely spread expressions:

$$P_{iron} = (P_h + P_e + P_{ad})A_m \pi D_m \tag{13.40}$$

where

A_m is the area where uniform flux density variation is plausible
D_m is the average diameter of that area

$$P_h = k_h f B_m^{\alpha};\quad P_e = \frac{2\pi\sigma_i d^2 f^2}{12\delta_i}\int_{2\pi}\left(\frac{dB}{d\theta_e}\right)^2 d\theta_e;\quad P_{ad} = \sqrt{2\pi}k_e f^{\frac{3}{2}}\int_{2\pi}\left(\frac{dB}{d\theta_e}\right)^{\frac{3}{2}} d\theta_e \tag{13.41}$$

Based on the variation of airgap flux density (13.39), the peak flux density and its time (linear) variation in the primary teeth and primary yoke are calculated; this way the primary core losses may by computed based on the multilayer analytical theory [11]. Finally, the flux density distribution versus time in the primary core during steady-state operation, at different speeds $U=\omega_1\tau/\pi$, may be calculated by 2D FEM. The results for the example in the previous paragraph ($F_x=250$ N, $U=6$ m/s) are shown in Figure 13.10 [10].

Though the on load primary core losses are larger than the ones under no load, the difference is less than 15%; also, they increase almost linearly with speed; finally, at 6 m/s they are 30 W, which

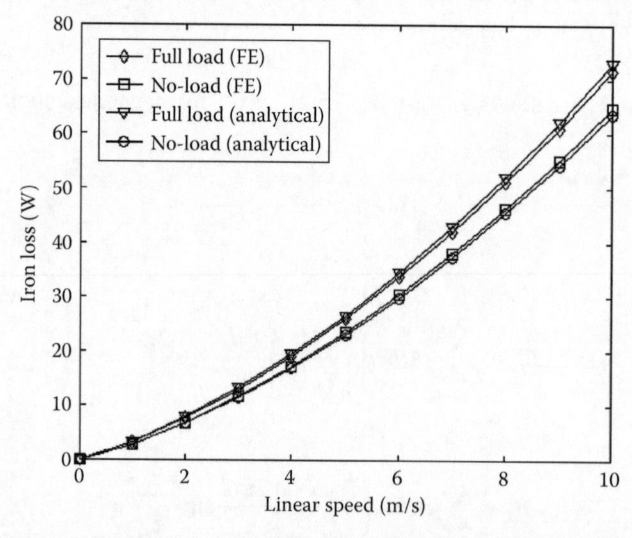

FIGURE 13.10 Primary core losses variation with speed. (After Amara, Y. et al., *IEEE Trans.*, MAG-41(4), 989, 2005.)

is about 2% of rated power ($F_xU = 1500$ W) ; the calculated efficiency [7] in the previous paragraph implies 6% of copper losses. So primary core loss under load is worthy of consideration in high efficiency T-LPMSMs. There are still the PM-secondary losses, due to slot openings and due to primary mmf space harmonics field. But the volume of the interior PM-secondary is in general less than 33% of the outer primary core volume. Thus, the PM-secondary losses are of less importance. A direct use of 2D FEM to calculate the flux density distribution and the core losses is recommended for the scope.

13.7 CONTROL OF T-LPMSMS

By T-LPMSM control, we mean here position, speed, or thrust control for motoring and regenerative braking of direct linear motion drives. Also, a kind of different control is needed for generators with imposed sinusoidal linear motion pulsations at given mechanical frequency such as with free-piston engine prime mover (stroke in the range of $70 \div 150$ mm, 30 Hz) or in wave energy conversion (strokes of $1 \div 3$ m and mechanical frequency of $1 \div 2$ Hz). Let us consider

- Field oriented control (FOC)
- Direct thrust and flux control (DTFC)

13.7.1 Field-Oriented Control

Due to the limited travel length (below 3 m in general) flux weakening is hardly necessary. Thus, unless an IPM secondary is used, $I_d^* = 0$ in FOC. For IPM secondary, exploiting maximum thrust current condition leads to

$$2I_d^{*2} + I_d^* \frac{\Psi_{PM}}{L_d - L_q} = 0; \quad L_d < L_q; \quad I_d^* < 0 \tag{13.42}$$

With L_d and L_q constant, for given stator current I_d^* is

$$I_d^* = \frac{-\Psi_{PM}(L_d - L_q)}{4} - \frac{1}{4}\sqrt{\left(\frac{\Psi_{PM}}{L_d - L_q}\right)^2 + 8I_S^{\,2}}; \quad I_d^* > -\frac{|\bar{I}_S|}{\sqrt{2}} \tag{13.43}$$

So, with controlled I_q^* and $|\bar{I}_S|$ required, I_d^* is easily calculated from (13.43); an expression that may be further simplified (see Figure 13.11); correction for magnetic saturation (with its cross-coupling effect) in IPM secondaries requires parameter online estimation and correction techniques, unless robust position control is applied.

For position tracking, a linear position sensor is necessary in general. Improving the position tracking may be approached many ways, but feed-forward or feedback disturbance observers [12] may be adequate, because in T-LPMSM not only temperature and magnetic saturation vary in time, but also the airgap, and, of course, the load. The generic FOC for T-LPMSMs, shown in Figure 13.12, is characterized by

- The inclusion of a maximum thrust/current, I_d^* referencer ($I_d^* = 0$ for the surface PM secondaries); the fast thrust response and the limited travel exclude, in general, flux weakening control (maximum thrust/flux control).
- A simplified online disturbance compensator based on the motion equation is included to quicken the position (and thrust) response in the presence of load force disturbance.

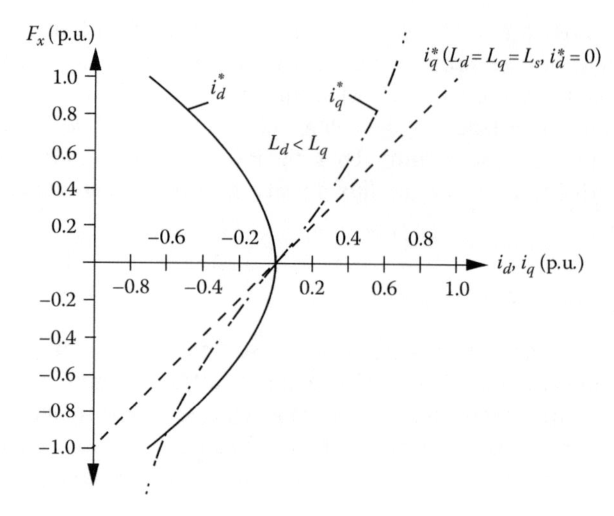

FIGURE 13.11 Maximum thrust/current conditions for IPM and SPM secondaries.

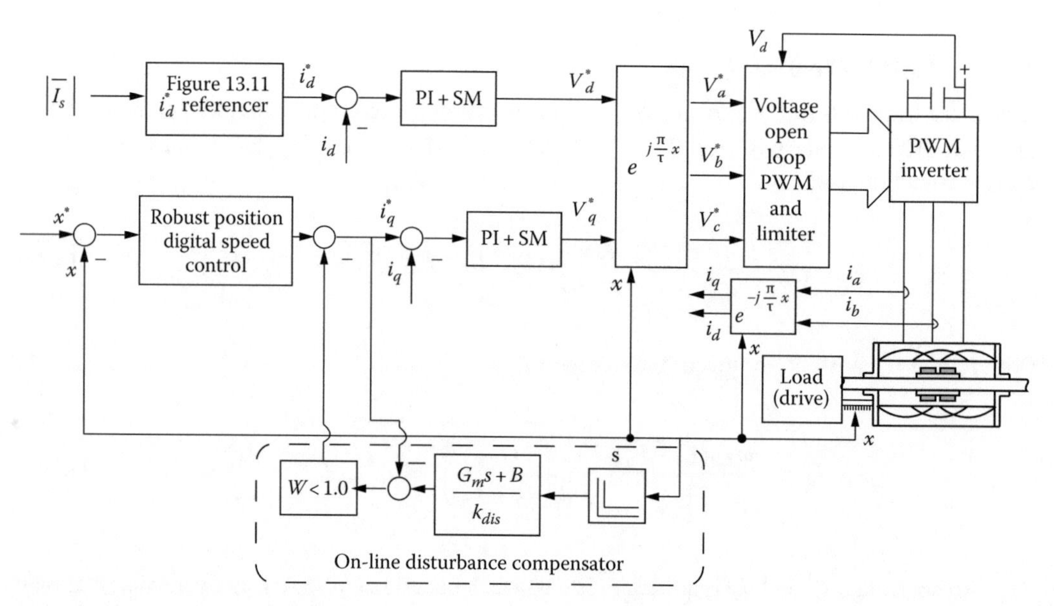

FIGURE 13.12 FOC of T-LPMSM.

- The PI + SM dc current regulators are considered sufficiently robust to allow—at rather low speeds—the elimination of motion voltages compensation; current limiters should be added, however.
- A reference voltage limiter is added, based on dc voltage (V_{dc}) knowledge, to secure controlled operation; the open loop PWM allows for rather sinusoidal voltages for T-LPMSM control.
- FOC may be used also for the control of an active suspension damper for cars or for a battery backup dc power source as generator. But, in this case, the position regulator is eliminated and direct current I_q^* control is operated based on the dc voltage and dc load current (Figure 13.13); the scheme works even for generator mode with no battery backup, but a capacitor filter, instead (Figure 13.13b).

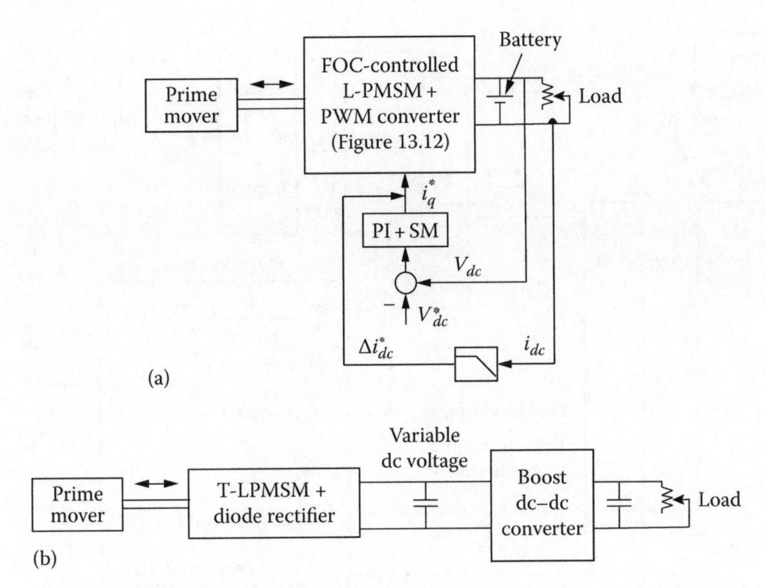

FIGURE 13.13 Generic generator T-LPMSM mode control for dc output with battery backup, (a) diode rectifier + boost dc–dc converter, (b).

- For a wave generator, the strong variations (with position) of terminal voltages might need an additional dc–dc converter for voltage boost and stabilization. In the latter case, it is possible to use a diode rectifier (instead of PWM converter) and thus "move" the output control to the boost dc–dc converter, to yield a constant dc output voltage source from the T-LPMSM as generator (Figure 13.13b)
- Direct active power and dc voltage control may replace FOC for generator mode with PWM converter.

13.7.2 Direct Thrust and Flux Control

When thrust control is needed, or for motion sensorless control, DTFC may be applied. As known, the cogging force may impede on control precision, vibration, and noise, and its attenuation by proper control may be achieved by a wide enough frequency band regulator of the thrust.

The cogging force versus position may be inserted as a table from FEM calculations; other force disturbance may be included also in the thrust control. However, in this case, a state observer is needed to estimate relative position $\hat{x}$ speed $\hat{u}$, flux $\hat{\psi}_S$ thrust $\hat{F}_x$, and dynamic force (Figure 13.14).

- The state observer estimates first the stator flux $\hat{\psi}_S$; only a simplified first-order delay filter is used, but a secondary order one may be put into service, to deal with the offset etc; a combined voltage–current observer may be used to work better at small speed, but the airgap (stator) irregularities would make the latter less robust. Ultimately, even the influence of airgap variation on inductance L_s may be estimated if FEM inductance/airgap curves are available.
- The concept of active flux ($\bar{\psi}_d^a$) is used here again [13]

$$\bar{\psi}_d^a = \hat{\bar{\psi}}_S - L_q \bar{I}_S = \left[\bar{\psi}_{PMd} + (L_d - L_q) I_d\right]\left(\cos\frac{\pi}{\tau}\hat{x} + j\sin\frac{\pi}{\tau}\hat{x}\right) \tag{13.44}$$

$\hat{\bar{\psi}}_d{}^a$ is aligned along PM axis and thus the position $\hat{x}$ reflects PM axis position with respect to phase a axis and thus $d\hat{x}/dt = \hat{u}$, the mover speed, both in steady state and during transients, in contrast to stator flux vector speed.

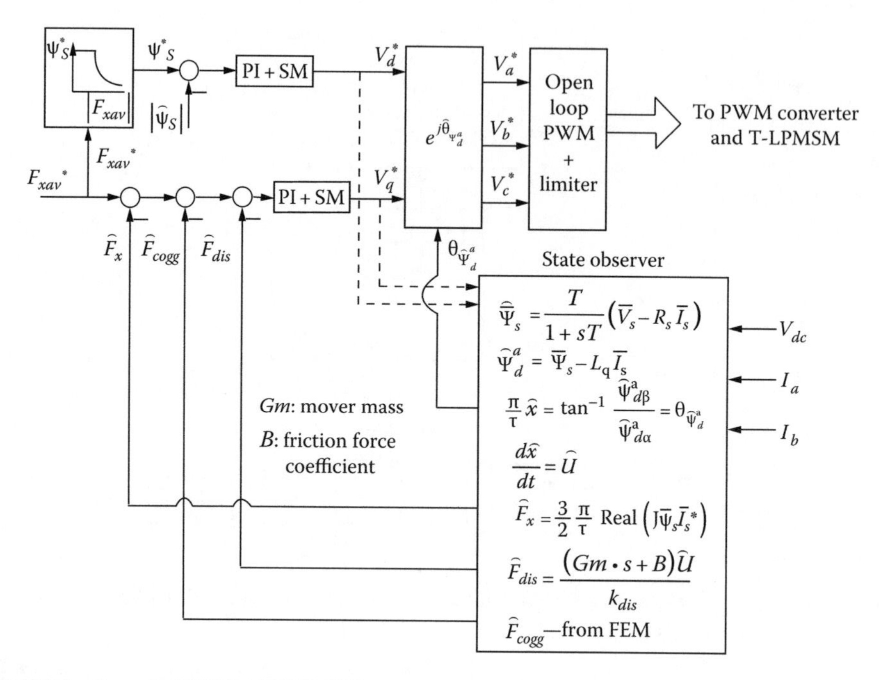

FIGURE 13.14 Generic DTFC of T-LPMSM.

- The speed $\hat{u}$ may be estimated from estimated position by a digital filter such as the PLL filter; at very low speeds, an additional speed estimation may be obtained through an observer based on motion equation and then added in a sharing mode.
- Dynamic force disturbance F_{dis} is included to speed up the thrust (and speed) response; a simple version of it is given in Figure 13.14.
- The FEM calculated function $F_{cog}(\hat{x})$ is stored in a table with, say, linear interpolation; another fast filter may be added here.
- To attenuate the total thrust pulsations, the thrust (and flux) regulators have to be fast and robust.
- For position sensorless control with DTFC, a position regulator is added; also, to "translate" the relative estimated position $\hat{\theta}_{\psi_d^a}$ into an absolute one, periodic proximity sensors along the way are needed with the knowledge of where the mover is at time zero; in many applications (such as robotic surgery), relative position control is, however, sufficient.
- The stator flux reference ψ_S^* is made a function of thrust to yield low enough losses at small loads.
- Open loop PWM of voltage with limiter provides for all available control effort at any time; flux weakening is not embedded in $\psi_S^*(F_x^*)$ referencer, but the voltage limiter does its job when needed (voltage limitation).

Cogging force attenuation in thrust via FOC of T-LPMSM [14,15] (Figure 13.15) shows low thrust ripple, as expected, and can be "transplanted" to F-LPMSM control.

13.8 DESIGN METHODOLOGY

Let us consider a T-LPMSM destined to pump the liquid in a limited diameter (60 mm) pipe. Trapezoidal current control is adopted as three low cost (Hall) proximity sensors are embedded in the short primary stack, for safe phase commutation; also, proximity sensors indicate braking toward end of travel.

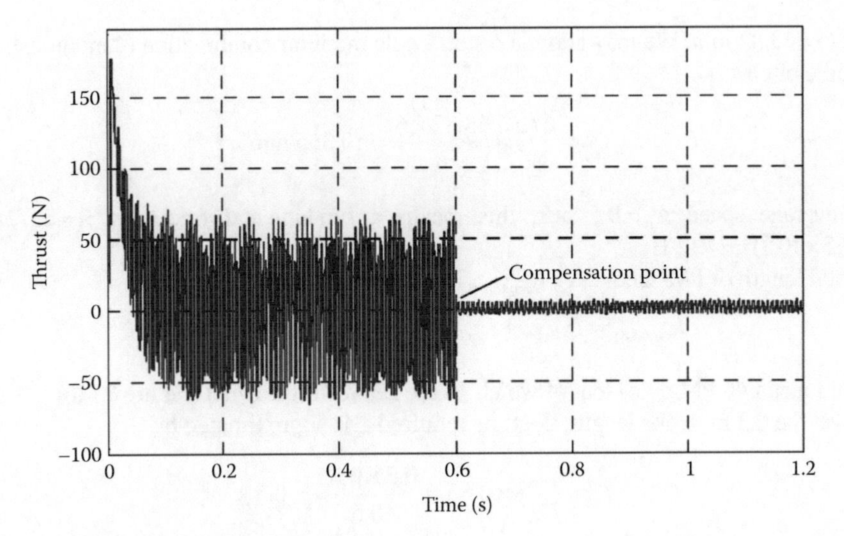

FIGURE 13.15 T-LPMSM no load thrust without and with cogging force compensator. (After Zhu, Y.-W. and Cho, Y.-H., *IEEE Trans.*, MAG-43(6), 2537, 2007.)

The complete specifications are as follows:

- Maximum outer (stator) primary diameter $D_{op}=60$ mm

Stroke length: $l_{stroke}=0.3$ m

- Average linear speed: $u_{av}=0.5$ m/s
- Mode of operation: oscillatory
- Thrust: 100 N continuous duty; 175 N peak thrust (for 33% duty cycle)
- $V_{dc}=300$ V

13.8.1 Design of Magnetic Circuit

To use the pipe walls for heat dissipation, an outer primary configuration of T-LPMSM is adopted. Also, a short primary is chosen to provide high efficiency; the long PM-secondary (stroke length+ primary length) is not excessive in terms of weight and cost. To simplify the construction, disk-shaped laminations are used to build the primary core; four radial slits will reduce the core losses to small values. This implies open slots in the primary. As the average speed is low, the pole pitch should be low to allow a reasonably thin primary yoke. On the other hand, as discussed in Ref. [3], a good ratio between airgap diameter D_{ip} and outer diameter of primary D_{oup} is around 0.6 and thus $D_{ip} \approx 0.56\ D_{oup}=0.56\times 60=33.6$ mm. Adopting a rather large peak thrust force density $f_{xk}=$ 1 N/cm² (for the small airgap diameter considered), we may calculate the approximate primary length $l_{primary}$:

$$l_{primary} = \frac{F_{xk}}{f_{xk}\pi D_{ip}} = \frac{175}{2\times\pi\times 0.036\times 10^4} = 0.166 \text{ m} \qquad (13.45)$$

Now a pole pitch has to be adopted. In general, as the airgap $g=1$ mm (at least) for high thrust density (without PM demagnetization) and the SPM radial depth $h_{PM}=4$ mm, the pole pitch has to be at least two times the magnetic airgap, $g + h_m=5$ mm, to avoid very large PM fringing through slots openings. Consequently, we choose a 2×6 teeth structure with a tooth pitch

$\tau = l_{primary}/6 = 13.82$ mm. We may adopt a 6 slot/8 pole modular combination (2 modules), and thus the PM pole pitch τ is

$$\tau = \frac{6\tau_s}{8} = \frac{13.82 \times 6}{8} = 10.365 \text{ mm} \tag{13.46}$$

For the average speed $u_{av} = 0.5$ m/s, this means a fundamental frequency $f_1 = u_{av}/2\tau_{PM} = 0.5/(2 \times 10.365 \times 10^{-3}) = 24.12$ Hz.

The total length of PM-secondary $l_{secondary}$ is

$$l_{secondary} = l_{primary} + l_{stack} = 2 \times 0.083 + 0.30 \approx 0.47 \text{ m} \tag{13.47}$$

This would mean 46 PM poles (out of which 16 are active all the time) (Figure 13.16).

To travel the 0.3 m stroke length, the time required t_s is approximated by

$$t_s \approx \frac{l_{stroke}}{U_{av}} \times 0.85 = \frac{0.3 \times 0.85}{0.5} = 0.5 \text{ s} \tag{13.48}$$

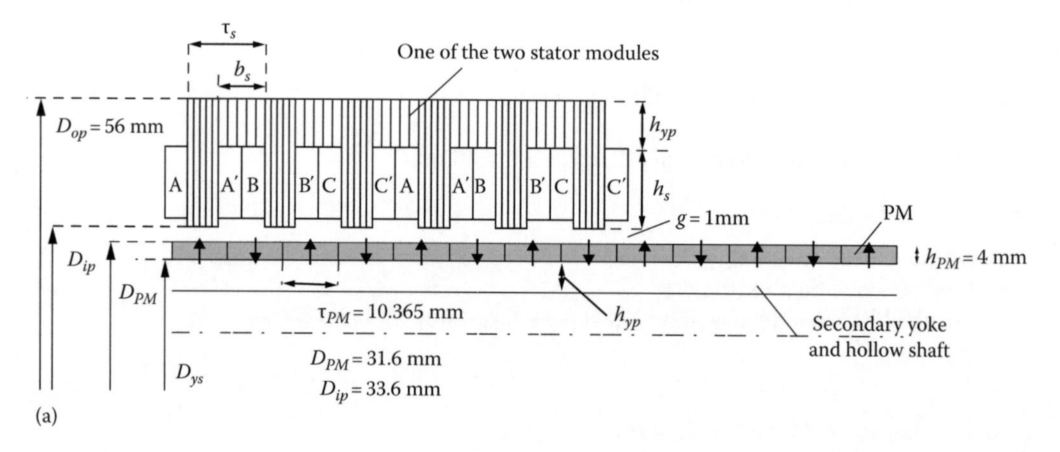

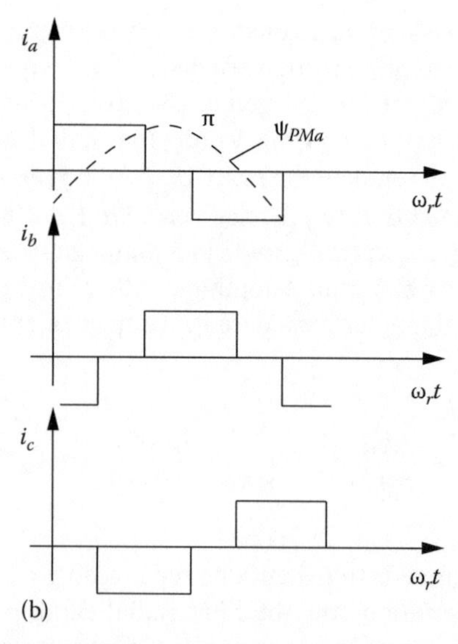

FIGURE 13.16 T-LPMSM: main geometry (a) ideal currents (b).

This corresponds to $t_s f_1 = 0.5 \times 24 = 8$ periods; as the acceleration to speed is considered rather fast, we may approximately consider that the T-LPMSM travels 8 periods under steady state for each stroke length.

13.8.2 Airgap Flux Density B_{gSPM}

According to (13.3)

$$B_{gSPM} = B_r\left(1 - \frac{(2gk_C + h_{PM})}{D_{ip}}\right)\frac{1}{1+\mu_{rec}\left(gk_C/\tau\right)\left(1-(2gk_C + h_{PM})/D_{ip}\right)}\frac{1}{(1+k_{fringe})(1+k_S)}$$

$$= 1.2\left(1 - \frac{2\times1\times1.25+4}{33.6}\right)\frac{1}{1+1.05\times(1\times1.25/10.365)(1-(2\times1\times1.25+4)/33.6)}\frac{1}{(1+0.1)(1+0.1)}$$

$$= 0.7284\,\text{T} \tag{13.49}$$

SmCo5 PMs, with $B_r \approx 1.2$ T, $\mu_{rec} = 1.05$ (linear demagnetization characteristic), are used to avoid corrosion effect, so typical in a pipeline.

The PM radial thickness $h_{PM} = 4$ mm and the PMs cover the whole PM pole span ($\tau_{PM} = \tau$); the PM fringing coefficient $k_{fringe} = 0.1$ and so the magnetic saturation coefficient $k_S = 0.1$ (despite the large total airgap ($g + h_{PM}$)), because the secondary yoke will be sized thin to reduce mover weight and cost). The Carter coefficient k_C was assigned a value of 1.25. The aforementioned analytical results may be easily verified by 2D FEM and corrected as required, to provide safe ground for the next steps in design.

13.8.3 Slot mmf for Peak Thrust

For trapezoidal current control, only two phases conduct at a time (except for phase commutation interval). In our case, that means 8 (of 12) primary coils are active (n_c turns/coil). Consequently, the thrust, written using the Lorenz force formula, is

$$F_{xk} = B_{gSPM} 8 n_c I \pi D_{ip} = 0.7284 \times 8 \times \pi \times 0.036 \cdot n_c \cdot I_p = 0.615 \cdot n_c \cdot I_p \tag{13.50}$$

For $F_{xk} = 175$ N it follows that

$$n_c \cdot I_p = \frac{175}{0.615} = 284.6\ \text{A} \cdot \text{turns/coil} \tag{13.51}$$

A peak current density may be adopted after calculating the RMS value corresponding to it:

$$I_{pRMS} = I_p\sqrt{\frac{2}{3}\times \text{duty cycle}} = I_p\sqrt{\frac{2}{3}\times 0.33} = 0.47 \cdot I_p \tag{13.52}$$

As the continuous duty cycle thrust is almost two times smaller, I_{pRMS} may be used for the machine design, to be valid for both situations. With current density $J_{Con}=8$ A/mm² for I_{pRMS}, the cross section of a slot A_{slot} (which hosts two coils) is

$$A_{slot} = \frac{2n_c I_p \cdot 0.47}{J_{Con} \cdot k_{fill}} = \frac{2\times 284.6\times 0.47}{8\times 0.55} = 60.8\ \text{mm}^2 \tag{13.53}$$

The slot filling factor $k_{fill}=0.55$ with preformed circular coils. But the slot width is only $b_s=\tau_s 0.55=13.82\times 0.55=7.6$ mm.

So the slot depth h_s is

$$h_s = \frac{A_{slotu}}{b_s}+1 = \frac{60.8}{7.6}+1 = 9\ \text{mm} \tag{13.54}$$

Now the primary yoke depth h_{yp} has to be verified first:

$$h_{yp} = \frac{(D_{op}-D_{ip})}{2} - h_s = \frac{(60-33.6)}{2} - 9 = 4.2\ \text{mm} \tag{13.55}$$

This may seem small but as the average diameter of the primary yoke is large, $D_{av\ yp}=D_{op}-h_{yp}=55.8$ mm, the area for yoke flux return is rather large. At no load B_{yp0} is

$$B_{ypo} \approx \frac{B_{gSPM}\cdot \tau\pi D_{ip}}{\pi D_{av\,ip} h_{yp}} = \frac{0.7284\times 10.35\times \not{\pi}\times 33.6}{\not{\pi}\times 55.8\times 4.2} = 1.082\ \text{T} \tag{13.56}$$

The primary peak current will produce a peak armature reaction flux density B_{ap}:

$$B_{ap} \approx \frac{\mu_0 n_c I_p}{2(g+h_{PM})} = \frac{1.256\times 10^{-6}\times 284.6}{2(1+4)\times 10^{-3}} = 35.74\times 10^{-3}\ \text{T} \tag{13.57}$$

This shows that the 4.2 mm yoke in the primary has to be kept that "large" —4.2 mm—mainly due to mechanical strength reasons. In the secondary, however, the yoke thickness h_{ys} is (from no load conditions)

$$h_{ys} \approx \frac{B_{gPM}\tau\pi D_{ip}}{B_{ys}\pi(D_{ip}-2h_{PM}-2g-h_{ys})} = \frac{0.7284\times 10.365\times \not{\pi}\times 33.6}{1.8\times \not{\pi}\times (33.6-10-h_{ys})} \tag{13.58}$$

So, even with $B_{ys}=1.8$ T the secondary yoke should be at least 7 mm thick. This leaves room for a hole (if any) in the solid back iron of secondary of only 9.6 mm.

13.8.4 Circuit Parameters

The low outer diameter of the primary (or of the pipe that holds the T-LPMSM) indicates that the tubular structure should, if used for such situations, target small thrust applications. We may now calculate the emf per phase:

$$E = 4n_c \cdot B_{gSPM}\cdot \pi \cdot D_{ip}\cdot U = 4\times 0.7284\times \pi\times 0.0336\times 0.5\times n_c = 0.4826\times n_c \tag{13.59}$$

The machine inductance $2L_S$ for two phases in series, with its two main components, is

$$2L_S = 2L_l + 2L_{m\,phase}. \tag{13.60}$$

The mutual inductances of the two phases in counter-series seem to cancel each other; this is why only $L_{m\,phase}$ is considered here ($k_q = 1$ (13.8)):

$$L_{m\,phase} = \frac{\mu_0 \times (4 \times n_c \times 0.867)^2 \times 0.010365 \times \pi \times 0.0336 \times (1+0.15)}{2 \times 8 \times 1.25 \times (1+4) \times 10^{-3}} = 0.7687 \times 10^{-6} n_c^{\,2} \tag{13.61}$$

From (13.11)

$$L_l = \mu_0 \pi (D_{ip} + h_s) 4 \cdot n_c^{\,2} \cdot \lambda_{ss} = \mu_0 \pi (0.0365 + 0.009) \times 4 \times 0.6158 \times n_c^{\,2} = 0.442 \times 10^{-6} \times n_c^{\,2} \tag{13.62}$$

With λ_{ss}

$$\lambda_{ss} = \frac{1}{b_s} + \frac{h_s - 1}{3b_s} = \frac{1}{7.6} + \frac{8}{3 \times 7.6} \approx 0.6158 \tag{13.63}$$

Finally, from (13.12)

$$R_s = \rho_{Co} \cdot \frac{l_c 4 n_c^{\,2} J_{Con}}{I_n n_c} = \frac{2.1 \times 10^{-8} \times \pi (0.0365 + 0.009) \times 4 \times 8 \times 10^6}{140} \times n_c = 6.857 \times 10^{-4} \times n_c^{\,2} \tag{13.64}$$

13.8.5 Number of Turns per Coil N_c (All Four Circular Shape Coils per Phase in Series)

The two phases in conduction equation are

$$V_{dc} = 2R_s i + 2E + 2L_s \frac{di}{dt} \tag{13.65}$$

In order to provide flat top current at $U = 0.5$ m/s for peak thrust, current chopping has to be allowed; if the hysteresis band of current control is 5% of peak current and the switching frequency in the PWM converter $fs = 10$ kHz Equation 13.65 becomes (for $V_{dc} = 300$ V)

$$300 = \left(2 \times 6.857 \times 10^{-4} \times 284 + 2 \times 0.4826 + 2 \times (0.7687 + 0.442) \times 10^{-6} \times 0.05 \times \frac{284}{10^{-4}} \right) \cdot n_c$$

$$= (0.38076 + 0.9652 + 0.34173) \cdot n_c \tag{13.66}$$

So the number of turns per coil $n_c = 178$ turns/coil.

The wire gauge is obtained from

$$d_{wire} = \sqrt{\frac{4}{\pi}\frac{I_n}{J_{Con}}} = \sqrt{\frac{4}{\pi}\frac{(n_c I_n / n_c)}{J_{Con}}} = \sqrt{\frac{4}{\pi}\frac{(274/2)/178}{8\times10^{-6}}} \approx 0.36 \text{ mm} \quad (13.67)$$

The rated current

$$I_n = (284/2)/n_c = (284/2)/178 = 0.8 \text{ A} \quad (13.68)$$

13.8.6 Copper Losses and Efficiency

The rated losses p_{Con} are

$$p_{Con} = 2R_s I_n^2 = 2\times6.857\times10^{-4}\times178^2\times0.8^2 = 27.8 \text{ W} \quad (13.69)$$

The electromagnetic P_{elm} power

$$P_{elm} = F_{xn}\cdot U = 100\times0.5 = 50 \text{ W} \quad (13.70)$$

So, neglecting other losses, the efficiency would be

$$\eta_n = \frac{P_{elm}}{P_{elm} + p_{Con}} = \frac{50}{50+27.8} = 0.6426 \quad (13.71)$$

The electric time constant

$$T_e = \frac{L_S}{R_S} = \frac{1.21\times10^{-6}\times n_c^2}{6.857\times10^{-4}\times n_c^2} = 1.765\times10^{-3} \text{ s} \quad (13.72)$$

This small time constant indicates that, for efficient (low band width) current chopping, at least 10 kHz switching frequency is required.

Note: The performance may seem acceptable at this low speed (0.5 m/s) but still other linear PM motors—with higher frequency (as those treated in the next chapter)—have to be investigated for small speeds, in tubular but especially in flat LPMSM configurations.

Once a preliminary design attempt has been made, an optimization design should follow; finally, 2D FEM key verifications or direct geometrical optimization refinements may be performed; all these tasks are, however, beyond our scope here.

13.9 GENERATOR DESIGN METHODOLOGY

Free-piston energy converters such as Stirling engines [22] may be used in series HEVs [16,17]. Also, wave energy conversion through T-LPMSMs, despite 0.5–1.0 m/s speeds and powers up to hundreds of kW for travel length of 3÷4 m or for smaller powers (10–15 kW), have been proposed recently [16–22]. Both flat and tubular LPMSMs have been investigated for the scope. Moreover, T-LPMSMs have been proposed for low frequency active dampers in cars [21].

Using two (three)-phase T-LPMSMs is natural for travels that go well above 6÷10 pole pitches (above 30 mm in general). In what follows the case of a linear wave—energy generator is approached due to its potential (7 kW mechanical power capacity for 1 m of sea shore around Japan, e.g., [20]).

The challenges are related to the low speed and super high thrust levels encountered. The tubular structure—with its ideally zero resultant radial force on the mover—makes it suitable for the scope. Let us consider a super large power (250 kW, 1 m/s), such as a case study, starting from Ref. [29], where, however, cogging force is investigated in a double-sided flat LPMSG.

The design has quite a few critical issues to tackle, such as small surface thrust density in N/cm^2 is not enough to build an economically viable solution; in our view, at least $6 \div 7$ N/cm^2, even $9 \div 10$ N/cm^2, would be more adequate. Such a high thrust density comes with two heavy "prices":

- A sizeable quantity of PMs and copper losses
- Large voltage regulation, which impedes the inverter kVA ratings

It is decided here, as a compromise, to choose a surface PM configuration (due to high thrust density) and low inductance.

The mechanical airgap for such an application should not go below $g=3$ mm. To avoid PM demagnetization and provide for large thrust, the radial thickness of the PMs should be $h_{PM} \geq 12$ mm. But the slot pitch should be large enough to allow a small enough PM flux fringing. For that, $\tau_s > 40$ mm. Let us consider $\tau_s = 60$ mm; in this case, a tentative primary frequency at $U_n = 1$ m/s is

$$f_{ni} \approx \frac{U_n}{2\tau} > \frac{1}{2 \times 0.06} \approx 8.33 \text{ Hz} \tag{13.73}$$

As $\tau \neq \tau_s$ this is only an orientative value.

This is rather low frequency especially for the three-phase PWM converter but still perhaps acceptable.

Let us consider $f_{xn} = 6$ N/cm^2; efficiency $\eta_n = 0.85$ (due to very small speed); the total thrust F_{xn} is

$$F_{xn} = \frac{(P_n/\eta_n)}{U_n} = \frac{250 \times 10^3/0.85}{1} = 294 \text{ kN} \tag{13.74}$$

Therefore, the total short primary outer stator area A_{airgap} is

$$A_{airgap} \approx \frac{F_{xn}}{f_{xn}} = \frac{294 \text{ kN}}{6 \times 10^4} = 4.9 \text{ m}^2 \tag{13.75}$$

With an airgap diameter $D_{ip} = 1.1$ m, the length of the primary $l_{primary}$ is

$$l_{primary} = \frac{A_{irgap}}{\pi D_{ip}} = \frac{4.9}{\pi \times 1.1} = 1.42 \text{ m} \tag{13.76}$$

Now we have to choose the number of short primary (stator) teeth:

$$N_p = \text{Integer}\left(\frac{l_{primary}}{\tau_s}\right) = \text{Integer}\left(\frac{1.42}{0.06}\right) \approx 24 \tag{13.77}$$

To reduce cogging force, we may choose two modules of 12 slots (teeth) with 14 PM poles (each). Therefore, we do have 24 teeth in the short primary (stator) that correspond to 28 PM poles (per same total length). So the PM pole pitch τ_{PM}, which in fact defines the primary frequency, is

$$\tau_{PM} = \tau_s \times \frac{24}{28} = 60 \times \frac{24}{28} = 51.43 \text{ mm} \tag{13.78}$$

To further reduce the cogging force, the PM span/PM pole pitch is optimized first; then the two primary halves of 12 teeth/14 PM poles may be shifted optimally by about $\tau_{PM}/6$, in general.

Still to be decided is if open slots may be used; here lies another compromise: open slots lead to large cogging force but semiclosed slots, which lead to smaller cogging force, experience large PM fringing (lower thrust) and higher leakage (and total) machine inductance.

Even if the frequency f_{nf}

$$f_{nf} = \frac{U_n}{2\tau_{PM}} = \frac{1}{2 \times 0.05143} = 9.72 \text{ Hz} \tag{13.79}$$

the armature reaction I_nL_s/ψ_{PM} ratio decides the power factor of the machine; ψ_{PM} is the PM flux linkage in a phase; with large force density I_nL_s increases, and thus a large L_s may not be tolerable. So, as the $h_{PM} + g = 12 + 3 = 15$ mm, it is felt here that, with a stator tooth width of $b_t = 35$ mm, and a slot width $b_s = \tau_s - b_t = 60–35 = 25$ mm, the PM span/PM pole pitch may be 40/51 = 0.77 and thus a good compromise may be achieved. Only 2D FEM key verifications could lead to an optimal solution here, based on a comprehensive fitting (objective) function. The design here serves only as a hopefully realistic start-up (Figure 13.17).

With strong PMs of the future ($B_r = 1.4$ T, $\mu_{rec} = 1.07$), already in the labs the airgap flux density produced by PMs may be calculated here ignoring the PM-secondary curvature $(D_{ip}) >> (g + h_{PM})$:

$$B_{gPM} = B_r \cdot \frac{h_{PM}}{h_{PM} + g} \cdot \frac{1}{(1 + k_{fringe})} \cdot \frac{1}{(1 + k_s)} \tag{13.80}$$

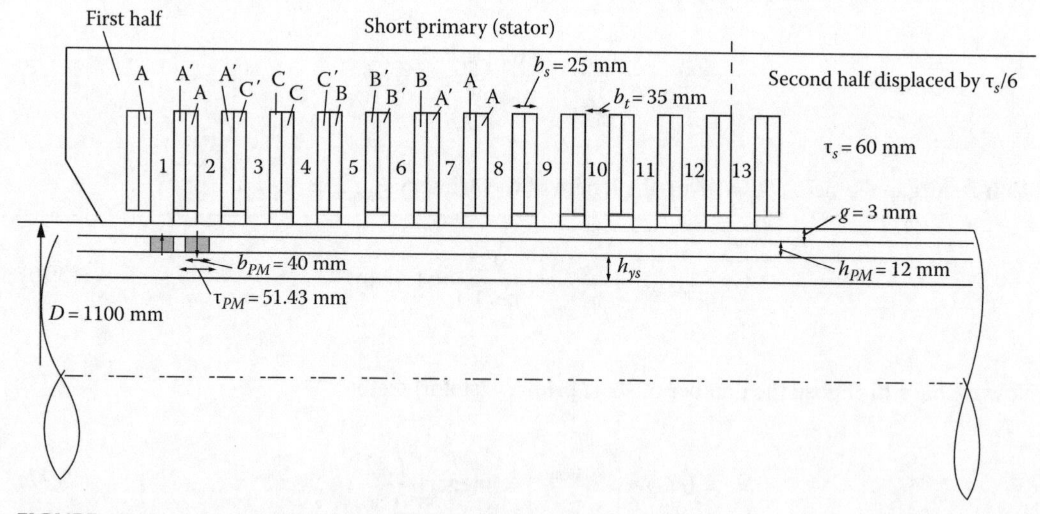

FIGURE 13.17 Large T-LPMSM generator geometry.

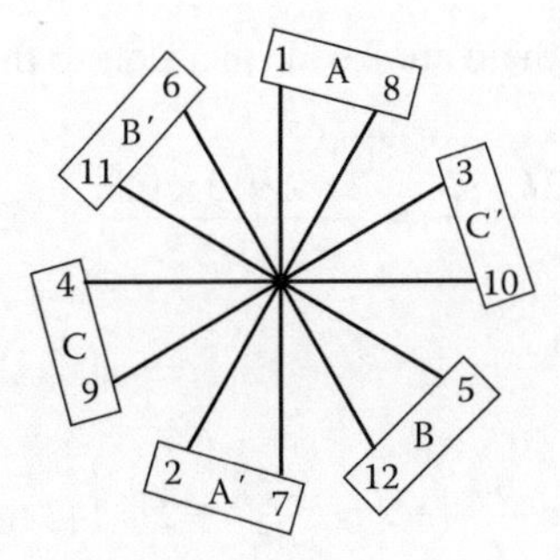

FIGURE 13.18 Star of emfs 12 coil/14 pole combination.

Magnetic saturation should not be very influential because the magnetic airgap (h_{PM} + g) is large, despite large primary mmf (to secure high thrust density); so with k_s=0.10 and with k_{fringe}=0.10 (open slots), the airgap average PM flux density is

$$B_{gPM} = 1.4\times\frac{12}{12+3}\times\frac{1}{1+0.1}\times\frac{1}{1+0.1} = 0.925\ \text{T} \tag{13.81}$$

More PM height (and weight) would produce higher B_{gPM} but, since the cost of PMs is high, it is not easy to justify the PM extra cost by a modest thrust (and efficiency) improvement. There are two neighboring coils per phase in series (in general) for 12 coils/14 poles module configuration (Figure 13.18).

The fundamental winding factor k_{w_1} for 12 coil/14 PM poles combination is $k_{w_1} = 0.933$ and can be easily inferred from Figure 13.18 (it is as for q=2!). Now the peak PM flux per one-phase coils is

$$\phi_{PM} \approx B_{gPM}\frac{b_{PM}}{2}\pi D_{ip}k_{w_1} \tag{13.82}$$

This is equivalent to calculating the emf per phase with the formula

$$E_{\max} = 16B_{gPM}\cdot\pi\cdot D_{ip}n_c\cdot k_{w_1}U = 16\times0.925\times\pi\times1.1\times0.933\times1\times n_c = 352\times n_c \tag{13.83}$$

In addition, the thrust F_{xn} is simply

$$F_{xn} = \frac{3}{2}B_{gPM}\pi D_{ip}\times16I_n\sqrt{2} = \frac{3}{2}\frac{\pi}{\tau}\psi_{PM}I_q;\quad I_q = I_n\sqrt{2};\quad I_d = 0 \tag{13.84}$$

Zero I_d control secures maximum thrust/current and also lagging power factor, as needed for the PWM converter boosting mode required to control power generating.

$$294\times10^3 = \frac{3}{2}\times0.925\times\pi\times1.1\times16\times0.933\times\sqrt{2}\times I_n n_c \tag{13.85}$$

So the mmf coil (RMS value) $I_n \cdot n_c$ is

$$I_n n_c \approx 2.915\times10^3\ \text{A turns (RMS)} \tag{13.86}$$

The slot width is only 25 mm and there are 2 coils in a slot; so the required area A_{slot} is

$$A_{slot} = \frac{2I_n \cdot n_c}{j_{Con} \cdot k_{fill}} = \frac{2\times 2.915\times 10^3}{3\times 0.55} = 3532 \text{ mm}^2 \tag{13.87}$$

So the slot height h_s is

$$h_s = \frac{A_{slot}}{b_s} + 2 = \frac{3532}{25} + 2 = 143 \text{ mm} \tag{13.88}$$

This is close to the acceptable limit, typical for large generators but as $h_s/b_s = 143/25 = 5.65 < (6 \div 7)$, the slot leakage inductance will be notable, though.

The copper losses for rated thrust p_{Con} are

$$p_{Con} = 3\rho_{Co}\frac{l_c 16 n_c^2 I_n^2 J_{Con}}{I n_c} = 3\times 2.1\times 10^{-8}\times \pi \times (1.1+0.143)\times 16\times 58.3\times 3\times 10^6 = 68.9 \text{ kW} \tag{13.89}$$

So, neglecting all other losses the efficiency would be

$$\eta_n \approx \frac{P_n}{P_n + p_{Con}} = \frac{250}{250+68.9} = 0.784 \tag{13.90}$$

This is not the expected 85%; some improvements may be obtained by decreasing further the current density and thus deepening the slots; but the gain will perhaps be up to 80%. It is possible to get better efficiency by adopting a smaller specific thrust f_{xn}, and thus for a larger diameter machine an even higher initial cost of the system is obtained. The machine inductance per phase with three active phases (sinusoidal (vector) current control) is

$$L_s = L_l + L_m \tag{13.91}$$

$$L_m = \frac{3}{2}\frac{4n_c^2 b_{PM}\pi D_{ip}\mu_0}{g + h_{PM}} = 1.256\times 10^{-6}\times \frac{3}{2}\times \frac{4\times 0.04\times \pi \times 1.1\times n_c^2}{(3+12)\times 10^{-3}} = 0.0694\times 10^{-3}\times n_c^2 \tag{13.92}$$

$$\begin{aligned} L_l &= 16\cdot \mu_0 n_c^2 \pi (D_{ip} + h_s)\lambda_{ss} = 16\times 1.256\times 10^{-6}\times 3.14\times (1.1+0.143)\times 1.96\times n_c^2 \\ &= 0.1537\times 10^{-3}\times n_c^2 \end{aligned} \tag{13.93}$$

$$\lambda_{ss} \approx \frac{h_s}{b_s} + \frac{b_{s0}}{b_s} = \frac{14}{3\times 25} + \frac{2}{25} = 1.96$$

$$L_s = (0.1537 + 0.0694)10^{-3}\times n_c^2 = 0.2231\times 10^{-3}\times n_c^2 \tag{13.94}$$

So

$$\frac{L_s I_n \sqrt{2}}{E_{\max}/\omega_r} = \frac{0.2231\times 10^{-3}\times n_c^2 \times I_n\sqrt{2}}{352\times n_c} = \frac{0.2231\times 10^{-3}\times \sqrt{2}\times 2915\times 2\times \pi \times 9.72}{352} = 0.159 \tag{13.95}$$

This is a rather small value, which means that the power factor is going to be very good; so the inverter kVA rating will be acceptable. Increasing the slot depth in order to increase efficiency, at the cost of a larger machine (primary), will also diminish the power factor; but since it is rather high, such a cost may be acceptable.

Note: The two design examples treat rather extreme specifications to test the capabilities of T-LPMSMs and are to be considered as preliminary sizing, start-up solutions.

13.9.1 Generator Control Design Aspects

The generator mode control depends heavily on the application. However, field oriented control (FOC) with pure I_q control is appropriate as it secures lagging power factor, and thus the PWM converter works as a forced commutation mode rectifier, which corresponds to the needed voltage boosting mode.

Figure 13.19a shows the autonomous generator mode with dc output load, where dc voltage control with current limiter may be appropriate to offer the reference current I_q^*; as the speed varies, the influence of I_q^* control on output power changes and thus some emf compensation in FOC is required. When the output dc voltage of T-LPMSM as generator is controlled from outside (Figure 13.19b), the control of the dc output current seems more adequate for T-LPMSG; as dc voltage is constant, or only slightly variable, the output power control is handled well through I_{dc}^* control.

As linear generators are characterized by oscillatory linear motion with large amplitude (which needs 6–10 pole travel length or more and thus imposes three-phase configurations), the power delivered over the travel length varies notably and the PWM converter has to be designed and controlled to handle the situation; a fixed dc voltage might not allow full energy extraction and thus an intermediary dc–dc converter with variable (controllable) boost ratio might be necessary.

In applications such as "active suspension damper" with a travel of about 0.2 m in car applications, the thrust control—through I_q^*—should be directly commanded from the suspension supervision system.

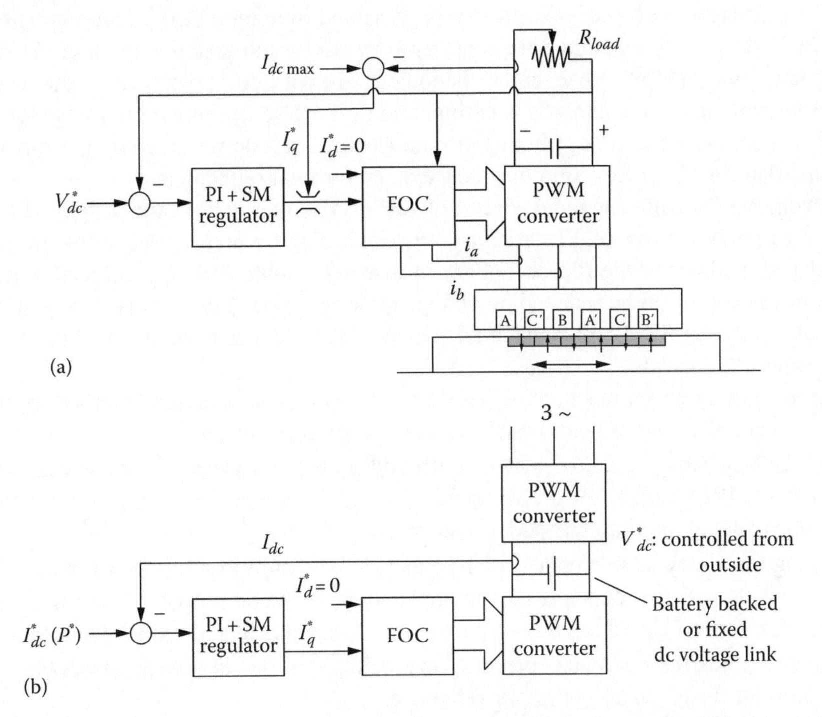

FIGURE 13.19 Typical generator control systems (a) for nonfixed dc link voltage (dc voltage control), (b) for fixed dc link voltage.

13.10 SUMMARY

- The tubular configuration for L-PMSMs is characterized by ideally zero radial force on the mover, low copper losses due to the circular shape of the coils, and better heat transmission from coils to primary core due to their contact along the entire coil length.
- Due to mechanical reasons, the tubular configuration is limited to a travel of at most $3 \div 4$ m.
- T-LPMSM applications range from free-piston thermal engine generators to wave energy converters and active suspension dampers for cars or linear position tracking from more than 0.3 m to 3 m travel length.
- The three-phase windings, supplied from off the shelf PWM converters used for rotary ac machines, are connected for $q=1$ slot/pole/phase or for fractionary q ($q \leq 0.5$) winding configurations to allow a small enough pole pitch and thus reduce the machine weight and initial cost; a travel of more than $6 \div 10$ pole pitches (PM pole pitches) justifies the use of T-LPMSMs.
- The primary iron core may be made of prefabricated soft magnetic composite (SMC) teeth (with two circular coils attached)—when the "slots" may be semiclosed (adequate for IPM secondaries)—or can be fabricated from disk-shaped laminations with three (four) radial slits, to reduce core losses in the yoke zone.
- Air-core three-phase windings (with circular coils) assembled: in $q=1$ or $q \leq 0.5$ winding configurations may be used when thrust pulsations are a sensitive issue, though the thrust/weight and efficiency tend to be smaller than those in iron core configurations.
- The fundamental winding factor of $q \leq 0.5$ windings is large, but their mmfs are rich in space harmonics and, consequently, the secondary (PM and yoke) eddy currents tend to be larger; however, as long as the fundamental frequency is below 50 Hz, their contribution to total losses is less than 5% in general.
- A technical 1D field theory is developed and shown to allow preliminary realistic performance assessment.
- A 2D multilayer analytical field theory is presented in notable detail and compared with 2D FEM results; slot openings are considered by conformal mapping method. Apart from magnetic saturation all other major phenomena are correctly portrayed by the analytical multilayer theory for notably less computation effort; so optimization analytical design may be approached this way; finally, around "analytical" optimum configuration a direct geometrical FEM optimization may be executed for final refinements.
- The cogging force may be reduced as for F-LPMSMs, mainly by optimizing the PM span/PM pole pitch and ($q \leq 0.5$) fractionary windings; also by making the whole primary of displaced modules or the PM-secondary of displaced poles. Still the end teeth drag force at zero current has to be reduced by dedicated design procedures; alternatively, it may be calculated by 2D FEM and considered a known disturbance, to be handled by the thrust (or current I_q) control.
- Core losses may be approached by 2D FEM (or a multilayer analytical method for magnetostatic field distribution, with widely accepted analytical formulae).
- The T-LPMSM may be field oriented controlled (FOC) when zero I_d^* (or maximum thrust/current for IPM secondary) control seems appropriate as flux weakening is not very probable due to small maximum speed (6 m/s or so).
- Cogging force may be treated as a disturbance, to be compensated through the I_q^* control.
- The generator control depends on the application; if the dc output voltage is controlled from other source, then a forced—commutation rectifier is field oriented controlled (FOC) in the sense that the reference current I_q^* is the output of the dc output current closed loop regulator for the given output power reference.
- It is also feasible to use for generator control a diode rectifier with a variable output—voltage and a constant dc output voltage dc–dc converter.

- Direct thrust and flux control (DTFC) is also used for T-LPMSMs, especially when thrust pulsations should be reduced drastically or when motion—sensorless control is performed.
- Two case studies for T-LPMSM design have been performed, both of small speed ($\leq$1 m/s) but one for small thrust (100 N) for motoring and one of 294 kN for generating. For the first case, the thrust density was 1 N/cm^2, while for the other it was 6 N/cm^2. At such small speeds, the efficiency is only above 64% in the first case and 78.8% for the second case, as a compromise between performance and T-LPMSM weight (and initial cost). New linear PM generator design efforts for ocean energy conversion are surfacing by the day [23].
- Linear magnetic gears have been incorporated recently into T-LPMSM for better performance at very low linear speed [24,25] or as magnetic screws [26].

REFERENCES

1. I. Boldea and S.A. Nasar, *Linear Electric Actuators and Generators*, Chapter 4, Cambridge University Press, Cambridge, U.K., 1997.
2. J. Wang, G.W. Jewell, and D. Howe, A general framework for the analysis and design of tubular linear PM machines, *IEEE Trans*., MAG-35(2), 1999, 1986–2000.
3. N. Bianchi, S. Bolognani, D.D. Corte, and F. Tonel, Tubular linear PM motors: An overall comparison, *IEEE Trans*., IA-39(2), 2003, 466–475.
4. E. Fornasiero, L. Albanti, N. Bianchi, and S. Bolognani, Consideration on selecting fractional-slot windings, *Record of IEEE-ECCE-2010*, Atlanta, GA, 2010, pp. 1376–1383.
5. N. Bianchi, M.D. Pre, G. Grezzoni, and S. Bolognani, Design considerations on fractional slot fault-tolerant synchronous motors, *IEEE Trans*., IA-42(4), 2006, 997–1006.
6. I. Boldea, *Electric Generators Handbook, Vol. 2: Variable Speed Generators*, Chapter 11, CRC Press, Taylor & Francis, New York, 2006.
7. J. Wang and D. Howe, Tubular modular PM machines equipped with quasi-Halbach magnetized magnets—Part I + II, *IEEE Trans*., MAG-41(9), 2005, 2470–2489.
8. N. Bianchi, Analytical computation of magnetic fields and thrust in a tubular PM linear servo motor, *Record of IEEE-IAS-2000*, Rome, Italy, 2000.
9. D.C.J. Krop, E.A. Lomonova, and A.J.A. Vandenput, Application of Schwarz-Cristoffel mapping to PM linear motor analysis, *IEEE Trans*., MAG-44(3), 2008, 352–359.
10. T.A. Driscoll, Schwarz-Cristoffel toolbox user's guide: Version 2.3, Department of Mathematical Sciences, University of Delaware, Newark, DE, 2005.
11. Y. Amara, J. Wang, and D. Howe, Stator iron losses of tubular PM motors, *IEEE Trans*., MAG-41(4), 2005, 989–995.
12. I. Boldea and S.A. Nasar, *Electric Drives*, 2nd edn., Chapter 7, CRC Press, Taylor & Francis, New York, 2006.
13. I. Boldea, C. Paicu, and G.D. Andreescu, Active flux concept in sensorless unitary ac drives, *IEEE Trans*., PE-23(5), 2008, 2612–2618.
14. H.-W. Kim, J.-W. Kim, S.-K. Sul, Thrust ripple free control of cylindrical linear synchronous motor using FEM, *Record of IEEE IAS-1996*, *Annual Meeting*, New Orleans, LA, vol. 1, 1996.
15. Y.-W. Zhu and Y.-H. Cho, Thrust ripple suppression of PM linear synchronous motor, *IEEE Trans*., MAG-43(6), 2007, 2537–2539.
16. W.R. Cawthorne, Optimization of brushless PM linear alternator for use with a linear internal combustion engine, PhD Thesis, West Virginia University, Morgantown, WV, 1989.
17. J. Lim, H.-Y. Choi, S.-K. Hong, D.-H. Cho, and H.-K. Jung, Development of tubular type linear generator for free-piston engine, *Record of ICEM-2006*, Chania, Greece, 2006.
18. V. Delli Colli, P. Cancelliere, F. Marignetti, R. Di Stefano, and M. Scarano, A tubular generator drive for wave energy conversion, *IEEE Trans*., IE-53(4), 2006, 1152–1159.
19. J. Faiz, M.E. Salari, and Gh. Shahgholian, Reduction of cogging force in linear permanent magnet generators, *IEEE Trans*., MAG-46(1), 2010, 135–140.
20. K. Hatakenaka, M. Sanada, and Sh. Morimoto, Output power improvement of PM-LSG for wave power generation by Halbach permanent magnet array, *Record of LDIA-2007*, 2007.
21. I. Boldea and S.A. Nasar, *Linear Motion Electromagnetic Devices*, Chapter 4, Taylor & Francis, New York, 2001.

22. P. Zheng, A. Chen, P. Thelin, W.M. Arshad, and Ch. Sadarangani, Research on a tubular longitudinal flux PM linear generator used for free-piston energy converter, *IEEE Trans.*, MAG-43(1), 2007, 447–449.
23. J. Prudell, M. Stoddard, E. Amon, T.K.A. Brekken, and A. von Jouanne, A PM tubular linear generator for ocean wave energy conversion, *IEEE Trans.*, IA-46(6), 2010, 2392–2400.
24. S. Niu, S.L. Ho, and W.N. Fu, Performance analysis of a novel magnetic-geared tubular linear PM machine, *IEEE Trans.*, MAG-47(10), 2011, 3598–3601.
25. R.C. Holehous, K. Atallah, and J. Wang, Design and realization of a linear magnetic gear, *IEEE Trans.*, MAG-47(10), 2011, 4171–4174.
26. J. Wang, K. Atalah, and W. Wang, Analysis of a magnetic screw for high force density linear electromagnetic actuators, *IEEE Trans.*, MAG-47(10), 2011, 4477–4480.

14 Multi-Pole Coil Three- or Two-Phase Linear PM Reluctance Motors

Linear electric motors discussed in previous chapters relied on primary coils that span a little more or less than a pole pitch. By pole pitch, we mean the half period of PM flux as the common attribute of many "novel typologies" with dedicated names.

In this chapter, we treat a variety of three- or two-phase linear PM motors that contain multi-pole nonoverlapping coils and are characterized by the following:

- Smaller copper weight per Newton of thrust and thus smaller copper losses; flat configurations are preferred, but tubular ones may be imagined also.
- Due to separate electric and magnetic circuit design, more armature mmf at small pole pitches (needed at low speeds) could be placed in the primary; thus, again, more thrust density is feasible.
- The higher thrust density is accompanied inevitably by higher $L_s \cdot i/\Psi_{PM}$ ratio and, consequently, lower power factor results (L_s—average inductance, Ψ_{PM}—PM phase flux linkage).
- The small pole pitch is a key condition for large force density, but this implies low airgaps, as these machines are essentially variable reluctance machines (with single- or double-magnetic anisotropy).
- Multi-pole coil three- or two-phase linear PM reluctance motors (MPC-LPMRMs) are basically single-phase machines that are assembled in three- or two-phase configurations: To reduce vibrations and noise, due to inevitable inequality of instantaneous thrust and normal force of each phase, besides longitudinal flux, transverse-flux configurations have been adopted.
- For larger pole pitch, variable reluctance secondary (stator) configurations could be considered, with PMs + ac windings primaries, especially for long travel (industrial, urban, or interurban transport). Most MPC-LPMRMs are, however, recommended for short travel (within 0.5–3 m), but at smaller airgap than F-LPMSMs or T-LPMSMs, as mentioned in previous chapters, for securing the claimed high thrust density (6 N/cm^2 at 0.5 m/s).
- The large normal force on the mover may be canceled by double-sided flat or tubular topologies.
- Application of MPC-LPMRMs has been tried for machining tables or for large stroke (0.5 m) low-frequency (1–5 Hz) vibrators for large mass movers.

14.1 FEW PRACTICAL TOPOLOGIES

14.1.1 Sawyer Linear PM Motor

We start, as we should, with the so-called Sawyer linear motor [1] (Figure 14.1), which is basically a two-phase PM primary and variable reluctance secondary (stator) topology.

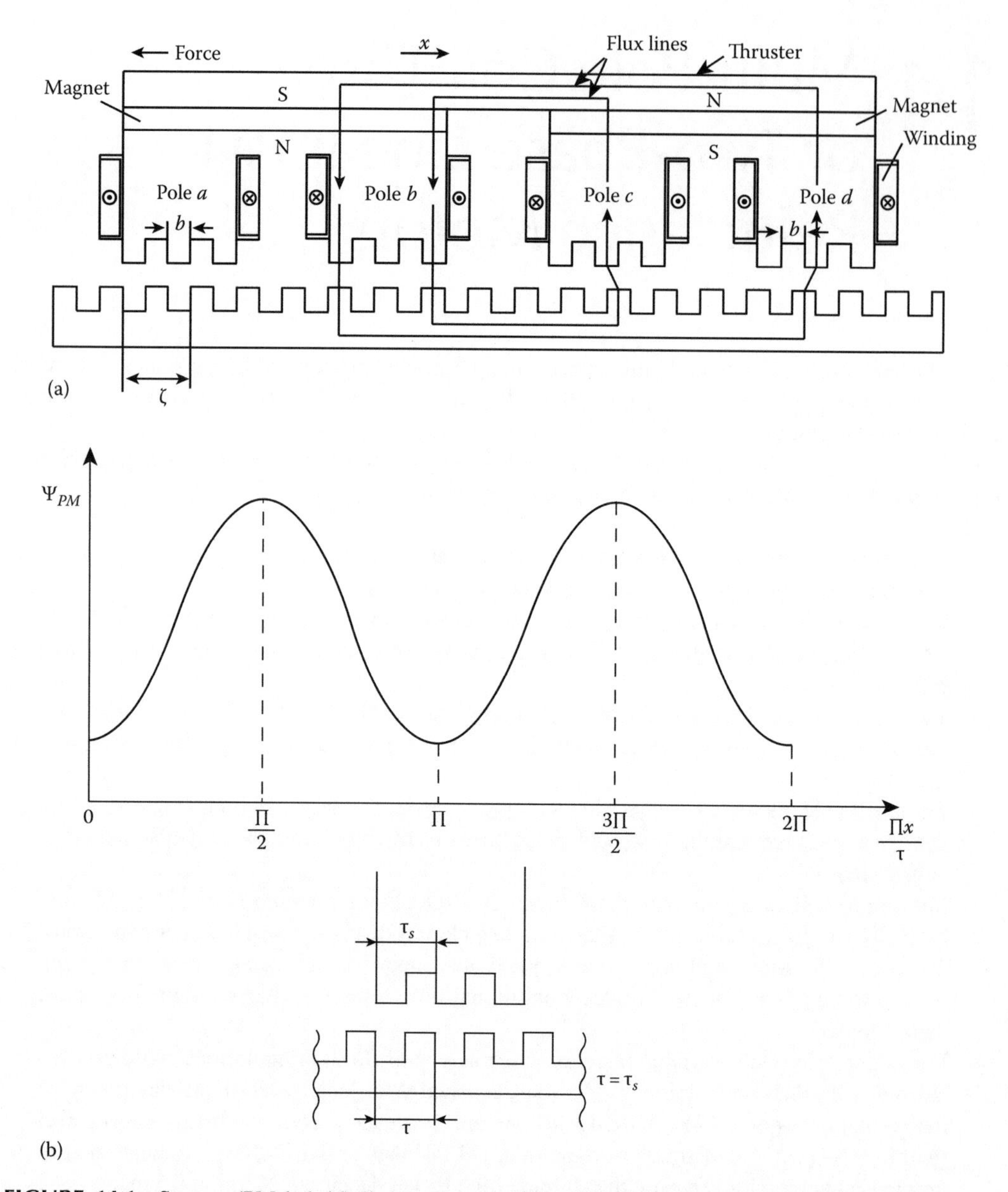

FIGURE 14.1 Sawyer (PM hybrid) linear stepper. (After Viorel, I.A. et al., Sawyer type linear motor dynamic modelling, *Proceedings of the International Conference on Electrical Machines (ICEM)*, Manchester, U.K., vol. 2, pp. 697–701, 1992.) (a) flux lines with coil A active and (b) PM flux in coil A versus secondary position.

The Sawyer motor is basically a pulsating homopolar PM flux reluctance machine: the PM flux in coil A (and in other coils) is unipolar, but it pulsates to produce emf and, thus, thrust; there is also a small pure reluctance force and a cogging, zero current, force.

The primary and secondary saliency for the four primary coils are phase shifted by $\pi/2$ or half slot pitch $\tau_s = \tau$ (primary and secondary slot pitches are equal to each other). So the current in coils A&A′ or B&B′ act together but, to get highest thrust available, the coil mmfs tend to double the PM flux in coil A (Figure 14.1a) and to cancel it in coil A′; the flux lines close through coils B and B′, which are passive at this position and thus do not produce any thrust (due to maximum and minimum reluctance positions, respectively).

So two coils are active and produce the whole thrust; besides, not so high thrust density, notable nonuniform normal force occurs and thus strong noise and vibration are present.

It is because of these reasons that we stop here the presentation of Sawyer linear PM motor, and the interested reader is advised to follow Refs. [3–5].

14.1.2 "Flux-Reversal" Configuration

In the search for higher thrust density, for the given volume and copper losses, a so-called PM flux-reversal single-phase module, three-phase linear PM reluctance configuration, was proposed [5,6]—Figure 14.2.

The PM flux-reversal configuration in Figure 14.2 [5] exhibits both phase A coils (A+A′) as active—thrust producing—and the PM flux derivative with position is theoretically twice the one for Sawyer motor, but it may be argued that the total maximum PM flux linkage in coil A is higher in the Sawyer linear motor because the PM flux fringing is larger in the flux-"reversal" configuration. These two phenomena do not conceal each other in the comparison and, thus, still the thrust density for equivalent geometry and copper weight and losses are at least 1.5 times higher for PM flux-reversal (switching) configuration (Figure 14.2).

With three phases, more uniform normal force distribution is obtained and thus lower noise and vibration are implicit. Experiments in [5] back up this phenomenological ascertaining.

Still a few more peculiarities of the flux-reversal (flux switch) topology are worth noticing:

- Because the PM flux lines go parallel to armature reaction (mmf) flux lines, there is hardly a danger of demagnetization; this merit leads, however, to a large machine inductance, which means a low power factor ($L_S i/\psi_{PM}$ is larger)—0.4 in Ref. [5]—and thus a higher kVA rating in the converter. So the "price" of high thrust density is a higher cost of the inverter.
- As much of the PM flux lines at zero current redistribute through primary core (inherent to large PM flux fringing), the normal force and cogging force tend to be smaller.
- Let us also note that $2\tau_s = \tau$ and thus as τ_s is lower limited (PM thickness has to be notably larger than the airgap), τ has to be larger than that in the Sawyer linear PM motor.

The flux-reversal configuration may be built with transverse flux when only one E shape core coil represents a phase, but then the secondary has the central region slotting shifted by $\tau/2$ with respect to lateral regions slotting (Figure 14.3, [5]).

While the primary E core may be made from laminations, the secondary variable reluctance has to be made from a soft magnetic composite (SMC) as its structure may not be executed with regular laminations (but it may be done with doubly skewed laminations—Figure 14.3b).

The experimental flux-reversal linear PM reluctance motor (two phase, Figure 14.3), with the geometry in Table 14.1 (after [5]), has been proved to produce a peak thrust density of 6 N/cm^2 in contrast to an equivalent Sawyer motor, which produced 50% of it (Figure 14.4).

Let us notice besides the very large thrust density (6 N/cm^2), the moderate efficiency and the low power factor (0.4); the latter aspect leads to the conclusion that the machine used very large mmf, which, however, did not demagnetize the PMs, due to large PM fringing (or parallel paths of mmf and PM flux lines); so the high thrust density was obtained at the cost of more PMs, and to reduce it, the airgap was extremely small (0.1 mm! for 2 kN of thrust). With F-LPMSMs (Chapter 11), almost similar performance may be obtained at a notably larger airgap (1 mm) with a surface (or Halbach) PM secondary.

However, for the flux-reversal linear PM reluctance machine, the secondary (stator) has a pure variable reluctance structure, and thus, when stray PM fields are not acceptable and the cost of the stator is sensitive (longer travel), the latter is preferable to F-LPMSMs.

A short primary mover means, however, that a flexible electric power cable (or a contactless power transmitter) is necessary.

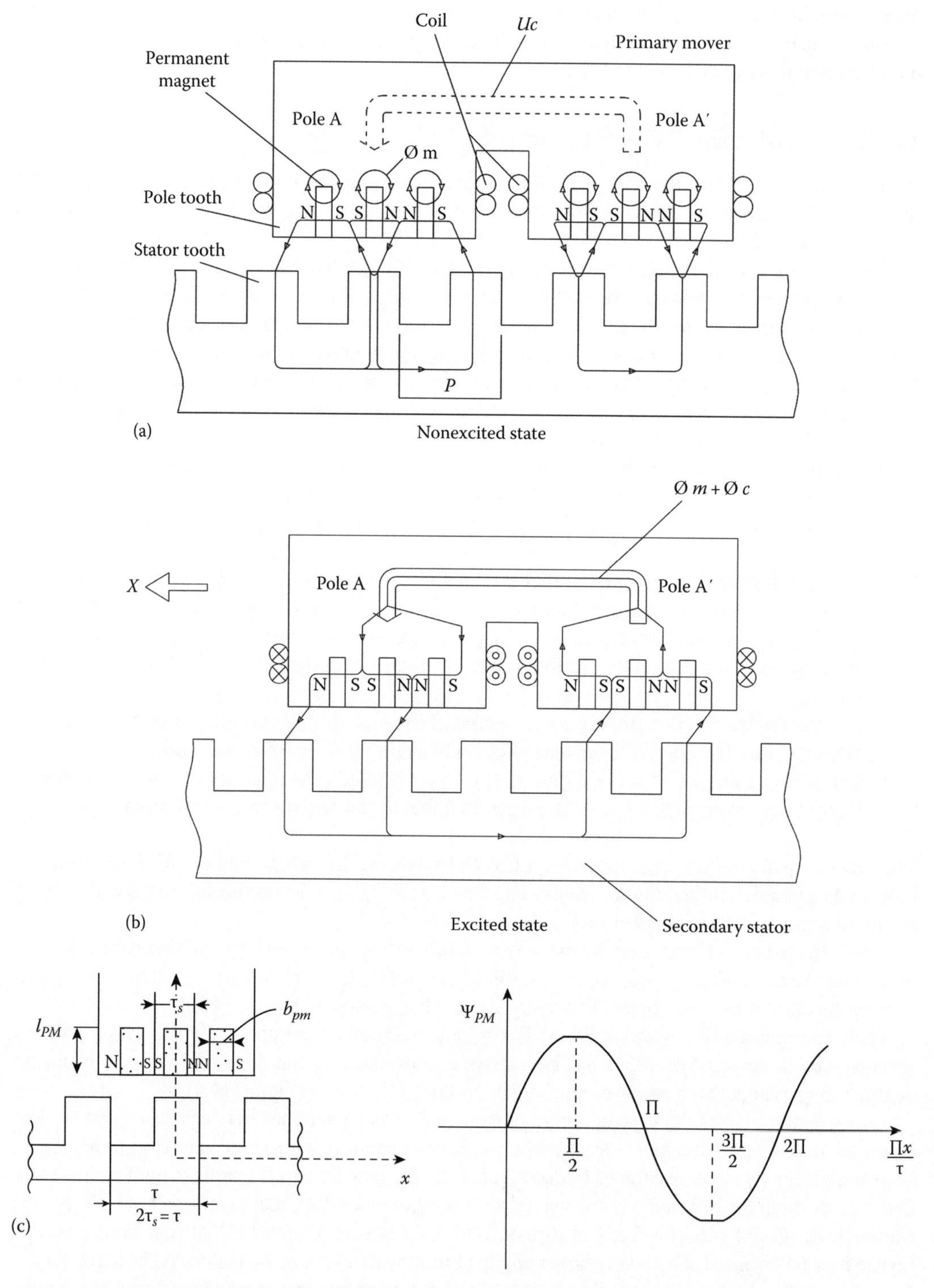

FIGURE 14.2 "Flux reversal" PM primary and variable reluctance secondary MPC-LPMRM. (a) PM flux lines at zero PM flux linkage in phase A position, (b) resultant flux lines with primary mmf at same position, and (c) PM flux linkage in phase A versus position. (After Karita, M. et al., High thrust density linear motor and its applications, *Record of LDIA-1995*, Nagasaki, Japan, pp. 183–186, 1995.)

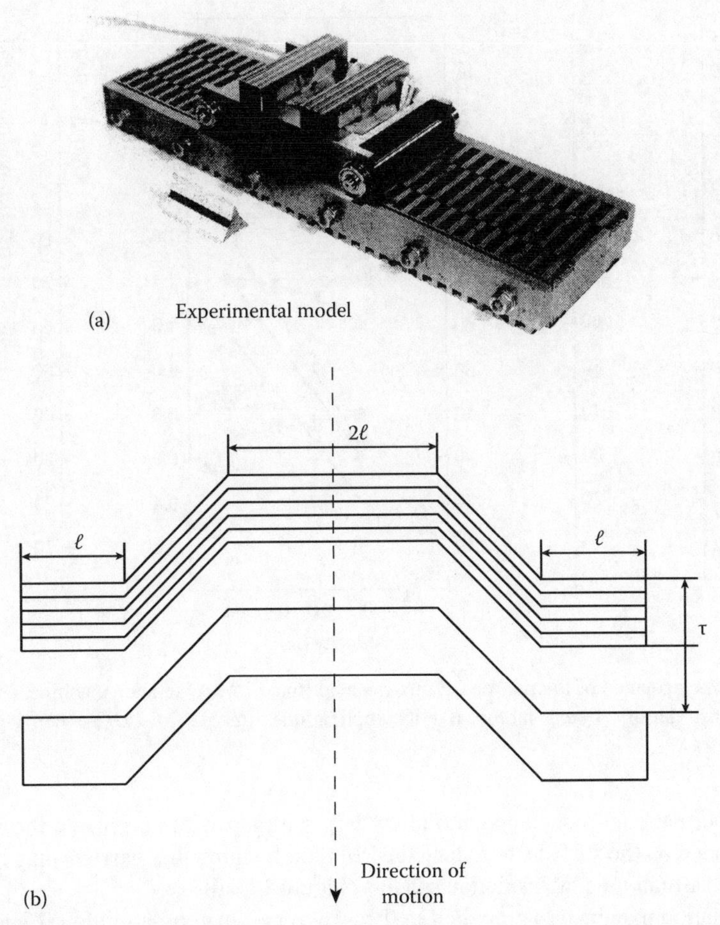

FIGURE 14.3 (a) Two-phase transverse-flux-reversal linear PM reluctance machine and (b) doubly skewed secondary. (After Karita, M. et al., High thrust density linear motor and its applications, *Record of LDIA-1995*, Nagasaki, Japan, pp. 183–186, 1995.)

TABLE 14.1
Dimensions of Initial and Final Shapes

Secondary Item	Value	Secondary Item	Value
Tooth pitch [mm]	5	Tooth pitch [mm]	10
Tooth width [mm]	3	Tooth width [mm]	4
Magnet thickness [mm]	2	Slot depth [mm]	6
Air gap [mm]	0.1		
Number of teeth/pole	4		
Gap area [cm^2]	13.68		
Coil turn	500		
Resistance [Ω]	3		
Inductance [mH]	40		
Permanent magnet	NdFeB		

Source: Karita, M. et al., High thrust density linear motor and its applications, *Record of LDIA-1995*, Nagasaki, Japan, pp. 183–186, 1995.

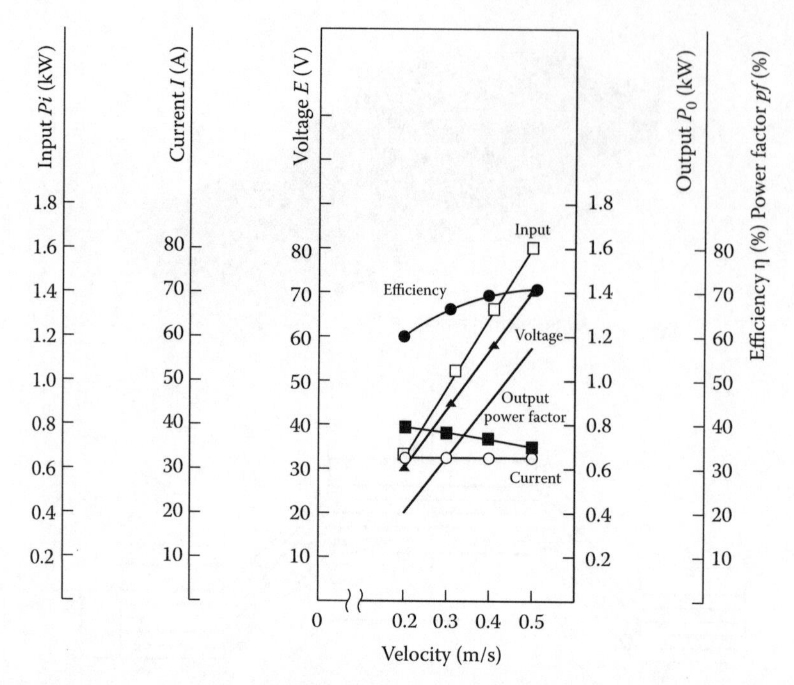

FIGURE 14.4 Performance of the two-phase flux-reversal linear PM reluctance machine. (After Karita, M. et al., High thrust density linear motor and its applications, *Record of LDIA-1995*, Nagasaki, Japan, pp. 183–186, 1995.)

To reduce the machine inductance and allow larger airgap with acceptable thrust density, the PMs may be placed on the surface; to reduce the PM flux, fringing flux barriers may be created into the longitudinal laminations by insulation spacers (Figure 14.5a).

A similar solution may be easier applied in the transverse-flux configuration (Figure 14.5b).

As evident from Figure 14.5, only half of the PMs (all of same polarity) are active; in the interior permanent magnet (IPM) structure (Figures 14.2 and 14.3) all PMs appear active at all times, but the fringing flux is very large and thus much of the advantage is lost. This loss may be compensated in IPM configurations by tall PMs (large l_{PM} in Figure 14.2b), such that even ferrite magnets may be used. On the other hand, the presence of insulation-made flux barriers in the primary core reduces the primary core weight and cost; the magnetic saturation is not heavy for surface permanent magnet (SPM) configuration as anyway the mmf is limited to the value that almost cancels the PM flux in the coil. It is thus feasible for larger airgap ($g=0.3$ mm) and $h_{PM}=1.5$ mm thick magnets, a flux barrier $\tau\text{-}b_{PM}=2$ mm wide and $\tau_s=5$ mm with $\tau=10$ mm, to obtain similar performance to IPMs, but for a larger power factor (0.6 or more).

Still, the SPMs are traveled by the mmf produced variable flux and thus eddy current occurs in the PMs; to reduce them, the PMs may be made of a few pieces in the transverse direction (motion direction is considered the longitudinal direction).

14.1.3 "Flux-Switching" Linear PM Reluctance Motors

Another form of IPM primary for the flux-reversal linear PM reluctance machine is the so-called flux-switching PM linear motor (Figure 14.6). Magnetic separation of phases may be adequate for good fault tolerance [7]. It is similar to flux-reversal machines.

In making a three-phase machine, without changing the slot pitch in the primary, we have to use the rules used for rotary flux-switching PMSMs, applied to the primary/secondary numbers of teeth combinations.

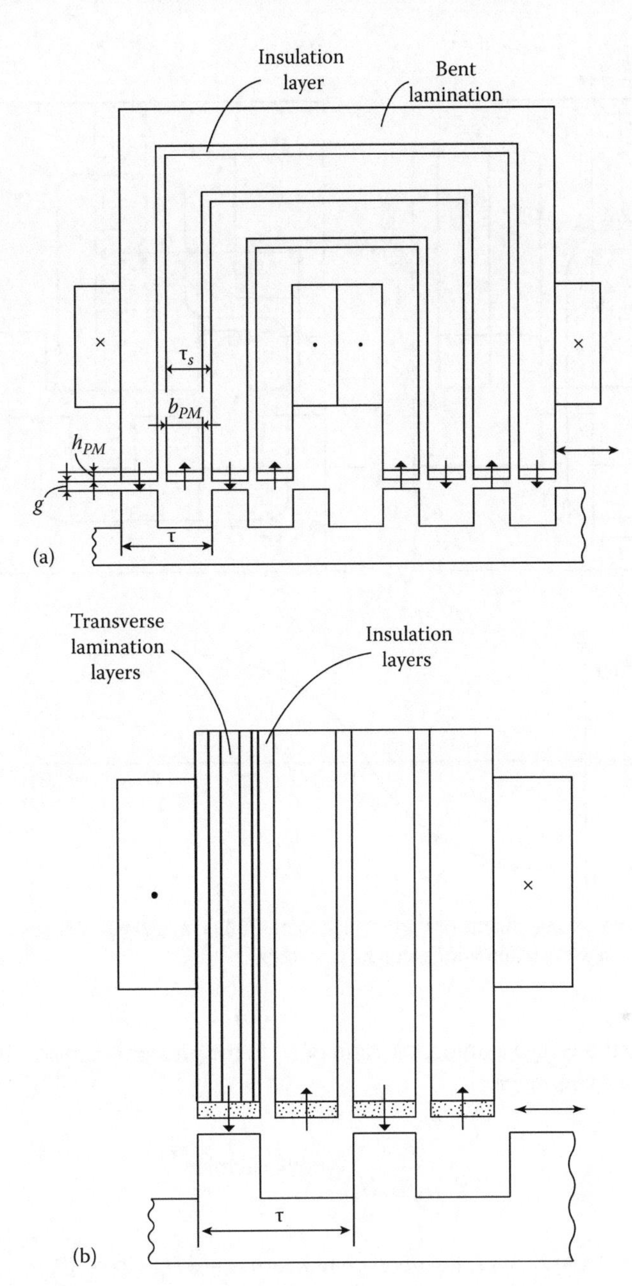

FIGURE 14.5 Surface PM "flux reversal" linear reluctance machine: (a) with longitudinal flux and (b) with transverse flux and flux barriers.

Let us consider six primary coils (for three phases), in general:

$$N_p = 3K; \quad K = 2,4,... \tag{14.1}$$

For the mover covering the same length, the number of secondary teeth N_s is

$$N_s = N_p(2n-1) \pm K \tag{14.2}$$

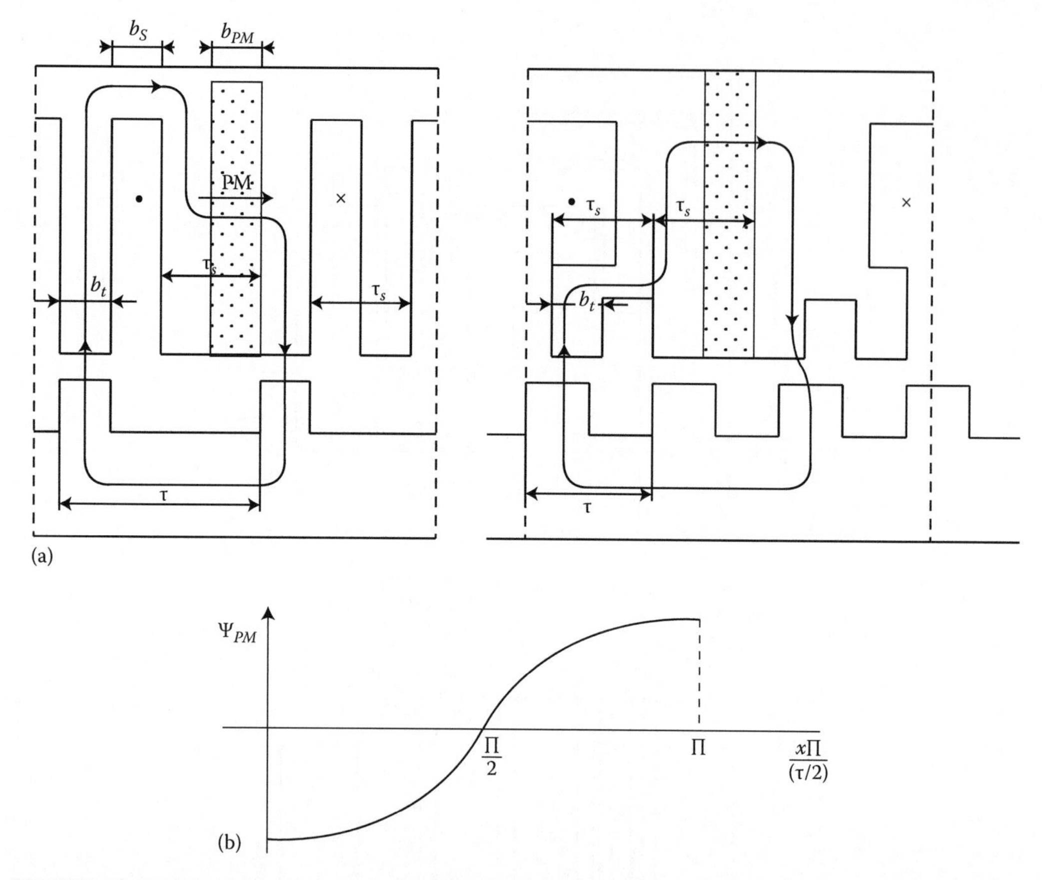

FIGURE 14.6 "Flux-switching" linear PM reluctance motor: (a) primary with 2 teeth/coil and with 4 teeth/coil and (b) PM flux linkage variation with secondary position.

To secure symmetric three-phase mmfs, the ratio of N_p to the greater common divisor (GCD) of N_p and N_s should be an even number:

$$\frac{N_p}{\text{GCD}(N_p,N_s)} = \text{even number} \tag{14.3}$$

For $2n=2$ teeth per coil and $N_p=12$, the good combinations are $N_s=10$ or 13.

For $2n=4$, 6 teeth per coil ($n=2$, 3), the best combinations (symmetric three-phase emfs) are as in Table 14.2.

These combinations lead to low cogging force. The 4 teeth/coil primary seems to be the one with maximum thrust per given geometry and copper losses.

The primary frequency f_p is

$$f_p = \frac{U}{2\tau} \tag{14.4}$$

The larger the number of secondary teeth for a given number of stator coil means lower pole pitch τ; that is, for given speed U, larger frequency in the primary. Therefore, it means higher core losses. Consequently, the total losses and maximum thrust, for given thrust and speed, have to be

TABLE 14.2
Primary/Secondary Teeth Combinations

n	$N_s = 3K$; $K = 2,4$	$N_r = N_p(2n - 1) \pm K$	$N_r = N_p(2n - 1) \pm 1$	$\frac{N_p}{GCD(N_p, N_s)}$
1	6	4, 8	5, [7]	3, 3, 6, 6
	12	8, 10, 12	[13]	3, 6, 3, 12
2	6	16, 20	17, [19]	3, 3, 6, 6
	12	32, 40	35, [37]	3, 3, 12, 12
3	6	28, 32	29, [31]	3, 3, 6, 6
	12	56, 64	59, [61]	3, 3, 12, 12

Source: Jin, M.-J. et al., *IEEE Trans.*, MAG-45(8), 3179, 2009.
$2n$ teeth per primary coil.
□, higher thrust density.

considered in an optimization design attempt together with the initial total cost (including the converter cost); higher frequency means lower power factor in general, in this large synchronous inductance machine.

This machine, in three-phase symmetric sinusoidal current control, is characterized by an inductance that is rather independent of mover position but is still dependent on current (on magnetic saturation) and is large.

14.1.4 Flux-Reversal PM-Secondary Linear Reluctance Motors

When the IPMs are placed on the secondary, especially in not-so-long travel applications (but with large thrust levels) (Figure 14.7), the multi-pole coils are used again and thus the copper losses per thrust and the thrust density are large; the PM flux concentration at low airgap (less than 0.5 mm) provides the claimed thrust density.

The transverse-flux version in Figure 14.7b may be preferred in some applications; the laminated structure in the PM secondary is, however, easier to build for the longitudinal flux configuration (Figure 14.7a).

14.1.5 Transverse-Flux Linear PM Reluctance Motors

Transverse-flux linear PM reluctance motors have been developed as counterparts of transverse flux permanent magnet synchronous motors (TF-PMSM) [8–11]. The configuration in Figure 14.8 refers to single-sided, passive variable reluctance secondary and PM plus ac coil primary. In Figure 14.8a, the PM flux concentration is complemented by additional interprimary, pole anti-fringing PMs placed there to notably reduce the main PM longitudinal flux fringing and, thus, create more thrust.

Similarly, in Figure 14.8b, the main PMs, placed on the primary pole with no PM flux concentration are complemented by additional anti-Halbach PMs to reduce main PM flux fringing and thus to increase thrust.

As visible in Figure 14.8, the pole pitch is the same in the primary and secondary, and thus the cogging thrust/phase is rather notable; however, in three (six)-phase configurations it may be reduced to less than 5%–7% of peak thrust.

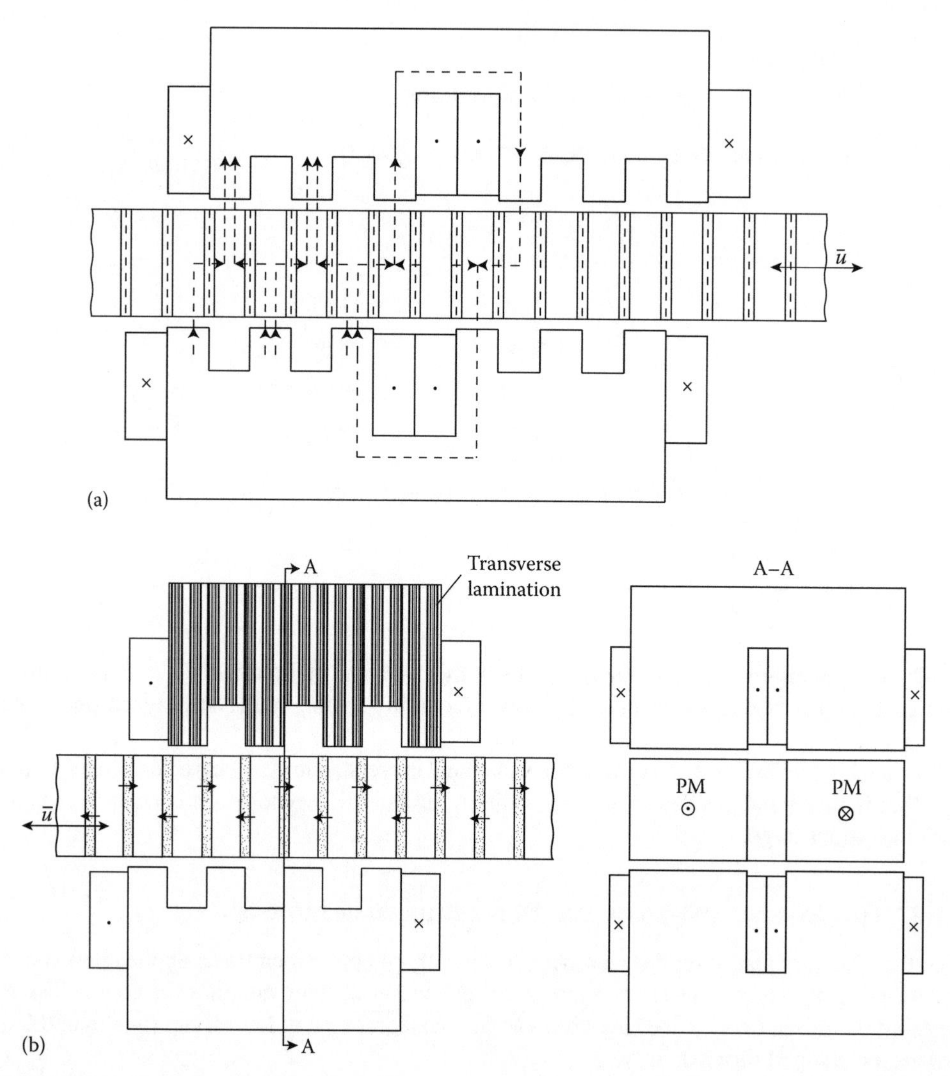

FIGURE 14.7 "Flux-reversal" IPM—secondary linear reluctance motor (one-phase section): (a) with longitudinal flux and (b) with transverse flux.

For large thrust, to reduce the transverse size, quite a few single-phase primary modules shifted between each other by 2π/m are required. This fragmentation of one-phase primary reduces to some extent the great merit of large thrust/copper losses.

The transverse flux linear permanent magnet reluctance machine (TF-LPMRM) may be applied even for large travel (urban and interurban transportation), mainly for propulsion, while levitation has to be complemented by dc controlled coils on board (since the normal to thrust ratio is in general less than 5–1). This acceptably low ratio is a merit if the normal force is always smaller than the mover primary weight, as it releases the stress on the linear bearings (or wheels) used to secure constant airgap without levitation control.

The fully active primary mover implies a feasible ac power cable or a friction type (or a contactless) electric power transmitter. However, the passive variable reluctance secondary (though laminated) is a special merit in terms of initial cost.

Note: As the linear homopolar synchronous motor (in Chapter 9), the fully active primary TF-LPMRMs seems adequate for passive guideway MAGLEVs, and a fully fair comparison between the two solutions is still due.

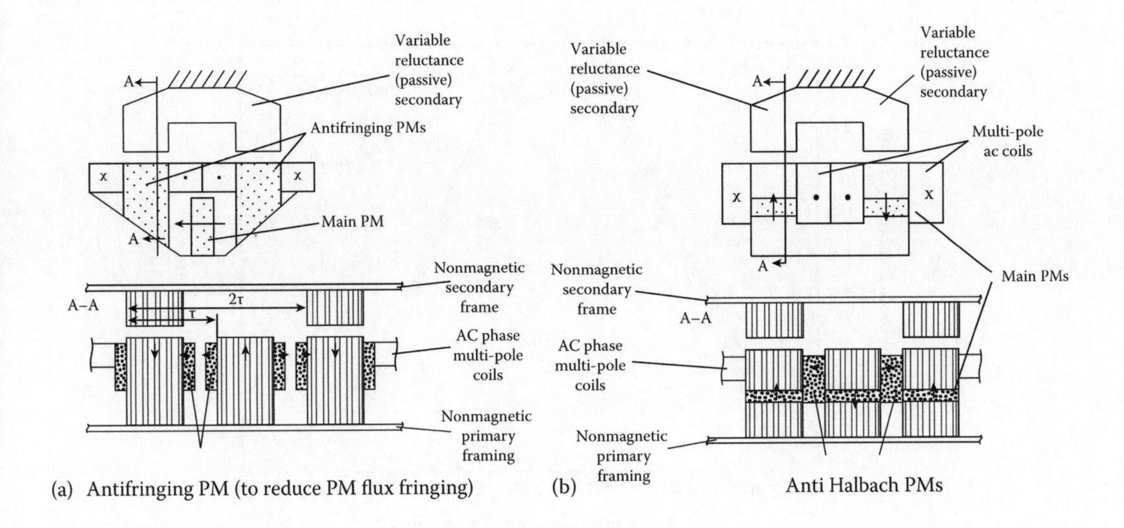

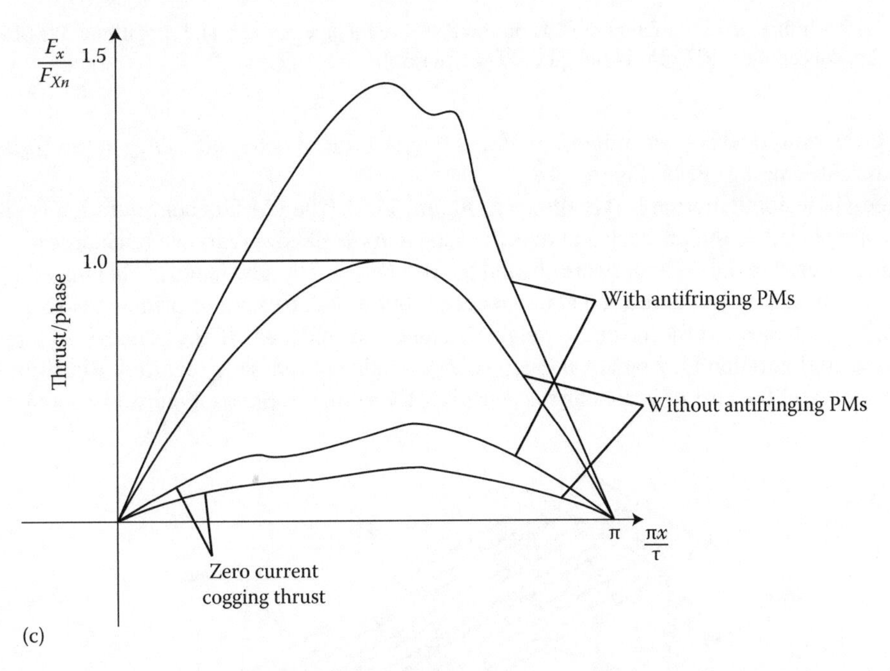

FIGURE 14.8 Transverse-flux linear PM reluctance motors (one-phase section): (a) flux concentration main PMs+ac coil primary, (b) main PMs+ac coil primary, and (c) thrust/phase versus mover position. (After Hoang, T.-K. et al., *IEEE Trans.*, MAG-46(10), 3795, 2010.)

In terms of how small the pole pitch can go in TF-LPMRMs, it all depends on the mechanically acceptable low airgap, as for all MPC-LPMRMs presented in this chapter, which are all single-phase modular multi-phase linear motors. The multi-pole coil windings lead also in TF-LPMRMs to large thrust/copper losses by the independent design of magnetic circuit and electric circuit. As for all other configurations in this chapter, PM flux fringing is the main cause of thrust reduction; consequently, a large mmf to produce large thrust is feasible (there is room for the coils) but, as the inductance is not very small, the ratio $L_S i_N/\psi_{PM}$ tends to be large and thus the power factor tends to be low; consequently, the inverter kVA rating is larger. Still the configuration with antifringing PMs increases the thrust per given mmf and thus reduces $L_S i_N/\psi_{PM}$. Thus, it increases the power factor,

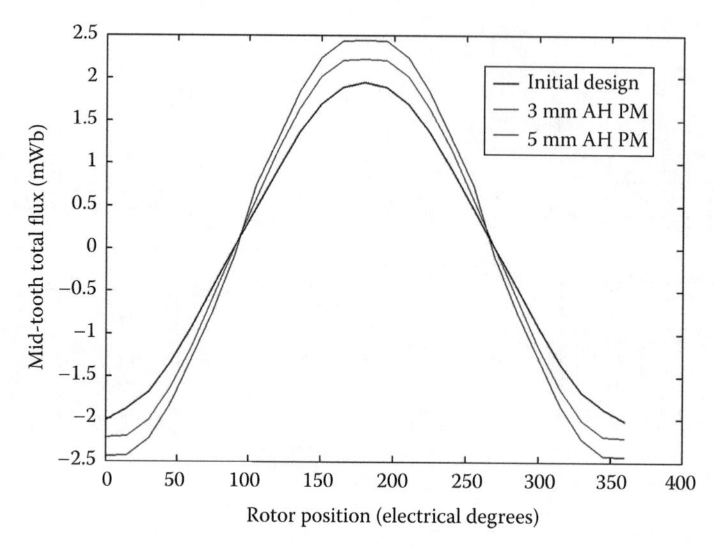

FIGURE 14.9 Influence of antifringing PMs on the PM flux linkage per coil and thrust in a TF-LPMRM with PM flux concentration (Figure 14.8b) quasi 3D FEM results.

and, for given mmf (and copper weight), produces larger thrust at better efficiency at the "price" of additional "antifringing" PMs (Figure 14.9).

A three-phase configuration [11] is shown in Figure 14.10. The PM flux concentration is visible. The PMs are placed on the primary and thus the secondary is passive (variable reluctance).

The three "legs" of the secondary are shifted by $2\tau/3$ ($t_p = \tau$). The advantage of "having all phases together" is "paid for" by more secondary material, but a better usage of primary E-shape core middle leg is a bonus; so the primary weight is somewhat reduced. If the primary is long, the three-phase configuration may be advantageous, as a single coil/phase is required. But if multiple units are used and the cost of secondary is sensitive, the configuration in Figure 14.8a and b may

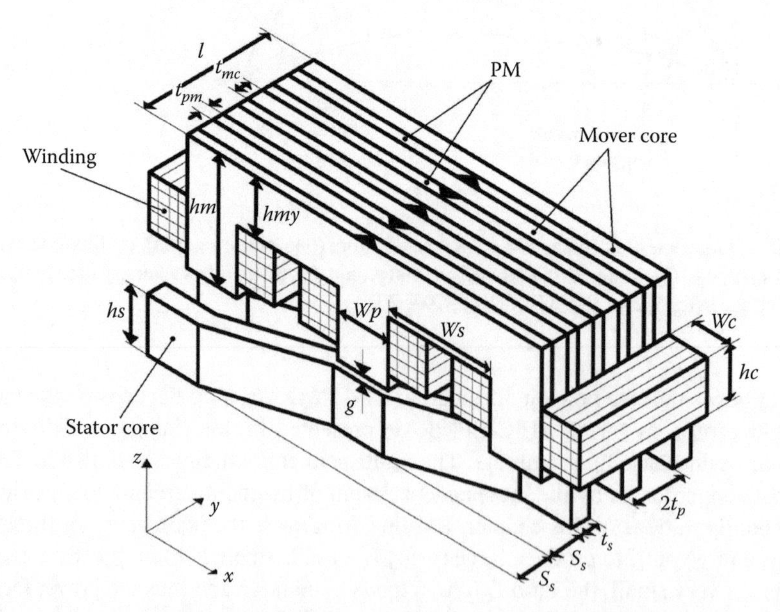

FIGURE 14.10 Three-phase transverse-flux linear PM reluctance motor. (After Hoang, T.-K. et al., *IEEE Trans.*, MAG-46(10), 3795, 2010.)

be more appropriate. For MAGLEVs, the 4(5)/1 normal force/thrust ratio typical for TF-LPMRMs means that additional levitation controlled electromagnets (to interact with the variable reluctance laminated secondary) are needed for a 1 m/s^2 acceleration vehicle, which requires a 10/1 or more normal force/thrust ratio.

Reducing cogging force is again not so easy with TF-LPMRMs as the primary and the secondary pole pitch counts are equal to each other.

14.1.6 Discussion

- Multi-pole coil linear PM reluctance motors in "flux-reversal," "transverse-flux," or "flux-switching" configurations are more or less similar: the PM flux changes polarity over secondary salient pole pitch in a multi-pole coil; the larger the number of poles, the smaller the pole pitch, and thus the higher the thrust for given total geometry and mmf, if and only if the PM fringing increase with pole pitch reduction is overcompensated by the increase in the number of poles.
- MPC-LPMRMs are essentially single-phase nonoverlapping coil modules assembled together to form three- or two-phase configurations required for travel length above 0.3 m in general.
- All PM+ac windings primaries, and thus MPC-LPMRM, share the "homopolar flux effect," which means that only 50% of the ideal PM flux becomes useful in thrust production, as they are locally single-phase ac synchronous motors in three- or two-phase configurations.
- The "separation" of magnetic from electric circuit design leads to more room for higher mmf per pole; but this way the $L_S I/\psi_{PM}$ ratio increases and thus the power factor decreases.
- Except for surface PM configuration (Figure 14.4b especially), the machine inductance Ls is large (in p.u.), which further decreases the power factor.
- A low power factor means a larger KVA rating PWM converter; now, it depends on the thrust and travel range to decide for the best compromise in a complex objective function optimization design.
- In industrial applications where a small airgap ($g \leq 0.5$ m) is acceptable and rather high thrust density (in N/cm^2 or in N/cm^3) is required, such machines are competitive as they allow low pole pitch; the maximum frequency is not very large, since the maximum speed is small (a few m/s) and thus core losses remain smaller than the copper losses.
- As any fair comparison between these configurations requires sound, though accessible, models, in the next sections we will deal with technical theories of two of the three types presented so far, and, whenever possible, 2D (3D) FEM and test results are brought in for validation.

14.2 TECHNICAL THEORY OF FLUX-REVERSAL IPM-PRIMARY LPMRM

Let us consider first the FR-LPMRM with IPM primary (Figure 14.2) redrawn here for convenience, with a few geometrical parameters highly visible.

The rather linear-slope flat-tapped appropriate PM flux linkage variation, with the mover position in the two coils of phase A (Figure 14.11), makes the technical theory (model) of this machine easy to develop; subsequent 2D (3D) FEM verifications and corrections are required, though. Therefore, in fact, only the maximum portion of PM flux linkage ψ_{PMp} has to be calculated. Apparently, the computation of ψ_{PMp} should not be very complicated, but the PM flux leakage (fringing) (Figure 14.1b) changes the picture. Moreover, the presence of coil current (in the magnetization direction) tends to inhibit the PM fringing flux and encourage the PM flux lines through airgap (Figure 14.2b). But in this situation of maximum flux through the airgap and iron, magnetic saturation becomes very heavy. So we may calculate ψ_{PMp} in the absence of current, as amended by the PM flux fringing; but remember that on load (high current) magnetic saturation intervenes.

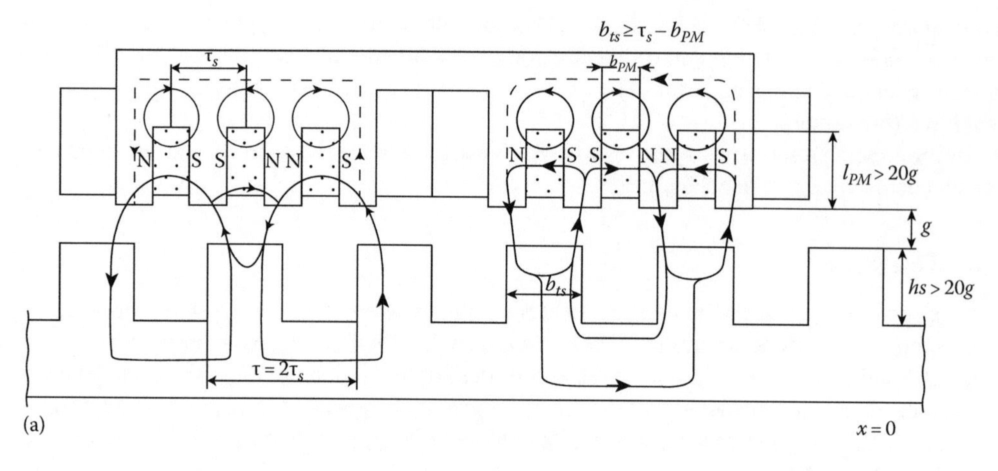

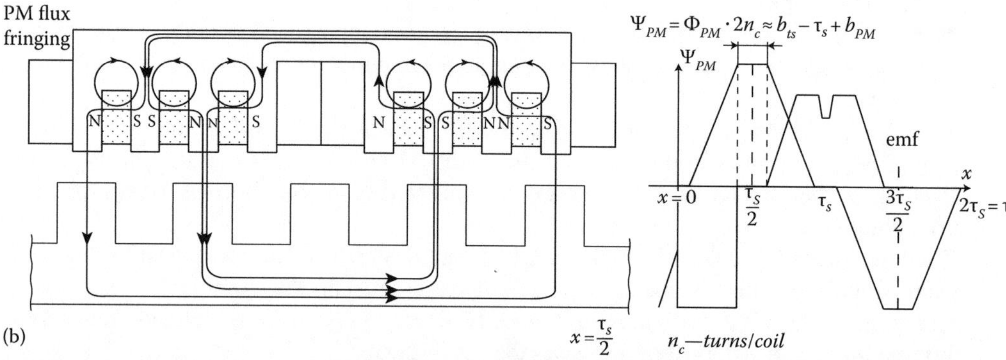

FIGURE 14.11 Flux reversal-LPMRM with IPM primary (one-phase module) and passive secondary: (a) zero PM flux position and (b) maximum PM flux position.

The peak PM flux linkage is simply

$$\psi_{PMp} = 2n_c n_p B_r l_{PM} L \frac{b_{PM}(\tau_s - b_{PM})/2 \cdot l_{PM}}{g + b_{PM}(\tau_s - b_{PM})/2 \cdot l_{PM}} \frac{1}{(1 + k_{fringe})(1 + k_s(I))} \tag{14.5}$$

In (14.5), the principle of magnetic reluctance ratio of PM and of airgap per half of stator tooth is used $(\tau_s - b_{PM})/(2 \cdot l_{PM})$.

A simplified analytical explanation of k_{fringe} in the presence of magnetic saturation ($k_s(I)$) is not easy to find, but, to reduce it, the airgap has to be very small. Even for zero airgap, $k_{fringe} > 1$; so we may take $k_{fringe} = 1.2$–1.5 for preliminary design and then verify it by 2D FEM calculations.

The machine phase inductance L_s includes the leakage inductance L_l and the main inductance L_m:

$$L_s = L_l + L_m(I) = L_l + \mu_0 \frac{(2 \cdot n_c)^2 \cdot 2b_{ts} \cdot L}{2g \cdot k_c \cdot (1 + k_s(I))} \tag{14.6}$$

L_l comprises the so-called slot leakage inductance (one part) and the end connection (three parts) leakage inductance; their expressions are rather straightforward for the two coils/phase in series.

A problem arrives if $L_m(I)$ varies also with the mover position; ideally, by design, it does not, because the number of mover teeth under the primary large pole is two at any time in Figure 14.11; small variations $L_m(I)$ are still possible.

But the cogging force is present as the position of the n_p ($n_p=3$ in Figure 14.11) PMs in the primary pole varies with mover position; its period is about τ_s.

The two coils/phase module resistance is also straightforward:

$$R_s = 2\rho_{co} \cdot l_c \cdot \frac{n_c}{I_n} j_{con} \tag{14.7}$$

I_n—rated current; j_{con}—design current density, with l_c—coil mean length.

To determine the thrust, with $\psi_{PM}(x)$ from Figure 14.1, we may calculate the emf per phase:

$$E_p = \frac{d\psi_{PM}}{dx} \cdot U \approx \frac{2\psi_{PM} \cdot U}{\tau_s - (b_{ts} + b_{pm} - \tau_s)} = k_{Ep} \cdot U \tag{14.8}$$

The rectangular (trapezoidal, rather) $d\psi_{PM}/dx$ dependence with zero-value zones suggests trapezoidal rather sinusoidal current control. So with two phases active at any time (except for commutation modes), the thrust F_x for a three-phase machine (three single-phase modules shifted by $2\,\tau_s/3$) is

$$F_x = 2 \cdot \frac{2\psi_{PMp}}{\tau_s - (b_{ts} + b_{pm} - \tau_s)} \cdot i = 2 \cdot k_{Ep} \cdot i \tag{14.9}$$

In (14.9), the machine main inductance variation with position ($L_m(x)$) is neglected.

Now the voltage equation with two phases active, and current chopping, is

$$V_{dc} = 2 \cdot R_s \cdot i + 2 \cdot E_p + 2 \cdot L_s \frac{di}{dt} \tag{14.10}$$

As L_s tends to be large, due to small airgap, heavy magnetic saturation may be justified to reduce the voltage drop, for given current, chopping hysteresis band Δi, and chopping frequency in the inverter and for a better inverter voltage ceiling utilization.

Though traditionally the maximum thrust is obtained when the primary current cancels the PM flux in the airgap—to avoid PM demagnetization (power factor 0.707)—in our case, we may use higher current for more thrust, because the PMs do not get demagnetized since the mmf field does not entirely cross the PMs (due to fringing). This explains why higher thrust density is feasible, at the "price" of lower "equivalent" power factor (0.4 in Ref. [5]). The lower power factor is reflected in (14.8) by a very large contribution of $2L_s di/dt$ in dc voltage, for trapezoidal chopped current control.

14.3 NUMERICAL EXAMPLE 14.1: FR-LPMRM DESIGN

Let us consider a three-phase FR-LPMRM (Figure 14.11) with the following data:

- Three IPMs per large primary pole
- PM height $l_{PM}=?$
- Airgap $g=0.12$ mm
- PM thickness $b_{PM}=2$ mm
- Primary slot pitch $\tau_s=5$ mm

- Primary tooth $b_{tp}=\tau_s - b_{PM}=5-2=3$ mm
- Secondary tooth $b_{ts}=5$ mm
- Secondary slot pitch $\tau=10$ mm
- PMs: NdFeB, Br = 1.15 T; $\mu_{rec}=1.05\,\mu_0$

To be calculated:

- The PM height l_{PM}, for $B_{gPM}=1$ T
- The thrust for trapezoidal current control and a stack width $l_{stack}=0.1$ m when the mmf airgap flux density of 1 T, if saturation is heavy $k_{sat}=1.0$
- The copper losses for the obtained thrust and $j_{con}=6$ A/mm^2
- The number of turns/coil and wire gauge at $U=0.5$ m/s and $V_{dc}=300$ V
- The voltage drop on inductance due to current chopping at 10 kHz with a hysteresis band of 5% of rated current
- Thrust density in N/cm^2

Solution

As we already inferred, the PM fringing is, in a way, reduced by magnetic saturation under machine load; let us suppose that with $k_{sat}=1$, the k_{fringe} is reduced to $k_{fringe}=0.5$. Consequently, under load, the peak flux in the two coils of a phase can be calculated from (14.5):

$$\psi_{PMp} = B_{gPM}\cdot n_p \cdot b_{tp}\cdot l_{stack} 2\cdot n_c \tag{14.11}$$

So

$$B_{gMP} = B_r\cdot\frac{l_{PM}}{b_{tp}}\cdot\frac{b_{PM}\cdot(\tau_s - b_{PM})/2\cdot l_{PM}}{g + b_{PM}\cdot(\tau_s - b_{PM})/2\cdot l_{PM}}\cdot\frac{1}{(1+k_{fringe})}\cdot\frac{1}{(1+k_s(I))} \tag{14.12}$$

$$1.0 = 1.15\cdot\frac{l_{PM}}{3}\cdot\frac{2\cdot(5-2)/2\cdot l_{PM}}{0.12+2\cdot(5-2)/2\cdot l_{PM}}\cdot\frac{1}{(1+0.5)}\cdot\frac{1}{(1+1)} \tag{14.13}$$

So $l_{PM}\approx 16$ mm.

The high degree of magnetic saturation was accepted in order to reduce the machine inductance (the airgap is very small).

So ψ_{PMp}, from (14.11), is

$$\psi_{PMp} = 1.0\cdot 3\cdot 3\cdot 10^{-3}\cdot 0.1\cdot 2\cdot n_c = 1.8\cdot 10^{-3} n_c \tag{14.14}$$

The emf (and thrust) coefficient in (14.8) and (14.9) k_{ep} is

$$k_{ep} = \frac{d\psi_{PMp}}{dx} = \frac{2\cdot 1.8\cdot 10^{-3} n_c}{(5-5+2-5)\cdot 10^{-3}} = 1.8\cdot n_c \tag{14.15}$$

Consequently, the thrust F_x is

$$F_x = 2 \cdot k_{ep} \cdot i = 3.6 \cdot n_c \cdot i \tag{14.16}$$

The mmf $n_c i$ per coil can be calculated for saturated conditions where the mmf airgap flux density is 1 T at $k_{sat} = 1$:

$$\frac{\mu_0 \cdot n_c \cdot i}{g(1+k_s)} = 1\text{T}; \quad n_c \cdot i = \frac{1 \cdot 0.12 \cdot 10^{-3}(1+1)}{1.256 \cdot 10^{-6}} = 238.8 \text{ of turns/coil} \tag{14.17}$$

So the thrust in such conditions would be

$$F_x = 3.6 \cdot 238.8 = 860 \text{ N} \tag{14.18}$$

The machine main inductance L_m is (from (14.6))

$$L_m = \frac{2 \cdot \mu_0 \cdot n_c^2 \cdot 2 \cdot b_{l_{stack}}}{2 \cdot g \cdot (1+k_s)} = \frac{2 \cdot 1.256 \cdot 10^{-6} \cdot (6 \cdot 10^{-3}) \cdot 0.1}{2 \cdot 0.12 \cdot (1+1) \cdot 10^{-3}} \cdot n_c^2 = 6.28 \cdot 10^{-6} n_c^2 \tag{14.19}$$

Assuming a leakage inductance $L_l = 0.2\ L_m$ (small airgap)

$$L_s = 1.2 \cdot L_m = 1.2 \cdot 6.28 \cdot 10^{-6} \cdot n_c^2 = 7.534 \cdot 10^{-6} \cdot n_c^2 \tag{14.20}$$

The copper losses P_{con} are (14.7)

$$P_{con} = 2 \cdot R_s \cdot i_n^2 = 2 \cdot 2 \cdot \rho_{co} \cdot l_{coil} \cdot \frac{n_c \cdot j_{con}}{i_n} i_n^2 = 4 \cdot 2.1 \cdot 10^{-8} \cdot 6 \cdot 10^6 \cdot 238.8 = 120.3552 \text{ W} \tag{14.21}$$

For a speed $U = 0.5$ m/s, the efficiency η is

$$\eta = \frac{F_x \cdot u}{F_x \cdot u + P_{con}} = \frac{860 \cdot 0.5}{430 + 120.355} = 0.78 \tag{14.22}$$

To calculate the number of turns, we make use of voltage equation (14.10); still $2 \cdot R_s \cdot i_n$ has to be calculated as a function of n_c (turns/coil):

$$2R_s i_n = 2 \cdot 2 \cdot \rho_{co} l_{coil} n_c j_{con} = 2 \cdot 2 \cdot 2.1 \cdot 10^{-8} \cdot 0.255 \cdot 6 \cdot 10^{-6} \cdot n_c = 2 \cdot 6.426 \cdot 10^{-2} \cdot n_c \tag{14.23}$$

$$l_{coil} \approx l_{stack} + 0.02 + 2 \times 3.5 \times b_{ts} = 2 \times 0.1 + 0.02 + 7 \times 0.005 = 0.255 \text{ m} \tag{14.24}$$

Now the voltage equation is used:

$$V_{dc} = 2R_s i_n + 2E_p + 2L_s \frac{di_n}{dt} \tag{14.25}$$

$$300 = n_c\left(2\cdot 6.426\cdot 10^{-2} + 2\cdot 1.8\cdot 0.5 + 2\cdot 7.534\cdot\frac{10^{-6}\cdot 0.05\cdot 238.8}{10^{-4}}\right)$$

$$= n_c(0.1285 + 1.8 + 1.8) \quad (14.26)$$

$$n_c = 80 \text{ turns/coil}$$

So the phase current

$$i_n = \frac{n_c i_n}{n_c} = \frac{238.8}{80} = 2.96 \text{ A} \quad (14.27)$$

The wire gauge is

$$d_{co} = \sqrt{\frac{4}{\pi}\frac{i_n}{j_{con}}} = \sqrt{\frac{4}{\pi}\frac{2.96}{6\cdot 10^6}} = 0.7938 \text{ mm} \quad (14.28)$$

Note: As seen from (14.26), the inductance voltage drop is still not very high (though already equal to emf) and thus the voltage ceiling utilization in the inverter is still acceptable.

So even higher current (and thrust) may be tolerated, provided the number of turns per coil n_c is reduced with a pertinent wire diameter increase.

The thrust density f_{tx} is

$$f_{tx} \approx \frac{F_x}{6L_{stack}\cdot 3.5\tau_s} = \frac{860 \text{ (N)}}{6\cdot 10 \text{ (cm)}\cdot 3.5\cdot 0.5 \text{ (cm)}} = 8.19 \text{ N/cm}^2 \quad (14.29)$$

This is a very large value for the low speed considered (0.5 m/s).

Note: Such a two-phase machine in Ref. [5] is also credited with very high thrust density; here a three-phase machine, working as in brushless dc (trapezoidal current) control mode, was investigated. 2D-FEM inquires to check the emf and thrust capability are still required, but this is beyond our scope here.

As the flux-switching LPMRM is very similar, in topology and modeling, to flux-reversal LPMRM, it will not be followed here further.

14.4 TRANSVERSE-FLUX LPMRM TECHNICAL THEORY

In order to secure a high thrust density, but still preserve a mildly large $L_q i_n/\psi_{PM}$ ratio (i.e., a not-so-small power factor), the so-called antifringing PM configuration in Figure 14.8b is considered. The one in Figure 14.8 contains IPMs and performs PM flux concentration and thus exhibits higher inductance L_q. We mention i_q control as it is in general performed (for motoring) to limit the i_d produced losses as the magnetic saliency (L_q/L_d ratio) is less than 1.3.

One more reason for reducing the PM fringing is that the value of ψ_{PMd} increases (Figure 14.9) and thus the thrust is obtained at smaller mmf (i.e., smaller copper losses and higher power factor); but, ultimately, the normal force increases notably. The passive, variable-reluctance, secondary stator is made of straight U-shaped transformer laminations, which makes it rather practical costwise.

A configuration of interest is redrawn here with a few geometrical parameters in display (Figure 14.12).

In general, $b_{ts}/g > 10$, $(\tau - b_{tp})/g > 3$; $1 \geq b_{tp}/\tau \geq 0.7$; $b_{ts} \approx b_{tp}$. The smaller the airgap, the better, but it may work with an airgap of 1 mm and more (perhaps up to 20 mm or so). There are three or six

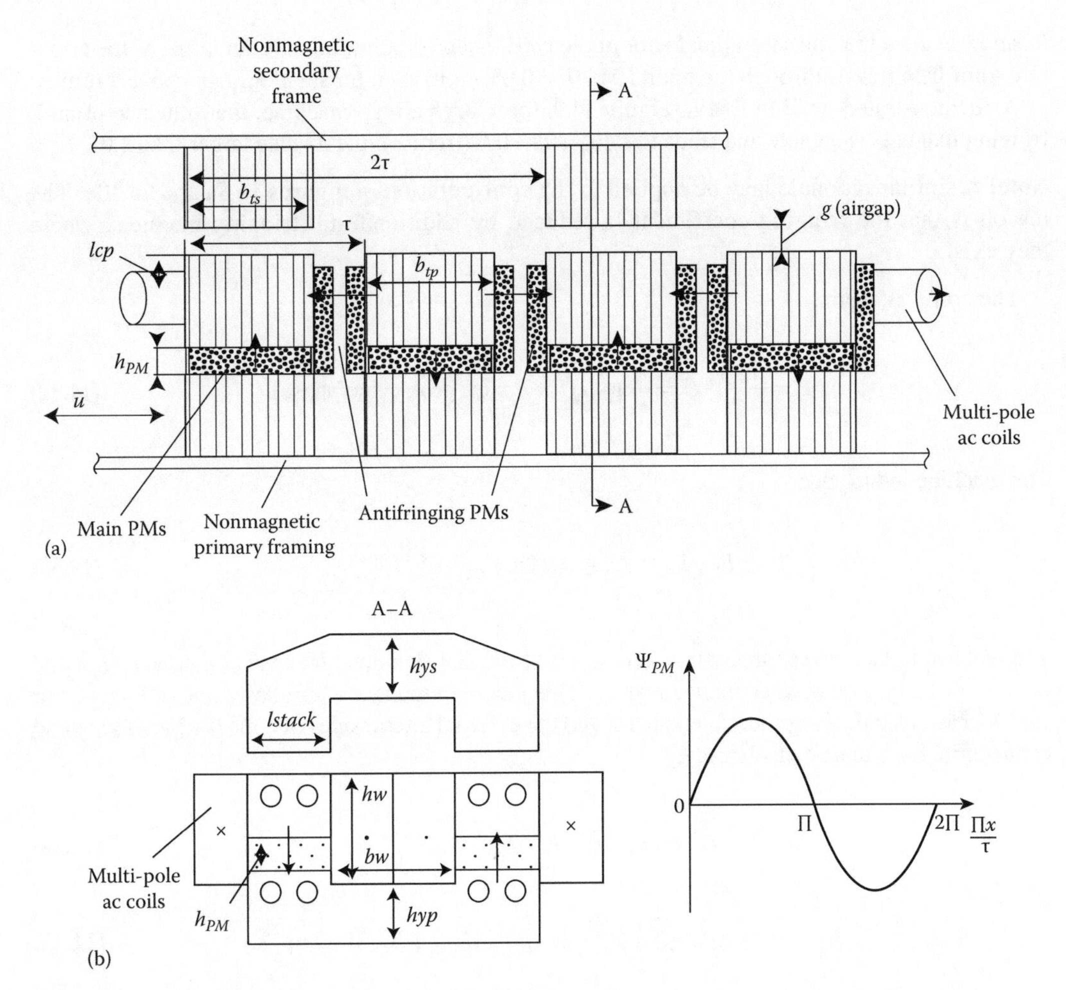

FIGURE 14.12 TF-LPMRM with antifringing additional PMs (one-phase module): (a) longitudinal view and (b) transverse cross section.

phases in a large TF-LPMRM to reduce cogging force, and the six units are phase-shifted along the direction of motion by $2\tau/m$; m is the number of phases.

A technical theory of TF-LPMRM is similar to the theory for FR-LPMRM, but the expression of total flux linkage in a phase $\psi_{PMs}(x)$, emf $E(x)$, machine phase inductance L_s and resistive R_s are somewhat different in the two cases. In this case, the mmf flux lines cross the main PMs and thus the design limiter is the zeroing of PM airgap flux by the peak phase current.

For TF-LPMRMs, the PM flux linkage in a coil tends to a sinusoidal space distribution more than to a trapezoidal one, and thus vector (field orientation) control is targeted.

Then, in a first approximation, mainly the peak value of PM flux linkage in the phase twin coils is required. As the PM flux contribution of antifringing PMs is not easy to calculate analytically, we consider their influence by allowing only a moderate main PM flux fringing:

$$\psi_{PMp} = B_{gPMi} \cdot p \cdot l_{stack} \cdot b_{ts} \cdot 2n_c \frac{1}{(1+k_{fringe}(l_{PMa}, h_{PMa}))(1+k_s)} \tag{14.30}$$

$$B_{gPMi} \approx B_r \frac{h_{PM}}{h_{PM} + g\mu_{rec}(\text{p.u.})}; \quad p\text{: pole pairs of primary} \tag{14.31}$$

It is now evident that the saturation factor in the investigated configuration is not large, as the armature mmf field travels through the main PMs (k_s=0.05 –0.10 even for small airgap: above 3 mm).

As demonstrated by 2D-FEM in Figure 14.9, for a large airgap machine, the influence of antifringing magnets is notable and leads to k_{fringe}=0.4–0.6 (from a typical value larger than 1.0).

Note: A similar rationale may be applied to the configuration in Figures 14.8a and 14.10e. The reason is that the fringing coefficient is reduced by additional antifringing magnets, where they exist.

The emf E is, again,

$$E = \frac{d\psi_{PM}}{dx}U = \frac{\pi}{\tau}\psi_{PMp}U = k_e n_c U \left(\text{peak value/phase}\right) \tag{14.32}$$

The machine inductance

$$L_r = L_l + L_m; \quad L_m \approx \mu_0 (2n_c)^2 \frac{pb_{ts}l_{stack}}{2(g + h_{PM})(1 + k_s)} \tag{14.33}$$

The leakage inductance expression is composed of the slot (window) leakage component (L_{ls}) and the interprimary core leakage inductance L_{la}. This latter component is similar to that of long power cables in air; also the longitudinal end parts and the external lateral sides of coils (a kind of long end connection) have to be considered, L_{le}:

$$L_{ls} = 2p\mu_0 b_{tp}(2n_c)^2\lambda_s; \quad \lambda_s \approx \frac{h_w}{3b_w} \tag{14.34}$$

$$L_{la} + L_{le} = \mu_0(2n_c)^2\left[2(\tau - b_{tp})p + 2\left(l_{stack} + b_w/2\right) + 2p\tau\right]\lambda_e \tag{14.35}$$

λ_e in general is lower than 0.15 (0.2), but it depends on coil height (h_w) per half width ($b_w/2$) ratio; the higher the ratio, the larger is λ_e. A more detailed modeling of leakage magnetic field and its inductance on TF-LPMRM (with 3D-FEM validation) is needed; but this is beyond our scope here.

The influence of longitudinal coil end connection $2(l_{stack}+b_w/2)$ in coil length is not very influential in the phase resistance and in the leakage inductance, as $2(l_{stack}+b_w/2)/p\tau<0.1$–$0.2$; this condition may be observed in sizing the length of each single-phase module; this may be the key merit of TF-LPMRM (rotary version): low copper losses per thrust are maintained.

The phase resistance R_s is

$$R_s = 2\rho_{co}l_{coil}\frac{j_{con}}{i_n}n_c; \quad \text{for two twin coils/phase module with the coil length } l_{coil} \tag{14.36}$$

$$l_{coil} = 2\left(2p\tau + l_{stack} + b_w/2\right) \tag{14.37}$$

The entire length of primary core phase module is about $2p\tau$.

Note: Core losses may be approached by analytical models (as done for SRM) [12] but care must be exercised because, being a single-phase machine, the TF-LPMRM experiences core losses both in primary and in the secondary. The 3D magnetic equivalent circuit model was used in Ref. [13] to determine the flux-density time variations in a few key regions of TF-LPMRM.

Analytical expressions, such as those already given in Chapters 11 and 12 for other LPMRMs, have been then used, where *dB/dt* is a key variable; no thorough test validation on core losses in TF-LPMRM is available yet. For small frequency and large thrust applications, even solid cores have been proposed [9,10], though the core losses are already large (even at less than 10 Hz). The thrust is reduced notably, with the main merit of lower core (and system) costs.

For a three-phase machine and sinusoidal emf and sinusoidal current, the average thrust is

$$F_{xav} = \frac{3}{2}\frac{\pi}{\tau}\left|\frac{d\psi_{dpx}}{dx}\right| i_{max} = \frac{3}{2}\frac{\pi}{\tau} k_e n_c i_{max} \cos\gamma, \quad i_{max} - \text{peak phase current} \tag{14.38}$$

where γ is the angle between the phase current and the emf; it is zero for pure i_q control; but the simultaneous propulsion and levitation control is not possible.

Concerning the normal force, for pure i_q control, the primary mmf produces magnetic flux density in the airgap B_{agq1}, which is phase shifted by 90° with respect to the ideal PM airgap flux density B_{gPMi1}; so the resultant flux density is

$$B_g(x,t) = B_{gPMi1}\cos\frac{\pi}{\tau}(x-\omega_r t) + B_{agq1}\sin\frac{\pi}{\tau}x\sin\omega_r t \tag{14.39}$$

So, on half the pole the armature reaction decreases the PM airgap flux density and it increases it along the other half. As position x varies in time, a corresponding time variation occurs. The PMs are producing a traveling field while the single-phase mmf produces a fixed magnetic field. So, as for the single-phase thrust, the normal force on each single-phase module produces a constant component and an ac component at $2\omega_r$. In the symmetric three-phase case, the ac components in both the thrust and the normal force compensate each other.

So the normal force per single-phase module is

$$(F_y)_{phase} \approx p\tau 2 l_{stack} \int_{-\frac{\tau}{2}}^{\frac{\tau}{2}} \frac{B_g^2(x,t)}{2\mu_0} dx \tag{14.40}$$

The average value of normal force for three-phase modules is

$$(F_{yav})_{phase} \approx \frac{\left(B_{gPMi1}^2 + B_{agg1}^2\right) b_{ts} p(2l_{stack})}{2\cdot 2\mu_0} \tag{14.41}$$

$$B_{agg1} \approx \frac{\mu_0 2 n_c i_{max}}{2\left(g + \frac{h_{PM}}{\mu_{rec}(\text{p.u.})}\right)(1+k_s)} \tag{14.42}$$

14.5 EXAMPLE 14.2: TF-LPMRM

The TF-LPMRM—intended for a suburban MAGLEV—as in Figure 14.12 has the following geometrical data:

- The U-shaped passive secondary width is 0.4 m ($l_w = l_{stack} = 0.133$ m).
- Airgap: $g = 6$ mm.

- Pole pitch: 60 mm.
- The primary pole span: $b_{tp}/\tau = 4/6$.
- The PM thickness: $h_{PM} = 12$ mm.
- The use of antifringing PMs leads to a fringing coefficient: $k_{fringe} = 0.5$.
- $B_r = 1.2$ T, $\mu_{rec}/\mu_0 = 1.05$ for the used PMs.
- The base propulsion force $F_{xavr} = 10$ kN.
- The normal force/thrust ratio $\geq 3/1$.
- The base speed $U_b = 10$ m/s.

To be calculated:

a. The PM flux density in the airgap at the aligned position.
b. For an mmf that produces $B_{agn}/B_{gPM} = 0.8$, the number of poles $2p$ per phase module; the number of phase modules (the phase modules may be attached separately to the bogie through a secondary (mechanical) suspension system).
c. The normal force/thrust ratio.
d. Copper losses for F_{xavr} and the efficiency at 10 m/s for the iron losses at 20% of copper losses.
e. Determine the power factor at base speed.

Solution

a. We may start with Equation 14.31 and determine the ideal PM airgap flux density B_{gPMi}:

$$B_{gPMi} = B_r \frac{h_{PM}}{h_{PM} + g\left(\mu_{rec}/\mu_0\right)} = 1.2 \frac{12}{12 + 6 \cdot 1.05} = 0.7868 \text{ T} \tag{14.43}$$

After the reduction due to PM flux fringing ($k_{fringe} = 0.5$, with antifringing PMs) and some magnetic saturation ($k_s = 0.05$), the active flux density (which manifests itself in the emf), B_{gPM}, is

$$B_{gPM} = \frac{B_{gPMi}}{(1 + k_{fringe})(1 + k_s)} = \frac{0.7869}{1.5(1 + 0.05)} \approx 0.5 \text{ T} \tag{14.44}$$

The peak PM flux in the two twin coils of one-phase module is (from 14.30)

$$\psi_{PMp} = B_{gPM} l_{stack} b_{ts} p \cdot 2n_c = 0.5 \cdot 0.133 \cdot 0.04 \cdot 2 \cdot pn_c = 5.32 \cdot 10^{-3} pn_c \tag{14.45}$$

b. For $B_{aq} = 0.8 \cdot B_{gMP} = 0.8 \cdot 0.5 = 0.4$ T the coil mmf peak value $n_c i_{max}$ is (14.42)

$$B_{aq} = 0.4 \approx \frac{\mu_0 2 n_c i_{max}}{2\left(g + \frac{h_{PM}}{\mu_{sec}(\text{p.u.})}\right)(1 + k_s)} = \frac{2 n_c i_{max} \cdot 1.256 \cdot 10^{-6}}{2\left(6 + \frac{12}{1.05}\right) \cdot 1.05 \cdot 10^{-3}} \tag{14.46}$$

So

$$n_c i_{max} = 5.828 \cdot 10^3 \text{ a turns per coil peak value} \tag{14.47}$$

With a current density $j_{con}=5$ A/mm^2 and $k_{fill}=0.6$ (preformed coils), the window area $l_w h_w$ is

$$l_w h_w = \frac{2n_c i_{max}}{\sqrt{2} j_{con} k_{fill}} = \frac{2 \cdot 5.828 \cdot 10^3}{\sqrt{2} \cdot 5 \cdot 10^6 \cdot 0.6} = 2.755 \cdot 10^{-3}\ \text{m}^2 \tag{14.48}$$

With $l_w=0.133$ m, the window height $h_w=2.755 \cdot 10^{-3}/0.133=0.207$ m; the "slot" aspect ratio of only $h_w/l_w=207/133=1.55$ should lead to an acceptable slot leakage inductance.

c. For the thrust, we still have to find the number of poles $2p$ per one-phase module(s) (14.38):

$$F_{xav} = 10 \cdot 10^3 = \frac{3}{2}\frac{\pi}{\tau}\psi_{PMp} i_{max} = \frac{3}{2}\frac{\pi}{0.06} \cdot 5.32 \cdot 10^{-3}\, p n_c i_{max} \tag{14.49}$$

So the number of poles $2p$ per one-phase module(s) is $2p=8$.

As the pole pitch $\tau=60$ mm, the total length of one-phase module $2p\tau=8 \cdot 60=480$ mm $=0.48$ m.

Such a length may be considered the only one needed for one phase. Again, the three-phase modules are properly phase shifted by $2\tau/3$ with respect to each other and fixed separately to the bogie.

The thrust density f_{xav} per total primary active area A_p is

$$f_{xav} = \frac{F_{xav}}{(12 p\tau l_{stack})} = \frac{10{,}000}{(6 \cdot 0.48 \cdot 0.133)} = 26{,}106\ \text{N/m}^2 = 2.6\ \text{N/cm}^2 \tag{14.50}$$

This is not a very large thrust density, but it is one that is obtained at reasonable copper losses.

Let us consider the normal force at zero current ($\overline{i_s} = \overline{j} \cdot i_q = 0$), from (14.41):

$$(F_{yav})_{phase} = 3\frac{B_{gPMi}^2}{2\mu_0} b_{ts} p \cdot 2l_{stack} = 3\frac{0.7868^2}{2 \cdot 1.256 \cdot 10^{-6}} 0.04 \cdot 4 \cdot 2 \cdot 0.133 = 31{,}465\ \text{N} \tag{14.51}$$

The ideal PM flux density B_{gPMi} is considered here in its flattop distribution. The ratio 3/1 of F_{xav}/F_{yav} is fulfilled. But this demonstrates that the TF-LPMRM is not capable alone to produce full levitation (which would need a 10/1 ratio, as proved with LHSMs or dc excited LSMs in the preceding chapters).

The main reason for this is the fact that the secondary laminated U-shaped cores occupy only 33% of primary length; this is enough for propulsion but not for full levitation; however, this secures lower secondary (stator) passive guideway costs.

d. The copper losses are (14.36) through (14.37)

$$p_{con} = 3 \cdot 2\rho_{co} l_{coil} j_{con} n_c i_n = 3 \cdot 2 \cdot 2.1 \cdot 10^{-8} \cdot 1.36 \cdot 5 \cdot 10^6 \frac{5820}{\sqrt{2}} = 3536\ \text{W} \tag{14.52}$$

$$l_{coil} = 2\left(\frac{8 \times 0.06 + 0.133 + 0.133}{2}\right) = 1.36\ \text{m} \tag{14.53}$$

So the ratio

$$\frac{F_{xav}}{p_{con}} = \frac{10{,}000}{3{,}536} = 2.827\ \text{N/W} \tag{14.54}$$

The efficiency η_b is

$$\eta_b = \frac{F_{xav}U_b}{F_{xav}U_b + p_{con} + p_{iron}} = \frac{10,000 \cdot 10}{10,000 \cdot 10 + 3,536(1+0.2)} = 0.959 \tag{14.55}$$

At 10 m/s, this is a rather large efficiency.

e. The power factor may be calculated using the vector diagram in Figure 14.3 (typical to all LSMs).

The frequency at base speed, f_b, is

$$f_b = \frac{\omega_b}{2\pi} = \frac{523.35}{(2\pi)} = 83.33 \text{ Hz} \tag{14.56}$$

The machine inductance components (14.33 through 14.35) are

$$L_m = \mu_0 (2n_c)^2 \frac{pb_{ts} l_{stack}}{2\left(g + \frac{h_{PM}}{\mu_{rec}(\text{p.u.})}\right)(1+k_s)} = \frac{1.256 \cdot 10^{-6} \cdot 4 \cdot 4 \cdot 0.04 \cdot 0.133}{2\left(6 + \frac{12}{1.05}\right)(1+0.05) \cdot 10^{-3}} n_c^2 = 2.921 \cdot 10^{-6} n_c^2 \tag{14.57}$$

$$L_{ls} = \mu_0 2 p b_{tp} (2n_c)^2 \lambda_s = 1.256 \cdot 10^{-6} \cdot 8 \cdot 0.04 \cdot 4 \cdot 0.5175 \cdot n_c^2 = 0.831 \cdot 10^{-6} n_c^2 \tag{14.58}$$

$$\lambda_s = \frac{h_w}{3b_w} = \frac{207}{3 \cdot 133} = 0.5175 \tag{14.59}$$

$$\begin{aligned} L_{la} + L_{le} &= \mu_0 (2n_c)^2 \left[2(\tau - b_{tp})p + 2\left(l_{stack} + b_w/2\right) + 2p\tau\right]\lambda_e \\ &= 1.256 \cdot 10^{-6} \cdot 4\left[2(0.06 - 0.04) \cdot 4 + 2\left(0.133 + 0.133/2\right) + 8 \cdot 0.06\right] \cdot 0.15 \\ &= 0.6326 \cdot 10^{-6} n_c^2 \end{aligned} \tag{14.60}$$

As there is no coupling between phases, the phase inductance

$$L_s = L_q = L_m + L_{ls} + L_{la} + L_{le} = (2.92 + 0.831 + 0.6326) \cdot 10^{-6} n_c^2 = 4.3836 \cdot 10^{-6} n_c^2 \text{ H} \tag{14.61}$$

The emf (14.32) is

$$E = \frac{\pi}{\tau} \psi_{PMp} U = \frac{\pi}{0.06} \cdot 5.32 \cdot 10^{-3} \cdot 4 \cdot n_c \cdot 10 = 11.136 \cdot n_c \tag{14.62}$$

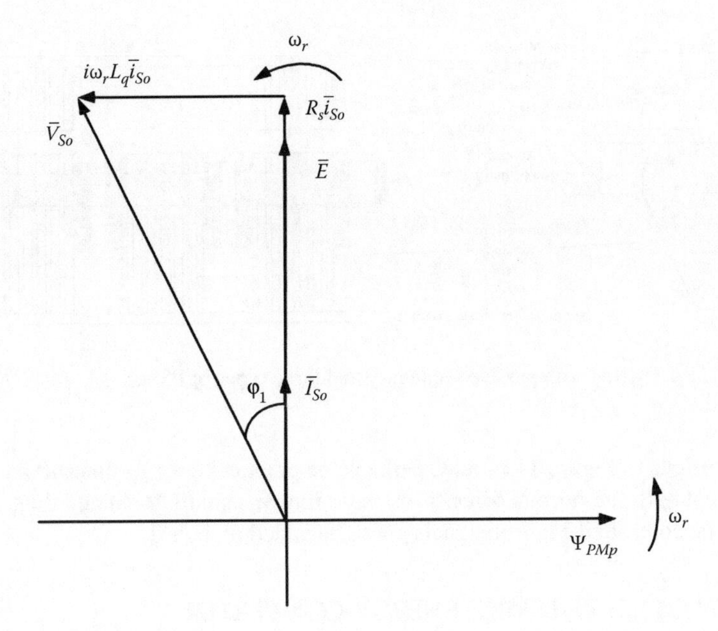

FIGURE 14.13 TF-LPMRM vector diagram (pure i_q mode).

Similarly (14.36),

$$R_s i_s = \frac{n_c p_{con}}{n_c i_{\max}} \frac{2}{3} = \frac{n_c \cdot 3536}{5820} \frac{2}{3} = 0.401 \cdot n_c \tag{14.63}$$

So

$$\omega_r L_q i_s = 523.33 \cdot 4.3836 \cdot 10^{-6} \cdot n_c \cdot 5820 = 13.351 \cdot n_c \tag{14.64}$$

So, from Figure 14.13,

$$\varphi_1 = \tan^{-1} \frac{\omega_r L_q i_s}{E + R_s i_s} = \tan^{-1} \frac{13.351 \times n_c}{(0.401 + 11.136) n_c} = \tan^{-1} 1.157 = 49°; \quad \cos \varphi_1 = 0.656 \tag{14.65}$$

Discussion

In spite of the mild thrust density f_{xav}=2.60 N/cm² the power factor is rather low, though the efficiency is very good. Even larger airgaps (in fact thicker PMs!) are required for better power factor, as efficiency is already very good. The placements of both PMs and the ac coils on the primary (the mover in our case) are "paid for" by roughly 50% PM utilization inherent limitation.

When the PMs are not placed on the primary, the PMs may all work, just to create four airgaps (instead of two); the ac primary has both transverse U and (in between) I-shaped cores; but the I cores lead to an "alarming" increase in the leakage inductance.

The IPM use on the primary allows PM flux concentration (Figure 14.8a), with additional PMs to reduce the machine inductance (Figure 14.14), and may lead to thrust densities of 10 N/cm² for the same geometry for similar power factor and efficiency (at 1.0 T useful PM maximum flux density in the U-shaped core of primary).

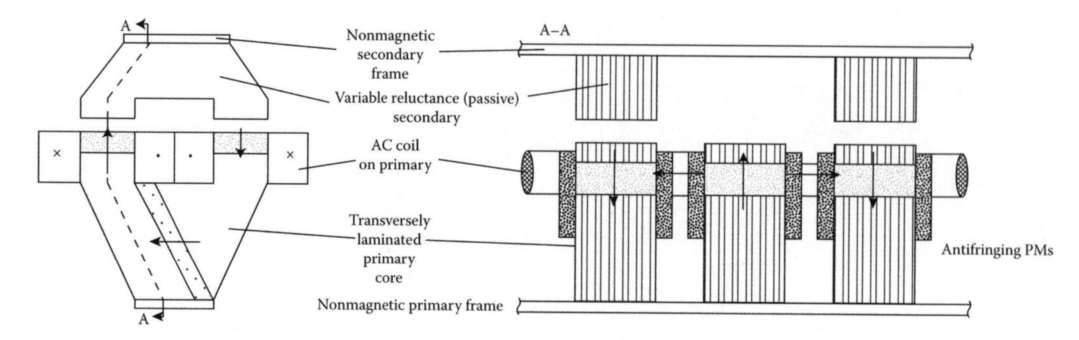

FIGURE 14.14 TF-LPMRM with passive secondary and three types of PMs.

The configurations in Figure 14.14 may, perhaps, be practical even for interurban MAGLEVs at 10 mm airgap; but still, the normal force/thrust ratio remains small (3–6) and thus, for full levitation, additional dc controlled electromagnets are to be added on board.

14.6 EXAMPLE 14.3: TF-LPMRG ENERGY CONVERTER

Let us consider a large, 250 kW, 1 m/s wave energy harvester and design (a) transverse flux linear permanent magnet reluctance generator (TF-LPMRG) generator for the scope; the primary is fixed and the passive secondary is the mover.

Solution

We will consider the configuration in Figure 14.14, but, to reduce the primary length, we will place the six phases (to reduce cogging force drastically) along a hexagon. This way the passive secondary of each phase placed on the mover (to avoid flexible electric power cable and "properly" insulate the primary from the corrosive environment, typical in wave energy harvesting) is shifted by $2\tau/3$ with respect to each other.

In addition, the fully active primary becomes shorter and thus the total length of secondary (primary length plus stroke length) is also reduced.

As the chosen configuration contains three types of PMs (visible in Figure 14.15), two of them for direct large PM flux production for a reasonable machine inductance, and one between the primary poles for antifringing, it may be feasible to produce 1.2 T of "useful" (emf producing) PM flux density in the airgap. In addition, at the limit, the phase mmf is allowed to produce a peak airgap flux density $B_{ag1}=1.0$ T. This means that the core laminations allow for higher saturation flux density (2.3 T at 20 kA/m). Despite its inherent large specific core losses, such material is chosen here to reduce the size and weight and, finally, core losses in the system.

Let us adopt (for the mild travel length of a few meters at most) a mechanical airgap $g=2.5$ mm and $SmCo_5$ PMs (corrosivity-resistant) with $B_r=1.2$ T and $\mu_{rec}=1.05$ (p.u.).

The pole pitch is better to be small but, to reduce PM flux fringing, $\tau/g\geq 20$, at least; so $\tau=20g\cdot 2.5=50$ mm. The stator (and mover) tooth length $b_{tp}\approx b_{ts}\approx 35$ mm; the interpole space in the primary $\tau-b_{tp}=50-35=15$ mm $=6g$, large enough to keep PM flux fringing within limits and leave room for primary pole mounting on the lateral nonmagnetic frame.

From now on, the design pursuit is similar to one in Example 14.2 but the PM flux concentration in the primary has to be considered. So the active (useful), emf producing,—PM flux density B_{gPM} is

$$B_{gPM}\approx\frac{B_r l_{PMc}}{l_{stack}\left(1+\dfrac{l_{PMc}(g\mu_{sec}+h_{PM})}{h_{PM}l_{stack}}\right)}\frac{1}{(1+k_{fringe})(1+k_s)} \tag{14.66}$$

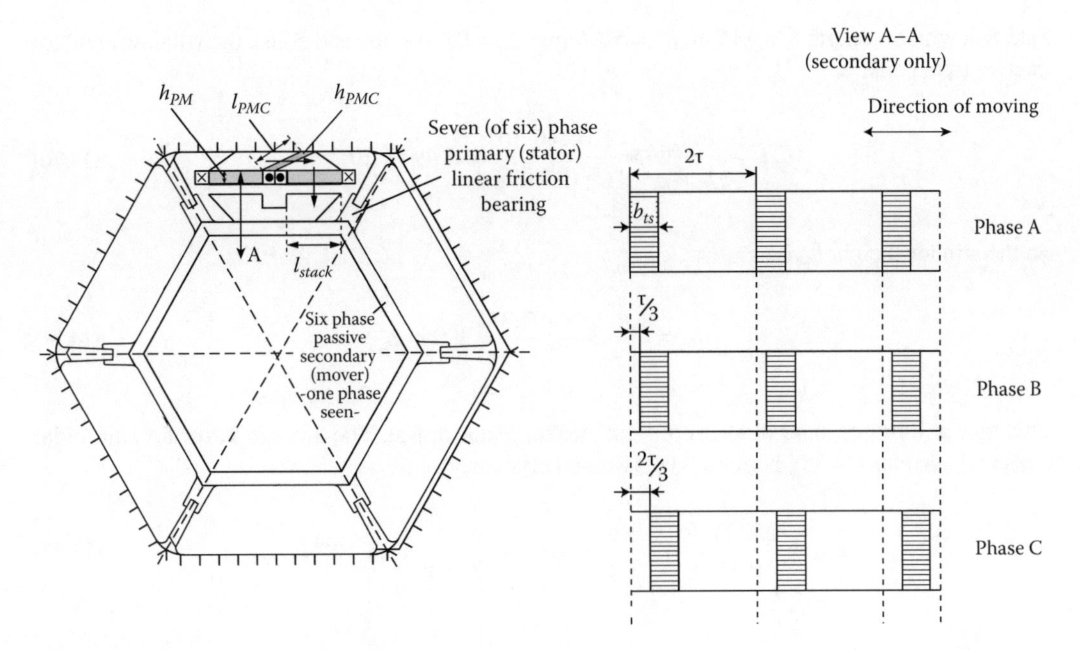

FIGURE 14.15 Hexagonal TF-LPMRG wave generator.

The contribution of both PMs, visible in Figure 14.15, is considered (14.66); the antifringing effect of the additional PMs (Figure 14.14) is considered through reducing the fringing coefficient $k_{fringe}=0.5$; here the antifringing effect of cutting the lamination corners is also considered [11].

To reduce the machine inductance, $h_{PM}=4g=10$ mm and the thickness of PM for flux concentration, $h_{PMC}=15$ mm! Thus the only unknown in (14.66) is the l_{PMC} (the length of PMs for flux concentration); if we choose the stack length l_{stack} ($l_{stack}=0.2$ m, for an external circle radius of secondary of roughly 1 m),

$$1.2=\frac{1.3\cdot l_{PMC}}{200\left(1+\dfrac{2\cdot l_{PMC}(2.5\cdot 1.05+10)}{15\cdot 200}\right)}\frac{1}{(1+0.5)(1+0.05)} \tag{14.67}$$

It may be proven that we obtain a negative l_{PMC} from (14.67), which is not practical. So let us eliminate the radial PMs (10 mm in thickness) and adopt $h_{PM}=0$ and increase the axial PMs thickness $h_{PMC}=25$ mm. Then applying (14.66) again, we get $l_{PMC}=418.5$ mm.

The PM flux in a phase, ψ_{PMp}, from (14.45), is

$$\psi_{PMp}=B_{gPM}l_{stack}b_{ts}p\cdot 2n_c=1.2\cdot 0.2\cdot 0.035\cdot 2\cdot pn_c=16.8\cdot 10^{-3}pn_c \tag{14.68}$$

The peak mmf per phase is already defined by $B_{agq1}=1$

$$B_{agq1}\approx 1.0\,\text{T}\approx\frac{2\mu_0 n_c i_{\max}}{2g+h_{PMC}\dfrac{l_{stack}}{l_{PMC}}}=\frac{2\cdot 1.256\cdot 10^{-6}\cdot n_c i_{\max}}{\left(2\cdot 2.5+25\dfrac{200}{418.5}\right)10^{-3}};\quad n_c i_{\max}=6746\cdot 10^3\ \text{A‡turns/coil} \tag{14.69}$$

Taking a window length $l_w=0.15$ m, $j_{con}=5$ A/mm², $k_{fill}=0.6$ (preformed coils), the window area, for the two twin coils, is

$$l_w h_w = \frac{2n_c i_{max}}{j_{con} k_{fill}} = \frac{2 \cdot 6.746 \cdot 10^3}{5 \cdot 10^6 \cdot 0.6} = 4.4977 \cdot 10^{-3}\ \text{m}^2 \tag{14.70}$$

So the window height h_w is

$$h_w = \frac{l_w h_w}{l_w} = \frac{4497.7}{150} = 300\ \text{mm} \tag{14.71}$$

This may still be accepted in terms of "slot" leakage inductance. The average thrust for sinusoidal emfs and currents (14.38), is (for a 0.96 assumed efficiency)

$$F_{xav} = \frac{250\ \text{kN}}{0.96} = 2 \cdot \frac{3}{2} \frac{\pi}{\tau} \psi_{PMp} i_{max} = \frac{6}{2} \frac{\pi}{0.05} 16.8 \cdot 10^{-3}\, p n_c i_{max} \tag{14.72}$$

So $p \approx 12$.

With a pole pitch $\tau=50$ mm, the primary length $l_p=2p\tau=24 \cdot 50=1200$ mm$=1.2$ m; this is an acceptable value for the application.

The copper losses p_{con} are (14.52)

$$p_{con} = 6 \cdot 2\rho_{co} l_{coil} j_{con} \left(\frac{n_c i_{max}}{\sqrt{2}} \right) = 6 \cdot 2 \cdot 2.1 \cdot 10^{-8} \cdot 3 \cdot 5 \cdot 10^6 \frac{6746}{\sqrt{2}} = 18{,}085\ \text{W} \tag{14.73}$$

$$l_{coil} \approx 2\left(2p\tau + \frac{3}{2} l_{stack} \right) = 2\left(1.2 + \frac{3}{2} \cdot 0.2 \right) = 3\ \text{m} \tag{14.74}$$

So, neglecting core losses—the frequency $f_n=U_n/(2\tau)=1/(2 \cdot 0.05)=10$ Hz—the efficiency η_n is

$$\eta_n = \frac{P_n}{P_n + p_{con}} = \frac{250 \cdot 10^3}{250 \cdot 10^3 + 18085} = 0.938! \tag{14.75}$$

The efficiency at this very low speed (1 m/s) is considered quite good although the design may be reiterated as $\eta_n=0.938<\eta_{assumed}=0.96$.

The design may be continued as in Example 14.2 to find the power factor but, as $B_{gq1}/B_{gPM}=1.0/1.2$, it means that $L_m i_{max}/\psi_{PMp} \approx 1.0/1.2$; with a leakage inductance $L_l \approx 0.3\, L_m$, we will obtain a power factor of about 0.62–0.65, even if the frequency is low ($f_n=10$ Hz). This is due to the armature reaction flux density B_{agq1}, accepted so large (1 T) to produce high thrust density.

Discussion

Comparing the design here with that for a tubular LPMSM (Section 13.9), for the same data, shows, for a similar size, that the TF-LPMRG "sports" a 0.938 efficiency versus only 0.784 for the tubular T-LPMSG, but the power factor is notably better with the latter. Also, the T-LPMSG was designed with a PM secondary mover placed along the entire stroke plus primary length, while the TF-LPMRG uses a passive variable reluctance rugged secondary (mover).

Note on MPC-LPMRMs: The so-called flux-reversal, flux-switching, and transverse-flux linear PM reluctance machines may produce either a trapezoidal or a close to sinusoidal emf and thus may be controlled as brushless dc or as brushless ac machines. As there are no notable differences in the control of MP-PLMRMs with respect to flat-LPMSMs or tubular LPMSM control treated in Chapters 12 and 13, we will not insist here on the subject. Also, most control methods applied to rotary PMSMs may be "transplanted" here for motoring and generating modes.

14.7 SUMMARY

- Besides linear PMSM with around unipole-span ac coils in distributed or fractionary (tooth-wound) windings, this chapter has dealt with multi-pole ac coils winding linear PM reluctance motors that are characterized by a kind of thrust multiplication ratio (equal to the number of pole pairs per coil); they have been called MPC-LPMRMs.
- MPC-LPMRMs, as any invention, bear different names such as flux-reversal [4,14,15], flux-switching [7], or transverse-flux [11] LPMRMs, but, in fact, the reversal of PM flux in the ac coil single-phase primary modules characterizes them all.
- Allowing for a bit of decoupling of magnetic circuit and electric circuit design, higher mmf for small pole pitch becomes feasible; consequently, larger thrust density is claimed.
- As all MPC-LPMRMs rely on some variable reluctance effect, their inductance flux drop $L_s i_n$ per PM flux linkage ψ_{PMp} is not very small, and thus the equivalent power factor decreases.
- PM flux fringing—PM flux "loss" due to flux fringing around airgap—reduces the ideal ψ_{PMp} by more than two times unless special measures are taken. This indicates that MPC-LPMRMs are PM intensive for the thrust they produce; still the specific thrust densities $f_{xt} \approx 6$–8 N/cm^2 are feasible in proper proportioned geometries (pole-pitch/airgap, primary interpole b_{sp}/airgap, PM thickness/pole pitch, etc.).
- The main advantage of MPC-LPMRMs, besides thrust density, is the rather low copper losses per Newton of thrust.
- All MPC-LPMRMs may be built with IPM or SPM secondaries and ac three or two or more phases primaries with variable reluctance cores; also, the primary may contain the PMs when the secondary becomes passive, with variable reluctance.
- Being essentially one-phase module three- or two-phase machines, both cores—of primary and of secondary—experience eddy current and hysteresis losses and so are the PMs (eddy currents); so all magnetic cores are laminated.
- The cogging force may be reduced by skewing (pole displacement), increase in the number of poles (to 6 for example), and, for flux-switching MPC-LPMRM, by a proper combination of the number of teeth of primary and secondary pole counts: $N_s = N_p(2n - 1) \pm k$; n—number of teeth per coil span.
- All MPC-LPMRMs may be built with transverse flux when the differences between them become insignificant.
- The single-phase modules assembled to make a three or two or more phase machines lead to thrust and normal force pulsations of each module at $2f_1$ (f_1—fundamental frequency); they cancel each other in the symmetric three-phase machine, but locally they are manifest and may produce noise and vibration. Also, the core losses in the primary and in the secondary core are notable due to this single phase topology.
- 2D and 3D FEM analyses are required to validate technical field theories, which are easy to use in preliminary design, though the large fringing flux makes them problematic in terms of precision.
- Design examples have shown the MPC-LPMRMs suitable even for speeds down to 0.5–1 m/s for acceptable efficiency and good enough power factor (in flux-reversal and flux-switching topologies) or for very good efficiency but for lower power factor in TF-LPMRMs.

- The normal force to thrust ratio is around 10/1 for FR and FS topologies and below 5/1 for TF topologies in general; so, basically, TF-LPMRMs cannot provide full levitation besides propulsion in industrial, urban, or interurban MAGLEV carriers.
- Thrust up to 279 kN has been proven feasible in Example 14.3 with TF-LPMRGs at 1 m/s for wave energy direct conversion at 93.8% efficiency, 10 Hz, and 0.6 power factor for an 8 N/cm^2 thrust density.
- The control of MPC-LPMRM(G)s is similar to that of BLDC or BLACPM motors or to flat and tubular-LPMSMs, as illustrated in Chapter 13.

REFERENCES

1. T. Kenjo, *Stepping Motors and Their Microprocessor Control*, Oxford University Press, Oxford, U.K., 1989.
2. I.A. Viorel, Z. Kovács, and L. Szabó, Sawyer type linear motor dynamic modelling, *Proceedings of the International Conference on Electrical Machines (ICEM)*, Manchester, U.K., vol. 2, 1992, pp. 697–701.
3. M. Sanada, S. Morimoto, and Y. Takeda, Vibration suppression linear pulse motors, *Record of IEEE-IAS-1994 Annual Meeting*, Denver, CO, vol. 1, 1994, pp. 517–522.
4. I. Boldea and S.A. Nasar, *Linear Motion Electromagnetic Devices*, Chapter 5, Taylor & Francis, New York, 2001.
5. M. Karita, M. Nakagawa, and M. Maeda, High thrust density linear motor and its applications, *Record of LDIA-1995*, Nagasaki, Japan, 1995, pp. 183–186.
6. I. Boldea, J. Zhang, and S.A. Nasar, Theoretical characterization of flux reversal machine in low speed servo-drives: The pole PM configuration, *IEEE Trans.*, IA-38(6), 2002, 1549–1557.
7. M.-J. Jin, C.-F. Wang, J.-X. Shen, and B. Xia, A modular PM flux-switching linear machine with fault-tolerant capability, *IEEE Trans.*, MAG-45(8), 2009, 3179–3186.
8. J. Chang, D.-H. Kang, J. Lee, and J. Hong, Development of transverse flux linear motor with PM excitation for direct drive applications, *IEEE Trans.*, MAG-41(5), 2005, 1936–1939.
9. J.-Y. Lee, J.-W. Kim, J.-H. Chang, S.-U. Chung, D.-H. Kang, and J.-P. Hong, Determination of parameters considering magnetic nonlinearity in solid core transverse flux linear motor for dynamic simulation, *IEEE Trans.*, MAG-44(6), 2008, 1566–1569.
10. J.-Y. Lee, J.-W. Kim, S.-R. Moon, J.-H. Chang, S.-U. Chung, D.-H. Kang, and J.-P. Hong, Dynamic characteristics analysis considering core losses in transverse flux linear machine with solid core, *IEEE Trans.*, MAG-45(3), 2009, 1776–1779.
11. T.-K. Hoang, D.-H. Kang, and J.-Y. Lee, Comparison between various designs of transverse flux linear motor in terms of thrust force and normal force, *IEEE Trans.*, MAG-46(10), 2010, 3795–3810.
12. L. Strete, L. Tutelea, I. Boldea, C. Martis, and I.A. Viorel, Optimal design of a rotary transverse flux motor (TFM) with PM in rotor, *Record of ICEM 2010*, Rome, Italy, 2010.
13. P.N. Materu and R. Krishnan, Estimation of switched reluctance motor losses, *IEEE Trans.*, IA-28(3), 1992, 668–679.
14. J.-Y. Lee, D.-H. Kang, J.-H. Jang, and J.-P. Hong, Rapid eddy current loss calculation for transverse flux linear motor, *IEEE-IAS 2006 Annual Meeting*, Tampa, FL, 2006, pp. 400–406.
15. S.-U. Chung, H.-J. Lee, and S.-M. Hwang, A novel design of LSM using FRM topology, *IEEE Trans.*, MAG-44(6), 2008, 1514–1517.

15 Plunger Solenoids and Their Control

15.1 INTRODUCTION

Plunger solenoids are electromagnetic actuators with 1(2) dc- or ac-fed coils, without or with permanent magnets, with limited (from a few tens of mm to 10–20 mm) linear motion of a soft magnetic core (Figure 15.1a) or PM mover (plunger Figure 15.1b) via attraction or even repulsive electromagnetic forces. Plunger solenoids are also called electromagnets and are used extensively in standard applications such as electric power switches (relays), door lockers etc., enjoying substantial worldwide markets.

Reference [1] presents a classical study of electromagnets.

Essentially, for plunger solenoids, on–off supply control is used as the forward motion is secured by the electromagnetic force while the backward motion is done by the mechanical spring to save energy [2].

To avoid the bouncing of the plunger as travel ends, a more advanced supply (and control) has to be provided. Bouncing is detrimental to the fixture that meets the plunger, in the sense of wearing, and should be limited.

In some applications, where the extreme positions should be maintained without current in the coil(s)—zero latching losses—the PM force has to overcompensate the mechanical spring force (Figure 15.2) [3].

For controlled positioning (with or without mechanical spring), advanced power control is needed, because, in general, the open loop system is statically and dynamically unstable (without PMs and without a mechanical spring).

Magnetic suspension in MAGLEVs is a particular case of controlled position solenoids, but, due to the interference of vehicle motion and the multiple degrees of freedom motions, the latter's design and control will be discussed in separate chapters. Also, in applications with constant frequency oscillatory linear motion, operation at resonance is sought to increase efficiency (and power factor) and thus reduce the power electronics kilovoltampere (kVA) ratings (and costs).

Linear oscillatory motors will thus be dealt with in a dedicated chapter.

Therefore, plunger solenoids treated here are restricted to nonoscillatory linear reciprocating motion when they close and open fluid flow (as in valve actuators) or electric current flow (as in electric contactors, etc.); also, small travel fast response is investigated with priority; position control will also be investigated for industrial small travel applications.

In view of the many topologies and applications, we will start here with the principles of PM-less plunger solenoids, then continue with their modeling, considering magnetic saturation and eddy currents; subsequently, a few design and control studies are developed in considerable detail, for configurations without and with PMs. Finally, a rather complete case study of FEM and dynamic modeling, optimal design, and position sensor-less control for PM twin-coil valve actuators is performed.

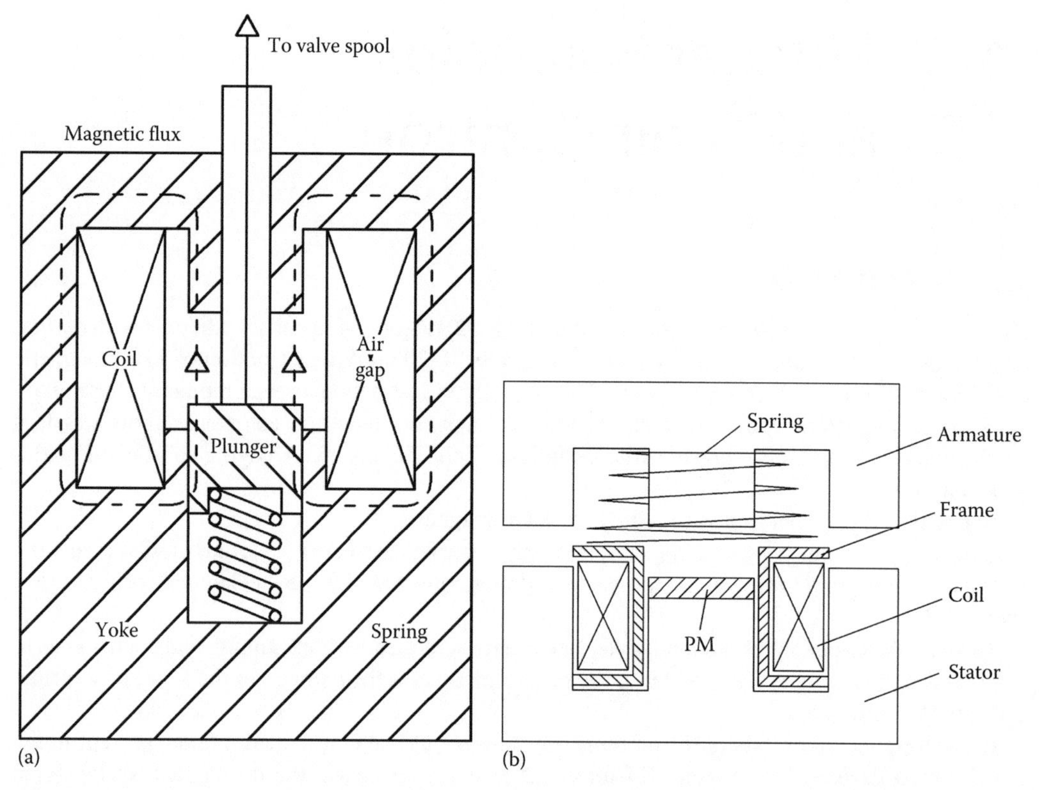

FIGURE 15.1 Cylindrical-shaped plunger solenoid: (a) without PM and (b) ac power switch with PMs in the stator.

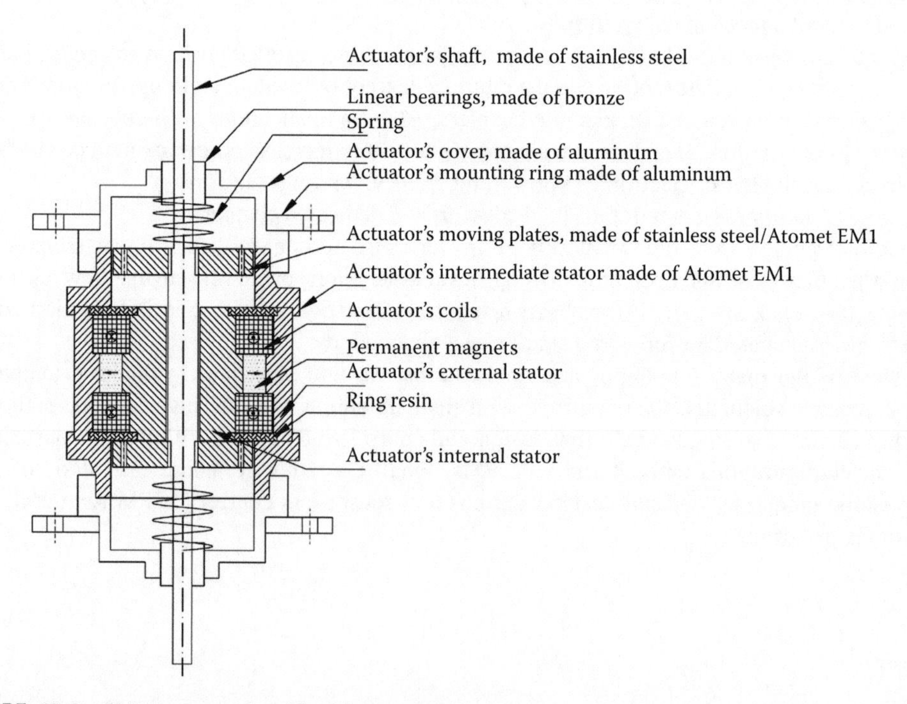

FIGURE 15.2 PM plunger solenoid with zero latching losses. (After Agarliţă, S., Linear single phase PM oscillatory machine: Design and control, PhD thesis, UPT.ro, 2009.)

15.2 PRINCIPLES

The PM-less plunger solenoids (Figure 15.1) include a cylindrical (in general) stator with a mild steel solid core or made of solid magnetic composite (SMC) with rather high permeability and small enough electrical conductivity ($\sigma_{el} < 1.5 \cdot 10^6\ \Omega^{-1}m^{-1}$), a multiturn coil and an SMC mover (plunger).

When the coil is voltage supplied, the current starts increasing and the plunger moves against the spring force until it closes the airgap (a residual resin-filled airgap is practical to avoid stacking).

When the voltage is turned off, the mechanical spring increases the airgap rather quickly. So far, no control is implied.

The attraction force between the plunger and the stator core may be calculated based on stored magnetic energy W_{mag} or coenergy W_{comag}:

$$F_x = -\left(\frac{\partial W_{mag}}{dx}\right)_{\psi=ct} = \left(\frac{(\partial W_{comag})}{dx}\right)_{i=ct} \quad (15.1)$$

with

$$W_{mag} = \in \int_0^{\psi} i d\psi; \quad W_{comag} = \in \int_0^{i_0} \psi di \quad (15.2)$$

where

ψ is the flux linkage
i is the current

The simple formulae (15.1) and (15.2) imply that no eddy currents are induced in the stator. In the absence of magnetic saturation, the dependence of total magnetic flux linkage of the coil flux linkage ψ on current i is linear and the slope depends on plunger position x (Figure 15.3).

The instantaneous force F_x is the same with both formulae in (15.1) and, for increased airgap x by Δx, the magnetic energy decreases while the coenergy increases (Figure 15.3) with airgap increase. Magnetic saturation produces a departure from the linearity of flux linkage $\psi(x, i)$ curve family (Figure 15.3d).

The average mechanical work and force F_{xav} over an energy cycle, in the presence of magnetic saturation (Figure 15.3d), is proportional, at constant current operation, to the area A_{OABCD}.

On the other hand, if the turn-off current coil energy is returned by power electronics to the source, its contribution is proportional to the area A_{OCBDO}, when losses are considered zero. So the energy conversion ratio per energy cycle η_{en} is

$$\eta_{en} = \frac{A_{OABCO}}{A_{OABCO} + A_{OCBDO}} \geq 0.5 \quad (15.3)$$

Apparently, magnetic saturation increases the energy conversion ratio per energy cycle; unfortunately, it does not do this for the same average thrust, for the same device, and for the same airgap excursion length (from x_{min} to x_{max}). However, higher η_{en} means a better utilization of the power electronic supply because magnetic saturation reduces the transient inductance of the coil. Consequently, the current rising and decaying are faster; that is, faster response in current and thus in force is obtained.

We will continue with a linear lump circuit model of the plunger solenoids, still ignoring magnetic saturation and eddy currents.

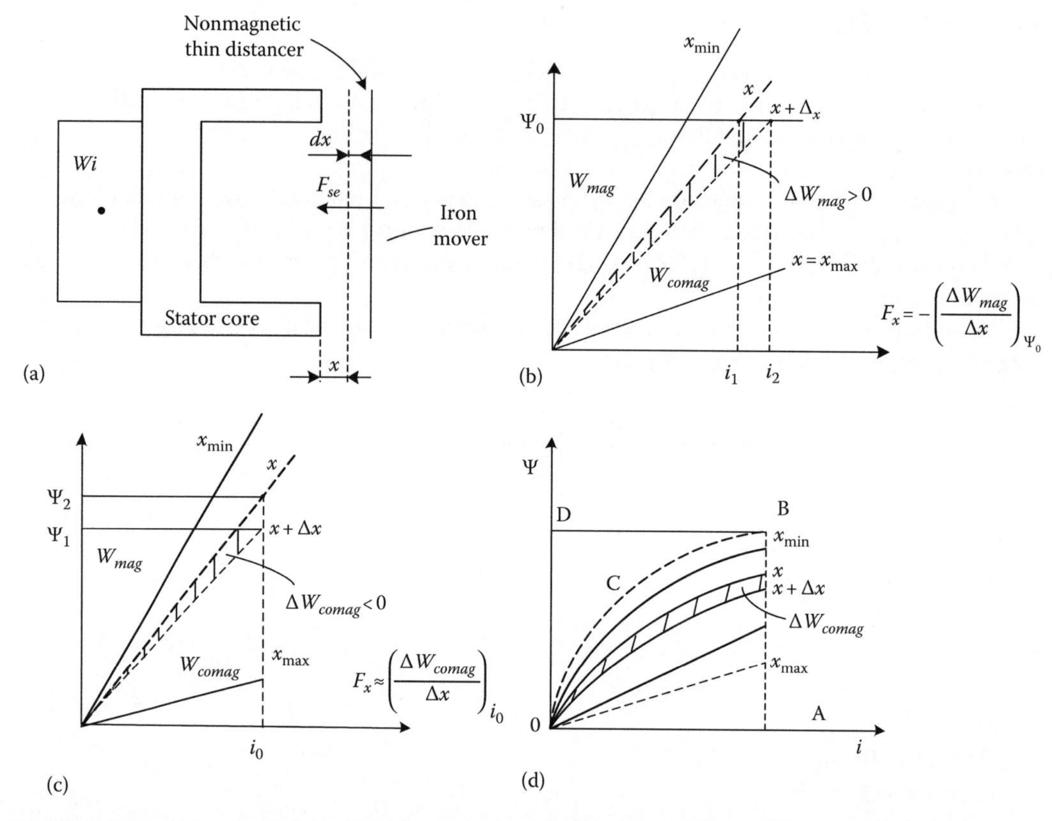

$$F_x = -\left(\frac{\Delta W_{mag}}{\Delta x}\right)_{\Psi_0}$$

$$F_x \approx \left(\frac{\Delta W_{comag}}{\Delta x}\right)_{i_0}$$

FIGURE 15.3 Instantaneous force from energy and coenergy differential (a) the basic topology of plunger solenoid; (b) constant flux magnetic energy increment ΔW_{mag}; (c) constant current coenergy increment ($\mu_{iron} = \infty$); (d) flux linkage/current/position curves family with magnetic saturation.

15.3 LINEAR CIRCUIT MODEL

A more detailed geometry of the plunger solenoid is shown in Figure 15.4.

During dc voltage turn-on (or off), both the current i and the plunger position x vary (from l_0 to l_1 and vice versa).

The mechanical spring is supposed to bring back the mover at $x_{min} = l_0$.

The electric circuit equation is straightforward:

$$V = Ri + \frac{d\psi}{dt}; \quad \psi = \left(L_l + L_m(x)\right)i \tag{15.4}$$

where

R is the coil resistance
L_l is the coil leakage inductance
L_m is the coil main inductance (airgap inductance when $\mu_{iron} = \infty$)

For $\mu_{iron} = \infty$, the main inductance L_m is approximately

$$L_m = \frac{\mu_0 A W^2}{l_1 - x} = \frac{a}{c + x}; \quad l_0 \le x \le l_1 \tag{15.5}$$

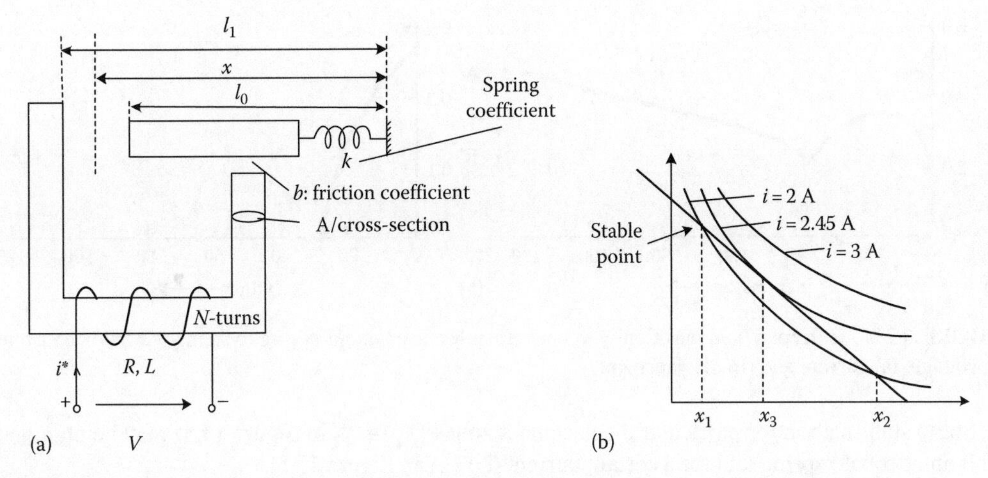

FIGURE 15.4 Basic plunger solenoid with one mechanical spring, (a) and equilibrium zone, (b).

with

$$a = -\mu_0 A W^2, \quad c = -l_1$$

On the other hand, the motion equation can be written as

$$M\frac{d^2x}{dt^2} + B\frac{dx}{dt} + K(x - l_0) = F_x = \frac{1}{2}i^2\frac{\partial L_m}{\partial x} = \frac{ai^2}{2(c+x)^2} \tag{15.6}$$

Equations 15.5 and 15.6 may be arranged into a state space format with either $(\psi, x, dx/dt)$ or $(i, x, dx/dt)$ as variables:

$$\begin{aligned}
&\frac{d\psi}{dt} = V - R\frac{\psi}{L_l + L_m(x)}; \quad i = \frac{\psi}{\left(L_l + L_m(x)\right)} \\
&\frac{dx}{dt} = U \\
&\frac{dU}{dt} = -\frac{1}{M}\left(BU + K(x - l_0) + \frac{a}{2(c+x)^2}\frac{\psi^2}{\left(L_l + L_m(x)\right)^2}\right)
\end{aligned} \tag{15.7}$$

As evident from (15.7), due to electromagnetic force complex formula, even in the absence of magnetic saturation and eddy currents, the third-order solenoid state space model is nonlinear.

Its stability may be approached by linearization: feedback linearization [4], linearization around an equilibrium point [5], asymptotically exact linearization [6], etc.

Linearization around an equilibrium point is the standard procedure used to yield insights into the general behavior, while feedback linearization is preferred when a feedback control law that transforms the nonlinear plant model into a linear one is targeted.

As we will dwell on dynamics and control later in this chapter, it suffices to say at this stage that, as known from Earnshaw's theorem, the plunger solenoid without a mechanical spring is statically and dynamically unstable, for constant voltage (or current) supply.

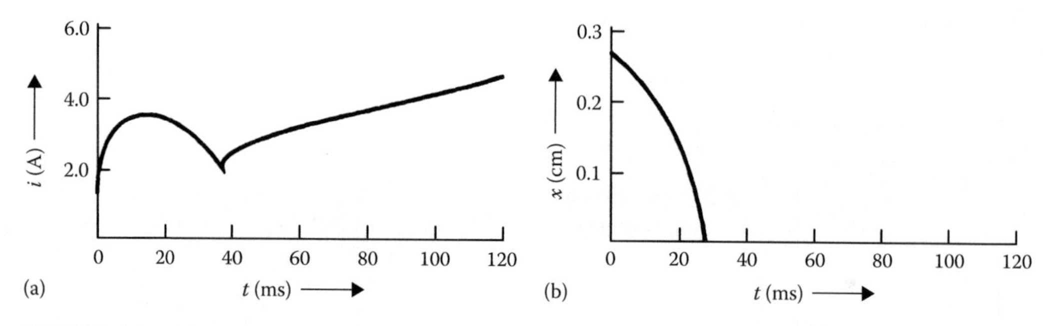

FIGURE 15.5 Current i and position x versus time for a solenoid-spring system, fed with constant dc voltage: (a) current and (b) displacement.

Static stabilization for particular displacement zones ($X_1 \rightarrow X_2$ in Figure 15.5) may be obtained with an adequate spring and for a certain current ($i=2$ A in Figure 15.5).

For larger current, the system is statically stable first at only one point (X_3) and then it becomes statically unstable. In reality, the current is not constant for constant voltage supply. Solving the state space model (15.7) through numerical methods in time for $M=10$ g, $B=0.0001$ N s/m, $R=1\ \Omega$, starting, $W_1=200$ turns, $A=1$ cm^2, $K=100$ N/m, $l_1=3$ cm, the current and position versus time in Figure 15.5 are obtained.

After the motion starts, the current increases first but then, due to motion start, the emf increases and as the inductance also increases, the current decreases; when the motion ends (after 28 ms; inductance is large [small airgap]), the current increases slowly toward V_0/R value, as expected. The results in Figure 15.5 are to be corrected by the influence of magnetic saturation and of eddy currents in the magnetic cores. During voltage turn-off, the skin effect in the cores delays the flux penetration and thus delays the occurrence of flux (force); magnetic saturation, fortunately, also reduces the inductance, increases the penetration depth, and thus has a "positive dynamic effect." During voltage turn-on, the current may decay quickly to zero (especially if the supply may apply not a zero but a negative voltage or a varistor is placed in series).

But, due to eddy currents, the delay in magnetic flux decay occurs, and thus the attraction force decay is slower. This phenomenon delays the spring retraction motion. Consequently, it is imperative to investigate quantitatively the influence of eddy currents and magnetic saturation.

15.4 EDDY CURRENTS AND MAGNETIC SATURATION

Flat active area solenoids may be made of laminations, though noise and vibration may be excessive, but for tubular topologies, which are notably more favorable (in terms of manufacturability), soft magnetic composite iron cores are the solution.

The cylindrical structure allows for approximate analytical eddy current solution, due to cylindrical symmetry, besides the use of FEM-transient methods.

We first investigate the eddy currents in a flat solid mild steel structure in order to grasp the essentials and then use a nonlinear circuit model with eddy currents for a quantitative thorough analysis of the plunger solenoid transients. The FEM inclusion of eddy currents is investigated later in the chapter for case studies of design and control.

For an infinite flat plate placed in the YX plane (Figure 15.6), a tangential uniform magnetic field H (along X), existing at time zero, is removed instantly.

The magnetic 1D eddy current field equation in iron is straightforward:

$$\frac{\partial^2 H(x,t)}{\partial x^2} = \mu_{iron}\sigma_{iron}\frac{\partial H(x,t)}{\partial t} \tag{15.8}$$

with σ_{iron} and μ_{iron} as electrical conductivity and magnetic permeability.

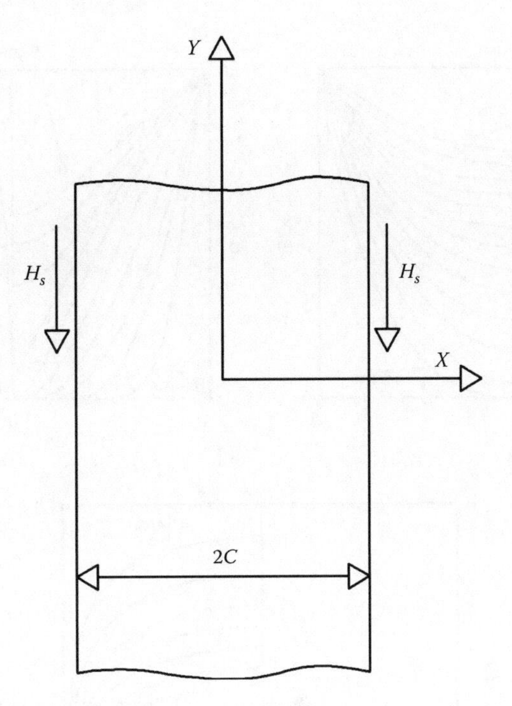

FIGURE 15.6 Flat solid iron plate.

Also,

$$H(\pm c,t) = 0 \quad \text{for } t > 0 \tag{15.9}$$

and

$$H(x,0) = H_0 \tag{15.10}$$

By using the separation of variables for the Fourier series method, the solution of (15.8) is

$$H(x,t) = \frac{4}{\pi} H_0 \sum_{\nu=1,3} e^{-\left(\frac{\nu^2\pi^2}{\mu_{iron}\sigma_{iron}c^2}\right)t} \sin\frac{\nu\pi x}{c} \tag{15.11}$$

The Fourier series converges quickly and thus, approximately, the average flux density $B_{av}(t)$, for constant permeability, is

$$B_{av}(t) \approx \frac{4}{\pi} B_e e^{-\left(\frac{4\pi^2}{3\mu_{iron}\sigma_{iron}c^2}\right)t}; \quad B_e = \mu_{iron} H_0 \tag{15.12}$$

But, if magnetic saturation is not neglected, μ_{iron} should be replaced by μ'_{iron} (transient permeability):

$$\mu^t_{iron} = \frac{dB}{dH} = \mu^t_{iron}(H) \tag{15.13}$$

Now the magnetization curve may be approximated by polynomials such as

$$H = \frac{a_1 - b_1 B^2 + c_1 B^3}{a_1' - b_1' B^2 + c_1' B^3} \tag{15.14}$$

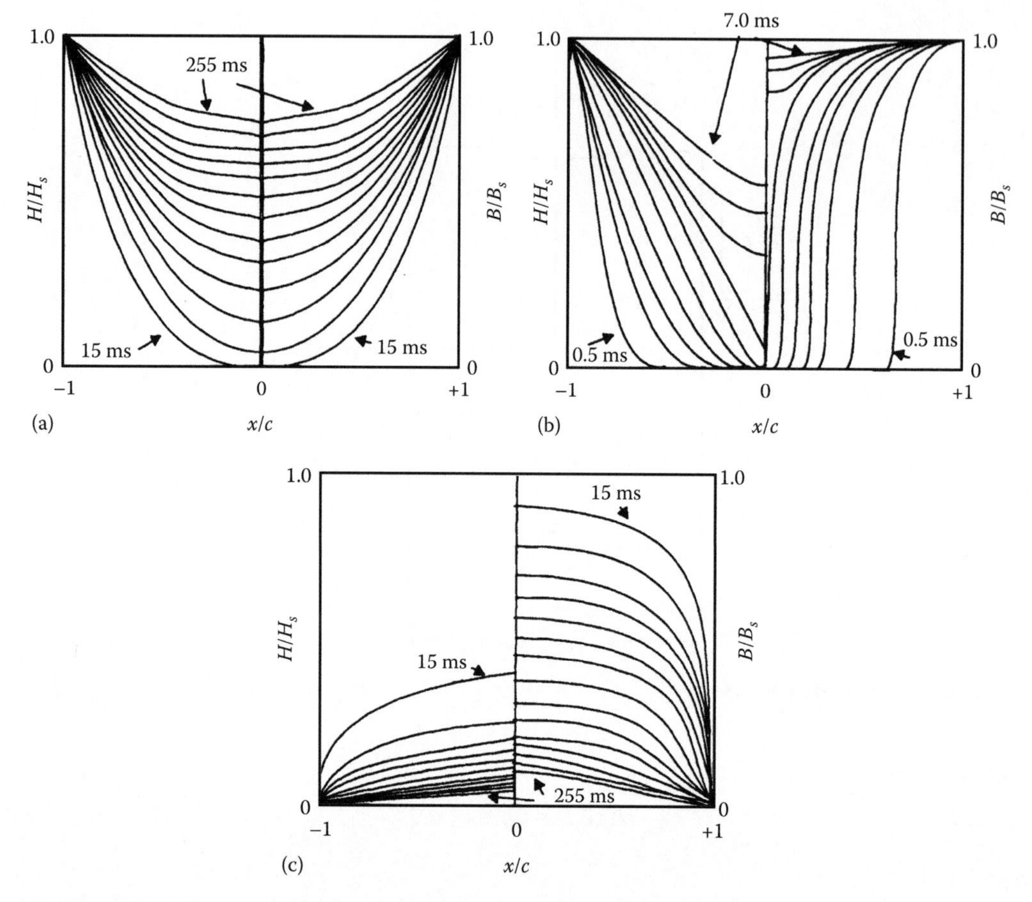

FIGURE 15.7 Field penetration for step magnetic field (mmf) change.

Let us consider $a_1 = 295, b_1 = 384, c_1 = 178, a_1' = 1, b_1' = 0.6, c_1' = 0.34$ for a 10 mm thick plate ($2c = 10$ mm) with $\sigma_i = 2.5 \cdot 10^6\ \Omega^{-1}\ \text{m}^{-1}$, and use FEM, for constant and variable permeability, for sudden magnetic field (mmf) application and removal [7]—Figure 15.7.

$$\begin{cases} (a)\ H_s = 224\ \text{A/m}, \quad B_s = 1.2\,\text{T}, \quad H_0 = 0\ (\text{field turn on}) \\ (b)\ H_s = 1590\ \ \text{A/m}, \quad B_s = 1.6\ \text{T}\ (\text{field turn on}) \\ (c)\ H_s = 0\ \text{A/m}, \quad B_0 = 1.6\,\text{T}, \quad H_0 = 1560\ \text{A/m}\ (\text{for field shut down}) \end{cases}$$

For the linear (nonsaturated) case, left side in Figure 15.7, the field builds up in 255 ms, while for heavy saturation (Figure 15.7b), it takes only 7.5 ms. The field decay (Figure 15.7c) is slower for the saturated core and it takes 200 ms. To corroborate these FEM results with the analytical expression in (15.2), we may define a magnetic inductance of eddy currents L_{eddy}, [8] besides the iron magnetic reluctance (resistance) R_{iron}:

$$R_{iron}\phi + L_{eddy}\frac{d\phi}{dt} = Wi \tag{15.15}$$

The corresponding time constant

$$T_{iron} = \frac{L_{eddy}}{R_{iron}}; \quad R_{iron} = \frac{l_p}{\mu_{iron} A_p} \tag{15.16}$$

A_p, l_p are plate cross section and field path length, respectively.

But T_{iron} should be consistent with the time dependence in (15.12):

$$L_{eddy} = \frac{3c^2 l_p \sigma_{Fe}}{4\pi^2 A_p} \; [\Omega^{-1}] \tag{15.17}$$

As a single value of permeability was considered in R_{iron} and L_{eddy}, its value does not appear in (15.17). In reality, to account approximately for local magnetic saturation variation during transients, both R_{iron} and L_{eddy} may be corrected [8] as

$$R^c_{iron} = \frac{l_i}{A(dB/dH)_{\phi_{ss}}}; \quad L^c_{eddy} = \frac{L_{eddy}}{S}; \quad S = \frac{(dB/dH)_{B_{av}}}{(dB/dH)_{B_s}} \tag{15.18}$$

S is called severity factor, and linearization is performed for small flux variation around Φ_{ss}. Notable improvements have been obtained this way, as seen in Figure 15.8 [8] for the example in Figure 15.7, for $S = 66.4$. Other eddy current delay considerations in the circuit model have been introduced with relative success [9,10].

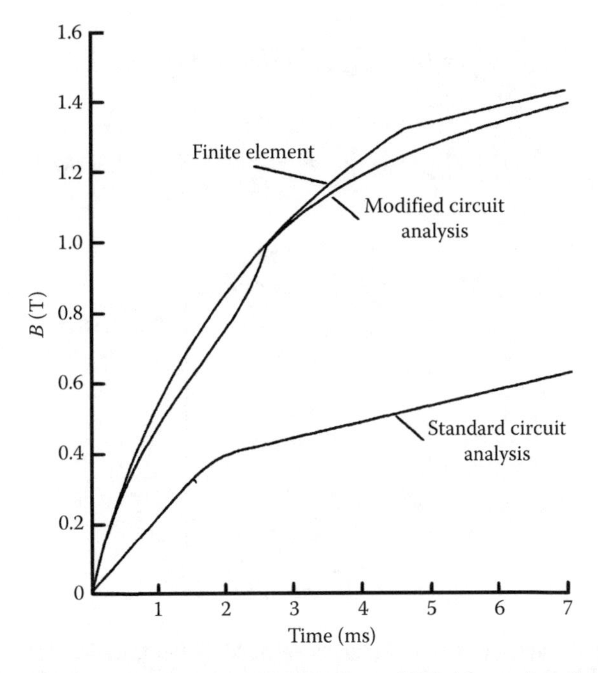

FIGURE 15.8 Transient response for a nonlinear flat plate to a 1592 A/m surface field. (After Caridies, J.P., Fast acting electromagnetic actuators, PhD thesis, Department of Mechanical Engineering, Pennsylvania State University, University Park, PA, 1982.)

Besides FEM—with eddy currents and magnetic saturation considered—the aforementioned nonlinear model may be used for optimal design and control system design in order to save precious computation time for reasonable relative precision.

15.5 DYNAMIC NONLINEAR MAGNETIC AND ELECTRIC CIRCUIT MODEL

A more realistic circuit model of the plunger solenoid not only accounts for magnetic saturation and eddy currents but also for the leakage and airgap magnetic fluxes Φ_l, Φ_g. FEM field distribution for a *C* core solenoid suggests that we may separate the coil mmf into a first part (αwi), which is responsible for the leakage flux, and a second one ($(1-\alpha)wi$), which determines essentially the airgap (force producing) flux.

Making use of the magnetic reluctances and magnetic inductances of stator (R_s, L_{ls}), mover (R_m, L_{lm}) cores, of airgap R_g, the dynamic magnetic circuit in Figure 15.9 is obtained. The reluctance R_{lm} refers to the flux paths produced by the total mmf.

The dynamic circuit model equations are rather straightforward:

$$\frac{d\phi_g}{dt} = \frac{\left[(1-\alpha)w \cdot i - (R_g + R_m + R_{lm})\phi_g + R_{lm}\phi_l\right]}{L_{lm}} \tag{15.19}$$

$$\frac{d\phi_l}{dt} = \frac{\left[\alpha w \cdot i - (R_s + R_{lm})\phi_l + R_{lm}\phi_g\right]}{L_{ls}} \tag{15.20}$$

We add the voltage and motion equation, for completing the model:

$$\frac{di}{dt} = \frac{1}{L_{endcore}}\left[V - Ri - \alpha w\frac{d\phi_g}{dt} - (1-\alpha)w\frac{d\phi_l}{dt}\right] \tag{15.21}$$

$$\frac{dx}{dt} = U; \quad L_{endcon}\text{: end turn inductance}$$

$$\frac{dU}{dt} = \frac{1}{M}\left[F_x - F_{load}(x,U)\right]; \quad F_x = \frac{1}{2}\phi_g^2\frac{\partial R_g}{\partial x}; \quad M\text{: mover mass} \tag{15.22}$$

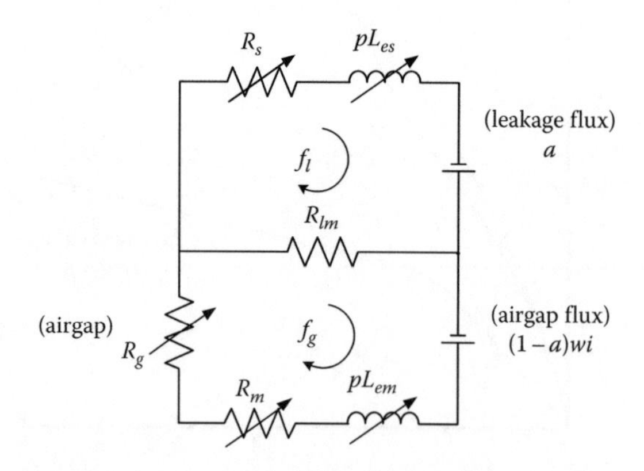

FIGURE 15.9 Dynamic magnetic circuit of plunger solenoid. (After Caridies, J.P., Fast acting electromagnetic actuators, PhD thesis, Department of Mechanical Engineering, Pennsylvania State University, University Park, PA, 1982.)

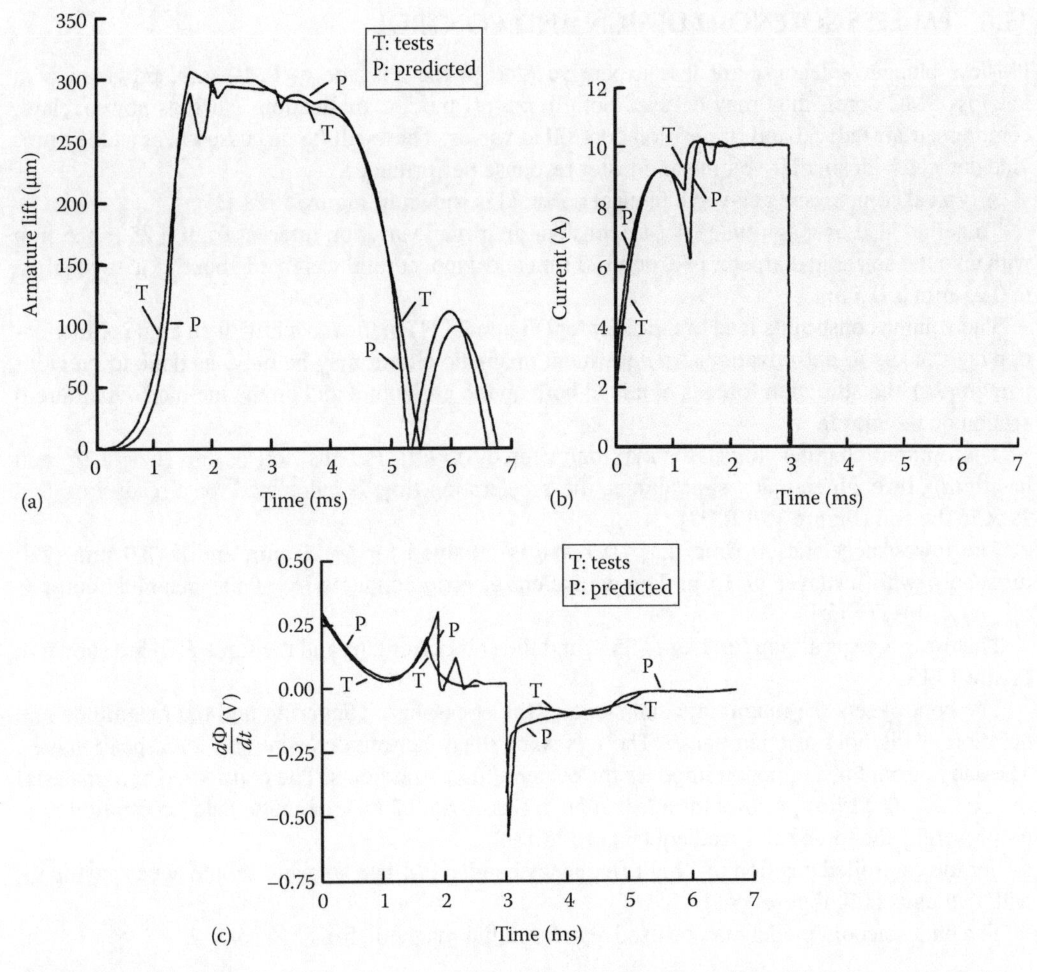

FIGURE 15.10 Dynamic transients: test versus predicted results: (a) position, (b) current, and (c) airgap flux time derivative. (After Caridies, J.P., Fast acting electromagnetic actuators, PhD thesis, Department of Mechanical Engineering, Pennsylvania State University, University Park, PA, 1982.)

In the attraction force expression (F_x in (15.22)), magnetic saturation has already been considered in calculating the airgap flux Φ_g, but it is ignored in the airgap magnetic reluctance R_g (which depends only on airgap).

Sample results obtained with this model [8] are shown in Figure 15.10.

In matching predicted to measured force versus displacement results, a residual airgap (of less than 20 μm) was added to account for contact surface imperfections; this may be necessary if a minimum nonmagnetic airgap coating is applied on contact surfaces.

Although the current decays quickly after voltage turn-off (Figure 15.10b), the plunger position takes reasonably more time until the first touchdown. In addition, in the absence of any control (or of strong damping), there is a notable plunger bouncing, which should be limited in order to prolong the life of the solenoid.

While the aforementioned analysis seems to establish the essentials of plunger solenoids, what follows describes a few implementations with their design and control issues for specific applications.

15.6 PM-LESS SOLENOID DESIGN AND CONTROL

PM-less plunger solenoids are less expensive (due to the absence of PMs) and, provided with low-loss SMC cores, they may be used both in on–off motion applications (such as power relays, compressed-air valves) and in position controlled valves. They will be discussed next as they provide devices with smaller volume and faster response performance.

A typical compressed gas-valve plunger solenoid is shown in Figure 15.11 [11].

To get an idea of magnitudes, let us consider an airgap variation interval from 0.25 to 0.5 mm with an outer solenoid diameter of around 27 mm and a mover total weight of about 15 g; travel time of 0.25 mm is 0.5 ms.

The volume constraints lead to the chamfer (Figure 15.11) in the mover, with an angle of inclination α as object to optimization. An equivalent magnetic circuit may be built (as done in previous paragraphs); the attraction force is obtained both on the horizontal and on the inclined (chamfered) section on the mover.

The optimal chamfer angle (for maximum force) $\alpha \approx 40°$. For the coil design (height (*b*) and length (*a*)), through complex simulations, the acceleration time is calculated for already installed 10 A in the coil (Figure 15.12) [11].

The lowest acceleration time ($t_{\min} \leq 0.5$ ms) is obtained for $b \approx 20$ mm and $a \approx 3.0$ mm (250 turns/coil) with a mover of 15 g. The equivalent electric conductivity of the magnetic cores is $\sigma_{iron} = 1.4 \cdot 10^6\ \Omega^{-1}\ \text{m}^{-1}$.

The thrust versus airgap for 15 and 25 A and the coil current $i(t)$ and the force $F_x(t)$ are shown in Figure 15.13.

The control sets the magnitude and slope of the open-phase triggering and the magnitude and duration of the hold of open phase. There is some delay between current and force peak values. The eddy current phenomenon impedes the current (force) gradients. The considered core material ($\sigma_{iron} = 1.41 \cdot 10^6\ \Omega^{-1}\text{m}^{-1}$) allows for a fast enough (less than 0.2 ms) magnetic field penetration time (as proved by the force/time gradient (Figure 15.13b)).

For the controlled position of a high-frequency band, a PM-less solenoid should act as push-pull, with two units [12], Figure 15.14.

The dual solenoid model may be used also for radial magnetic bearings control.

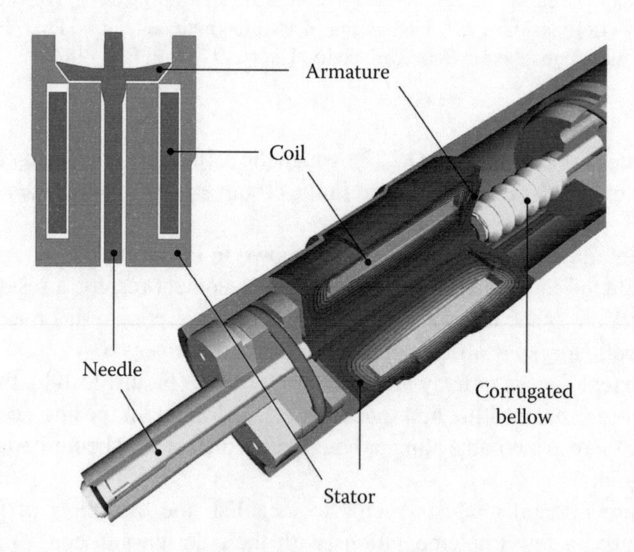

FIGURE 15.11 PM-less solenoid as on–off valve actuator. (After Albert, J. et al., *IEEE Trans.*, MAG-45(3), 1741, 2009.)

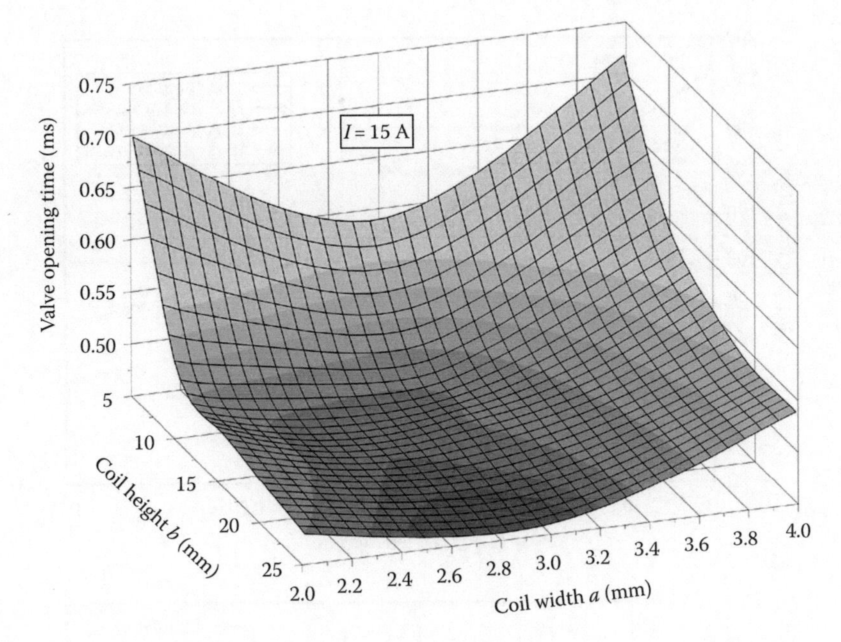

FIGURE 15.12 Opening time (for 0.25 mm travel) variation with coil length a and height b. (After Albert, J. et al., *IEEE Trans.*, MAG-45(3), 1741, 2009.)

The dynamic model (with $\mu_{iron} = ct$ and no eddy currents) is rather straightforward (based on the linear model in Section 15.3):

$$
\begin{aligned}
\frac{dx}{dt} &= U \\
\frac{dU}{dt} &= \frac{\beta_0 i_2^2}{2M(d-x)^2} - \frac{\beta_0 i_1^2}{2M(d+x)^2} - F_u U - F_c \operatorname{sgn}(U) \\
\frac{di_1}{dt} &= \frac{(v_1 - R_1 i_1)(d+x)}{\beta_0} + \frac{i_1 U}{d+x} \\
\frac{di_2}{dt} &= \frac{(v_2 - R_2 i_2)(d-x)}{\beta_0} - \frac{i_2 U}{d-x}
\end{aligned}
\qquad (15.23)
$$

An example for $d = 3.85 \cdot 10^{-3}$ m, $\beta_0 = 4.4 \cdot 10^{-4}$ N m²/A², $R_1 = R_2 = 100\ \Omega$, $M = 0.015$ kg, $F_u = 18.5$ N s/m, $F_c = 1.3$ N is illustrated here. The two coils are turned on in shifts (or, for faster response, differential control of the two solenoids by $v_1 = v_0 + \Delta v$, $v_2 = v_0 - \Delta v$ may be performed). For robust position control response of the nonlinear system of (15.23), quite a few strategies may be applied.

Gain scheduling [5] switching PI, on–off and zero vibration on–off control [12] (Figure 15.15), or sliding mode control [13] may be tried.

For zero vibration on–off control [12], overshooting and bouncing are not visible but the response quickness is not particularly high (tens of milliseconds).

15.6.1 Bouncing Reduction

Now in single solenoid on–off control, for opening classic valves (contacts), the plunger bouncing against the stator (valve) chair has to be attenuated. Even for a simplified control scheme

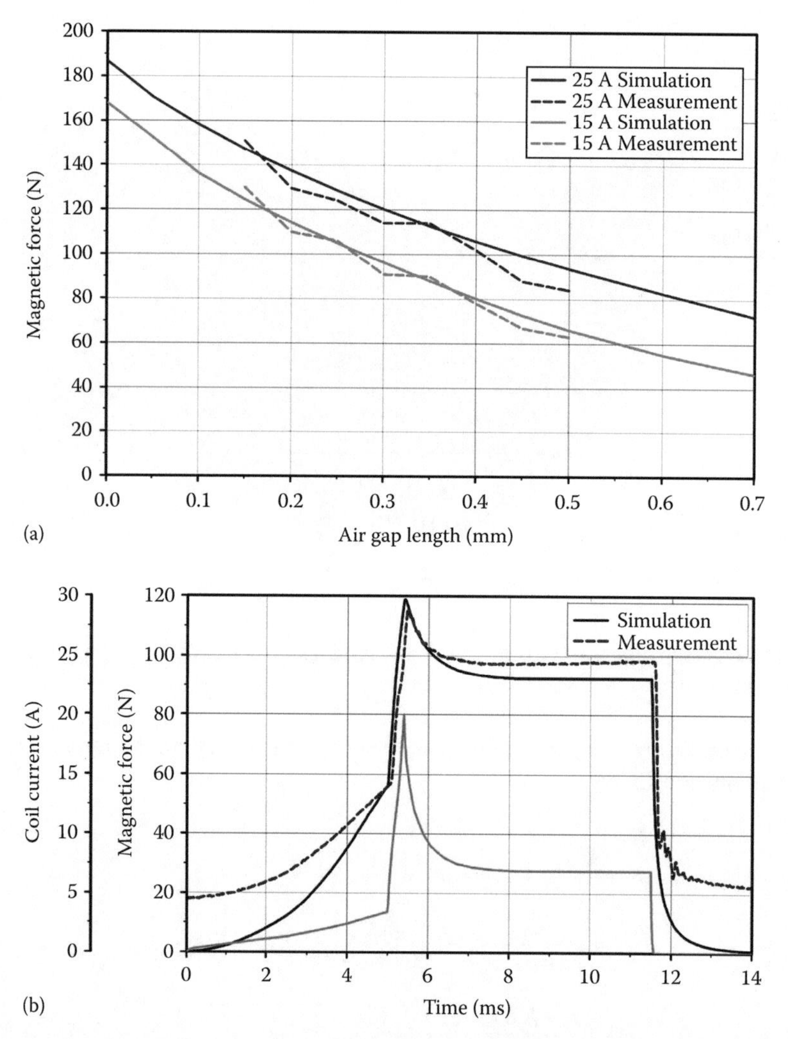

FIGURE 15.13 Calculated and measured: (a) force/airgap and (b) current and force versus displacement (for standard power amplifier supply). (After Albert, J. et al., *IEEE Trans.*, MAG-45(3), 1741, 2009.)

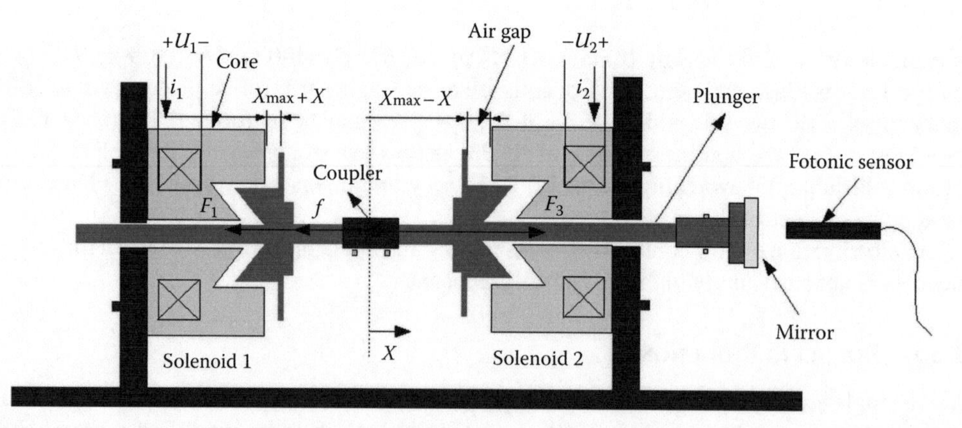

FIGURE 15.14 Dual PM-less solenoid. (After Yu, L. and Chang, T.N., *IEEE Trans.*, IE-57(7), 2519, 2010. With permission.)

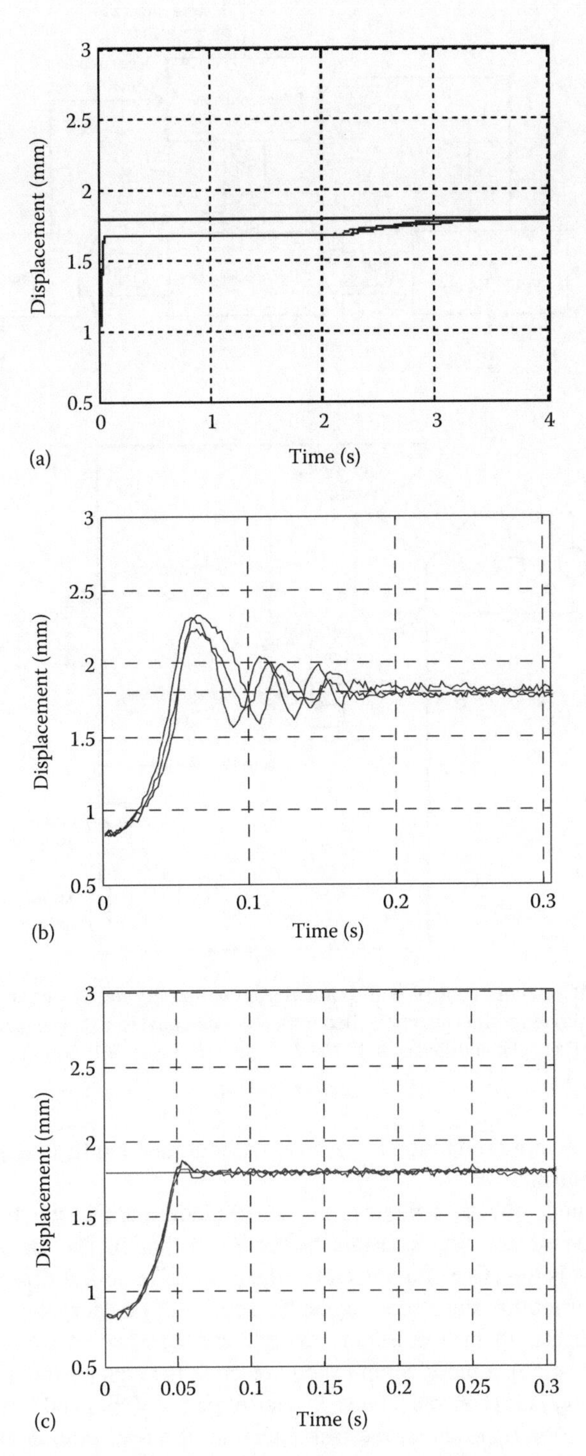

FIGURE 15.15 Experimental position step response: (a) with switching PI control, (b) with on–off control, and (c) with zero vibration on–off control (set point: 1.8 mm; initial point: 0.85 mm). (After Yu, L. and Chang, T.N., *IEEE Trans.*, IE-57(7), 2519, 2010.)

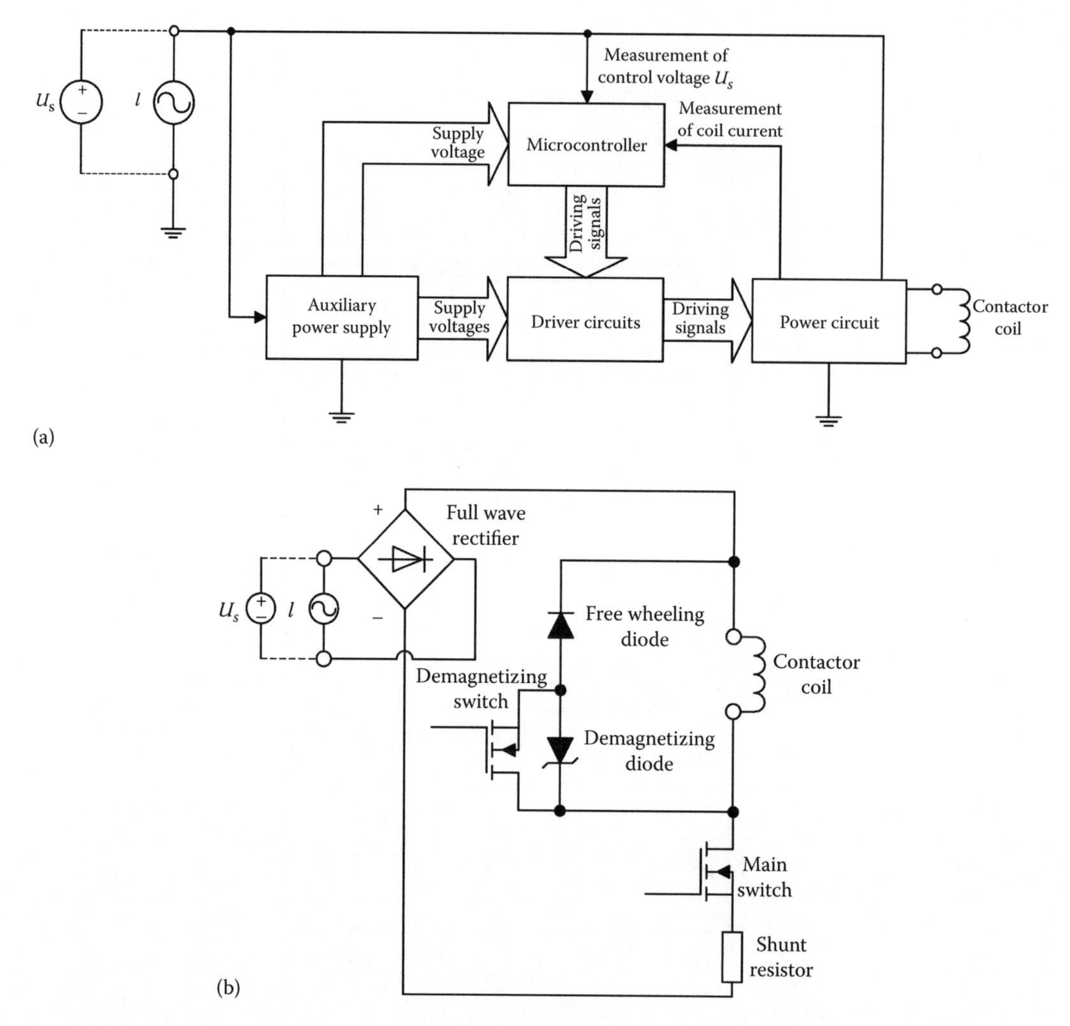

FIGURE 15.16 On–off solenoid control with bouncing reduction (a) and power supply (b). The dc power source (direct or with a diode rectifier) is controlled with a dc–dc converter that in turn, controls the coil current. (After de Morales, P.M.S.D. and Perin, A.J., *IEEE Trans.*, IE-55(2), 861, 2008.)

(Figure 15.16) [14], bouncing reduction is possible without position feedback (or estimation) and without close-loop control.

The freewheeling diode, the demagnetization (Zener) diode, and the switch allow for fast current decay (reduction of the electric time constant by the Zener diode). The main power switch is off either when the current is too large or at the end of the turn-off period. A drop in the current profile occurs when the mover motion starts; consequently, a current first decay, below an experimentally learned threshold, indicates that the motion is about to end (the contactor closes). For an ac power source (without rectification), a more complicated detection through current profile is still feasible [14]. The holding current is further reduced to I_{hold} (also found by trial and error).

The bouncing reduction depends on the fast detection of mover motion and, consequently, on the validity of current samples (which should be ignored close to zero voltage crossing, to avoid large error).

This limitation suggests that, perhaps, reduced bouncing with faster motion (even position) detection (estimation) is the next solution for better performance.

15.6.2 FEM Direct Geometric Optimization Design

Quite a few FEM direct geometric optimization designs of PM-less solenoids have been introduced (such as level-set [15] or geometric algorithms [16]); as only a few geometrical variables are used for optimization, both magnetic saturation and eddy currents may be considered (Figure 15.17, after [16]).

15.7 PM PLUNGER SOLENOID

15.7.1 PM Shielding Solenoids

When no latching (at zero current) is required, the PM may be used for shielding—that is, for flux deviation—in order to produce higher force for the given geometry and mmf (losses) [17], Figure 15.18.

The PMs shielding and the stepped airgap lead to larger force and thus, for constant current, the force does not vary much with airgap (Figure 15.18b); this indicates favorable conditions for position close-loop control. The voltage-on and voltage-off transients (Figure 15.18c and d) are very similar to those in Figure 15.10, derived for PM-less plunger solenoids.

The flux $\psi(t, x)$ and force $F(t, x)$ variations are calculated by FEM [17]. The fast closing of the contact (Figure 15.18c) is achieved but, again, eddy currents delay the displacement decrease during voltage turn-off (Figure 15.18d), when the motion is determined mainly by the mechanical spring ($K = 1700$ N/m, mover mass 0.77 g!).

15.7.2 PM-Assisted Solenoid Power Breaker

Single (Figure 15.19a) and dual-coil (Figure 15.19b) PM solenoids [19] have been proposed for fast response power circuit breakers.

The main asset of PM assistance in plunger solenoids is to provide holding force at zero current, which leads to lower energy consumption, when the holding time is rather large in relative values.

Typical, practical control strategies for the single and dual coil PM solenoids are shown in Figure 15.20.

Loop I (Figure 15.20a) refers to "making loop" while loop II refers to the "breaking loop."

For ac voltage supply, the initial phase voltage may vary randomly; the initial angle of voltage has a strong influence on the displacement/time, which is, however, in the millisecond range (Figure 15.21a).

The dual coil PM solenoid (Figure 15.21b) [19] designed for a 3200 A/400 V circuit breaker shows 22 ms making time, 3 ms trigger time, and 13 ms breaking time ($C_a = 10{,}000\ \mu$F, $C_b = 1{,}000\ \mu$F). For PM solenoids, the PM holding force at travel end is larger than the mechanical spring force at one end for the single coil and at two ends for the dual coil.

Another dual coil PM solenoid for small excursion—submillimeter range (20–30 μm) —application was proposed for surface machining, micro-optics, metal molds, light-enhanced films, and axial active magnetic bearings [20]; accelerations up to 100 times the gravitational acceleration (forces of 100 N for a plunger mass of 12–15 g) are required.

While the force/volume capability of PM plunger solenoids in various high dynamics applications seems to be competitive, the optimization design and the better loss assessment and robust control are subjects of paramount importance in the effort to extend controlled PM solenoid markets.

For a more elaborate case study, we will investigate now a twin-coil (single inverter control) PM solenoid, intended for valve actuating, in terms of analytical, FEM, circuit model, optimization design, and close-loop sensorless control.

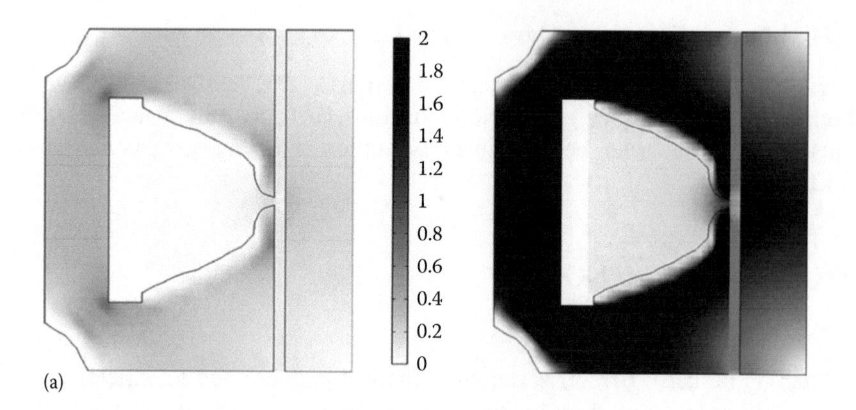

Comparisons of the Best and the Worst Solution after 10 Simulations (Case I)

	Magnetic Force (N/m)	Optimized Topology	Evaluation Number
Best solution	1207.9		531
Worst solution	1143.9		572
Average value	1199.7	—	630.7

(b)

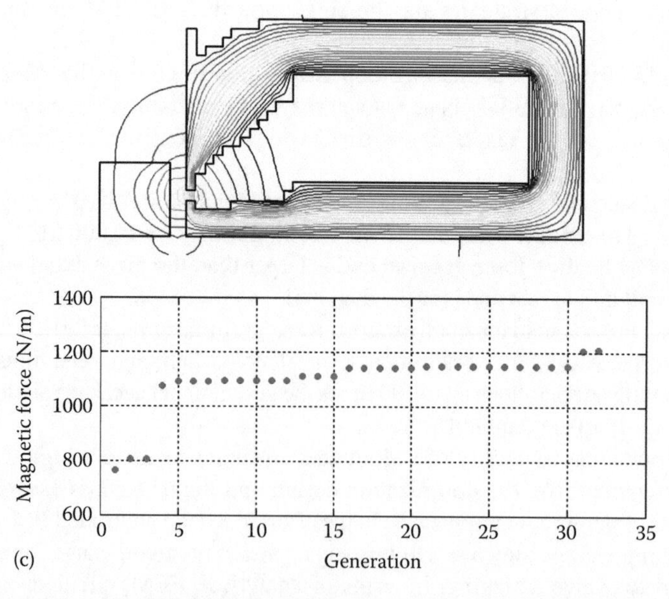

FIGURE 15.17 Optimal topology FEM design, (a), level-set method, (b). (After Park, S. and Min, S., *IEEE Trans.*, MAG-46(2), 618, 2010.) (c) GA optimization.

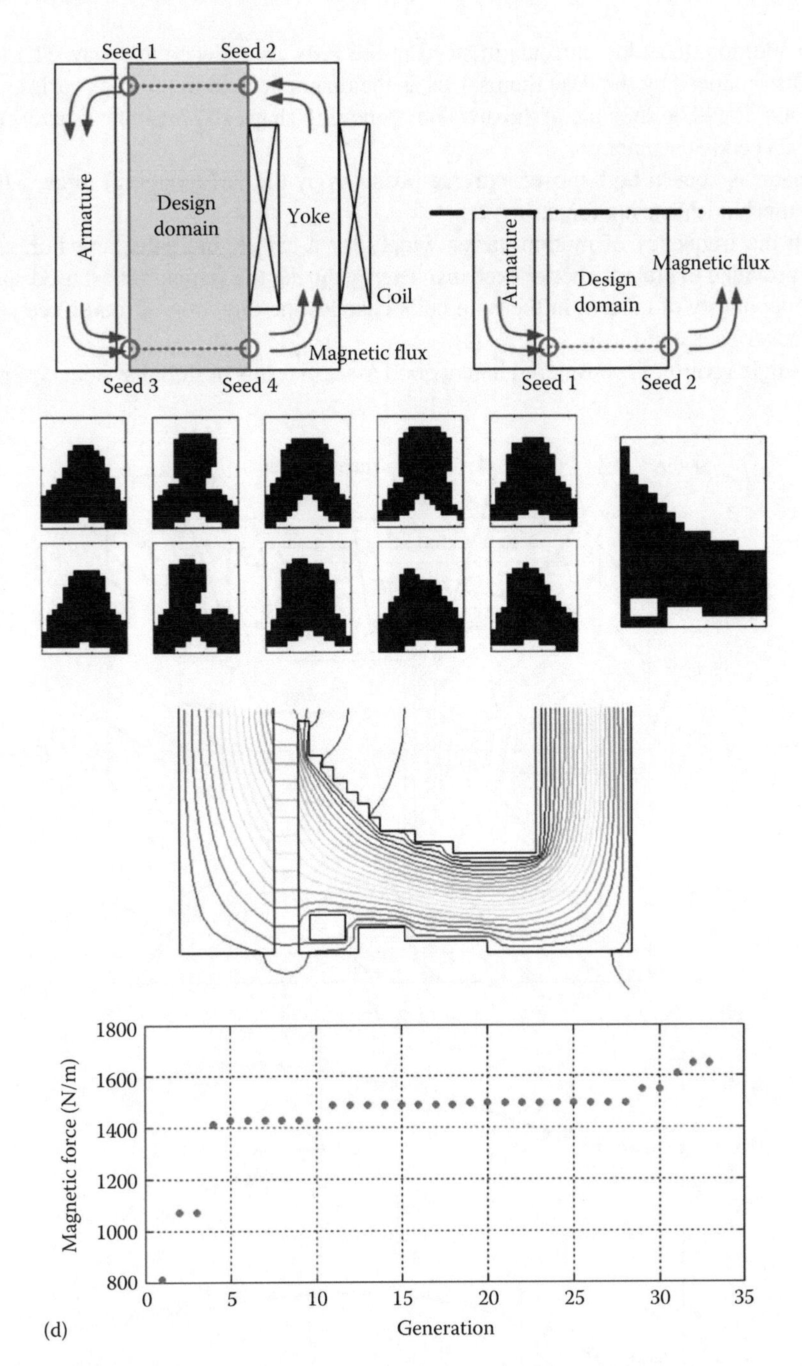

FIGURE 15.17 (continued) (d) GA optimization. (After Choi, J.S. and Yoo, J., *IEEE Trans.*, MAG-45(5), 2276, 2009.)

15.8 CASE STUDY: PM TWIN-COIL VALVE ACTUATORS

15.8.1 Topology and Principle

The PM twin-coil valve actuator investigated here is shown in Figure 15.22.

The PM twin-coil actuator is controlled by a single inverter, and thus the twin-coil mmf produced magnetic field does not cross the PMs; consequently, the danger of demagnetization is reduced. In addition, the eddy currents induced in the PMs are reduced this way. There are some eddy currents produced by the PMs themselves as the magnetic field in the PMs varies a little with mover position; it will be shown that the division of the ring-shape PM into a few sectors drastically reduces the PM eddy currents.

The actuator is kept in both mover extreme positions by the PM (cogging) force, which has to surpass the mechanical spring force.

Although the frequency of motion varies widely for a car engine valve actuator, mechanical springs are provided to globally better exchange energy during the vehicle's most used speed range. Changing the polarity of current in the twin coils changes the direction of total force and quickly drives the mover back and forth.

Using a single inverter is considered here a good asset in terms of initial system cost reduction.

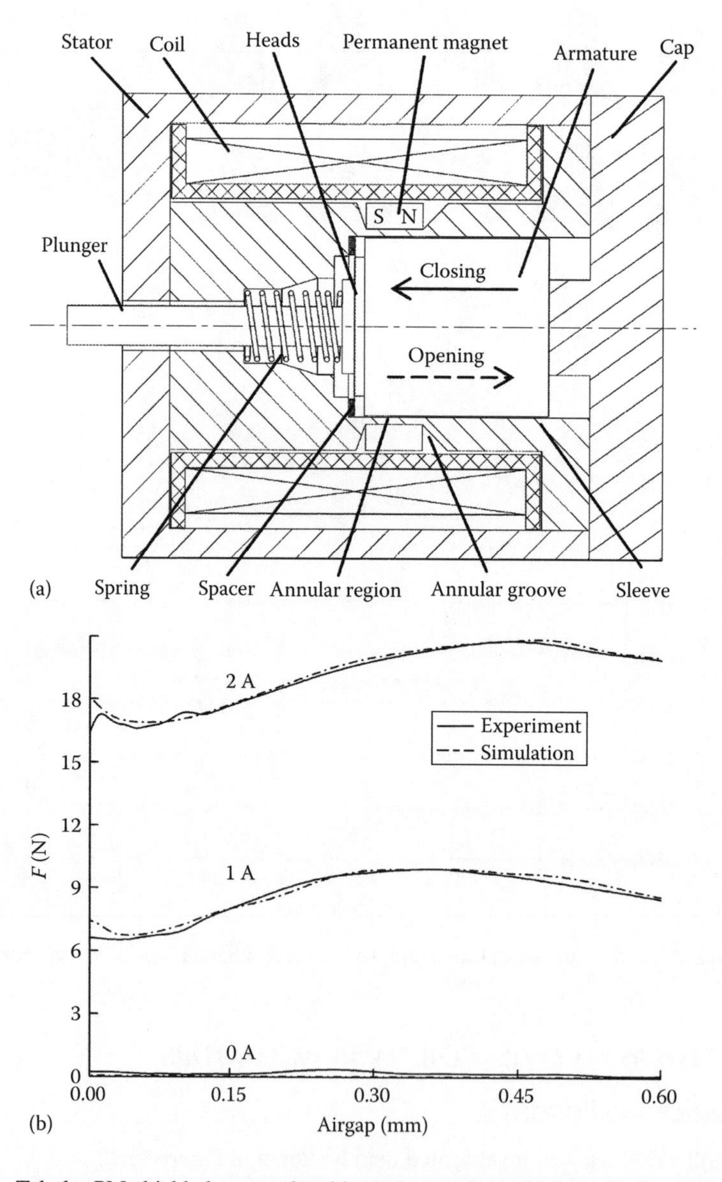

FIGURE 15.18 Tubular PM-shield plunger solenoid: (a) the topology, (b) force displacement.

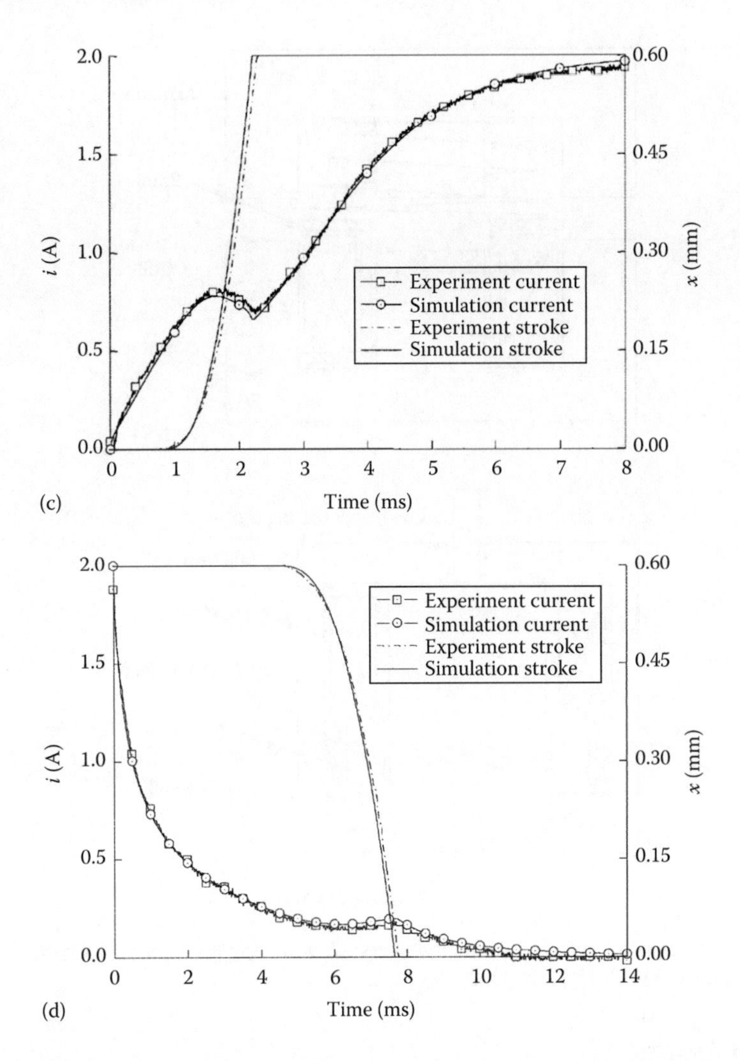

FIGURE 15.18 (continued) Tubular PM-shield plunger solenoid: (c) current and displacement versus time during contact closing, and (d) current and displacement versus time during contact opening. (After Man, J. et al., *IEEE Trans.*, MAG-46(12), 4030, 2010.)

A variable valve car transmission needs essentially

- Zero holding current
- ±4 mm travel, within 3–4 ms, enough for 6 krpm engine speed
- Low (soft) landing velocity: less than 0.3 m/s at 6 krpm engine speed
- Maximum valve speed: 3.5 m/s
- Peak valve-mover acceleration: 5000 m/s^2
- Package (valve) volume ≤0.2 dm^3
- Up to 600 N developed thrust
- No more than 3 kVA peak power for an 8 intake + 8 exhaust valve engine
- Low electrical time constant to facilitate fast force dynamics

These are very demanding specifications unmet yet in industrial applications by plunger solenoid valve actuators (without or with PMs).

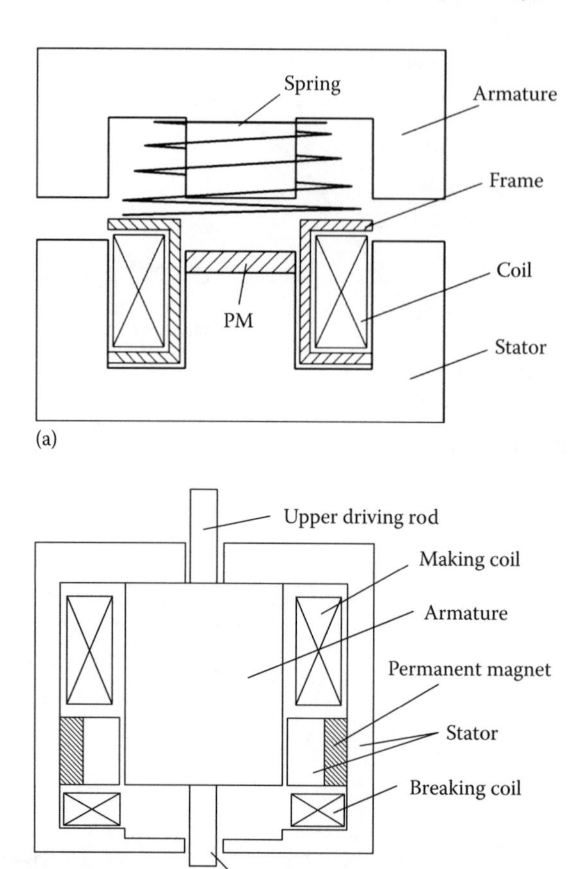

FIGURE 15.19 PM-assisted solenoid power breakers (a) with single coil and (b) with dual coil. (After Fuang, S., *IEEE Trans.*, MAG-45(7), 2990, 2009.)

15.8.2 FEM Analysis

As the volume constraints are very challenging, magnetic saturation will be heavy, and thus only 2D FEM may produce practical useful results. The Atomet EM1 soft magnetic composite core losses will be computed after magnetostatic FEM field variation calculations.

They are proven to be modest in comparison with copper losses (Atomet M1 reaches 1.5 T at 20 kA/m), for the dimensions marked in Figure 15.23.

The FEM calculated PM cogging force, total force, total coil flux, and the transient inductance are given in Figure 15.24a through d.

The forces have been calculated using the weighted stress tensor volume integral.

An acceptable agreement of measured and calculated forces is visible in Figure 15.24. Further on, the actuator cores have been divided into eight regions with different flux-density time variation patterns; then these $\pm B_{max}$ variations were used to calculate the core losses here:

$$p_{core} = \rho_{A-EM1} \cdot \text{vol} \cdot p_{100\,\text{Hz},1\,\text{T}} \cdot \frac{f}{100} \cdot B_{max}^2 \tag{15.24}$$

with $\rho_{A\text{-}EM1} = 7100$ kg/m^3, $p_{100\,\text{Hz},\,1\,\text{T}} = 20$ W/kg.

Finally, adding different current values of all eight volume contributions, the results in Figure 15.25 are obtained.

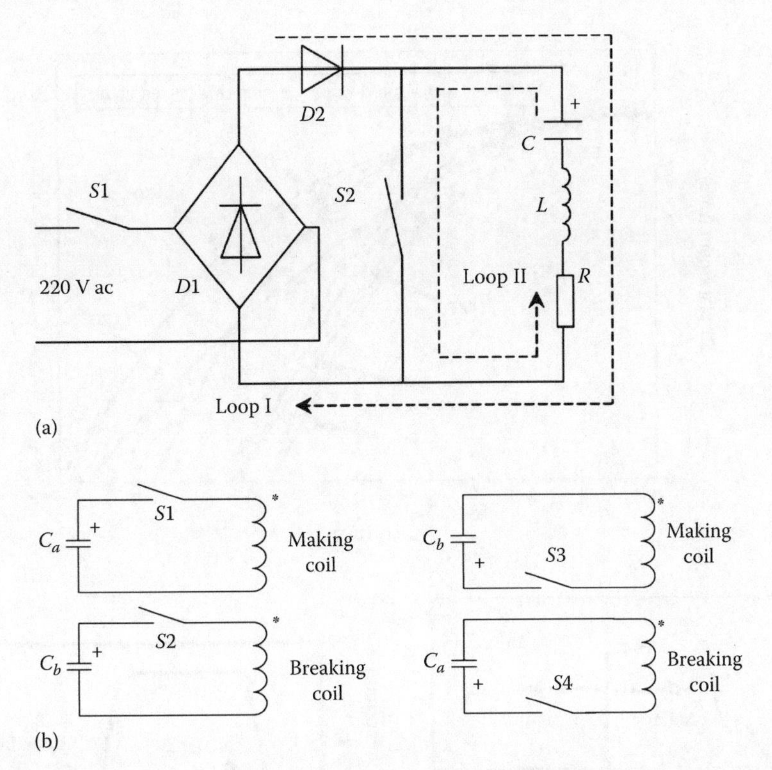

FIGURE 15.20 Control circuit system. (After Fuang, S. et al., *IEEE Trans.*, MAG-45(10), 4566, 2009.) (a) single-coil PM solenoid and (b) dual coil PM solenoid (making and braking control).

It is notable that p_{core} < 1.6 W; if we add the 3 W PM eddy current losses (the ring-shaped PM is made of eight sectors), we get 4.6 W, which is rather small in comparison with 115 W copper losses at 21 A (RMS) sinusoidal current.

Thermal and mechanical FEM analyses have also been performed. The temperature evolution at maximum current is shown in Figure 15.26a; mechanical stress results are visible in Figure 15.26b.

While mechanically the stress distribution is within bounds, for a temperature-safe operation, $SmCo_5$ PMs are to be used.

15.8.3 Direct Geometrical FEM Optimization Design

Grid search and three variations of Hooke–Jeeves optimization method have been used to optimize the design of PM twin-coil valve actuator.

First, the objective function and its constraints are placed:

- $F_{P\max} > 300$ N—PM force
- $K_f = F_{\max}/F_{P\max} > 1.8$—force coefficient
- $a_{\max} \geq 5000$ m/s² (maximum acceleration of the mover)
- $p_{copper} < 120$ W—copper losses
- Actuator volume: vol < 0.2 L

Now the variable vector is chosen: r_0—shaft radius, r_1—interior shaft average radius, r_2—intermediate stator average radius, r_3—external stator interior radius, r_4—external stator external radius, l_{PM}—PM height (axial), l_k—axial air zone height between coils and stator core; l_c—coil height (axial); l_m—mover plate height (axial); h—intermediate stator external height.

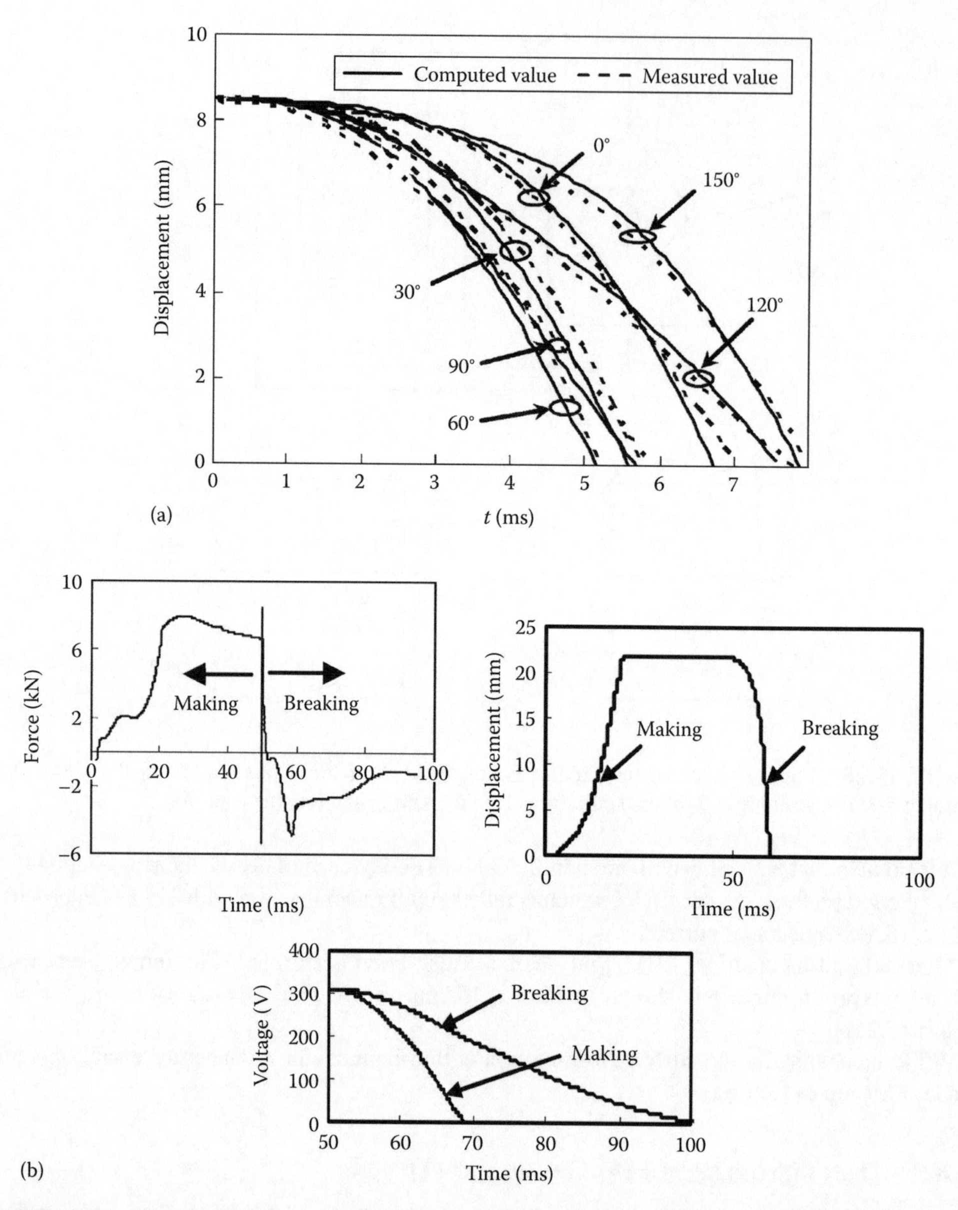

FIGURE 15.21 Displacement/time (single-coil PM solenoid), (a); force, displacement, voltage versus time, dual coil PM solenoid dynamics, (b). (After Fuang, S. et al., *IEEE Trans.*, MAG-45(10), 4566, 2009.)

About 2300 situations have been run by the grid search method, with the best results as follows: F_{PM}=343.92 N, F_{max}=702.01 N, a_{max}=5204.7 m/s^2, p_{copper}=115.9 W, volume=0.16316 dm^3, for r_0=3 mm, r_1=14 mm, r_2=19 mm, r_3=24 mm, x_{max}=8 mm, l_m=6.68 mm, h=5 mm, l_k=2 mm; a total computer time of 3 h was used to draw the actuator geometry for the direct use in the grid search optimization. For the Hooke–Jeeves method (min, max, and modified), the objective function F_{ob} has been defined as

$$F_{ob} = p_{copper} \times (p_f + p_a + p_v + 1) \tag{15.25}$$

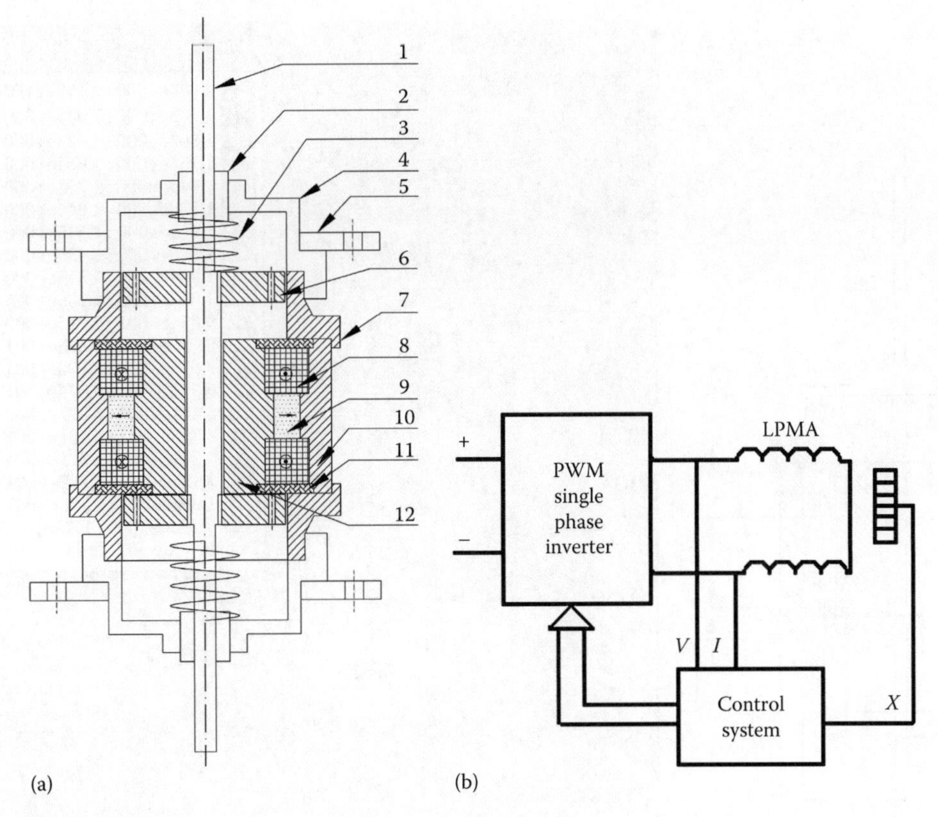

FIGURE 15.22 PM twin-coil valve actuator: (a) the topology and (b) basic control scheme. Where (1) shaft (stainless steel); (2) linear bearing (bronze); (3) springs; (4) aluminum cover; (5) mounting ring; (6) mover plates (Atomet EM1); (7) intermediate stator (Atomet EM1); (8) twin coils; (9) permanent magnets; (10) outer core; (11) insulator; and (12) inner core. (After Agarliță, S., Linear single phase PM oscillatory machine: Design and control, PhD thesis, UPT.ro, 2009.)

with p_f—force penalty factor, p_a—acceleration penalty factor, p_v—volume penalty factor, and

$$
\begin{aligned}
p_f &= K_f\left(\frac{F_{PM\min}}{F_{PM}}\right)^2 \quad \text{if } F_{PM\min} > F_{PM}, \quad \text{otherwise } p_f = 0 \\
p_a &= K_a\left(\frac{a_{\min}}{a}\right)^2 \quad \text{if } a_{\min} > a, \quad \text{otherwise } p_a = 0 \\
p_v &= K_v\left(\frac{V_T}{V_{T\max}}\right)^2 \quad \text{if } V_T > V_{T\max}, \quad \text{otherwise } p_v = 0
\end{aligned}
\tag{15.26}
$$

with

$$K_f = K_a = 5 \quad \text{and} \quad K_v = 1 \tag{15.27}$$

Sample evolutions of acceleration and volume with i iteration number (Figure 15.27) show that all four optimization methods lead more or less to similar results. Table 15.1 confirms this trend.

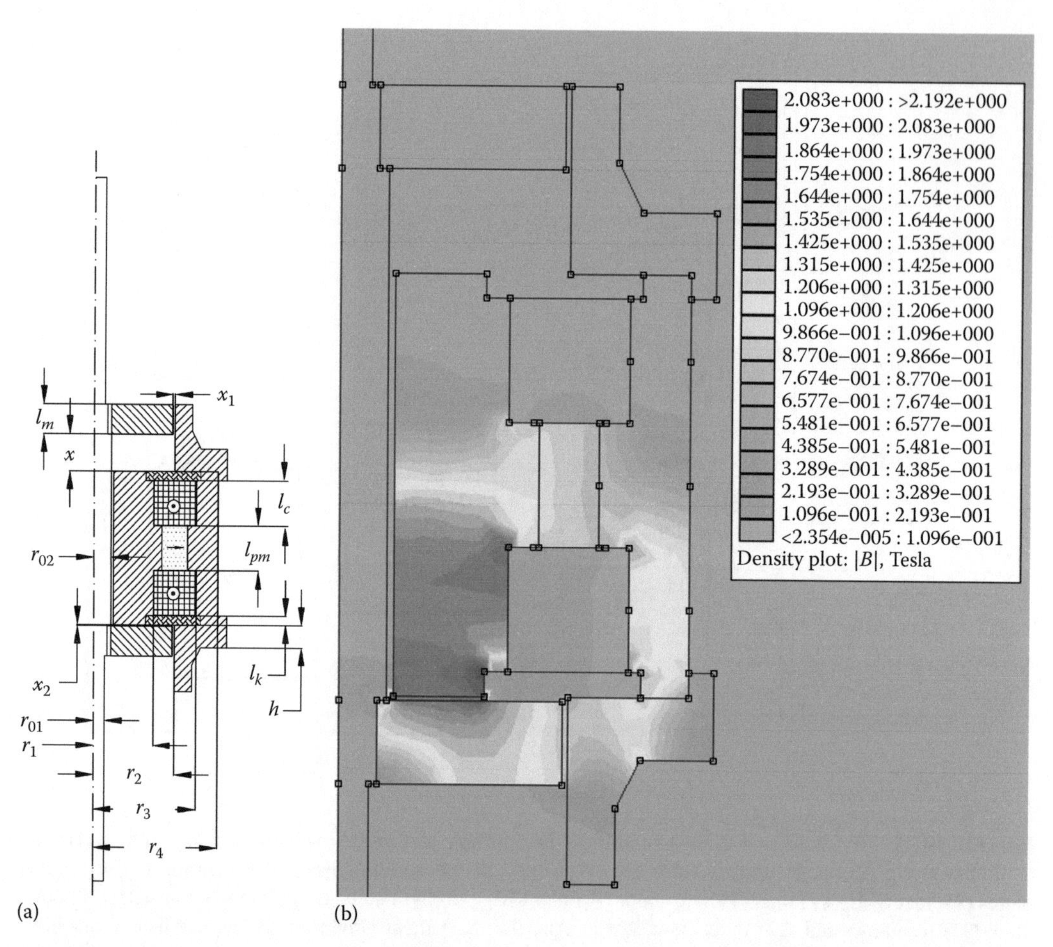

FIGURE 15.23 Geometrical parameters, (a) and flux density distribution for $I=25$ A, valve position, (b). The geometrical parameter values are $r_{01}=5$ mm, $r_{02}=8$ mm, $r_1=14$ mm, $r_2=14$ mm, $r_3=24$ mm, $r_4=29$ mm, $y=8$ mm, value max displacement $x_1=0.5$ mm, $x_2=0.2$ mm, $l_m=6.68$ mm, $l_c=12$ mm, $l_{pm}=8$ mm, $l_k=2$ mm, $h=5$ mm. (After Agarliță, S., Linear single phase PM oscillatory machine: Design and control, PhD thesis, UPT.ro, 2009.)

But the computation time with modified Hooke–Jeeves method decreases to less than 25 min (from 3 h and 5 min for the grid-search method). Hooke–Jeeves method should be performed from a few randomly different points to make sure that a global optimum design is obtained.

15.8.4 FEM-Assisted Circuit Model and Open-Loop Dynamics

The intricate influence of mover displacement and current on total coils flux, total force, and on inductance makes the use of FEM-obtained results curve fitting a must in building a realistic circuit model for the actuator dynamics and control.

As an example, the total flux variation with position x and current y may be curve fitted by

$$\phi = C_1x^3 + C_2x^2y + C_3xy^2 + C_4y^3 + C_5x^2 + C_6xy + C_7y^2 + C_8x + C_9y + C_{10} \tag{15.28}$$

The least squares method is used in the process of $C_1 \cdots C_{10}$ calculation from FEM results (of Figure 15.24b).

The system equations are straightforward:

$$V = Ri + \frac{d\phi(i,x)}{dt}$$

$$\frac{dx}{dt} = U; \quad U\text{: speed} \tag{15.29}$$

$$M\frac{dU}{dt} = F(i,x) - K \cdot x - F_c - F_{load}; \quad F_c\text{: friction force}$$

A MATLAB® Simulink® code is generated to simulate the system (15.29) —Figure 15.28a and b.

Simulated and measured open-loop dynamics results (for given voltage and frequency) are shown in Figure 15.29.

Similar verifications have been done for various given current amplitude and frequency.

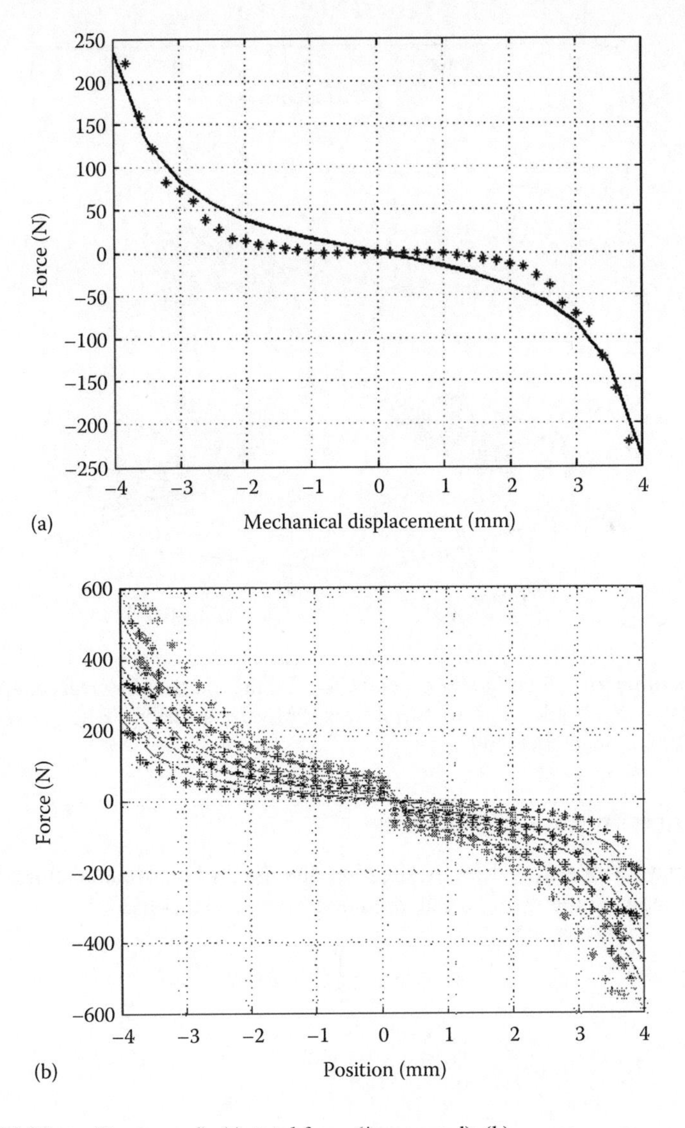

FIGURE 15.24 PM force (*measured), (a), total force (*measured), (b).

(continued)

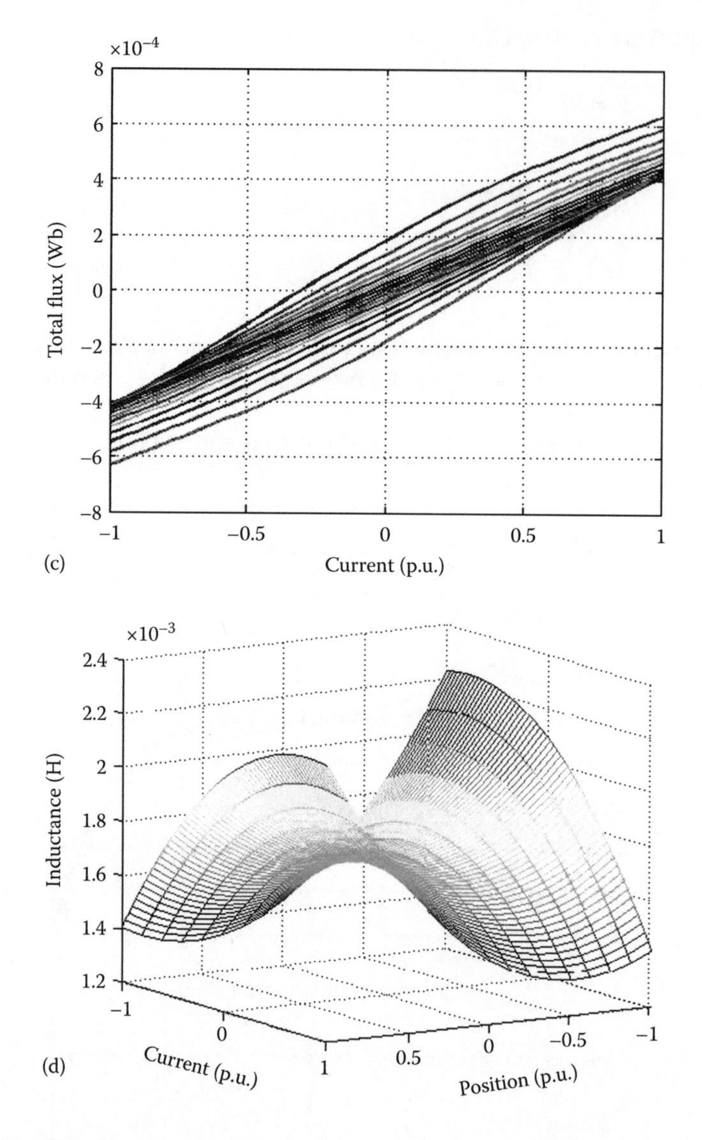

FIGURE 15.24 (continued) Total flux in the coils, (c), and the transient inductance, (d), total flux versus position and current. (After Boldea, I. et al., Novel linear PM valve actuator. FE design and dynamic model, *Record of LDIA-2007*, Lille, France, 2007.)

15.8.5 FEM-Assisted Position Estimator

As the current i may be measured directly, the total flux Φ may be estimated either from the voltage equation or by processing the extra flux Φ_e obtained from search coils:

$$\phi_m = \phi_{m0} + \int (V - Ri)dt$$
$$\phi_e = \phi_{e0} + \int V_e dt \tag{15.30}$$

The FEM-computed $\Phi(x, i)$ curves are now "inverted" to obtain $x(\Phi, i)$ by adequate curve fitting with a function similar to (15.28).

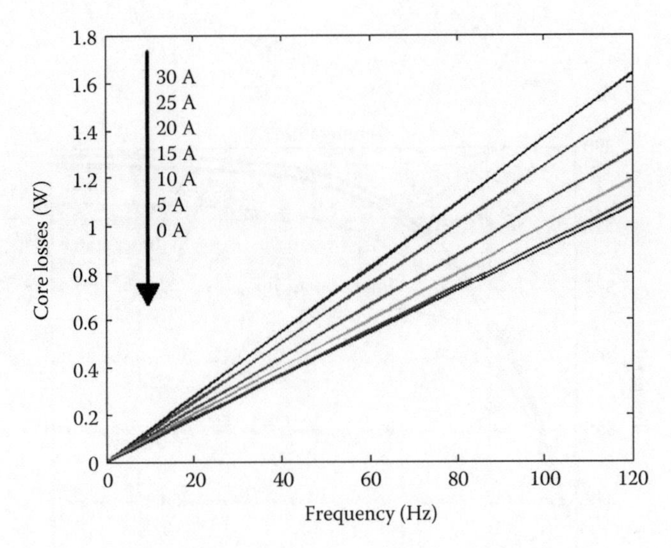

FIGURE 15.25 Core losses versus motion frequency and different currents. (After Boldea, I. et al., Novel linear PM valve actuator. FE design and dynamic model, *Record of LDIA-2007*, Lille, France, 2007.)

Consequently, with known flux Φ, and i from the $x(\Phi, i)$ function, the position is "estimated" on line.

The integral in (15.30) has to be reset; instead, a second-order filter replaces the integral:

$$H(s) = \frac{s}{s^2 + 2\xi\omega_0 s + \omega_0^2} \quad (15.31)$$

Reasonable phase and amplitude precision results are obtained from 1 Hz upward.

As expected, the usage of search coils (coil flux is Φ_e) avoids the Ri term integration and thus is more robust to temperature and current errors.

Typical estimation results, at 30 Hz, are given in Figure 15.30 (after [23]).

The gain in precision with search coils should be seriously considered.

15.8.6 Close-Loop Position Sensor and Sensorless Control

A PI + RCG (relay constant gain) position with an interior current controller is used for close-loop robust position sensorless control (Figure 15.31).

$$\begin{aligned} I^* &= \varepsilon_x\left[K_{px}\left(1+\frac{1}{sT_{ix}}\right)+K_{RCGx}\right]; \quad K_{px,i} \approx 2\cdot K_{RCGx,i} \\ V^* &= \varepsilon_i\left[K_{pi}\left(1+\frac{1}{sT_{ii}}\right)+K_{RCGi}\right]; \end{aligned} \quad (15.32)$$

The RCG behaves as a primitive sliding mode control contribution to deliver robustness.

When the position sensor is used, the results of position control are quite good at 40 Hz. (Figure 15.32).

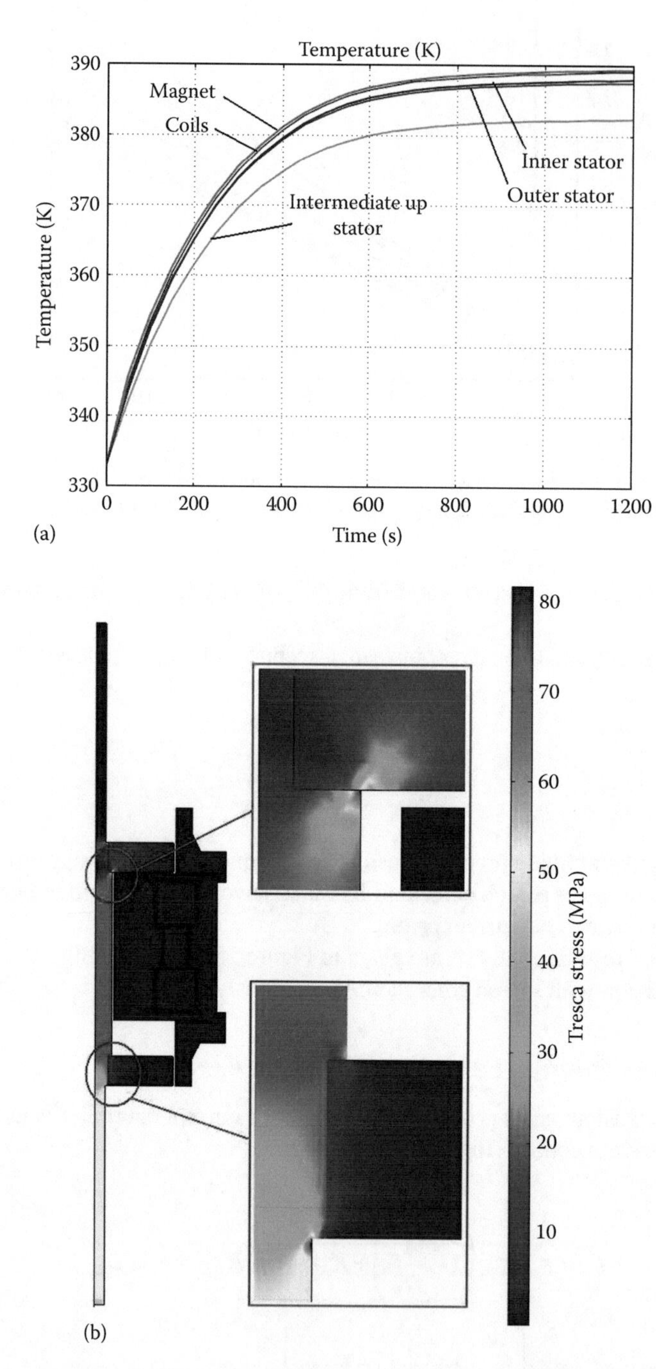

FIGURE 15.26 Temperature evolution at max current (a) and T_{resco} stress distribution (b) from 2D FEM. (After Boldea, I. et al., Electromagnetic, thermal and mechanical design of Linear PM valve actuator laboratory model, *Record of OPTIM-2008*, Brasov, Romania, vol. 2, 2008.)

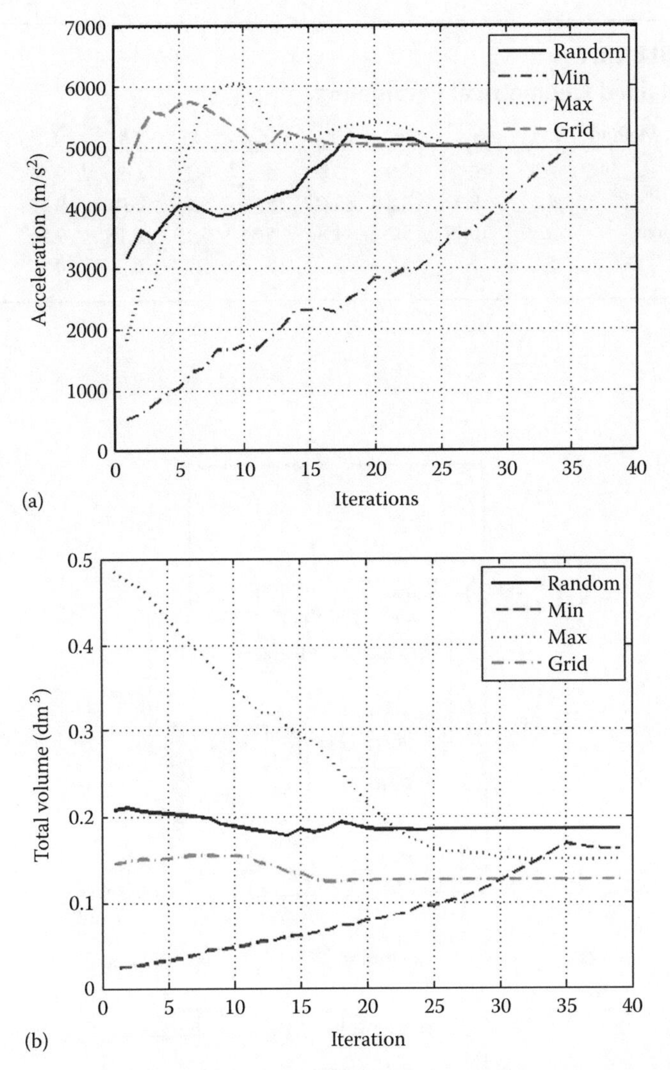

FIGURE 15.27 Acceleration (a) and volume (b) evolution during optimization process. (After Agarlită, S., Linear single phase PM oscillatory machine: Design and control, PhD thesis, UPT.ro, 2009.)

Comparatively, sensorless control loop for 1 Hz (Figure 15.33a), at 40 Hz (Figure 15.33b) and at 80 Hz (Figure 15.33c), look very reasonable.

Note on soft-landing control: The rather acceptable precision (0.05 mm error) close-loop sensorless position control suggests that improvements in the control may lead to active landing speed limitation, that is, bouncing limitation. The independence of machine inductance of airgap—and its rather small value—should be an asset in fast current control, once the eddy currents in the Atomet-EM1 (or other SMC) cores are fairly small.

Although the PM twin-coil actuator was intended for valve actuators, it may be considered for other applications as well. It merely indicates a comprehensive method of design and control approach for fast action plunger solenoids.

TABLE 15.1
Obtained Geometrical Parameters

Opt. Method	r_0	r_{10}	r_{21}	r_{32}	r_{43}	l_m	h	l_c	l_{pm}
Min	4.5	10	4.9	5.8	6	7.1	6.2	9.66	11
Max	4.6	9.7	4.84	5.62	5.38	6.65	5.41	9.22	11.2
Random	4.36	10.17	5.2	4.8	5.36	7.19	6.16	9.56	10.8
Grid	5	9	5	5	5	6.7	6	10	10

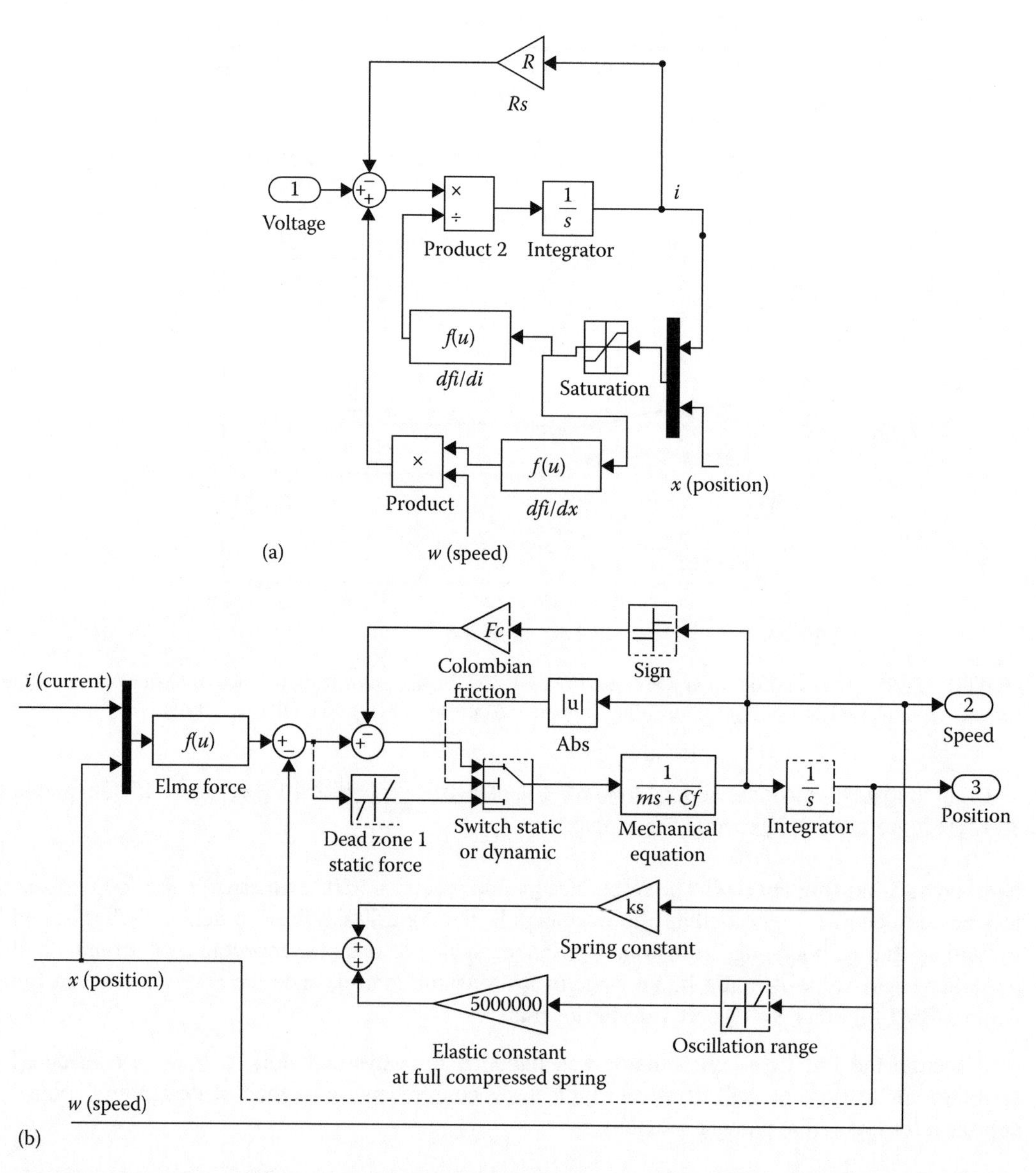

FIGURE 15.28 Electrical (a) and mechanical (b) parts of actuator model. (After Agarlită, S., Linear single phase PM oscillatory machine: Design and control, PhD thesis, UPT.ro, 2009.)

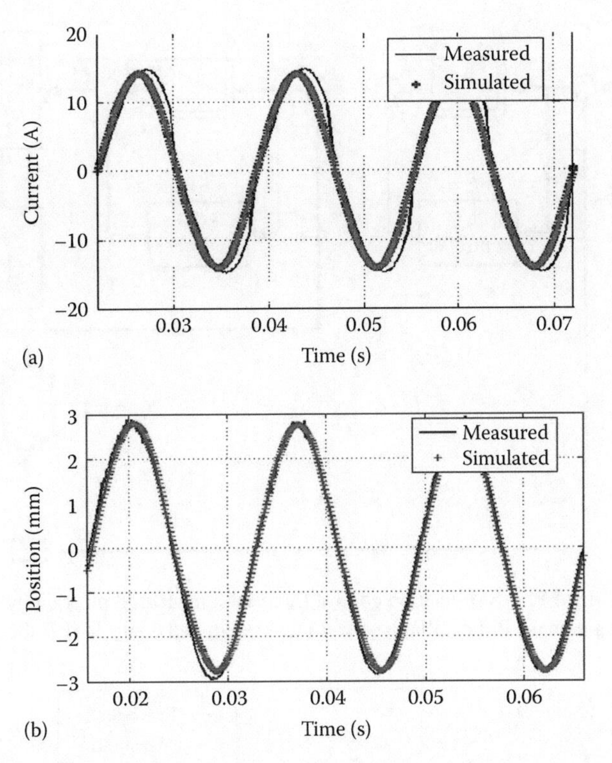

FIGURE 15.29 Calculated and measured current (a) and position (b) for 10 V dc (After Agarlită, S., Linear single phase PM oscillatory machine: Design and control, PhD thesis, UPT.ro, 2009.)

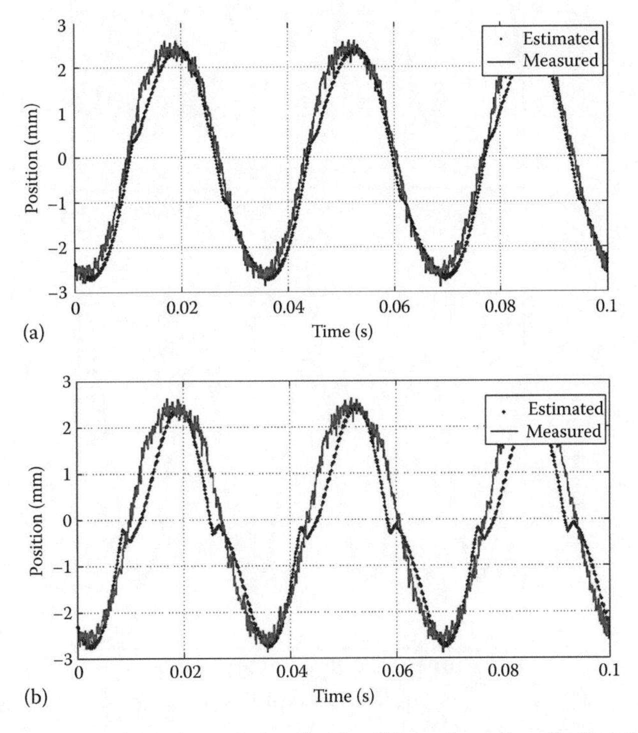

FIGURE 15.30 Estimated position using search coils (a) and from the main coils directly (b), at 30 Hz. (After Agarlită, S., Linear single phase PM oscillatory machine: Design and control, PhD thesis, UPT.ro, 2009.)

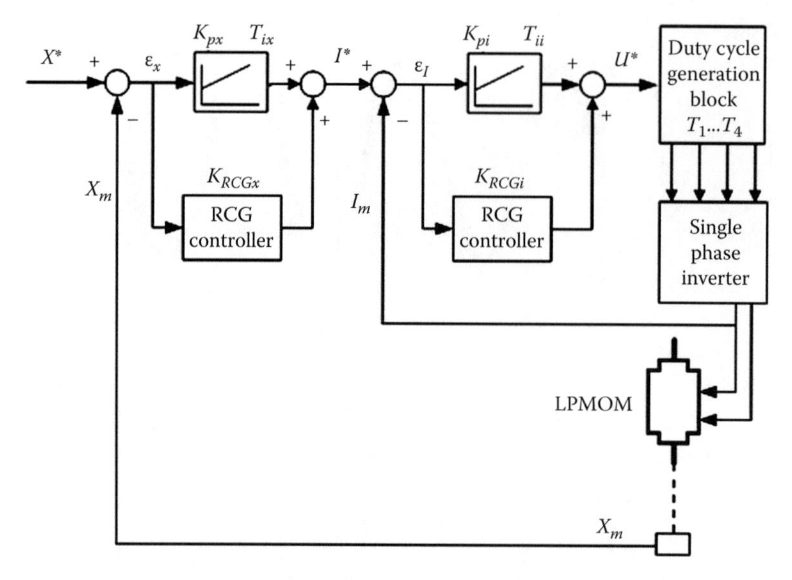

FIGURE 15.31 Combined PI + SM position control (here with measured position by a laser device). (After Agarliță, S., Linear single phase PM oscillatory machine: Design and control, PhD thesis, UPT.ro, 2009.)

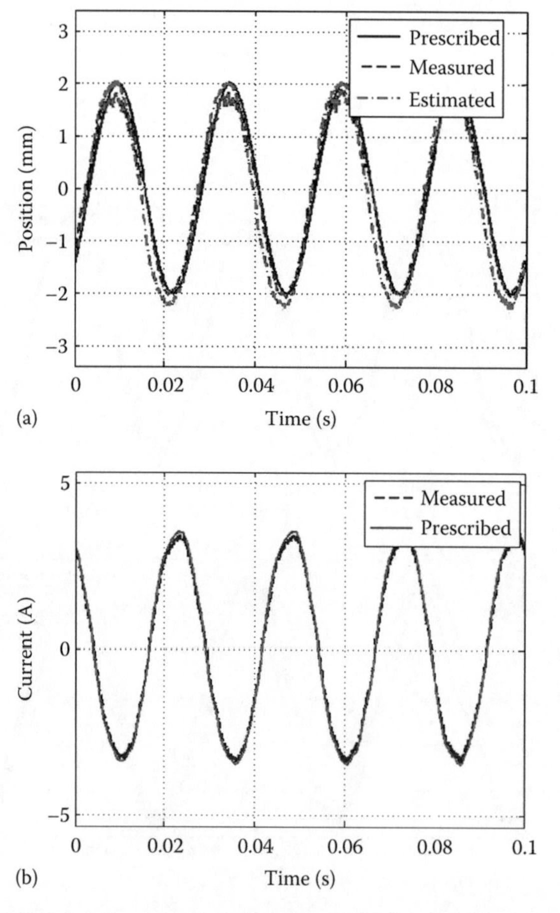

FIGURE 15.32 Close-loop position-sensor control: (a) position and (b) current. (After Agarliță, S., Linear single phase PM oscillatory machine: Design and control, PhD thesis, UPT.ro, 2009.)

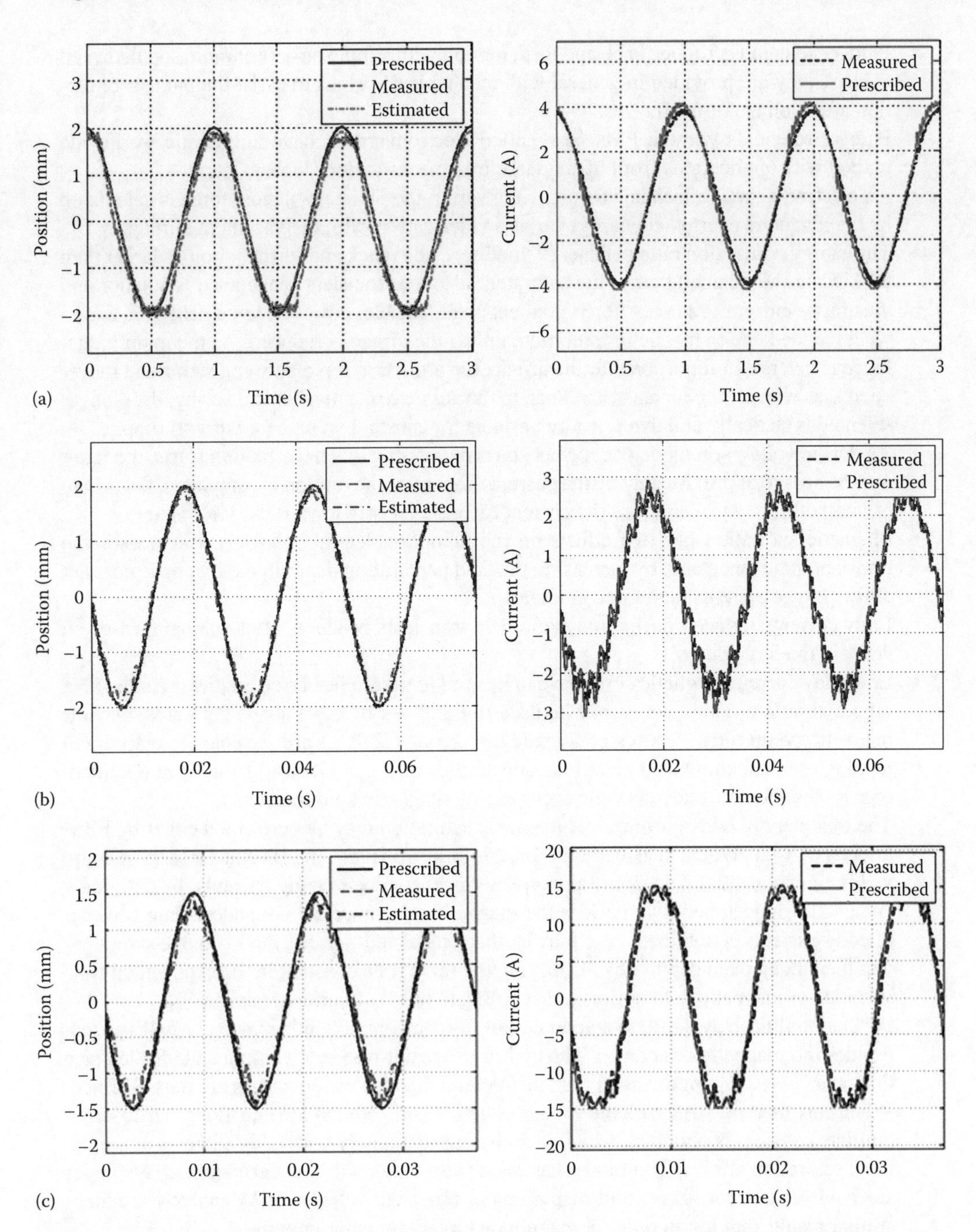

FIGURE 15.33 Close-loop position-sensorless control (a) at 1 Hz, (b) at 40 Hz, and (c) 80 Hz. (After Agarliță, S., Linear single phase PM oscillatory machine: Design and control, PhD thesis, UPT.ro, 2009.)

15.9 SUMMARY

- Plunger solenoids are single (dual) coil electromagnetic devices without or with PMs that provide linear motion based on electromagnetic forces, for travels from 20–40 μm to 10 (20) mm.

- Plunger solenoids produce back and forth motion with simple on–off or more sophisticated control; they are provided in general with mechanical springs to assist the plunger retraction after voltage turn-off.
- Plunger solenoids without PMs, also called electromagnets, have a dynamic worldwide market with applications from lifting large iron pieces to door lockers, etc.
- An electromagnetic-mechanical energy converter, the plunger solenoid force is calculated by the magnetic energy (coenergy) variation (derivative) with displacement (airgap).
- The energy conversion ratio per energy (motion) cycle (back and forth motion) is larger than 50% when the magnetic cores are saturated; also, the transient inductance is smaller and thus faster current gradients (force gradients) are feasible; however, force density is somewhat reduced due to magnetic saturation, unless the airgap is reduced. As the main inductance is inversely proportional to the airgap, the attraction force between stator and mover decreases with airgap increase and thus, in the absence of a mechanical spring, the plunger solenoid is statically and dynamically unstable for constant voltage (or current) supply.
- As during voltage-on mode the current starts decreasing when the motion starts, the latter may be sensed approximately by the current drop; after the motion is completed, the airgap is small (almost zero) and thus the current increases slowly toward the V/R value.
- Magnetic saturation has two effects on the plunger solenoid behavior: it decreases the transient inductance and, by increasing the field penetration depth in the magnetic cores, it allows larger currents (and force gradients).
- Eddy currents induced during magnetic field transients produce, after voltage turn-off, a delay in the force decay.
- Low eddy current magnetic cores have to be used to secure fast force gradients (in the kHz range); thin lamination cores may be used in flat C + I or E + I shape flat cores, but soft magnetic composites (ferrites or Somaloy or Atomet ET1...) are feasible for cylindrical topologies when equivalent electric conductivities of $\sigma_{iron} = 1.4 \cdot 10^6\ \Omega^{-1}\ m^{-1}$ are obtained; thus acceptably low eddy currents occur during magnetic field transients.
- The treatment of eddy currents and magnetic saturation may be performed either by FEM directly or by analytical field methods; for an infinite iron plate, the 1D Fourier series method yields an exponential field decay along plate thickness due to eddy currents; by defining a "magnetic" inductance L_{eddy} besides the magnetic reluctance, an equivalent time constant of eddy currents is obtained; L_{eddy} may be then calculated—even with magnetic saturation variation along plate depth—by identifying the same time constant in the exponential time variation of magnetic field in time as according to the 1D Fourier series method.
- For the nonlinear dynamic magnetic circuit, the "magnetic" inductance concept is used, besides the magnetic reluctance. The coil mmf is split into $\alpha\ W_i$, responsible for leakage flux, and $(1 - \alpha)\ W_i$, for the main flux; differential flux/current equations are thus obtained. When corroborating this with the voltage equation and with motion equations, a fifth-order nonlinear system is obtained, which includes approximately both eddy currents and magnetic saturation effects. The model has been fully validated in experiments on a plunger solenoid for current, force, and displacement transients. Direct FEM analysis produces similar results, but for an order of magnitude larger computation times.
- To speed up the response for retraction motion mode, dual-coil topologies have been proposed for compressed air valves, with 0.3–0.5 mm maximum excursion length, where displacement time should be less than a few milliseconds for a mover mass, which turns to be around 15 g. Zero vibration on–off or sliding mode control of the current in the two coils provides for such a fast motion control.
- Soft landing of the mover on its "chair" is necessary to secure a long life for the valve device. Passive bouncing reduction may be obtained if the turn-off is triggered, with motion sensing (by current decay with voltage on), quick enough and if the current turn-off is fast (by using a Zener diode or a varistor).

- Fast response solenoid optimization design has been attempted by direct geometric methods using FEM. Level set and genetic algorithms have provided notable improvements in terms of force/volume and force/copper losses.
- PMs have been lately added to solenoids for
  - Flux shielding (to deviate the magnetic flux for better flux density) when no latching force at zero current is required.
  - Flux assistance when zero current latching is required; the PM force at travel end(s) has to surpass the mechanical spring force.
- For dual-coil PM solenoids, capacitors are used to charge and, respectively, discharge the current in the two coils quickly. For a 3200 A/400 V circuit breaker, 22 ms making time with 3 ms trigger and 13 ms breaking time are obtained; this is considered competitive performance.
- In a special topology dual-coil PM solenoid, sub-millimeter travel control (20–30 μm) is obtained for 100 N force and 12.5 g mover; this configuration is suitable for micro-optics, metal molds for light-enhanced films or for axial active magnetic bearings [20].
- In a case study, a twin-coil PM valve actuator is rather fully investigated—theory and experiments—with respect to the following:
  - FEM analysis of a preliminary analytical design.
  - Direct 2D FEM optimization design by grid search and modified Hooke–Jeeves methods.
  - Eddy current losses have been calculated for magnetostatic conditions in eight regions by FEM.
  - A dynamic circuit model, and code, with force and flux versus position and current curve-fitted polynomial from FEM results, has been developed and validated in tests under constant voltage and frequency operation. Mechanical springs are added (with resonance at the most used frequency) though the frequency varies from zero to 90 (100) Hz; 3 ms travel time over 8 mm is provided, with a maximum acceleration of 5000 m/s^2, and a device volume of less than 0.17 L for a 50 g mover.
  - Although the device works stably with the given voltage (current) and frequency (with current controller), close-loop position control with a faster position sensor or sensorless produces better results.
  - Position estimation from search coil voltage filtering for flux (Φ_e) estimation from 1 to 90 Hz, with FEM data assistance (by the mover $x = f$(flux, current) function, with current also measured), has shown satisfactory results with position errors below 0.05 mm over the entire frequency range.
- The aforementioned case study is just a sample methodology of investigation for fast response solenoids; dynamic R&D efforts in the area are expected in the near future. The rather complete FEM dynamic model of a plunger solenoid (with magnetic saturation, hysteresis, and eddy currents included) [24,25] or more complete models [26,27] are indications of the trends for the near future.

REFERENCES

1. H.C. Roters, *Electromagnetic Devices*, John Wiley & Sons, London, U.K., 1941.
2. S. Fang, H. Lin, and S.L. Ho, Magnetic field analysis and dynamic characteristic prediction of ac PM contactor, *IEEE Trans.*, MAG-45(7), 2009, 2990–2995.
3. S. Agarlită, Linear single phase PM oscillatory machine: Design and control, PhD thesis, UPT.ro, 2009.
4. J.D. Lindau and C.R. Knopse, Feedback linearization of an active magnetic bearing with voltage control, *IEEE Trans.*, CST-10(1), 2002, 21–31.
5. A. Forrai, T. Ueda, and T. Yamara, Electromagnetic actuator control: A linear parameter varying (LPV) approach, *IEEE Trans.*, IE-54(3), 2007, 1430–1441.
6. L. Li, T. Shinshi, and A. Shimokohbe, Asymptotically exact linearization for active magnetic bearings actuators in voltage control configuration, *IEEE Trans.*, CST-11(2), 2003, 185–195.

7. J.P. Caridies and S.R. Tunes, Fast acting electromagnetic actuators—Computer model development and verification, *Record SAE Trans.*, 91, 1982, 820–822.
8. J.P. Caridies, Fast acting electromagnetic actuators, PhD thesis, Department of Mechanical Engineering, Pennsylvania State University, University Park, PA, 1982.
9. T. Kajima, Dynamic model of the plunger type solenoids at deenergizing states, *IEEE Trans.*, MAG-31(3), 1995, 2315–2323.
10. M. Piron, T.J.E. Miller, D. Jonel, P. Sangha, G. Reid, and J. Coles, Rapid computer-aided design method for fast acting solenoid actuator, *Record of IEEE-IAS-1998 Annual Meeting*, St. Louis, MO, 1998.
11. J. Albert, Banucu, W. Hafla, and W.M. Reecker, Simulation based development of a valve actuator for alternative drives using BEM-FEM code, *IEEE Trans.*, MAG-45(3), 2009, 1741–1747.
12. L. Yu and T.N. Chang, Zero vibration on-off position control of dual solenoid actuator, *IEEE Trans.*, IE-57(7), 2010, 2519–5226.
13. V. Utkin, J. Guldner, and J. Shi, Sliding mode control in electromechanical systems, Taylor & Francis, London, U.K., 1999.
14. P.M.S.D. de Morales and A.J. Perin, An electronic control unit for reducing contact bounce in electromagnetic contactors, *IEEE Trans.*, IE-55(2), 2008, 861–869.
15. S. Park and S. Min, Design of magnetic actuator with nonlinear ferromagnetic materials using level-set based topology optimization, *IEEE Trans.*, MAG-46(2), 2010, 618–621.
16. J.S. Choi and J. Yoo, Structural topology optimization of magnetic actuators using genetic algorithms and on-off sensitivity, *IEEE Trans.*, MAG-45(5), 2009, 2276–2279.
17. J. Man, F. Ding, Q. Li, and J. Da, Novel high speed electromagnetic actuators with PM shielding for high pressure applications, *IEEE Trans.*, MAG-46(12), 2010, 4030–4033.
18. S. Fuang, Magnetic field analysis and dynamic characteristics prediction of ac PM contactor, *IEEE Trans.*, MAG-45(7), 2009, 2990–2995.
19. S. Fuang, H. Lin, S.L. Ho, X. Wang, P. Jin, and H. Lin, Characteristic analysis of simulation of PM actuator with a new control method for air circuit breaker, *IEEE Trans.*, MAG-45(10), 2009, 4566–4569.
20. D. Wu, X. Xie, and S. Zhou, Design of a normal stress electromagnetic fast linear actuator, *IEEE Trans.*, MAG-46(4), 2010, 1007–1014.
21. I. Boldea, S.C. Agarliță, L. Tutelea, and F. Marignetti, Novel linear PM valve actuator. FE design and dynamic model, *Record of LDIA-2007*, Lille, France, 2007.
22. I. Boldea, S.C. Agarliță, F. Marignetti, and L. Tutelea, Electromagnetic, thermal and mechanical design of Linear PM valve actuator laboratory model, *Record of OPTIM-2008*, Brasov, Romania, vol. 2, 2008 (IEEExplore).
23. S. Agarliță, I. Boldea, F. Marignetti, and L. Tutelea, Position sensorless control of a linear interior PM oscillatory machine, with experiments, *Record of OPTIM 2010*, Brasov, Romania, 2010, pp. 689–695 (IEEExplore).
24. O. Bottauscio, M. Chiampi, and A. Manzin, Advanced model for dynamic analysis of electromechanical devices, *IEEE Trans.*, MAG-41(1), 2005, 36–46.
25. H.S. Choi, S.H. Lee, Y.S. Kim, K.T. Kim, and I.H. Park, Implementation of virtual work principle in virtual airgap, *IEEE Trans.*, MAG-44(6), 2008, 1286–1289.
26. M.A. Batdorff and J.H. Lumkes, High fidelity magnetic equivalent circuit model for an axisymmetric electromagnetic actuator, *IEEE Trans.*, MAG-45(8), 2009, 3064–3073.
27. Y. Kawase, T. Ota, and N. Nakamura, Dynamic analysis of solenoid valve taking into account discharge current of capacitor using FEM, *Record of LDIA-1998*, Tokyo, Japan, 1998, pp. 379–390.

16 Linear DC PM Brushless Motors

16.1 INTRODUCTION

The linear dc homopolar-PM motor has been developed from the voice-coil speaker-microphone linear electromagnetic device; however, in most cases, it uses a flat (rather than tubular) topology and is applied for progressive (rather than oscillatory) linear motion; also, the travel may reach 100 mm or more.

As it is based on Lorenz force, to change direction of motion the polarity of the current is reversed [1–3]. The main asset is the simplicity of the control with full force starting from any position when supplied from a 4 quadrant dc–dc converter, for a rather long (100 mm or even more), but limited, travel. Short coil mover with long PM or long coil stator with short PM-mover combinations may be used, depending on the application.

The copper losses per N of thrust are not small as the coil is in the air, to avoid any cogging (zero current) force and thus allow for precision control.

Recording pens, optical disk drives, focusing systems in digital video cameras, x–y chart recorders, etc., are typical applications of linear dc PM brushless motors [4–9].

Linear dc PM brush motors have also been proposed [1], but today these are of mere historical interest; consequently, only brushless topologies are considered.

Finally, as tubular configurations will be investigated for linear oscillatory machines in a subsequent chapter, only flat configurations will be treated in this chapter.

Design issues of iron-core-backed or air-core-backed PM topologies are studied in some detail; then iron losses and geometrical optimization design are illustrated; finally, sliding mode precision control is investigated, as applied in focusing a digital video camera.

16.2 TOPOLOGY ASPECTS

Two basic topologies are shown in Figure 16.1a and b [3,8].

The single-coil mover in Figure 16.1a—intended for travels above 10 mm—allows for a larger force density than the dual-coil mover with PM in air (Figure 16.1b), but the latter has a larger force/weight, so crucial in some applications, when the mover has a small weight (1–2 g). For larger force per volume, single-coil dual-core configurations may be used (Figure 16.2).

As the thin magnetic fixture may be replaced even with a nonmagnetic rugged one, the inductance of the coil is rather small and thus the force variation with mover position is very small, because the core magnetic saturation under load is not large, provided the PM flux does produce a reasonable flux density in the core. Also, the iron losses are to be smaller as they are produced mainly by the local variation of the core magnetic flux density due to armature current.

Out of the four coil sides, this time two are active (only one side is active in Figure 16.1); this justifies the claims of larger force density and lower copper losses per N of force. Lateral nonmagnetic plates (Figure 16.2b) may provide better PM stiffness over PM length up to 0.1–0.15 m travel or even more.

To double the thrust for the single coil, the PM thickness has to be doubled (roughly), but the core weight is only slightly larger (20% or so) while the coil weight is also only slightly increased (20% or so). The configuration in Figure 16.2 shows further potential for improvements in linear dc PM brushless topologies.

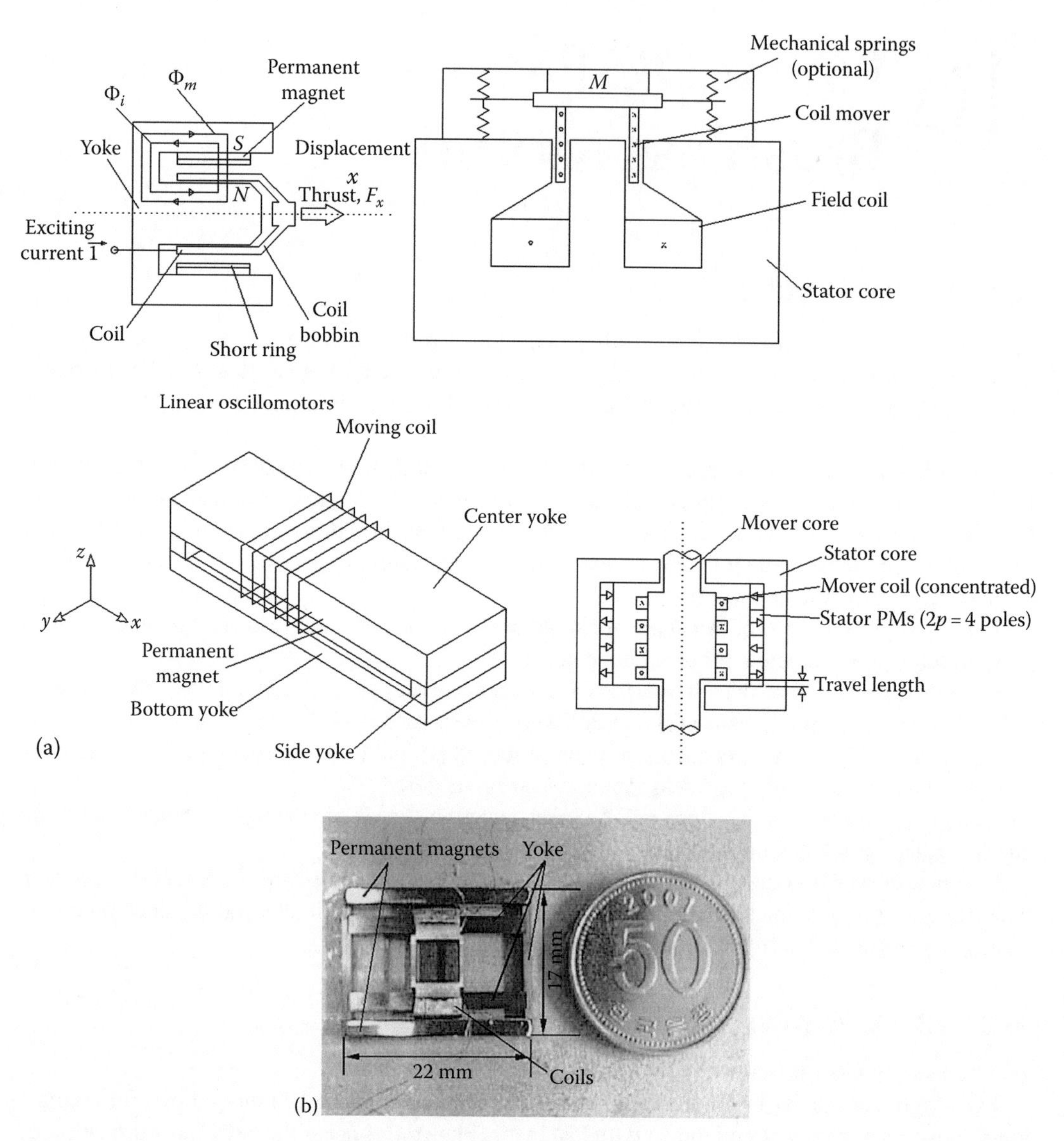

FIGURE 16.1 Basic topologies: (a) with one-coil mover and iron cores, (b) with PM in air and dual-coil mover.

16.3 PRINCIPLE AND ANALYTICAL MODELING

Let us consider here the analytical model of the configuration in Figure 16.1a, illustrated in more detail in Figure 16.3.

With N turns per coil, w_c, active coil length and B_{PM}, *PM* flux density, considered uniform here along x (direction of motion), the Lorenz force on the coil F_x is

$$F_x = K_f \cdot i; \quad K_f = B_{PM} \cdot N \cdot w_c \tag{16.1}$$

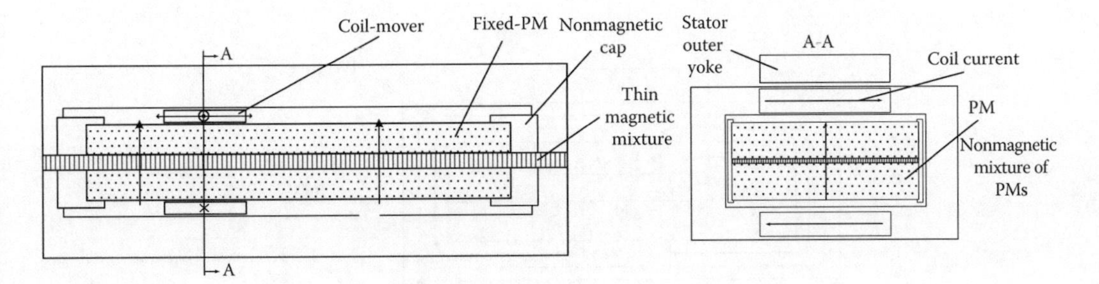

FIGURE 16.2 Single-coil dual-core configuration.

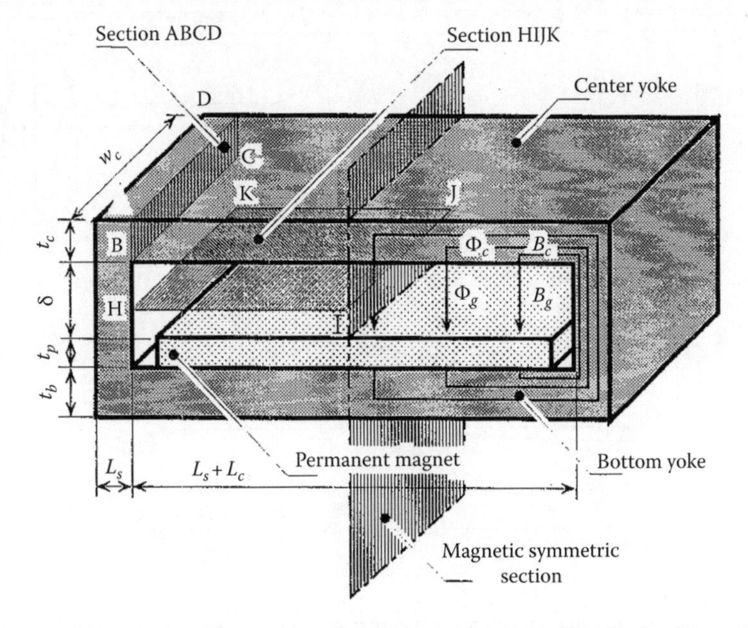

FIGURE 16.3 Geometry and flux paths of linear dc PM mover. (After Wakiwaka, H. et al., *IEEE Trans.*, MAG-32(5), 5073, 1996.)

At center point, the PM flux and the armature (coil mmf) flux lines and flux density distribution are given in Figure 16.4.

As visible in Figure 16.4, the armature reaction increases the core flux on the right side of mover and decreases it on the left side, much like in a standard rotary dc brush motor (transversal armature reaction).

Now, with the mover at extreme position, the coil (armature) flux stays the same (within the core) while the PM fringing (leakage) at its ends leads to an even smaller force. So K_f is expected to vary with mover position due to magnetic saturation (Figure 16.4c) because the field of the coil mmf closes mainly within the magnetic core.

This reality would become evident with 3D FEM (even with 2D FEM) analysis, but an approximate analysis, valid for preliminary design, would be in order here.

So let us consider the center position when the flux produced in the core by the coil is

$$\phi = B_a \cdot w_c \cdot t_c; \quad B_a = \frac{Ni\mu(B_c)}{2(L_s + L_c + l_s + t_p + g + t_c)} \tag{16.2}$$

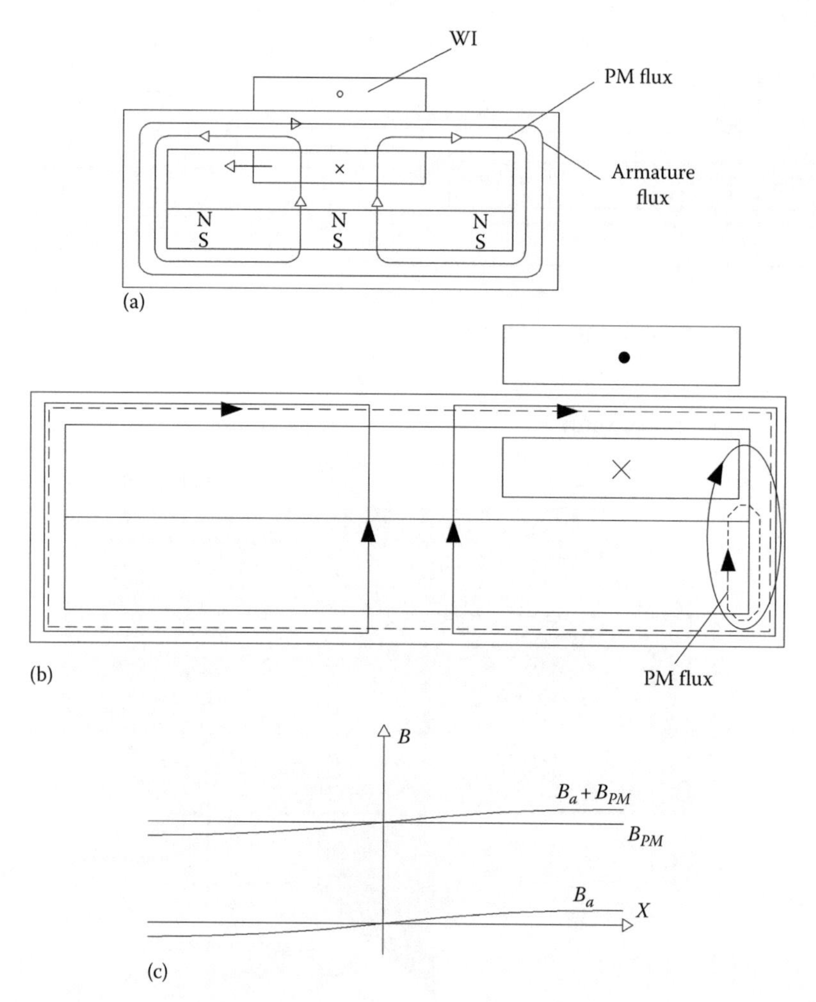

FIGURE 16.4 Flux lines and flux densities at (a) center position, (b) extreme position, and (c) force versus position at given current. (After Boldea, I. and Nasar, S.A., *Linear Motion Electromagnetic Devices*, Taylor & Francis, New York, 2001.)

On the other hand, the PM flux in the core right corner ϕ_c is

$$\phi_c = B_{c\max} \cdot t_c \cdot w_c = B_a \cdot w_c \cdot t_c + B_{PM} \frac{(L_s + L_c)}{2} w_c \tag{16.3}$$

$B_{c\max}$ in the core is limited by the core material $B(H)$ curve and may be assessed by combining (16.2) and (16.3):

$$\phi_{c\max} = w_c \cdot t_c \cdot B_{c\max} = \left[\frac{\mu \cdot (B_c) \cdot N \cdot i}{2(L_s + L_c + l_s + l_p + g + t_c)} + B_{PM} \frac{(L_s + L_c)}{2 \cdot t_c} \right] \cdot w_c \cdot t_c \tag{16.4}$$

Equation 16.4 may be solved iteratively, if first

$$B_{PM} \approx B_r \cdot \frac{t_p}{t_p + g} \cdot \frac{1}{1 + K_{fringe}} \tag{16.5}$$

The fringe factor K_{fringe} depends on mover position to some extent (for start $K_{fringe}=0.1–0.2$).

Equation 16.4 suggests that, for given $B_{c\max}$, for more force, more core area ($w_c t_c$) and more L_s+L_c (total length) are required. Conversely, if the length of travel L_s increases, by specifications, then, for more force, the core area ($w_c t_c$) should also increase.

The force per watt of copper losses is also to be considered. The ideal coil thickness ($g=h_c+2g_w$) has to be smaller or equal to *PM* thickness to secure the highest force density:

$$h_c \approx g = \frac{N \cdot i_n}{L_c \cdot j_{con} \cdot K_{fill}} \tag{16.6}$$

K_{fill} is the copper fill factor
j_{con} is the design current density

The coil length l_{coil} is

$$l_{coil} \approx 2(w_c + g) \tag{16.7}$$

and thus the coil resistance R_c is

$$R_c = \frac{1}{\sigma_{co}} \frac{l_{coil}}{i_n} j_{con} \cdot N \tag{16.8}$$

σ_{co} is the copper electrical conductivity.

So the design copper losses p_{con} can be written as

$$p_{con} = \frac{2}{\sigma_{con}} \left(w_c + \frac{N \cdot i_n}{L_c \cdot j_{con} \cdot K_{fill}} \right) j_{con} \cdot N \cdot i_n \tag{16.9}$$

Finally, the design force/watt of copper losses is

$$f_x = \frac{F_{xn}}{p_{con}} = \frac{B_{PM} \cdot \sigma_{co}}{2\left(1 + \dfrac{N \cdot i_n}{w_c \cdot L_c \cdot j_{con} \cdot K_{fill}}\right) j_{con}} \tag{16.10}$$

Now the *PM* flux in the coil Ψ_{PM} becomes

$$\psi_{PM} = N \cdot B_{PM} \cdot w_c \cdot x \cdot \frac{1}{1 + K_{fringe}(x)} \tag{16.11}$$

With $K_{fringe} \approx 0.1–0.2 \approx$ const, $x=0$ in the center position.

So the emf is

$$E = -\frac{d\psi_{PM}}{dx} \cdot \frac{dx}{dt} \approx \frac{-N \cdot B_{PM} \cdot w_c}{1 + K_{fringe}(x)} \cdot u; \quad u \text{ is the speed} \tag{16.12}$$

Finally, the coil inductance is

$$L_c \approx (1 + K_l) \frac{\mu(i) \cdot N^2 \cdot w_c \cdot t_c}{2(L_s + L_c + l_s + t_p + g + t_c)} \tag{16.13}$$

The core permeability $\mu(i)$ is an average one; in an analytical model, it may be a weighted average of three values in three key points of the core, while the actual magnetization curve $B(H)$ of the core has to be considered. The factor K_l refers to the leakage inductance, which, in any case, is less than 0.2 even for, say, a soft magnetic composite core; not so for an air-core device.

Note on core losses: An analytical model of core losses may be obtained by refining the previously mentioned analytical model via the magnetic equivalent circuit model, where quite a few zones of uniform flux density variations have to be identified.

Then, by knowing the flux density variation in time, for the given linear motion profiles, the core loss may be calculated, by standard expressions, in each region; then they are added together.

Alternatively, 3D FEM may be used in magnetostatic mode, and then with the flux density/time variations in each finite element, the total core losses are calculated. For a tubular configuration, 2D FEM may be used directly, considering eddy current losses and even hysteresis losses [5].

16.4 GEOMETRICAL OPTIMIZATION DESIGN BY FEM

For the configuration in Figure 16.1a, detailed, for the geometrical variables, in Figure 16.5, the optimization variables are obtained as L_1, L_2, L_3, L_4, and L_5. Two inequality constraints in mm [7] are visible in Figure 16.5:

$$\frac{2}{36} - 8 \cdot 8 \cdot L_3 - \frac{L_7}{2} \geq 0.5 \qquad 2 \cdot L_3 + L_5 \leq 36 \tag{16.14}$$

A few equality constraints (specifications) are also given [7]:

- Current density $j_{con} = 24.5$ A/mm^2
- Coil volume = 280 mm^3
- Mechanical gap $G_1 = 0.4$ mm (on one side of the coil)
- $G_2 = 0.5$ mm—gap at the end of travel

With a given initial variable vector $[L_1, L_2, L_3, L_4, L_5]^T$, a 2D FEM (at least), and an optimization method, to change L_i in order to get maximum thrust, the optimization design is completed.

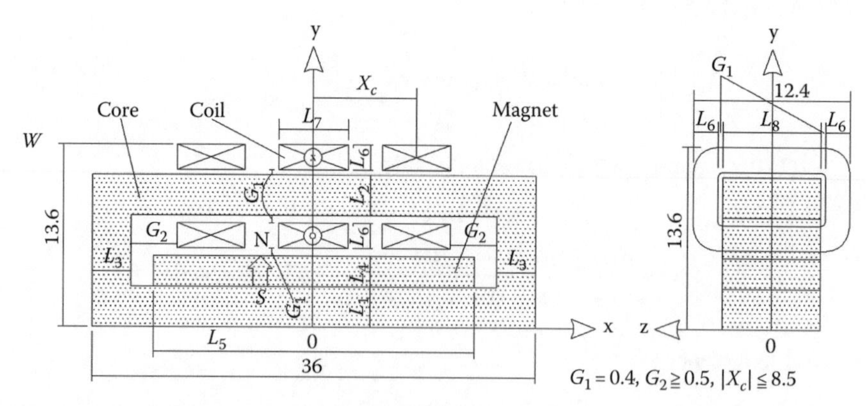

FIGURE 16.5 Definition of geometrical design variables. (After Boldea, I. and Nasar, S.A., *Linear Motion Electromagnetic Devices*, Taylor & Francis, New York, 2001.)

We may augment the objective function by adding, say, temperature constraints, based on a simplified thermal model, etc.

In our case, with maximum force as the objective function, but also with given coil volume and current density, the copper losses are fixed. Therefore, in fact we are seeking here maximum thrust/watt of copper losses.

In Ref. [7], the Rosenbrock optimization method was used for the scope. The initial and final variable vectors are shown in Table 16.1 and the results are given in Table 16.2 [7].

The nonuniform distribution of force versus position is proven by FEM (Figure 16.6) but is not severe (Table 16.1).

During the 39 iterations (for a 0.01 mm convergence criterion), the variable vector was changed by a given step of 10% of initial values and the thrust was calculated by 2D FEM. If the thrust increased with respect to the previous step, the variables were changed by

$$\Delta L_j^{(k+1)} = \alpha L_j^{(k)}; \quad \alpha > 1.5$$

$$\text{otherwise,} \tag{16.15}$$

$$\Delta L_j^{(k+1)} = -\beta L_j^{(k)}; \quad \beta < 0.75$$

TABLE 16.1
Dimensions of Initial and Final Shapes

Dimensions	Initial Value (mm)	Optimal Value (mm)
L_1	3.8	4.19
L_2	3.4	3.70
L_3	3.6	4.27
L_4	2.4	2.76
L_5	26.0	26.0
L_6	1.4	26.22
L_7	6.0	9.44
L_8	8.8	9.85

Source: After Takahashi, S., Optimal design of linear dc motor using FEM, *Record of LDIA-1995*, Nagasaki, Japan, pp. 331–334, 1995.

TABLE 16.2
Turns of Coil

	Thrust (N)		
Position of Coil (mm)	Initial Shape	Optimal Shape	Ratio Increase (%)
−8.5	0.82	0.99	20
0	0.88	1.13	29
8.5	0.86	1.05	22

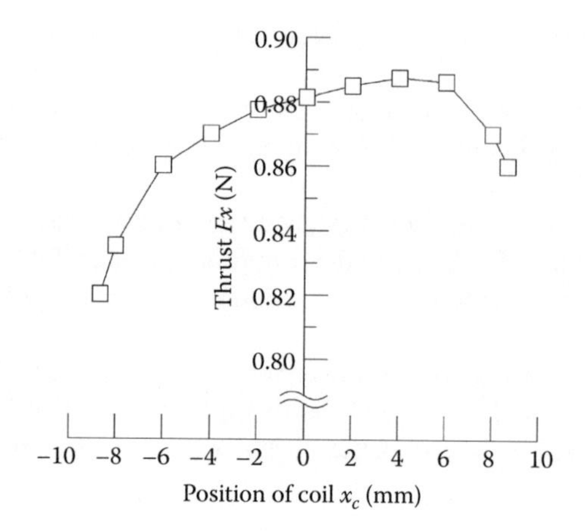

FIGURE 16.6 Initial force versus position. (After Takahashi, S., Optimal design of linear dc motor using FEM, *Record of LDIA-1995*, Nagasaki, Japan, pp. 331–334, 1995.)

A 20% improvement in the thrust was obtained to reach 0.23 N/W [7], which is still modest, but the fixed current density was 24.5 A/mm² and the coil volume was 280 mm³.

16.5 AIR-CORE CONFIGURATION DESIGN ASPECTS

For an optical disk drive application, the PM-in-air configuration (Figure 16.1b) was adopted [8]. A virtual path magnetic circuit model of the machine proved to compare acceptably with 3D FEM results (Figure 16.7), except for small airgap values. Even for this air-core machine, the PM flux density in the coil area is about 0.3 T.

The experimental results (Figure 16.8) for the mini linear dc PM motor of Table 16.3 [8] show a wide frequency response band, suitable for the application.

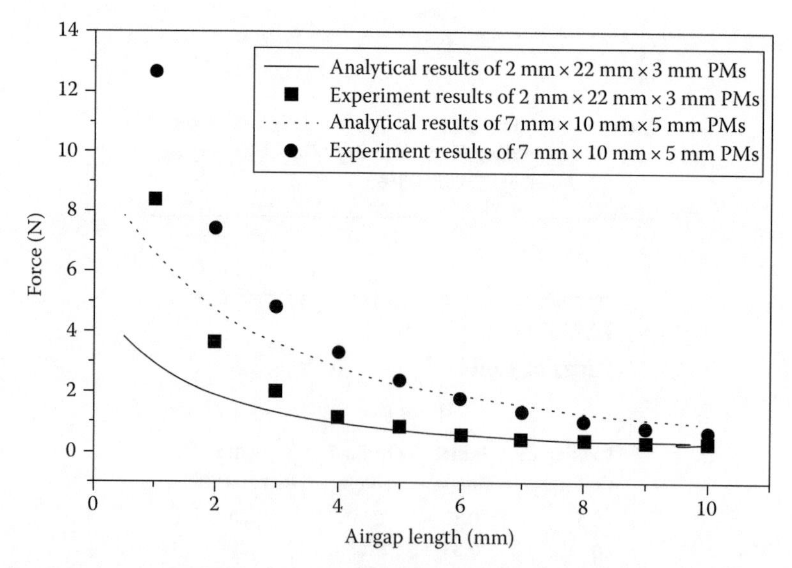

FIGURE 16.7 Force versus airgap length between the two PMs. (After Park, J.H. et al., *IEEE Trans.*, MAG-39(5), 3337, 2003.)

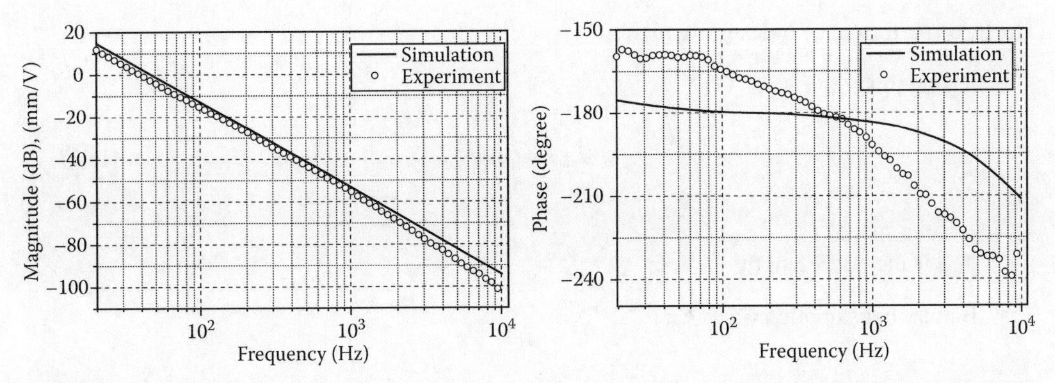

FIGURE 16.8 Frequency response of the prototype. (After Park, J.H. et al., *IEEE Trans.*, MAG-39(5), 3337, 2003.)

TABLE 16.3
Discretization Data

Number of elements	3.456
Number of nodes	1.741
Number of iterations of Rosenbrock's method	39
CPU time (s)	201

16.6 DESIGN FOR GIVEN DYNAMICS SPECIFICATIONS

For the iron-core linear dc brushless PM motor, in a small power application (a digital video camera focuser), a more complicated optimal design should include not only the energy conversion ratio and the battery consumption but also the rising time (travel time for given travel length).

If the space is restricted to volume ($l_x \times l_y \times l_z$), Figure 16.9, we may end up with a few main variables [6]:

$$\gamma = \frac{l_m}{l_w} = \frac{PM \text{ thickness}}{\text{coil thickness}}; \quad N = \text{round}\left(\frac{l_w}{\phi}\right) \cdot \text{round}\left(\frac{l_p}{\phi}\right) \tag{16.16}$$

l_p is the coil length
ϕ is the coil conductor diameter

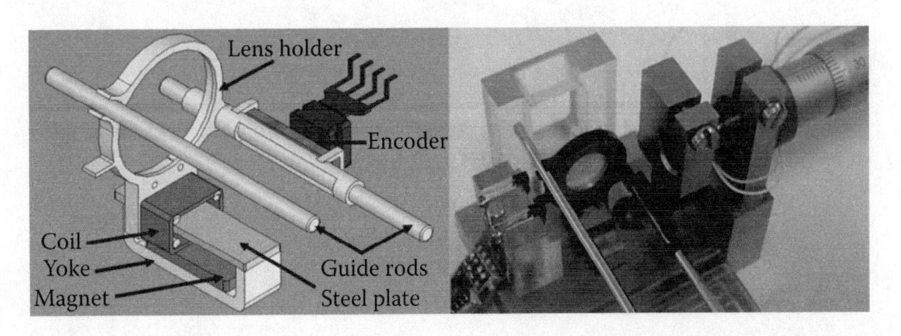

FIGURE 16.9 Linear dc PM focuser for digital video camera. (After Yu, H.C. and Liu, T.S., *IEEE Trans.*, MAG-43(11), 4048, 2007.)

The three performance indexes are then

- Rising time t_r:

$$d_{\max} = \int_0^{t_r} u dt \tag{16.17}$$

u is the speed
$d_{\max}$ is the travel length

- Battery consumption (in J) E_0:

$$E_0(\gamma,\phi) = \int_0^{t_r} i(t)v(t)dt \tag{16.18}$$

- Energy efficiency:

$$\eta_e(\gamma,\phi) = \frac{E_0 - \int_0^{t_r} i^2(t)\cdot R(\gamma,\phi)dt}{E_0} \tag{16.19}$$

All variables in (16.17) through (16.19) are functions of γ and ϕ; to use them in the optimization design, we have to solve for current i and speed u from voltage and motion equations:

$$\frac{di}{dt} = \frac{v(t) - i(t)\cdot R(\gamma,\phi) - K_u(\gamma,\phi)\cdot u(t)}{L(\gamma,\phi)} \tag{16.20}$$

$$\frac{du}{dt} = \frac{K_u(\gamma,\phi)\cdot i(t) - F_{load} - B\cdot u(t)}{M} \tag{16.21}$$

where
L is the coil inductance
M is the mover mass
K_u is the emf (and force) constant (see Equation 16.11):

$$K_u(\gamma,\phi) = \frac{d\Psi_{PM}}{dx} = \frac{F_x}{i} = K_f \tag{16.22}$$

If the force F_x is calculated by 2D FEM, then $K_u(\gamma, \phi)$ may be determined from (16.22); also, the inductance L may be calculated from FEM, while the resistance R_c is

$$R_c = \frac{4}{\sigma_{co}}\cdot\frac{N\cdot l_{coil}(\gamma)}{\pi\phi^2} \tag{16.23}$$

With the given voltage, R, L, K_f, K_u, B_m, $d_{\max}$, and M and F_{load}, the three performance indexes may be calculated after solving for U and i in (16.20 and 16.21).

For $M=2$ g, $v=3$ V, $i_{max}=0.03$ A, $d_{max}=5.2$ mm, $B_m=0.005$ N s/m, and $F_{load}=0.05$ gw (gram weight), to compromise the three performance indexes finally [6], $\gamma=2.5$, $\phi=0.07$ mm, for $\eta_e=4.9\%$ (energy efficiency!), $E_0=1.3$ mJ, and rising time $t_r=44$ ms [6]. This rising time is considered much better (6 times smaller) than that obtained with commercial rotary-stepper-plus transmission actuator. To fully characterize the design, $R_c=32.8\ \Omega$, $L=1.2$ mH, and $K_u=K_f=42.3$ gw/A (measured) [6].

Though the energy efficiency η_e seems low, we have to consider it in comparison with existing solutions for such a small device.

16.7 CLOSE-LOOP POSITION CONTROL FOR A DIGITAL VIDEO CAMERA FOCUSER

We are extending here the digital video camera focuser (Figure 16.9) analysis by dealing with the close-loop position control (open loop dynamics were investigated in the previous paragraph).

At this time, the voltage and motion equations parameters are as given previously.

The movable part is composed of the lens holder, the coil, and a linear magnetic strip mounted on a side tube. The stator includes the PMs, the yoke, a steel plate, and two motion-guiding pads. The position encoder includes the magnetoresistive (MR) sensor and a linear magnetic strip with 0.88 mm pole pitch; the final position estimation precision is around 3 μm. In addition, speed estimation is calculated from position feedback.

Two position control methods are used for comparison:

- PID control
- Sliding mode control

The PID control contains two cascaded loops, one for position and one for speed (Figure 16.10).

For sliding mode control, a sliding mode functional $S(e)$, $e=d^*-d$ (position error) should be defined [9]:

$$S(e)=\ddot{e}+c_2\dot{e}+c_3e \tag{16.24}$$

If the sliding mode poles are $p_{1,2}=-\alpha\pm i\beta$, then

$$c_2=2\alpha,\quad c_3=(\alpha+\beta)^2 \tag{16.25}$$

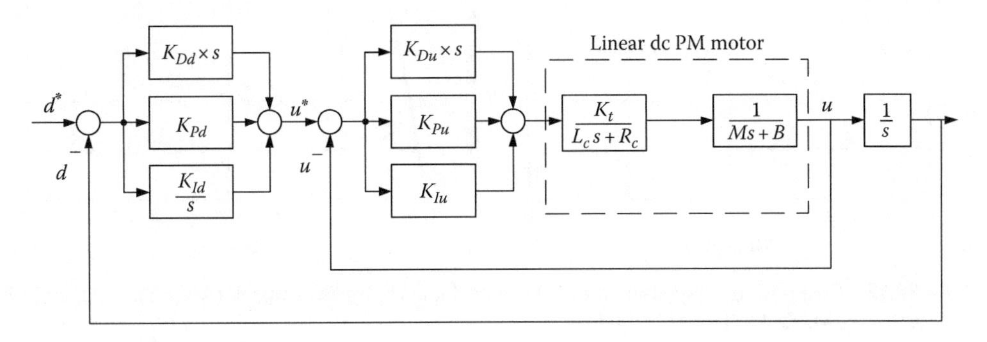

FIGURE 16.10 PID position control.

To be stable, the system has to stay close to the sliding mode surface $S(e)=0$, but first to reach the sliding mode condition (surface)

$$S\dot{S} < -\sigma(S); \quad \sigma > 0 \tag{16.26}$$

To verify the stability and define a control law, the Lyapunov criterion is used, after a Lyapunov (energy) function V is chosen [9]:

$$V = S^2 \tag{16.27}$$

But

$$\dot{V} = \frac{dV}{dt} = 2S\dot{S} < -2\sigma(s) = -2\sigma\sqrt{V} < 0 \tag{16.28}$$

As $V>0$, $\dot{V}<0$ and thus the control will be asymmetrically stable if (16.26) is satisfied.

The control law may be defined, based on (16.24), as

$$U = -Q_{sat}(s) - L_3^{-1}c_2\ddot{e} - L_3^{-1}c_3\dot{e} \tag{16.29}$$

With $Q>0$, the maximum boundary of system states and disturbance is

$$Sat(S) = \begin{matrix} sign(S); & if\,|S| > \varepsilon \\ S/\varepsilon; & if\,|S| \leq \varepsilon \end{matrix} \tag{16.30}$$

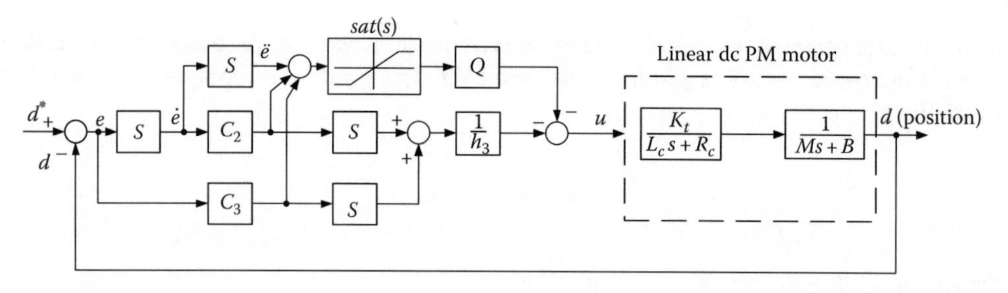

FIGURE 16.11 Sliding mode position control.

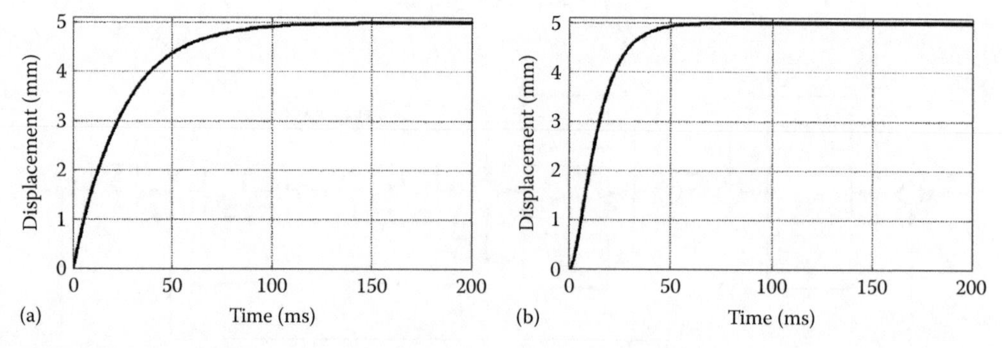

FIGURE 16.12 Step position response: (a) in PID control and (b) by SM control. (After Yu, H.C. and Liu, T.S., *IEEE Trans.*, MAG-43(11), 4048, 2007.)

The Sat function was used, instead of Sign function, to provide for low chattering response around the target. The close-loop control scheme is illustrated in Figure 16.11. The parameter Q may be adjusted based on minimization of error energy function $E_n(e) = \int e^2 dt$.

It has to be noticed that

- There is no cascaded control (as for PID control).
- Quite a few position error time derivatives are required, but, as position is measured, the two derivatives may be obtained by adequate observers; also if an inertia accelerometer were made available, it would smooth notably the $\dot{e}$ and $\ddot{e}$ observers (as done in MAGLEVs).

Sample results for the case in point, for PID and SM control, are given in Figure 16.12.

Both focusing time (from 120 to 60 ms) and steady-state precision (from 21 to 7 μm) are reduced for SM in comparison with PID control, as expected (SM control is known to provide fast and robust response with small overshooting).

16.8 SUMMARY

- The linear dc homopolar brushless PM motor has been developed from the voice-coil speaker microphone, but it develops control for progressive linear motion rather than oscillatory motion.
- It is based on the Lorenz force and is essentially a single-phase linear synchronous brushless PM motor with limited travel.
- The practical travel length goes from a few mm to 100 mm or more.
- Flat configurations with air core and iron core are typical for the linear dc brushless PM motor, while tubular ones are used in most speaker-microphone applications.
- The Lorenz force sign changes with current polarity.
- For limited travel, no brushes are required and the single (dual)-coil mover is fed from a flexible power cable.
- The PM-mover configuration is feasible, but then the copper Joule loss in the stator coils is larger.
- Iron-core configurations show higher force/volume while air-core configurations show slightly better force/weight but, also, higher frequency response band (due to lower electric time constant and inductance).
- For iron-core flat configurations, analytical field model for circuit model parameters, and performance including magnetic saturation, produce satisfactory results for preliminary design purposes.
- Secondary effects such as force reduction at travel ends may be ascertained properly with FEM analysis.
- Force density and force/copper losses are two conflicting performance indexes that have to be reconciled in a proper design.
- Core losses may be calculated by analytical or numerical field modes and are important when the duty cycle of back and forth motion is large.
- For air-core topologies, virtual magnetic circuit models have been proven acceptable (validated by 3D FEM), except for a small distance between PMs (when two or three of them are provided); notable computer time is saved this way for optimization design.
- Direct FEM geometrical optimization design has been successfully proven on linear dc brushless PM motors with 0.8 N force for 17 mm travel length at 0.23 N/W of copper losses as the objective function.

- Direct FEM geometrical optimization may be exercised with some sophisticated performance indexes such as rising time, travel time per travel length, total energy consumption per rising time, and energy efficiency for the same. For a 2 g coil mover, a rising time of 44 ms, at total energy consumption of 1.3 mJ and energy efficiency of 4.9%, was obtained in view of focusing a digital video camera.
- For the same application with given circuit parameters $R=32.8\ \Omega$, $L=1.2$ mH, $K_u=K_f=$ 42.3 gw/A, and $M=1.8$ g, both cascaded PID position and speed control and sliding mode (SM) position control have been investigated for a magnetostriction position sensor (5 μm in precision); SM control reduced the settling time for a 5 mm travel at 60 ms for a steady-state positioning precision of 7 μm, adequate for design video camera focusing; the existing rotary microstepper needs 400 ms for the same job.
- Due to ruggedness, simplicity, and its rather long travel length, the linear dc brushless PM motor will find more and more applications when high-frequency band motion (in the hundreds of Hz range) is needed; the hot spots in R&D will perhaps be geometrical FEM, optimization design, and robust high-precision fast positioning control (For more, see Refs. [10,11]).

REFERENCES

1. S.A. Nasar and I. Boldea, *Linear Motion Electric Machines*, John Wiley, New York, 1976.
2. A. Basak, *Permanent Magnet Linear Dc Motors*, Oxford University Press, 1996.
3. I. Boldea and S.A. Nasar, *Linear Motion Electromagnetic Devices*, Taylor & Francis, New York, 2001, Chapter 7.
4. H. Wakiwaka, H. Yajima, S. Senoh, and H. Yamada, Simplified thrust limit equation of linear Dc motor, *IEEE Trans.*, MAG-32(5), 1996, 5073–5075.
5. M. Hippner, H. Yamada, and T. Mizuno, Iron losses in linear dc motor, *IEEE Trans.*, MAG-38(5), 1999, 3715–3717.
6. H.-C. Yu, T.-Y. Lee, S.-J. Wang, M.-L. Lai, J.-J. Ju, D.-R. Huang, and S.-K. Lin, Design of a voice coil motor used in the focusing system of a digital video camera, *IEEE Trans.*, MAG-41(10), 2005, 3979–3981.
7. S. Takahashi, Optimal design of linear dc motor using FEM, *Record of LDIA-1995*, Nagasaki, Japan, pp. 331–334.
8. J.H. Park, Y.S. Baek, and Y.P. Park, Design and analysis of mini-linear actuator for optical disk drive, *IEEE Trans.*, MAG-39(5), 2003, 3337–3339.
9. H.C. Yu and T.S. Liu, Output feedback sliding mode control for a linear focusing actuator in digital video cameras, *IEEE Trans.*, MAG-43(11), 2007, 4048–4050.
10. P. Fang, F. Ding, Q. Li, and Y. Li, High response electromagnetic actuator with twisting axis for gravure systems, *IEEE Trans.*, MAG-45(1), 2009, 172–175.
11. C.-S. Liu, P.-D. Lin, S.-S. Ke, Y.-H. Chang, and J.-B. Horng, Design and characterization of miniature auto focusing voice coil motor actuator for cell-phone camera applications, *IEEE Trans.*, MAG-45(1), 2009, 155–159.

17 Resonant Linear Oscillatory Single-Phase PM Motors/Generators

17.1 INTRODUCTION

Fixed frequency linear motion applications [1] such as compressors, pumps, vibrators, as well as speakers/microphones use linear oscillatory single-phase PM motors (LOMs), while short-stroke (up to 20 mm in general) applications use linear single-phase PM generators (LOGs) such as Stirling engines or other linear piston engines due to

- Elimination of rotary to linear motion mechanical transmission
- Simplicity and ruggedness of their topology
- Rather high efficiency and high force density, especially with mechanical springs and operation at resonance frequency (mechanical eigenfrequency = electrical frequency)
- Simplicity (single-phase PWM converter) for close-loop control

Note: Other linear generators with larger stroke (20 mm and more) have been investigated in 2(3) phase configurations in previous chapters. Here only the single-phase ones are treated.

The LOMs (LOGs) may be classified into three main categories:

1. With coil mover
2. With PM mover
3. With iron mover

Flat and tubular, single-phase PM topologies are feasible, but the tubular ones are favored, as the blessing of circularity leads to better volume usage and better heat transmission. As expected, each category exhibits a plethora of different topologies, with a few being actually practical.

This chapter will deal with the tubular coil-mover LOM (LOG) both for large power (25 kW—say for series hybrid electric vehicles) and for very small power microphones and vibrators of mobile phones.

Then the PM-mover topology in its tubular shape will be treated in both single- and multiple-pole (and stator coil) variants for refrigerator-like compressors. In addition, the flat PM-mover LOM (LOG) in reversal flux configurations with PM flux concentration will be investigated due to its high-performance potential.

Finally, a tubular and a flat iron-mover LOM (LOG) will be covered, from topology through modeling, FEM-analysis dynamics, control, and optimal design with case studies.

17.2 COIL-MOVER LOMs (LOGs)

Single-coil (or homopolar)-mover LOMs (Figure 17.1a) are typical for strokes up to 100 mm for high-rated forces: hundreds of Newtons to kN at frequencies below 6–70 Hz and for much lower forces, in the sub-mm range, for microspeakers in mobile phones, etc.

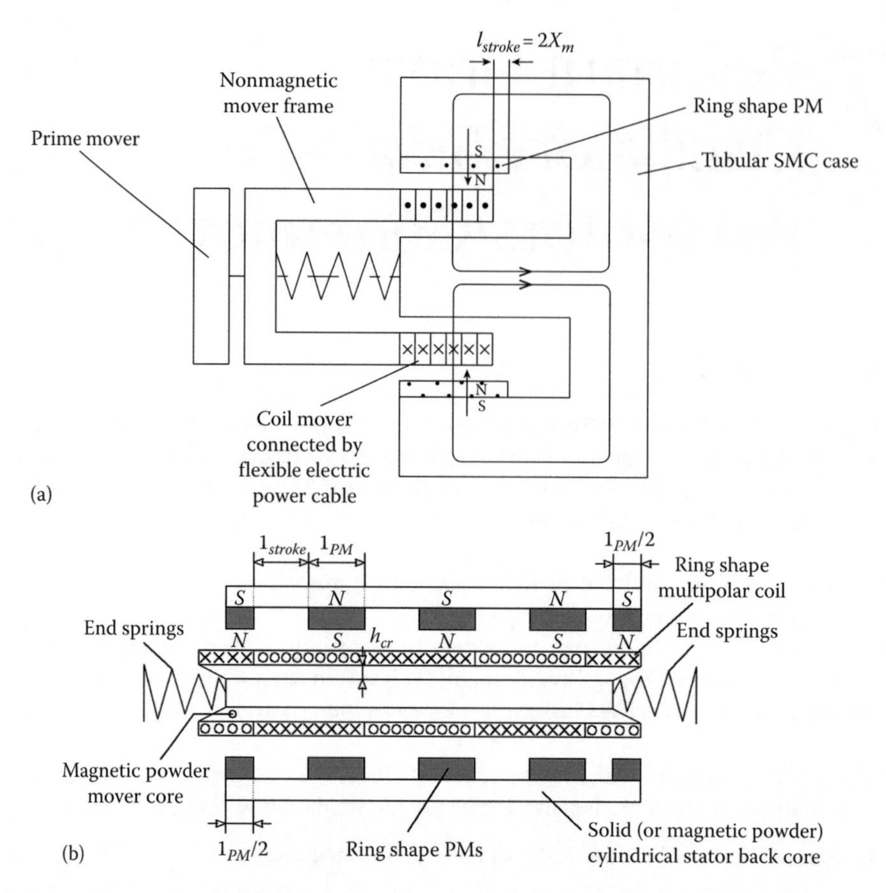

FIGURE 17.1 Coil-mover LOM(G), (a) single-coil–single PM, (b) multiple-coil–multiple PM.

The multiple-coil topology (Figure 17.1b) should be preferable when the weight of the core (and total weight) is critical and thus the multi-PM poles lead to lower PM core flux. The flexible electric power cable (or even slip rings and brushes) to transmit or collect the electric input (output) power is a demerit of the coil-mover solution; however, the mass M of the mover is reduced, so that the mechanical spring constant K, for the given resonance frequency, is smaller; the cost of the spring is reduced:

$$2\pi f_m = \sqrt{\frac{K}{M}} \tag{17.1}$$

In Figure 17.1, the coil length is longer than the PM length by the stroke length, l_{stroke}:

$$l_{stroke} \approx 2X_m \quad X_m\text{: Motion amplitude} \tag{17.2}$$

This leads to lower PM weight but to more mover weight and copper losses. The shorter coil–longer PM by (l_{stroke}) combination (Figure 17.1a) may also be adopted for the multiple-coil topology (Figure 17.1b).

But larger PMs mean more costs and thicker stator core and weight.

Depending on the key constraints of application, one or the other option is used in the design.

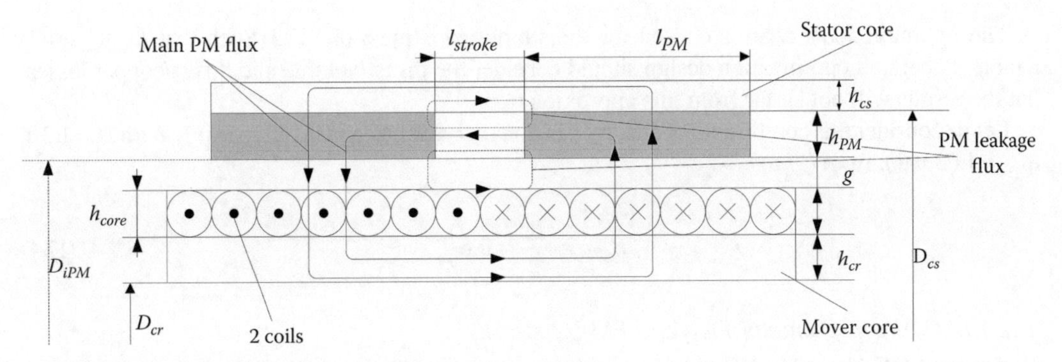

FIGURE 17.2 Two-coil mover two-PM pole stator LOM (G).

The air coil(s) leads to smaller inductance and thus larger power factor, but it also implies lower thrust density; the fragility of the coil mover, which has to be enforced with a nonmagnetic frame (siluminum), may also be a problem as the electromagnetic force is exerted directly on the conductors of the coil(s).

In an effort to reconcile these conflicting requirements, for large power devices (kW range and more), a two or more pole topology is selected; this time the inner magnetic core may be attached to the mover to reinforce it mechanically, in spite of its additional mover weight (Figure 17.2).

Additional axially magnetized PMs may be added to produce more flux density in the coils, but also to reduce outer core flux (and radial thickness).

The dual (multiple)-coil-mover LOM may also be manufactured for smaller diameter and larger axial length geometry.

As the average PM airgap flux density is around $0.5 \times B_r$ at best, LOM(G), due to the air-core coils' mild magnetic saturation in the cores, can be kept under control, without excessive core volume, but the thrust density remains moderate.

17.2.1 Four-Coil-Mover LOM: Modeling and Design by Example

Let us consider the configuration in Figure 17.2 (but with four PMs and four coils) and develop an analytical model for it, and then use this model for preliminary electromagnetic design as a generator for an electrical output of 20 kW, 220 V_{RMS}, $f_1 = 60$ Hz, stroke length $l_{stroke} = 30$ mm, and for harmonic (sinusoidal) motion: $x = l_{stroke}/2\sin(2\pi f_1 t)$.

To limit the space required for presentation, we will first define the objectives and then proceed, in parallel, with the modeling and the numerical example: airgap PM flux density, airgap diameter and PM sizing, emf, thrust, inductance, resistance and core losses, core sizing and core loss, voltage equation and phasor diagram, number of turns (total), efficiency, power factor and its compensation by a capacitor, mover weight, and mechanical spring rigidity constant K.

17.2.1.1 Airgap: PM Flux Density

Making use of the Ampere law, while accounting for the axial PM and for the flux fringing and for magnetic saturation by equivalent coefficients, $K_{aPM} < 0.3$, $K_{fringe} < 0.2$, $K_{sat} < 0.1$ (large total airgap), the airgap PM flux density B_{gPM} is

$$B_{gPM} = \frac{B_r h_{PM} \mu_{rec}(1 + K_{aPM})}{h_{PM} + (g + l_{coil})\mu_{rec}(1 + K_{fringe})(1 + K_{sat})} \tag{17.3}$$

In Equation 17.3, the curvature of the tubular structure is neglected as the thrust, power, and airgap diameter are notable.

The optimum h_{PM}/g ratio is chosen for the simplified expression (17.3) such that $h_{PM}\mu_{rec}/g \approx 1$; a more elaborated optimization design should consider the thrust/volume and thrust/copper losses, but the results will not be far from this approximation.

Let us consider this condition met and, by FEM, $K_{aPM}=0.20$, $K_{sat}=0.1$, $K_{fringe}=0.1$. With $B_r=1.2$ T ($\mu_{rec}=1.05$ p.u.), from (17.3)

$$B_{gPM} \approx \frac{B_r}{2} = 0.6\ \text{T} \tag{17.4}$$

17.2.1.1.1 Airgap Diameter D_{iPM} and PM Length l_{PM}

To size the LOG, the thrust has to be known by assuming an efficiency $\eta_n \approx 0.93$ and calculating the average speed U_{av}:

$$\begin{aligned} U_{av} &= 2l_{stroke}f_1 = 2\cdot 0.03\cdot 60 = 3.6\ \text{m/s} \\ F_{xav} &= \frac{P_n}{U_{av}\eta_n} = \frac{20\cdot 1^3}{3.6\cdot 0.93} = 5973.7\ \text{N} \end{aligned} \tag{17.5}$$

A moderate flux density $f_{xn}=2$ N/cm² is adopted. Consequently, the total active PM area A_{PM} is

$$A_{PM} = \frac{F_{xav}}{f_{xr}} = \frac{5973.7}{2\cdot 10^4} = 0.2986\ \text{m}^2 \tag{17.6}$$

Now the problem is to discriminate between PM length and airgap diameter D_{iPM}. It is evident that the PM pole length (l_{PM}) should be longer than stroke length, in order to limit the additional copper losses in the inactive core length ($2pl_{stroke}$); $2p$—coils (PMs); in our case $p=2$.

Let us consider $l_{PM}=2l_{stroke}=2\cdot 0.03=0.06$ m. For 4 active poles, the airgap diameter D_{iPM} is approximately

$$D_{iPM} \approx \frac{A_{PM}}{2pl_{PM}\pi} = \frac{0.2986}{4\cdot 6\cdot 10^{-2}\cdot \pi} = 0.396\ \text{m} \tag{17.7}$$

But the average thrust (Lorenz force) F_{xav} is

$$F_{xav} \approx B_{gPM}\frac{l_{PM}}{l_{PM}+l_{stroke}}\cdot \pi\cdot D_{avc}\cdot 2p;\ \ n_c\cdot I_{av};\ \ I_{av} = I_n\sqrt{2}\cdot\frac{2}{\pi} \tag{17.8}$$

n_c is the turns per coil (there are $2p$ coils) and each coil is $l_{PM}+l_{stroke}$ in length. But for a given current density $j_{con}=6$ A/m² and coil filling factor $K_{fill}=0.8$ (the coil may be made of multiple thin copper foils),

$$\frac{n_c I_{av}}{l_{PM}+l_{stroke}} = j_{con}\cdot h_{coil}\cdot k_{fill}\cdot n_c\cdot I_{av} = 0.09\cdot 6\cdot 10^6\cdot h_{coil}\cdot 0.8 \tag{17.9}$$

So

$$h_{coil} = \frac{n_c\cdot I_{av}}{0.432\cdot 10^6} \tag{17.10}$$

Also,

$$D_{avc} \approx D_{iPM} - g - h_{coil}; \quad g = 1.5 \text{ mm} \tag{17.11}$$

From (17.8) through (17.11), we may calculate simultaneously $h_{coil} \approx (h_{PM} + g)/\mu_{rec}$ and $n_c I_{av}$: $n_c I_{av} \approx 3017$ A turns/coil and $h_{coil} = 7.0 \cdot 10^{-3}$ m.

So the PM height $h_{PM} = h_{coil} \cdot \mu_{rec} + g = 7 \cdot 1.05 + 1.5 = 8.85 \text{ mm} = 8.85 \cdot 10^{-3}$ m; $D_{avc} = D_{iPM} - (g + h_{coil}) = 0.396 - (1.5 + 7.0) \cdot 10^{-3} \approx 0.386$ m.

The total resistance R_s is

$$R_s = 2p \cdot \rho_{co} \frac{\pi D_{avc} n_c^2 j_{con}}{n_c I_{av}} = \frac{4 \cdot 2.0 \cdot 10^{-8} \cdot \pi \cdot 0.386 \cdot 6 \cdot 10^6 \cdot n_c^2}{3017} = 2.603 \cdot 10^{-4} n_c^2 \tag{17.12}$$

So the copper losses p_{con} are

$$p_{con} \approx R_s I_{av}^2 = 2.60 \cdot 10^{-4} \cdot 3017^2 = 1755 \text{ W} \tag{17.13}$$

With iron and mechanical losses neglected, the copper losses would lead to an "ideal" efficiency of 0.92. As this is not far away from the assigned value of 0.93, and we are dealing here with a preliminary design, the obtained geometry holds; otherwise, the design should be redone from (17.5) with a lower (0.9) value of assigned efficiency.

17.2.1.2 Inductance

The inductance comprises essentially only the airgap component:

$$L_s \approx \frac{1}{8} 2p\mu_0 n_c^2 \pi D_{avc} \frac{(l_{PM} + l_{stroke})}{\frac{h_{PM}}{\mu_{rec}} + g + h_{coil}} = \frac{1}{8} \cdot 4 \cdot 1.256 \cdot 10^{-6} \cdot \pi \cdot 0.386 \cdot \frac{0.09}{\left(\frac{8.85}{1.05} + 1.5 + 7.0\right) \cdot 10^{-3}} n_c^2$$

$$= 4.046 \cdot 10^{-6} n_c^2 \tag{17.14}$$

The emf (E)

$$E(t) = B_{gPM} \cdot U(t) \cdot \pi \cdot D_{avc} \cdot 2p \cdot n_c \frac{l_{PM}}{l_{PM} + l_{stroke}} \tag{17.15}$$

The speed is

$$U(t) = \omega_1 x_m \cos \omega_1 t = 2\pi \cdot 50 \cdot 0.015 \cdot \cos \omega_1 t = 4.71 \cdot \cos \omega_1 t \tag{17.16}$$

$$E(t) = 0.6 \cdot 4.71 \cdot \pi \cdot 0.386 \cdot 4 \cdot \frac{0.06}{0.09} \cdot n_c \cdot \cos \omega_1 t = 9.134 \cdot n_c \cdot \cos \omega_1 t \tag{17.17}$$

In (17.15), a PM flux linkage in the coil linear variation with position is supposed; consequently, the harmonic motion leads to a sinusoidal emf. In reality, at the ends of PMs, both the emf and thrust are somewhat reduced; these effects have to be calculated by 2D FEM and accounted for by fudge factors in the force and emf expressions.

17.2.1.3 Core Sizing and Core Losses

As the PM airgap flux density is $B_{gPM}=0.6$ T, the peak "armature reaction" is

$$B_{agpeak} \approx B_{gPM}\frac{\omega_1 L_s I_{RMS}}{E_{1RMS}} = \frac{.6\cdot 100\cdot\pi\cdot 4.096\cdot 10^{-6} n_c^2 I_{RMS}}{\left(9.130\cdot n_c/\sqrt{2}\right)} \approx 0.395 \text{ T} \tag{17.18}$$

The average PM flux in the cores is given, however, by the B_{gPM}, as magnetic saturation is kept under control with $B_{cs}=B_{cm}\approx 1.5$ T. For $2p=4$ PM poles not 1/2 but approximately 2/3 of PM pole flux travels the coils axially.

The stator outer core depth h_{cs} is

$$h_{cs} \approx \frac{2}{3}\frac{B_{gPM}\cdot l_{PM}\cdot\pi(D_{iPM}+h_{PM})}{B_{cs}\pi\left(D_{iPM}+2h_{PM}+(h_{cs})\right)} \tag{17.19}$$

From (17.19), with $B_{gPM}=0.6$T, $l_{PM}=0.06$ m, $D_{iPM}=0.396$ mm, $h_{PM}=8.55$ mm, $h_{cs}\approx 16$ mm.

In a similar way, the thickness of the mover core is $h_{cm}=18$ mm.

Now the iron weight—for both stator and mover—is

$$\begin{aligned} G_{ironsm} &= 2p(l_{PM}+l_{stroke})\cdot\pi(D_{iPM}+2h_{PM}+h_{cs})\cdot h_{cs}\cdot\gamma_{iron} \\ &= 4\cdot 0.09\cdot\pi(0.396+2\cdot 0.00855+0.016)\cdot 0.016\cdot 7200 \\ &= 55 \text{ kg} \end{aligned} \tag{17.20}$$

$$\begin{aligned} G_{irons} &= 2p(l_{PM}+l_{stroke})\cdot\pi\left(D_{iPM}-2(g+h_{coil})-h_{cm}\right)\cdot h_{cm}\cdot\gamma_{iron} \\ &= 4\cdot 0.09\cdot\pi\left(0.396+2(1.5+7.0)\cdot 10^{-3}-0.018\right)\cdot 0.018\cdot 7200 \\ &= 52.84 \text{ kg} \end{aligned} \tag{17.21}$$

Note: the mover iron weight seems large and may be "moved" to an interior stator, by splitting the airgap in two and making the coil-mover structure of nonmagnetic (siluminum) materials; its mechanical ruggedness has then to be considered carefully.

With SMC cores, the P_{iron} at 1.5 T/50 Hz may be considered to be 4 W/kg, so the total core losses may be as high as 4 W/kg · 110 kg = 440 W. This is nontrivial and should be considered in efficiency calculations and in the thermal design. A lower core loss material would be beneficial, provided its cost is affordable for the application.

17.2.1.4 The Phasor Diagram

The coil-mover LOM(G) is in fact a single-phase synchronous machine where the sinusoidal emf is produced by the harmonic motion, while the PM flux linkage per coil varies linearly with mover position; the inductance is independent of mover position as the machine exhibits air coils.

Under steady state, at resonance, $\omega_1=\omega_m=2\pi\sqrt{k/M}$, the mechanical spring "moves" the mover back and forth, with small mechanical losses, while the electromagnetic (Lorenz) force produces electromagnetic power for the load.

The decreases in force (and emf) coefficients $Ke(x)$, $(F_x=K_e i)$, toward the ends of stroke, are in general less than 10%, and thus a sinusoidal emf may be used for the preliminary design. Also, the

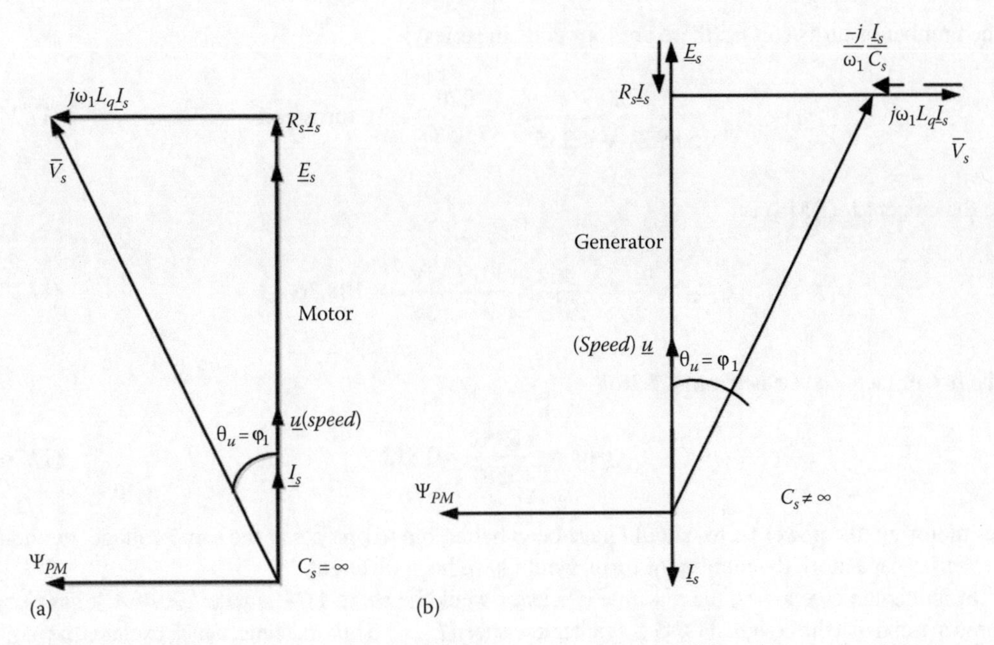

FIGURE 17.3 Coil-mover LOM(G); the phasor diagram for resonance conditions (a) motoring, (b) generating.

cogging force is practically zero. Consequently, the standard phasor diagram may be used, based on steady-state voltage equation in complex variable terms (Figure 17.3)

$$\underline{V_s} = R_s \underline{I_s} + j\omega_1 L_s \underline{I_s} - j\frac{1}{\omega_1 C_s}\underline{I_s} + K_e \underline{U} \tag{17.22}$$

$\underline{U}$ is speed phasor; $\underline{E_s} = K_e \underline{U}$

For resonance conditions $\omega_1 = \omega_m$ and the current is in phase with emf (and speed), to produce maximum power/current. In this case, however, the power factor is lagging (for motoring) and the voltage regulation is notable for generating.

A series capacitor (C_s) may be used to compensate (or overcompensate) the inductance L_s, to extract more power for a given machine at given output voltage in generating and provide for stability in autonomous operation, as shown later in this chapter when stability and control aspects are treated. The series capacitor C_s may be used in the motoring also to only partially compensate for the inductance L_s and thus improve the power factor and increase starting current for faster self-synchronization at direct connection to the power grid; care must be exercised in this case to avoid PM demagnetization during C_s-assisted direct starting; ultimately, C_s may be shorted during direct starting.

17.2.1.5 The Number of Turns per Coil n_c

From the phasor diagram

$$(E_s - R_s I_s)^2 + \omega_1^2 L_s^2 I_s^2 = V_s^2; \quad I_s = I_{RMS} = I_{av} \cdot \left(\frac{\pi}{2\sqrt{2}}\right) \tag{17.23}$$

Making use of already calculated parameters,

$$E_s = 2.2835 \cdot \frac{n_c}{\sqrt{2}}; \quad R_s I_s = 2.603 \cdot 10^{-4} n_c (n_c I_s)$$

$$\omega_1 L_s I_s = 100 \cdot \pi \cdot 4.046 \cdot 10^{-6} n_c (n_c I_s), \; Vs = 220\, V(RMS)$$

The number of turns/coil n_c (there are four coils in series) is

$$n_c \frac{220}{\sqrt{5.8952^2 + 4.235^2}} = \frac{220}{7.2587} \approx 31 \text{ turns/coil} \tag{17.24}$$

So the current I_1 (RMS) is

$$I_{1RMS} = \frac{n_c I_{av}}{n_c} \cdot \frac{\pi}{2\sqrt{2}} = \frac{3017}{31} \cdot \frac{\pi}{2\sqrt{2}} = 108.36 \text{ A} \tag{17.25}$$

The power factor cos φ is [from (17.26)]

$$\cos\varphi = \frac{5.8952}{7.2587} = 0.812 \tag{17.26}$$

For motoring, the power factor should have been better, but to operate at the same voltage, (without any series capacitor), the number of turns would have been different.

Including the core losses, the machine efficiency would be about 1.6% smaller (90.4%). It has to be borne in mind that the designed LOG is a rather low-speed $U_{av} \approx 3.6$ m/s machine, which explains the moderate efficiency, but the elimination of mechanical transmission and the increased reliability might pay off.

The copper weight in the mover can be easily calculated and, after admitting additional mover mass for framing and from the prime mover (or load machine: compressor for motoring), the mechanical spring rigidity K may be calculated, at resonance frequency.

In an attempt to reduce copper losses and eliminate the electric power flexible cable, the PMs (longer than coils) may be placed on the mover but then core weights will be larger.

Note: Though we may follow here other aspects of LOM(LOG), we will exercise them over other topologies, to spread more evenly the new knowledge over this chapter.

17.2.2 Integrated Microspeakers and Receivers

Microspeakers and receivers are required for advanced mobile phones that integrate laptop PC functions, web searching, MP3 song players, etc. [2–4].

High-quality sound, broader frequency range, and reduced size and costs are paramount for microspeakers and receivers. Harmonic distortion affects the quality of sound and is defined by the ratio between the sum of sound power of higher harmonics to the power of fundamental frequency.

Harmonic distortion is produced by the diaphragm asymmetry, input source voltage harmonics, and uneven magnetic field distribution. The latter also influences the sound pressure level (SPL). These are all related to the microspeaker/receiver design.

But for an acoustical analysis, the sound power radiated, which vibrates in a mean rms surface with speed of ($\dot{x}^2(f)$), is

$$W_{rad}(f) = \rho c S_{rad} \sigma_{rad}(f) < \dot{x}^2(f) >, [\text{W}] \tag{17.27}$$

where

- ρ is air density
- c is the sound speed in air
- f is the radiation frequency
- S_{rad} is the diaphragm area
- σ_{rad} is the radiation efficiency

For a monopole source type diaphragm, [5]

$$\sigma_{rad}(f) = \frac{K^2a^2}{1+K^2a^2} \tag{17.28}$$

$K = 2\pi f/c$ is the wave number, and a is the diaphragm radius. The sound pressure level (SPL) at distance d from the source $L_p(d)$ is

$$L_p(d) = L_w - 20\log\left(\frac{d}{d_0}\right) - 8; \text{ [dB]} \tag{17.29}$$

L_w is the sound power level, $d_0 = 1$ m:

$$L_w(d) = 10\log\left(\frac{W_{rad}}{10^{-12}}\right); \text{ [dB]} \tag{17.30}$$

Finally, the total harmonic distortion (THD) in % is [5]

$$\text{TWD} = \frac{\text{Sum of power in harmonics } (\nu \geq 2)}{\text{Total power}} \tag{17.31}$$

The microspeaker and the dynamic receiver in the cellular phones may be separate units, but, to reduce space, the integration of the two in one piece seems a practical solution (Figure 17.4) [4]. Consequently, only the integrated solution is detailed here.

The two-coil motors are both placed (Figure 17.4c) in the airgap between the three permanent magnets and exposed to large PM flux densities (0.35–0.45 T), in spite of the microdimensions at play for cellular phones, due to PM flux concentration.

A 2D FEM analysis, given the cylindrical structure, provides the instrument to calculate with acceptable precision the field distribution, emf, and thrust.

The examples in Figure 17.5a and b show flux distribution [4]; the flux density and thrust versus coil position (height h_2 in mm) are illustrated in Figure 17.5c. They are typical of the efficacy of 2D FEM for the scope.

For $h_1 = 1.1$ mm, the PM flux in the coil is most symmetric, as it should. The average PM flux density over the coil height pulsates inevitably with the coil position, when the latter vibrates. The force variation with coil position is a clear indication of this phenomenon.

The average PM flux density (ac component) contributes to the pure sinusoidal SPL, with sinusoidal current. The harmonics lead to harmonics distortion.

PM height h_2, coil position h_1, and central to the outer membrane diameter ratio, coil height and total airgap may be considered in an optimization design, where the sound pressure should be maximized and THD and volume should be constraint.

To obtain sound power $W_{rad}(f)$, however, the speed of coil $\dot{x}(f)$ should be calculated. Consequently, a dynamic model is required. The latter includes the voltage and motion equations:

$$\frac{di}{dt} = \frac{V - iR - K_e(x)\dot{x}}{L} \tag{17.32}$$

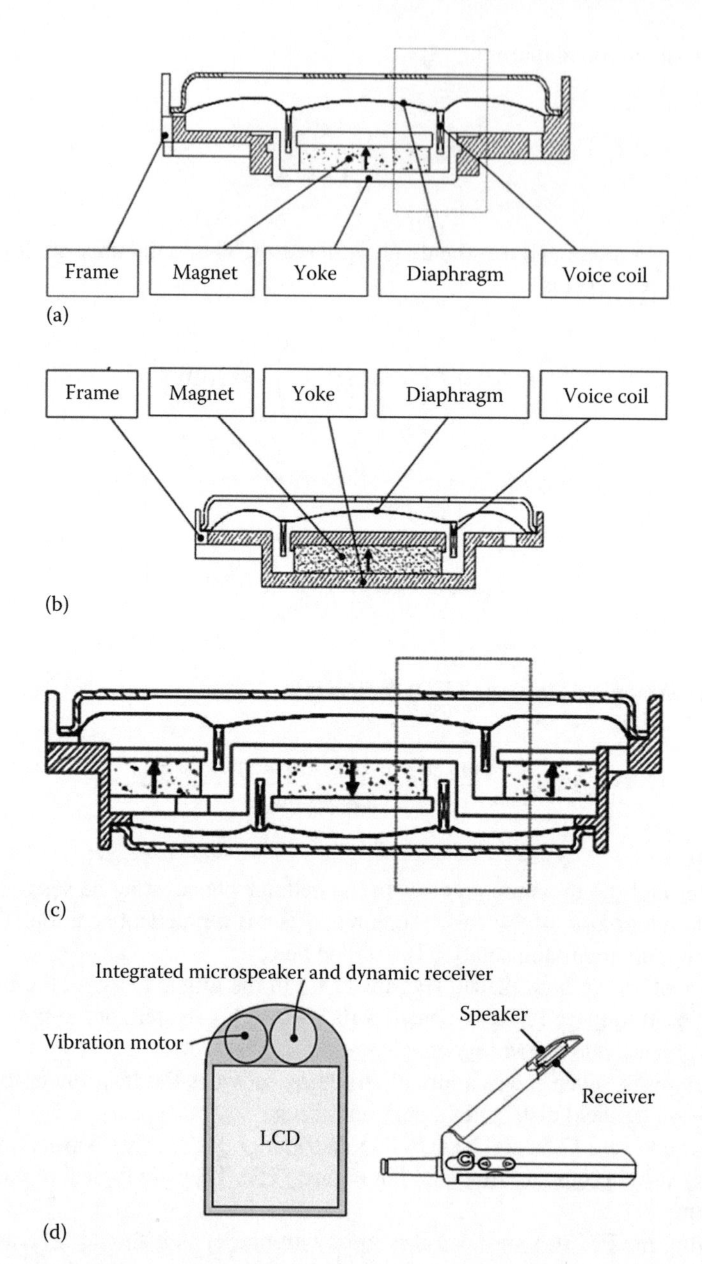

FIGURE 17.4 (a) Microspeaker alone, (b) receiver alone, (c) integrated microspeaker and dynamic receiver, and (d) layout of integrated device. (After Hwang, S.M. et al., *IEEE Trans.*, MAG-39(5), 3259, 2003.)

$$\dot{x} = \frac{K_e(x)i - C\dot{x} - Kx}{M} \tag{17.33}$$

C is the dynamic factor
K is the spring factor
M is the mover mass
R, L are the coil resistance and inductance
V is the supply (or input) voltage

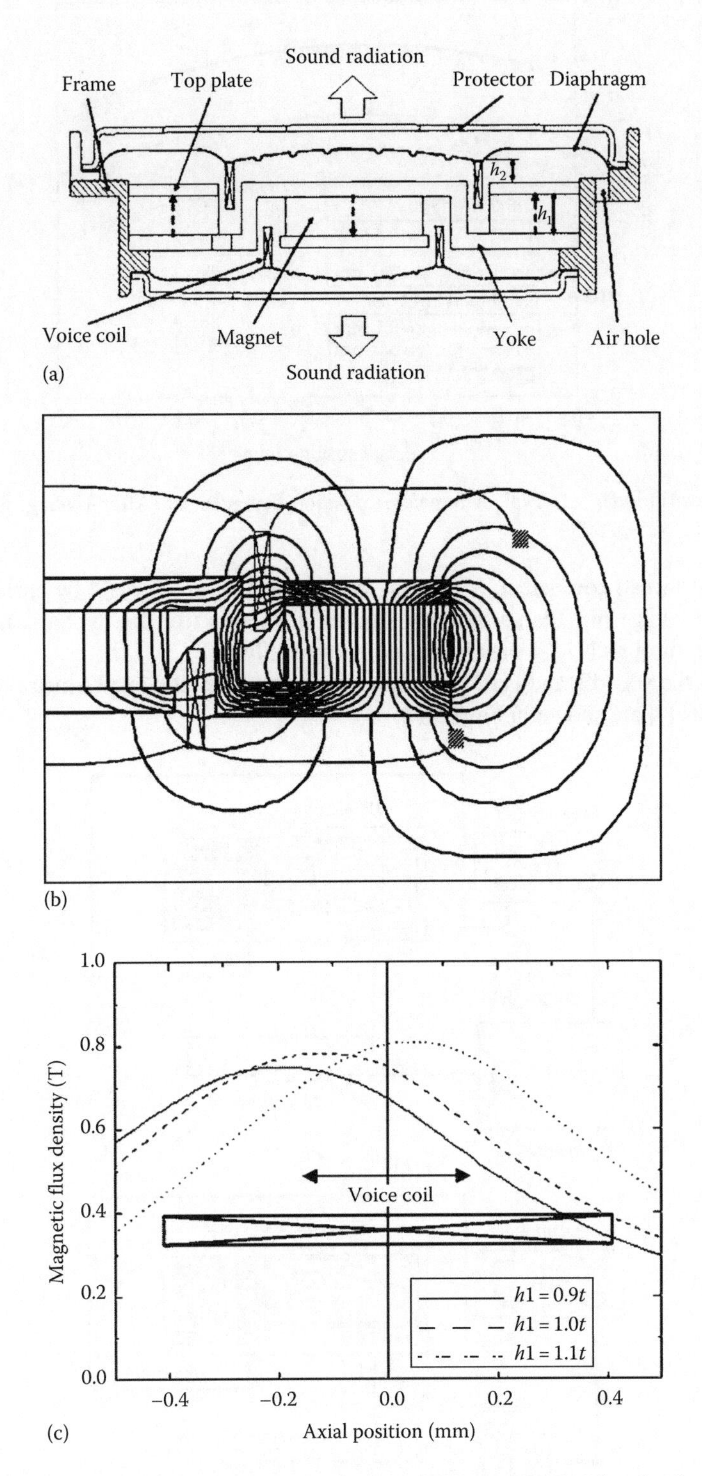

FIGURE 17.5 (a) Integrated device components, (b) flux distribution, and (c) PM flux density in the coil airgap.

(*continued*)

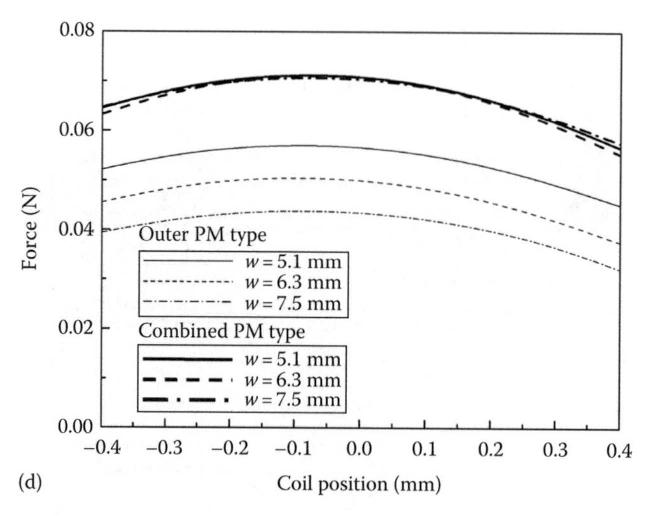

FIGURE 17.5 (continued) (d) Typical force/coil position dependence. (After Hwang, S.M. et al., *IEEE Trans.*, MAG-39(5), 3259, 2003.)

The emf (and thrust) coefficient, $K_e(x)$, may be obtained from 2D FEM by curve fitting. Either with sinusoidal voltage (for the speaker) or sinusoidal motion (for the dynamic receiver), model equations (17.32) through (17.33) may be solved numerically.

Subsequently, the SPL (dB) and THD (%) may be calculated. They also may be measured (Figure 17.6). Typical results [4] are shown in Figure 17.7.

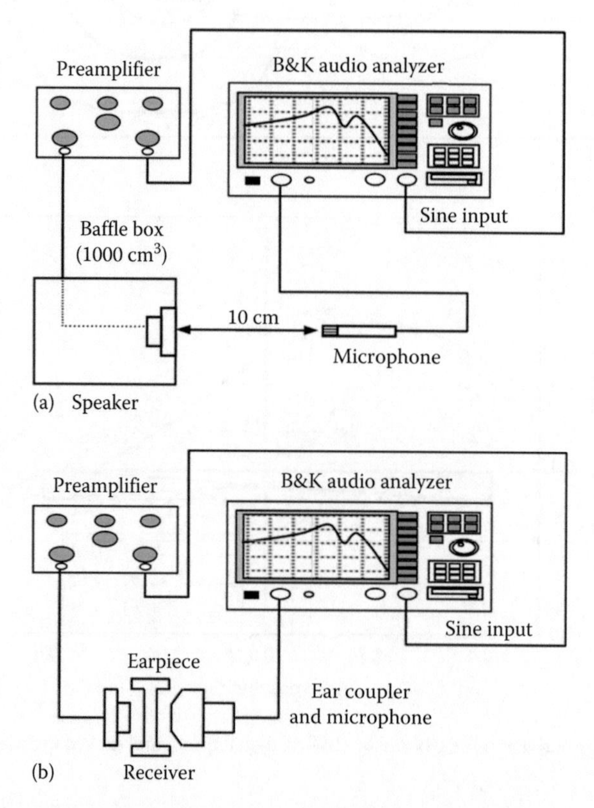

FIGURE 17.6 Test rig for: (a) speaker SPL and (b) receiver SPL. (After Hwang, S.M. et al., *IEEE Trans.*, MAG-39(5), 3259, 2003.)

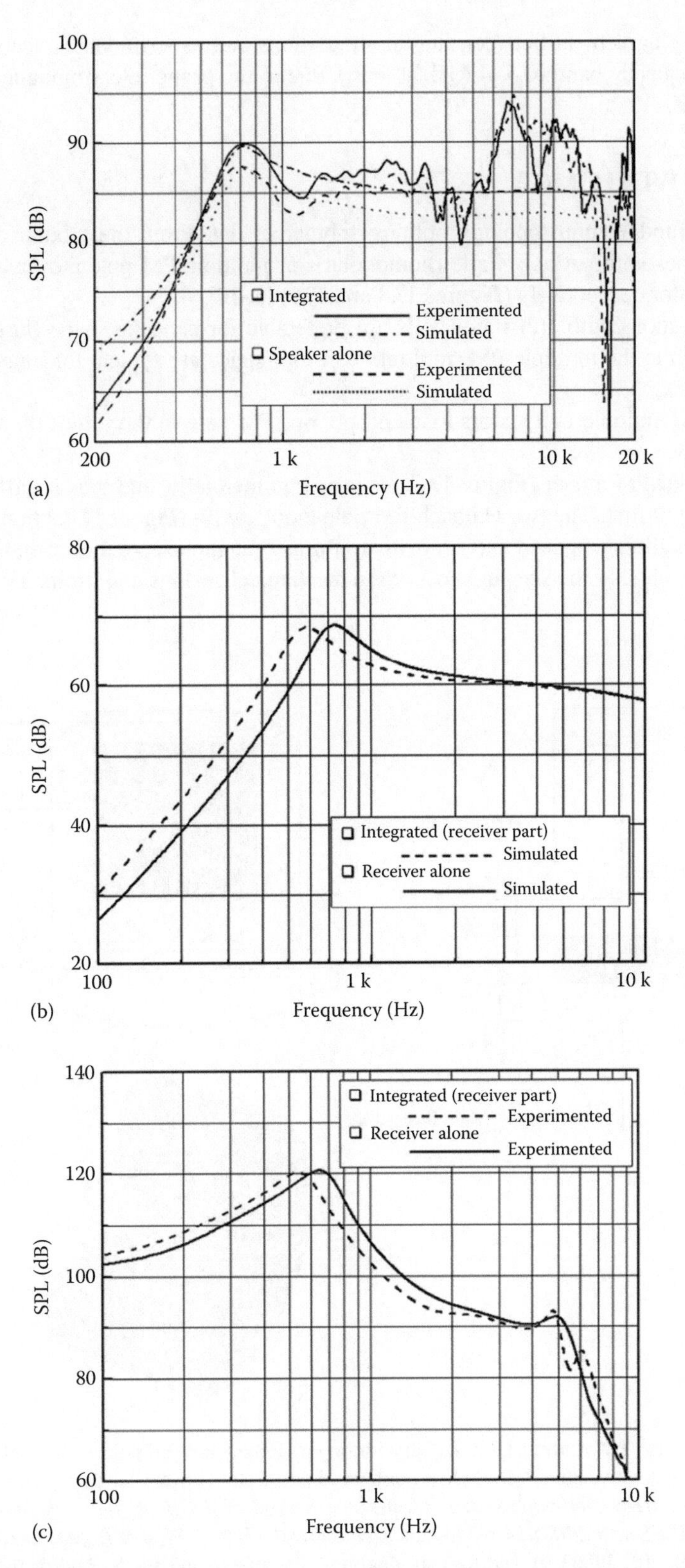

FIGURE 17.7 Sound pressure for speakers (a), receiver in open air (b), in an auspice (c). (After Hwang, S.M. et al., *IEEE Trans.*, MAG-39(5), 3259, 2003.)

A satisfactory agreement between digital simulations and experiments is accompanied by the visible large frequency band (0.5–4 kHz at least), due to the proper electromagnetic design of the coil-mover LOM.

17.3 PM-MOVER LOM(G)

There are two fundamental topology options: tubular or flat. From the tubular option, we present here only ones with surface single (homopolar)- or multiple-PM pole mover, with single- and multiple-coil stator, respectively (Figures 17.8 and 17.9) [6–10].

The configurations with 1(2) stator coils are preferable for pancake shape (large diameter and short length), while the multiple-PM, multiple-coil topologies are typical for longer with smaller-diameter volumes.

In single- and multiple-coil stators (one coil per one PM pole in the rotor), the PM flux linkage switches polarity.

The homopolar PM mover (Figure 17.8a) is used commercially, and thus it will be investigated here in some detail first. The two Halbach PM pole topology [9] (Figure 17.8c) that claims a reduction of mover weight is supposed to contain also a thinner magnetic core; however, the attached core weight increases notably the size and cost of the mechanical springs and limits the eigenfrequency to 50–60 Hz.

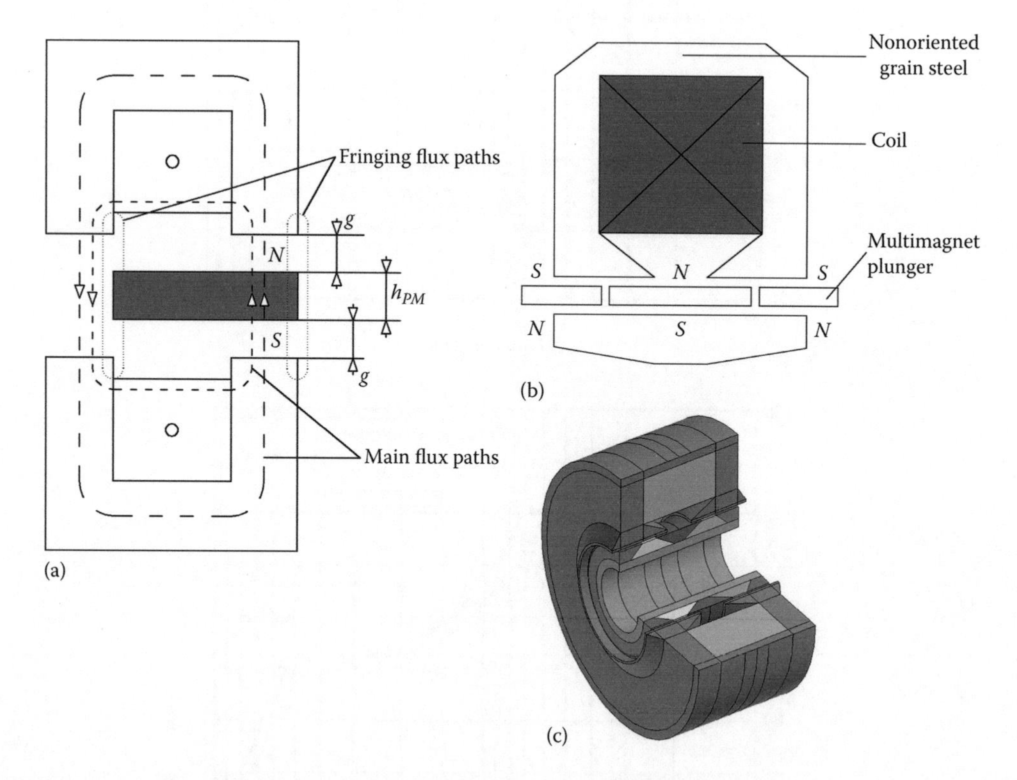

FIGURE 17.8 Tubular PM-mover LOM(G)s, (a) with single (homopolar) PM pole mover and stator coil. (After Wang, J. et al., Comparative study of winding configurations of short-stroke, single phase tubular permanent magnet motor for refrigeration applications, *Conference Record of the IEEE Industry Applications Society Annual Meeting* (*IEEE-IAS'2007*), New Orleans, LA, 2007; Redlich, R.W. and Berchowitz, D.H., *Proc. Inst. Mech. Eng.*, 199(3), 203, 1985.) (b) and (c) with dual-pole PM mover and single- and dual-stator coil. (After Ibrahim, T. et al., Analysis of a short-stroke, single-phase tubular PM actuator for reciprocating compressors, *6th International Symposium on Linear Drives for Industrial Applications* (*LDIA2007*), Lillie, France, 2007.)

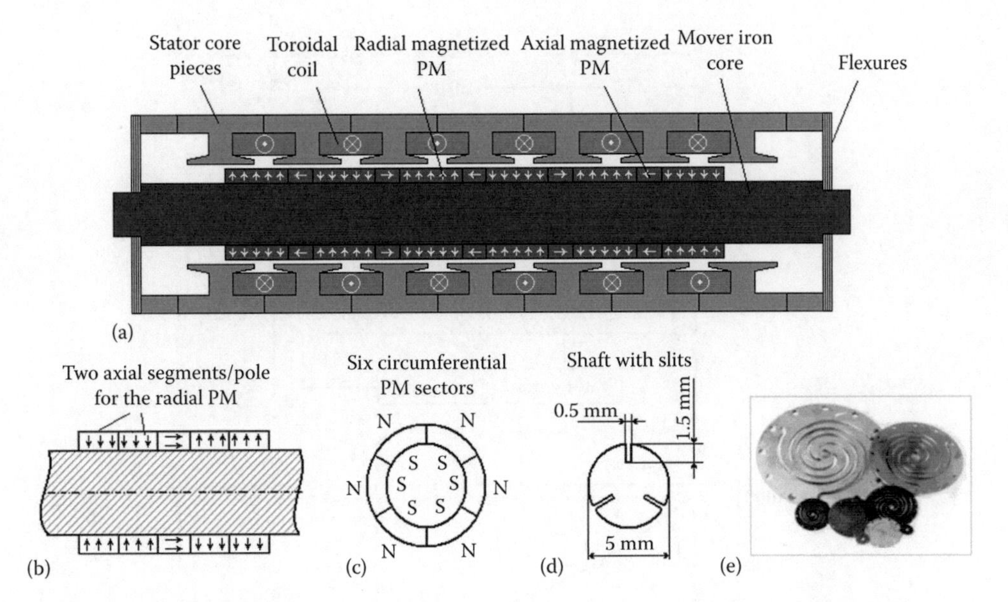

FIGURE 17.9 LOM(G): Tubular multiple PM-pole mover and multiple-coil stator topology: (a) cross section, (b) two PM poles, (c) radial PMs, (d) slits, and (e) flexures.

Although less rugged (mechanically), the "pure" PM mover with nonmagnetic lightweight (siluminum or resin) framing is considered here further as it allows a reasonable mechanical spring for high enough eigenfrequencies (up to 300 Hz for 50 W devices).

The cogging force of yet another variant of the single-stator-coil three-pole PM mover is shown in Figure 17.10; it is apparently capable of producing a linear cogging force, sufficient to replace the mechanical spring if 25% of the ideal maximum stroke length is "sacrificed" [11].

The tubular configurations are credited with the blessing of the circularity, and using "flexures" (Figure 17.9) as mechanical springs, both the resonance conditions and the linear bearing functions are performed.

On the other hand, a flat but double-sided topology, eventually with PM flux concentration, with square-like stator coil shape may allow a more rugged PM mover and a higher force (power) density. A so-called flux reversal of such topology is shown in Figure 17.11.

The double-sided structure provides for ideally zero normal force on the mover (when placed central in the airgap) and all PM utilization all the time.

The rather large PM flux leakage and fringing are inevitable. This situation is also typical for tubular single-coil single-PM-mover configurations (Figure 17.8a).

The stroke length in all PM-mover LOMs is rather small (less than 20 mm in general); the theoretical maximum stroke length corresponds to the motion length that leads to maximum positive maximum negative PM flux in the stator coil(s)—or the pole pitch $\tau = x_{max} = l_{stroke}$.

In what follows, we will treat in some detail three PM-mover LOM topologies: the tubular homopolar PM mover, the tubular multiple-PM pole mover with multiple stator coils, and the flat double-sided flux-reversal PM-mover LOM(G).

17.3.1 Tubular Homopolar LOM(G)

A low-power high-frequency ($f_e = f_m = 300$ Hz) tubular homopolar PM-mover LOM is illustrated in Figure 17.12, where the stator is the interior part (to reduce copper losses) and the cores are made of SMC (somaloy 700 or better).

To thoroughly investigate such a configuration, considering radial and axial eccentricities, a 3D electromagnetic field model is needed. This was done [12] with remarkable success.

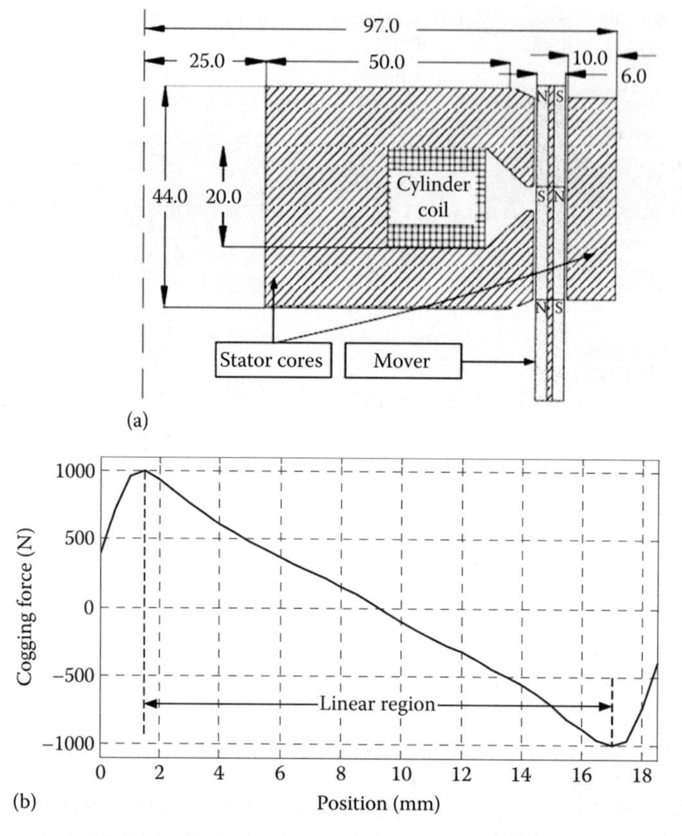

FIGURE 17.10 (a) Potentially spring-less tubular PM-mover LOM(G) and (b) with its cogging force ("magnetic" spring). (After Boldea, I. et al., Springless resonant linear oscillatory PM motors? *International Symposium on Linear Drives for Industrial Applications* (*LDIA2011*), Eindhoven, the Netherlands, 2011.)

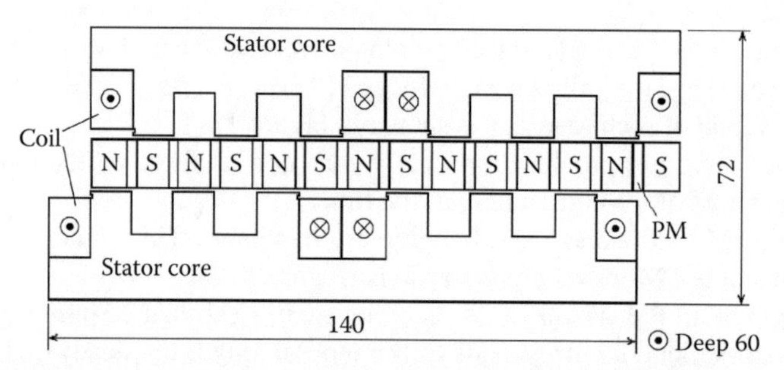

FIGURE 17.11 Flat flux-reversal PM-mover LOM(G) with PM flux concentration (double sided). (After Pompermaier, C. et al., Analytical and 3D FEM modeling of a tubular linear motor taking into account radial forces due to eccentricity, *IEEE International Electric Machines and Drives Conference* (*IEMDC'09*), Miami, FL, 2009.)

However, to investigate the machine dynamics, with reasonable computation effort, a practical magnetic circuit model may be used (Figure 17.13a), with flux lines illustrated in Figure 17.13b. To account for PM flux fringing and leakage—which is paramount—the magnetic circuit should be modified with mover position a few times.

As an example, the magnetic reluctances of flux lines 1 and 6 (in Figure 17.13) are

$$R_{line1} \approx \frac{g+t}{2\mu_0 t_m (2r_a + g)} \tag{17.34}$$

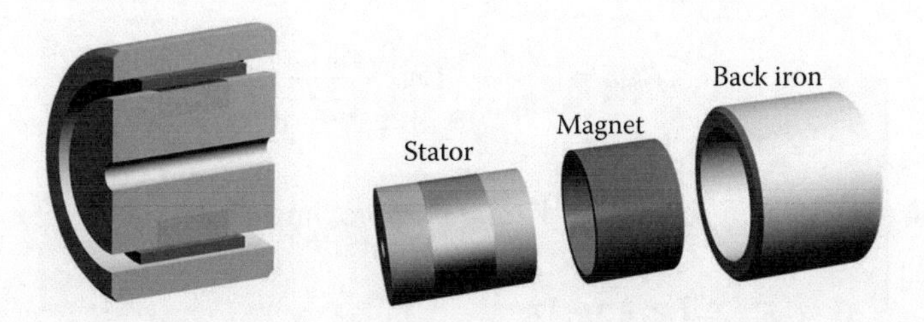

FIGURE 17.12 Tubular homopolar PM-mover LOM with interior single-coil stator. (After Pompermaier, C. et al., *IEEE Trans.*, IE-59(3), 1389, 2012.)

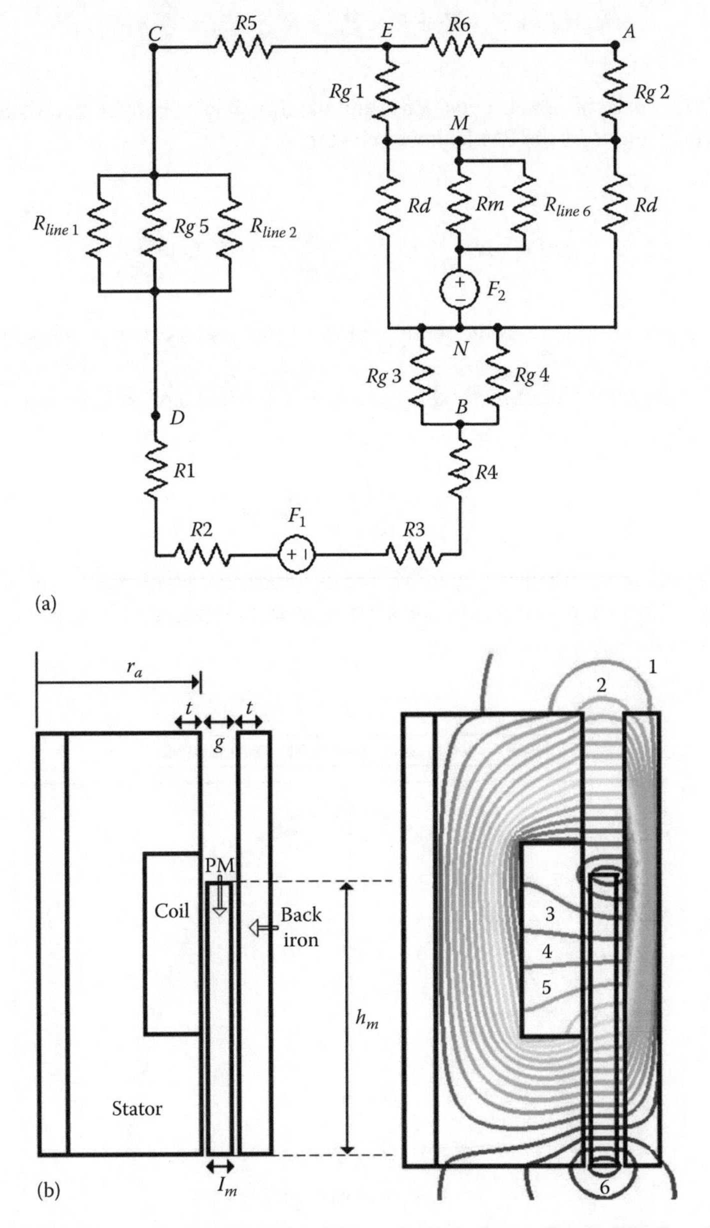

FIGURE 17.13 Magnetic circuit, (a) fringing (1,2) and (b) leakage (3,4,5) PM flux line tubs. (After Pompermaier, C. et al., *IEEE Trans.*, IE-59(3), 1389, 2012.)

$$R_{line6} \approx \frac{1}{2\pi\mu_0 h_m} \log\left(\frac{r_a + l_m}{r_a}\right) \tag{17.35}$$

The airgap flux Φ (traversing R_{g5} in Figure 17.13a) is approximately

$$\Phi = \frac{F_{mm1} - \left(R_d \parallel 2R_b\right)F_{mm2}}{R_a + R_{g1} \parallel (R_6 + R_{g2}) + R_m \parallel \left(R_g/2\right) + R_{g3} \parallel R_{g4}}; \quad F_{wm1} = Ni; \quad F_{mm2} = l_m \frac{B_r}{\mu_0} \tag{17.36}$$

$$R_a = R_1 + R_2 + R_3 + R_4 + R_{g5} + R_{g6}; \quad R_b = \frac{R_a}{2} + R_m \tag{17.37}$$

The nonlinear $B(H)$ curve of SMs is considered and the flux Φ is calculated iteratively. R_m is the PM magnetic reluctance and R_d is the PM leakage reluctance.

The *emf* is

$$emf = -N\frac{d\Phi}{dt} = -N\frac{d\Phi}{dx}\frac{dx}{dt} = -NK_e(x)\frac{dx}{dt} \tag{17.38}$$

Typically, emf results obtained via the analytical 2D FEM and by experiments are compared in Figure 17.14.

Therefore, the magnetic circuit method is reliable to calculate emf, and, consequently, the interaction thrust F_x also:

$$F_x = N\frac{d\Phi}{dx}i = NK_e(x)i \tag{17.39}$$

The cogging force has to be calculated by FEM, and its variation with displacement (in Figure 17.15a) suggests that it is trying to expel the mover outside the airgap (it is negative for negative displacement):

Thus, it opposes the spring action.

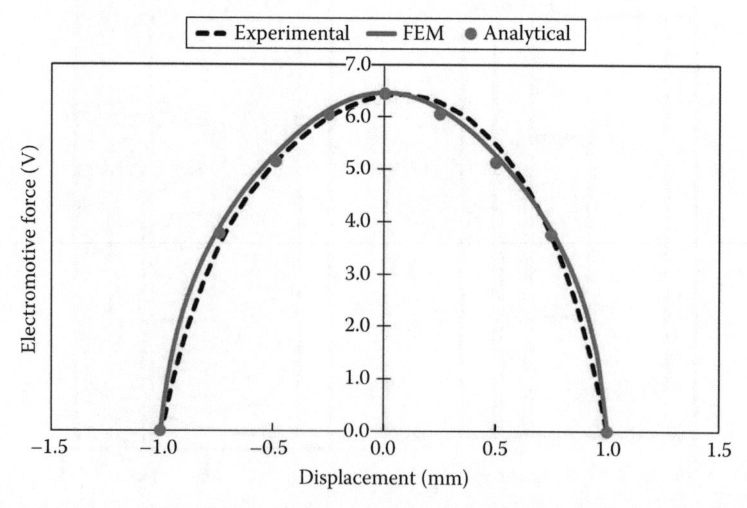

FIGURE 17.14 Emf: computed versus experimental results. (After Pompermaier, C. et al., *IEEE Trans.*, IE-59(3), 1389, 2012.)

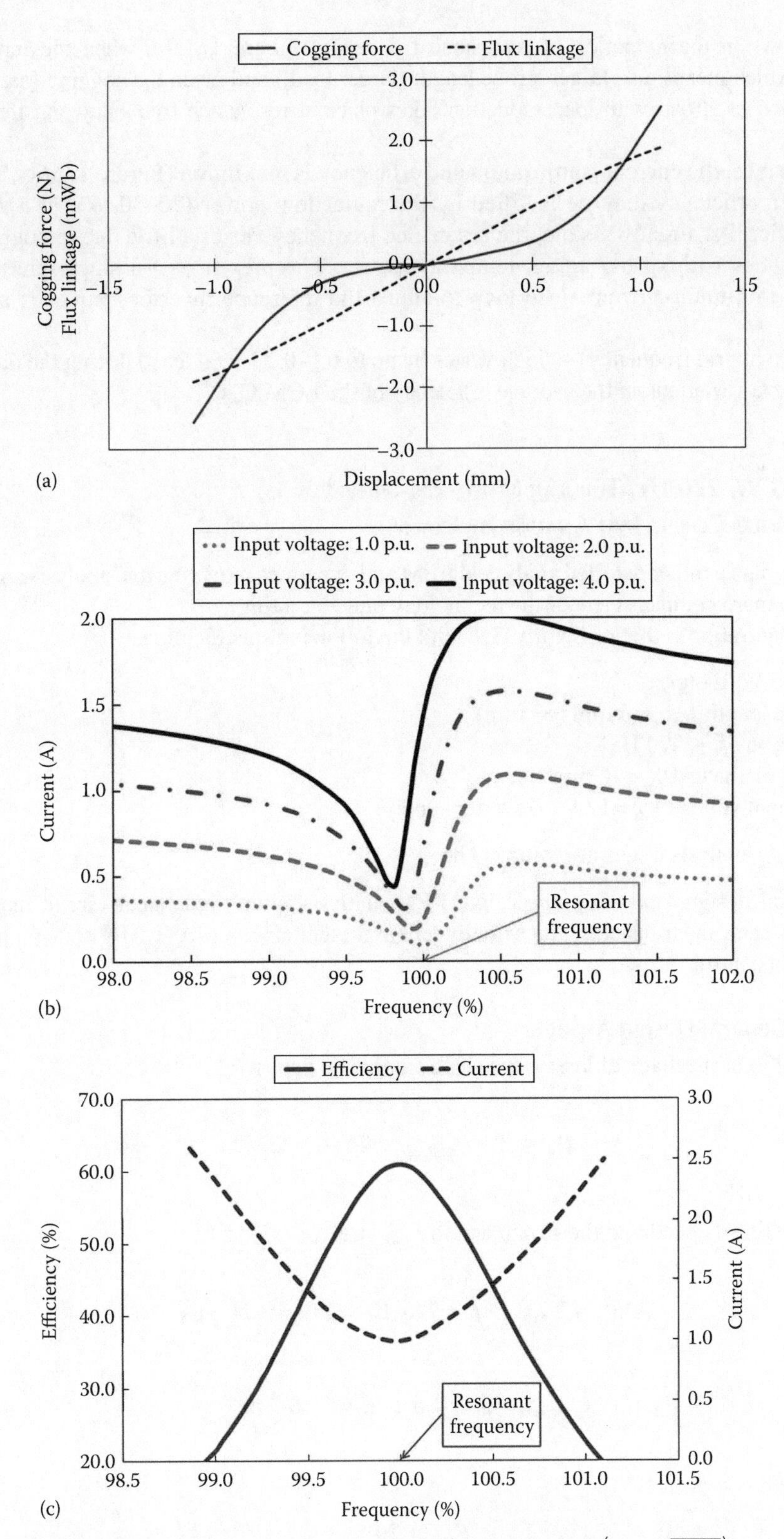

FIGURE 17.15 (a) Cogging force, (b) current versus resonance frequency $\left(f_m = \sqrt{K/M}\right)$, and (c) expecting efficiency and current. (After Pompermaier, C. et al., *IEEE Trans.*, IE-59(3), 1389, 2012.)

This shows in the reduction of resonance frequency (Figure 17.15b), when the input voltage increases (which means also larger stroke length [larger load]) and when the cogging force is larger. The reduction of stroke with load reduction takes place at resonance frequency and thus at good efficiency.

At resonance, the current is minimum and efficiency is maximum (Figure 17.15c). The rather modest 60% efficiency may be justified by the rather low power (20–30 W), in a very small volume device. As already visible, the resonance frequency varies a little due to cogging force, but it also does with spring aging, temperature, etc. The presence of a single-phase inverter and a slow minimum-current close loop to adjust the reference inverter frequency solves this problem.

Operation at grid frequency—which varies by up to 0.1–0.2 Hz (at least) during the day—would lead to notable variations in the average efficiency of the LOM(G)s.

17.3.2 25 W, 270 Hz, Tubular Multi-PM-Mover Multi-Coil LOM: Analysis by Example

In this paragraph a rather detailed analytical FEM and dynamics modeling and analysis are unfolded to provide a more complex view of the technology under scrutiny.

The configuration is that of Figure 17.9, with the following specifications:

- $P_n = 25$ W (motor)
- Stroke length $l_{stroke} = 6$ mm (±3 mm)
- Frequency $f_1 \leq 270$ Hz
- Outer diameter $D_{OS} < 16$ mm
- DC input voltage $V_{dc} = 12$ V_{dc} (inverter supply)

The following analysis issues are followed here:

- General design, optimization design, FEM analysis, simplified linear circuit model for steady state and transients, and a nonlinear circuit model with MATLAB® code for dynamics and control

17.3.2.1 General Design Aspects

With 1 W (4%) of mechanical losses, the electromagnetic power P_{en} is

$$P_{en} = P_n + P_{mechloss} = 25 + 1 = 26 \text{ W} \tag{17.40}$$

The average linear speed and the stroke length l_{stroke} are

$$u_{av} = 2 \cdot l_{stroke} \cdot f_1 = 2 \cdot 6 \cdot 10^{-3} \cdot 270 = 3.24 \text{ m/s} \tag{17.41}$$

$$l_{stroke} = 2 \cdot 3 = 6 \text{ mm} = 6 \cdot 10^{-3} \text{ m} \tag{17.42}$$

The average thrust is

$$F_{xav} = \frac{P_{en}}{u_{av}} = \frac{26}{3.24} = 8.02 \text{ N} \tag{17.43}$$

Choosing an average thrust surface density $f_s = 0.64$ N/cm^2 and a stator bore diameter $D_{is} \approx 0.5$ $D_{os} = 8$ mm, the required stator length L_s is

$$L_s = \frac{F_{av}}{\pi \cdot D_{is} \cdot f_s} \approx 50 \text{ mm} \tag{17.44}$$

The airgap PM flux density B_{agv} has two components:

$$B_{gav} = B_{PMr} + B_{PMa} \tag{17.45}$$

which correspond to radially and, respectively, axially magnetized PMs.

Approximately

$$B_{PMr} = B_r \frac{h_{PM}}{(h_{PM} + g\mu_{rec})(1 + k_s)(1 + k_{fr})} \tag{17.46}$$

$$B_{PMa} = B_r \frac{1}{\left(\frac{2g}{l_{PM}} + \frac{\tau}{2h_{PM}}\right)(1 + k_{fa})} \tag{17.47}$$

With $B_r = 1.2$ T, $\mu_{rec} = 1.07$ p.u., $g = 0.3$ mm, $h_{pm} = 1.2$ mm, $k_s = 0.1$ (saturation factor), $k_{fr} = 0.2$ (fringing factor for radial PMs), and $k_{fa} = 2$ (fringing factor for axially magnetized PMs), we get $B_{PMa} = 0.124$ T, $B_{PMr} = 0.85$ T. The ratio $B_{PMa}/B_{PMr} = 0.124/0.85 \approx 0.15 = 15\%$ represents the additional thrust proportion brought by the axially magnetized PMs.

Skipping design details, for $V_n = 12$ Vdc, we end up with the following PM-LOM parameters: Emf (peak value) $\approx 0.49 \cdot w_c$. (w_c, turns per coil; $N = 6$ coils, only 5 coils are fully active). The two end coils experience only homopolar PM flux variations, so they "account" to one fully active coil. In such conditions, the number of turns per coil $w_c = 20$ turns and rated current is 5.2A.

The inductance is $L_s \approx 0.28 \cdot 10^{-6} w_c^2 = 0.112 \cdot 10^{-3}$ H, and the resistance $R_s = 9.52 \cdot 10^{-4} w_c^2 = 0.3772\,\Omega$.

The electric time constant $T_e = L_s/R_s = 0.3$ ms, which implies 20 kHz or more PWM switching frequency. The copper losses $p_{con} = 6.868$ W ($J_{co} = 20$ A/mm^2) and the iron losses $p_{iron} = 2.47$ W (Somaloy 54 W/kg at 1 T and 270 Hz) lead to efficiency $\eta_n = 0.7075$ (an improved 74% efficiency will be obtained by optimization design and then validated by FEM when the mover and the stator core losses will be checked). A rather complete circuit model including the mover losses (attributed in the stator) is shown in Figure 17.16; R_{PM} is the PM eddy current resistance

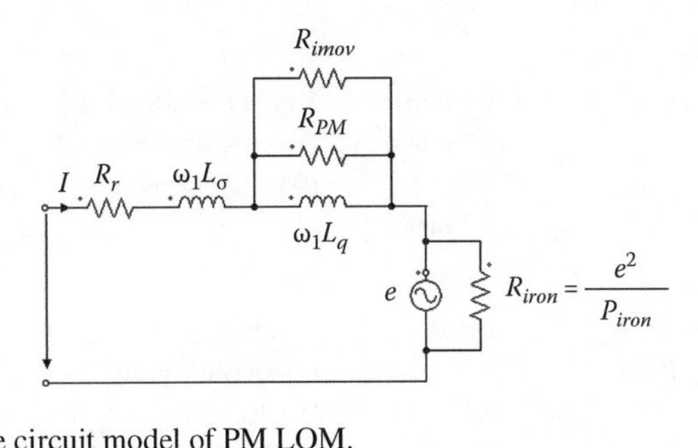

FIGURE 17.16 Complete circuit model of PM LOM.

and R_{imov} is the rotor solid back iron resistance. After key FEM validations and corrections of the analytical model used earlier, in the general design, by fudge factors, optimization design proceeds.

17.3.2.2 Optimization Methodology by Example

Optimization design methodology and MATLAB code include a few distinct stages as seen from the input file summarized in Table 17.1.

There are imposed "primary dimensions," "technical requests," "technical limitations," initial values of "optimization variables and limits for them," "objective function coefficients," "step size," and other specifications such as stator winding and rotor temperature limitations. An improved "Hooke–Jeeves" optimization method is used and the objective function is

$$F_{ob} = c_1 \frac{p_{co} + p_{iron}}{p_{loss0}} + c_2 \frac{m_{mover}}{m_{mover0}} \tag{17.48}$$

where

p_{loss0} are the electric losses
m_{mover0} is the weight of the mover, both from the initial design
c_1, c_2 are proportionality coefficients chosen by trial and error (Table 17.1)

The objective function is a compromise between electric losses and mover's weight as the latter influences the mechanical spring size (weight) and cost.

TABLE 17.1
Given Parameters in the Input File of the Optimization Design Code

Parameter	Value	Parameter	Value
Technical requests			
Rated power, P_n	25 W	Oscill. magnitude, x_n	3 mm
Rated frequency, f_n	270 Hz	Rated voltage, V_{dc}	12 V
Primary dimensions			
Outside diameter, D_{os}	16 mm	Stator piece length	7 mm
Poles	6		
Initial values of optimization variables			
Stator inner diameter, D_{si}	8 mm	Tooth width, w_{st}	2 mm
PM height, h_{PM}	1.2 mm	Radial PM width, w_{pmr}	5 mm
Slot mouth, sMs	1.5 mm		
Technological limitations			
Airgap, g	0.3 mm	Slot filling factor, k_{fill}	0.6
Teeth tip height, h_{s4}	0.3 mm	Winding temperature, T	120 C
Wedge place height, h_{s3}	0.5 mm	PM temperature, T_{PM}	105 C
Yoke height, h_{sc}	1.5 mm		
Other specifications			
Losses coefficient, c_1	1	Step rate	2
Mover weight coeff., c_2	0	Current duty cycle	2/3
Initial step, d_1	0.2	Iron losses factor, k_{pfe}	1.13
Final step, d_2	0.1	Mechanical losses, P_{mec}	1 W

TABLE 17.2
Computed Parameters from Output File Optimization Design Code

Parameter	Value	Parameter	Value
Electrical parameters			
Rated current, In	4.65 A	Winding inductance, L_s	156 μH
Rated coil mmf	97.7 A	Copper losses, *Pcu*	6.02 W
PM linkage flux	97 μWb	Iron losses, *Pfe*	1.74 W
Winding resistance, R_s	0.42 Ω	Efficiency	74%
Constructive dimensions			
Stator inner diameter, D_{si}	8 mm	Tooth width, w_{st}	2 mm
PM height, h_{PM}	1 mm	Radial PM width, w_{pmr}	5.9 mm
Slot mouth, *sMs*	1.1 mm	Turns per coil, w_1	21
Weights (g)			
One stator core piece	5	One axial mag. PM	0.2
Total stator core	35.3	Total PMs	6.32
One coil	1.62	Stator	45
Total cupper	9.7	Mover, m_m	15.7
Mover iron	9.4	Total	60.7
One radial mag. PM	0.9		

The Hooke–Jeeves optimization algorithm is a pattern search algorithm using two kinds of moves: exploratory moves and pattern moves. Five optimized variables (Table 17.1.) are grouped in a vector, and each element is modified with a given step in the exploratory moves. The objective function gradient is computed and after that, the algorithm starts pattern moves until the objective function reaches a minimum. From that point on, another search move is started. If there is not a smaller value for the objective function, then the step is decreased in ratio of step rate. The algorithm is stopped after the final step d_2 (Table 17.2) is reached. At each step, the optimization variables are limited within their range.

As the motor is supposed to be cooled together with the linear piston compressor, the 120° winding temperature is used only to calculate the copper losses.

A sample output file is summarized in Table 17.2, where we can see a 4% improvement in the efficiency from 70% in the initial design, to 74%, with a total motor weight of 60 g without mechanical springs.

The number of iterations needed for a given progress ratio in the optimization was around 10 and computation CPU time was 9.35 s (on a dual core 2.4 GHz processor, Windows XP, MATLAB 6.5). The same optimization design was run for f_1=500 Hz; a very tempting efficiency η_n=0.82% for 58 g of active materials and the same 6(±3) mm stroke length was obtained. If a mechanical spring for this eigenfrequency (500 Hz) can be built (mover weight ≈11 g), then the 500 Hz solution might be preferable.

17.3.2.3 FEM Analysis

A 2D FEM analysis of the tubular configuration has been performed to validate PM flux density distribution, emf, thrust, and rotor losses. The results on PM airgap flux density distribution with position for three mover positions (x=0 means mover in the middle position), Figure 17.17, shows rather high levels, as required. Saturation "hot-spot" places have been kept also within limits to limit core losses (stator: Somaloy core; the mover back iron: mild solid steel). The 0.83 T average of PM airgap flux density value from the initial design is also confirmed.

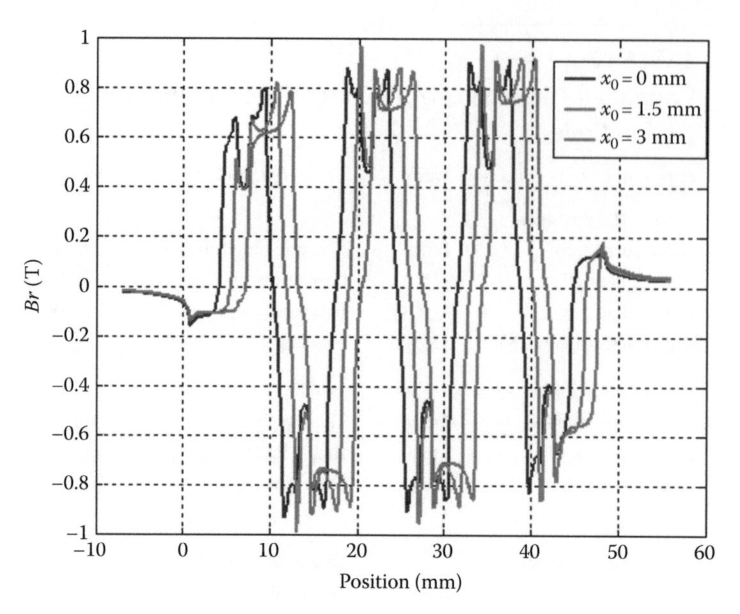

FIGURE 17.17 Airgap radial PM flux density distribution in the middle of the airgap.

The total force at constant mmf versus position, shown in Figure 17.18, proves that the PM LOM is capable of producing the required average thrust of 8 N. It also shows that there is a cogging force (zero current force) that manifests itself as a virtual nonlinear spring.

The thrust decreases toward excursion's end, but in reality, as the mmf is in phase with the speed (sinusoidal (harmonic) motion—imposed mainly by the mechanical springs) and with the current, the thrust has to go anyway down to zero at excursion's end. The average electromagnetic power ($F_x(x) \cdot u(x)$) for sinusoidal motion, constant current, is shown in Figure 17.19 for both 270 and 500 Hz PM-LOM operation.

Typical FEM emf waveform for one turn/coil is shown in Figure 17.20, for harmonic motion at 270 Hz, with its frequency spectrum.

The FEM calculated inductance is shown in Figure 17.21. It depends on mover position and on mmf (load) but, in fact, on local magnetic saturation. Its average value (at rated 102 A turns/coil) is about equal to the 0.11 mH analytically calculated value ($w_c = 20$ turns/coil) in the initial design.

To check the level of mover eddy current losses, the latter have been investigated by FEM at standstill for the rated (270 Hz) frequency of the sinusoidal rated stator current (102 A turns/coil). Their distribution is shown in Figure 17.22a ($\rho_{Fe} = 10\ \rho_{Co}$, $\rho_{PM} = 1.2 \cdot 10^{-6}\ \Omega m$, $\mu_{r\ iron} = 730$ *p.u.*), and the eddy current density variation along mover outer surface is evident in Figure 17.22b. The mover eddy current losses are (at 270 Hz) 16.4 mW in PM and 33.5 mW in the mover solid back iron. These values are rather small with respect to the total core losses of 2.4 W considered in the analytical design model, but still worthy of consideration as they are about 2% of all core losses.

Note: At 25 W, (500 Hz), the mover eddy current losses are slightly smaller than for 270 Hz because the stator current (the main source of these losses) is notably smaller (because efficiency is better).

17.3.2.4 Simplified Linear Circuit Model for Steady State and Transients

For sinusoidal current and sinusoidal motion (and emf), a simplified circuit model can be developed considering the single-phase linear PMSM with a cage-less PM mover.

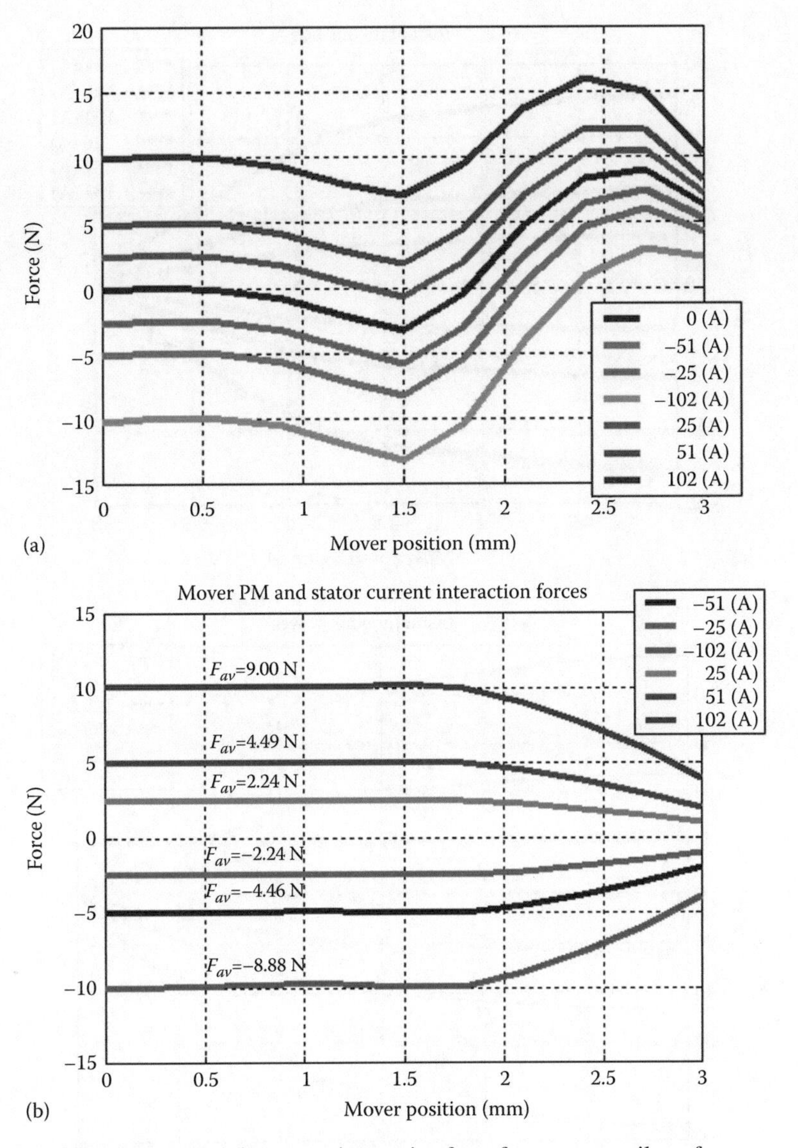

FIGURE 17.18 (a) Total thrust and (b) current-interaction force for constant coil mmf.

Thus, for sinusoidal steady state, at resonance $\left(\omega_1 = \omega_{res};\ \omega_{res} = \sqrt{k_{spring}/m_m}\right)$, the PM-LOM equations are

$$\underline{V}_1 = R_s\underline{I}_1 + j\omega_1 L_s \underline{I}_1 + \underline{E}_1$$
$$\underline{E}_1 = k_e \underline{U}_1 \tag{17.49}$$

$$F_x = k_e I_1 \tag{17.50}$$

For transients we consider that the average load is proportional to speed:

$$F_{load}(t) \approx C_{load} U(t) \tag{17.51}$$

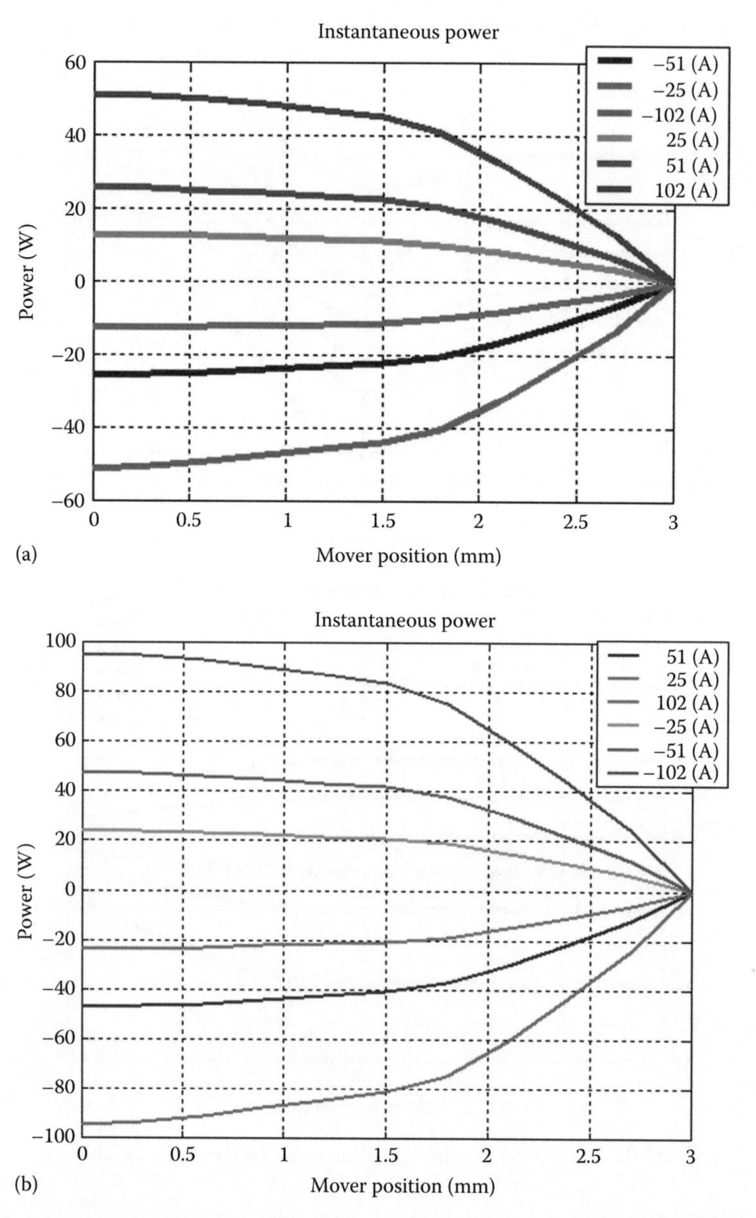

FIGURE 17.19 Average electromagnetic power versus position at constant mmf per coil (in A) (a) $f_1 = 270$ Hz, (b) $f_1 = 500$ Hz.

We may rewrite now the voltage and motion equations for transients:

$$V_1(t) = R_s i(t) + L_s \frac{di}{dt} + k_E u(t) \tag{17.52}$$

$$m\frac{du(t)}{dt} = k_E i(t) - k_{spring} x(t) - C_{load} u(t) \tag{17.53}$$

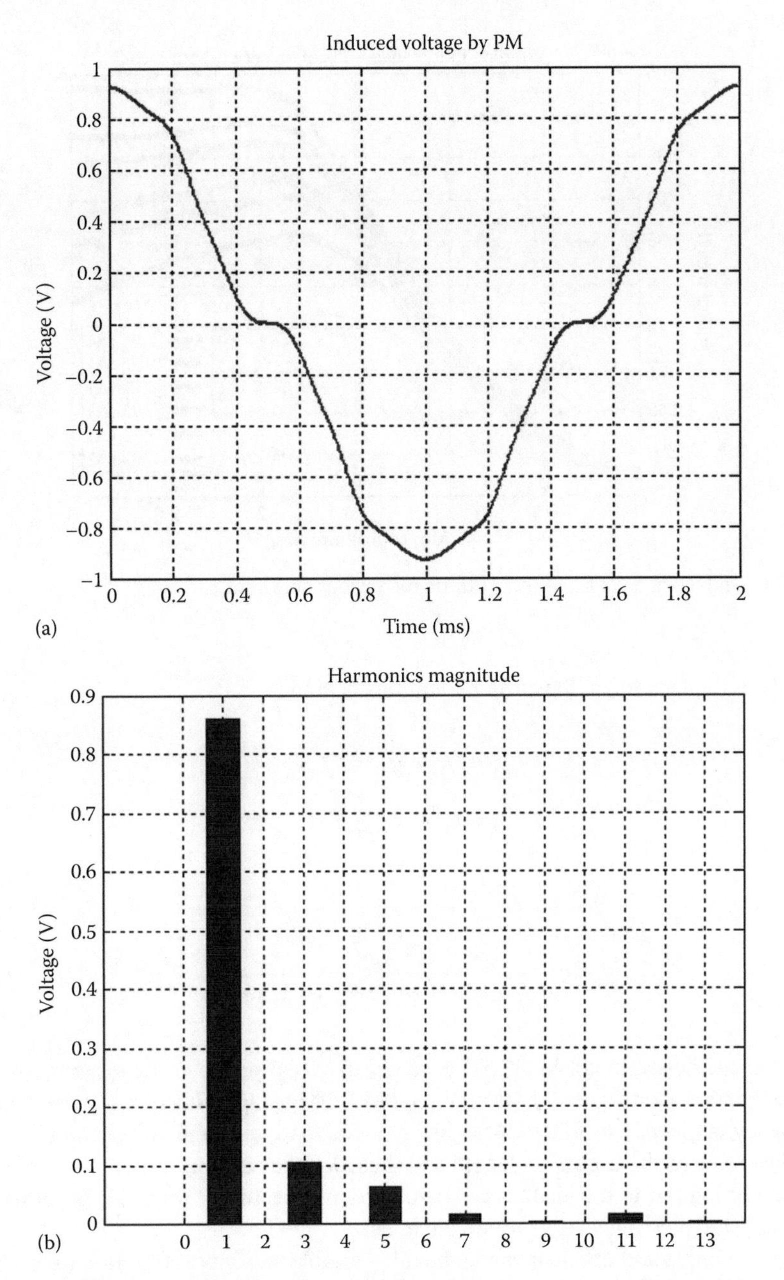

FIGURE 17.20 Emf for one turn/coil versus (a) time for harmonic motion and (b) its harmonic spectrum.

The cogging force is neglected in the transient model here. With sinusoidal voltage and steady-state motion

$$\underline{V_1} = V_1\sqrt{2}e^{j\omega_1 t}; \quad \underline{U_1} = j\omega_1 \underline{X}_1 \tag{17.54}$$

$$\underline{I_1} = \frac{\underline{X}_1\left(k - \omega_1^2 m + j\omega_1 C_{load}\right)}{k_E} = \underline{U_1}\left(\frac{C_{load}}{k_E}\right)_{\omega_1=\omega_{res}} \tag{17.55}$$

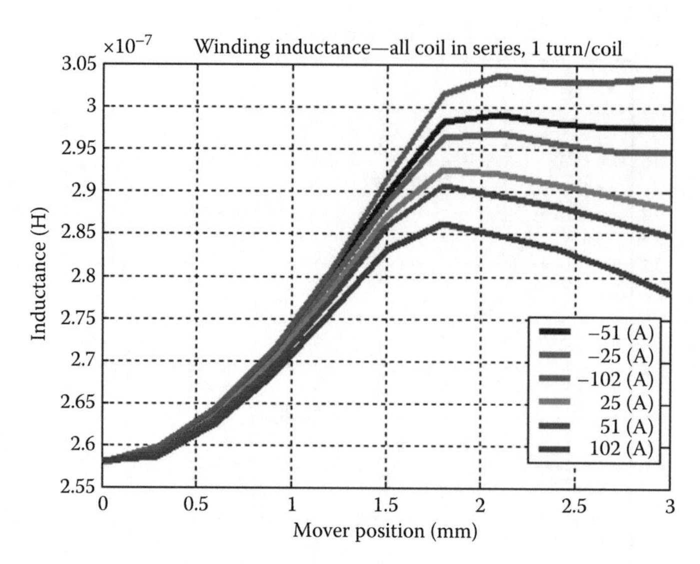

FIGURE 17.21 Inductance versus mover position for various coil mmf's.

For transients in Laplace form, the earlier equalities become

$$\tilde{V}_1 = (R_s + sL_s)\tilde{i} + k_E s\tilde{x}$$
$$ms^2\tilde{x} = k_E\tilde{i} - k_{spring}\tilde{x} - C_{load}s\tilde{x} \tag{17.56}$$

or

$$\tilde{x} = \frac{k_E\tilde{V}}{(R_s + sL_s)(s^2m + sC_{load} + k_E) + sk_E^2} \tag{17.57}$$

All the roots of the denominator in (17.57) have negative real parts, so the response to any voltage input is always stable. Consequently, starting the PM LOM by applying directly the voltage at resonance frequency is the way to follow. Also, for reduced load, only the voltage amplitude has to be reduced: it just reduces the excursion length and thus also the current.

This would be similar to the ideal case when, in rotary machines, we could hypothetically modify continuously the pole pitch (l_{stroke} = pole pitch, here).

This seems a simple and efficient way to handle variable capacity refrigeration (or other loads).

This simplified dynamic model has served only as basis for insights and for steady-state performance calculation under load.

To illustrate its usefulness, we show in Figure 17.23 the displacement and sinusoidal current magnitudes (Figure 17.23a and b) and the output average power versus frequency (Figure 17.23c).

The beneficial effect of resonance conditions is evident. The presence of the inverter can help to follow on line the eventual slow variations of spring eigenfrequency due to temperature and mechanical aging, by just hunting slowly for minimum current at given load.

17.3.2.5 Nonlinear Circuit Model and MATLAB® Code with Digital Simulation Results

The analytical model for steady state and transients in the previous paragraph did not account precisely for actual thrust/current/position or cogging force/position curves. Therefore, especially when

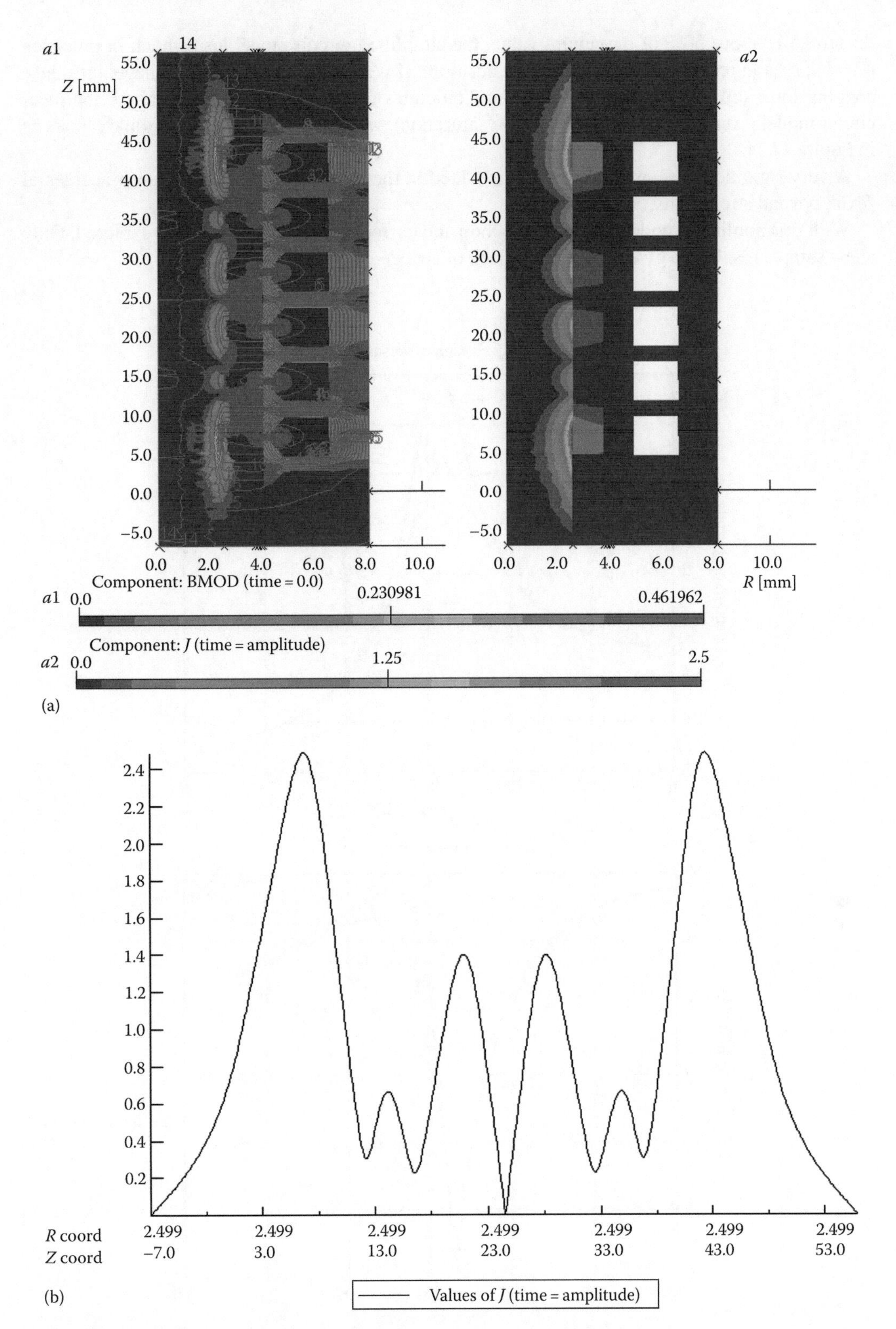

FIGURE 17.22 Mover eddy currents distribution at standstill with ac rated current (a) and current density variation along mover outer surface (b).

the stroke is above 50% of maximum value, the simplified circuit model loses much in precision ($F_x = k_e i$; k_e is in reality a function of mover position). If we import $k_e(x,t)$, the emf shape, introduce cogging force (all from FEM), and add also Coulomb's friction influence, we obtain a nonlinear circuit model (even with neglected magnetic saturation), which, in MATLAB–Simulink®, looks as in Figure 17.24.

A very rigid fictitious spring (material) is added in the model to "reflect" the mover accidental "exit" beyond stroke ends.

With this nonlinear model, voltage open-loop and current close-loop controls are explored. Only a few sample results are given here (due to lack of space).

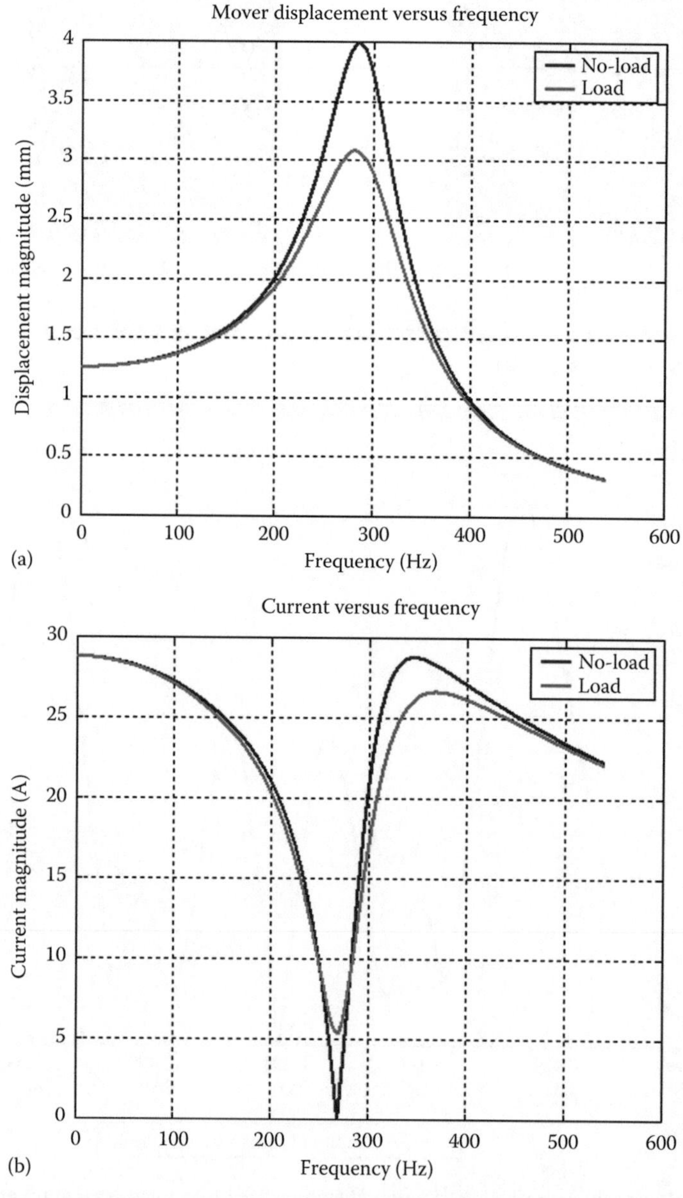

FIGURE 17.23 Sinusoidal voltage and motion steady-state characteristics versus frequency (a) displacement, (b) current.

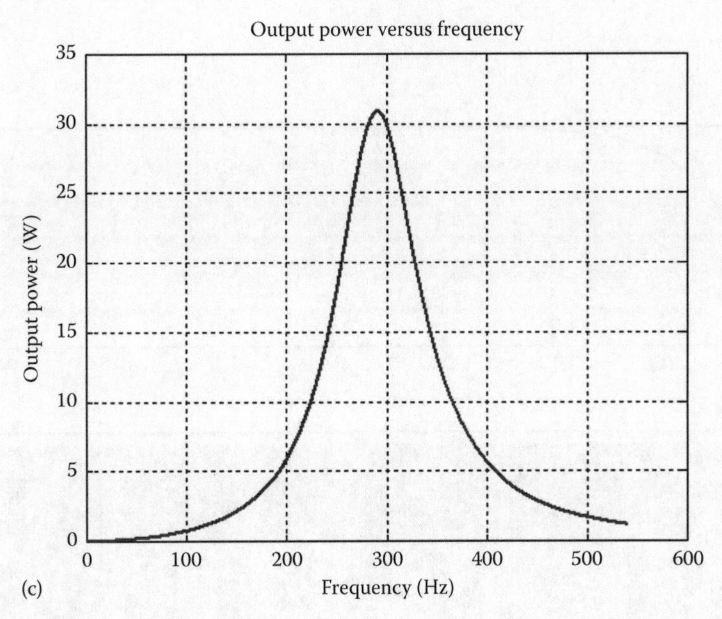

FIGURE 17.23 (continued) Sinusoidal voltage and motion steady-state characteristics versus frequency (c) output average power.

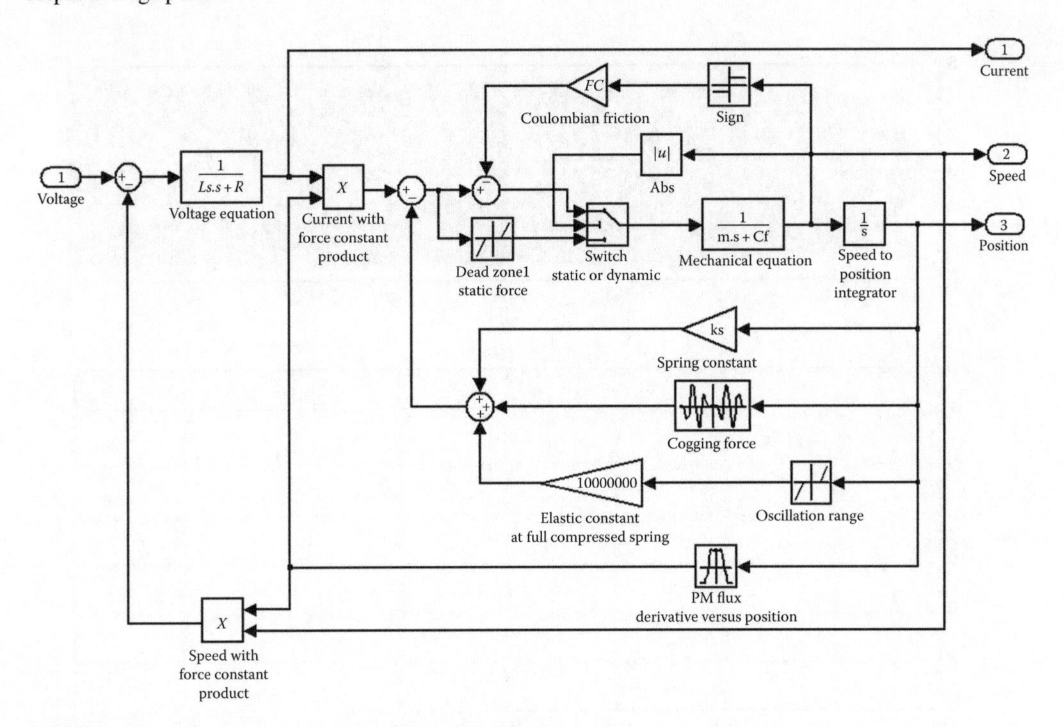

FIGURE 17.24 Proposed nonlinear model of PM-LOM.

Direct starting and operation under load at constant (resonance) frequency with open-loop gradual voltage application is shown in Figure 17.25a through d (voltage amplitude, current waveform, instantaneous speed, and average power).

The MATLAB–Simulink solver automatically chooses the sampling time to keep the relative errors smaller than 10^{-4}. The PWM signal is generated by a modified sigma-delta modulator that contains a zero-order hold with 2 μs sampling time in its local close loop.

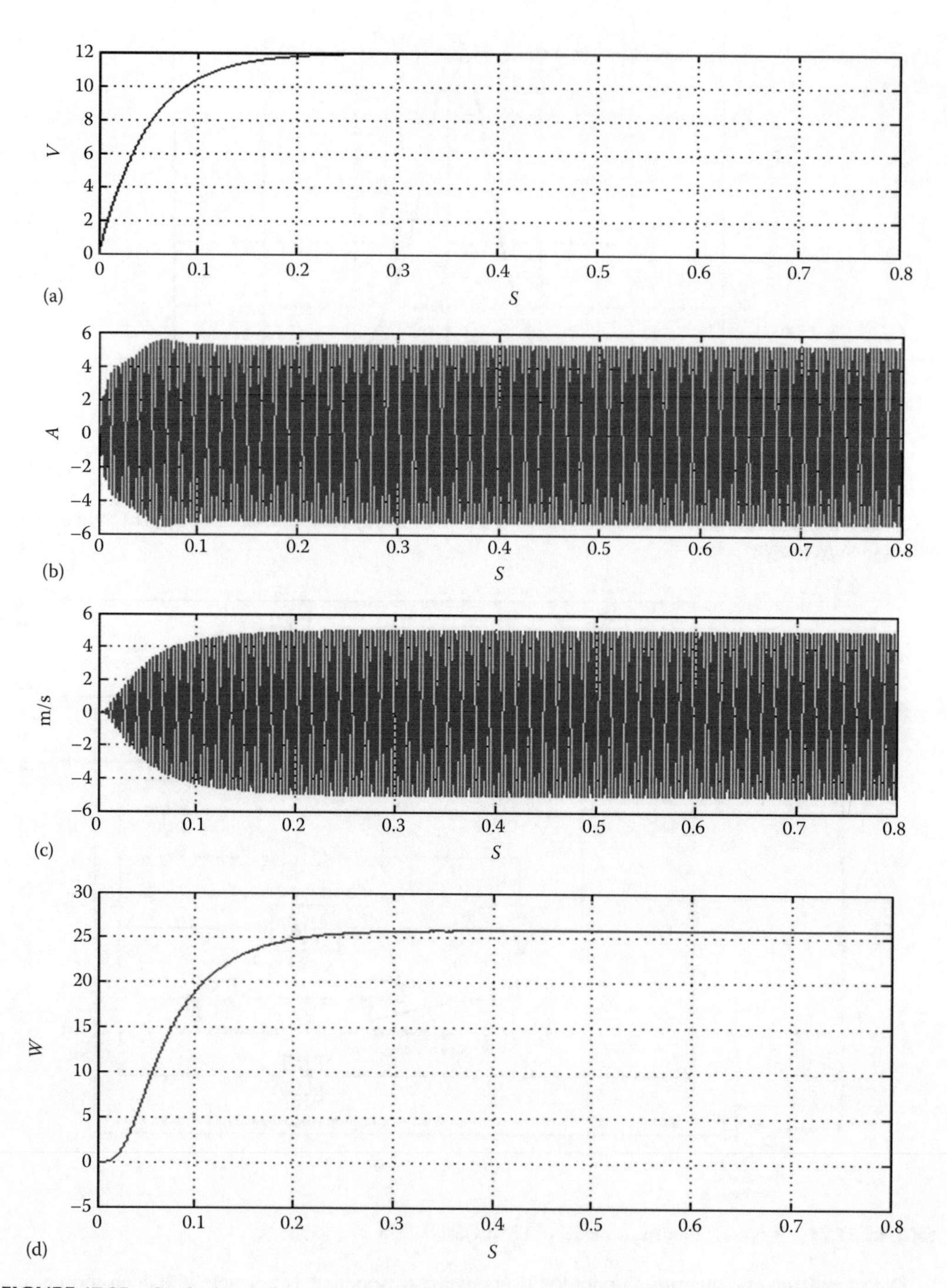

FIGURE 17.25 Gradual open-loop voltage application for start at 270 Hz: (a) voltage amplitude, (b) current, (c) speed, and (d) average power versus time.

The control objective of the linear motor is to obtain maximum efficiency for compressor drives or to precision tracking if it is used as an actuator.

Sinusoidal close current loop control at mechanical resonance frequency, for maximum efficiency, was also investigated, with satisfactory results for voltage, current, speed, and average power (Figure 17.26).

A zoom on steady state for voltage, current, and speed would show very clearly the resonance conditions (the current is almost in phase with speed). The current control loop is of on-off type with the application of $\pm V_{dc}$ (dc voltage). The sampling time for zero order hold is set at 20 μs in this case to limit the maximum switching frequency at 25 kHz and keep the current ripple within an acceptable range (Figure 17.26b). The current control is robust to electrical parameters uncertainly. The output power variation is practically not sensitive to a large winding resistance and inductance variation. It has an acceptable sensitivity with respect to permanent magnet flux as

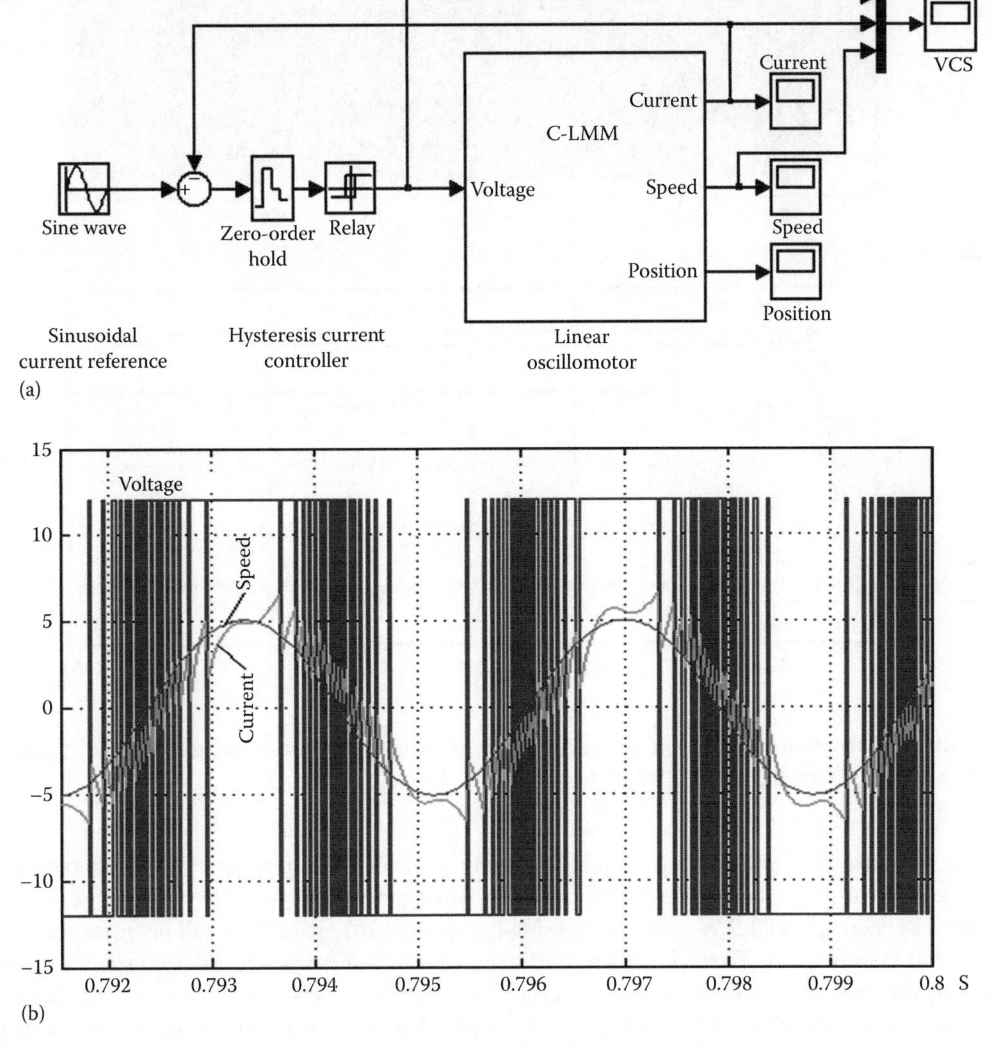

FIGURE 17.26 Close-loop sinusoidal current control: (a) genetic control, (b) current and speed versus time (zoom).

(continued)

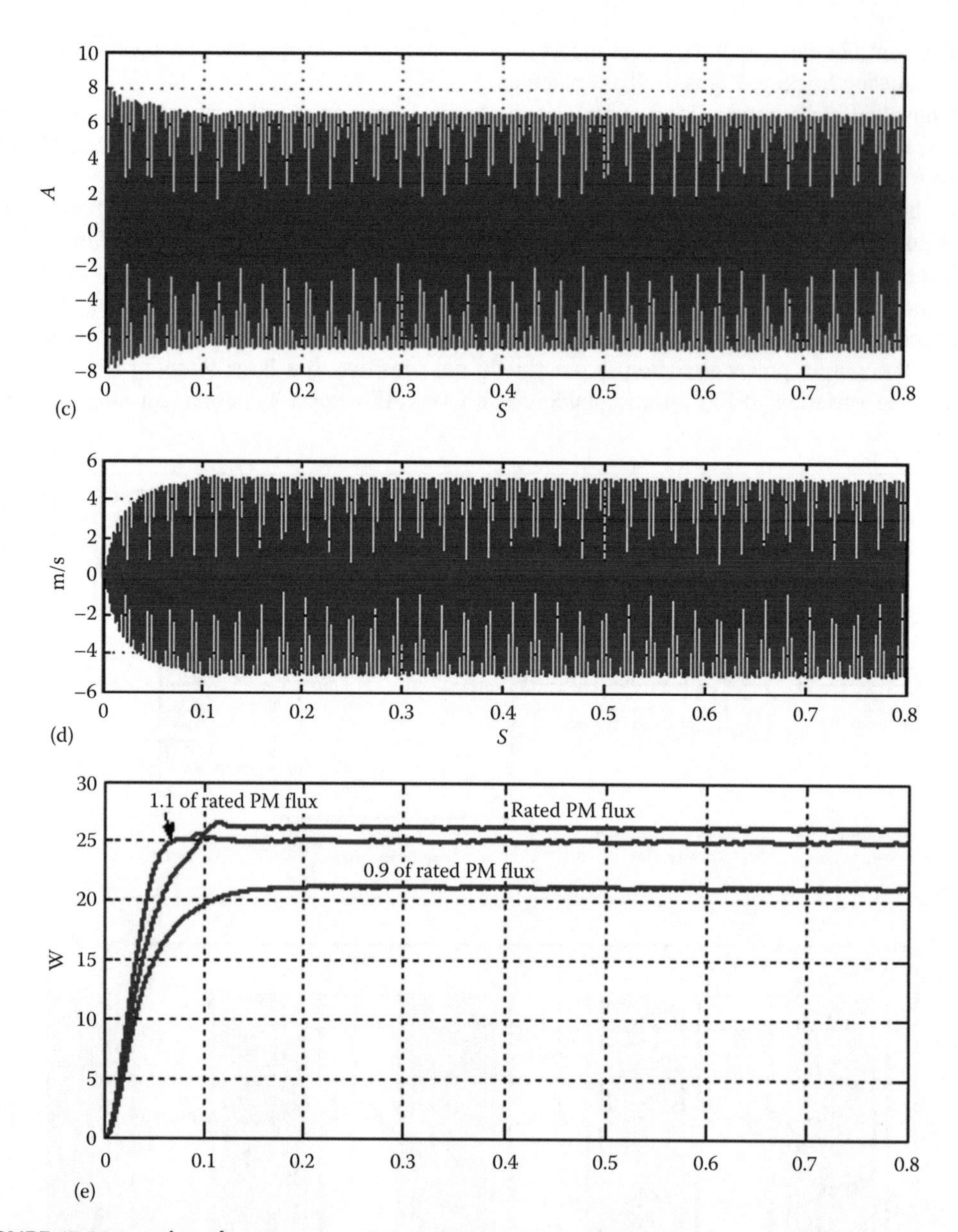

FIGURE 17.26 (continued) Close-loop sinusoidal current control: (c), (d) current and speed, and (e) output average power in W, during acceleration speed-proportional load.

shown in Figure 17.26e. There is a notable sensitivity to mechanical parameters (mover total mass and mechanical spring constant). For example, increasing the spring rigidity by 10% the output power will decrease at 13.5 W, while decreasing the spring stiffness at 90% will decrease the output power to only 7.7 W. To track the small changes in resonant frequency, the frequency synthesized by the inverter is changed in order to drive the phase angle between current and emf to zero (a simple voltage model would provide good emf estimation because the frequency is rather large). This aspect is beyond our scope here.

Comparing Figures 17.25 and 17.26, we may infer that much faster starting is achieved with close-loop current control (from 0.25 to 0.1 s), though in both cases (open-loop and close-loop current control) successful synchronization under full load is obtained. To modify the refrigeration capacity, the reference voltage (respectively, current) is changed. Close-loop current control is to be preferred as it is less dependent on machine parameters.

17.3.3 Double-Sided Flat PM-Mover LOM

In this paragraph, two built prototypes are investigated: one with surface and one with interior PM mover [13,14]. The scope is to model and characterize them thoroughly both for steady state and for dynamic operation.

The prototypes' configurations are shown in Figure 17.27. The mover is sliding on linear bearings and the kinetic energy is recovered by two mechanical springs.

The main dimensions of the prototypes are

- Stroke length: 10 mm
- Teeth width: 5 mm
- Airgap: 1 mm
- Permanent magnet thickness: 2 mm
- Permanent magnet material is NdFeB with B_r = 1.13–1.18 T, H_c = 844–900 kA/m
- Number of turns/coil: 285
- Wire diameter: 0.45 mm
- Mover mass: 2 kg

17.3.3.1 State-Space Model of the Linear Machine

The model of the linear machine is based on voltage and mechanical differential equations:

$$V = RI + \frac{d\Psi}{dt} \tag{17.58}$$

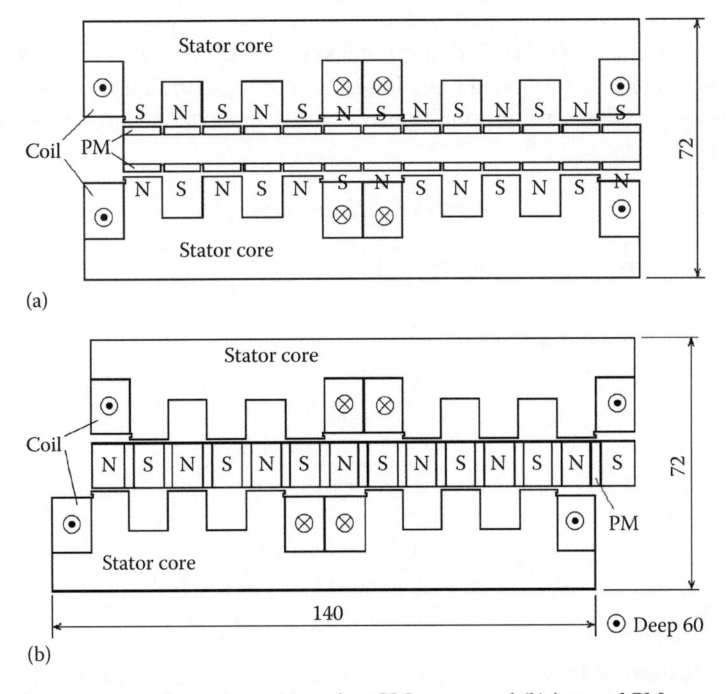

FIGURE 17.27 Prototypes configurations: (a) surface PM mover and (b) internal PM mover.

$$m\frac{du}{dt} = F_{em} - F_{cg} - F_{load} - F_f - F_s, \tag{17.59}$$

where

V is the armature voltage
R is the resistance
I is the current
Ψ is the total linkage flux
m is the mover mass
u is the mover speed
F_{em} is the electromagnetic force
F_{cg} is the cogging force
F_f is the friction force
F_s is the spring force
F_{load} is the load force

If the core saturation is low, the superposition principle is operational and the total linkage flux is the sum of permanent magnet flux and the flux produced by current. Moreover, if the inductance variation with mover position is negligible, then the voltage equation becomes

$$V = RI + L\frac{dI}{dt} + \frac{\partial \Psi_{PM}}{\partial x}\cdot\frac{dx}{dt} \tag{17.60}$$

The electromagnetic force depends on the total flux derivative versus mover position and armature current. The electromagnetic force is given in (17.61) for a constant inductance:

$$F_{em} = \frac{\partial \Psi_{pm}}{\partial x}\cdot I \tag{17.61}$$

The linear oscillomotor has a strongly nonlinear behavior, which will be illustrated later.

The linearization model presented in Equations 17.62 considers the load force proportional to the speed, while viscous friction force is proportional to c_f (friction coefficient), and the flux derivative versus position is equal k_f:

$$\begin{aligned} \tilde{V}_1 &= (R_s + sL_s)\tilde{i} + k_f s\tilde{x} \\ m_t s^2\tilde{x} + c_f s\tilde{x} + k\tilde{x} &= k_f\tilde{i} \end{aligned} \tag{17.62}$$

The analytical solution of the linear model is

$$\tilde{x} = \frac{k_f\tilde{V}_1}{(R_s + sL_s)(s^2 m_t + sc_f + k) + sk_f^2} \tag{17.63}$$

$$\tilde{i} = \tilde{V}_1\frac{(s^2 m_t + sc_f + k)}{(R_s + sL_s)(s^2 m_t + sc_f + k) + sk_f^2} \tag{17.64}$$

The poles locus of transfer function is shown in Figure 17.28 for R_s=9 Ω, L_s=0.22 H, m_t=2 kg, k=146 N/mm, k_f=100 Wb/m, and the loading coefficient c_f varies between 0 and 500 [kg/s].

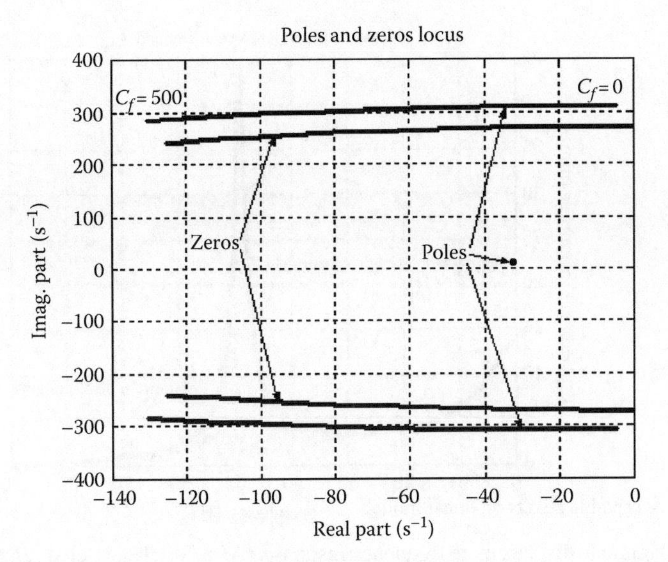

FIGURE 17.28 Poles and zeros locus of transfer functions. (After Tutelea, L. et al., *IEEE Trans.*, IE-55(2), 492, 2008.)

All poles have negative real part, even for zero friction coefficients (Figure 17.28), so the system is stable and its response is finite for finite inputs. The current magnitude versus frequency is shown in Figure 17.29 and the displacement magnitude of the mover in Figure 17.30, for 180 V voltage magnitude. The maximum magnitude of the mover displacement is given by the imaginary part of the poles of transfer function (electromagnetic resonance frequency). The electromagnetic resonance frequency is 49.16 Hz for no load and 48.82 Hz for a load with an equivalent friction coefficient of $c_f = 166$ kg/s. The current minimum magnitude depends on the imaginary part of zeros and only mildly on the poles of the current transfer function (17.64). In general, the minimum current is reached for a little smaller frequency than the mechanical resonance frequency, $f_0 = 43$ Hz, which depends on load.

The maximum current magnitude is reached for a little larger frequency than electromechanical resonance frequency, as it is shown in Figure 17.29.

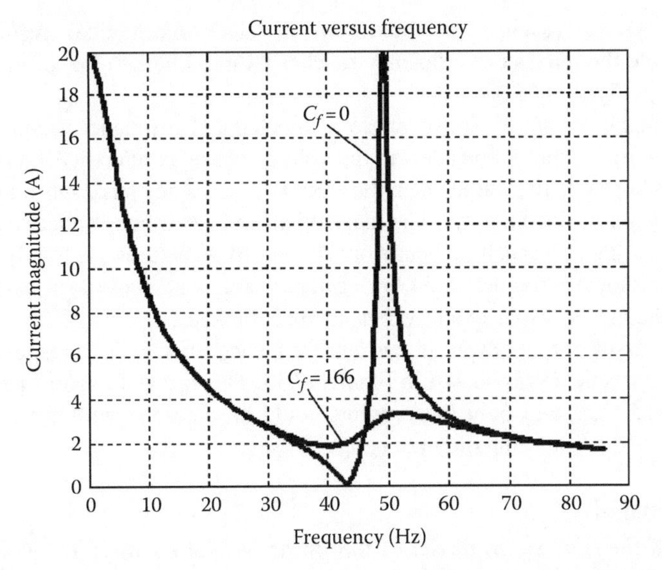

FIGURE 17.29 Current magnitude frequency response. (After Tutelea, L. et al., *IEEE Trans.*, IE-55(2), 492, 2008.)

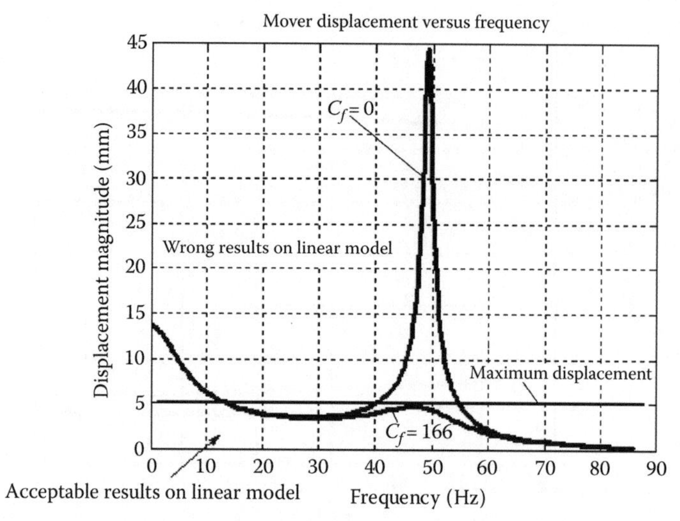

FIGURE 17.30 Magnitude displacement frequency response. (After Tutelea, L. et al., *IEEE Trans.*, IE-55(2), 492, 2008.)

The loading friction coefficient, $c_f = 166$, was chosen to produce the maximum mechanical power at mechanical resonance frequency. The voltage magnitude $V_1 = 180$ V was chosen to keep the displacement magnitude closer and under its maximum value of 5 mm for loading conditions, Figure 17.30. For this voltage, the no-load displacement is very large. In fact, for the real machine, the maximum mechanical available displacement is 8 mm and the force is changing sign at 5 mm. The linear model is approximately correct only for mover displacements smaller than 5 mm. However, the model shows that it is necessary to reduce the voltage magnitude about nine times for the no-load regime.

The mechanical power (for harmonic oscillation) is

$$P_{mec} = \frac{1}{2} c_f (\omega X_m)^2 \tag{17.65}$$

The mechanical power reaches its maximum at electromechanical resonance frequency (Figure 17.31), while the maximum efficiency reaches its maximum value around the mechanical resonant frequency (Figure 17.32).

The current and mover speed are in phase at mechanical resonance frequency as shown in Figure 17.33. The current phase, considering the voltage phase as reference, changes from a large negative value to a large positive value when the electrical frequency passes through the mechanical resonant frequency, for no-load regime. The mechanical and electromechanical resonance frequencies are points of extreme for machine behavior. The machine features are totally different at these frequencies as it is shown in Figure 17.34, for current versus load coefficient, and, respectively, in Figure 17.35, for the mover displacement versus load coefficient.

The efficiency versus output power at constant frequency (mechanical, respectively, electromechanical resonance frequency) is shown in Figure 17.36. The curves in solid line are for displacement smaller than 5 mm, and only these points could be obtained with the real machine. The efficiency is computed considering only the copper losses.

17.3.3.2 FEM Analysis

The parameters of the state-space model of the linear oscillatory machine could be determined by tests on an existing machine, but this procedure does not work during the design process. The FEM analysis could be used for this scope. The FEM setup and magnetic field lines are shown

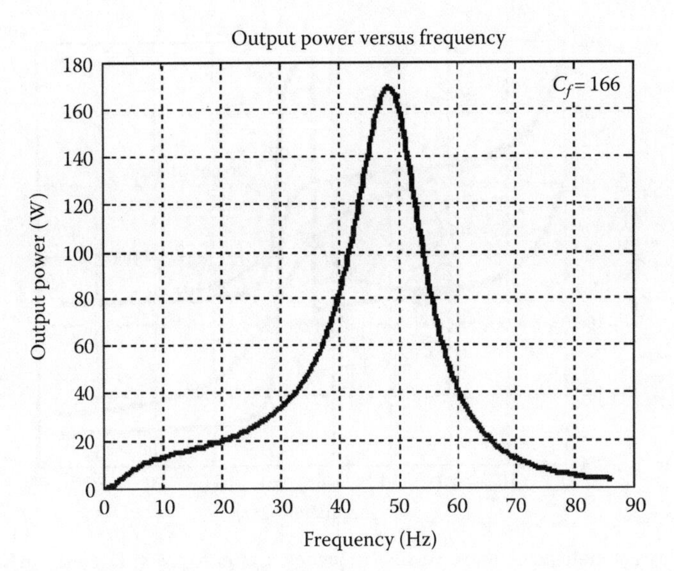

FIGURE 17.31 Output power at constant load coefficient. (After Tutelea, L. et al., *IEEE Trans.*, IE-55(2), 492, 2008.)

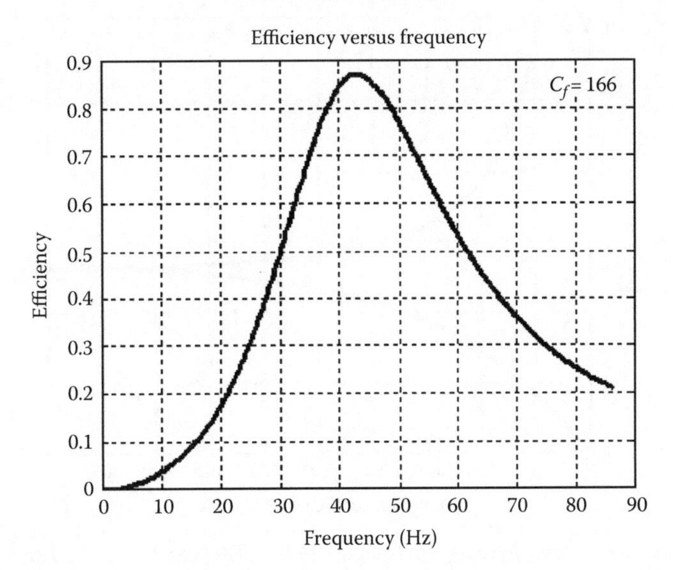

FIGURE 17.32 Efficiency at constant load coefficient. (After Tutelea, L. et al., *IEEE Trans.*, IE-55(2), 492, 2008.)

in Figure 17.37 for the interior PM flux concentration linear oscillatory machine (Figure 17.27b). Moreover, the FEM analysis could provide a piece of information otherwise difficult to measure: the airgap flux density shown in Figure 17.38. The FEM analysis is validated by the measured thrust (Figure 17.39) and inductance (Figure 17.40).

The FEM results are in good agreement with standstill tests, as shown in Figure 17.39.

The FEM inductance does not contain the overhang leakage inductance, so it is below the measured inductance with about 20 mH. Also for small current, the FEM inductance is constant compared with measured inductance. The core magnetic saturation curve is linearized around zero current in order to improve the algorithm convergence. The agreement between FEM and test results is satisfactory, considering the measurement errors. The FEM analysis could produce the parameters for the linear state model and also for the nonlinear model via look-up tables.

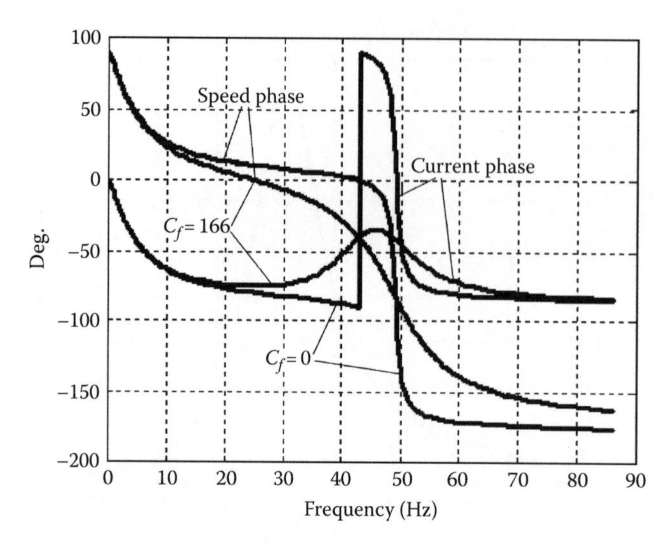

FIGURE 17.33 Current and speed phase versus frequency. (After Tutelea, L. et al., *IEEE Trans.*, IE-55(2), 492, 2008.)

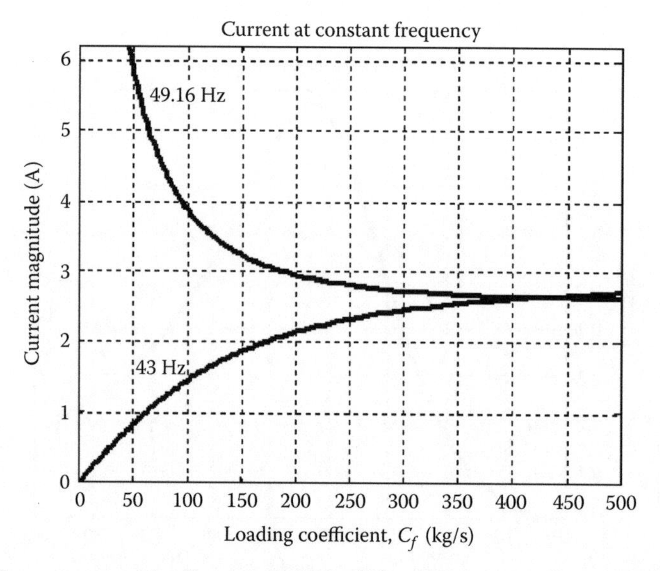

FIGURE 17.34 Current versus loading coefficient. (After Tutelea, L. et al., *IEEE Trans.*, IE-55(2), 492, 2008.)

17.3.3.3 Nonlinear Model

The nonlinearities are produced by the dependence of flux derivative on position and by magnetic saturation and cogging force.

The permanent magnet flux derivative depends on the mover position and could be available in the model via a table. A trapezoidal shape may be considered, as shown in Figure 17.41.

The cogging force dependence on the mover position is introduced in the MATLAB–Simulink model also as a look-up table from test results, after noise filtering (Figure 17.42).

The friction force could be divided into viscous friction force F_{vf} and Coulomb friction force F_{Cf}, which is the other nonlinearity source, especially in small oscillations around equilibrium position:

$$F_f = F_{Cf} + F_{vf}$$

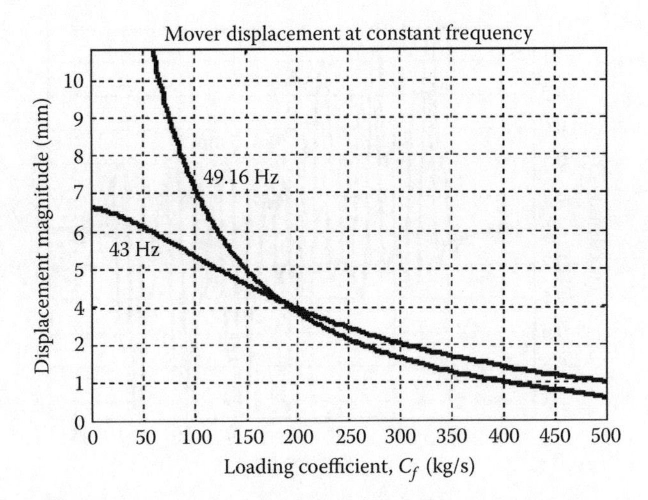

FIGURE 17.35 Displacement magnitude versus loading coefficient. (After Tutelea, L. et al., *IEEE Trans.*, IE-55(2), 492, 2008.)

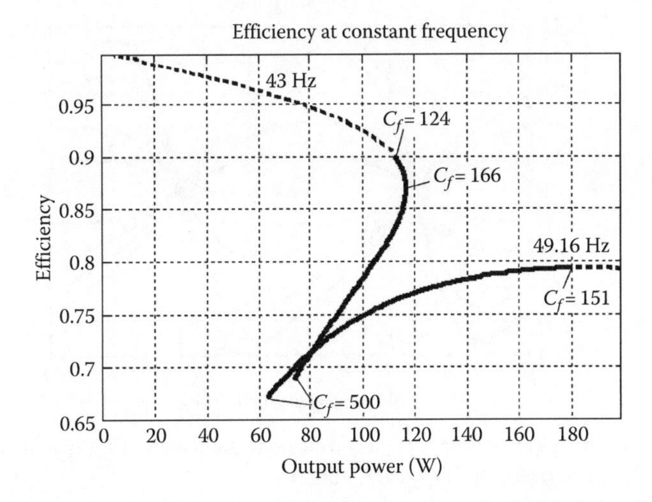

FIGURE 17.36 Efficiency versus output power. (After Tutelea, L. et al., *IEEE Trans.*, IE-55(2), 492, 2008.)

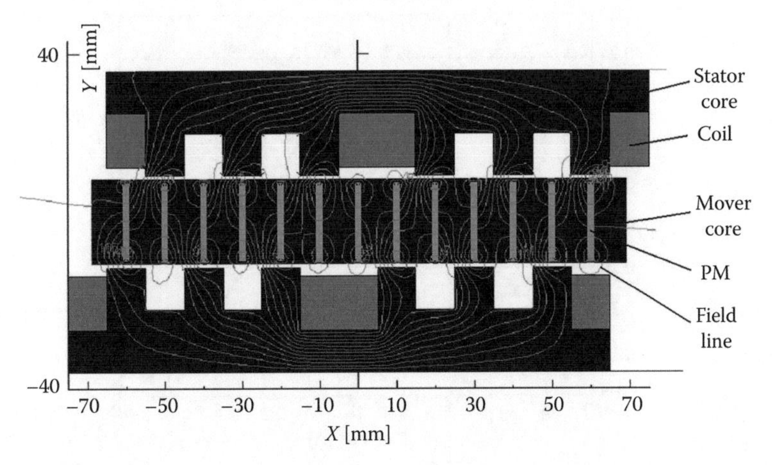

FIGURE 17.37 FEM setup and magnetic field lines for mover at maximum force position and 1 A/coil current. (After Tutelea, L. et al., *IEEE Trans.*, IE-55(2), 492, 2008.)

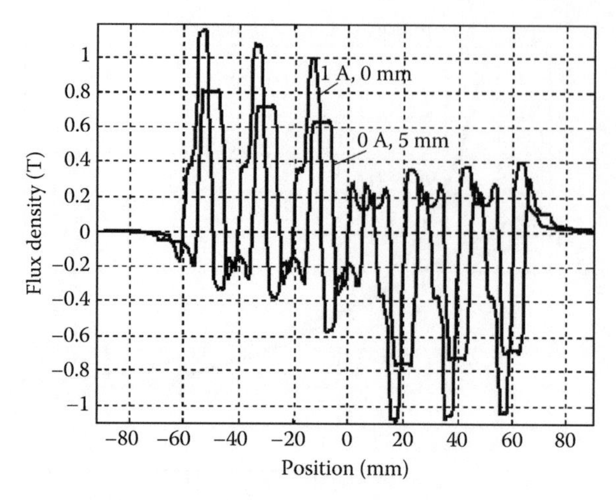

FIGURE 17.38 Air gap flux density by FEM. (After Tutelea, L. et al., *IEEE Trans.*, IE-55(2), 492, 2008.)

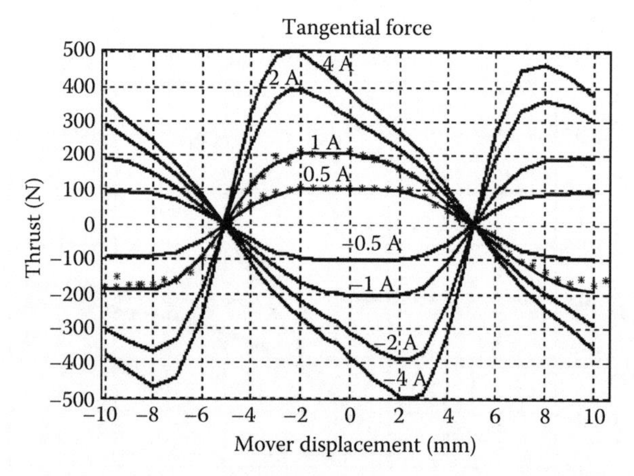

FIGURE 17.39 Thrust versus mover position and coil current; -solid line—FEM simulation, * standstill test. (After Tutelea, L. et al., *IEEE Trans.*, IE-55(2), 492, 2008.)

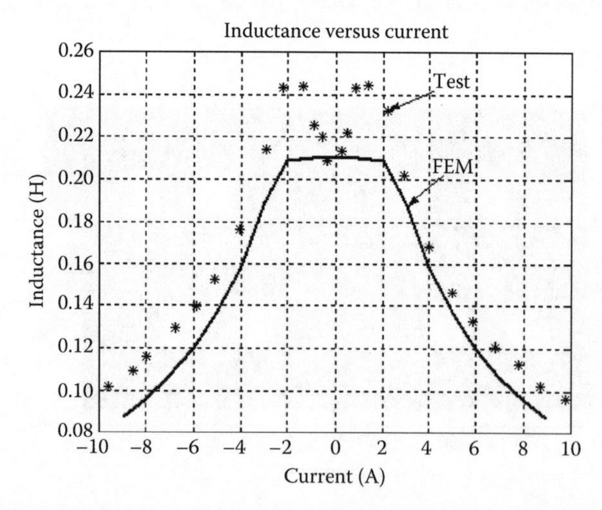

FIGURE 17.40 Machine inductance versus current—two parallel paths. (After Tutelea, L. et al., *IEEE Trans.*, IE-55(2), 492, 2008.)

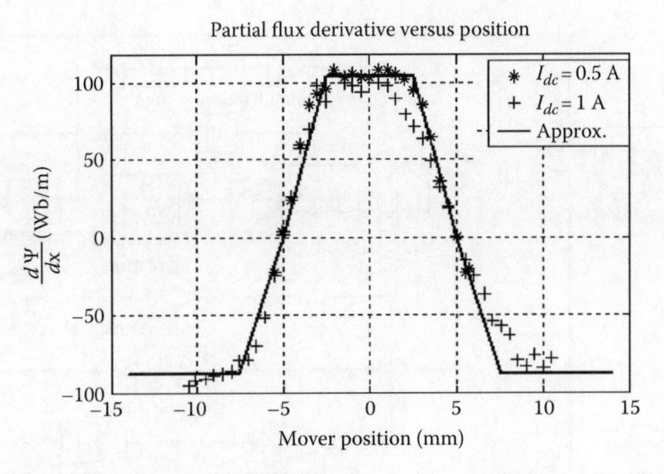

FIGURE 17.41 Flux derivative distribution. (After Tutelea, L. et al., *IEEE Trans.*, IE-55(2), 492, 2008.)

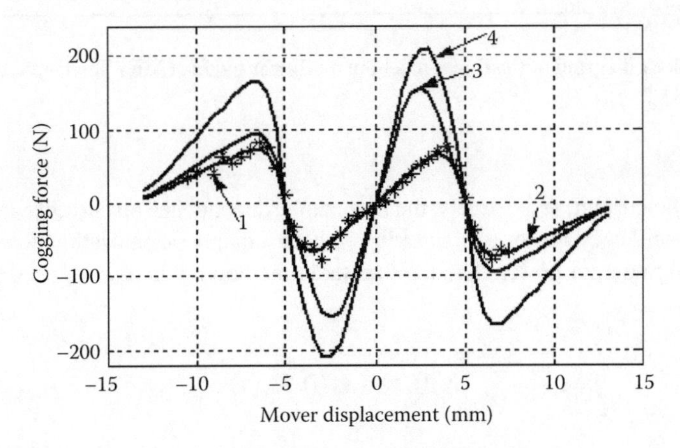

FIGURE 17.42 Cogging force: (1) measured points for FCPM, (2) approx. curves for FCPM, (3) approx. curves for SPM, (4) sum of 2 and 3. (After Tutelea, L. et al., *IEEE Trans.*, IE-55(2), 492, 2008.)

$$F_f = \begin{cases} F_C sign(u) + k_{vf} u & \text{for } u \neq 0 \\ \min\left(\left|F_{rez}\right|, F_c\right) \cdot sign(F_{rez}) & \text{for } u = 0 \end{cases} \tag{17.66}$$

where F_{rez} is the sum of electromagnetic force and mechanical spring force.

The mechanical spring force is assumed to vary linearly with mover displacement, except for the accidental situation when the mover hits the frame, or the spring is fully compressed, when the elastic force suddenly increases.

Finally, the nonlinear model of the linear machine is shown in Figure 17.43, considering constant parameters as R, coil resistance; L, coil inductance; k_f, force coefficient (maximum value of flux derivative versus position); m, mover mass; k_s, spring constant; F_c, Coulombian friction force; c_f, viscous friction coefficient; and x_{max}, maximum mechanical stroke. Moreover, the model contains two distributions: flux derivative versus position and cogging force versus position already presented in Figures 17.41 and 17.42, respectively. Core losses have been neglected.

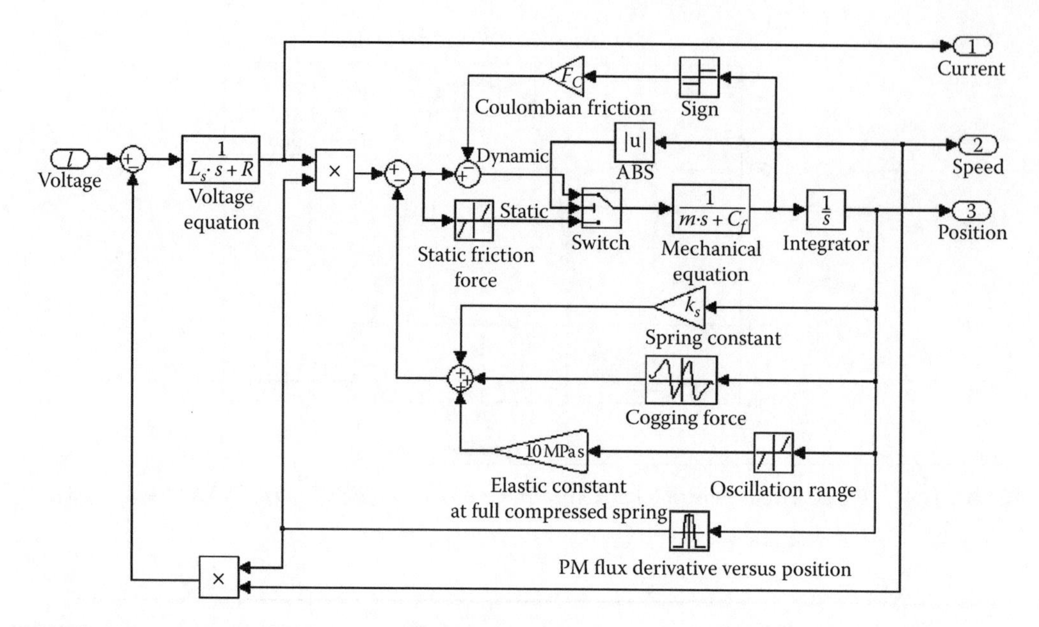

FIGURE 17.43 Block diagram of the linear machine nonlinear model. (After Tutelea, L. et al., *IEEE Trans.*, IE-55(2), 492, 2008.)

The force coefficient and, respectively, the inductance dependence on current cannot be neglected when the linear oscillatory machine is working in heavy magnetic saturation conditions, as in our prototype for a large output power. The look-up table method could also be used in this case. The force coefficient is

$$K_f(i,x) = k_{if}(i) \cdot k_{xf}(x) \tag{17.67}$$

The force coefficient dependence on current (from FEM), k_{if}, is shown in Figure 17.44, while the force dependence coefficient versus mover displacement k_{xf} has the same variation as was shown in Figure 17.41, except for the magnitude that is unity.

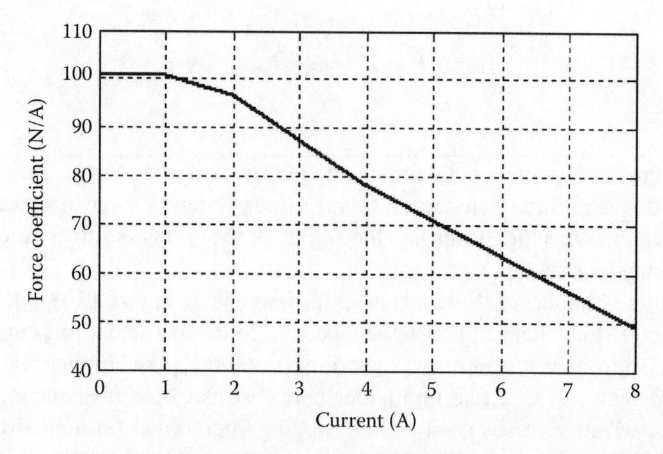

FIGURE 17.44 Force coefficient versus current. (After Tutelea, L. et al., *IEEE Trans.*, IE-55(2), 492, 2008.)

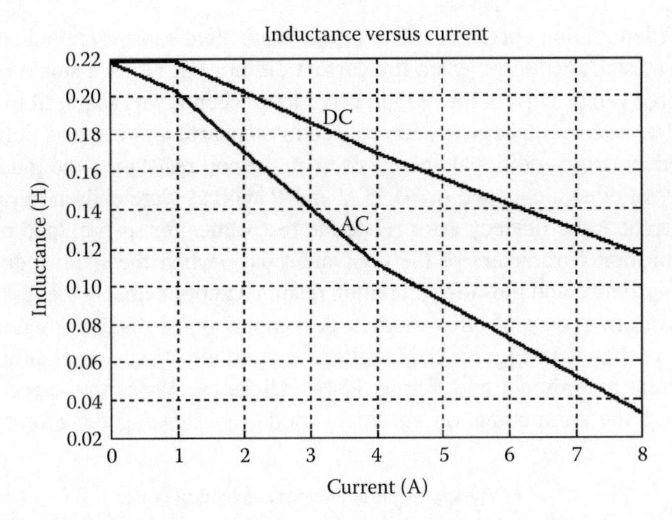

FIGURE 17.45 DC and AC inductance versus current. (After Tutelea, L. et al., *IEEE Trans.*, IE-55(2), 492, 2008.)

The ac inductance, computed from FEM results (17.68), is used in the look-up table and it is shown in Figure 17.45:

$$L_{ack} = \frac{\Psi(I_{k+1}) - \Psi(I_k)}{I_{k+1} - I_k} + 0.02$$

$$I_{ak} = \frac{I_{k+1} + I_k}{2} \tag{17.68}$$

where the 0.02 H term is the end coil leakage inductance.

The block diagram of the nonlinear model, considering the force coefficient and inductance dependence on current, is shown in Figure 17.46. A larger current was observed in the test results than in the simulation results. This is produced by a construction particularity, as a screw bolt was

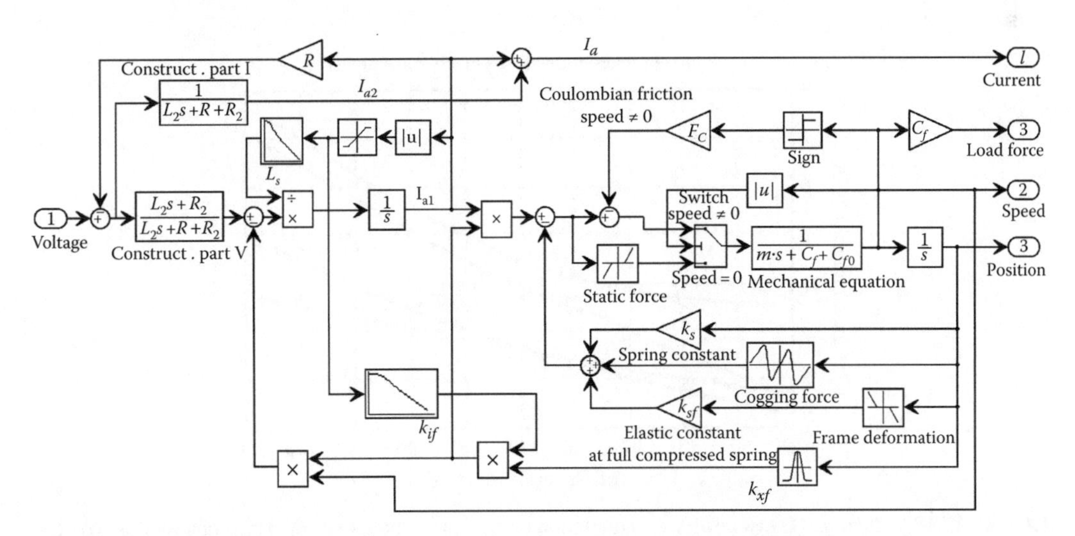

FIGURE 17.46 Block diagram considering inductance and force coefficient dependence on current. (After Tutelea, L. et al., *IEEE Trans.*, IE-55(2), 492, 2008.)

used to tighten the lamination core. The effect of this equivalent short-circuited cage was observed also in the dc decay standstill tests, when the current did not fall along a single exponential wave. In addition, the force versus current hodograph had a loop when it was acquired in standstill tests at low frequency (1 Hz) ac current. The cage introduced in this particular construction is increasing the current and is producing the additional voltage drop on the coil resistance and inductance leakage.

The parameters of equivalent cage L_2=0.35 H and R_2=40 Ω were chosen in order to minimize the simulation current and efficiency error related to test values for several load points (45–230 W output power). Constant parameters of the equivalent cage when the main inductance has large variations is a compromise and thus the simulation results do not fit closely the test results; however, they are better than in the case of other models. A comparison of results produced with different models and tests is shown in Figure 17.47, voltage magnitude versus mechanical output power; Figure 17.48, current magnitude; and Figure 17.49, efficiency, where the curve 1 represents the test results, curve 2 the simulations on nonlinear model considering the prototype construction

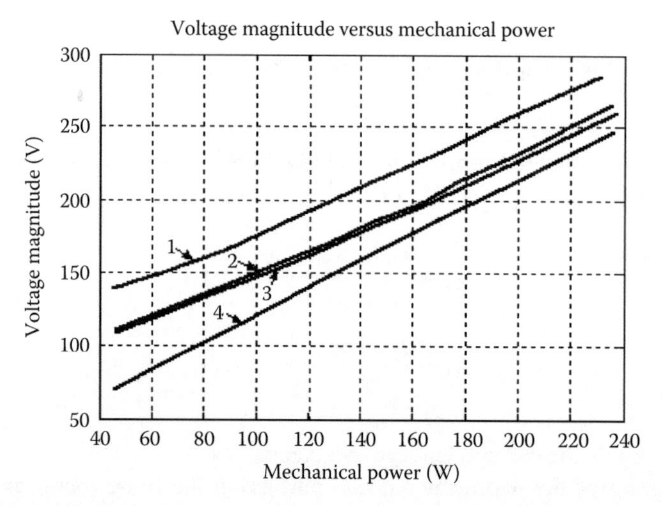

FIGURE 17.47 Supplied voltage: (1) test results, (2) simulation on nonlinear model considering construction particularity, (3) simulation on nonlinear model, and (4) simulation on linear model. (After Tutelea, L. et al., *IEEE Trans.*, IE-55(2), 492, 2008.)

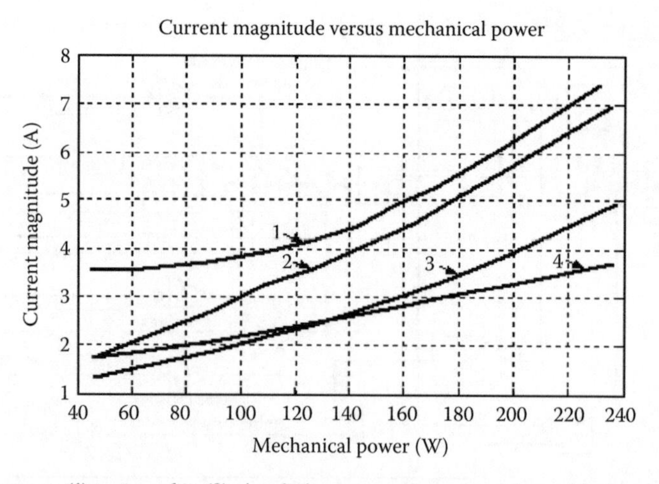

FIGURE 17.48 Current: (1) test results, (2) simulation on nonlinear model considering construction particularity, (3) simulation on nonlinear model, and (4) simulation on linear model. (After Tutelea, L. et al., *IEEE Trans.*, IE-55(2), 492, 2008.)

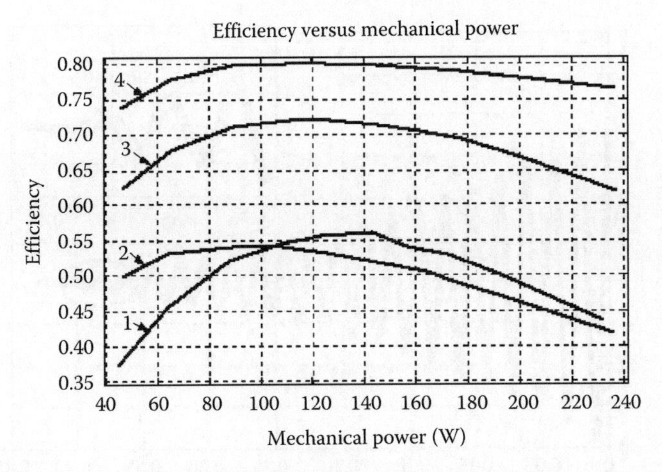

FIGURE 17.49 Efficiency: (1) test results, (2) simulation on nonlinear model considering construction particularity, (3) simulation on nonlinear model, and (4) simulation on linear model. (After Tutelea, L. et al., *IEEE Trans.*, IE-55(2), 492, 2008.)

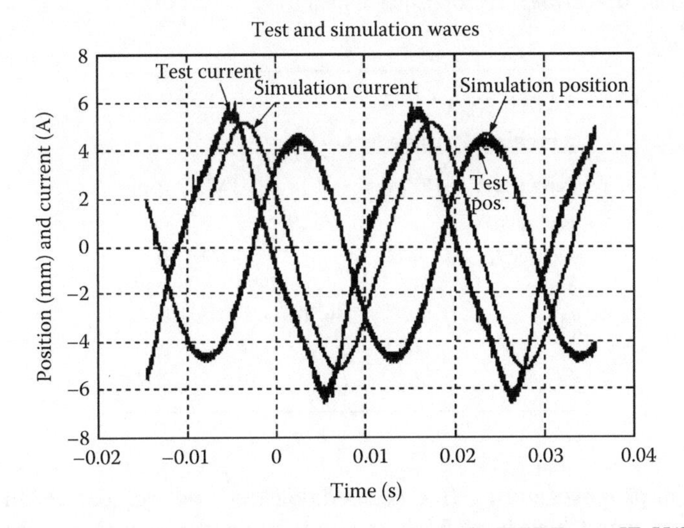

FIGURE 17.50 Current and position waves. (After Tutelea, L. et al., *IEEE Trans.*, IE-55(2), 492, 2008.)

particularity, curve 3 the simulations on nonlinear model, and curve 4 the simulations on linear model. The efficiency could increase by around 15%–20% by eliminating the construction problem.

The current and position waves in dynamic simulation and test results are shown in Figure 17.50.

The simulation base times were shifted to overlap the simulation position and measured position. The current–speed (position) phase is very sensitive to frequency around the resonance frequency (Figure 17.33), so a small error in the model parameters (resonance frequency) could produce a large error in the current–speed phase. Moreover, it is the short-circuit cage, which was not fully modeled and which should be eliminated in a future prototype.

Free deceleration of back-to-back coupled machines simulation is in good agreement with test results (Figure 17.51).

17.3.3.4 Parameters Estimation

Some of the model parameters, such as the resistance, could be measured directly while the inductance is computed from dc current decay tests and the flux derivative with mover position is computed from standstill thrust measurement.

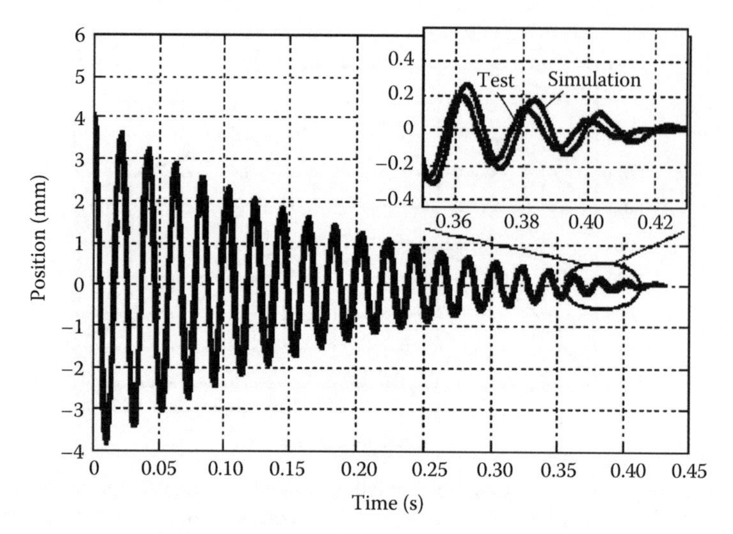

FIGURE 17.51 Free oscillation test and simulation for coupled machines particularity (short-circuit cage). (After Tutelea, L. et al., *IEEE Trans.*, IE-55(2), 492, 2008.)

TABLE 17.3
Machine Parameters

Parameters	FCPM		SPM	Unit
Resistance		9		Ω
Inductance	0.22		0.125	H
k_f	104		90	Wb/m
k_s		146,000		N/m
Fc	2.35		3	N
Cf	17		16	N s/m

Free oscillation of mover during free deceleration test was recorded to find the mechanical resonance frequency for each machine and also for the two machines, when mechanically coupled back to back as in dynamic tests. The Coulomb friction force, F_c, and viscous coefficient, c_f, are adjusted in order to produce the same mover displacement in the simulation as in recorded data that were measured by a laser-based position transducer. The machine parameters are shown in Table 17.3.

17.3.3.5 Further Performance Improvements

The FEM results on thrust and inductance are in good agreement with the test results. Therefore, we conclude that we can use the FEM to design a better oscillatory machine. By reducing the airgap from 1 to 0.4 mm, and by increasing the PM height from 2 to 3 mm and the stator back core from 10 to 15 mm, the airgap flux density is increased (Figure 17.52) due to real flux concentration. Consequently, the thrust versus current is increased more than three times as it is observed in Figure 17.53. The inductance does not increase much (Figure 17.54), so the voltage drop on it does not increase notably. In conclusion, the power density and efficiency could be dramatically improved.

Note: This paragraph, starting from experiments, is indicative of the various problems encountered in modeling an LOM to fit test results.

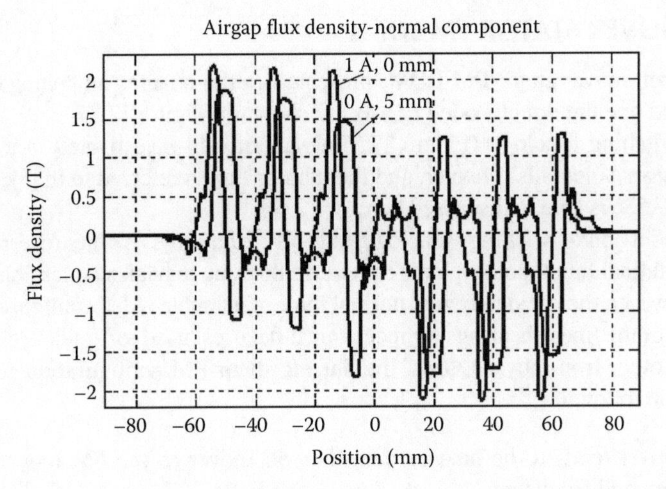

FIGURE 17.52 Airgap flux density on improved design. (After Tutelea, L. et al., *IEEE Trans.*, IE-55(2), 492, 2008.)

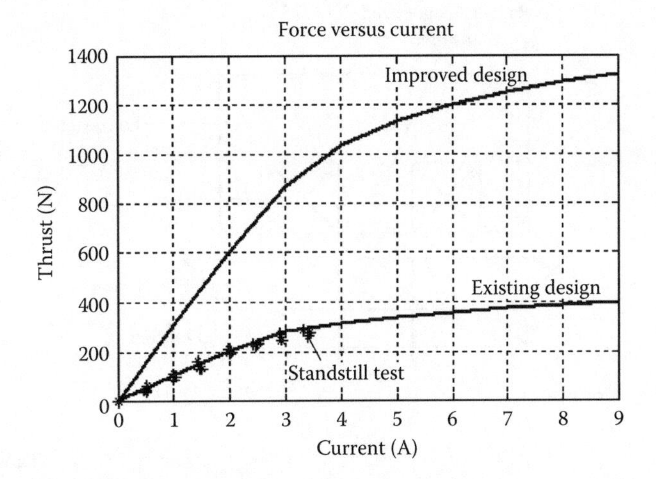

FIGURE 17.53 Thrust for existing and improved designs. Solid line- FEM, * standstill test, 2 parallel path. (After Tutelea, L. et al., *IEEE Trans.*, IE-55(2), 492, 2008.)

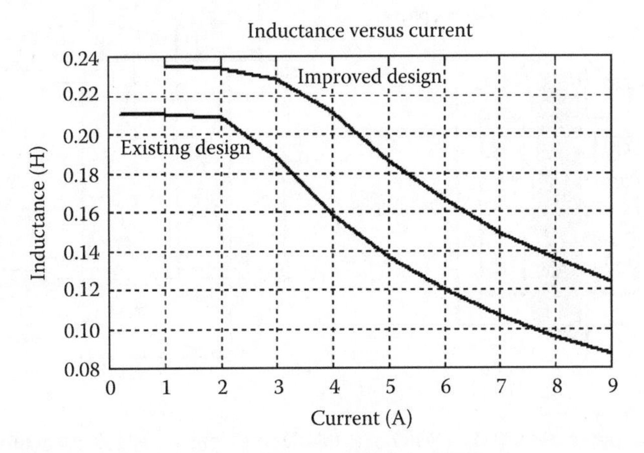

FIGURE 17.54 Inductance for existing and improved design, FEM at zero mover displacement. (After Tutelea, L. et al., *IEEE Trans.*, IE-55(2), 492, 2008.)

17.4 IRON-MOVER STATOR PM LOMs

Aside from the iron-mover stator PM LOM, presented in the chapter on "plunger solenoids," two typical tubular and one flat double-sided LOMs are shown in Figure 17.55.

As the flux-switching topology (Figure 17.55a) was already investigated in a slightly different shape in the "Plunger Solenoids" chapter, and the tubular flux is referred to in [8], only the configuration in Figure 17.55c is briefly discussed here.

It may be called a "flux-switching" topology, but this is the same as "flux reversal." Figure 17.55c illustrates a longitudinal flux topology, but a transverse flux one may also be feasible in a single-sided configuration. However, the net ideally zero normal force of a double-sided configuration is paramount in easing the task of the linear bearings (or mechanical flexures) used to "guide" the linear motion.

Together with other iron-mover LOMs, the flat FR stator PM configuration shows a few peculiarities such as the following:

- The iron mover tends to be heavier than the coil mover or the PM mover for the same thrust, stroke, and frequency.
- The airgap variations lead to notably uncompensated normal force (due to the small airgap).

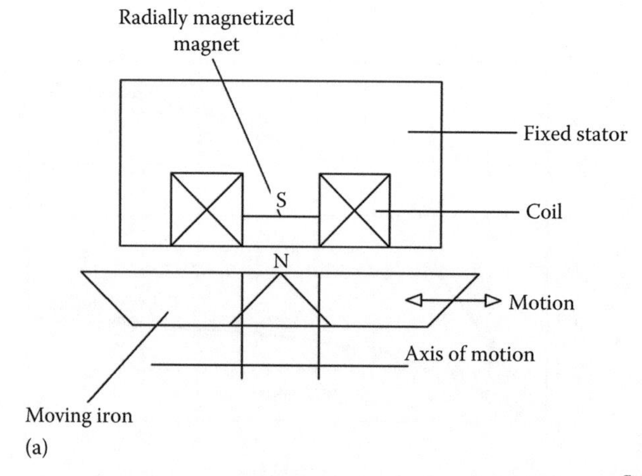

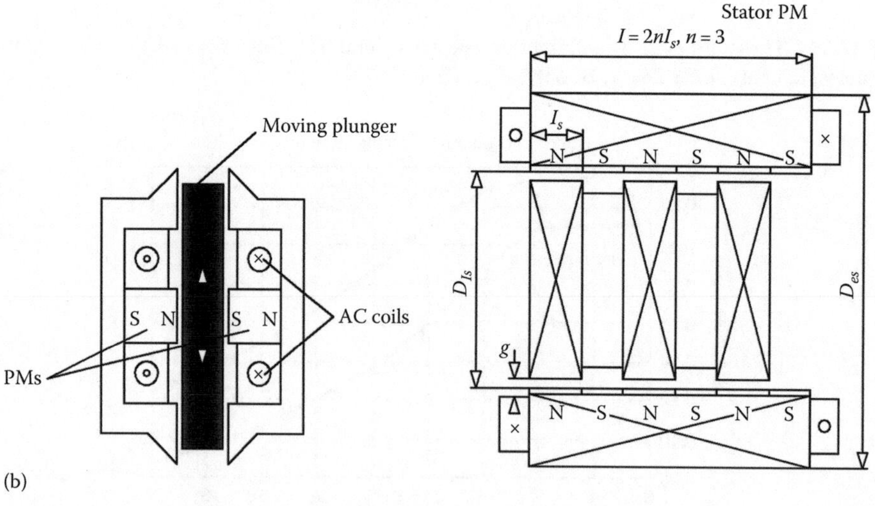

FIGURE 17.55 Iron-mover stator PM LOM(G)s (a) tubular with pulsed-PM flux, (b) tubular with flux reversal, (c) flat with PM flux concentration (double sided). (After Redlich, R.W. and Berchowitz, D.H., *Proc Inst. Mech. Eng.*, 199(3), 203, 1985.)

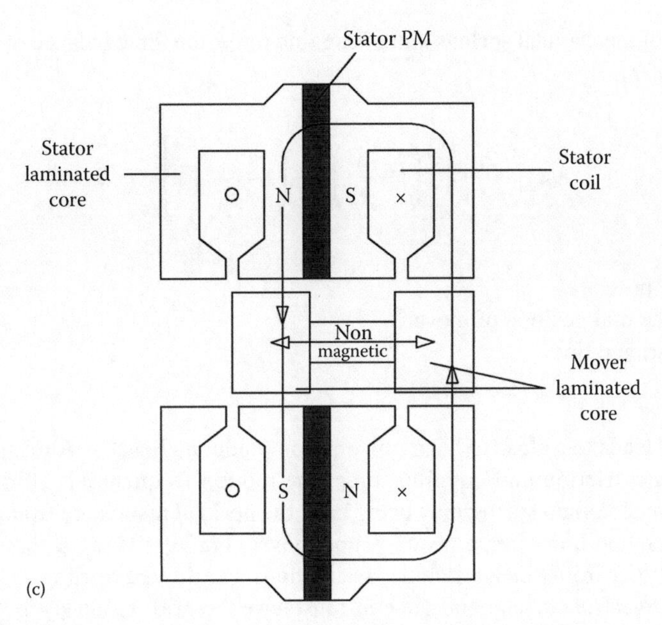

FIGURE 17.55 (continued) Iron-mover stator PM LOM(G)s (c) flat with PM flux concentration (double sided). (After Redlich, R.W. and Berchowitz, D.H., *Proc Inst. Mech. Eng.*, 199(3), 203, 1985.)

- There is a strong cogging force, which in general may not be used as a "magnetic" spring due to its insufficient level and "wrong" direction.
- The rather large inductance leads to lower power factor, slower current response, and stronger danger of PM demagnetization.
- Fringing and leakage PM flux are important, and optimal geometrical design should be used to reduce them.
- The force (and emf) coefficient $K_e(x)$ depends notably on mover position, so the emf departs from a sinusoidal waveform and so does the current for sinusoidal input voltage; all this might lead to some vibration and noise.

17.5 LINEAR OSCILLATORY GENERATOR CONTROL

LOGs driven by a spark-ignited gasoline linear engine or by a Stirling free-piston engine may operate stably in stand-alone mode based on the prime mover stabilizing means.

For small powers, a Stirling free-piston engine may, by its controller, function stably at the power grid and deliver controlled power at the grid frequency [15–18].

Reference [15] introduces a model for the two-cylinder internal combustion linear engine (ICLE) by using the balance of forces:

$$P_L(x)A_b - P_R(x) - F(x) = m_t\ddot{x} \tag{17.69}$$

P_L is the instantaneous pressure in the left cylinder
P_R is the instantaneous pressure in the right cylinder
A_b is the bore area
$F(x)$ is the electromagnetic (LOG) force and mechanical (spring, if any) force
m_t is the total mover mass

Under no-load and ideal (frictionless) conditions, no spring is required to sustain the mover motion: gas compressing and expending forces will do it, and thus, the natural frequency of the engine is reached.

In the absence of mechanical springs and of cogging force, the force balance of ICLE, at no load, can be written [16,17]

$$A_B P_1 \left(\frac{2r}{r+1}\right)^n \left[\left(1+\frac{x}{x_n}\right)^{-n} - \left(1-\frac{x}{x_m}\right)^{-n}\right] = m_t \ddot{x} \tag{17.70}$$

P_1 is the intake pressure
$x_m = l_{stroke}/2$ is the mid position of mover
r is the combustion ratio
n is a constant

Equation 17.70 leads to a close to harmonic motion (under no load). As during load and in real conditions (nonzero friction and cogging force) the motion frequency is slightly reduced, the introduction of a mechanical spring may bring back the no-load resonance frequency conditions.

For a Stirling engine linear force piston prime mover (Figure 17.56a), a stability analysis was performed in [18,19]. Stability means damped oscillations. For this the Stirling engine displacer and power pistons, the electric current, and the emf must have the grid frequency ω. Stability is based on the inequality

$$W_d \le W_c \tag{17.71}$$

where W_c is the total energy removed from the shaft (by the LOG) over one cycle of ω:

$$W_c = P_c \frac{2\pi}{\omega} \tag{17.72}$$

and W_d is energy generated over one cycle of ω (by the Stirling engine)

$$W_d = P_d \frac{2\pi}{\omega} \tag{17.73}$$

A typical result for a Stirling engine and an LOG with a series capacitor C_s is shown in Figure 17.56b [19]. For a few output power levels (2, 10, 20, 25, 30 kW), stability gradually grows from 55–60 to (55–71) Hz frequencies.

A trade-off between a stronger PM and a larger tuning capacitor C_s, based on minimum initial cost, may be performed.

The tuning capacitor C_s does a few things:

- Increases the rate of power removal from the prime mover P_c, such that it is faster than the rate of power generated by the engine, P_d.
- It may reduce the danger of PM demagnetization during larger loads; not so under sudden short circuit.
- Improves the LOG power factor.

All three requirements are important, but the first two demand higher priority in sizing the tuning capacitor. An additional mechanical spring would change the picture of stability mostly for

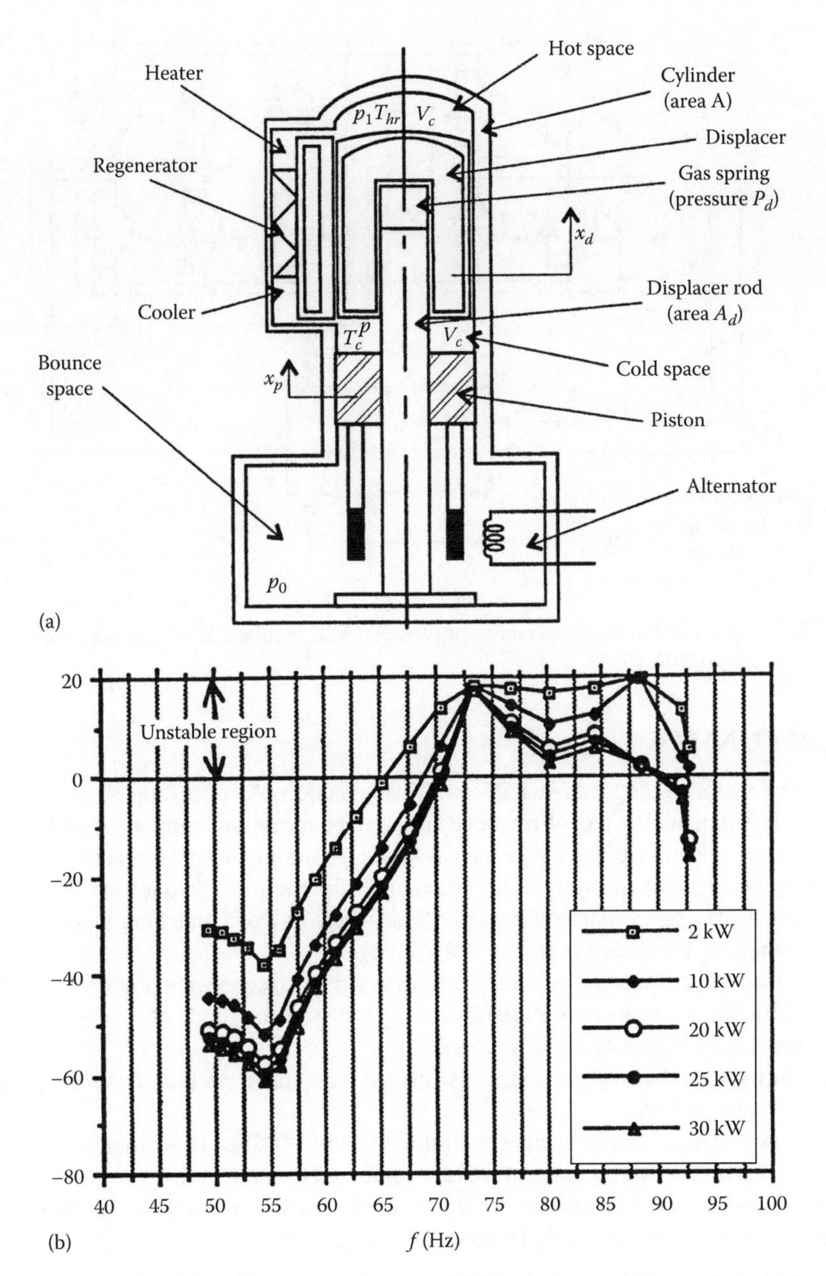

FIGURE 17.56 (a) Stirling engine linear prime mover and (b) Bode plots stability analysis. (After Tutelea, L. et al., *IEEE Trans.*, IE-55(2), 492, 2008.)

the better, but the resonance frequency may be more rigid, to follow and secure maxim LOG efficiency.

However, if a converter is used, and a strong mechanical spring "governs" the motion, the system may be close-loop stabilized either for grid or for autonomous operation (Figure 17.57).

As expected, the dual converter may operate bidirectionally and under open- or close-loop control, either in generator or in motor mode.

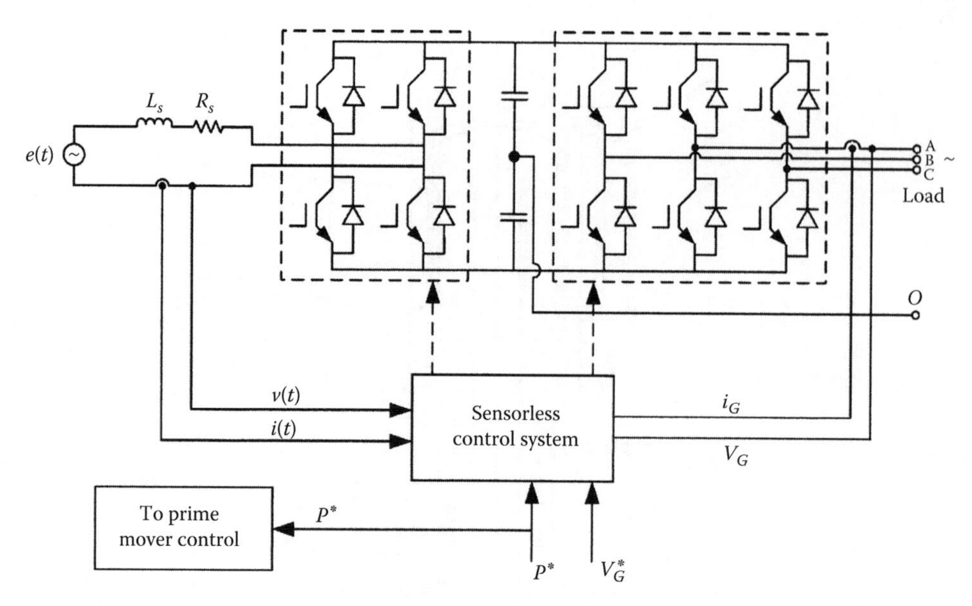

FIGURE 17.57 LOG with bidirectional converter control. (After Redlich, R.W. and Berchowitz, D.H., *Proc Inst. Mech. Eng.*, 199(3), 203, 1985.)

17.6 LOM CONTROL

The control for motoring depends heavily on the application. As operation at resonance is most desirable, only a small adjustment of frequency to produce minimum current would be required.

Also, the larger the voltage, the larger the stroke length for larger load force, in general.

As a strong mechanical spring is used "to control" the motion, the question arises if position close-loop control may be eliminated for applications where its exact matching is not critical. Let us consider reciprocating compressors (Figure 17.58) [20].

The current amplitude versus stroke characteristics at a few frequencies (Figure 17.59) [19] shows that, say at 50 Hz, the stroke increases almost linearly with the amplitude of current after the pumping point (right side in Figure 17.59) is reached.

This should indicate that current control is sufficient and thus, position control for vapor compressors may be eliminated.

A B-spline neural network + PI current controller (Figure 17.60) is shown to produce good results with the phase angle (θ) of the sinusoidal current command as input [19].

Even when the current reference is increased, to increase stroke, the response in current, position, and compressor pressure are all stable (Figure 17.61), after [19].

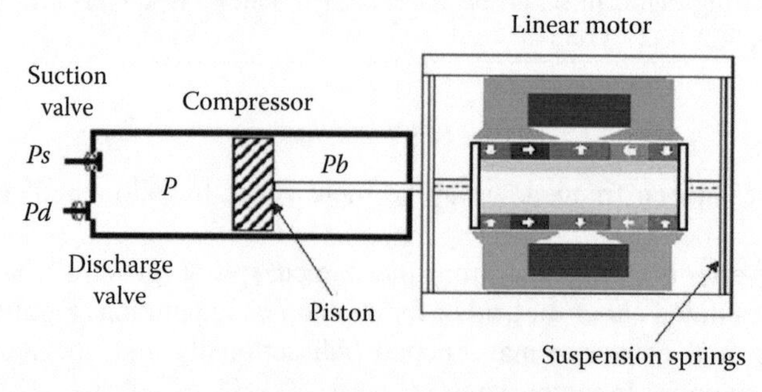

FIGURE 17.58 Linear compressor LOM. (After Boldea, I. and Nasar, S.A., *Linear Motion Actuators and Generators*, Chapter 8, Cambridge University Press, Cambridge, U.K., 1997.)

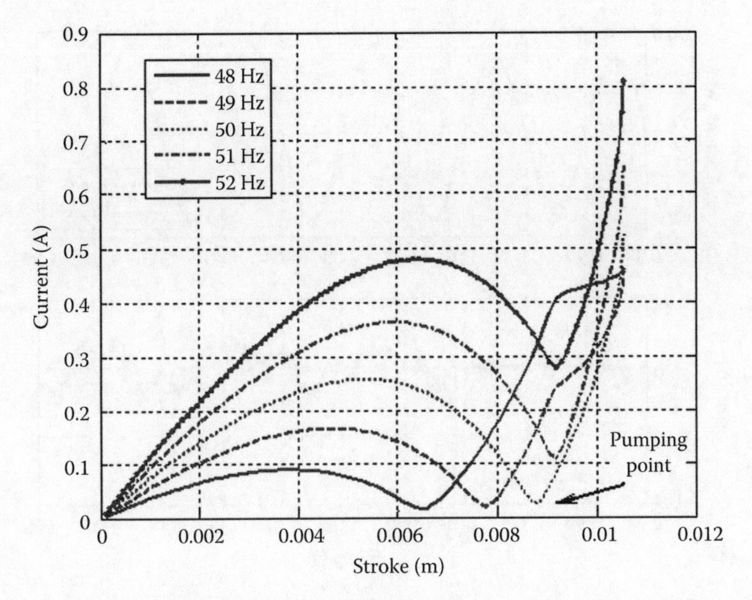

FIGURE 17.59 Current versus l_{stroke} curves for linear compressor. (After Lim, Z. et al., A hybrid current controller for linear reciprocating vapor compressors, *Conference Record of the IEEE Industry Applications Society Annual Meeting* (*IEEE-IAS'2007*), New Orleans, LA, 2007.)

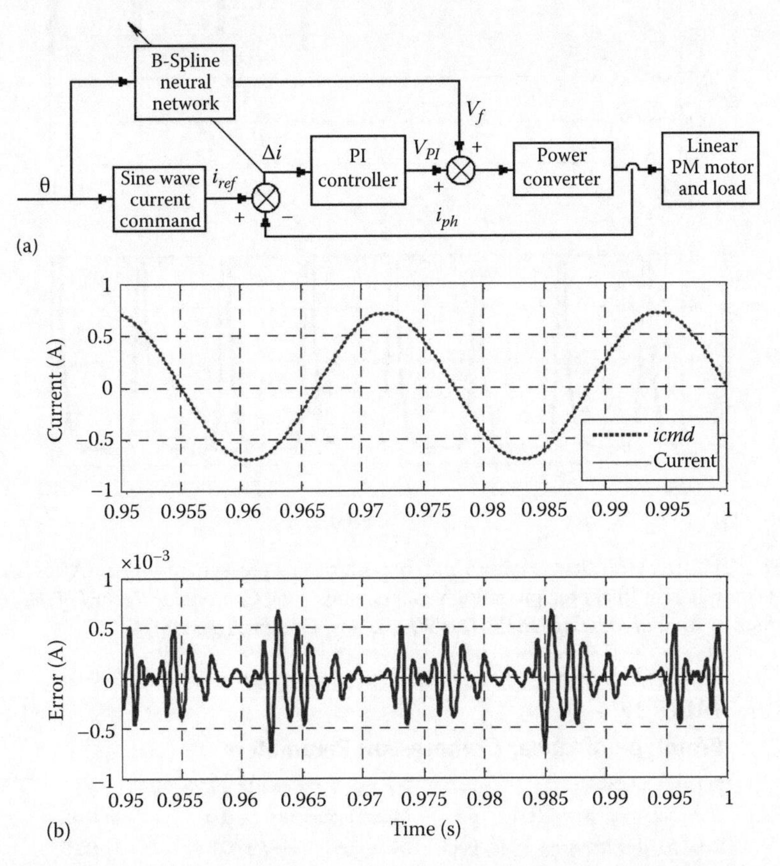

FIGURE 17.60 (a) BSNN current controller for LOM and (b) typical results. (After Lim, Z. et al., A hybrid current controller for linear reciprocating vapor compressors, *Conference Record of the IEEE Industry Applications Society Annual Meeting* (*IEEE-IAS'2007*), New Orleans, LA, 2007.)

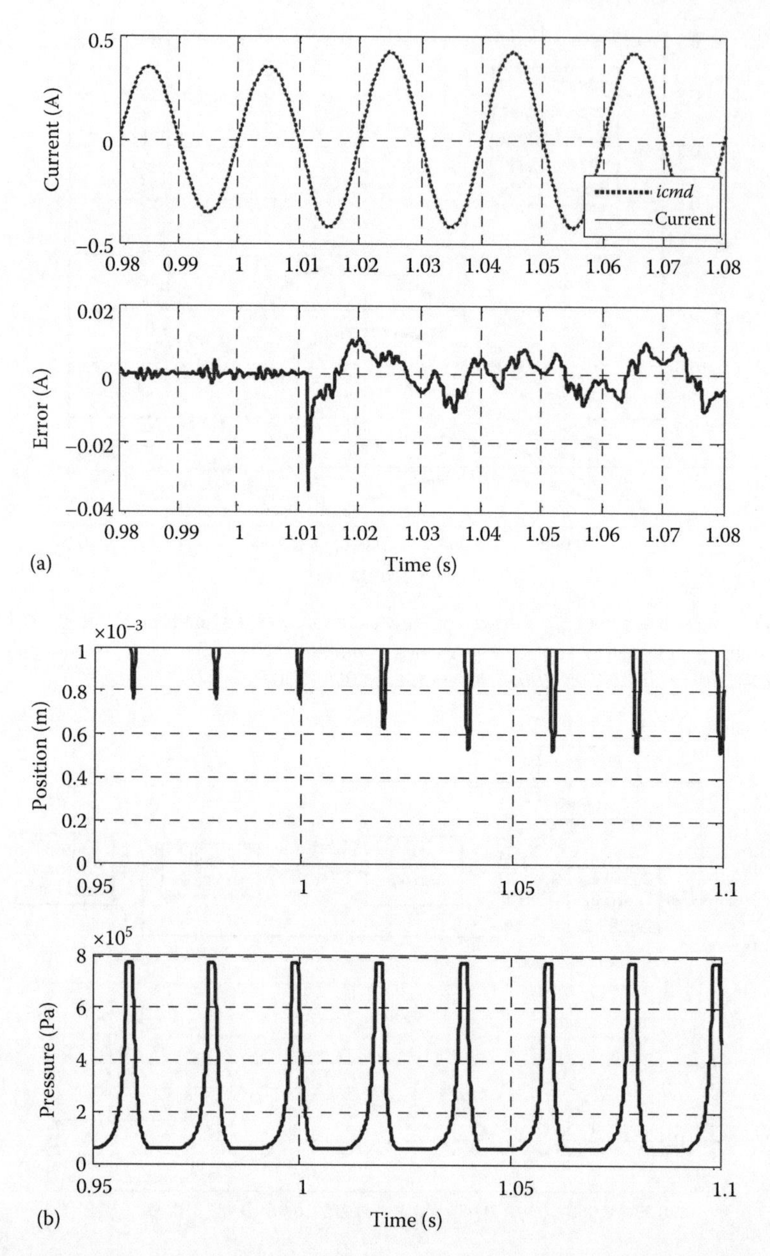

FIGURE 17.61 (a) Current transient response and (b) position and pressure response. (After Lim, Z. et al., A hybrid current controller for linear reciprocating vapor compressors, *Conference Record of the IEEE Industry Applications Society Annual Meeting* (*IEEE-IAS'2007*), New Orleans, LA, 2007.)

TABLE 17.4
Prototype of Linear Compressor: Parameters

Supply voltage V (V_{rms})	230	Average force constant KTa (N/A)	92
Nominal frequency f (Hz)	50	Motor inductance Le (H)	0.45
Total moving mass m (kg)	0.43	Motor resistance Re (Ω)	02.5
Spring constant K (kN/m)	29.8	Viscous damping B (N s/m)	1.10
Rated stroke U (mm)	10.5		

The PI current controller alone was shown incapable of small enough current regulation errors. Typical data of the prototype in [19] are given in Table 17.4.

Other robust current controllers (such as sliding mode + PI) may also apply even for a smaller computation effort, as shown in Chapter 15, dedicated to "plunger solenoids."

17.7 SUMMARY

- Single-phase PM linear oscillatory synchronous motors/generators are preferred for quite a few applications with a stroke of up to 60 mm or so, while operating at constant frequency f_1 and variable voltage.
- Mechanical spring (for vapor compressor load) or the prime mover (linear internal combustion engine or Stirling engine) is capable of harmonic linear motion at resonance ($f_m = f_1$).
- This way the efficiency is high because the electromagnetic force deals only with the load, while the mover motion is controlled mechanically (by the mechanical spring or by the prime mover (for generator mode).
- The stroke length l_{stroke} corresponds to two motion amplitudes x_{max} ($l_{stroke} = 2x_{max}$):

$$x(t) = x_{max}\sin(2\pi f_1 t);\, f_1 = f_m \tag{17.74}$$

- The stroke may be assimilated to the electric machine pole pitch τ; consequently, when the load decreases, the stroke (pole pitch) is reduced by reducing the voltage at resonance frequency $f_1 = f_m$. So at reduced loads the excursion length and speed are reduced by reducing a "virtual" pole pitch in the single-phase PM linear synchronous machine; but modifying pole pitch is, in terms of efficiency, as good as modifying frequency to reduce speed; this explains why the efficiency at reduced loads remains large. This is a special merit of LOM(s).
- There are tubular [9] and flat topologies [21–23] for LOM(G)s; the tubular ones enjoy "the blessing" of circularity and thus show ideally zero radial (normal) force and, consequently, the mechanical flexures (or linear bearings) are "saved" very large normal forces; for flat structures, double-sided configurations are required to preserve this asset.
- The heavier the mover of LOM(G)s, the larger the mechanical spring for given eigenfrequency $f_m = f_1$; air-coil movers and PM movers tend to result in lower mover weight and thus imply smaller (less expensive) mechanical springs, but they are mechanically more fragile as their framing should be nonmagnetic with low electrical conductivity (siluminum, Kevlar, etc.)

  There are also iron movers to consider, though their weight tends to be larger for given stroke, frequency, and power, but they are rugged mechanically and the PMs on the stator are better protected mechanically and thermally.
- The coil movers require a mechanically flexible electric power cable, which might be a liability for some applications.
- As even a small departure from resonance frequency leads to notable reduction in LOM(G) efficiency, operation with an inverter, to trace minimum current in correcting the reference frequency, is required.
- LOM(G)s lack the rotary to linear mechanical transmission and could compete with standard rotary motor drives for LOMs if their efficiency is high and their total cost is competitive at equal or better reliability, which is the case with PM LOM(G)s.
- Air-coil-mover LOMs may be built in the tenth of kW range for, say, series hybrid electric vehicle generators driven by dual linear piston gas engines, and down to small powers in microspeakers (or recorders for mobile phones, etc.).

The coil in air with surface PMs on the stator leads to a small electric time constant, which provides acceptable power factor at large powers and fast current (force) response in low-power applications.

- A 20 kW, 60 mm stroke, 60 Hz LOG is proven to show 90% efficiency at 0.8 lagging power factor in a 110 kg stator and 30 kg mover device (average speed is only 3.6 m/s, which explains the not so small weight/kW).
- For microspeakers/receivers with high-quality sound and broad frequency range (kHz), reduced size and cost are paramount; radiated sound power, dependent on linear speed, geometry, etc., are related to sound pressure level (SPL); harmonic distortion, then, defines the sound quality globally.

Once the diaphragm speed is given, the integration of microspeaker and dynamic receiver LOM(G), with PM flux concentration, was proven to produce competitive performance in a smaller volume device.

To avoid the necessity of a mechanically flexible electric power cable, the PM-mover topologies have been introduced, especially for vapor compressors.

- The tubular homopolar PM-mover, single-coil, C-core stator topology, first introduced by Redlich in the 1980s has become commercial, and others with multiple heteropolar PM mover and multiple-coil stators have been proposed in the last two decades.
- The tubular homopolar PM mover (Redlich) topology, though it leads to a moderate airgap PM flux density, overall leads to a low-weight device of pancake shape with a good efficiency.
- The multiple heteropolar PM mover with multiple stator topology is proven in this chapter to produce 25 W at 270 Hz, 6 mm stroke at 74% efficiency within a device whose total weight is around 60 g (mover weight: 16 g). An increase in size would jump the efficiency to 80%–85%; a two-pole PM-mover device was proven to produce 42 W at 89% efficiency for a 10 mm stroke at 50 Hz, however, for 866 g total weight (230 g mover), which demands a strong (large) mechanical spring [24].
- A double-sided flat flux-reversal LOM is introduced and analyzed in detail with experiments to illustrate various techniques in its characterization; PM flux concentration in the mover is performed to increase force/total weight and efficiency, but the PM flux concentration leads to a high p.u. inductance and thus moderate power factor (limited frequency).
- Iron-mover LOMs have rugged movers but suffer from high inductance p.u. syndrome; they are suitable when safety (reliability) is critical for a long life (such as in vapor compressors, Stirling engine driven LOGs for deep space missions).
- LOGs may produce sustained harmonic motion by the prime mover, even without mechanical springs, by using wisely a centralizing large linear cogging force obtained by expert design [11].
- Both a two-piston linear ICE and the Stirling (free-piston) engine have been proven capable of stable motion; mechanical springs may be added to help stability and so is a tuning series capacitor in the LOG.
- The ultimate control of LOGs (and LOMs) is close-loop control, which needs, for bidirectional power flow, a dual (back to back) inverter.
- For LOM, the control depends on the application; for vapor compressors, the robust current control was proved sufficient (no position control is needed), beyond the pumping point of the compressor.
- Overall, resonance single-phase PM linear motors/generators prove to be highly competitive devices for applications with strokes from under 1 to 60 mm (or so) and stroke length at frequencies from 40–60 to 300 Hz and more (at very low power). Strong R&D and industrial efforts are expected with LOM(G)s, especially for vapor compressors [20,22], Stirling engine generators, and microspeakers/dynamic receivers.

REFERENCES

1. I. Boldea and S.A. Nasar, *Linear Electromagnetic Devices*, Chapter 7, Taylor & Francis Group, New York, 2001.
2. S.M. Hwang, H.J. Lee, K.S. Hong, B.S. Kang, and G.Y. Hwang, New development of combined PM type microspeakers used for cellular phones, *IEEE Trans.*, MAG-41(5), 2005, 2000–2003.
3. S.M. Hwang, K.S. Hong, H.J. Lee, J.H. Kim, and S.K. Jeung, Reduction of dynamic distortion in dual magnetic type microspeakers, *IEEE Trans.*, 40(4), 2004, 3054–3056.
4. S.M. Hwang, H.J. Lee, J.H. Kim, G.Y. Hwang, W.Y. Lee, and B.S. Kang, New developments of integrated microspeakers and dynamic receivers used for cellular phones, *IEEE Trans.*, MAG-39(5), 2003, 3259–3261.
5. F.F. Mazda, *Electronic Instruments and Measurement Techniques*, Cambridge University Press, Cambridge, U.K., 1987.
6. T. Mizuno, M. Kaway, F. Tsuchiya, M. Kosugi, and H. Yamada, An examination for increasing the motor constant of a cylindrical moving magnet-type linear actuator, *IEEE Trans.*, MAG-41(10), 2005, 3976–3978.
7. J. Wang, D. Howe, and Z. Lin, Comparative study of winding configurations of short-stroke, single phase tubular permanent magnet motor for refrigeration applications, *Conference Record of the IEEE Industry Applications Society Annual Meeting* (*IEEE-IAS'2007*), New Orleans, LA, 2007.
8. R.W. Redlich and D.H. Berchowitz, Linear dynamics of free piston Stirling engine, *Proc. Inst. Mech. Eng.*, March 1985, 199(3), 203–213.
9. T. Ibrahim, J. Wang, and D. Howe, Analysis of a short-stroke, single-phase tubular PM actuator for reciprocating compressors, *International Symposium on Linear Drives for Industrial Applications* (*LDIA2007*), Lillie, France, 2007.
10. I. Boldea, *Variable Speed Generators*, Chapter 12, CRC-Press, Taylor & Francis Group, New York, 2006.
11. I. Boldea, S. Agarlita, and L. Tutelea, Springless resonant linear oscillatory PM motors? *International Symposium on Linear Drives for Industrial Applications* (*LDIA2011*), Eindhoven, the Netherlands, 2011.
12. C. Pompermaier, F.J. Kalluf, M.V. Ferreira da Luz, and M. Sadowski, Analytical and 3D FEM modeling of a tubular linear motor taking into account radial forces due to eccentricity, *IEEE International Electric Machines and Drives Conference (IEMDC'09)*, Miami, FL, 2009.
13. L. Tutelea, M. Ch. Kim, M. Topor, J. Lee, and I. Boldea, Linear PM oscillatory machine: Comprehensive modeling for transients with validation by experiments, *IEEE Trans.*, IE-55(2), 2008, 492–500.
14. L. Tutelea, M.C. Kim, T.H. Kim, J. Lee, and I. Boldea, Development of a flux concentration type linear oscillatory machine, *IEEE Trans.*, MAG-40(4), 2004, 2092–2094.
15. C. Pompermaier, K.F.J. Haddad, A. Zambonetti, M.V. Ferreira da Luz, and I. Boldea, Small linear PM oscillatory motor: Magnetic circuit modeling corrected by axisymmetric 2-D FEM and experimental characterization, *IEEE Trans.*, IE-59(3), 2012, 1389–1396.
16. N. Clark, S. Nandkumar, and P. Famouri, Fundamental analysis of a linear two-cylinder ICE, *Record of SAE International FUEL and Lubricants Meeting and Exposition*, San Francisco, CA, 1998.
17. A. Cosic, I. Lindback, W.M. Arshad, and P. Thelis, Application of a free-piston generator in a series hybrid vehicle, *Proceedings of the Fourth International Symposium on Linear Drives for Industry Applications* (*LDIA'2003*), Birmingham, U.K., 2003, pp. 541–544.
18. Z.X. Fu, S.A. Nasar, and M. Rosswurm, Stability analysis of free piston Stirling engine power generation system, *27th IECECE Conference*, San Diego, CA, August 3–7, 1992.
19. I. Boldea and S.A. Nasar, *Linear Motion Actuators and Generators*, Chapter 8, Cambridge University Press, Cambridge, U.K., 1997.
20. Z. Lim, J. Wang, and D. Howe, A hybrid current controller for linear reciprocating vapor compressors, *Conference Record of the IEEE Industry Applications Society Annual Meeting* (*IEEE-IAS'2007*), New Orleans, LA, 2007.
21. I. Boldea, S.A. Nasar, B. Penswich, B. Ross, and R. Olan, New linear reciprocating machine with stationary PM, *Conference Record of the IEEE Industry Applications Society Annual Meeting* (*IEEE-IAS'1996*), New Orleans, LA, 1996, vol. 2, pp. 825–829.
22. J. Lee and I. Boldea, Linear reciprocating flux reversal PM machine, US patent No. 6538.349, 25.03.2003
23. L. Tutelea, M.C. Kim, T.M. Kim, J. Lee, and I. Boldea, A set of experiments to more fully characterize linear PM oscillatory machines, *IEEE Trans.*, MAG-41(10), 2005, 4009–4011.
24. Z.Q. Zhu and X. Chen, Analysis of an E-core interior PM linear oscillating actuator, *IEEE Trans.*, MAG-45(10), 2009, 4384–4387.

18 Multiaxis Linear PM Motor Drives

By multiaxis linear PM motor drives (M-LPMDs), we mean here small power devices that produce either long stroke *xy* (planar) motion with small rotation, without or with active *z* levitation (vertical) motion—along axis *z*—control, or *x*, *y*, *z* (3D) controlled short displacements with precision in the micro- to nanometer range.

Typical applications are pick and place machines, inspection systems, and fabrication of small objects (when nanometer precision is necessary).

Such applications have been and could be "covered" by using multiple rotary PM electric motors with adequate mechanical transmissions such as lead screws, which, however, suffer from lost motion, stick-slip, and wind-up.

To prevent backlash and dry friction, regardless of travel length, inchworm-like-clamping, inertial sliding/walking, flexure, and levitated mechanisms have been proposed. The latter two combined are related directly to multiaxis linear PM motor drives.

As long stroke (up to hundreds of mm) *x-y* motion applications require distinct M-LPMD topologies, we will treat them separately.

18.1 LARGE *x–y* (PLANAR) MOTION PM DRIVE TOPOLOGIES

Two- or three-phase flat linear PMSMs are expected to be used for efficient large (up to tens and hundreds of mm) travel along *x–y* axes. The dilemma about using fixed ac coils primary with PM mover (Figure 18.1) or fixed PM secondary with contactless ac power-fed primary mover [1,2] (Figure 18.2) seems to be over, with the former as winner, due to the elimination of ac power feeding (even if contactless at 100 kHz or so).

Both in Figures 18.1 and 18.2, the ac windings are basically three phase, with tooth-wound *x* and *y* coils, but in one case they are placed in a plastic (air) core, while in the other they are "locked" in an iron core.

The PM secondary is similar in both cases and is characterized by N-S pole sequencing along *x* and *y* axes.

As expected, more sophisticated PM arrays have been proposed to increase the force per watt of copper losses and to reduce the "cross talk" between the *x* and *y* coils via PMs (through emfs) [3,4], Figure 18.3.

The second dilemma, air-core or iron-core ac coils, also seems to be solved in favor of air-core coils to reduce cogging force and linearize the system (no magnetic saturation) with total weight reductions, in spite of smaller force density.

The straight-through long coils in Figures 18.1 and 18.2 also tend to be replaced by *x-y* small, multiple coils placed at 90° [3] or in zigzag shape (Figure 18.4) [8,9].

The rectangular ac coils and PM shapes are subject to geometrical optimization design for various objective functions, from minimum temperature for given performance and size to minimum "cross talk" via emf [3], or minimum levitation (vertical) force pulsations.

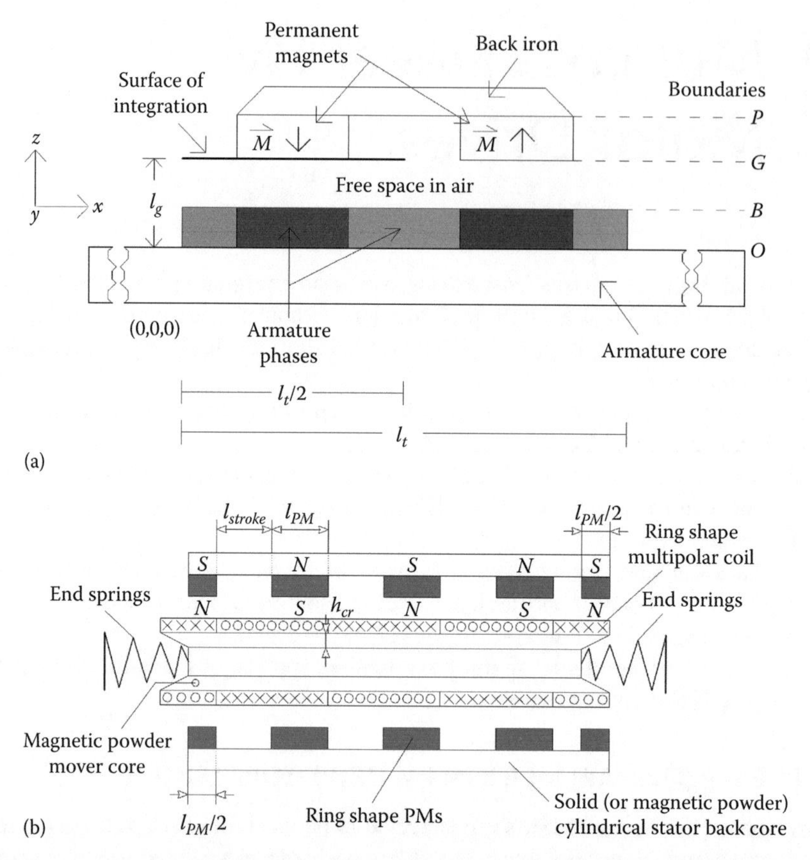

FIGURE 18.1 (a) Fixed ac-fed primary with PM-mover configuration of planar-motion PM motor drive and (b) tubular. (After da Silveira, M.A. et al., *IEEE Trans.*, MAG-41(10), 4006, 2005.)

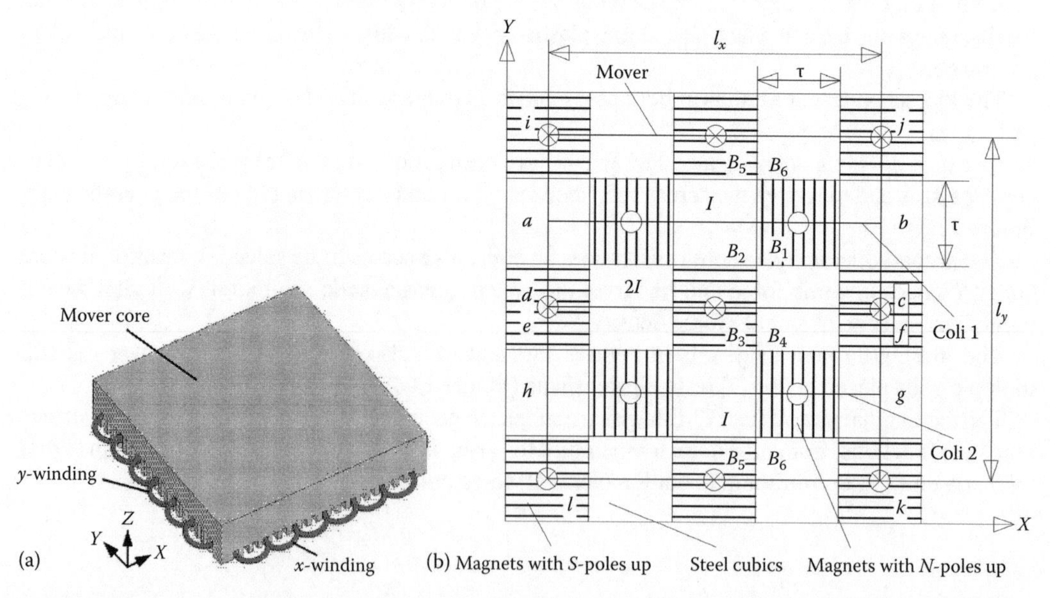

FIGURE 18.2 (a) Iron-core ac-fed primary mover and (b) with fixed PM array. (After Cao, J. et al., *IEEE Trans.*, MAG-41(6), 2156, 2005.)

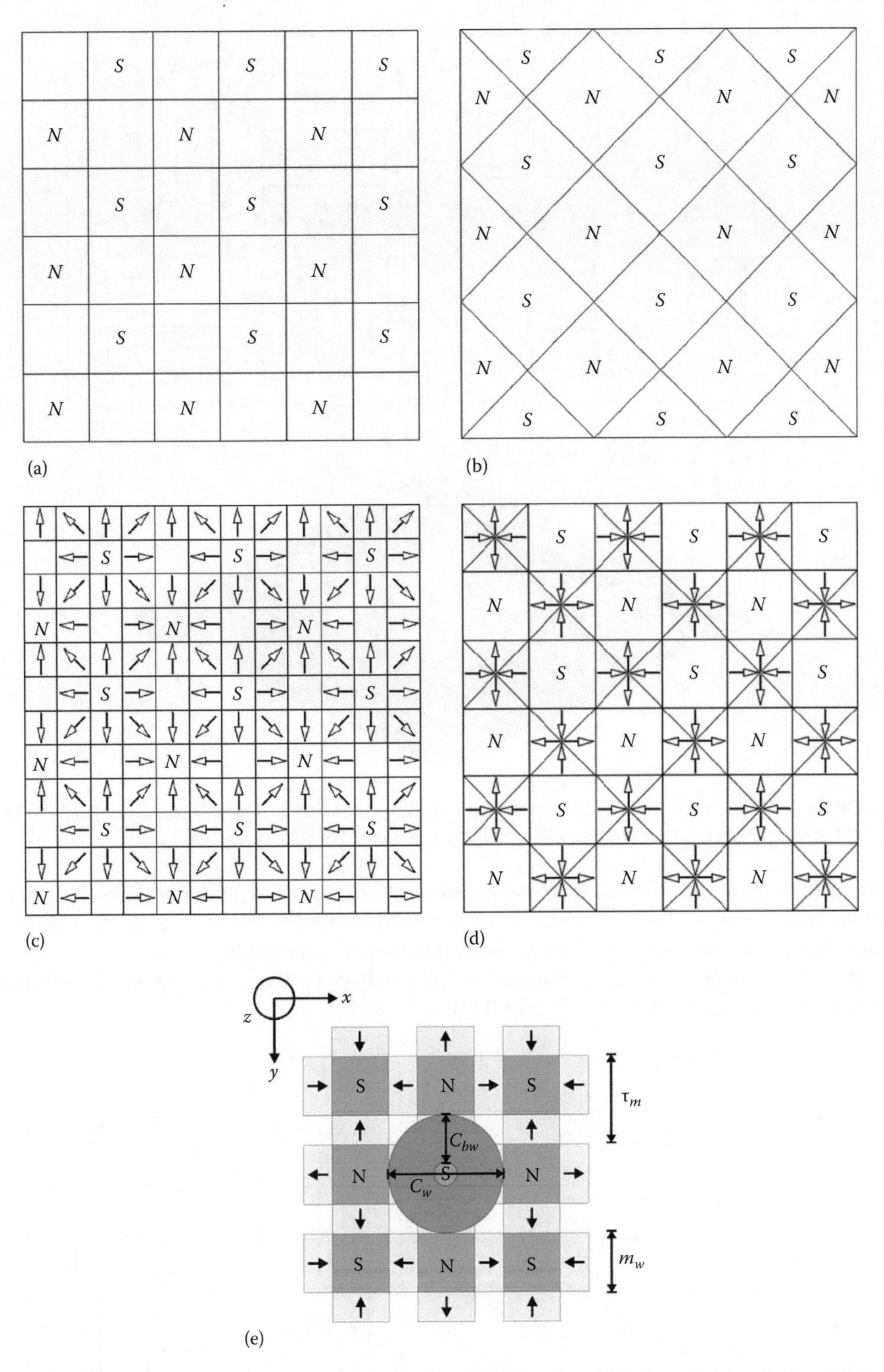

FIGURE 18.3 PM arrays: (a) Asakawa. (After Asakawa, Y., Two dimensional positioning device, US Patent 4626749, December 1986.) (b) Chitayat. (After Chitayat, T., Two axis motor with high density magnetic platen, US Patent 5777402, July 1999.) (c) Trumper. (After Trumper, D.L. et al., Magnetic arrays, US Patent 5631618, May, 1877.) (d) Jung. (After Cho, H.-S. and Jung, H.K., *IEEE Trans. Ener. Conver.*, 17(4), 492, 2002.) (e) Vanderput. (After de Boeij, J. et al., *IEEE Trans.*, MAG-45(6), 1118, 2008.)

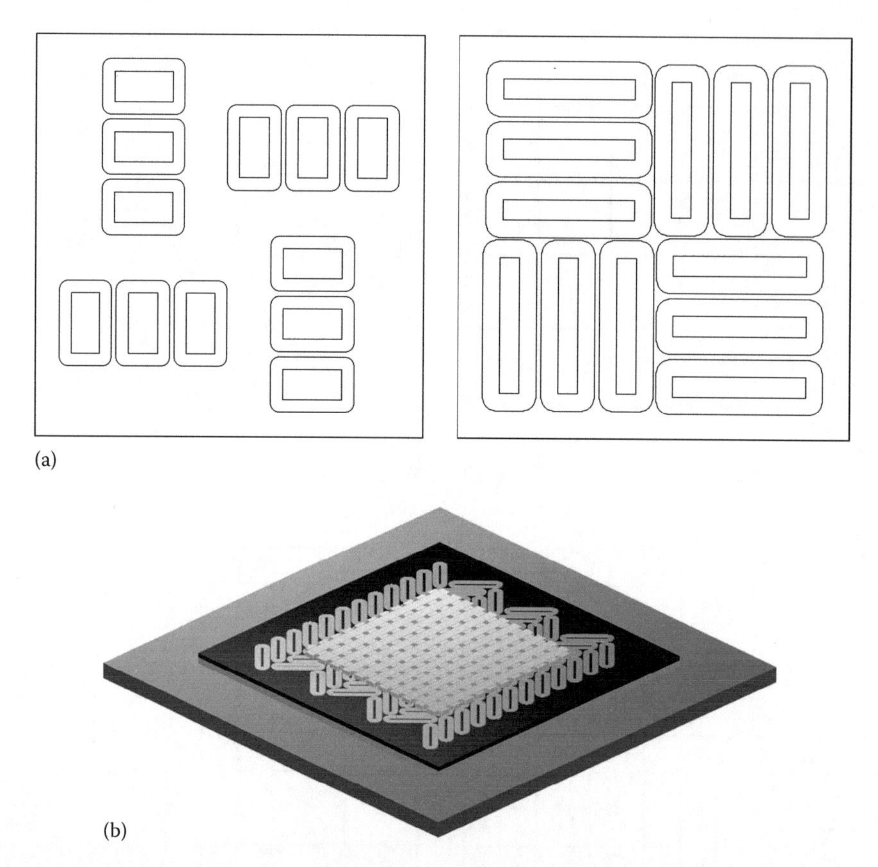

FIGURE 18.4 AC coil arrays: (a) at 90°. (After Cho, H.-S. and Jung, H.K. *IEEE Trans. Ener. Conver.*, 17(4), 492, 2002.) (b) in zigzag. (After Jansen, J.W. et al., *IEEE Trans.*, MAG-43(1), 15, 2007.)

Another dilemma that seemed to have been solved is that between mechanical (or air) bearings and magnetic levitation. Today MAGLEV is preferred due to many advantages, the first being the elimination of "dry friction," which allows better precision in positioning.

For small travel (mm range) x, y, z motion and small φ, Φ, θ control, custom-shaped planar MAGLEV stages may be built [10] (Figure 18.5).

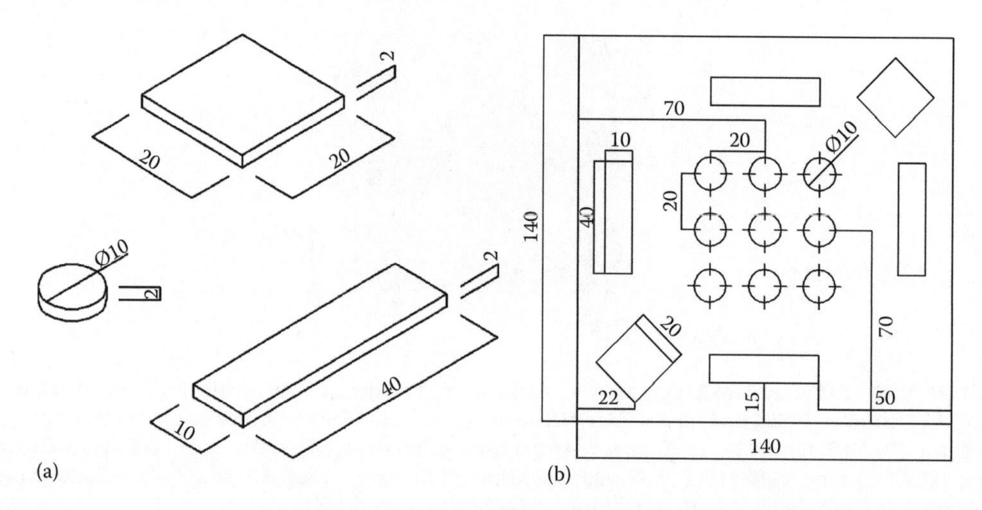

FIGURE 18.5 Small ranges 6 DOF planar MAGLEV: (a) the planar configuration and (b) the circular ac coil.

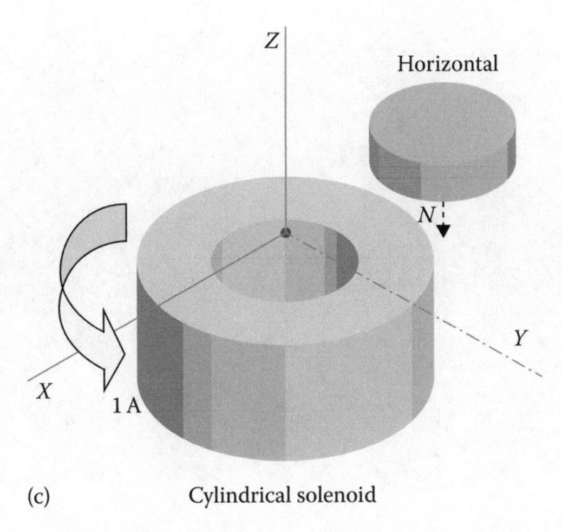

FIGURE 18.5 (continued) Small ranges 6 DOF planar MAGLEV: (c) cylindrical solenoid. (After Lai, Y.-C. et al., *IEEE Trans.*, MAG 43(6), 2600, 2007.)

While all coils in Figure 18.5 are circular and identical, the PMs have three different shapes (circular, rectangular, quadratic) and sizes, to handle better the 3D translational and the 3D rotational small range motions; each coil is controlled through a dedicated single-phase inverter.

18.2 MODELING OF LARGE TRAVEL PLANAR LINEAR PM DRIVES WITH RECTANGULAR AC COILS

Let us consider the PM-mover planar air-core PM motor device (Figure 18.4b) presented here with only one coil in the PM array (Figure 18.6) [8].

The Halbach configuration in Figure 18.6 has the particularity that the coil length (along 0*y*) is not related to the PM pole pitch.

The surface magnetic charge method (Figure 18.6b) [8] leads to the magnetic flux density of the cuboidal PM in the translator coordinates, which has three components B_x^{PM}, B_y^{PM}, B_z^{PM}:

$$\begin{aligned}
B_x^{PM} &= \frac{B_r}{4\pi}\sum_{i=0}^{1}\sum_{j=0}^{1}\sum_{k=0}^{1}(-1)^{i+j+k}\log(R-T) \\
B_y^{PM} &= \frac{B_r}{4\pi}\sum_{i=0}^{1}\sum_{j=0}^{1}\sum_{k=0}^{1}(-1)^{i+j+k}\log(R-S) \\
B_z^{PM} &= \frac{B_r}{4\pi}\sum_{i=0}^{1}\sum_{j=0}^{1}\sum_{k=0}^{1}(-1)^{i+j+k}\text{atan2}\left(\frac{ST}{RU}\right)
\end{aligned} \tag{18.1}$$

with

$$R=\sqrt{S^2+T^2-U^2};\quad S=x^{PM}-q_x^{PM}-(-1)^i a$$

$$T=y^{PM}-q_y^{PM}-(-1)^j\times b;\quad U=z^{PM}-q_z^{PM}-(-1)^k\times c$$

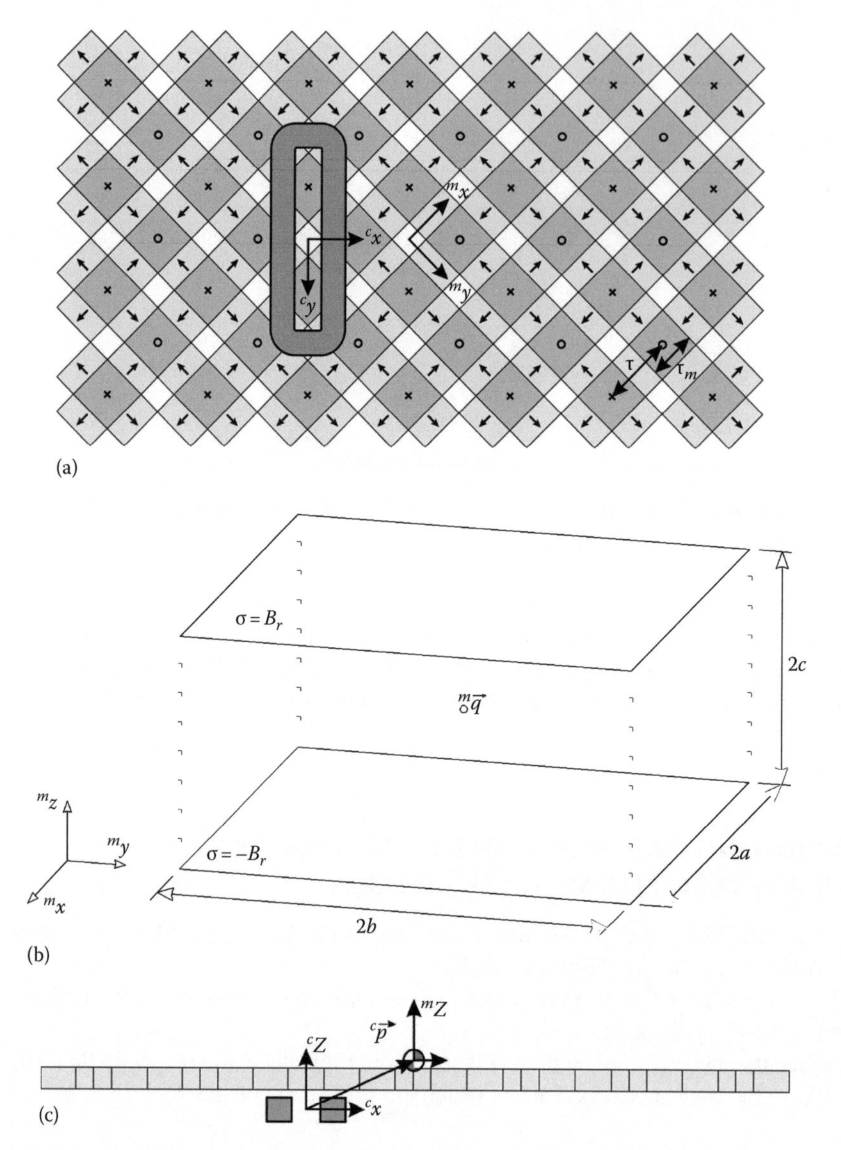

FIGURE 18.6 (a) View from below Halbach PM array and ac coil, (b) side view, and (c) and surface magnetic charge model of PM. (After Jansen, J.W. et al., *IEEE Trans.*, MAG-43(1), 15, 2007.)

and atan(2) is the $\tan^{-1}$ function in 4 quadrants; q_y^{PM}, q_x^{PM}, q_z^{PM} is the location of PM center and x^{PM}, y^{PM}, z^{PM} are the coordinates of the point where B_x^{PM}, B_y^{PM}, B_z^{PM} are calculated in PM coordinates.

For an array of PMs, we have to add each PM contribution, accounting for its specific center position.

The force and the torque can be calculated using the Lorenz force principle. Using the coordinate transformation, the mover can be calculated in a 6 DOF reference with respect to the stator, and thus the force and torque on the PM are [8]

$$\vec{F}^{PM} = -\int_{Vcoil} {}^{m}R_c\, {}^{c}\vec{j}\left({}^{c}\vec{x}\right) \times \vec{B}^{PM}\left({}^{m}R_c\left({}^{c}\vec{x} - {}^{c}\vec{p}\right)\right) d^{c}v \tag{18.2}$$

$$\vec{T}^{PM} = \int_{Vcoil} \left({}^{m}R_c\left({}^{c}\vec{x} - {}^{c}\vec{p}\right)\right) \times \left({}^{m}R_c\, {}^{c}\vec{j}\left({}^{c}\vec{x}\right) \times \vec{B}^{PM}\left({}^{m}R_c\left({}^{c}\vec{x} - {}^{c}\vec{p}\right)\right)\right) d^{c}v \tag{18.3}$$

The coordinates ${}^{c}\vec{x}$ and ${}^{c}\vec{p}$ appear in Figure 18.6a and $\left|{}^{m}R_{c}\right| = \left[{}^{c}R_{m}\right]^{T}$; also,

$$\left|{}^{c}R_{m}\right| = Rot\left({}^{c}y, {}^{m}\theta\right) \cdot Rot\left({}^{c}x, {}^{m}\phi\right) Rot\left({}^{c}z, {}^{m}\psi\right)$$

$$Rot\left({}^{c}y^{m}\theta\right) = \begin{vmatrix} \cos^{m}\theta & 0 & \sin^{m}\theta \\ 0 & 1 & 0 \\ -\sin^{m}\theta & 0 & \cos^{m}\theta \end{vmatrix}$$

$$Rot\left({}^{c}x^{m}\psi\right) = \begin{vmatrix} 1 & 0 & 0 \\ 0 & \cos^{m}\psi & -\sin^{m}\psi \\ 0 & \sin^{m}\psi & \cos^{m}\psi \end{vmatrix} \tag{18.4}$$

$$Rot\left({}^{c}z^{m}\theta\right) = \begin{vmatrix} \cos^{m}\phi & -\sin^{m}\phi & 0 \\ \sin^{m}\phi & \cos^{m}\phi & 0 \\ 0 & 0 & 1 \end{vmatrix}$$

${}^{m}{}_{\theta}$, ${}^{m}{}_{\psi}$, and ${}^{m}{}_{\Phi}$ are the rotations angles about the ${}^{c}x$, ${}^{c}y$, ${}^{c}z$ axes.

The volume integral is calculated over the coil volume, numerically, by using a 3D mesh; in each mesh element, the 3D trapezoidal rule may be used to do the integral.

For 5×5 poles (85 PMs) and 3 coils, the computation time is reduced to 20–30 s in comparison with 40 min required for 3D FEM analysis of the same case [8].

For design optimization, even the surface magnetic charge model may be too time consuming; the harmonic model may be used to further reduce the computation time [8] to, perhaps, 1 s.

Test result comparisons with surface magnetic charge and harmonics model do show good agreements of surface magnetic charge model, while the harmonics model cannot predict the end effects, as it supposes an infinitely long mover (Figure 18.7a and b) [8].

A more refined Halbach PM-mover configuration [9], analyzed with the Fourier series (harmonics) model, has led to optimal vertical/horizontal PM sections τ_m/τ in the Halbach array that further reduces the z component of PM flux density pulsations and increases the vertical flux density for more levitation force (${}^{c}F_{z}$), in comparison with results in [8].

In a further effort to reduce the "cross talk" between neighboring coils for the configuration in Figure 18.8, where groups of three coils are planted at 90° with the PM array of Figure 18.3d [3], by using the harmonics model again, an almost zero emf induced in a y direction coil, when an x direction coil is activated, is obtained by geometrical optimization of coil dimensions [3] (Figure 18.8). This is obtained with sinusoidal symmetric three-phase currents control in the three coil groups.

An elaborated optimization study of planar linear PM motor drives is developed in [4], which includes geometrical optimization and parametric search sequences to reduce computation effort for minimum copper losses for given temperature rise constraints. Only the fundamental from the harmonics model is used, to save time for the optimization task, while the higher harmonics are, anyhow, in the range of manufacturing tolerances.

18.3 PLANAR LINEAR PM MOTOR MICRON POSITIONING CONTROL FOR MILLIMETER RANGE TRAVEL

The configuration in Figure 18.9a and b exhibits

- Stator with four singular rectangular shape coils
- x-y flexure stage
- 2 PM (N-S) movers attached to the flexures

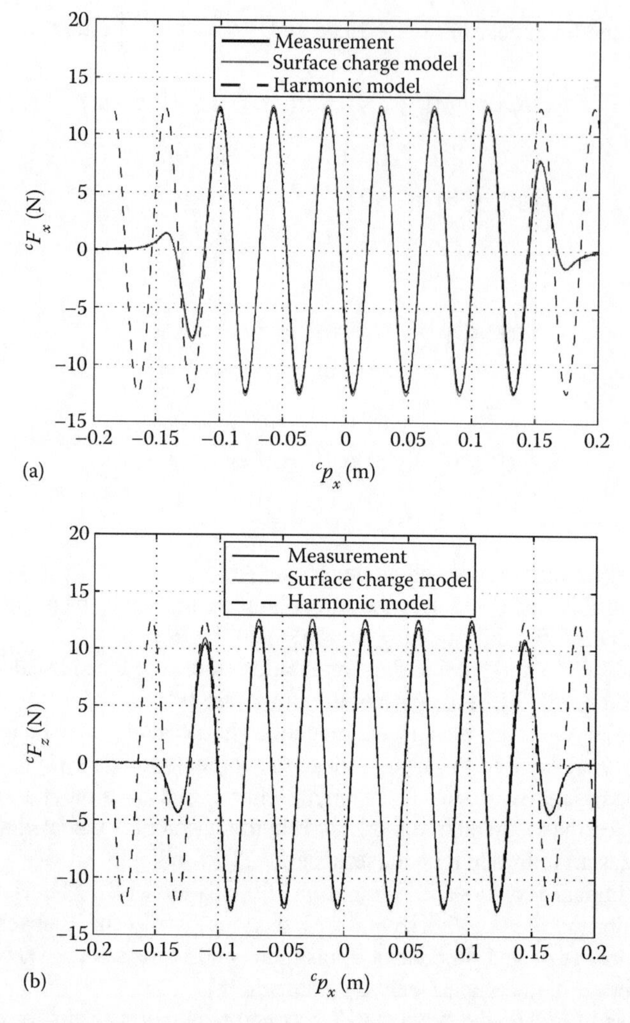

FIGURE 18.7 (a) Distribution of force along x and (b) force along z, at 1 mm airgap. (After Jansen, J.W. et al., *IEEE Trans.*, MAG-43(1), 15, 2007.)

- Optical feedback position sensors
- Airgap, unusually small (50 μm)

The individual coils plus PM pairs operate as voice–coil single-phase actuators because their travel length is much smaller than the PM or coil lengths along the directions of motion x and y.

The guiding mechanism of x y flexure is visible in Figure 18.9b, after [11].

For the case in point, x-y maximum displacement is ±0.5 mm and maximum rotation is 5° [11]. As the motion implies two forces f_x^m, f_y^m and a torque τ^m and we do have four force (thrust) sources $f_{x1}, f_{x2}, f_{x3}, f_{x4}$, some form of allocation is needed to handle this redundancy.

Because the design of the control system is targeted, a simplified model of the system is used.

As the displacements (stroke) along x and y are small, with respect to coil and PM geometry, the Lorentz force formula becomes simple (Figure 18.10):

$$f_{act} = 2NLBi = K_I i \tag{18.5}$$

N is the turns per coil.

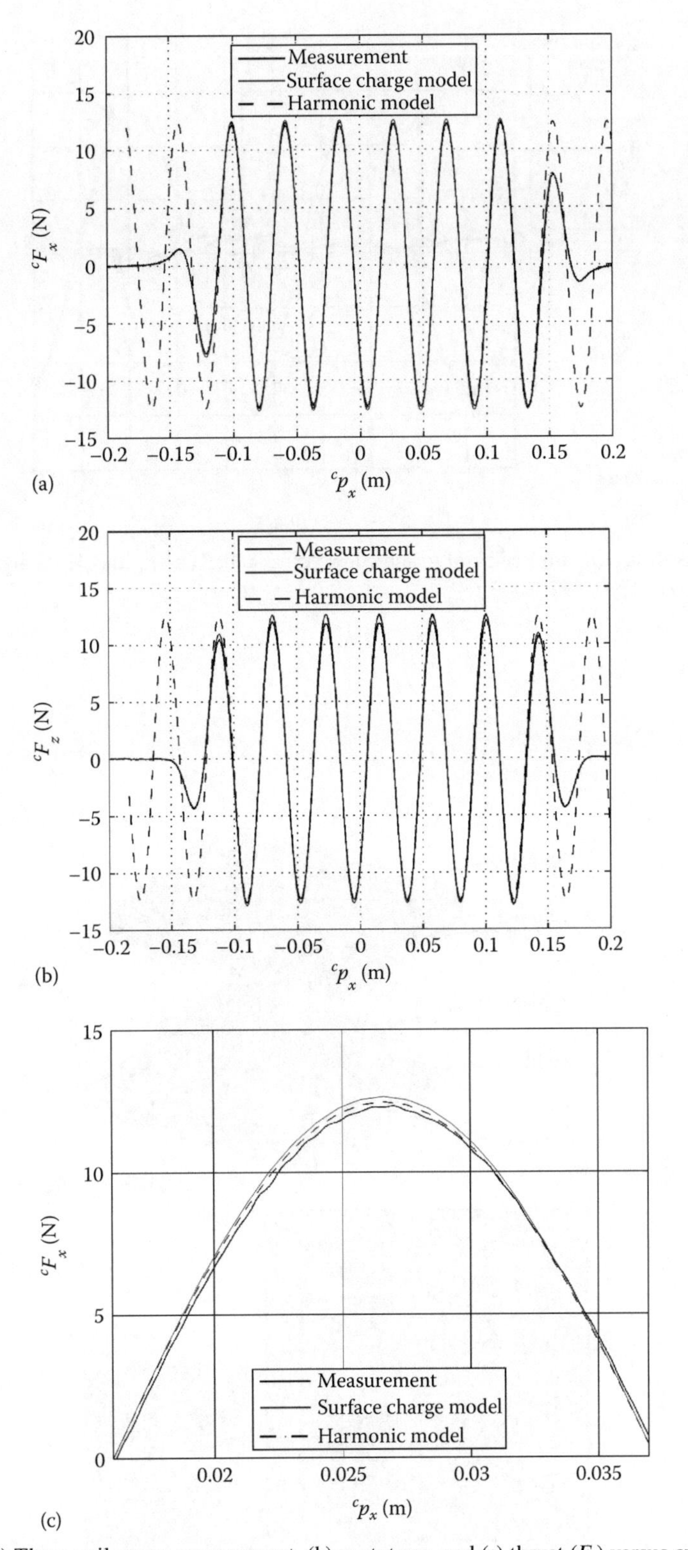

FIGURE 18.8 (a) Three coil group arrangement, (b) prototype, and (c) thrust (F_x) versus current.

(*continued*)

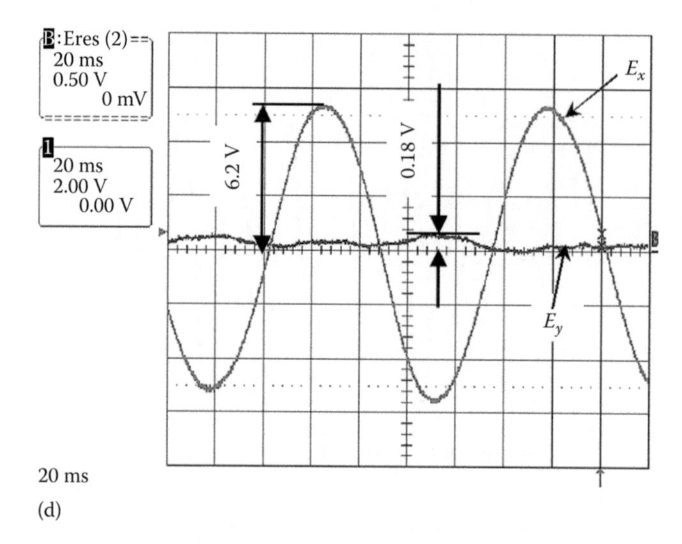

FIGURE 18.8 (continued) (d) Back emf to show low "cross-talk." (After Cho, H.-S. and Jung, H.K., *IEEE Trans. Ener. Conver.*, 17(4), 492, 2002.)

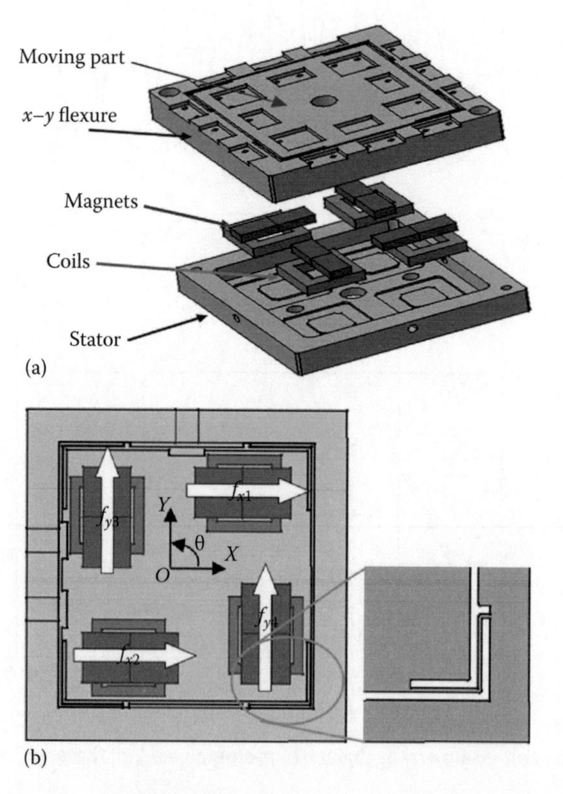

FIGURE 18.9 (a) Four ($2x+2y$) linear PM configuration and (b) with x–y flexure-PM mover. (After Chen, M.-Y. et al., *IEEE Trans.*, MAG 57(1), 96, 2010.)

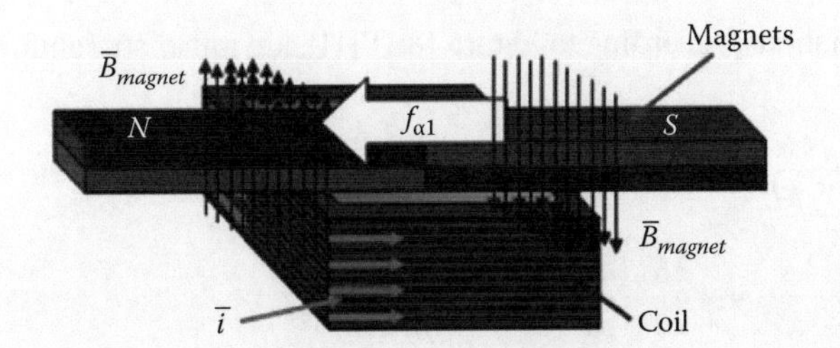

FIGURE 18.10 Voice–coil PM motor. (After Lai, Y.-C. et al., *IEEE Trans.*, MAG 43(6), 2600, 2007.)

Only two sides of the coil (of length L, each) are active. The force allocation for the four single-phase linear voice–coil motors, to produce the required $f_x{}^m$, $f_y{}^m$, τ^m motion, based on Figure 18.11 [11], where the flexure is also modeled, is

$$\begin{vmatrix} f_x^m \\ f_y^m \\ \tau_m \end{vmatrix} = \begin{vmatrix} 1 & 1 & 0 & 0 \\ 0 & 0 & 1 & 1 \\ -d_1 & d_2 & -d_3 & d_y \end{vmatrix} \cdot \begin{vmatrix} f_x^1 \\ f_x^2 \\ f_x^3 \\ f_x^4 \end{vmatrix} \tag{18.6}$$

Due to redundancy, an inverse transformation of (18.6) leads to the force allocation as presented in (18.7) [11]:

$$\begin{vmatrix} f_{x1}^x \\ f_{x2}^x \\ f_{x3}^x \\ f_{x4}^x \end{vmatrix} = \begin{vmatrix} \frac{d_2}{d_1+d_2} & 0 & \frac{-1}{d_1+d_2} \\ \frac{d_1}{d_1+d_2} & 0 & \frac{1}{d_1+d_2} \\ 0 & \frac{d_4}{d_3+d_4} & \frac{-1}{d_3+d_4} \\ 0 & \frac{d_3}{d_3+d_4} & \frac{1}{d_3+d_4} \end{vmatrix} \cdot \begin{vmatrix} f_x^m \\ f_y^m \\ \tau^m \end{vmatrix} \tag{18.7}$$

The flexure (which replaces the linear bearings) is modeled by four springs (K_x, K_y) and four dampers (b_x, b_y) (Figure 18.11).

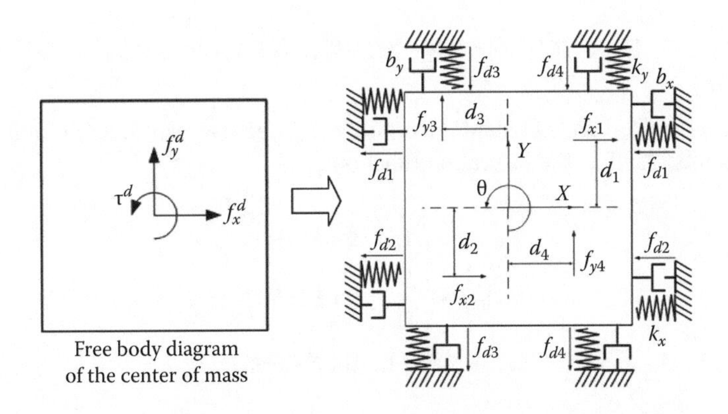

FIGURE 18.11 Forces on the mover. (After Chen, M.-Y. et al., *IEEE Trans.*, MAG 57(1), 96, 2010.)

The equations of motions, according to Figure 18.11 [11], are rather straightforward (with $\sin\theta \approx \theta$):

$$\ddot{x} + \frac{4b_x}{M}\dot{x} + \frac{4K_x}{M}x - \frac{2K_x(d_1 - d_2)}{M}\theta = \frac{K_i}{M}(i_1 + i_2)$$

$$\ddot{y} + \frac{4b_y}{M}\dot{y} + \frac{4K_y}{M}y - \frac{2K_y(d_3 - d_4)}{M}\theta = \frac{K_i}{M}(i_3 + i_4) \tag{18.8}$$

$$\ddot{\theta} + \frac{4b_\theta}{J}\dot{\theta} - \frac{2K_x(d_1^2 - d_2^2) + 2K_y(d_3^2 - d_4^2)}{J}\theta - \frac{2K_x(d_1 - d_2)}{J}x - \frac{2K_y(d_3 - d_4)}{J}y$$

$$= -\frac{K_i}{M}(d_1 i_1 - d_2 i_2 + d_3 i_3 - d_4 i_4)$$

With $|X| = |x, \dot{x}, y, \dot{y}, \theta, \dot{\theta}|$ as state variables in (18.8) and $|Y| = |x, y, \theta|$ as output vector, the state space form of (18.8) is straightforward:

$$|\dot{X}| = |A||X| + |B||u|$$

$$|Y| = |C||X| \tag{18.9}$$

From now on, the control system may be designed in quite a few ways.

From (18.8) it is evident that if $d_1 = d_2$ and $d_3 = d_4$, the motions along x and y are decoupled from rotation and thus a three single-input single-output system is obtained.

In reality, manufacturing inaccuracies lead to coupling effects, and thus disturbance terms should be added.

For sliding mode control, a sliding surface $S = 0$ is added:

$$S = \dot{E} + \lambda E$$

$$\lambda = Diag(\lambda_1, \lambda_2, \lambda_3) \tag{18.10}$$

$$[E] = [x_1 - x_{1d}, y_1 - y_{1d}, \theta_1 - \theta_{1d}]^T$$

x_{1d}, y_{1d}, θ_{1d} are initial values and x_{1d}, y_{1d}, θ_{1d} are the desired values of variables.

If a system parameter online estimator is added, its output is used in the control command U_{AS} [11]:

$$U_{AS} = \hat{B}^{-1}\left[-Ax + \ddot{x}_d - \lambda\dot{E} - \hat{W}_{cons} - KS - N \cdot sat(S)\right] \tag{18.11}$$

with $\hat{B}$ matrix as estimated and constant uncertainty $\hat{W}_{const}$ also as estimated; K and N are diagonal 3rd matrix constants; "Sat" is the saturation function:

$$sat(s) = \begin{cases} 1, & \text{for } S > \varepsilon \\ \dfrac{S}{|\varepsilon|}, & \text{for } \varepsilon \ge S > -\varepsilon \\ -1, & \text{for } S < -\varepsilon \end{cases} \tag{18.12}$$

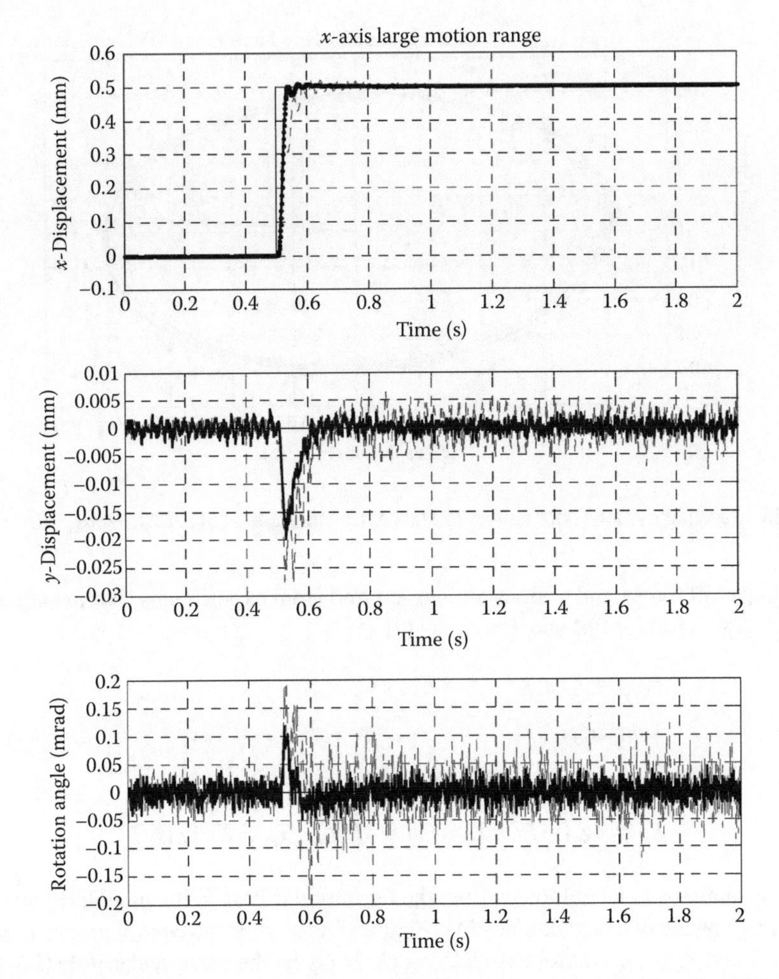

FIGURE 18.12 Large, ±0.5 mm, transient along x: thick line, with parameter estimation; thin line, without parameter estimation. (After Chen, M.-Y. et al., *IEEE Trans.*, MAG 57(1), 96, 2010.)

The Lyapunov stability method may be considered to yield λ and K and N diagonal matrix constants λ_1, λ_2, λ_3, $K1$, K_2, K_3, N_1, N_2, N_3.

A ±0.5 mm transient along axis x is presented in Figure 18.12 [11].

The error is below 5 μm with parameter adaptation (thick lines), which is quite satisfactory.

Circular motion (±0.2 mm at 0.25 Hz) illustrated in Figure 18.13 [12] shows an rms error of about 20 μm (thick line).

With a better resolution position reader, even a ±150 nm steady-state error was proved [11].

These results show clearly the robustness of adaptive sliding mode control for planar motion of 1 mm range amplitude with submicron steady-state positioning precision.

18.4 SIX DOF CONTROL OF A MAGLEV STAGE

Similarly good results for x, y, z, (levitation) control with dedicated voice–coil linear motors are shown in Figure 18.14 [10] for each motion of the 6 DOF.

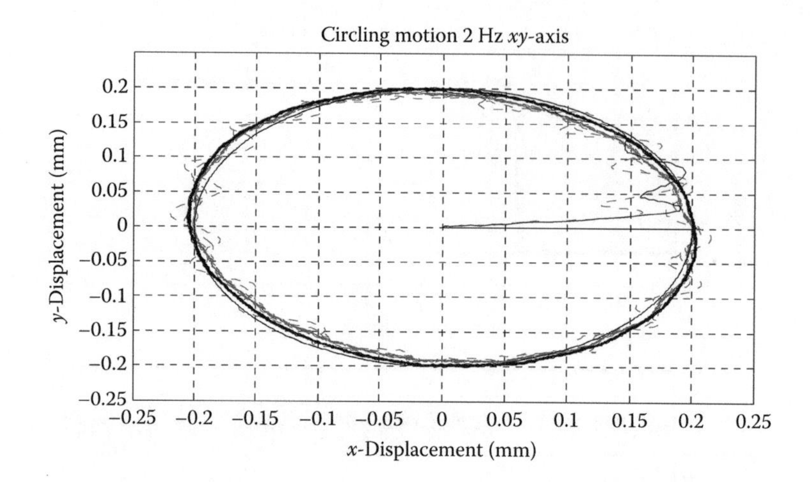

FIGURE 18.13 Cycling motion: ±0.2 mm at 25 Hz. (After Verma, S. et al., *IEEE Trans.*, MAG-41(5), 1159, 2005.)

For simplicity, all forces and torques are curve fitted from experiments. For example, the levitation force $F_{air,z}$, of a circular PM and I_z coil, is [10]

$$\begin{aligned} &F_{air,z} = f_z(z)I_z; \quad f_z \approx C_1 z + C_2 \\ &\textit{Similarly:} \\ &F_z = 5f_z(z)I_z, \quad F_x = 5f_r(r)I_z \\ &F_{rr} = 2g_r(r)I_r, \quad T_{ez} = 2h_r(r)I_{rz}I_z, \quad T_{er} = 2f_z(z)I_{rr}I_z \end{aligned} \tag{18.13}$$

F_z and F_x are levitation and lateral forces from the five circular PMs in the middle (Figure 18.14a); F_{rr} is the lateral force by the two rectangular PMs (Figure 18.14b), T_{ez} is the control torque along the *z*-axis (Figure 18.14c), and T_{er} is the control torque along *Ox* or *Oy* by the two circular PMs (Figure 18.14d).

The overall model equations are now straightforward (Figure 18.14) [10]:

$$\begin{aligned} M\ddot{x} &= F_r\cos\theta - F_{xr} \\ M\ddot{y} &= F_r\sin\theta - F_{ry} \\ M\ddot{z} &= F_z - mg \\ J_{xx}\ddot{\varphi} + \dot{\phi}\dot{\theta}\left(J_{zz} - J_{yy}\right) &= m_z B_y - T_{ex} \\ J_{yy}\ddot{\phi} + \dot{\varphi}\dot{\theta}\left(J_{xx} - J_{zz}\right) &= m_z B_x - T_{ey} \\ J_{zz}\ddot{\theta} + \dot{\varphi}\dot{\phi}\left(J_{yy} - J_{xx}\right) &= \left[\left(F_r\sin\theta - F_{ry}\right)x - \left(F_r\cos\theta - F_{rx}\right)y\right] - T_{ez} \end{aligned} \tag{18.14}$$

M is the total mass carrier, and J_{xx}, J_{yy}, J_{zz} are the inertias. B_y, B_x are the flux densities; $r = \sqrt{x^2 + y^2}$; $J_{xy} = J_{yz} = J_{zx} = 0$. Around steady-state points, we may also assume $\dot{\theta} = \dot{\varphi} = \dot{\phi} = 0$, which will further simplify the model (18.14).

The voice–coil actuator current transients have been neglected in comparison with the mechanical transients; that is, current control is very fast; this is quite practical in an air-core machine.

Typical results on a prototype, with only PID control on all axes (Table 18.1), are shown in Figure 18.15 [10].

Satisfactory results are visible in Figure 18.15, but a more robust control should improve the performance further.

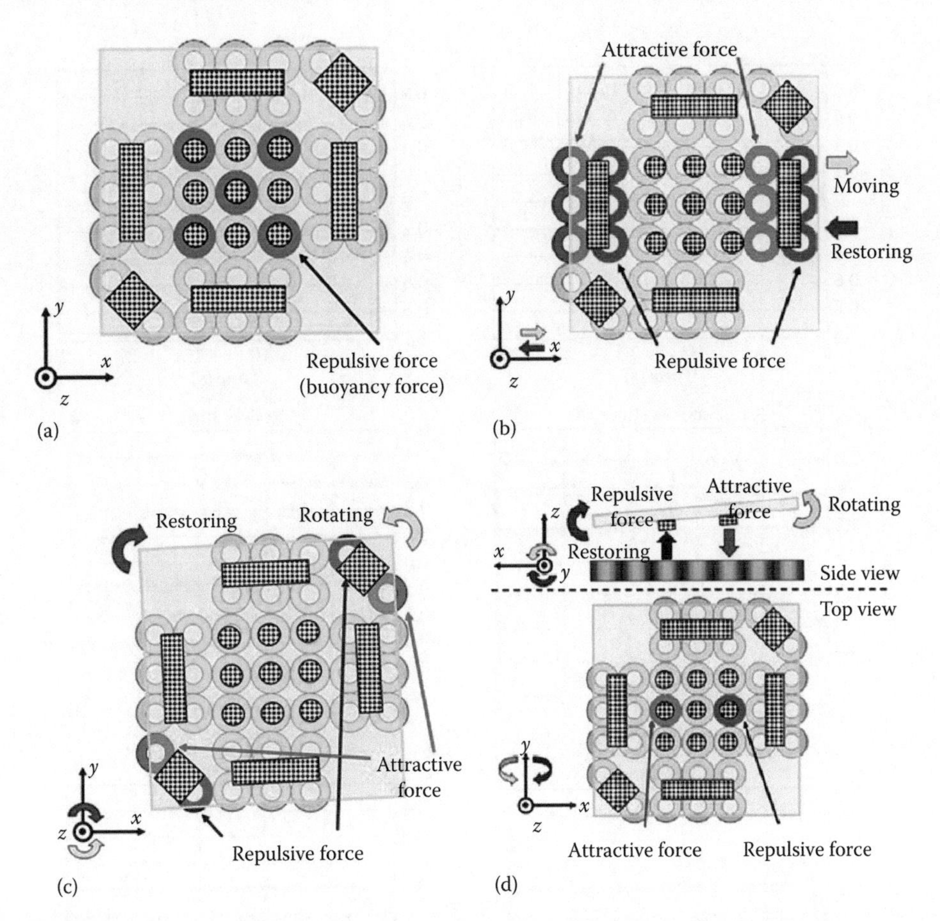

FIGURE 18.14 Six DOF driver of planar MAGLEV stage: (a) z direction (levitation), (b) x direction, (c) z-rotation, and (d) y-axis rotation. (After Lai, Y.-C. et al., *IEEE Trans.*, MAG 43(6), 2600, 2007.)

TABLE 18.1
Experimental PID Parameters

	K_P	K_I	K_D
x	0.073	0	0.0032
y	0.073	0	0.0032
z	0.01	0.16	0.00074
φ	0.075	0	0.04
ϕ	0.075	0	0.04
θ	0.1	0	0.003

18.5 MULTIAXIS NANOMETER-POSITIONING MAGLEV STAGE

A nanometer-positioning MAGLEV stage system for photolithography or semiconductor manufacturing, microscopic scanning, or fabrication of nanostructures and microscale rapid prototyping may be built with only six voice–coil PM motors (Figure 18.16) [12]: three vertical (with one PM mover) and three horizontal (with two PM movers).

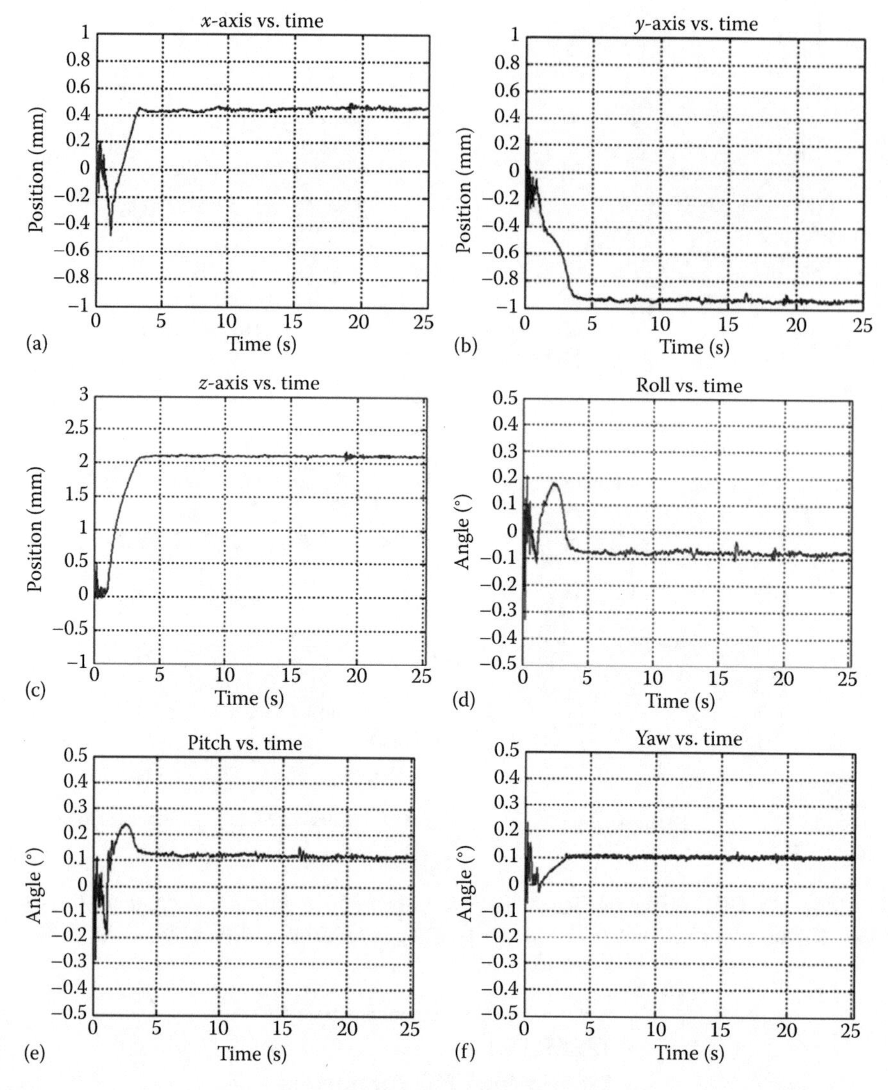

FIGURE 18.15 Experimental results: (a) x direction, (b) y direction, (c) z direction (levitation), (d) x rotation, (e) y rotation, and (f) z rotation. (After Lai, Y.-C. et al., *IEEE Trans.*, MAG 43(6), 2600, 2007.)

The z- and x-axis (axial) forces of individual voice air-coil actuators are now independent and stem from the Lorenz force equation [12]:

$$F_z(z) = \int (\bar{j} \times \bar{B}) dv = \left(\frac{J\mu_0 M}{4\pi}\right) \cdot \int_{c-\frac{w}{2}}^{c+\frac{w}{2}} \int_{z-\frac{h}{2}}^{z+\frac{h}{2}} \int_0^{2\pi} \left[\int_0^R \int_0^{2\pi} \frac{\rho d\theta d\rho}{\sqrt{\left(z-\frac{d}{2}\right)^2 + r^2 + \rho^2 - 2r\rho\cos(\theta-\phi)}} - \int_0^R \int_0^{2\pi} \frac{\rho d\theta d\rho}{\sqrt{\left(z+\frac{d}{2}\right)^2 + r^2 + \rho^2 - 2r\rho\cos(\theta-\phi)}} \right] \quad (18.15)$$

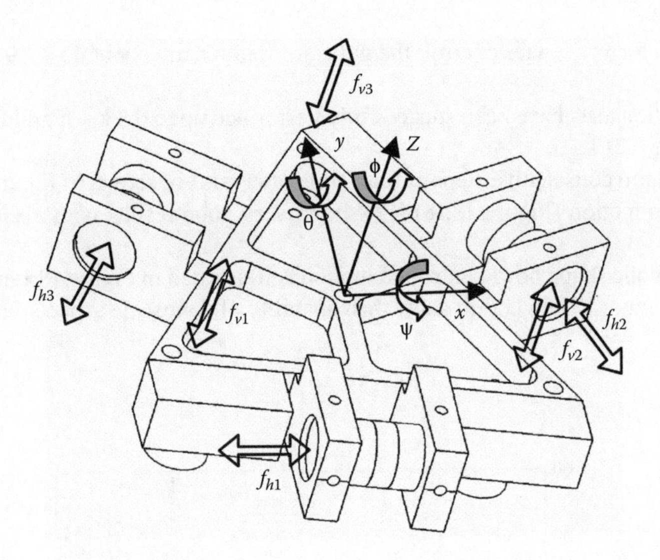

FIGURE 18.16 Six DOF MAGLEV for nanopositioning. (After Verma, S. et al., *IEEE Trans.*, MAG-41(5), 1159, 2005.)

where
- R is the radius
- d the magnet height
- w the coil width
- c the average radius of coil
- z the height of coil center

r, Φ, z represent the coordinates of the small volume element of coil; ρ, θ are coordinates of the small surface element of the magnet at the top and bottom surfaces in cylindrical coordinates.

As the platen is magnetically levitated, it may be modeled by a pure mass M and thus

$$M\frac{d^2x}{dt^2} = F; \quad F\text{: modal force} \tag{18.16}$$

So,

$$\frac{x(s)}{F(s)} = \frac{1}{Ms^2} \tag{18.17}$$

Similarly for platen rotation,

$$\frac{\theta(s)}{T_e(s)} = \frac{1}{Js^2} \tag{18.18}$$

Let us consider $M=0.21$ kg and J_{xx}, J_{yy}, J_{zz}: 133, 122, $236\cdot10^{-6}$ kg·m² [12].

The control system is designed with a lead-lag filter with a damping ratio $\xi=0.7$, phase margin $\gamma=50°$, and crossover frequency $f_c=48$ Hz [12]:

$$G(s) = \frac{K(s+130)(s+8)}{s(s+1130)} \tag{18.19}$$

$K = 6.8 \cdot 10^{-4}$ N/m, for x, y, z axes control; the gains for rotation are about 42.4, 39.05, 75.33 N for Ψ, θ, Φ, respectively.

Load perturbation tests have been successfully performed up to 0.3 kg of additional mass (above the platen mass of 0.21 kg).

In addition, 15 nm consecutive steps at 0.5 s time intervals are shown in Figure 18.17 [12].

Circular 50 nm motion (Figure 18.18) [12] shows very good results with a ±10 nm steady-state error.

Finally, a parabolic shape bowl controlled motion is illustrated in Figure 18.19a and 3D impeller-shape motion in Figure 18.19b [12] to show the capability of nanoscale aspects fabrication.

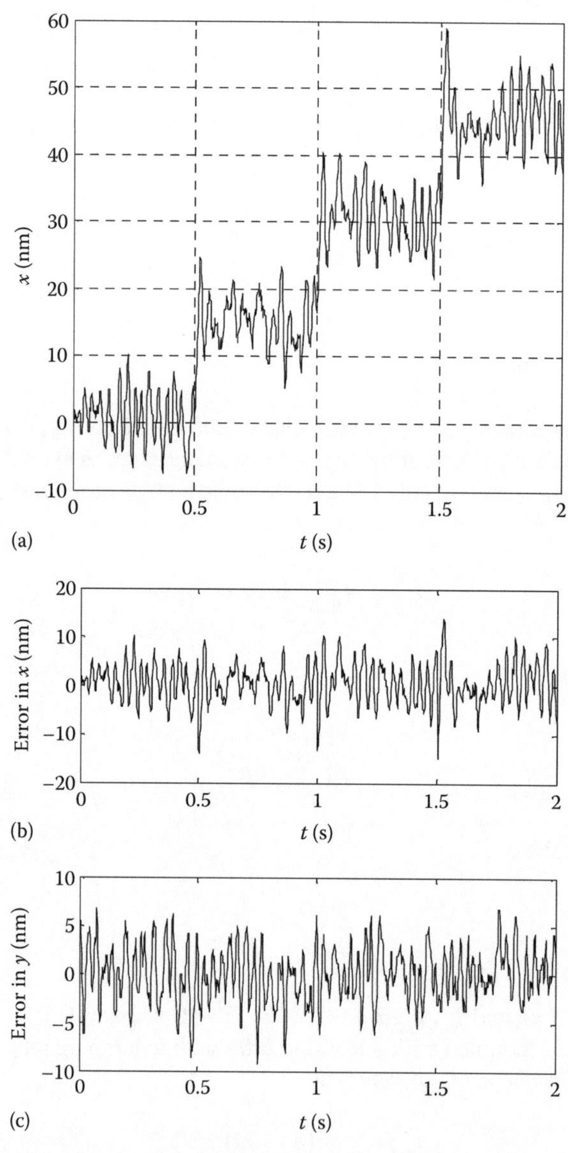

FIGURE 18.17 (a) 15 nm steps response, (b) error in x, and (c) and error in y. (After Verma, S. et al., *IEEE Trans.*, MAG-41(5), 1159, 2005.)

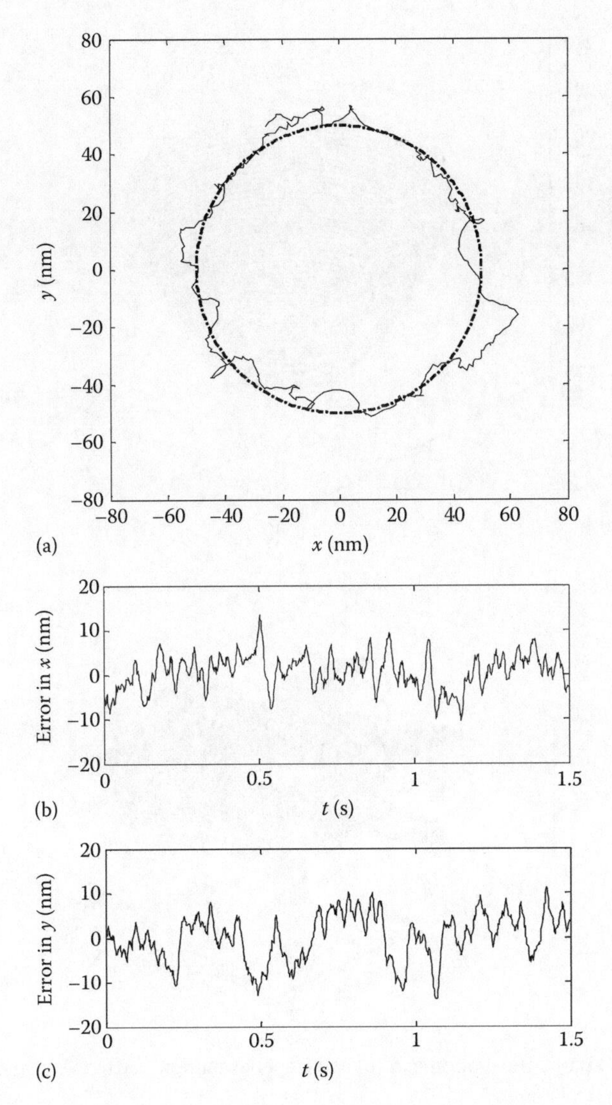

FIGURE 18.18 (a) Circular 50 nm motion, (b) with x, and (c) y errors. (After Verma, S. et al., *IEEE Trans.*, MAG-41(5), 1159, 2005.)

18.6 SUMMARY

- Multiaxis motion is required for many industrial applications such as photolithography, semiconductor wafer manufacturing, macroscopic scanning, fabrication of microstructures, and microscale rapid prototyping.
- 2 DOF planar-motion applications with long travel (tens and hundreds of mm) may be approached with special multiphase flat linear PM motor drives, with magnetic levitation to eliminate dry friction and backlash of systems using rotary actuators and rotary to linear transmissions.
- Fixed ac coil primary with primary mover configurations are preferred as they eliminate the ac power transmission to primary (even in contactless realization).
- The PM array makes use of various Halbach originated configurations to increase levitation and propulsion forces per watt of copper losses in air-core coil primaries.

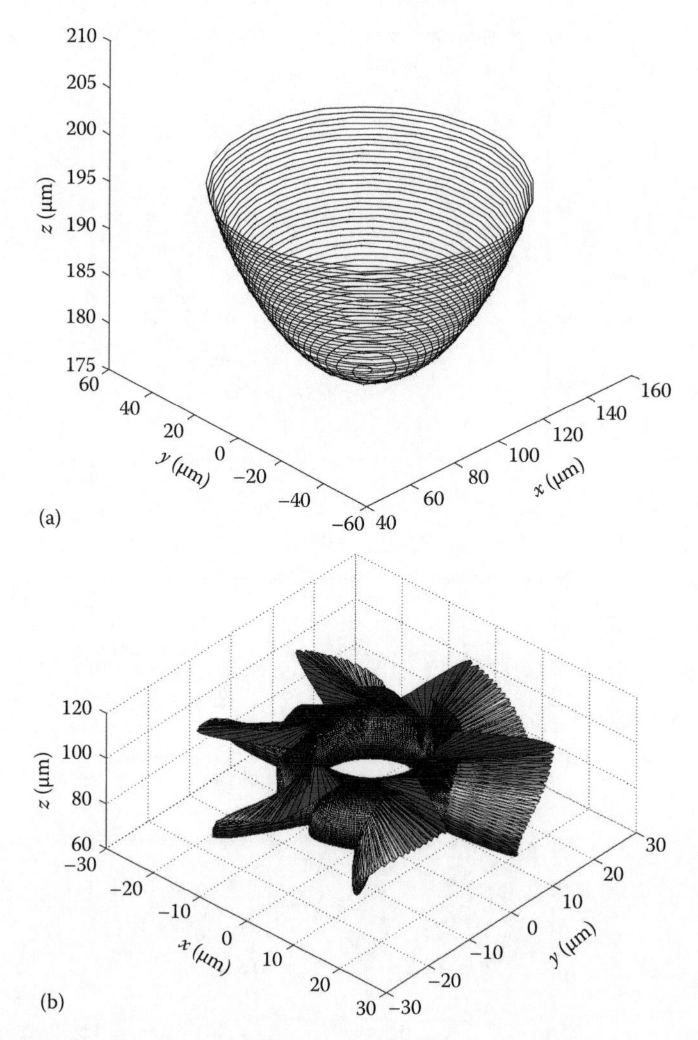

FIGURE 18.19 (a) 3D parabolic motion and (b) and impeller motion. (After Verma, S. et al., *IEEE Trans.*, MAG-41(5), 1159, 2005.)

- The 3(2) phase ac coil primaries may be built with three coil units or multiple all-through nonoverlapping three-phase coils along *Ox* and *Oy* axes.
- For short travel (mm range) 2 DOF or 6 DOF motion, voice–coil (single-phase) PM actuators are used; each is controlled by its own inverter.
- For electromagnetic modeling, besides 3D FEM, which is computer time consuming, the surface magnetic charge model and the harmonics (Fourier series) models have been introduced. The surface magnetic charge model is used to reproduce correctly even the end effects—due to limited number of coil and PMs—while it reduces the computation time by 100 times in comparison with 3D analysis.
- Though incapable of modeling the end effects, the harmonics model (even the fundamental one) is used for optimal design and for control system design due to its low computation time.
- For planar motion of small range, the combination of mechanical flexures with PM array mover to provide bearing function at 50 μm airgap has been proven very practical when the singular rectangular coil interacts with 2 pole PM mover, along *Ox* and *Oy* units.

Micrometer-range position precision has been demonstrated with adaptive sliding mode control, both for unidirectional and circular motions.

- For similar (mm range) planar-motion MAGLEV stages, similar configurations have been proved to yield very good results, with only PID control along all axes (x, y, x, Ψ, θ, Φ).
- A nanometer precision positioning multiaxis MAGLEV stage made with 3+3 single-coil PM linear motor drives has also been built and tested successfully [12]. Its capability of x, y, z and 6 DOF motion in the submillimeter excursion length with nanometer precision makes it suitable for nanostructures fabrication.
- With only lead-lag filter control, better position sensors, and robust control, even better performance may be forecasted for such systems.
- Strong future R&D and industrial efforts are expected in multiaxis linear PM motor drives in the near future due to their potential, especially as MAGLEV stages, to enhance productivity in high-precision positioning fabrication.

REFERENCES

1. M.A. da Silveira, A.F.F. Fihlo, and R.P, Homrich, Evaluation of the normal force of a planar actuator, *IEEE Trans.*, MAG-41(10), 2005, 4006–4008.
2. J. Cao, Y. Zhu, J. Wang, W. Yin, and G. Duan, A novel synchronous permanent magnet planar motor and its model for control applications, *IEEE Trans.*, MAG-41(6), 2005, 2156–2163.
3. H.-S. Cho and H.K. Jung, Analysis and design of synchronous PM planar motors, *IEEE Trans. Ener. Conver.*, 17(4), 2002, 492–499.
4. J. de Boeij, E.A. Lomonova, IEEE Trans. vol. MAG, and A.J.A. Vandenput, Optimization of contactless planar actuator with manipulator, *IEEE Trans.*, MAG-45(6), 2008, 1118–1121.
5. Y. Asakawa, Two dimensional positioning device, US Patent 4626749, December 1986.
6. T. Chitayat, Two axis motor with high density magnetic platen, US Patent 5777402, July 1999.
7. D.L. Trumper, M.J. Kim, and M.E. Williams, Magnetic arrays, US Patent 5631618, May 1877.
8. J.W. Jansen, C.M.M. van Lierop, E.A. Lomonova, and A.J.A. Vandenput, Modeling of magnetically levitated planar actuators with moving magnets, *IEEE Trans.*, MAG-43(1), 2007, 15–25.
9. W. Min, M. Zhang, Y. Zhu, B. Chen, G. Duan, J. Hu, and W. Yin, Analysis and optimization of a new 2D magnet array for planar motor, *IEEE Trans.*, MAG 46(5), 2010, 1167–1171.
10. Y.-C. Lai, I.-L. Lee, and J.-Y. Yen, Design and servo control of a single-deck planar MAGLEV stage, *IEEE Trans.*, MAG 43(6), 2007, 2600–2602.
11. M.-Y. Chen, H.-H. Huang, and S.-K. Hung, A new design of a submicropositioner utilizing electromagnetic actuators and flexure mechanism, *IEEE Trans.*, MAG 57(1), 2010, 96–105.
12. S. Verma, W.-J. Kim, and W.-J. Shakir, Multiaxis MAGLEV nanopositioner for precision manufacturing and manipulation applications, *IEEE Trans.*, MAG-41(5), 2005, 1159–1161.

19 Attraction Force (Electromagnetic) Levitation Systems

Attraction levitation systems (ALS) use attraction current–controlled electromagnetic force to control the airgap between fix and mobile mild magnetic cores. Applications for ALS range from magnetic bearings and vibrating tables to magnetically levitated vehicles (MAGLEVs) [1–3].

For MAGLEVs, the primary is placed on board and contains permanent magnet (PM)-less or PM-assisted dc controlled current electromagnets (solenoids) and a solid mild iron fix secondary (track).

On the other hand, for active magnetic bearings and for vibrating tables, the secondary is the mover and is made, in general, from a laminated mild steel core or of a soft magnetic composite (SMC).

In this chapter, we will concentrate on ALS for MAGLEVs, as active magnetic bearing (or bearing-less rotary electric motor) is a field of R&D in itself that deserves a separate treatment.

The following main aspects of ALS for MAGLEVs are treated in what follows:

- Competitive topologies
- Simplified analytical model (with no end effect) for design
- End effect analytical modeling
- ALS design methodology and sample performance for MAGLEVs
- ALS circuit model for dynamics and control
- Open-loop transfer functions of the linearized system
- State-space feedback control
- Linear parameter varying control
- Sliding mode control
- Zero power control
- Control system performance by example
- Collision avoidance
- Average control power
- Ride comfort

19.1 COMPETITIVE TOPOLOGIES

The most obvious topology of ALS for MAGLEVs comprises *C*-core (laminated) electromagnets on board of vehicle and *C*-core (made of solid mild steel) of secondary (track)—Figure 19.1.

The two *C*-cores of same width L_w allow for a self-centering guidance force when the *C*-core electromagnets are off-centric; this effect may help the guidance system, which may also be performed with *C*-core electromagnets.

PMs may be added to the dc controlled electromagnets either in the *C*-core (Figure 19.2a) or in an *E*-core configuration with dual airgap (Figure 19.2b). Also, the double *C*-core structure leads to lower eddy currents in the solid iron secondary during vehicle motion and controlled dynamics.

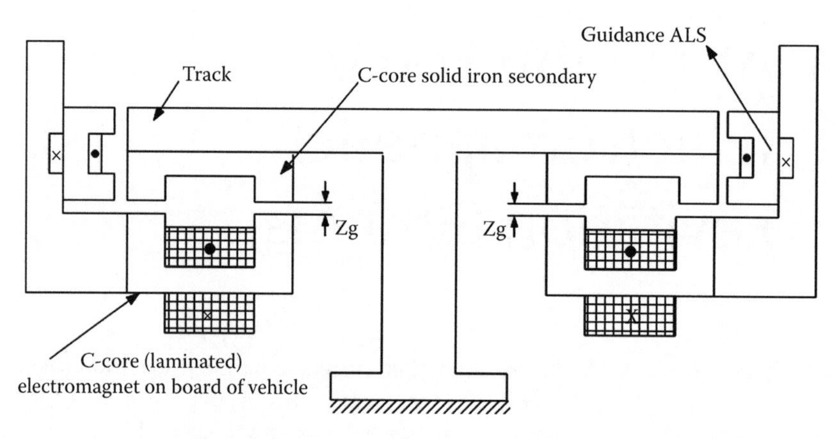

FIGURE 19.1 *C*-core ALS for MAGLEVs.

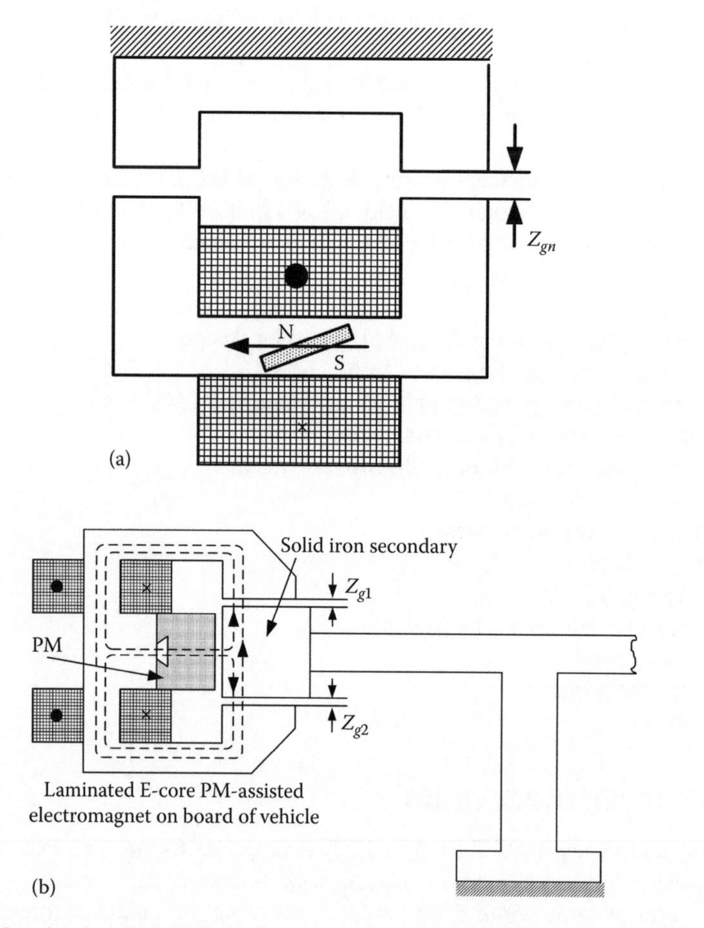

FIGURE 19.2 PM-assisted typical ALS: (a) *C*-core electromagnet and (b) *E*-core dual electromagnet.

The PM may be designed to provide for rated airgap Z_{gn} the entire weight of the vehicle (loaded), with the coil current control used for dynamics and control purposes.

In the *E*-core topology (Figure 19.2), the two twin coils may be potentially controlled together for levitation control only, or separately if both levitation and guidance control is necessary. Rotated by 90°, the configuration in Figure 19.2b may be used for axial active magnetic bearings also.

As the cost of PMs tends to be high and they experience large (or fast) magnetic flux variations, notable eddy currents are induced in them, apparently their use for MAGLEVs may not be the first choice, despite a drastic reduction in control power.

The solid iron secondary core, chosen for economic reasons, leads to eddy currents induced in it due to

- Magnetic flux variation during airgap dynamic control (control eddy currents)
- Longitudinal end effect at medium and high speeds of vehicle (end effect eddy currents)

In general, however, for MAGLEVs, the frequency band of ALS is less than 25 Hz. Thus, the control eddy currents should not be very large, but, in active magnetic bearings, the frequency band is notably larger and, thus, they should be accounted for.

On the other hand, the end effect eddy currents are related to the limited length of dc electromagnets traveling at a speed U with respect to the solid core secondary track [4–7].

It is a kind of dc linear eddy current brake in the sense that eddy currents are induced at the entry and exit ends; they decay over a certain length. If the speed is large, the decay length may be a good part of the electromagnet length and thus notable flux density and normal (attraction) force F_n reduction occurs. The eddy currents induced by motion also produce a drag force F_d; the drag force has to be accounted for in the propulsion system design for high speeds.

We will first develop a simplified analytical field model without eddy currents and then treat the latter in a subsequent paragraph.

19.2 SIMPLIFIED ANALYTICAL MODEL

As the core width b_i=40–60 (80) mm and the airgap Z_g in the MAGLEVs vary from 15 (20) to 8 (10) mm, respectively, in general in interurban and urban applications, the fringing flux (Figure 19.3) has to be considered.

Approximately, the distribution of airgap flux density in the airgap $B(z)$ varies as

$$B(y) = B_0; \quad \text{for } 0 \le |y| \le \frac{b_i}{2} \tag{19.1}$$

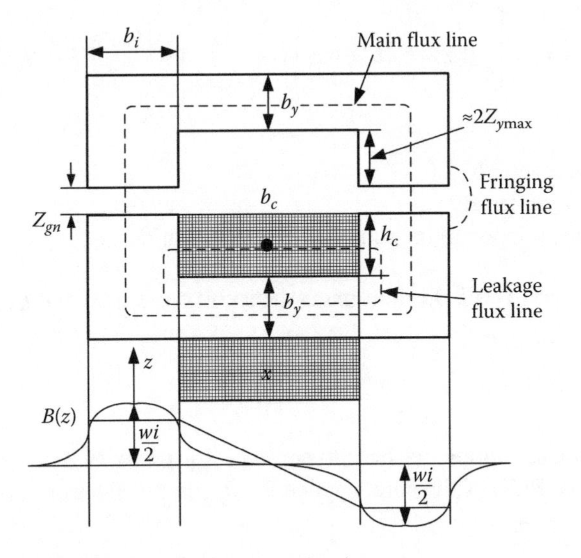

FIGURE 19.3 Airgap distribution along 0y (transverse) axis.

$$B(y) = B_0 \cdot \exp\left(\frac{-2\left(|z| - \frac{b_i}{2}\right)}{g_0}\right); \quad \text{for } |y| \geq \frac{b_i}{2}$$

So the average track core flux density B_c is

$$B_c \cong B_0\left(1 + \frac{2g_0}{b_i}\right); \quad B_0 = \frac{\mu_0 Wi}{2g_0\left(1 + K_s\left(B_{av}\right)\right)} \tag{19.2}$$

where g_0 is the rated airgap.

In the electromagnet C core, there is also an additional leakage flux density whose average value in the core is B_{av}:

$$B_{av} \approx \frac{\mu_0 Wi}{2b_c}; \quad b_c >> 2g \tag{19.3}$$

Only approximately, and because $b_c \gg 2g$, we may consider that the average iron flux density in the electromagnetic core B_{iav} is (if $b_i \approx b_{y1} = b_{y2}$)

$$B_{iav} \approx B_c + B_{av} \tag{19.4}$$

This simplified expression could serve in sizing the core thickness (or width $b_i = b_{y1} = b_{y2}$).

The attraction levitation force F_{L0} is

$$F_{L0} = \frac{2LB_0^2\left(1 + \frac{2g_0}{b_i}\right)^2 \cdot b_i}{2\mu_0} \tag{19.5}$$

In addition, the electromagnet weight G_0 yields

$$G_0 = 2L\left[\left(2\left(b_i + h_c\right)b_i + b_c b_i\right)\gamma_i + \frac{Wi}{J_{con}}\left(1 + \frac{(b_i + h_c)\pi}{2L}\right)\gamma_{co}\right] \tag{19.6}$$

where

L is the electromagnet length

J_{co} is the design current density

γ_i, γ_{co} are the iron and copper (aluminum sheet) mass densities

With a filling factor $K_{fill} = 0.6–0.7$, the aluminum-sheet coil cross section $b_c h_c$ is

$$b_c h_c = \frac{Wi}{J_{con} K_{fill}} \tag{19.7}$$

To account for magnetic saturation, we have to consider Equations 19.2 through 19.4 together with the magnetization curve $B(H)$ of the core, applied for B_{iav} in the electromagnet and B_c in the core of the track.

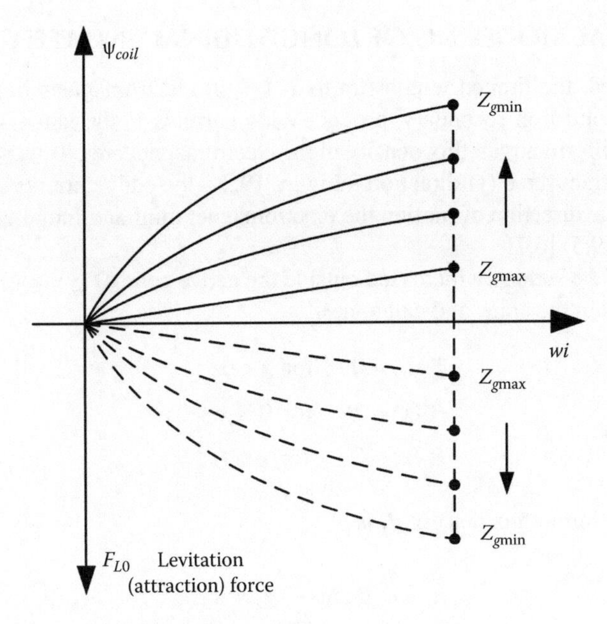

FIGURE 19.4 Coil flux linkage $\Psi(i, g)$ and attraction force F_{L0} (i, g).

Alternatively, a refined magnetic circuit may be developed for the scope. Typical curves of $\Psi_c(i, Z_g)$ may be obtained (as done for solenoids), together with F_{L0} (i, Z_g)—Figure 19.4.

Magnetic saturation influence is small at higher airgap (say at standstill airgap: $Z_{g\max} = 2Z_{g\ rated}$), but it may be important at smaller airgaps (say $0.5Z_{g\ rated}$). In general, during dynamic control of airgap, the latter varies by ±20%–25%, as a compromise between energy consumption and collision avoidance with the track.

The Ψ_c (i,g) and F_{L0} (i,g) curve families may be better calculated by finite element method (FEM); 2D (or even 3D) FEM is required for the scope.

Curve fitting Ψc (i,g) and F_{L0} (i,g) functions by simplified polynomial (or exponential) functions will provide reliable data for dynamics and control studies.

To avoid direct touch between the ALS and the track, a thin nonmagnetic–nonconducting coating is stuck to the electromagnet surface.

A sample approximation is

$$\Psi_c \approx \frac{ci - di^2}{aZ_g + b} + L_l i; \quad L_l\text{—leakage inductance} \tag{19.8}$$

$$F_{L_0} = K_f \cdot \left(\frac{ci - di^2}{aZ_g + b}\right)^2 \tag{19.9}$$

The voltage and motion equations are added to complete the model:

$$V = Ri + \frac{d\Psi_c}{dt}; \quad M\frac{dZ_g}{dt} = -F_{L_0} + M_{g_g} + F_{ext} \tag{19.10}$$

19.3 ANALYTICAL MODELING OF LONGITUDINAL END EFFECT

As already mentioned, the limited length (up to 1–1.5 m) electromagnets in motion at speed U, with respect to the solid iron secondary, produce eddy currents in the latter. These eddy currents contribute to a nonuniform airgap flux density in the electromagnet area ($0 \le x \le L$) and a drag force F_d. While along the transverse (y) direction—Figure 19.3—the eddy currents are assumed to vary sinusoidally, along the direction of motion the electromagnet mmf and initial airgap flux density is rectangular (Figure 19.5) [6,7].

The airgap should be suddenly increased outside the active zone ($0 \ge x \ge L$) to portray properly the field distributions in the entry and exit zones:

$$\begin{aligned} B_0(x) &\approx 0 \quad \text{for } x < 0 \\ B_0(x) &= B_0 \quad \text{for } 0 \le x \le L \\ B_0(x) &\approx 0 \quad \text{for } x > L \end{aligned} \tag{19.11}$$

The amplitude of the initial flux density B_1 is

$$B_1 \approx \frac{4}{\pi} B_0 \sin\frac{\pi}{2C} a_x = K_0 B_0 \tag{19.12}$$

$2C = 2a_x = (b_i + 2g)$, if the two C-core legs are equal in width: b_i.

The eddy currents have in reality two components: one along $0y$ and the other along $0x$ (Figure 19.6).

Using Ampere's and Faraday's laws, the reaction field of eddy currents, H_y, satisfies the following equation:

$$\frac{\partial^2 H_z}{\partial x^2} + \frac{\partial^2 H_y}{\partial^2 y} - \mu_0 \sigma_{ie} U \frac{d_i}{g_e} \frac{\partial H_z}{\partial x} = 0; \quad \frac{\partial^2 H_z}{\partial^2 y} = -\left(\frac{\pi}{2C}\right)^2 H_z \tag{19.13}$$

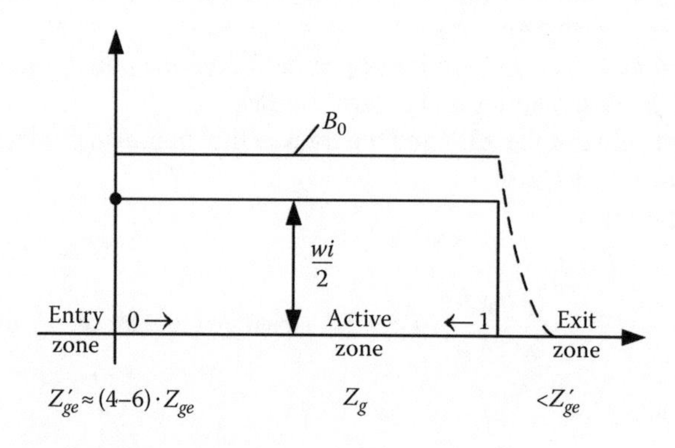

FIGURE 19.5 Longitudinal model.

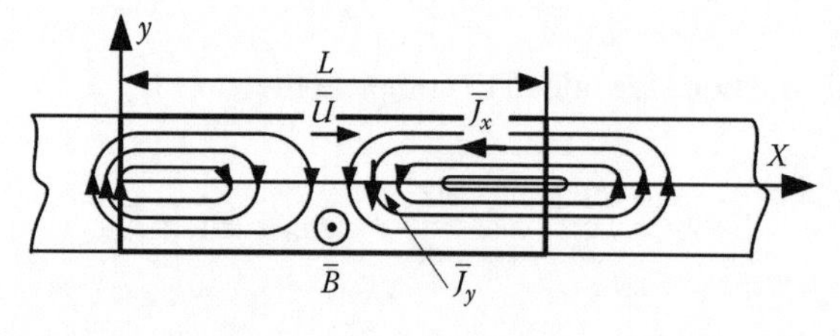

FIGURE 19.6 Theoretical eddy current density lines.

d_i is the reaction field penetration depth in the solid iron secondary (track):

$$\frac{\partial^2 H_z}{\partial x^2} - \mu_0 \sigma_{ie} \frac{d_i}{g_e} \frac{\partial H_z}{\partial x} - \left(\frac{\pi}{2C}\right)^2 H_z = 0 \tag{19.14}$$

$$\underline{\gamma}_{12} = \frac{C_u}{2} \pm \sqrt{\left(\frac{C_u}{2}\right)^2 + \left(\frac{\pi}{2C}\right)^2}; \quad C_u = \frac{\mu_0 \sigma_{ie} d_i U}{g_e} \tag{19.15}$$

Again, $2C = b_i + 2g$ for C-shape cores.

Therefore, the solution to (19.13) with (19.15) is

$$H_{r0}(x,y) = A_0 e^{\gamma_1 x} \cos\frac{\pi}{2C} y, \quad x \le 0$$

$$H_{r1}(x,y) = \left[A_1 e^{\gamma_1 (x-L)} + B_1 e^{\gamma_2 x}\right] \cos\frac{\pi}{2C} y; \quad 0 \le x \le L \tag{19.16}$$

$$H_{r2}(x,y) = B_2 e^{\gamma_2 (x-L)} \cos\frac{\pi}{2C} y; \quad x \ge L$$

For the low speeds ($U < 10$ m/s in general) $C_u/2 << \pi/2C$ and thus $|\underline{\gamma}_1| = |\underline{\gamma}_2| \approx \pi/2C$, and consequently, the end effect is low because $H_r(x,z)$ attenuates quickly along the active electromagnet length.

At high speeds, $C_u/2 >> \pi/2C$ and thus $\gamma_1 >> |\underline{\gamma}_2|$, and the end effect is notable.

The integration constants A_0, A_1, B_1, B_2 are to be obtained from boundary conditions: continuity of resultant flux density $B = B_{1z} + \mu_0 H_r$ and of $\partial H_r/\partial x$ at $x = 0, L$:

$$\underline{B}_1 = \frac{K_0 B_0}{(\underline{\gamma}_2/\underline{\gamma}_1) - 1}; \quad \underline{A}_0 = B_0 K_0 + \underline{B}_1 + \underline{A}_1 e^{-\underline{\gamma}_1 L}$$

$$\underline{A}_1 = \frac{-K_0 B_0}{(\underline{\gamma}_1/\underline{\gamma}_2) - 1}; \quad \underline{B}_2 = B_0 K_0 + \underline{A}_1 + \underline{B}_1 e^{\underline{\gamma}_2 L} \tag{19.17}$$

At very high speed, B_1 may approach $-K_0 B_0$ and thus $\underline{A}_1$ approaches zero.

The harmonics in transverse (0y) direction may be considered by $K_{yv} = (4/\pi v)\sin \pi v (b_i + 2g)/4C$ instead of K_0.

For preliminary calculations, the fundamental along 0y (19.17) suffices.

In Equation 19.13, the reaction field penetration depth is not considered; here we may consider it to be equal to the C-core leg height $h_c \approx d_i$, because of the very strong transverse edge effect coefficient K_t (decreased electric conductivity) and due to longitudinal, J_i, component of current density J_0.

As the resultant flux density in the airgap is

$$B_{rez} \approx B_0 + \mu_0 H_r(x,y) \tag{19.18}$$

and H_r decreases with x ($0 \le x \le L$), the B_{rez} will gradually increase with x from a small value to a larger than B_0 value, due to eddy currents field reaction.

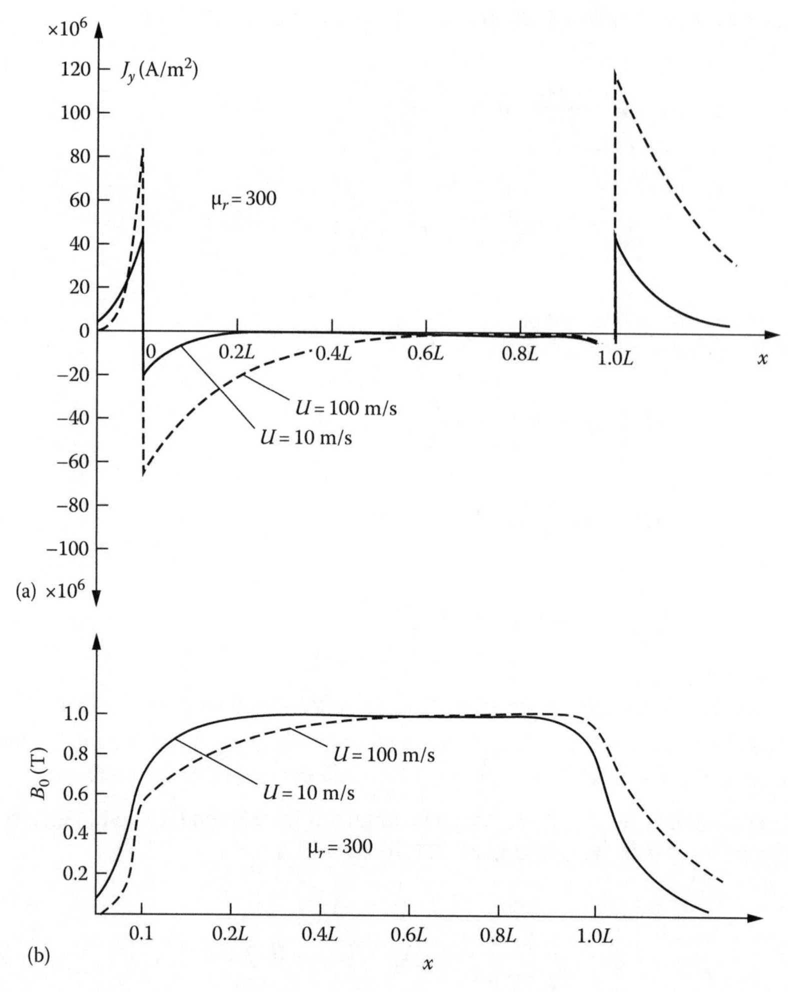

FIGURE 19.7 Airgap flux density and eddy current density distribution along 0*X* due to longitudinal end effect. (After Matsumura, F. and Yamada, S., *Elec. Eng. Jpn.*, 94(6), 50, 1974.)

For $b_i = 0.04$ m, $B_0 = 1$ T, $B_i = 0.12$ m, $z_g = 0.015$ m, $d_i = 0.025$ m, $L = 1$ m, $\sigma_i = 3.52 \times 10^6$ $(\Omega\ \text{m})^{-1}$, the distribution of airgap flux density and of eddy current density along x is shown in Figure 19.7 at $U_1 = 10$ m/s and $U_2 = 100$ m/s.

The attraction (levitation) and drag forces F_l, F_d are

$$F_l \approx \frac{2}{\mu_0} \int_0^{\frac{b_i}{2}+g_0} dy \int_0^L \left[B_{1z}(y) + \mu_0 H_r(x,y) \right]^2 dx \tag{19.19}$$

$$F_d \approx \frac{1}{\sigma_i U} \int_{-c}^{c} \int_0^{\infty} d_i \left(J_x^{\ 2} + J_y^{\ 2} \right) dy dx \tag{19.20}$$

$$J_x d_i = g_e \frac{\partial H_r}{\partial y}; \quad J_y d_i = g_e \frac{\partial H_r}{\partial x} \tag{19.21}$$

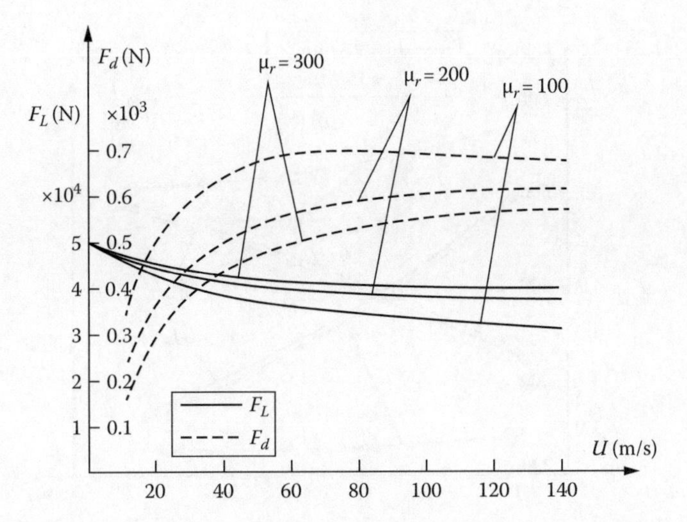

FIGURE 19.8 Levitation, F_l, and drag, F_d, forces versus speed. (After Boldea, I. and Nasar, S.A., *Linear Motion Electromagnetic Systems*, John Wiley & Sons, New York, 1985.)

For the example in Figure 19.7, the variation of levitation (attraction) force and of drag force, for different equivalent iron permeability, is shown in Figure 19.8.

A notable but not catastrophic decay of levitation force of 25% at 100 m/s and drag force F_d = 17% of acceleration force at 1 m/s² (0.7×10^3 N), for a mass corresponding to the levitation force 4×10^4 N (4×10^3 kg), are observed.

The ratio $(F_l/F_d)_{100\,\text{m/s}} \approx 4\times10^4/0.7\times10^3 = 57.1$, which may be called the goodness factor of levitation, is considered good for high-speed MAGLEVs.

19.4 PRELIMINARY DESIGN METHODOLOGY

Based on an analytical or FEM-derived model, the design of the ALS starts with the sizing of the dc controlled electromagnets, capable of producing the rated and maximum attraction levitation forces F_{Lr} and $F_{L\max}$ for a required F_n/electromagnet weight ratio and for minimum copper losses in the electromagnet for that attraction force (copper loss/normal force).

The two criteria are conflicting, and this is why one was considered a constraint:

$$\frac{F_L(\text{N})}{g_g G_0(\text{kg})} > K_{lev}; \quad K_{lev} = 7\text{–}15 \tag{19.22}$$

The higher values of K_{lev} correspond to forced cooling (j_{con} > (6–8) A/mm²), while the smaller values correspond to natural cooling (j = 3–3.5 A/mm²).

Higher current density j_{con} means not only a smaller coil but also larger copper losses per developed levitation force:

$$\min\left(\frac{P_{co}}{F_L}\right) \text{ is desired} \tag{19.23}$$

In addition, the guideway solid iron secondary (for levitation) has to be limited in weight (cost), and thus the total *C*-core electromagnet (and secondary) width is $2b_i + h_c < 0.3\text{–}0.35$ m (there are two such secondary cores, one on each side of MAGLEVs). In a simplified form, this leaves room, for a given rated airgap Z_g to investigate the electromagnet length L versus width b_i (*C*-core leg width).

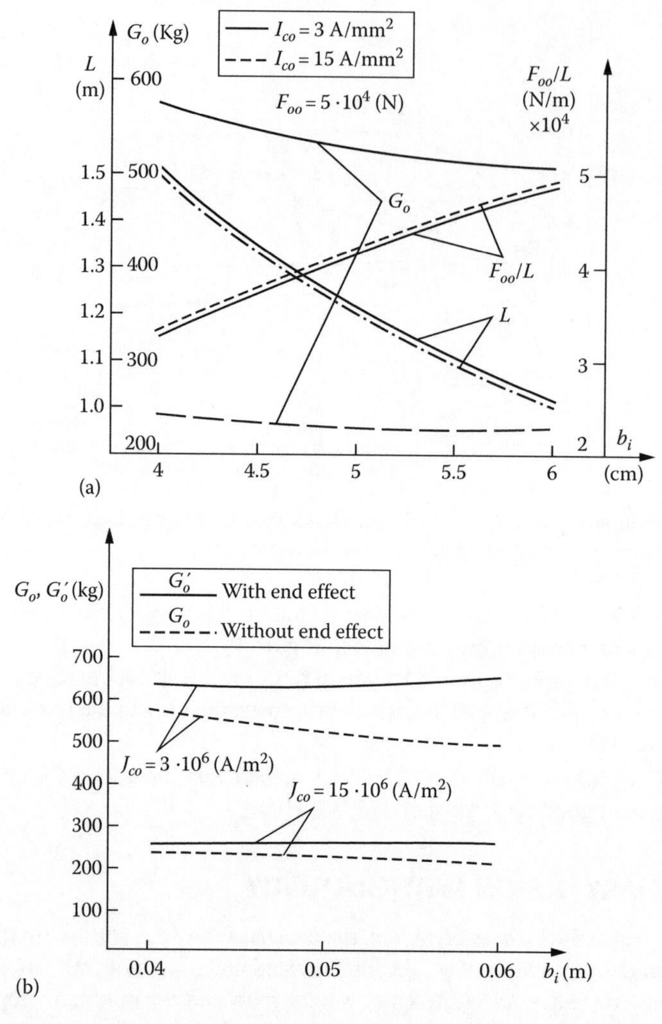

FIGURE 19.9 Electromagnet mass to length L, levitation force/unit length, for $j_{con}=3$ A/mm² and $j_{con}=15$ A/mm², versus C-core leg width d_i (a); recalculated mass with longitudinal end effect at 100 m/s (b). (After Matsumura, F. and Yamada, S., *Elec. Eng. Jap.*, 94(6), 50, 1974.)

Let us continue the numerical example of Figures 19.7 and 19.8 and obtain, by using the simplified model in the previous paragraph, the results in Figure 19.9.

A visible increase in electromagnet weight is required to maintain the levitation force $F_n=5\times10^4$ N, to counteract the longitudinal end effect at $v=100$ m/s.

When the C-core leg width is reduced, the cost of the track is reduced and also the electromagnet length L is increased; so the end effect influence is reduced. Values of L around 1.5 m are typical for high-speed MAGLEVs.

19.5 DYNAMIC MODELING OF ALS CONTROL

The single electromagnet guideway subsystem is presented in Figure 19.10.

It has only one degree of motion freedom along z (vertical) direction.

A MAGLEV involves multiple motions. But using secondary and tertiary mechanical suspension systems between the electromagnets and the bogie and between the latter and the passenger cabin, enough decoupling is obtained to allow for separate (decentralized) robust control of each electromagnet.

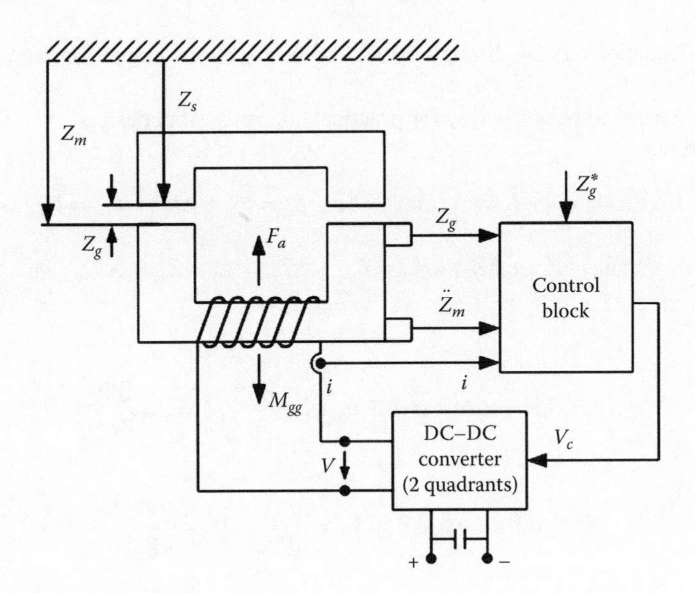

FIGURE 19.10 Single electromagnet–guideway system.

For stable levitation, the airgap Z_g is maintained close to the reference value Z_g*, within 20%–25% dynamically, in the presence of quite a few perturbations like external forces F_{ext} (from additional weight or wind forces) and guideway irregularities (from Z_s), such that a reasonable degree of ride comfort is secured for reasonable specific energy consumption (average kW/ton and peak kVA/ton).

To choose the input vector of variables, let us introduce here again the complete set of single electromagnet–guideway system equations:

$$V(t) = Ri(t) + \dot{\psi}(t)$$

$$\Psi(t) = L(Z_g, i) * i(t)$$

$$W_m(t) = \int_0^{\Psi} i d\Psi; \quad W_{com} = i\Psi - W_m(t) \tag{19.24}$$

$$F_L = -\frac{\partial W_m}{\partial Zg}\bigg|_{\Psi=ct} = \frac{\partial W_{com}}{\partial Z_g}\bigg|_{i=ct}$$

$$M\ddot{Z}_m = M_g - F_L + F_{ext}; \quad Z_m = Z_s + Z_g \tag{19.25}$$

It should be noticed that the motion equation now refers to the absolute acceleration $\ddot{Z}_m$ and not to $\ddot{Z}_g$, because the guideway irregularities (visible in Z_s) are accounted for.

Using the approximations in (19.8) and (19.9),

$$\Psi = \Psi_{Z_g} + L_l i; \quad \Psi_{Z_g} = \frac{ci - di^2}{az_g + b} \tag{19.26}$$

$$F_e \approx K_x \cdot \Psi_{Z_g}{}^2 \tag{19.27}$$

L_l is the leakage inductance of the electromagnet, considered independent of current i and position Z_g.

As visible in Equations 19.24 through 19.27, the single electromagnet guideway subsystem is nonlinear.

Linearization is used to pave the way for practical control system design.

For linearization,

$$V = V_0 + \Delta V;\quad \psi = \psi_0 + \Delta\psi;\quad i = i_0 + \Delta i;\quad F_L = F_{L_0} + \Delta F_L;\quad F_{ext} = F_{ext_0} + \Delta F_{ext}$$

$$Z_q = Z_{g_0} + \Delta Z_g;\quad Z_m = Z_{m_0} + \Delta Z_m;\quad Z_s = Z_{s_0} + \Delta Z_s \tag{19.28}$$

Finally,

$$\Delta V = R\Delta i + \alpha_i \Delta\dot{i} + \alpha_g \Delta\dot{Z}_g;\quad \alpha_i = \left.\frac{\partial\Psi}{\partial i}\right|_0;\quad \alpha_g = \left.\frac{\partial\Psi}{\partial Z_g}\right|_0$$

$$\Delta F_L = \beta_i \Delta i + \beta_g \Delta Z_g;\quad \beta_i = \left.\frac{\partial F_L}{\partial i}\right|_0;\quad \beta_g = \left.\frac{\partial F_L}{\partial Z_g}\right|_0 \tag{19.29}$$

$$M\Delta\ddot{Z}_m = -\Delta F_L + \Delta F_{ext}$$

$$\Delta Z_m = \Delta Z_s + \Delta Z_g$$

□$Z_s(t)$—for guideway irregularities—is considered given.

The coefficients α_i, α_g, β_i, β_g are calculated at the steady-state point characterized by V_0, i_0, Ψ_0, F_{L0}, F_{ext0}, Z_{g0}, Z_{m0}, and they are supposed to vary notably with the linearization "point."

The structural diagram of the linearized model (19.28) and (19.29) is portrayed in Figure 19.11a, with Δi as the electrical variable.

A similar linearization is possible with $\Delta\Psi$ instead of Δi as the electrical variable (Figure 19.11b) with

$$\alpha_\psi = \left.\frac{\partial i}{\partial\Psi}\right|_0;\quad \beta_\Psi = \left.\frac{\partial F_L}{\partial\Psi}\right|_0;\quad \alpha_{gx} = \left.\frac{\partial i}{\partial Z_g}\right|_0;\quad \beta_{gx} = \left.\frac{\partial F_L}{\partial Z_g}\right|_0 \tag{19.30}$$

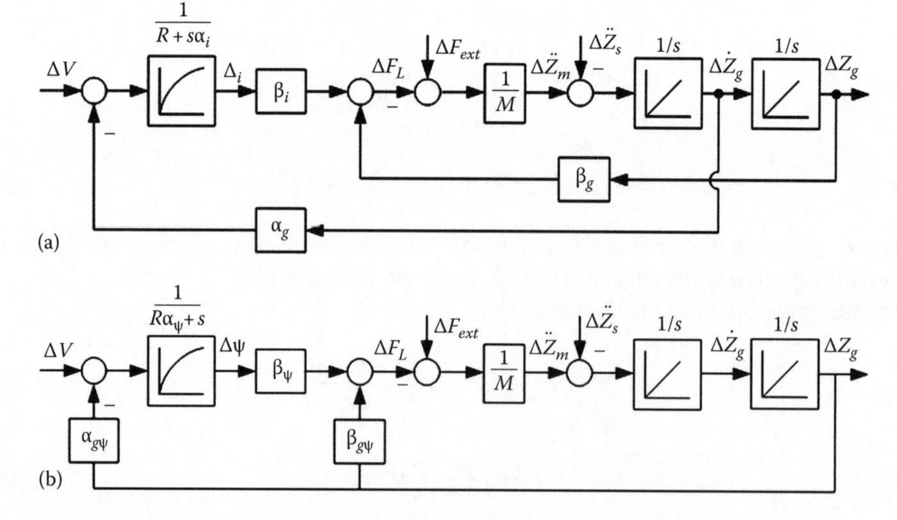

FIGURE 19.11 Linearized system with (a) Δi and (b) $\Delta\Psi$ as the electrical variable.

Again, the coefficients α_Ψ, β_Ψ, $\alpha_{g\Psi}$, $\beta_{g\Psi}$ depend on the linearization point. It has been demonstrated that, being related to magnetic flux, which determines the levitation force irrespective of airgap Z_{g0}, they should vary less with Z_{g0} in comparison with the coefficients for Δi as electric variable [11].

The problem is that, in general, the current may be measured (and controlled) easier than the flux linkage Ψ, especially at steady state (dc).

As the dynamic errors, introduced by linearization, depend heavily on the linearization point (Z_{g0}, F_{L0}, i_0...), especially with Δi as variable, when the linear control system is adopted, the latter is to be adaptive for MAGLEVs, where the airgap varies, at least at vehicle lift off and landing, by its rated value.

The structural diagrams suggest three possible state vectors:

$$\underline{\Delta X_1} = \begin{vmatrix} \Delta Z_g \\ \Delta \dot{Z}_g \\ \Delta i \end{vmatrix}; \quad \underline{\Delta X_2} = \begin{vmatrix} \Delta Z_g \\ \Delta \dot{Z}_g \\ \Delta \ddot{Z}_m \end{vmatrix}; \quad \underline{\Delta X_3} = \begin{vmatrix} \Delta Z_g \\ \Delta \dot{Z}_g \\ \Delta \Psi \end{vmatrix} \tag{19.31}$$

The output vectors for the three situations are

$$\begin{vmatrix} \Delta Z_g \\ \Delta i \end{vmatrix} = \begin{vmatrix} 1 & 0 & 0 \\ 0 & 0 & 1 \end{vmatrix} \underline{\Delta X_{1,2,3}}$$

$$\begin{vmatrix} \Delta Z_g \\ \Delta \dot{Z}_g \end{vmatrix} = \begin{vmatrix} 1 & 0 & 0 \\ 0 & 0 & 1 \end{vmatrix} \underline{\Delta X_{2,1,3}} \tag{19.32}$$

$$\begin{vmatrix} \Delta Z_g \\ \Delta \Psi \end{vmatrix} = \begin{vmatrix} 1 & 0 & 0 \\ 0 & 0 & 1 \end{vmatrix} \underline{\Delta X_{3,2,1}}$$

From the structural diagrams in Figures 19.11a and b, we get state equations:

$$\Delta \dot{X}_1 = A_1 \Delta X_1 + B_1 \Delta V + C_1 \begin{vmatrix} \Delta F_{ext} \\ \Delta \ddot{Z}_s \end{vmatrix}$$

$$\Delta \dot{X}_2 = A_2 \Delta X_2 + B_2 \Delta V + C_2 \begin{vmatrix} \Delta F_{ext} \\ \Delta \dot{F}_{ext} \\ \Delta \ddot{Z}_s \end{vmatrix}$$

$$\Delta \dot{X}_3 = A_3 \Delta X_3 + B_3 \Delta V + C_3 \begin{vmatrix} \Delta F_{ext} \\ \Delta \ddot{Z}_s \end{vmatrix} \tag{19.33}$$

The state variables have to be measured or some of them are to be estimated. In our case, Z_g may be measured by dedicated sensors; $\ddot{Z}_m$ may be measured in general by an inertial sensor (ADx105 of analog devices).

The flux is not easily measurable, especially at steady state (dc), but Hall sensors with low temperature sensitivity could solve the problem.

However, $\dot{Z}_g$ (even $\ddot{Z}_g$) have to be estimated through an observer. Also, the flux Ψ may be observed if the $\Psi(i, Z_g)$ FEM curve family is available (tabled or curve-fitted) and i and Z_g are measured directly. This way, even the flux at given current is observed.

Now if we calculate the transfer function between the airgap deviation $\Delta Z_g(s)$ and the input voltage $\Delta V(s)$, for zero external force and guideway irregularity perturbations ($\Delta F_{ext}(s)=0$, $\Delta Z_s(s)=0$) is of the form

$$\frac{\Delta Z_g(s)}{\Delta V(s)} = \frac{-\Delta V}{(R+s\alpha_i)\left((s^2 M/\beta_i)+\beta_g\right)+\alpha_g s} \tag{19.33a}$$

It may be demonstrated that some roots of the denominator of Equation 19.33 have positive real parts at any linearization point (by the signs of α_i, β_i, α_g, β_g), and thus the open-loop voltage-fed system is not stable; even for controlled constant current, the system would be unstable.

The decrease in the levitation force with airgap increase is a strong phenomenological explanation for this.

As the system is statically and dynamically unstable, its forced stabilization is required.

19.6 STATE FEEDBACK CONTROL OF ALS

Let us control the ALS linearized model with $\Delta \underline{X}_2 = \left[\Delta Z_g, \Delta \dot{Z}_g, \Delta \ddot{Z}_m\right]^T$ as the state vector with the perturbation vector $\Delta \underline{Z}_2 = \left[\Delta F_{ext}, \Delta \dot{F}_{ext}, \Delta \ddot{Z}_s\right]^T$ and the output $\Delta \underline{Y}_2 = \left[\Delta Z_g, \Delta \ddot{Z}_m\right]^T$:

$$\Delta \dot{\underline{X}}_2 = A\Delta \underline{X}_2 + \underline{B}\Delta \underline{Y}_2 + \underline{P}_2 \Delta \underline{Z}_2 \tag{19.34}$$

From (19.28 and 19.29),

$$\underline{A} = \begin{vmatrix} 0 & 1 & 0 \\ 0 & 0 & 1 \\ -\beta_g/M & -(\beta_g - \beta_i\alpha_g/\alpha_i)/M & -R/\alpha_i \end{vmatrix}; \quad B = \begin{vmatrix} 0 \\ 0 \\ -\beta_i/M\alpha_i \end{vmatrix}; \quad \underline{P_2} = \begin{bmatrix} 0 & 0 & 0 \\ 0 & 0 & -1 \\ R/M\alpha_i & 1/M & 0 \end{bmatrix} \tag{19.35}$$

Though the control system should be adaptive, we hereby deal with it for one linearization point, only. The procedure may be repeated for a few critical situations and then the control law may use adaptable coefficients:

$$\Delta V = K_g \Delta Z_g + K_u \Delta \dot{Z}_g + K_a \Delta \ddot{Z}_m = \underline{K}^T \Delta \underline{X_2} \tag{19.36}$$

such that integral squared criterion is minimum:

$$I(\Delta V) = \frac{1}{2}\int_0^\infty \left[q_g^2 \Delta Z_g^2(t) + q_a^2 \Delta \ddot{Z}_m(t) + \Delta u^2(t)\right] dt \tag{19.37}$$

The control coefficients K_g, K_u, K_a will depend on q_g and q_a in (19.37).

The optimum criterion (19.37) aim to

- Reduce the airgap deviations (ΔZ_g) to avoid electromagnet–guideway collisions
- Reduce absolute acceleration ($\dot{Z}_m$) for ride comfort
- Reduce control power (by ΔV)

The control law is obtained from the known equation:

$$\underline{K}^T = \begin{bmatrix} K_g & K_u & K_a \end{bmatrix}^T = -\underline{R}^{-1} B_2^T \underline{P} = -b' \begin{bmatrix} P_{13} & P_{23} & P_{33} \end{bmatrix}^T \quad (19.38)$$

$$\underline{R} = [1], \quad b' = \frac{-\beta_i}{M\alpha_i}, \quad \underline{P} = \begin{bmatrix} p_{11} & p_{12} & p_{13} \\ p_{12} & p_{22} & p_{23} \\ p_{13} & p_{23} & p_{33} \end{bmatrix} \quad (19.39)$$

Matrix $\underline{P}$ is the solution of the Riccati equation:

$$\underline{P}\,\underline{A}_2 + \underline{A}_2^T \underline{P} - \underline{P} \cdot \underline{B} \cdot \underline{B}^T \underline{P} = -\underline{Q}, \quad \underline{Q} = \text{diag}\left[q_g^{\,2}, 0, q_a^{\,2} \right] \quad (19.40)$$

With known $\underline{Q}$, K^T (control law) may be calculated.

To determine $\underline{Q}$, Equation 19.40 is decomposed into six equations to find $p_{11}, p_{12}, p_{13}, p_{22}, p_{23}, p_{33}$. Then, using (19.38) and (19.39) and eliminating p_{12}, three equations with three unknowns are obtained:

$$K_g^{\,2} + \frac{2a_1}{b} K_g = q_g^{\,2} \quad (19.41)$$

$$K_a^{\,2} + \frac{2a_3}{b'} K_a + \frac{2}{b'} K_u = q_a^{\,2}$$

$$-\frac{2a_3}{b'} K_g - \frac{2a_1}{b'} K_a - 2K_g K_a + \frac{2a_2}{b'} K_u + K_u^{\,2} = 0$$

with

$$a_1 = \frac{-\beta_g R}{M\alpha_i}; \quad a_2 = \frac{-\beta_g}{M} \frac{T_d R}{\alpha_i}; \quad a_3 = -\frac{R}{\alpha_i}; \quad T_d = \frac{R}{\alpha_i} + \frac{\beta_i \alpha_g}{\beta_g R}$$

Through an algebraic stability criterion,

$$K_g > 2K_{g0}\left(K_a - K_{a0}\right)\left(K_u - K_{u0}\right) + \left(\frac{K_g - K_{g0}}{b'}\right) \geq 0 \quad (19.42)$$

where

$$K_{g0} = \frac{\beta_g R}{M\alpha_i b'}; \quad K_{u0} = \frac{-\beta_g R T_d}{\alpha_i M b'}; \quad K_{a0} = -\frac{R}{b'\alpha_i} \tag{19.43}$$

Also, the amplifiers have limits: $K_g \le K_{g\max}$; $K_u < K_{u\max}$; and $K_a < K_{u\max}$.

The system (19.41) has to have real solutions: K_g, K_a, K_u and thus q_g and q_a have to be chosen carefully in a domain. As the vehicle weight varies from unloaded mass (M_1) to loaded mass M_2, the q_a, q_g domain has to yield practical solutions for both situations.

For control implementations, we should notice that Z_g and $\ddot{Z}_m$ are measurable, but $\dot{Z}_g$ is not.

So an observer for $\hat{\dot{Z}}_g$ is required.

The control scheme with a state observer is illustrated in Figure 19.12 [11].

The $\hat{\dot{Z}}_g$ observer in Figure 19.12 solves the problem of system behavior with respect to guideway irregularities.

It implies only the choice of $S_{ob} = -\beta_0$ pole since

$$\hat{\dot{Z}}_g(1) = \frac{\ddot{Z}_m(1)}{s+\beta_0} + \frac{s\beta_0}{s+\beta_0}\dot{Z}_g(s) \tag{19.44}$$

For an ALS with the data $M = 200$ kg, $Z_{g0} = 10^{-2}$ m, $T = \alpha_i/R = 0.053$ s, $R = 1.532\ \Omega$, $\beta_i = 236$ N A^{-1}, $-\beta_g = 9\times10^5$ N m^{-1}, results as in Figure 19.13 are obtained.

It is evident that without the adopted $\hat{\dot{Z}}_g$ observer the system cannot tolerate force and track irregularity perturbations.

To further improve the performance, an integral of airgap error may be added to the control law to secure zero steady-state airgap error. Also an additional interior current loop may be added to limit the current and quicken the response.

A two-quadrant dc–dc converter ($\pm V_{\max}$) is needed for quick enough control of MAGLEVs.

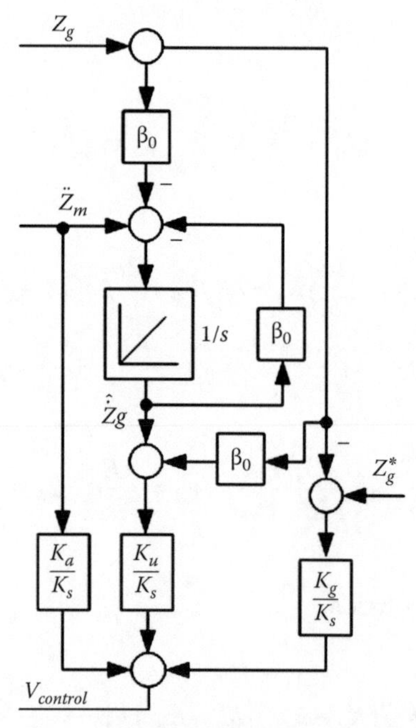

FIGURE 19.12 Control system with $\hat{\dot{Z}}_g$ observer (K_s: dc–dc converter gain).

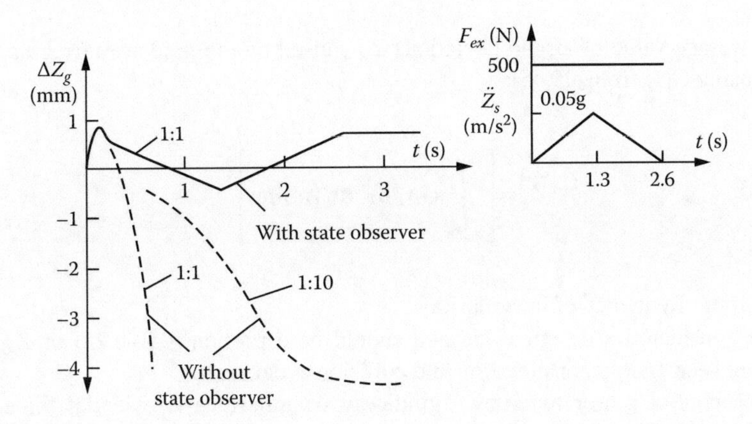

FIGURE 19.13 Control system response to external force F_{ext} and $\ddot{Z}_m$ (track irregularities) perturbations. (After Matsumura, F. and Yamada, S., *Elec. Eng. Jap.*, 94(6), 50, 1974.)

19.7 CONTROL SYSTEM PERFORMANCE ASSESSMENT

The control system performance for MAGLEVs is related to

- Electromagnet–guideway collision avoidance
- Average control power
- Ride comfort

For active magnetic bearings, in some applications, the maximum (or average) dynamic deviation of the airgap is limited to a small fraction of the rated airgap (when applied for machining purposes); ride comfort is not considered for active magnetic bearings but noise and vibration are.

To assess this performance, the following transfer functions of the close-loop system are required:

$$G_1(s) = \frac{Z_g(s)}{Z_s(s)} = \frac{-\left[s^3 + \left(\frac{R}{\alpha_i} + \frac{\beta_i K_s K_a}{M\alpha_i}s^2 + \frac{K_s \beta_i}{M\alpha_i}s\right)\right]}{D(s)} \quad (19.45)$$

$$G_2(s) = \frac{Z_g(s)}{F_{ext}(s)} = \frac{\frac{1}{M}(s + R/K_i)}{D(s)} \quad (19.46)$$

$$G_3(s) = \frac{Z_g(s)}{V(s)} = \frac{-\beta_i}{MK_i D(s)} \quad (19.47)$$

with

$$D(s) = s^3 + \left(\frac{R}{K_i} + \frac{\beta_i K_s K_a}{M\alpha_i}\right)s^2 + \left(\frac{\beta_i K_s K_a}{M\alpha_i} - \frac{(-\beta_j)L_l}{M\alpha_i}\right)s + \frac{\beta_i K_s K_g}{M\alpha_i} + \frac{\beta_g R}{M\alpha_i} \quad (19.48)$$

The guideway (track) irregularities may be described by the power spectral density (PSD):

$$\phi_z(\omega):$$

$$\Phi_z(\omega) = \frac{A_g U}{\omega^2}; \quad U\text{: vehicle speed}; \quad A_g\text{: guideway constant} \quad (19.49)$$

The squared average value of airgap deviations $\overline{Z}_g$ caused by the guideway irregularities may be expressed by using $G(1)$ from (19.45):

$$\overline{Z_g} = \left[\int_{0.2*\pi}^{\omega_0} \left| G_1(j\omega) \right|^2 \Phi_z(\omega) d\omega \right]^{1/2} \tag{19.50}$$

ω_0 is the maximum frequency of irregularities.

$\overline{Z}_g$ may be considered a design value and should be a portion (0.2–0.25) of Z_{g0} (for a good compromise between energy consumption and collision avoidance).

The average control power for known guideway irregularities is essential for a competitive system, and, for assessment, it is required first to calculate the voltage and current squared average deviations $\overline{\Delta V}$ and $\overline{\Delta i}$:

$$\overline{\Delta V} = \left[\int_{0.2*\pi}^{\omega_0} \left| \frac{G_1(j\omega)}{G_3(j\omega)} \right|^2 \Phi_z(\omega) d\omega \right]^{1/2} \tag{19.51}$$

$$\overline{\Delta i} = \left[\int_{0.2*\pi}^{\omega_0} \frac{\left| G_1(j\omega) * G_3^{-1}(j\omega) + (\alpha_i - L_l)\beta_i * j\omega \right|}{R + j\omega\alpha_i} \Phi_z(\omega) d\omega \right]^{1/2} \tag{19.52}$$

Thus, the average control power P_{con} is

$$P_{con} = \left(V_0 + \overline{\Delta V}\right)\left(I_0 + \overline{\Delta i}\right) \tag{19.53}$$

Finally, the ride comfort may be calculated using the power spectral density concept applied for absolute accelerations, $\phi_{\ddot{Z}_m}(j\omega)$:

$$\Phi_{\ddot{Z}_m}(\omega) = \omega^4 \left| G_1(j\omega) + 1 \right|^2 \Phi_z(\omega) \tag{19.54}$$

The human comfort level when exposed to a spectrum of accelerations of various frequencies has been quantified based on dedicated studies.

The "Janeway curve" is by now an accepted standard.

To meet the ride comfort standard, the $\Phi_{\ddot{Z}_m}(\omega)$ curve has to be below "Janeway curve" at all frequencies of interest.

The higher the speed, the more difficult it is to meet the "Janeway criterion."

Even if it would be possible to meet the "Janeway criterion" at high speeds (say at $U_{max} = 100$ m/s), the energy consumption in the ALS might render the solution unpractical.

Consequently, a secondary (and a tertiary) mechanical suspension system is used on MAGLEVs.

A typical passive system is shown in Figure 19.14.

The secondary suspension system equations are

$$M\ddot{Z}_M = Mg - K\left(Z_M - Z_S - Z_g - l_0\right) - \beta\left(\dot{Z}_M - \dot{Z}_m\right) \tag{19.55}$$

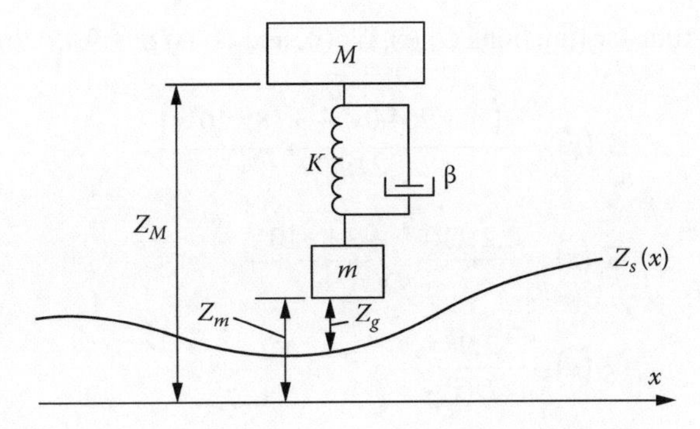

FIGURE 19.14 Passive secondary suspension system. (After Matsumura, F. and Yamada, S., *Elec. Eng. Jpn.*, 94(6), 50, 1974.)

where

l_0 is the distance between unsprung (ALS) mass m and sprung mass M

K and β are the spring rigidity and damping coefficients when they are unloaded:

$$Mg = K\left(Z_{M_0} - Z_{g_0} - l_0\right) \tag{19.56}$$

where

Z_{M0} is the rated height of mass M

Z_{g0} is the rated height of mass m

For the mass m, with no spring (the ALS),

$$m\ddot{Z}_m = mg + K\left(Z_M - Z_S - Z_g - l_0\right) + \beta\left(\dot{Z}_M - \dot{Z}_m\right) - F_L + F_{ext} \tag{19.57}$$

In this case, the problem may be solved as up to now but with $\overline{Z_g} = \left(0.15 - 0.25\right)Z_{g_0}$; higher track irregularities may be tolerated or smaller average control power is obtained due to the presence of the secondary suspension system.

The ride comfort will be checked for $\ddot{Z}_m$; in general, for a wise use of secondary system $m/M < 0.2$; a condition that is easily fulfilled by ALSs.

19.8 CONTROL PERFORMANCE EXAMPLE

Let us suppose a MAGLEV of 50 ton at 100 m/s with rated airgap $Z_{g0} = 1.5 \times 10^{-2}$ m that uses electromagnets that are 1.5 m in length (L) with C = core leg width $b_i = 0.04$ m for a current density $j_{con} = 15$ A/mm^2 (forced cooling).

Ten electromagnets are used, so $F_{L0} = 5 \times 10^4$ N with $W_{i0} = 3.58 \times 10^4$ A-turns, $V_0 = 200$ V, $I_0 = 200$ A, electromagnet weight $m = 300$ kg; coil resistance $R = 1\ \Omega$ and the rated inductance L (saturation neglected) is $L_0 = \propto_i = 0.1$ H, $\beta_i = 1490$ N/A, $\beta_g = -1984 \times 10^7$ N/m; $L = \Psi/i = (2.5 + 1.43/2\ g)10^{-3}$(H).

Based on the control system design in Section 19.6, for $q_a = 308$ s^2/m, $q_g = 1.65 \times 10^5$ V/m, we obtain the state feedback control law coefficients:

$$K_g = 1.8 \times 10^5 \text{ V/m},\ K_u = 1.07 \times 10^4 \text{ V s/m},\ K_a = 340.0 \text{ V s}^2\text{/m}$$

With these values, the transfer functions $G_1(s)$, $G_2(s)$, and $G_3(s)$ of (19.45) through (19.48) are

$$G_1(s) \approx \frac{-\left(s^3 + 964.0s^2 + 3.18\times10^4 s\right)}{D(s)}$$

$$G_2(s) \approx \frac{0.2\times10^{-3}s + 2.1\times10^{-3}}{D(s)} \tag{19.58}$$

$$G_3(s) \approx -\frac{2.98}{D(s)}$$

$$D(s) \approx s^3 + 964s^2 + 3.18\times10^4 s + 4.96\times10^5$$

With welded guideway, the constant A_g (in (19.49)) is $A_g = 1.5\times10^{-6}$ m.

From (19.50), the squared average airgap deviation $\overline{Z_g} = 4.6\times10^{-3}\ \text{m} < Z_{g0}/3$ ($f_0 = \omega_0/2\pi = 25$ Hz). Therefore, the MAGLEV avoids the collision with the guideway at 100 m/s. However, the squared average voltage and current deviations $\bar{\sigma}_V$ and $\bar{\sigma}_i$ are huge:

$$\bar{\sigma}_V = 4690\ \text{V};\quad \bar{\sigma}_i = 670\ \text{A}$$

The average control power is inadmissibly high, and thus a mechanical secondary system is needed.

To do so, we have to reconsider the whole control system design for $m = 300$ kg (only the electromagnet); finally, with $q_a' = 2.73$ V s²/m, $q_g' = 5.8\times10^4$ V/m, $K_g' = 7.4\times10^4$, $K_u' = 5600$, $K_a' = 14.8$, $\overline{Z_g}' = 4.65\times10^{-3}$ m$<Z_{g0})/3$, we obtain $\bar{\sigma}_V' = 630.8$ V, $\bar{\sigma}_i' = 246.8$ A.

This time the voltage deviation is about 3/1 and current deviation is 1.2/1. Such values are almost acceptable, but, if possible, they should be further reduced by the cabin's additional suspension system contribution.

The ride comfort has been calculated (19.49) for both situations, and the results are shown in Figure 19.15.

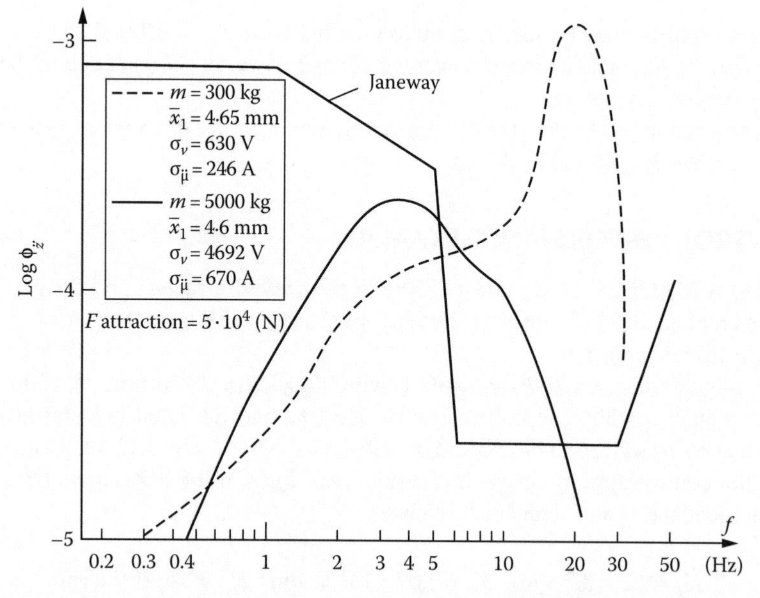

FIGURE 19.15 Ride comfort $m = 300$ kg, $M = 5000$ kg. (After Matsumura, F. and Yamada, S., *Elec. Eng. Jpn.*, 94(6), 50, 1974.)

The secondary system was not introduced in the data in Figure 19.15; only the mass M was decreased to $m = 300$ kg.

So a secondary (and tertiary) mechanical system is responsible to provide ride comfort for $f_0 > 5$ Hz.

Vehicle lifting (from Z_{gmax} to Z_{g0}) and touchdown should be controlled by a special sequence, as the linear control system discussed so far may not be stable for such large airgap deviations.

This and other reasons raise the problem of alternative, more robust, control of ALS.

19.9 VEHICLE LIFTING AT STANDSTILL

For a 4 ton MAGLEV [9], similar to the previous, the state feedback control system was used for $M = 1000$ kg, $Z_{g0} = 0.01$ m, $R = 0.7\ \Omega$, $i_0 = 47$ A, $V_0 = 33$ V, $\Psi_0 = \alpha_i\, i_0 = 5.20$ Wb, $K_g = 5000.0$, $K_u = 3263$, $K_a = 110.0$, $\omega_{0s} = 25$ rad/s, and $\xi = 0.45$.

An additional flux linkage feedback is added to increase stability limits at large airgaps. A search coil voltage with a first-order large time constant delay is used for flux feedback Ψ_t (Figure 19.16).

Experimental results [9] (Figure 19.17) picturing the vehicle stable lifting (4 ALSs of 1 ton each) from 20 to 12.5 mm, with reasonable voltage and current dynamic profiles, tend to substantiate the benefic role of the flux feedback and of the airgap deviation PI loop, added to provide for ideally zero steady-state airgap deviation error.

The steady-state estimated average levitation control power is about 1.45 kW/ton, while the measured one was 1.54 kW/ton at standstill. As expected, naturally cooled electromagnets have

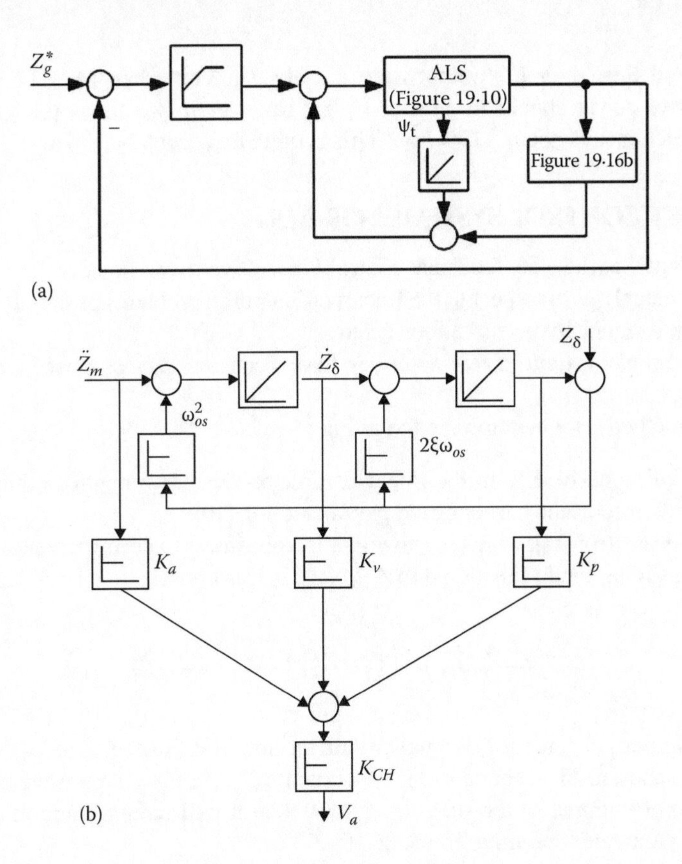

FIGURE 19.16 State feedback control system with additional flux feedback (a); observer (for $\hat{\dot{Z}}_g$) (b).

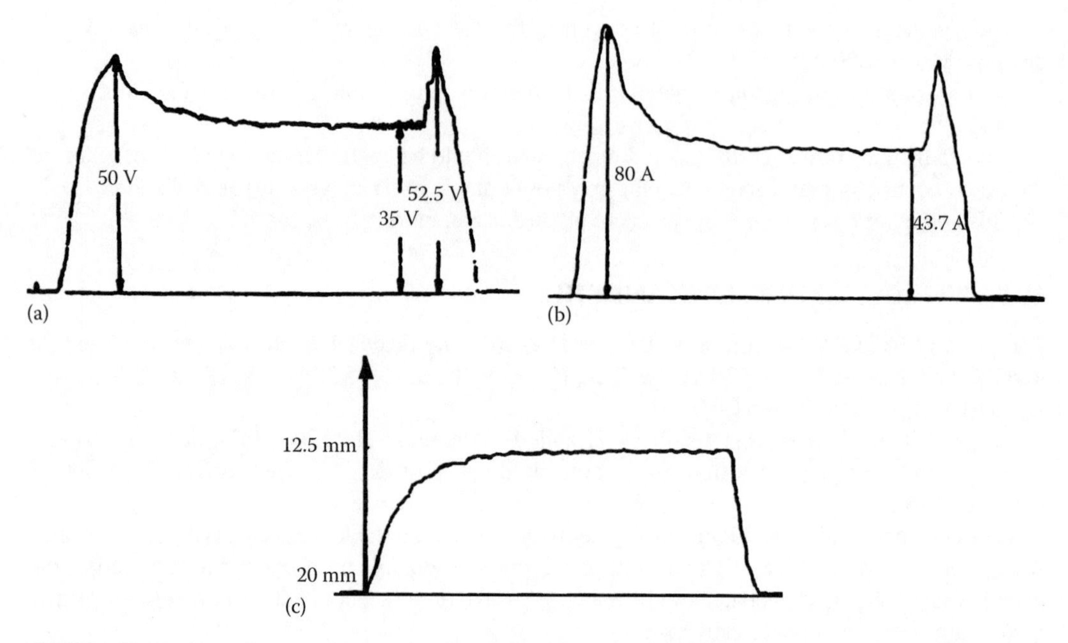

FIGURE 19.17 Experimental lifting of a 4 ton MAGLEV at standstill (4×1 ton ALSs): (a) input voltage (single quadrant dc–dc converter), (b) coil current, and (c) airgap (from 20 to 12.5 mm). (After Boldea, I. et al., *IEEE Trans.*, VT-37(4), 213, 1988.)

been designed and thus their lift/weight ratio is only 7/1. For urban MAGLEVs ($U \leq 20$ m/s), the average control power should in general be not more than two times the steady-state value (at standstill), which means about 3 kVA/ton. This is quite an acceptable value.

19.10 ROBUST CONTROL SYSTEMS FOR ALSs

It is by now evident that the state feedback control is sensitive to the linearization point and, when the airgap varies notably with respect to the linearization point, problems of stability in the presence of force and track irregularity perturbations occur.

To solve this problem, quite a few solutions have been recently proposed (mostly for active magnetic bearings).

Among such methods, we mention the following:

- Gain scheduling derived from the linear parameter-dependent control initiated from the controllers defined at the corner of the parameter box [10]
- Sliding mode control [11], which is known for its robustness to perturbations and parameter detuning; a sliding mode functional $\sigma(Z_g, \dot{Z}_g, \ddot{Z}_g)$ is first chosen:

$$\sigma\left(Z_g, \dot{Z}_m, \ddot{Z}_m\right) = \left(Z_g^* - Z_g\right) + T_s\hat{\dot{Z}}_g + T_s T_{sm}\ddot{Z}_m \tag{19.59}$$

T_s is first to be imposed; in general T_s is smaller than the smallest time constant of the system. Then, T_{sm} has also to be chosen. The second-order functional $\sigma(Z_g, \dot{Z}_g, \ddot{Z}_g) = 0$ represents a surface. If the system is capable of jumping to the $\sigma(Z_g, \dot{Z}_g, \ddot{Z}_g) = 0$, then it will remain there under well-defined perturbation and parameter detuning limits.

The stability of SM control system is a problem in itself and may be treated in many ways but preferably (yet) by the Lyapunov method [11,12].

The control law is rather simple and may take the form [11]

$$V(t)=\begin{cases}+V_{\max}; & \sigma\left(Z_g,\dot{Z}_g\right)>h\\ K_R\sigma(t)+\dfrac{1}{T_i}\displaystyle\int\sigma dt; & |\sigma|<h\\ -V_{\max}; & \sigma\left(Z_g,\dot{Z}_g\right)<-h\end{cases}\tag{19.60}$$

Here, $\pm V_{\max}$ are the positive and negative maximum output voltages of the dc–dc converter that supplies the ALS.

Close to the target ($|\sigma|<h$), a PI controller over the SM functional is used to reduce chattering around the target. The parameters K_R and T_i have to be determined in the design process and have to be estimated as done for state feedback control systems.

But the PI system, which acts close to the target in corroboration with the term $T_sT_{sm}\ddot{Z}_m$—via $\ddot{Z}_s$, Equation 19.54, introduces a serious perturbation due to track irregularities [11].

To eliminate this perturbation, $T_{sm}=0$ in (19.59), which becomes

$$\sigma_s\left(Z_g,\dot{Z}_g\right)=\overline{Z}_g^{*}-Z_{g0}+T_s\dot{Z}_g\tag{19.61}$$

But now, to reduce the perturbation influence, the SM control is cascaded with an interior fast (bipositional) current loop.

Typical results with such a system are shown in Figure 19.18 [11].

The airgap varies in Figure 19.18a from 0.03 to 0.01 m and, respectively, to 0.005 m. The response time is about 0.2 s. The perturbation rejection looks very good, and this is due to the fast current (bipositional) loop response; $\ddot{Z}_s$ is followed strictly by $\ddot{Z}_m$ toward its rejection.

The fast current and force responses (Figure 19.18b) are evident. The high-frequency oscillations in the current loop are visible in force also, but they are limited and may be handled by standard insulate gate bipolar transistor (IGBT) dc–dc converters.

Moreover, the current fast (PD) current loop reduces the system to one of second order.

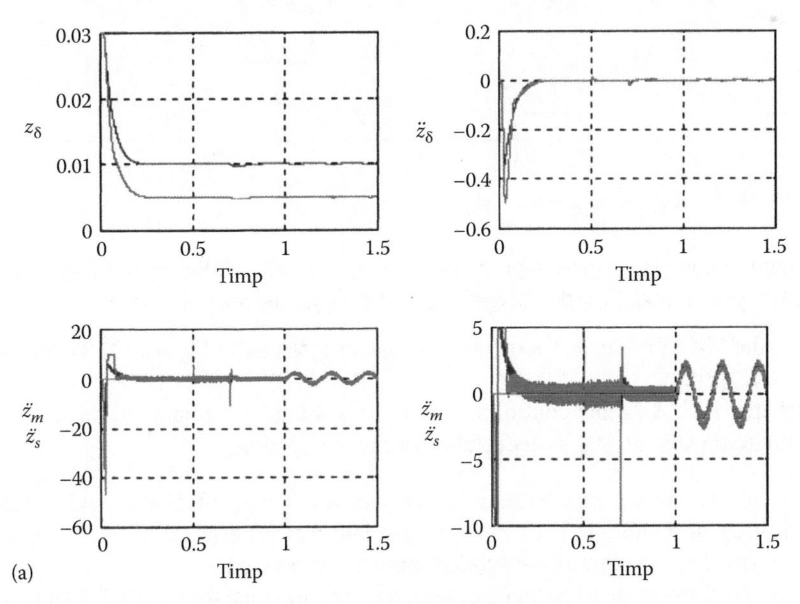

FIGURE 19.18 SM + current control: (a) airgap, $\dot{Z}_g\ddot{Z}_m\ddot{Z}_s$ with details (zoom).

(continued)

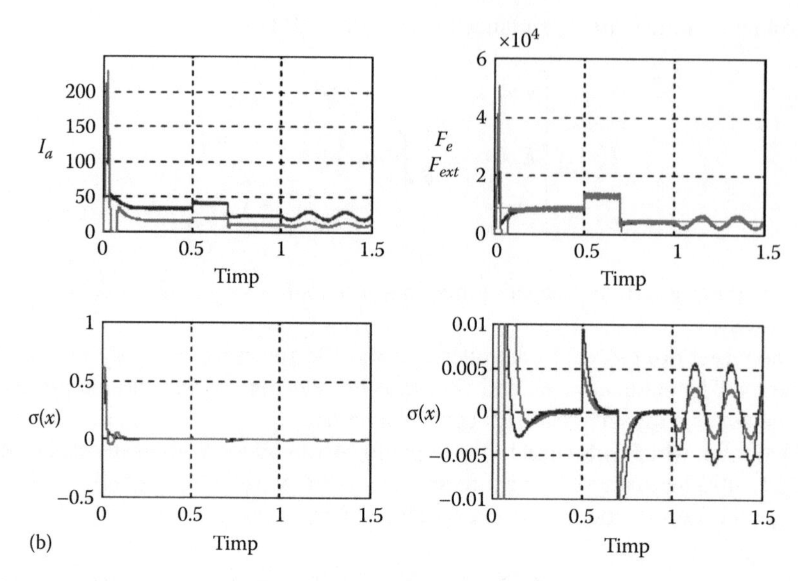

FIGURE 19.18 (continued) SM + current control: (b) current, external force, $\sigma(Z_g, \dot{Z}_g)$ with details (zoom). (After Trica, Al., *Electromagnetic Suspension Systems*, University Politehnica Timisoara Publishers, 2009.)

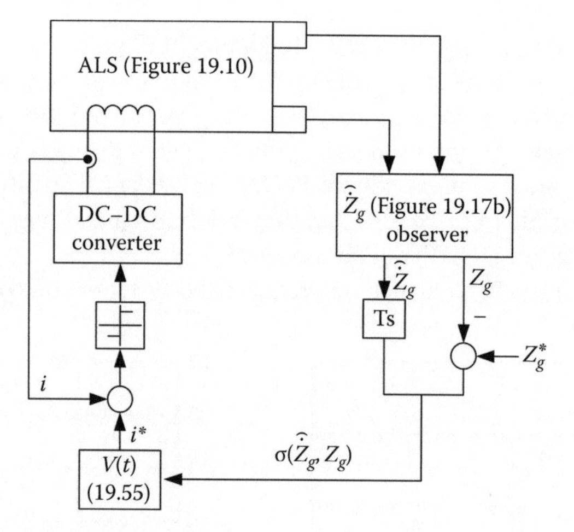

FIGURE 19.19 SM + PI + current control system.

Note: Though the absolute acceleration $\ddot{Z}_m$ disappears from the sliding mode functional (19.61), it remains in the system model for the design of the sliding mode controllers.

Therefore, the SM + PI + current loop control system looks as in Figure 19.19 (For more on SM control, see Ref. [12]).

It appears that such a simple control system is very robust. It remains to be seen if the rather increased maximum voltage V_{max} is acceptable for the application.

- State feedback control with current change rate feedback of dual-sided ASL system (for linearization of levitation force with current and airgap) applied to an active magnetic bearing was shown to produce sensorless airgap control [13].
- The dual ALS for suspension together with one for guidance for MAGLEV transportation industrial platform is proportional integral derivative (PID)-fuzzy adaptive controlled differentially in a master–slave solution with $I_0 \pm i$ [14] (Figure 19.20a and b).

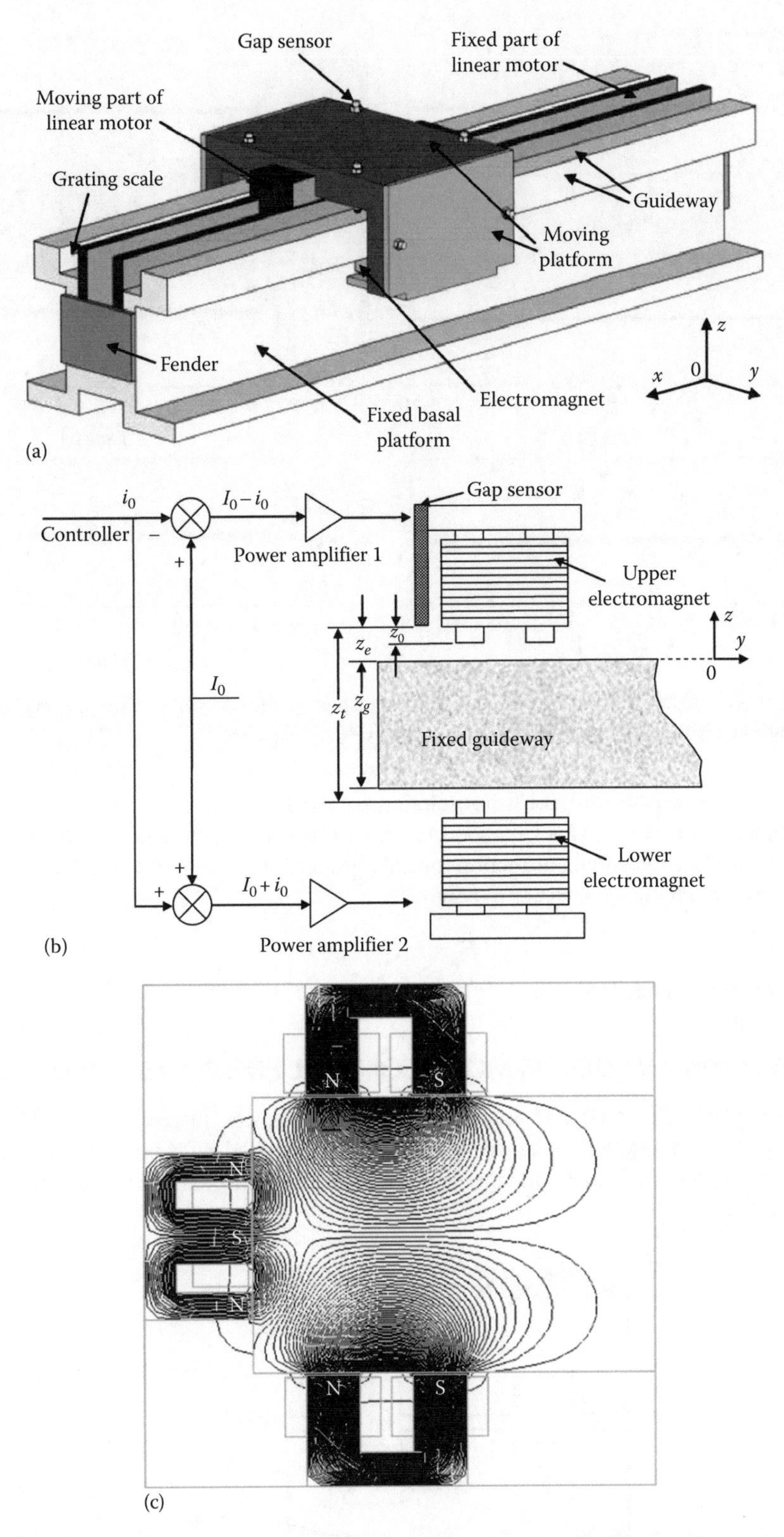

FIGURE 19.20 MAGLEV industrial transport platform: (a) framework, (b) control system for levitation (vertical motion), (c) optimized geometry for levitation (vertical motion) and guidance (lateral motion).

(*continued*)

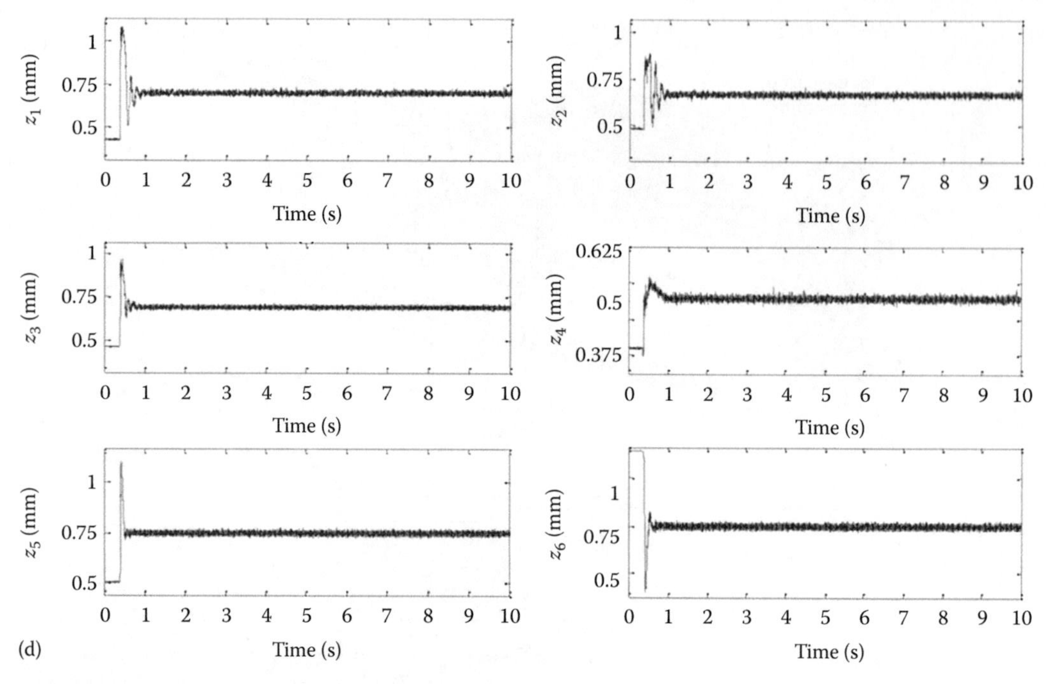

FIGURE 19.20 (continued) MAGLEV industrial transport platform: (d) six airgaps evolution during MAGLEV motion. (After Li, L. et al., *IEEE Trans.*, MAG-40, 3512, 2004.)

Airgap eddy current sensors with 0.1 μm resolution are used.

Another dual PM-assisted ALS (intended for axial active magnetic bearings Figure 19.2b) has been shown to produce a resultant attraction force F_L proportional to current and displacement from middle position, which makes it easier to control:

$$F_L = K_i I + K_x X \tag{19.61a}$$

over the entire airgap range [15].

19.11 ZERO POWER SLIDING MODE CONTROL FOR PM-ASSISTED ALSs

As shown in Figure 19.2a and b, PM-assisted ALS may also be considered (Figure 19.21).

The control of such a system is similar to that of dc controlled ALS presented so far, but here quite a different approach is taken for the description of the four operation modes to be handled, integrated by ALS:

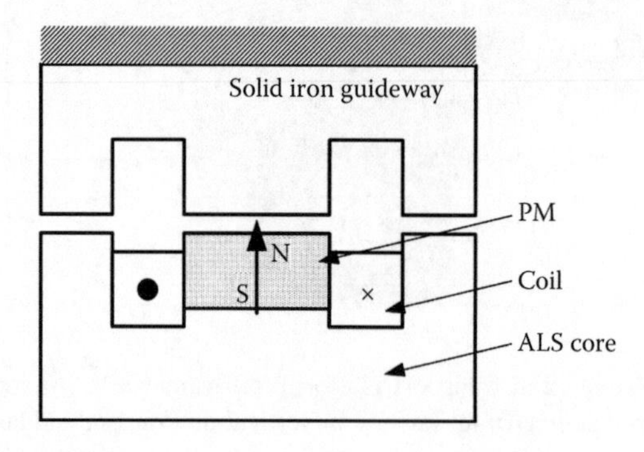

FIGURE 19.21 PM-assisted ALS.

- Smooth takeoff with switching to zero power control afterward.
- Zero control power levitation: The steady-state control coil current is maintained to zero even with payload variation, but the targeted airgap will now vary such that the PM alone may produce the whole levitation force.
- Guideway collision avoidance control takes over and controls the system at minimal airgap.
- Soft landing is provided by planned airgap versus time reference tracking control.

For zero power (current) control, sliding mode control is adopted for robustness, but first one new row is added in the state-space model, which contains $\int \Delta i dt$[16]:

$$\begin{vmatrix} \Delta Z_g \\ \Delta \dot{Z}_g \\ \int \Delta i dt \\ \Delta i \end{vmatrix} = \begin{vmatrix} 0 & 1 & 0 & 0 \\ a_{210} + \Delta a_{21} & 0 & 0 & a_{240} + \Delta a_{24} \\ 0 & 0 & 0 & 1 \\ 0 & a_{420} + \Delta a_{42} & 0 & a_{440} + \Delta a_{44} \end{vmatrix} \begin{vmatrix} \Delta Z_g \\ \Delta \dot{Z}_g \\ \int \Delta i dt \\ \Delta i \end{vmatrix} + \begin{vmatrix} 0 \\ 0 \\ 0 \\ b \end{vmatrix} \Delta V + \begin{vmatrix} 0 \\ d \\ 0 \\ 0 \end{vmatrix} F_{ext} + \begin{vmatrix} 0 \\ 1 \\ 0 \\ 0 \end{vmatrix} \ddot{Z}_s \tag{19.62}$$

$$|A| = |A_0| + |\Delta A|$$

and

$$a_{210} = \frac{1}{M}\frac{\partial F_L}{\partial Z_g}\bigg|_0, \quad a_{240} = \frac{1}{M}\frac{\partial F_L}{\partial i}\bigg|_0, \quad a_{420} = \frac{-w_1 \partial \Phi}{\partial Z_g}\bigg|_0, \quad a_{440} = \frac{-R}{L_0}, \quad b = \frac{1}{L_0}, \quad d = \frac{1}{M} \tag{19.63}$$

where

L_0 is the coil inductance at rated airgap
M is the ALS mass
R is the coil resistance
$\ddot{Z}_m$ is the track irregularities acceleration

The similarity with previous paragraphs (19.6) is strong; just a new row is added.

A sliding hyperplane σ, its reaching law $\dot{\sigma}$, and the control law ΔV are defined as [16]

$$\sigma = \underline{C} \cdot \underline{X}; \quad \underline{C} = [C_1 C_2 C_1 K, 1]; \quad \underline{X} = \left[\Delta Z_g, \Delta \dot{Z}_g \int \Delta i dt, \Delta i\right]^T \tag{19.64}$$

with

$$\dot{\sigma} = -p\sigma - q sign(\sigma) \tag{19.65}$$

and

$$\Delta V = \left(\underline{C} b^{-1}\right)\left(-\underline{C} A_0 \underline{X} - p\sigma - q sign(\sigma)\right) \tag{19.66}$$

If the system characteristic on the hyperplane is chosen as [16]

$$CF(s) = (s + t\omega_n)\left(s^2 + 2\xi\omega_n s + \omega_n^2\right) \tag{19.67}$$

then the parameters of the control law C_1, C_2, and K may be calculated as

$$C_1 = \frac{2\xi t\omega_n^{\ 2} + \omega_n^{\ 2} + a_{210}}{a_{240}}$$

$$C_2 = \frac{\left(t\omega_n + 2\xi\omega_n + t\omega_n^{\ 3}\right)/a_{210}}{a_{240}} \tag{19.68}$$

$$K = \frac{-t\omega_n^{\ 3}}{a_{210} \cdot c_1}$$

For stability, not only $p>0$ and $q>0$ in (19.65) [17], but also the external force is estimated as

$$\hat{F}_{ext} = M \cdot a_{210} \cdot K \int \Delta i dt \tag{19.69}$$

From the existence condition of the sliding hyperplane, $\sigma\dot{\sigma} < 0$, the values of p and q are obtained as

$$p > sup\left|\frac{C\Delta A + \ddot{Z}_s}{\sigma}\right| > 0$$

$$q = \left|c_2 a_{210} K \int \Delta i dt\right| > 0 \tag{19.70}$$

On a 25 kg empty vehicle in Ref. [16], a 12 kg payload is added and the airgap response is shown in Figure 19.22.

It is now evident that the steady-state current is driven to zero after the airgap is reduced from 9.00 to 6.5 mm (such that the PMs alone can handle the entire larger levitation force).

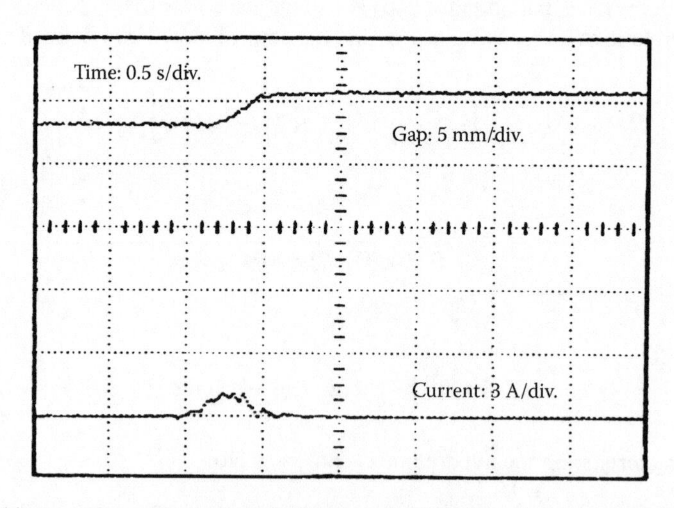

FIGURE 19.22 Airgap and current response for 12 kg payload addition (to 25 kg dead weight).

For vehicle takeoff and soft landing, as well as operation at minimum airgap safety rides, the airgap tracking mode should be initiated.

This time the state space model gets an additional state of $\int \Delta Z_g{}' dt$ (in fact, it gets the equality $\left(\int \Delta Z_g{}' dt\right)' = \Delta Z_g{}'$); and thus the current will be automatically adjusted to a nonzero value such as

$$\Delta i - T\int \Delta Z_g{}' dt \to 0 \tag{19.71}$$

The state space vector is similar to that mentioned previously, that is, $\underline{X} = \left[\int \Delta Z_g{}' dt, \Delta Z_g, \Delta \dot{Z}_g, \Delta i\right]$, and so are the sliding hyperplanes $\sigma', \dot{\sigma}'$ and $\Delta V'$. The design of the controller is similar to the one for zero current control, and the two modes may be integrated into a single control system.

Reference [16] shows a typical planned soft takeoff and landing (Figure 19.23).

Also Figure 19.23c shows minimum (safe) airgap tracking control for maximum overload (additional 25 kg for 25 kg dead weight).

The control performance is evaluated as done previously in this chapter with respect to rms airgap variation, average control power, and ride comfort ((19.45) through (19.48)), only to find

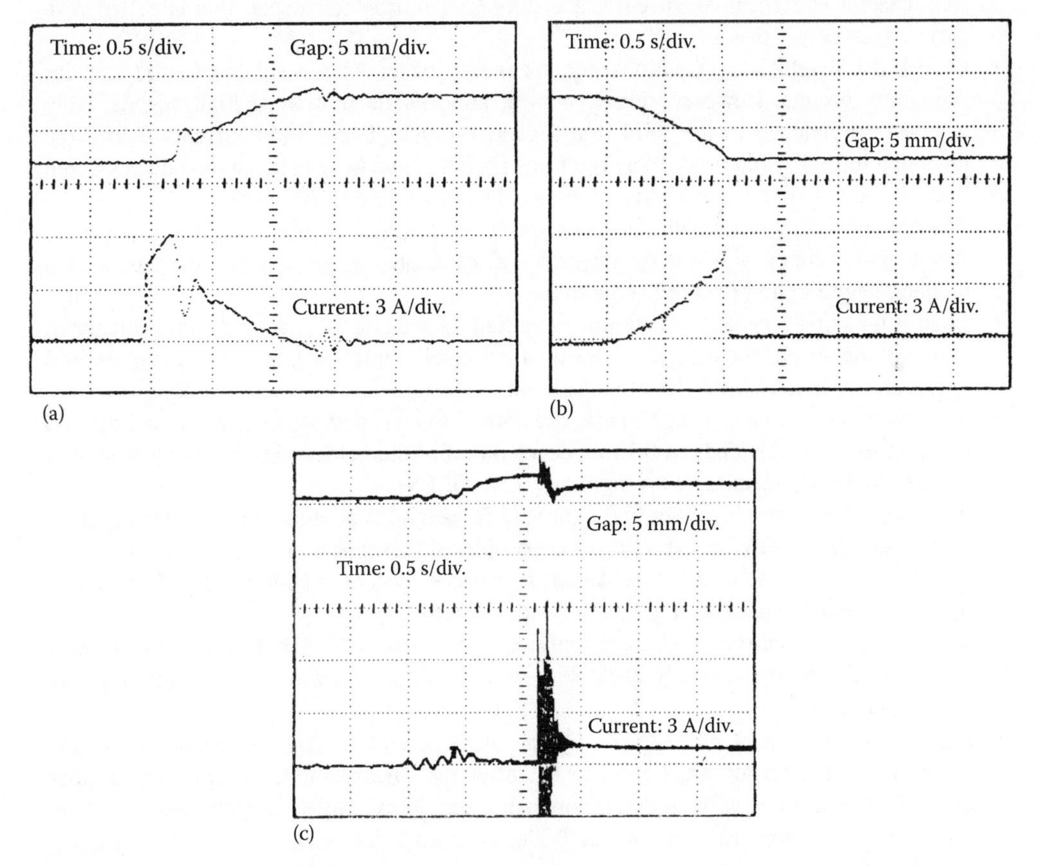

FIGURE 19.23 (a) Takeoff from 14 to 9 mm, (b) soft landing, and (c) overload (25 kg payload) operation at minimum airgap (5 mm). (After Tzeng, Y.K. and Wang, T.C., Analysis and experimental results of a MAGLEV transportation system using controlled—PM electromagnets with decentralized robust control strategy, *Record of LDIA-1995*, Nagasaki, Japan, pp. 109–113, 1995.)

that $\bar{Z}_g/Z_{g0}=1/7$ and $P_{con}=0.32$ kW/ton for a full-size 400 km/h MAGLEV [18]. Compared to at least 3 kW/ton for a dc-controlled electromagnet ALS, it seems some spectacular progress. But the cost of the PMs for high-speed MAGLEVs should not be underestimated; however, for MAGLEV industrial platforms, the zero power control via sliding modes seems to us quite practical.

19.12 SUMMARY

- ALS provide stable equilibrium of an electromagnet below a mild steel body, at a reference distance by active control of the current in the electromagnet; an ALS uses electromagnetic attraction force as expressed in Maxwell tensor.
- ALS applications range from magnetic bearings through industrial platforms to vehicular MAGLEVs (at medium and high speed).
- The primary electromagnet may be provided also with a PM when the current control in the electric coil serves mostly for airgap control.
- For PM-assisted ALS, it is feasible to operate at variable airgaps when the payload increases such that only the PMs provide the steady-state levitation force when the steady-state current is zero; nonzero current occurs only during transients; this is called zero current (power) control.
- In MAGLEV vehicle applications, the secondary of an ALS is made of solid iron for economic reasons. However, during control and vehicle motion at high speeds, eddy currents are induced in the solid iron secondary/track. These eddy currents reduce the levitation force F_{L0} and produce a drag force F_d; both have to be calculated and considered in the design of ALS for medium-/high-speed MAGLEVs (20–100 m/s).
- The design of ALS starts with the specifications of F_L/ALS weight, rated voltage, current, speed, and secondary width L_W interval and calculates essentially the electromagnets number, length L, and C-core leg width b_i.
- Then, after a first geometry is obtained, longitudinal end effects on F_L and F_d are calculated at maximum speed; the design is corrected iteratively until the value of F_L is the desired one, with end longitudinal effects considered.
- It has been shown that even at 100 m/s, the normal force is reduced only by about 25% and the drag force corresponds to 0.1 m/s deceleration of the vehicle due to longitudinal end effects. So ALSs are suitable for high-speed MAGLEVs.
- The copper losses per levitation force (or vehicle weight) is another performance criterion to be checked; 3–4 kW/ton at 100 m/s is considered reasonable and ALSs are capable of that, but only if a secondary mechanical suspension system is added; for full comfort a tertiary (cabin) mechanical suspension system is used.
- The state space equations of ALS are nonlinear; after linearization they prove to correspond to a statically and dynamically unstable system; so stabilization has to be forced on it by close-loop control.
- State feedback control with Z_g, $\hat{\dot{Z}}_g$, $\ddot{Z}_m$ (absolute acceleration) as the state vector and Z_g, $\ddot{Z}_m$ as output vector has been proven, after the linearization, to allow for smaller coefficients variation with linearization point conditions. So an airgap sensor and an absolute acceleration sensor are needed. The track irregularities are represented by $\dot{Z}_s(t)$, $Z_s+Z_g=Z_m$.
- State feedback control may be designed using an optimization criterion that secures electromagnet–guideway collision avoidance, ride comfort, and limited control power.
- As $\dot{Z}_g$ cannot be measured, an observer is needed; particular observer properties are necessary to handle track irregularity perturbations.

- Dedicated transfer functions (19.45 through 19.48) are used to calculate performance indices:
  - rms airgap deviation error: $\overline{Z}_g$
  - Voltage and current rms deviations: $\overline{\Delta V}, \overline{\Delta i}$

Ride comfort: $\phi_{\ddot{Z}_m}(\omega)$

- By a numerical example it is proven that even without a mechanical suspension system ALS may provide $\overline{Z}_g/Z_{g0} < 0.33$ (to avoid collisions with the track) but at huge voltage $\overline{\Delta V}/V_0 = 20$ and large current rations $\overline{\Delta i}/i_0 > 3$ at 100 m/s. Consequently, secondary and tertiary mechanical suspension systems are needed to handle the control at reasonable control power and good ride comfort above 5 Hz, in 100 m/s (or so) MAGLEVs.
- The vehicle lifting (take off) and landing requires special control as state feedback control cannot handle such large variations of airgap. More robust control to handle both running and lifting and landing of the vehicle is needed.
- Sliding mode control, plus a PI control close to the target and an interior fast (bi-positional) current loop, has been proven capable of handling most operation modes of a medium-speed (15–20 m/s) MAGLEV at less than 2.5 kW/ton control power and airgap variation from 20 mm (idle position) to the rated 10 mm.
- Also state feedback control plus a flux feedback was shown to be proper for robust control of ALS.
- Finally, sliding mode zero power (current) control for running and SM nonzero current control for takeoff and landing with airgap planned trajectory tracking have been proven to be adequate for PM-assisted ALSs. A calculated spectacular 0.36 kW/ton control power at 100 km/h suggests that PM-assisted ALSs should be given more consideration in the future, especially for an industrial platform.
- At the other end of the scale, for axial/magnetic bearings, a PM-assisted ALS has been proven, even by PID + SM + current control, capable for large axial force perturbation rejection for speed transients of ±18 krpm [19].
- Even the placement of a high-temperature superconductor to produce constant-width-airgap attraction force has been proven to be adequate for simplified control ALSs.
- It is estimated that ALS will spread more in the future due to low energy consumption, low noise and vibration in transportation of people or in industry, and for industrial MAGLEV platforms [20].

REFERENCES

1. H. Kemper, Suspension systems through electromagnetic forces, a possible approach to basically new transportation technologies, *ETZ*, 4, 1933, 391–395 (in German).
2. H. Kemper, Electrical railroad vehicles magnetically guided, *ETZ-A*, 1, 1953, 11–14 (in German).
3. G. Bohn, P. Romstedt, W. Rothmayer, and P. Schwarzler, A contribution to magnetic levitation technology, *Proceedings of the Fourth ICEC*, Eindhoven, the Netherlands, 1972, pp. 202–208.
4. R.M. Borcherts and L.C. Davis, Lift and drag forces for the attraction electromagnetic systems, Ford Motor Company Scientific Research Staff Report, 1976.
5. S. Yamamura and T. Ito, Analysis of speed characteristics of the attractive electromagnetic levitation vehicles, *Elec. Eng. Jap.*, 95(162), 1975, 84–89.
6. I. Boldea and S.A. Nasar, *Linear Motion Electromagnetic Systems*, Chapter 10, John Wiley & Sons, New York, 1985.
7. I. Boldea, Optimal design of attraction levitation magnets including the end effect, *EME* (now *EPCS*) *J.*, 6, 1981, 57–66.
8. F. Matsumura and S. Yamada, A method to control the suspension system utilizing magnetic attraction force, *Elec. Eng. Jpn.*, 94(6), 1974, 50–57.

9. I. Boldea, A. Trica, G. Papusoiu, and S.A. Nasar, Field tests on MAGLEV with passive guideway linear inductor motor transportation system, *IEEE Trans.*, VT-37(4), 1988, 213–219.
10. A. Forrai, T. Ueda, and T. Yumura, Electromagnetic actuator control: A linear parameter—Varying (LPV) approach, *IEEE Trans.*, 1E-54(3), 2007, 1440–1441.
11. Al. Trica, *Electromagnetic Suspension Systems*, University Politehnica Timisoara Publishers, 2009 (in Romanian).
12. V. Utkin, J. Guldner, and J. Shi, *Sliding Mode Control in Electromechanical Systems*, 2nd edn., CRC Press, Taylor & Francis Group, New York, 2009.
13. L. Li, T. Shinshi, and A. Shimokohbe, State feedback control of active magnetic bearing based on current change rate alone, *IEEE Trans.*, MAG-40, 2004, 3512–3517.
14. J.A. Duan, H.B. Zhou, and N.P. Guo, Electromagnetic design of a novel linear MAGLEV transportation platform with finite element analysis, *IEEE Trans.*, MAG-47(1), 2011, 260–263.
15. D. Wu, X. Xie, and S. Zhou, Design of a normal stress electromagnetic fast linear actuator, *IEEE Trans.*, MAG-46(4), 2010, 1007–1014.
16. Y.K. Tzeng and T.C. Wang, Analysis and experimental results of a MAGLEV transportation system using controlled—PM electromagnets with decentralized robust control strategy, *Record of LDIA-1995*, Nagasaki, Japan, 1995, pp. 109–113.
17. W. Gao and J.C. Hung, Variable structure control of linear system: A new approach, *IEEE Trans.*, 1E-40(1), 1993, 45–55.
18. Y.K. Tzeng and T.C. Wang, Dynamic analysis of the MAGLEV system using controlled—PM electromagnets with robust zero power control strategy, *Record of IEEE Intermag 1995*, San Antonio, TX, 1995.
19. K. Hijikata, M. Takemoto, S. Ogasawara, A. Chiba, and T. Fukao, Behavior of a novel thrust magnetic bearing with a cylindrical rotor on high speed rotation, *IEEE Trans.*, MAG-45(10), 2009, 4617–4620.
20. M. Ghodsi, T. Ueno, and T. Higuchi, Improvement of magnetic circuit in levitation system using HTS and soft magnetic material, *IEEE Trans.*, MAG-41(10), 2005, 4003–4005.

20 Repulsive Force Levitation Systems

Repulsive forces (RFLS) occur between two current carrying conductors of positive polarity or between a moving current carrying coil and a short-circuited fix electric circuit (conducting sheet or shorted coil) and also between a moving permanent magnet (PM) and a fix short-circuited electric circuit or between two PMs of opposite polarity. The repulsive (electrodynamic) force, in contrast to attraction force (Chapter 19), is such that it repels the other side and thus tends to be statically stable. Not so dynamically, but the damping of the motion oscillations is simpler than for ALSs. Also, the distance (gap) between the mover and the guideway can be 10 times larger than in ALSs for the same speed (for MAGLEVs, $U_{max} > 100$ m/s). But to obtain high levitation force RFLS per weight for a net mechanical airgap of around 100 mm, superconducting dc electromagnets on board of vehicle are required. A row of such superconducting magnets of opposite polarity on board of MAGLEVs interacts through motion-induced currents within an aluminum sheet or ladder secondary to levitate the vehicle. Strong PMs may be used instead (at room temperature), once their remanence flux density would go above 2 T or so. The interaction of levitation, guidance, and propulsion in RFLS MAGLEVs has been introduced to save energy and reduce initial system cost.

For magnetic bearings, PMs with active motion damping by controlled coils may be used.

The larger gaps for MAGLEVs and the simpler control for magnetic bearings make RFLS attractive in applications. A 550 km/h MAGLEV with RFLS has been demonstrated already and RFLS bearings are extensively tested for special applications.

To summarize the RFLS technology, this chapter deals with

- RFLS competitive technologies
- SC field distribution
- Levitation and drag forces for sheet secondary
- Levitation and drag forces for ladder secondary
- RFLS dynamics, stability, and active damping
- PM—ac coil RFLS
- PM-PM magnetic bearings

20.1 SUPERCONDUCTING COIL RFLS: COMPETITIVE TECHNOLOGIES

RFLS for MAGLEVs may be embedded in quite a few technologies. We select here a few, deemed as fully representative, [1,6]:

- With conductive sheet track (secondary)—Figure 20.1a
- With conductive ladder track (secondary)—Figure 20.1b
- Normal-flux type (Figure 20.1a,b)
- Null-flux integrated levitation, guidance and propulsion, Figure 20.2
- Coil + PM magnetic bearing (Figure 20.3a) [7]
- PM-PM magnetic bearing (Figure 20.3b) [8]

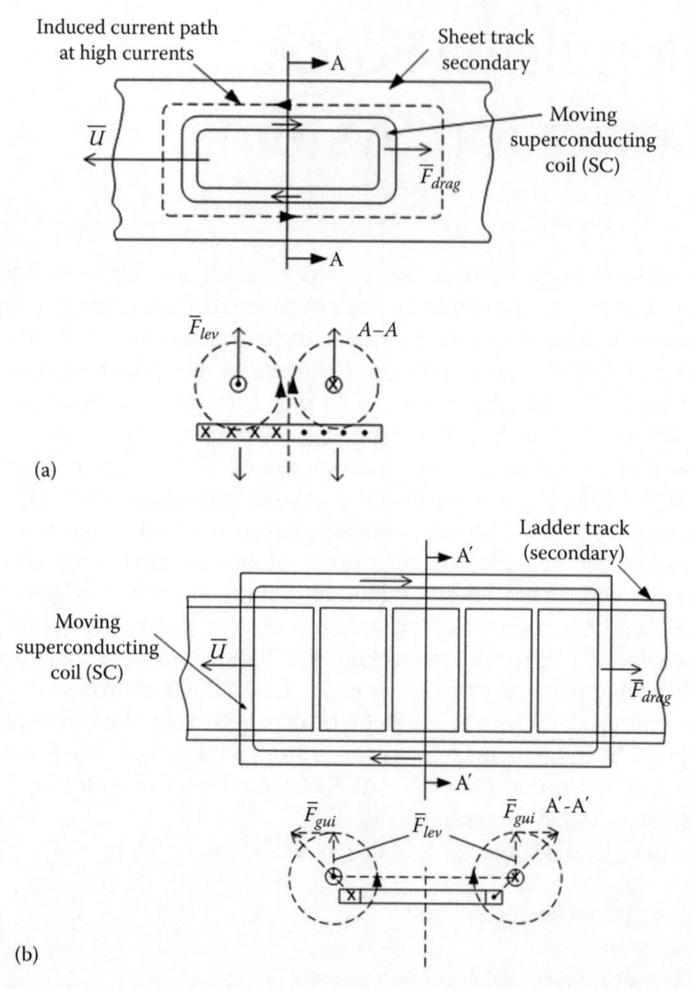

FIGURE 20.1 Normal-flux SC RFLS (only one SC is shown, from a row of NSNS coils): (a) with sheet track and (b) with ladder track.

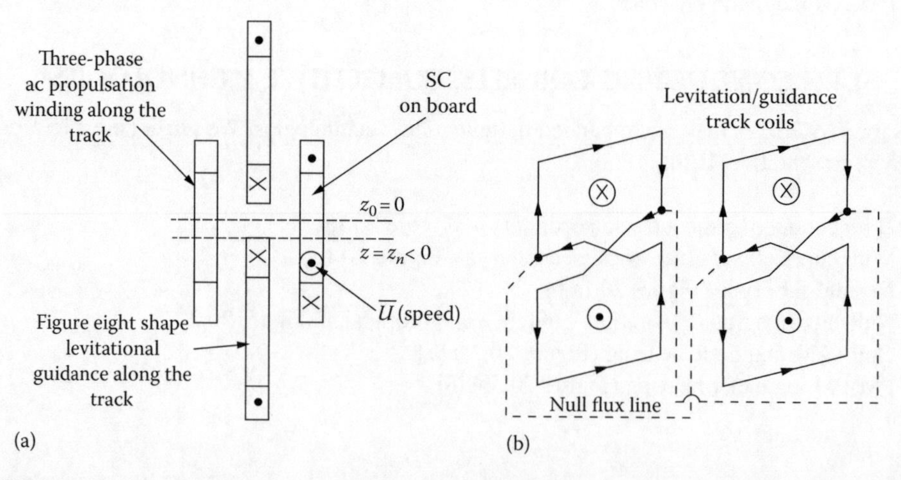

FIGURE 20.2 Null-flux SC-RFLS: (a) transverse cross section and (b) lateral view of light-shape levitation/guidance coils along the track.

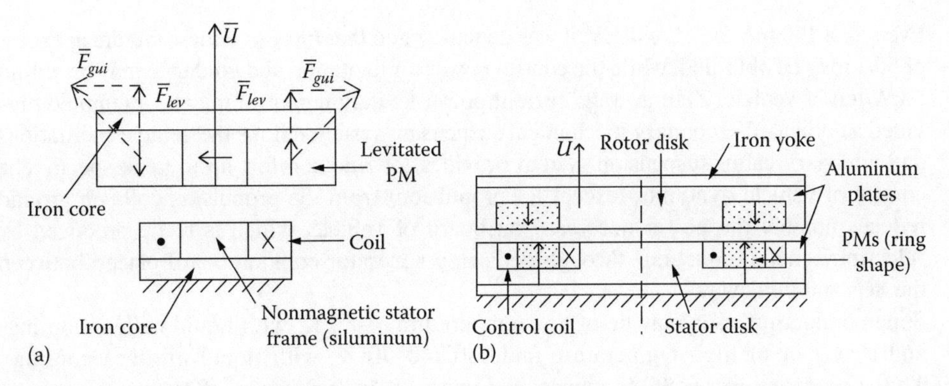

FIGURE 20.3 PM-RFLS as magnetic bearing: (a) coil-PM topology and (b) PM + PM + control coil topology.

The competitive technologies in Figures 20.1 through 20.3 may be characterized as follows:

- The normal-flux topologies with sheet or ladder track are considered as basic because they reveal the principles and fundamental characteristics: the levitation and eddy currents drag forces, F_{lev}, F_{drag} versus speed for various net airgap (levitation height) values in the 50–150 mm range. A levitation goodness factor for forces at peak speed may be defined as

$$G_F = \left(\frac{F_{lev}}{F_{drag}} \right)_{U_{max}} \tag{20.1}$$

  The typical system dynamics behavior characterized by static stability but undamped oscillations occurs in the some conditions.
- The ladder track is used to increase force goodness factor G_F but levitation and drag force pulsations occur, due to ladder transverse currents.
- The null-flux systems (Figure 20.2) have been introduced to increase the force goodness factor (which has a notable impact on propulsion power) and to lower the speed at which the peak value of drag force occurs and its value. The neighboring propulsion system ac armature coils may provide repulsive-attractive vertical (and guidance) forces by adequate vector control with controlled i_d, for damping the levitation vertical or lateral oscillations. Controlled i_q is applied for propulsion. A secondary mechanical system is all that is required for meeting the Janeway ride comfort curve (standard).

  Note: In Chapter 8, the integrated propulsion/levitation/guidance SC-MAGLEV topology has been presented. The eight-shape stator track coils produce propulsion/levitation/guidance by proper connections and control on the two sides of the vehicle. This topology is not covered here.

- Repulsive magnetic PM bearings may be built with an air coil, with controlled current, that repels a moving PM (or vice versa). As the air-coil mmf (and losses) tends to be notable, a dual PM-PM bearing may be built where the control coils (connected together) may be controlled to regulate the system dynamics during rotor motion as well as the takeoff and soft landing at standstill. For this configuration, if the airgap target is not to be tracked with very high precision, the zero (current) power control described in Chapter 19 may be applied as well.
- A power levitation goodness factor for levitation and guidance may be defined as

$$G_p = \frac{F_{lev} \cdot U}{F_{drag} \cdot U + P_{control}} \tag{20.2}$$

  $P_{control}$ is the control power

- Even for a 130 m/s SC MAGLEV, it was demonstrated that force goodness factors in excess of 50/1 may be obtained, while the control power for levitation and guidance may be within 5 kW/ton of vehicle. Zero average current/power levitation control may be attempted provided a dedicated secondary mechanical suspension system damps the vehicle oscillations and a tertiary cabin suspension system provides for ride comfort up to peak speed. The control of vehicle dynamics (except for propulsion) from the propulsion coils on ground reduces notably the power necessary on board of vehicle, which is to be produced by "electromagnetic induction" through dedicated generator coils on board placed between the SCs and guideway.
- Superconducting coils may be of low temperature (4–10 K, with liquid helium cooling) and strong or of high temperature and soft (40–70 K, with liquid nitrogen cooling). The appearance of an SC is shown in Figure 20.4a (for basics of superconductivity, see Chapter 1).

In essence, Figure 20.4a comprises the superconducting-wire coil, liquid helium (nitrogen) dewars, fiber glass evacuated insulation, supports, and electrical connections. The helium (nitrogen) dewar is isolated from the environment by a vacuum zone with stainless steel walls around it.

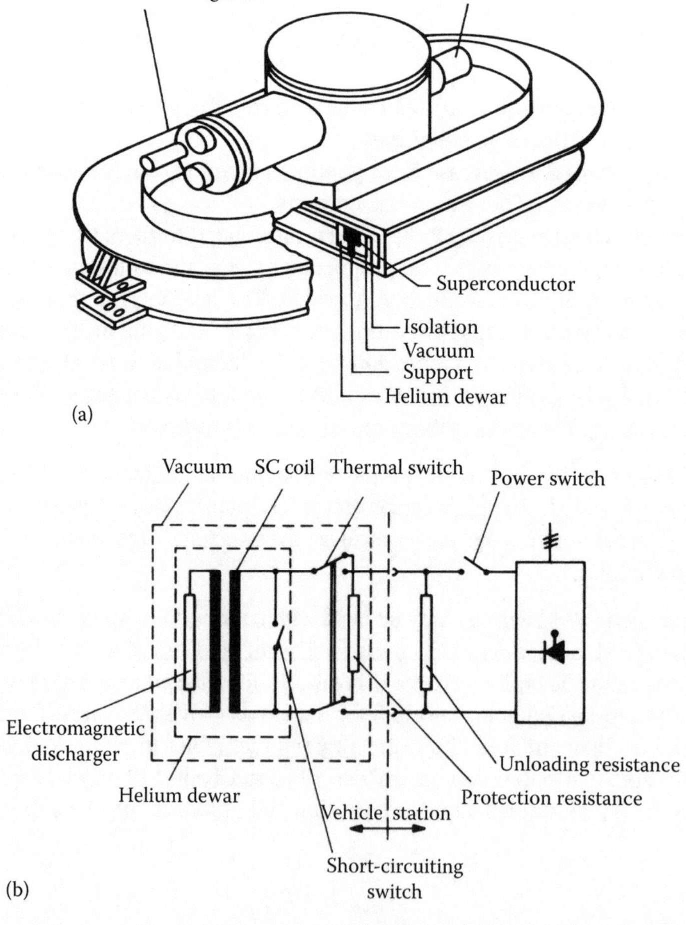

FIGURE 20.4 Superconducting coil: (a) general view and (b) electric power circuit and supply. (After Boldea, I. and Nasar, S.A., *Linear Motion Electromagnetic Systems*, John Wiley Interscience, New York, 1985.)

For strong SCs, the liquid nitrogen is first poured; after its evaporation, liquid helium is poured until the desired (4–10 K) temperature is reached. Only after that, the electric current is flown into the SC coil gradually; finally, the coil is short-circuited inside the low-temperature structure (Figure 20.4b). The SC may be replenished periodically (after 1 day or more).

20.2 SHEET SECONDARY (TRACK) NORMAL-FLUX RFLS

The superconducting coils have large mmf (above (3–4) 10^5 A turns) as they are supposed to produce in the guideway coils plane NSNS flux densities in the range of at least 0.2–0.3 T for a vertical (lateral) distance of a 0.2–0.3 m (center to center) between the two.

For start, let us consider a coil in air producing a magnetic field in point P (Figure 20.5a). As shown in Chapter 8, by using Biot–Sawart law, via Neumann formula, the magnetic flux density components B_x, B_y, B_z in point P may be obtained, as illustrated in Figure 20.5b.

The normal (along levitation direction) B_z component for one coil is far from a sinusoid along the vehicle motion direction. However, B_z varies closely to a sinusoid along the lateral direction. A row of SCs may produce, by proper length/width pitching and levitation heights, the desired flux density distribution along all three directions: x, y, z.

The SC field distribution can then be investigated analytically allowing for

- Cosinusoidal distribution along oy
- Fourier decomposition of longitudinal distribution (along ox) with the retention of fundamental, third, and fifth harmonic
- Dependence along oz (vertical) direction results from the Laplace equation and is essentially exponential for each longitudinal harmonic:

$$\frac{\partial^2 H_{x0}}{\partial x^2} + \frac{\partial^2 H_{y0}}{\partial y^2} + \frac{\partial^2 H_{z0}}{\partial z^2} = 0 \tag{20.3}$$

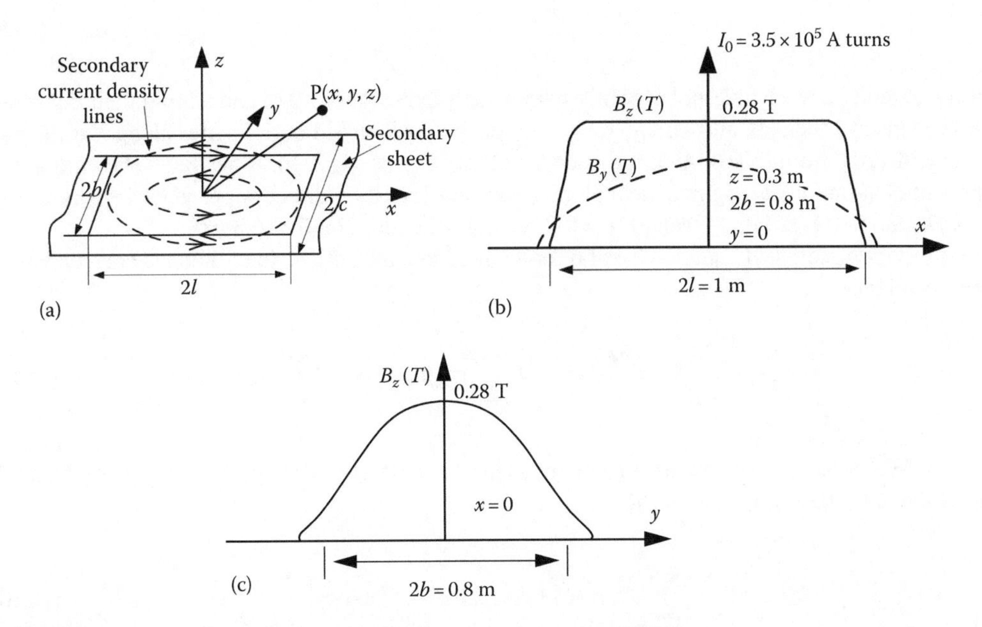

FIGURE 20.5 SC: (a) rectangular coils, (b) B_z and B_y versus x, and (c) B_z versus y.

with the solutions

$$H_{x0}(x,y,z)=\sum_{\nu=1,3,5} C_{\nu 0}\cdot j\cdot\frac{\pi\nu}{L_\nu}\cdot e^{-\gamma_{\nu 0}(z-z_0)}\cdot\cos\left(\frac{\pi}{2c}y\right)\cdot e^{j\frac{\pi\nu}{L_\nu}x} \tag{20.4}$$

$$H_{y0}(x,y,z)=\sum_{\nu=1,3,5} C_{\nu 0}\cdot\frac{\pi}{2c}\cdot e^{-\gamma_{\nu 0}(z-z_0)}\cdot\sin\left(\frac{\pi}{2c}y\right)\cdot e^{j\frac{\pi\nu}{L_\nu}x} \tag{20.5}$$

$$H_{z0}(x,y,z)=\sum_{\nu=1,3,5} C_{\nu 0}\cdot\gamma_{\nu 0}\cdot e^{-\gamma_{\nu 0}(z-z_0)}\cdot\cos\left(\frac{\pi}{2c}y\right)\cdot e^{j\frac{\pi\nu}{L_\nu}x} \tag{20.6}$$

$$\gamma_{\nu 0}=\sqrt{\left(\frac{\pi}{2c}\right)^2+\left(\frac{\pi\nu}{L_\nu}\right)^2} \tag{20.7}$$

To find the integration constants $C_{\nu 0}$, we may use another expression of $B_{z0}=\mu_0 H_{z0}$ obtained from the Neumann formula (for a rectangular coil) [6]:

$$B_{z0}=\mu_0 H_{x0}=\frac{\mu_0 I_0}{4\pi}\left\{\begin{aligned}&\frac{y+b}{(y+b)^2+z^2}\left[\frac{x+L/2}{\left[(x+L/2)^2+(y+b)^2+z^2\right]^{1/2}}-\frac{x-L/2}{\left[(x-L/2)^2+(y+b)^2+z^2\right]^{1/2}}\right]\\&-\frac{y-b}{(y-b)^2+z^2}\left[\frac{x+L/2}{\left[(x+L/2)^2+(y-b)^2+z^2\right]^{1/2}}-\frac{x-L/2}{\left[(x-L/2)^2+(y-b)^2+z^2\right]^{1/2}}\right]\end{aligned}\right\} \tag{20.8}$$

Putting together a few NSNS polarity SCs with a pitch $L\nu>L$ ($L\nu\approx 1.2\,\mathrm{L}$) and adding their contribution for a given z (height), and at, say, $y=0$, after decomposition in Fourier series along X direction, we may identify from (20.6) the integration constants $C_0\nu$ ($\nu=1, 3, 5$). It has been shown that the exponential variation of magnetic field along z (vertical/levitation direction) produces errors within (2%–3%) for $z=0.1$–0.3 m, SC length $L=1$–3 m, and SC width $2b=0.5$–0.8 m.

A similar Poisson field equation is valid for the reaction magnetic field of induced currents in the sheet secondary:

$$\frac{\partial^2 H_{rx}}{\partial x^2}+\frac{\partial^2 H_{ry}}{\partial y^2}+\frac{\partial^2 H_{rz}}{\partial z^2}=-jU\frac{\pi\nu}{L_\nu}\sigma_{Al}\mu_0 H_{rz}=jU\frac{\pi\nu}{L_\nu}\sigma_{Al}\mu_0 H_{z0} \tag{20.9}$$

If in (20.9) the SC field variation in the sheet secondary is neglected, $\partial^2 H_{rz}/\partial z^2=0$; ($d_{AL}=20$–$25$ mm), the solution of (20.9) simplifies to [6]

$$H_{rz}(x,y,z)=\sum_{\nu=1,3,5} C_{\nu\nu}\cdot H_{z0}^{\nu}(x,y,z_0)\cdot e^{-\gamma_\nu(z-z_0)}\cdot\cos\left(\frac{\pi}{2c}y\right)\cdot e^{j\frac{\pi\nu}{L_\nu}x} \tag{20.10}$$

$$H_{rx}(x,y,z)=\sum_{\nu=1,3,5} C_{\nu\nu}\cdot H_{x0}^{\nu}\left(x,y,z_0\right)\cdot e^{-\gamma_\nu(z-z_0)}\cdot\cos\left(\frac{\pi}{2c}y\right)\cdot e^{j\frac{\pi\nu}{L_\nu}x} \tag{20.11}$$

$$H_{y0}(x,y,z)=\sum_{\nu=1,3,5} C_{\nu\nu}\cdot H_{y0}^{\nu}\left(x,y,z_0\right)\cdot e^{-\gamma_\nu(z-z_0)}\cdot\sin\left(\frac{\pi}{2c}y\right)\cdot e^{j\frac{\pi\nu}{L_\nu}x} \tag{20.12}$$

with

$$\left|\gamma_\nu\right|^2=\gamma_{\nu0}^2+jU\frac{\pi\nu}{L_\nu}\mu_0\sigma_{Al};\quad C_{\nu\nu}=\frac{-jU\nu\pi\mu_0\sigma_{Al}}{L\cdot\gamma_\nu^2} \tag{20.12a}$$

but $H_{z0}^{\nu}, H_{y0}^{\nu}, H_{x0}^{\nu}$ are available, (20.4) through (20.6), and thus H_{rz}, H_{rx}, H_{ry} may be calculated.

The x, y secondary current density lines may simply be calculated as

$$d_{A2}\bar{J}=rot\hat{H}_r;\quad d_{Ai}\bar{J}_x=\frac{\partial H_{ry}}{\partial y}-\frac{\partial H_{rz}}{\partial z};\quad d_{Ai}\bar{J}_y=\frac{\partial H_{rz}}{\partial z}-\frac{\partial H_{rx}}{\partial x} \tag{20.13}$$

For $L=2.7$ m, $L_\nu=3$ m, $I_0=3\times10^5$ A turns, $2b=0.5$ m, $z_0=0.25$ m, $2c=0.7$ m, $d_{Al}=25$ mm at $U=10$ m/s, and at $U=100$ m/s, the variation of induced current density components in the secondary sheet along the direction of motion is as shown in Figure 20.6.

As the secondary sheet thickness is $d_{Al}=25$ mm, it seems (from Figure 20.6) that the variation of current density along its depth, at high speeds, may not be neglected.

This will be true even for ladder secondary when, to reduce skin effect, a transposed wire conductor has to be used.

Further on, the levitation and drag forces F_L and F_d (Figure 20.7) are

$$F_L(u)=\frac{\mu_0}{2}\cdot N\cdot Re\left[\int_0^{L_\nu}\int_{-c}^{c}\int_{z0}^{z0+d_{Al}}(J_x\cdot H_{y0}^{*}-J_y\cdot H_{x0}^{*})dx\cdot dy\cdot dy\right]$$

$$=\frac{\mu_0 NL_\nu c}{2}\sum_{\nu=1,3,5}C_{\nu0}^2\cdot\gamma_{\nu0}^2\cdot Re\left[C_{\nu\nu}\frac{\gamma_\nu-\gamma_{\nu0}}{\gamma_\nu}\left(1-e^{-\gamma_\nu\cdot d_{Al}}\right)\right] \tag{20.14}$$

$$F_d(u)=\frac{\mu_0}{2}\cdot N\cdot Re\left[\int_0^{L_\nu}\int_{-c}^{c}\int_{z0}^{z0+d_{Al}}\left(J_y\cdot H_{z0}^{*}\right)dx\cdot dy\cdot dy\right]$$

$$=\frac{\mu_0 Nc\pi}{2}\sum_{\nu=1,3,5}\nu\cdot C_{\nu0}^2\cdot\gamma_{\nu0}\cdot Re\left[j\cdot C_{\nu\nu}\frac{\gamma_\nu-\gamma_{\nu0}}{\gamma_\nu}\left(1-e^{-\gamma_\nu\cdot d_{Al}}\right)\right] \tag{20.14a}$$

An attractive method to calculate the levitation force is to start with image levitation force F_{Li}, which is produced by the SC in interaction with its image (Figure 20.8):

$$F_{Li}=-\left(\frac{\partial W_m}{\partial z}\right)_{I_0=ct.}=I_0^2\left(\frac{\partial M_L}{\partial z}\right)_{z=2z_0} \tag{20.15}$$

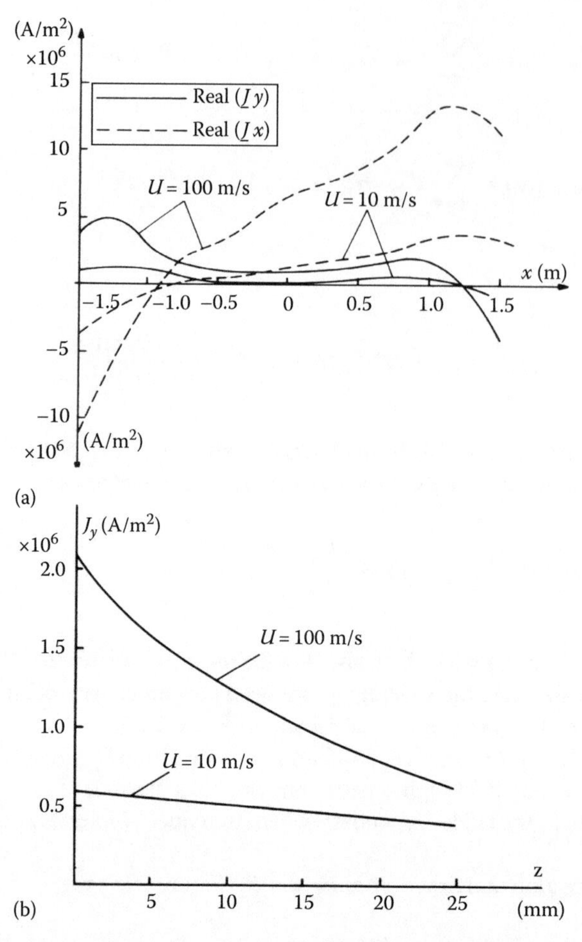

FIGURE 20.6 Secondary current density components variation along motion direction, (a) and the variation of J_y along secondary sheet depth, (b) at 10 m/s and at 100 m/s. (After Boldea, I. and Nasar, S.A., *Linear Motion Electromagnetic Systems*, John Wiley Interscience, New York, 1985.)

$$M_L = \mu_0 \left\{ \begin{array}{l} L\left\{ \ln\left[\dfrac{L}{2z_0} + \left(1+\left(\dfrac{L}{2z_0}\right)^2\right)^{1/2} \right] - \dfrac{2z_0}{L} - \left(1+\left(\dfrac{2z_0}{L}\right)^2\right)^{1/2} \right\} \\ +2b\left\{ \ln\left[\dfrac{b}{z_0} + \left(1+\left(\dfrac{b}{z_0}\right)^2\right) \right]^2 - \left(1+\left(\dfrac{z_0}{b}\right)^2\right)^{1/2} + \dfrac{z_0}{b} \right\} - \\ -L\left\{ \ln\left[\dfrac{L}{(4b^2+4z_0^2)^{1/2}} + \left(1+\dfrac{L^2}{4b^2+4z_0^2}\right)^{1/2} \right] - \left(1+\dfrac{4b^2+4z_0^2}{L^2}\right)^{1/2} + \dfrac{\left[4b^2+4z_0^2\right]^{1/2}}{L} \right\} \\ -2b\left\{ \ln\left[\dfrac{2b}{\left[L^2+4z_0^2\right]^{1/2}} + \left(1+\dfrac{4b^2}{L^2+4z_0^2}\right)^{1/2} \right] - \left(1+\dfrac{L^2+4z_0^2}{4b^2}\right)^{1/2} + \dfrac{\left[L^2+4z_0^2\right]^{1/2}}{2b} \right\} \end{array} \right\} \quad (20.16)$$

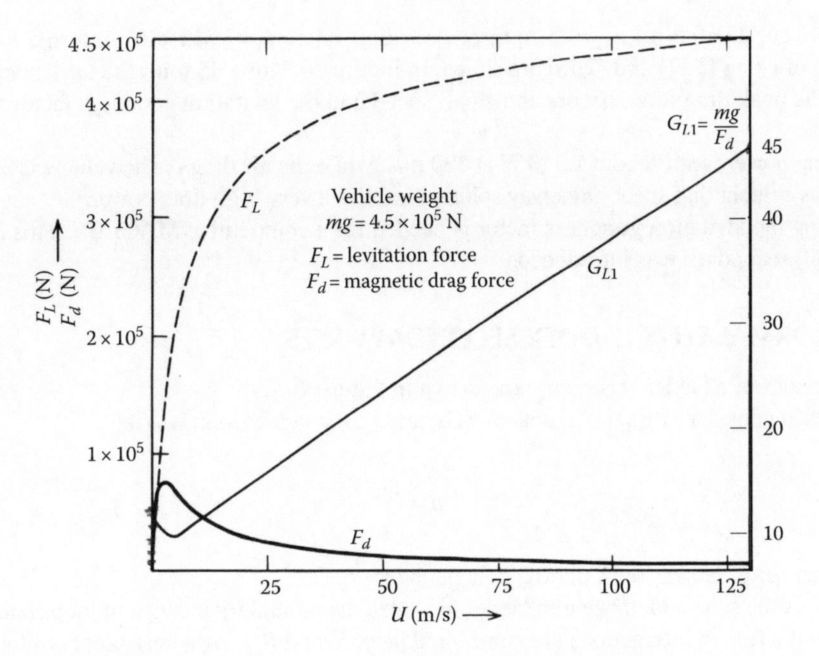

FIGURE 20.7 Levitation F_L, drag force F_d, and levitation goodness force factor G_{L1}. (After Nasar, S.A. and Boldea, I., *Linear Motion Electric Machines*, John Wiley Interscience, New York, 1976.)

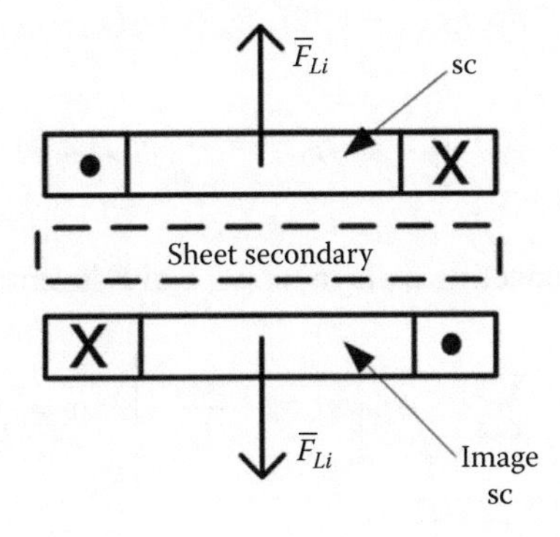

FIGURE 20.8 Image SC.

The variation of levitation and drag force with speed may be approximated as [6]

$$F_L(v) = F_{Li}\left[1 - \frac{1}{\left[1 + (U/U_0)^2\right]^{n1}}\right] \tag{20.17}$$

$$F_d(u) = \frac{U_0}{U}F_L; \quad U \approx \frac{2}{\mu_0 \sigma_{AL} d_{AL}} \tag{20.18}$$

Analytical 3D or FEM-3D methods may be used to calculate $F_L(u)$, $F_d(u)$, and then calibrate Equations 2.17 and 2.18 as practical approximations.

For $L=3$ m, $2b=0.5$ m, $d_{AL}=25$ mm, $z_0=0.3$ m, $n_1=0.2$, $I_0=2.315\times10^5$ A turns, $F_{Li}=55$ ton, the results of using (2.17) and (2.18) are shown in Figure 20.7 for a 45 ton vehicle. The results look good, as the peak drag force occurs at a small speed and the levitation goodness factor at 130 m/s is about 45.

The drag power is still about 1.3 MW at 130 m/s, while the air drag of the vehicle is 3 MW.

It is thus evident that sheet guideway solution implies a very high drag power.

Doubling the levitation goodness factor is needed for a competitive MAGLEV. This is how the ladder (coil) secondary was introduced.

20.3 NORMAL-FLUX LADDER SECONDARY RFLS

The schematics of a ladder secondary are shown in Figure 20.9.

Let us still consider l_T, the pole pitch of SCs, and l_R, the ladder loop length:

$$n=\frac{l_T}{l_R} \tag{20.19}$$

n is the member of ladder loops per SC pole pitch ($n>1$).

Also, L_c is the self-inductance of a ladder loop, M_{oj} the mutual inductance of loops o and j. Only the right and left loop interactions are considered here. R_l and R_t are the resistances of longitudinal and transversal sides of the loops.

So the equation of loop o in Figure 20.9 is

$$\begin{aligned} &L_c\frac{d}{dt}(i_o(x_o,t))+2R_ei_o(x_o,t)+R_t[2i_o(x_o,t)-i_o(x_o-l_R,t)-i_o(x_o+l_R,t)\\ &-M_{cj}\frac{d}{dt}(i_o(x_o+l_R,t)+i_o(x_o-l_R,t)]=-\frac{d}{dt}\psi_{sc}(x_o,t)=E(x_o,t) \end{aligned} \tag{20.20}$$

where ψ_{sc} is the flux of superconducting coils in loop o and E its emf:

$$E\approx\sum_{\nu=1,3,5}E_\nu\cos\left[\nu\left(\omega t+\frac{\pi x_o}{l_T}\right)\right];\quad \omega=\frac{\pi u}{l_T} \tag{20.21}$$

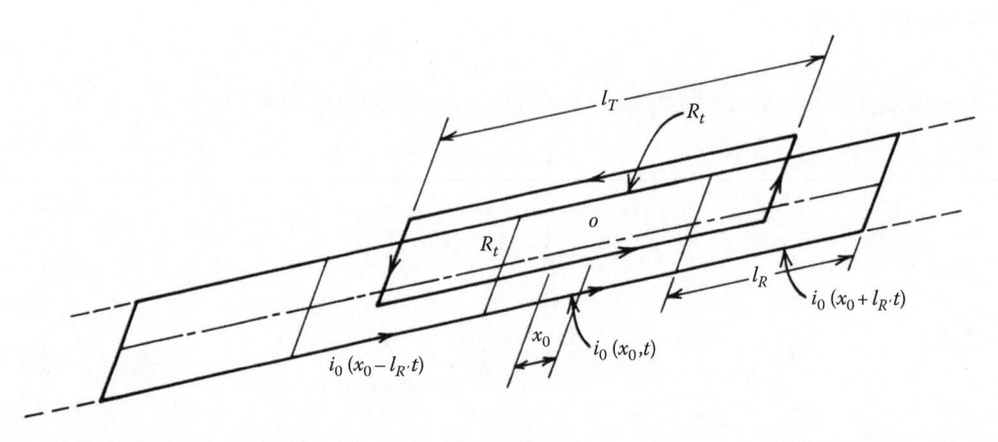

FIGURE 20.9 Ladder secondary schematics. (After Nasar, S.A. and Boldea, I., *Linear Motion Electric Machines*, John Wiley Interscience, New York, 1976.)

So the induced current in the loop o writes

$$i_o(x_o,t) \approx \sum_{\nu=1,3,5} i_\nu \cos\left(\nu\left(\omega t + \frac{\pi x_o}{l_T}\right) - \varphi_k\right) \tag{20.22}$$

Finally,

$$i_o(x_o,t) = \sum_{\nu=1,3,5} \frac{E_\nu\left(\nu(\omega t + (\pi x_o/l_T)) - \Phi_k\right)}{\left\{\left[2R_l + 2R_t\left(1-\cos\nu\pi/n\right)\right]^2 + \nu^2\omega^2\left(L_c - 2M_{c1}\cos\nu\pi/n\right)^2\right\}^{1/2}} \tag{20.23}$$

with

$$\tan\Phi_k = \omega T_{ov} \tag{20.24}$$

and

$$T_{ov} = \frac{\nu\left(L_c - 2M_{c1}\cos\nu\pi/n\right)}{2\left[R_l + R_t\left(1-\cos\nu\pi/n\right)\right]} = \nu\frac{L_{ce}}{R_{ce}} \tag{20.25}$$

Now the levitation and drag forces for the superconducting coil are

$$F_L(t) = \sum_{i=1}^{n} I_o i_o(m\,l_R,t)\frac{\partial M_m}{\partial z} \tag{20.26}$$

$$F_d(t) = \sum_{i=1}^{n} I_o i_o(m\,l_R,t)\frac{\partial M_m}{\partial x} \tag{20.27}$$

where M_m is the mutual inductance between the SC and m the track loop ($m>R$ for each SC). As expected, the discrete structure of the ladder secondary leads to levitation and drag force pulsations. They should be reduced by proper design; also the peak drag force should be forced to occur at low speeds to facilitate vehicle starting.

For $l_T = 3.2$ m, L=2.2 m (SC length), $2b$=0.5 m, the harmonics of E are very small. For n=2 and N-count SCs the forces are obtained from (20.23) to (20.27):

$$F_L \approx NM_L I_o^2 \frac{\omega^2 L_{ce}}{R_{ce}^2 + \omega^2 L_{ce}^2}\frac{\partial M}{\partial z} \tag{20.28}$$

$$F_d \approx NM_L I_o^2 \frac{\pi}{L_T}\frac{\omega R_{ce}}{R_{ce}^2 + \omega L_{ce}^2} \tag{20.29}$$

with M_L—the mutual inductance between an SC and its image (2.15). For $n=2$ and $\nu=1$,

$$T_{01}=\frac{L_{ce}}{R_{ce}}=\frac{L_c}{2(R_t+R_l)} \tag{20.30}$$

So

$$F_L\approx\frac{F_{Li}U^2}{U^2+\left(l_T/\pi T_{01}\right)^2};\quad F_d\approx F_L\left(\frac{M_L}{-\partial M/\partial z}\right)\frac{1}{T_{01}}\frac{F_{Li}U^2}{U^2+\left(l_T/\pi T_{01}\right)^2} \tag{20.31}$$

For $N=12$ SCs, and the aforementioned data, $T_{01}=0.15$ s, $I_0=2.3\times10^5$, when the vehicle has 45 ton results as in Figure 20.10 [5] are obtained $\left(-M_L/(\partial M/\partial z)=0.134\text{ m}\right)$.

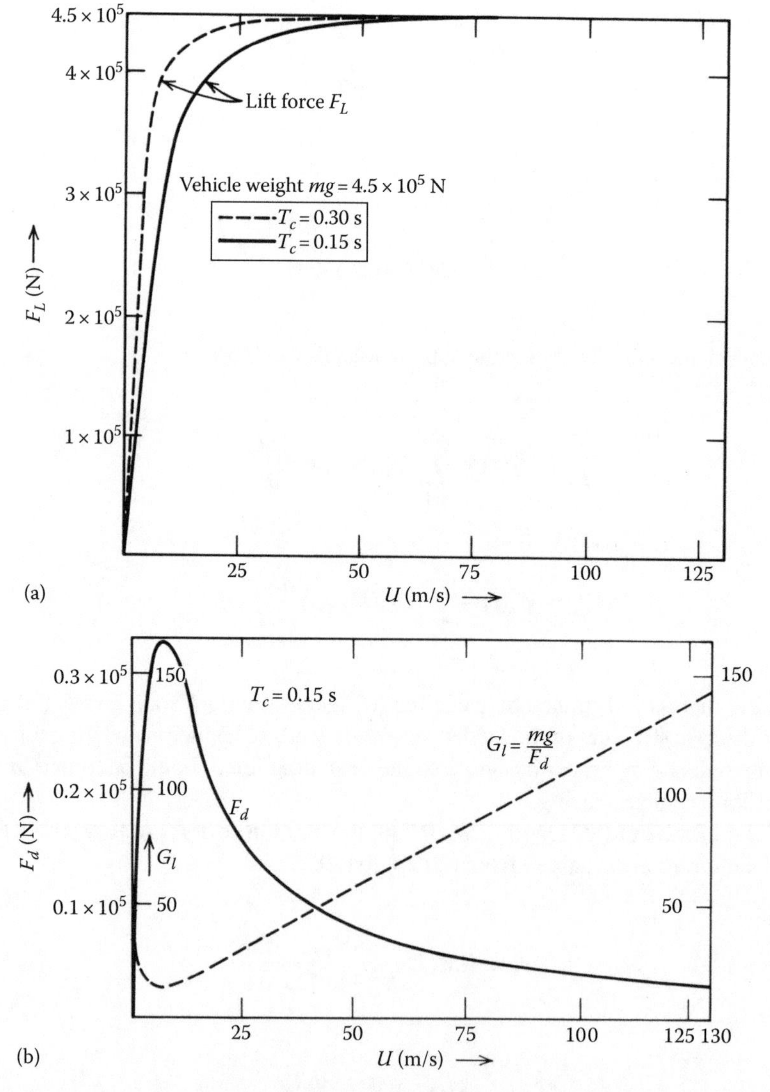

FIGURE 20.10 Ladder secondary normal-flux RFLS: (a) levitation force and (b) drag force F_d and goodness factor. (After Nasar, S.A. and Boldea, I., *Linear Motion Electric Machines*, John Wiley Interscience, New York, 1976.)

It is clear that the larger the time constant T_{01}, the faster the image levitation force is obtained, however, at increased ladder secondary costs. For a realistic $T_{01}=0.15$ s, the image force is obtained around 200 km/h. The peak value of drag force is only 0.32×10^5 N more than two times lower than for the sheet secondary. Also, the levitation goodness factor reaches a staggering value of 142 at 130 m/s. The force pulsations are small as E_1 has small time harmonics, by proper design. The drag power is much smaller than for sheet secondary, but track cost may not be smaller as the manufacturing costs of the ladder secondary tend to be larger.

For even better performance, the null-flux system was introduced.

20.4 NULL-FLUX RFLS

The null-flux system may be realized with horizontal or with vertical coils; the vertical coil solution is easy to build and handle by the vehicle. The figure-eight-shape secondary is a typical realization of this solution (Figure 20.11).

When the SC is vertically in the horizontal axis of the track coils, the SC flux in the latter is zero. But, as soon as the SC (the vehicle) falls by z_0 from that position, the flux is not zero anymore. A net levitation force is produced by the interaction of the SC field with the track-induced currents. Let us denote by M_1 and M_2 the mutual inductances between the SC and the two parts (upper and lower part) of the eight-shape coils in the track.

Then, proceeding as for the ladder secondary (with $n=2$) we may obtain the levitation and drag forces per vehicle as

$$F_{Ld} \approx -N(M_1 - M_2)I_o^2 \frac{\omega^2 L_{cd}}{R_{cd}^2 + \omega^2 L_{cd}^2} \frac{\partial}{\partial z}(M_1 - M_2) \tag{20.32}$$

$$F_{dd} \approx N(M_1 - M_2)^2 I_o^2 \frac{\omega R_{cd}}{R_{cd} + \omega^2 L_{cd}^2}; \quad T_{cd} = L_{cd}/R_{cd} \tag{20.33}$$

R_{cd} is the resistance of upper (or lower, if equal to each other) track coil and L_{cd} is the inductance of the same, considering the influence of neighboring track coils.

Taking the same example ($l_T = 3.2$ m, $L=2.2$ m, $2b=0.5$ m) with $z_0 = -2\times10^{-2}$ m, we may obtain

$$\frac{M_1 - M_2}{\partial M_2/\partial z - \partial M_1/\partial z} \approx 0.1 \tag{20.34}$$

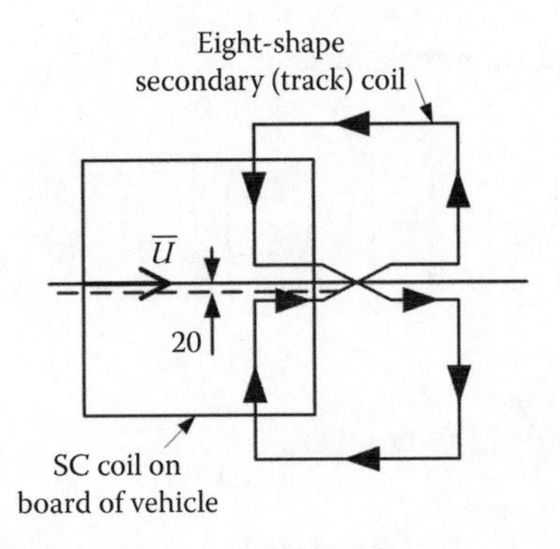

FIGURE 20.11 Vertical null-flux RFLS.

Following (20.31)

$$F_{Ld} \approx -F_{Li}\frac{U^2}{U^2+\left(l_T/\pi T_{cd}\right)^2};\quad F_{dd} \approx F_{Li}\left(\frac{M_1-M_2}{\frac{\partial}{\partial z}\left(M_1-M_2\right)}\right)\frac{1}{T_{cd}}\frac{U^2}{U^2+\left(l_T/\pi T_{cd}\right)^2} \tag{20.35}$$

This way, the levitation stiffness S_z is

$$S_z = \frac{\partial F_L/\partial z}{F_L} \tag{20.36}$$

and may be calculated as $S_z=12$ (it was about 8.0 for the sheet secondary); the peak drag force, for the 45 ton levitation force is now $F_{dk} = 0.228\times10^5\,\text{N} < 0.32\times10^5\,\text{N}$ of the ladder secondary (so $G_L>200$!). The levitation goodness factor is thus better. But to obtain the required levitation force, the SC mmf has to be notably larger than for the normal-flux systems.

Note: As shown in Chapter 8, the vertical eight-shape track coils connected to form a nonoverlapping three-phase winding ($q=1/2$ slots/pole/phase), and connected across the vehicle to secure guidance higher stiffness, can provide integrated propulsion levitation/guidance.

All done so far in this chapter has paved the way with fields and forces fundamental knowledge to approach this rather complex solution; but it is beyond our scope to follow it farther.

20.5 DYNAMICS OF RFLS

To illustrate the fundamentals of RFLS dynamics, we still use the normal-flux topology, as it is simpler to evaluate the essentials.

Let us consider the SCs of a bogie, on one side, and use the small deviation theory:

$$z(t) = z_0 + z_1(t) \tag{20.37}$$

The track irregularities are neglected and no other suspension system is considered; the vertical motion equation is

$$m\ddot{z}_1 = -NI_0^2\left\{1-\left[1+\left(\frac{U}{U_0}\right)^2\right]^{-n_1}\right\}\frac{\partial M}{\partial z} - mg \tag{20.38}$$

Notice the $\partial M/\partial z<0$ with I_0 and U as constants:

$$\ddot{z}_1 = -\frac{1}{m}NI_0^2\left\{1-\left[1+\left(\frac{U}{U_0}\right)^2\right]^{-n_1}\right\}\left(\frac{\partial^2 M_L}{\partial z^2}\right)_{2z_0}\cdot z_1 \tag{20.39}$$

In (20.39), it was assumed that $F_{Lo} = mg$.

The solution of (20.39) is evidently sinusoidal:

$$z_1(t) = A\cos\left(\gamma t + \Phi_z\right) \tag{20.40}$$

At high speeds, with

$$\gamma = \left[-g \left(\frac{\partial^2 M_L / \partial z^2}{\partial M_L / \partial z} \right)_{2z_0} \right]^{1/2} \; ; \quad \frac{\partial^2 M_L}{\partial z^2} > 0$$

the system stiffness S_z (20.36) is

$$S_z = \frac{\partial^2 M_L / \partial z^2}{\left| \partial M_L / \partial z \right|} > 0 \tag{20.41}$$

For $z_0 = 0.3\ m$ and the sheet secondary numerical example in Section 20.2 $S_z = 8.0$, $\gamma = 8.75$ rad, $f_0 = \gamma / 2\pi \approx 1.4$ Hz.

The oscillations are not attenuated, and thus the system is dynamically unstable while statically it is stable, as $S_z > 0$.

This rationale is valid mainly at high speeds. It has been proved that at same speeds [9] there is some positive damping for the ladder secondary (not so for sheet secondary).

20.6 DAMPING RFLS OSCILLATIONS

The oscillations in an RFLS may be damped in three main ways:

- Passive electric dampers (PED) (Figure 20.12)
- Active electric dampers (AED) (Figure 20.13)
- Secondary suspension system (SSS) (Figure 20.14)
- Combined passive electric and secondary suspension system (PED + SSS) (Figure 20.15)

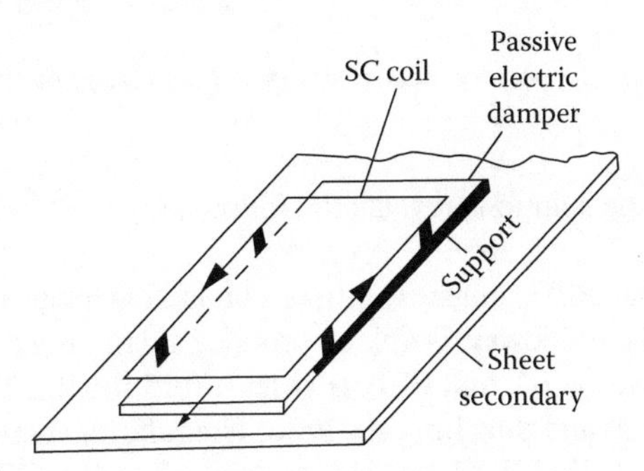

FIGURE 20.12 Passive electric damper sheet. (After Nasar, S.A. and Boldea, I., *Linear Motion Electric Machines*, John Wiley Interscience, New York, 1976.)

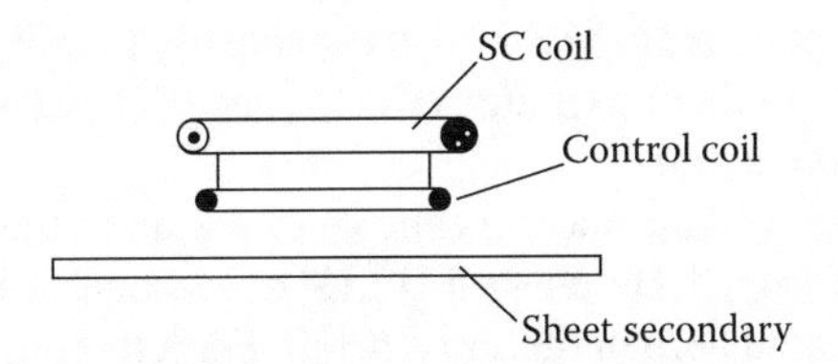

FIGURE 20.13 Active electric damping coil. (After Nasar, S.A. and Boldea, I., *Linear Motion Electric Machines*, John Wiley Interscience, New York, 1976.)

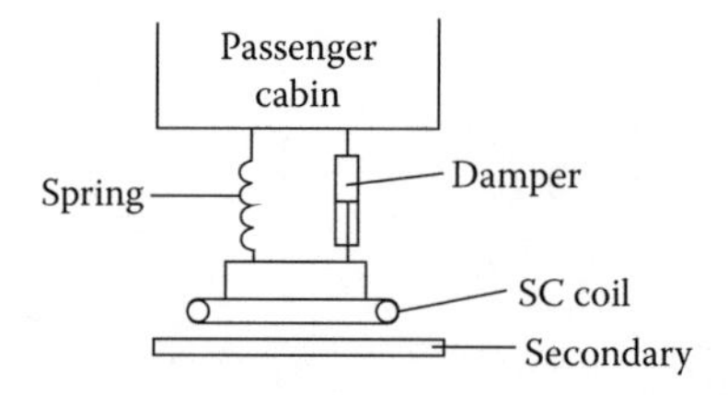

FIGURE 20.14 Secondary suspension system (SSS).

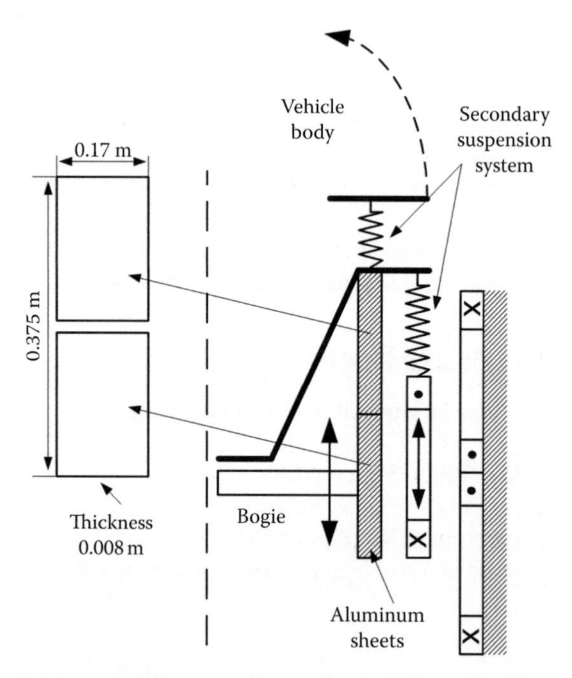

FIGURE 20.15 Combined SSS and passive electric damper sheet placed on the bogie.

A short characterization of the solutions reveals the following:

- It has been shown that PED, consisting of a conducting plate (or shorted coil) placed between the SC and the guideway, is able to produce a damping time constant $\tau < 1$ s only if the distance between the SC and PED is at least 0.15 m [10]. This leaves 0.02–0.05 m vehicle-guideway airgap and thus imposes lower irregularity (higher costs) track, to avoid collisions. Also, below 2 Hz PED is not able to damp the SC oscillations.
- The active electric damper is supposed to be capable of damping the MAGLEV oscillations and assure enough ride comfort, but at notable energy consumption on board; this solution will be detailed later.
- The active (controlled) pneumatic SSS is also capable of damping oscillations and providing full ride comfort up to the maximum vehicle speed [11]; being fully mechanical, it is not followed here in more detail.
- The PED + SSS (passive) system may be capable of damping the vertical oscillations and providing ride comfort from 1 Hz onward [12,13] at reasonable initial system costs. This solution will also be treated here in the same detail. For a thorough synthesis on dynamics and stability of RFLS, see Ref. [13].

20.6.1 Active Electric Damper

Let us start with levitation force approximate linearized expression for a normal-flux RFLS [6]:

$$F_L \approx F_{Li} \cdot \frac{\omega^2\tau^2}{1+\omega^2\tau^2}\left[1-2\gamma_{10}z_1(t)-\frac{1}{\omega^2\tau}\cdot\frac{(1-\omega^2\tau^2)}{(1+\omega^2\tau^2)}\dot{z}_1(t)\right] \quad (20.42)$$

$\tau = L_e/R_e$ equivalent electric time constant for sheet or ladder guideway coils, F_{Li}—image levitation force and $\omega = \pi \cdot U/L_v$ is related to vehicle speed; z_1—vertical airgap deviation from linearization value z_0; $\gamma_{10} = \pi/L_v$.

If we add now a control coil fixed below the SC, at z_c height, the motion equation of the SC is

$$m\ddot{z}_1 = F_L - mg - 2I_c(t)I_0\left(\frac{\partial M'_{Lc}}{\partial z}\right)_{z_0+z_c} \cdot \frac{\omega^2\tau^2}{1+\omega^2\tau^2} \quad (20.43)$$

$I_c(t)$ is the control coil current and M'_{Lc} is its mutual inductance with SC coil image, with respect to the secondary (track). Other image interactions are neglected.

To damp the vertical oscillations, the control current is proposed to be

$$I_c(t) = \alpha_c\dot{z}_1(t) - kz_1(t) \quad (20.44)$$

This control law of current presupposes that the value of $z_1(t)$ may be measured and $\hat{\dot{z}}_1(t)$ may be estimated (as done for attraction force levitation systems, by an observer, with z_1 and $\ddot{z}_m$ measured (the latter by an inertial accelerometer).

Introducing (20.44) in (20.43), one obtains

$$m\ddot{z}_1(t) + \frac{\omega^2\tau^2}{1+\omega^2\tau^2}\left[2F_{Li}\gamma_{10} - 2K_zI_0\left(\frac{\partial M'_{Lc}}{\partial z}\right)_{z_0+z_c}\right]\cdot z_1(t)$$
$$+\left[\frac{\gamma_{10}}{\omega^2\tau}F_{Li}\frac{(1-\omega^2\tau^2)}{(1+\omega^2\tau^2)} + 2\alpha_vI_0\left(\frac{\partial M'_{Lc}}{\partial z}\right)_{z_0+z_c}\right]\dot{z}_1(t) = 0 \quad (20.45)$$

The solution of (20.46) should be a damped oscillation—for stable operation—with a time constant τ_d and a frequency ω_0:

$$z_1(t) = z_0e^{-t/\tau_d}\cos\omega_0t \quad (20.46)$$

with

$$\tau_d = \frac{2\cdot m}{\frac{\omega^2\tau^2}{1+\omega^2\tau^2}\left[\frac{\gamma_{10}}{\omega^2\tau}F_{Li}\frac{(1-\omega^2\tau^2)}{(1+\omega^2\tau^2)} + 2\alpha_vI_0\left(\frac{\partial M'_{Lc}}{\partial z}\right)_{z_0+z_c}\right]} \quad (20.47)$$

and

$$\omega_0 = \left\{ \frac{\omega^2\tau^2\left[2\gamma_{10}F_{Li} - 2K_z I_0\left(\partial M'_{Lc}/\partial z\right)_{z_0+z_c}\right]}{m\left(1+m^2\tau^2\right)} \right\}^{1/2} \tag{20.48}$$

With $\omega_0^2 < 0$ there will be no oscillations, but this case implies small τ_d values (heavy secondary) or high control current. Let us consider the previous numerical example in this chapter for which we adapt $\omega_0 = 3.6$ rad/s and $\tau_d = 0.4$ s, $z_c = 0.1$ m at u=100 m/s for $I_0 = 3\times10^5$ A turns; $z_0 = 0.3$ m, $L_v = 3$ m, 2b=0.5 m.

Finally, from (20.46) to (20.47) we calculate $\alpha_v = 1.33\times10^4$ (A s/m); $K_z = 1.16\times10^5$ (A/m).

The position $z_1(t)$ and acceleration response $z_1(t)$ for a sudden (step) change in position of $z_i = 0.03$ m are shown in Figure 20.16:

$$z_1(t) = z_i\left(\cos\omega_0 t + \frac{\sin\omega_0 t}{\tau_d\omega_0}\right)e^{-t/\tau_d} \tag{20.49}$$

It must be seen that the response is periodic and attenuated. The frequency response not only to vertical position modifications but to external perturbations such as wind gusts or track irregularities at various speeds is required before deciding that AED is suitable for a practical system.

Here we investigate only the response to guideway irregularities as it is related to ride comfort.

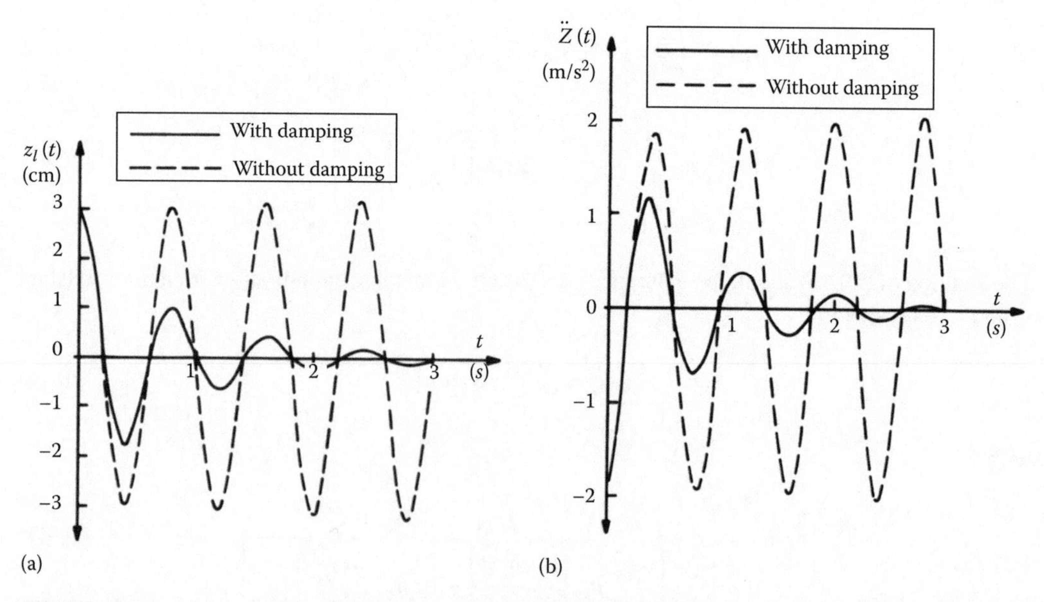

FIGURE 20.16 Active electric damper response to step change reference position of $z_i = 0.03$ m at 100 m/s: (a) vertical position response and (b) vertical acceleration response. (After Nasar, S.A. and Boldea, I., *Linear Motion Electric Machines*, John Wiley Interscience, New York, 1976.)

20.6.2 AED Response to Guideway Irregularities

The guideway irregularities may be considered sinusoidal (as done for attractive force levitation systems):

$$z_0 = z_g \sin \omega t \tag{20.50}$$

where z_g is

$$z_g = \left[\int_{\Omega_i}^{\infty} \psi(\Omega') d\Omega' \right]^{1/2}; \quad \psi(\Omega) = \frac{A}{\Omega'^2}; \quad \Omega' = 2\pi f/U \tag{20.51}$$

with f as the frequency of oscillations.

Detailing from z_g in (20.51) yields

$$z_g = \left[\int_{0.2}^{f} \frac{AU}{2\pi f^2} df \right]^{1/2} \tag{20.52}$$

If this perturbation is introduced in Equation 20.45, its solution becomes

$$z_1(t) = z_a \sin(\omega t + \varphi_i) + A_1 e^{-t/\tau_d} \cos(\omega_0 t + \varphi_0) \tag{20.53}$$

$$z_a = \frac{z_g \left(\omega_0^2 + 1/\tau_d^2\right)}{\sin \varphi_i \left|\omega^2\right| \tan \varphi_i + 2\omega/\tau_d - \omega_0^2 - 1/\tau_d^2} \tag{20.54}$$

$$\varphi_i = \tan^{-1} \frac{2\omega/\tau_d}{\omega^2 - \left(\omega_0^2 + 1/\tau_d^2\right)} \tag{20.55}$$

With zero initial conditions A_1 and φ_0 may be calculated:

$$\tan \varphi_0 = \frac{\omega \cos \varphi_i + (1/\tau_d) \sin \varphi_i}{\omega_0 \sin \varphi_i} \tag{20.56}$$

$$A_1 = -\frac{z_a \sin \varphi_i}{\sin \varphi_0} \tag{20.57}$$

Knowing $z_1(t)$ we can calculate $\dot{z}_1(t)$ and $\ddot{z}_1(t)$ for different vehicle speeds u and track irregularities frequencies f, for given guideway constant A. The maximum value of $\ddot{z}_1(t)$, a_m may be calculated and represented as a function of frequency f. Typical results for our ongoing example, together with the Janeway curve, are shown in Figure 20.17a. It appears that the AED is capable of providing alone the required ride comfort at 100 m/s ($A = 1.5 \times 10^6$ m) at control coil mmf as in Figure 20.17b (in p.u. values). A peak value of 3.5% of SC mmf (in effect, 16,500 A turns) is required for providing full ride comfort at 100 m/s.

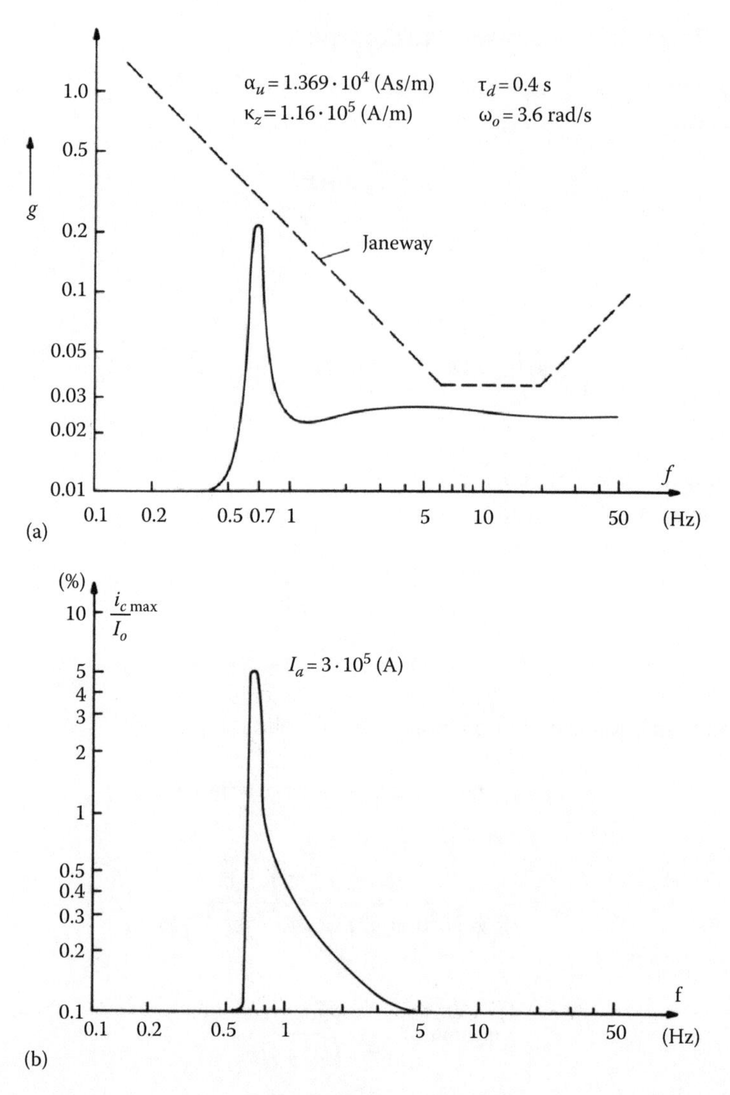

FIGURE 20.17 AED response to track irregularities at 100 m/s: (a) acceleration response and (b) control coil mmf requirement. (After Nasar, S.A. and Boldea, I., *Linear Motion Electric Machines*, John Wiley Interscience, New York, 1976.)

This amounts to about a maximum of 5 kW/ton, which is about in the range of attractive force levitation systems at 100 m/s.

This simplified analysis of AED may serve as a preliminary design basis, as for an actual system a multitude of other factors have to be included [13].

20.6.3 PED + SSS Dampers

As alluded to in Figure 20.15, a combination of an aluminum double sheet (one upper, one lower, 0.008 m thick), placed on the MAGLEV bogie, and the SCs fixed to the bogie and to passenger cabin by a mechanical suspension system, may be able to produce stable operation for the vertical motion (levitation) in an eight-shape (null-flux) RFLS with vertical SCs [12].

FEM computation results for an assumed vertical motion of SC at 10 Hz for an amplitude of 0.02 m have led to a vertical (levitation) force as in Figure 20.18a [12] for a full-size MAGLEV.

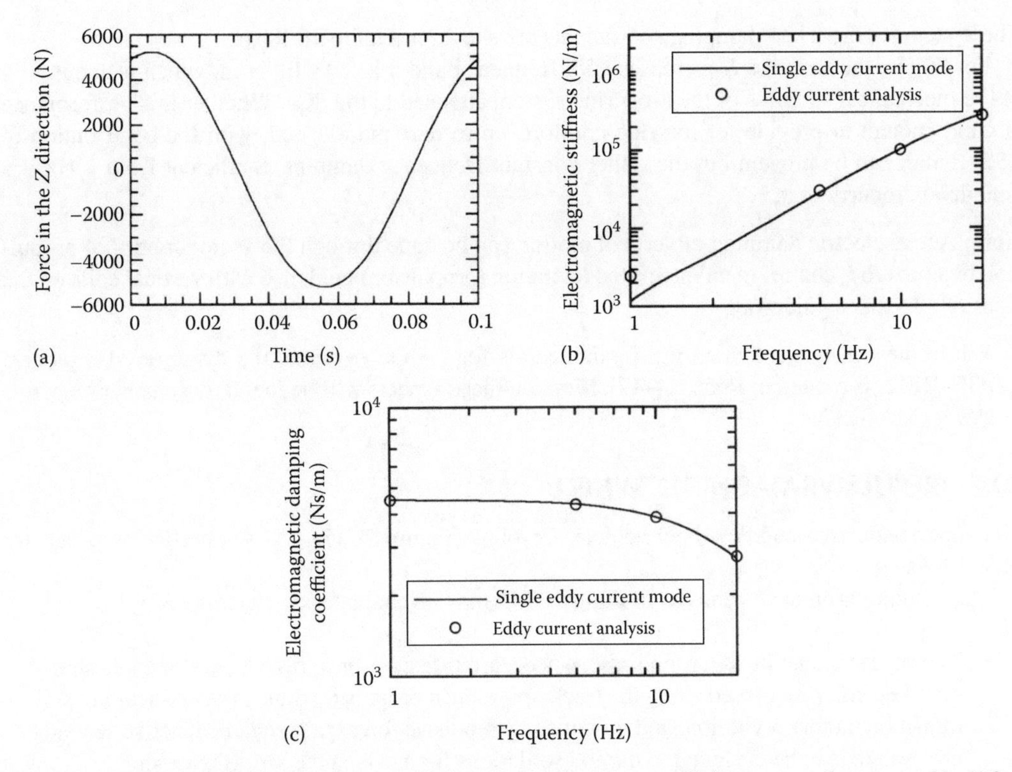

FIGURE 20.18 AED+SSS system (a), stiffness *K* (b), damping coefficient versus frequency (c). (After Miikura, S. et al., Electromagnetic stiffness and damping effects in the secondary suspension of a superconducting MAGLEV vehicle, *Record of LDIA-1995*, Nagasaki, Japan, pp. 97–100, 1995.)

The force seems rather sinusoidal and thus a single eddy current mode (harmonic) suffices. Therefore, for the bogie-attached aluminum sheet, the circuit equation is

$$L\frac{di}{dt}+Ri=E;\quad E=\frac{\partial M}{\partial z}\cdot I_0\cdot \dot{Z}_{sc} \tag{20.58}$$

In (20.58), the SC mmf was considered constant.

The vertical force F_z between the SC and bogie-attached aluminum sheet is

$$F_z=-\frac{\partial M}{\partial z}iI_0 \tag{20.59}$$

As F_z from FEM varies sinusoidally in time, it may be represented in complex numbers:

$$F_z=\left(\frac{\partial M}{\partial z}\right)^2\frac{I_0^2}{R}\frac{j\omega}{j\omega\tau+1}\approx\left(k+j\omega c\right)z_c \tag{20.60}$$

The time constant τ results from fitting (20.60), with the FEM force F_z in Figure 20.18a as

$$\left|F_z\right|=\frac{\alpha\omega z_{max}}{\sqrt{\omega^2\tau^2+1}};\quad z_{max}\text{ — SC motion amplitude} \tag{20.61}$$

The system stiffness and damping coefficients are shown in Figure 20.18b,c.

For 20 Hz, the stiffness K increases with frequency and it is 2.7×10^5 N/m, which is about 30% of the mechanical stiffness of the air spring system attached to the SCs. Whether this performance is good enough to provide for the ride comfort, up to maximum speed, with the contribution of SSS, remains to be proven; but the rather constant frequency damping coefficient from 1 Hz is a beneficial property of it.

Note: Active electric damping of vertical motion can be done through the vector control of propulsion by nonzero i_d control in an integrated levitation (propulsion) guidance with vertical coils on the sides of a U-shape guideway.

While previously we treated the fundamentals for 1 DOF systems, the dynamic of a passive 5 DOF–RFLS is treated in Refs. [14–17]. More on these aspects will be found in a chapter on active guideway MAGLEVs.

20.7 REPULSIVE MAGNETIC WHEEL

The superconducting paddle wheel vehicle, Cryobus (Figure 20.17) [6], was earlier proposed for MAGLEVs [18].

The configuration of Cryobus as in Figure 20.19 may be characterized as follows:

- The superconducting heteropolar SC wheels are rotated over a triple ladder or triple short-circuited coil raw, placed along the track, to produce propulsion (much like a linear air-coil induction motor), levitation, and guidance by repulsive forces; a mechanical active secondary system is to be designed to damp oscillations for a safe and comfortable ride.

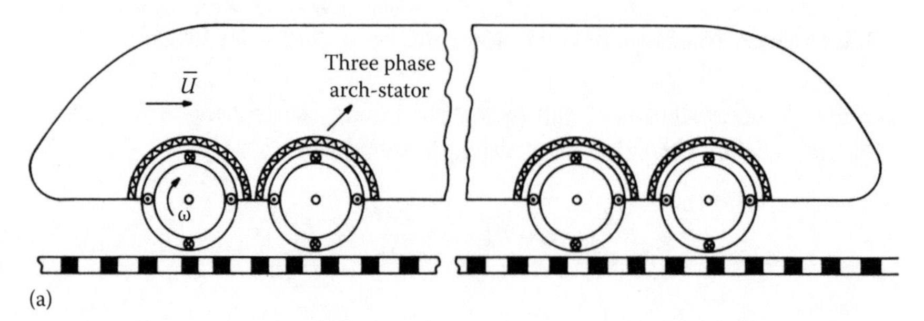

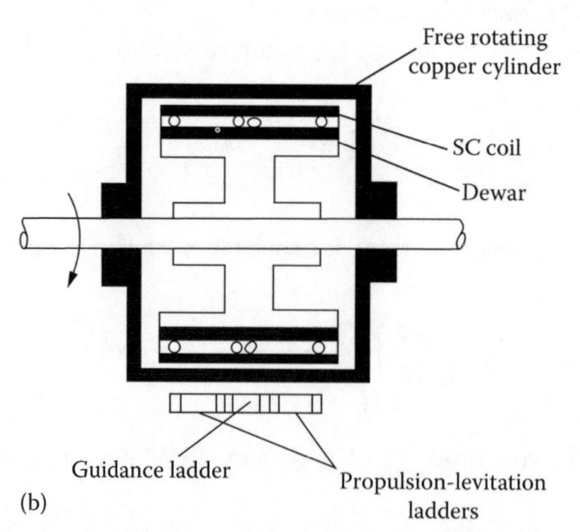

FIGURE 20.19 Cryobus longitudinal view, (a) and cross-section of the paddle wheel, (b). (After Nasar, S.A. and Boldea, I., *Linear Motion Electric Machines*, John Wiley Interscience, New York, 1976.)

- The magnetic wheels produce propulsion/levitation/guidance as long as ac is produced in the guideway; it means that by spinning the wheel at, say, nonzero slip frequency the MAGLEV might develop enough force to lift the vehicle at zero speed. To provide lift force at zero propulsion force, we simply rotate half of the wheels in one direction and the rest in the opposite direction. So, theoretically the takeoff may take place from standstill.
- When the wheel peripheral speed U_{wp} is larger than the vehicle speed U, the propulsion force is positive while for $U_{wp} < U$ regenerative braking is produced.
- As only the lower half of the wheel interacts with the guideway, the upper half may interact with arch-shape synchronous motor ac stator windings, to "paddle" the magnetic wheel, and save space and hardware. With $2p = 8$ poles, 4 poles may be used to "paddle" the wheel.
- The free rotating copper cylinder, Figure 20.17b, may be used to damp the oscillations in levitation (as PED).
- As the magnetic coupling between the circular wheel and the flat guideway is weaker than for flat SCs, the overall efficiency of Cryobus is expected to be lower than for flat-coil RFLS–MAGLEVs. However, when used to integrate proportional levitation guidance functions, the Cryobus may attain a total power efficiency up to 65% and thus compete with electrodynamic flat-coil active guideway MAGLEVs in a simpler topology (see Ref. [6], Chapter 12, for preliminary design).
- The Cryobus needs all power to be transferred on board of vehicle as the track is passive, and thus it may prove to be an overall cost/performance solution worth reconsideration.
- Recently, the superconducting wheel has been proposed to be replaced by a heteropolar PM wheel [19,20]. The same concept was proposed to compensate the dynamic end effect in high-speed LIMs. If the PM wheel could duplicate the performance of the SC dual-axial-pole wheel (Figure 20.17b) with a higher than 1.5 T remanent flux density, then this latter solution might stand a practical application challenge, even if flying at a smaller height (0.04–0.06 m).
- In general, the replacement of SCs by strong PMs is so far less practical because of the lower weight of SCs at 0.15–0.2 m rated gap for given lift force. But stronger PMs of the future at smaller gap might reverse the situation soon [21,22].

20.8 COIL-PM REPULSIVE FORCE LEVITATION SYSTEM

The experimental setup in Figure 20.20 [23] may be applied directly to an industrial MAGLEV platform or to an axial bearing of an air-core PM synchronous motor/generator inertial battery. In contrast to attraction force levitation systems, it allows long (up to 20 mm or more) excursion length.

To analyze the dynamics of this 1 DOF system with vertical position tracking control, the motion equation including coil force, gravity, friction, and external force disturbance is used:

$$F_L - m \cdot g - B \cdot \dot{x} - F_{ext} = m \cdot \ddot{x} \tag{20.62}$$

where

x is the vertical distance between PM and air coil
m is the PM mass
B is the friction coefficient
F_L is the levitation (repulsive) force of coil current
F_{ext} is the external force disturbance

If $V(t)$ is the control effort, the levitation force may be approximated as follows [23]:

$$F_L \cong \frac{V(t)}{a(x+b)^4} \tag{20.63}$$

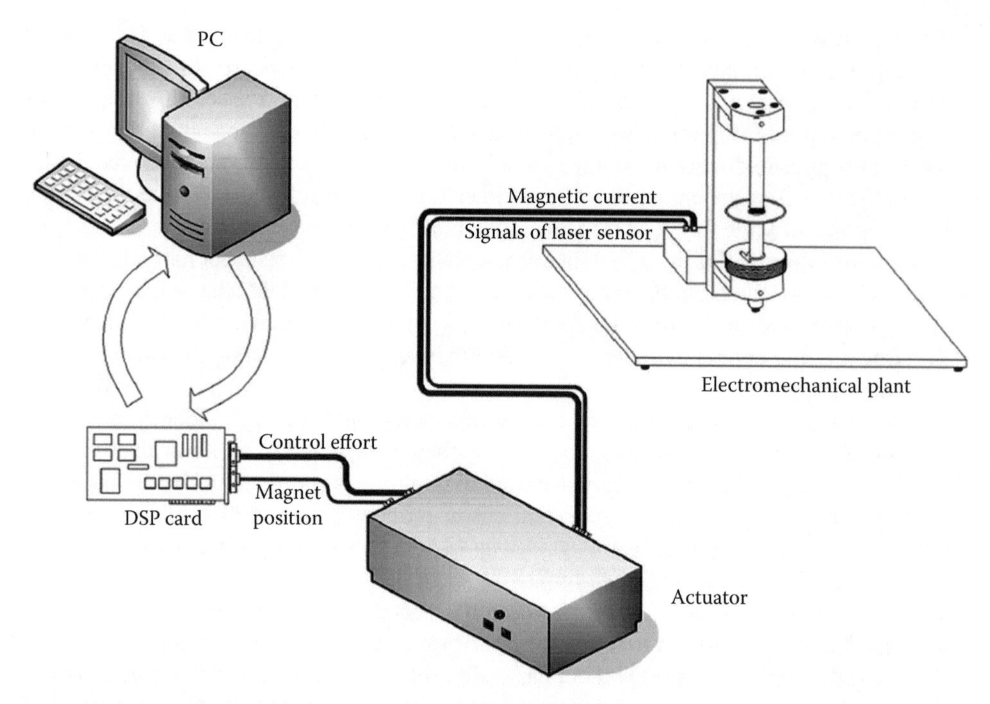

FIGURE 20.20 Coil-PM repulsive force levitation system. (After Slotine, J.J.E. and Li, W., *Applied Nonlinear Control*, Prentice-Hall, Englewood Cliffs, NJ, 1991.)

With and, after separating the time varying uncertainties, from lumped uncertainties $L(x,t)$, Equations 20.62 and 20.63 may be put in the form

$$\ddot{x}(t) = f_n(\underline{x},t) + G_n(\underline{x},t)\cdot V(t) + d_n(\underline{x},t) + L(x,t) \tag{20.64}$$

with

$$L(\underline{x},t) = \Delta f + \Delta GV(t) + \Delta d < \delta \tag{20.64a}$$

For robust control, the sliding mode control approach is corroborated here with a radial basis function network (RBFN), to estimate online the uncertainties in the system dynamics.

The sliding mode functional $S(t)$ is standard:

$$S(t) = \dot{e}(t) + \lambda_1 e(t) + \lambda_2 \int_0^t e(\tau)d\tau; \quad e(t) = x^* - x \tag{20.65}$$

The values of λ_1 and λ_2 represent the desired dynamics and may be calculated from given rise time and overshooting as in a second-order system. The globally asymptotic stability may be guaranteed (based on the Lyapunov method) with a control law of the form

$$V(t) = G_n^{-1}(x,t)\left[-f_n(\underline{x},t) - d_n(\underline{x},t) + \ddot{x}_m - \lambda_1\dot{e}(t) - \lambda_2 e(t) - \delta\cdot sign\cdot S(t)\right] \tag{20.66}$$

Following the approach with the RBFN estimator, for $m=0.121$ kg, $B=2.69$, $a=1.65$, $b=6.2$, $\lambda_1=10$, $\delta=10$, $\lambda_2=30$, results as in Figure 20.21 and Table 20.1 [24] have been obtained.

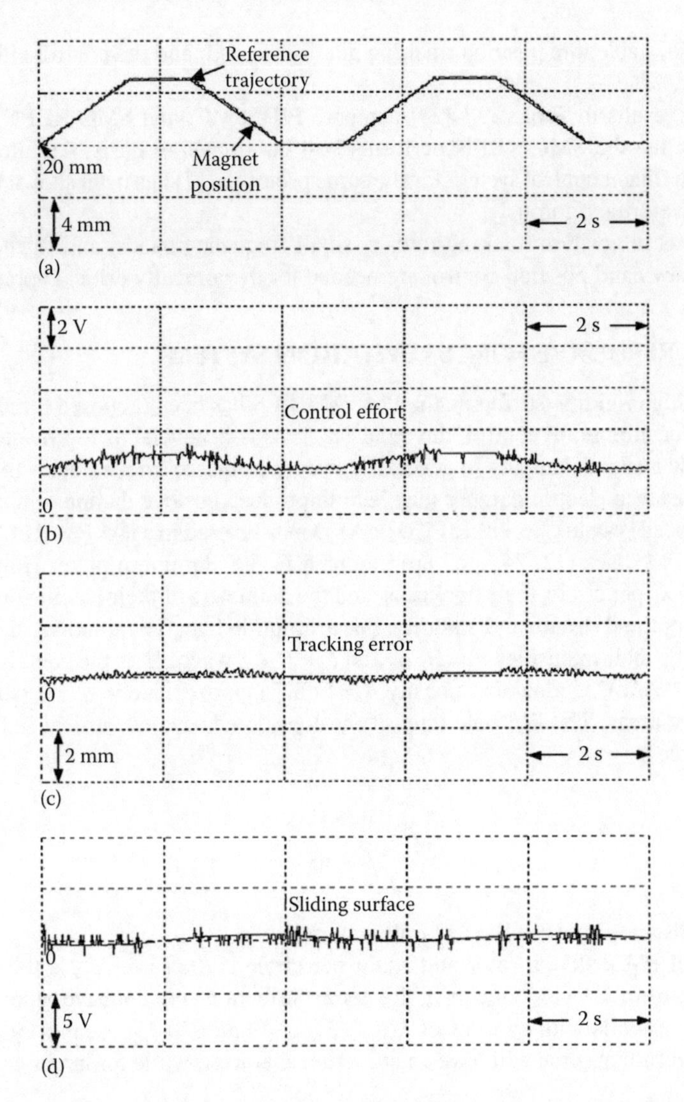

FIGURE 20.21 Experimental results for trapezoidal airgap command: (a) reference and actual position, (b) control effort, (c) tracking error, and (d) sliding surface. (After F.J. Lin et al., *IEEE Trans.*, MAG-54(3), 2007, 1752.)

TABLE 20.1
Control Performance

	Performance Indexes					
	Sum of Squared Tracking Error (mm^2) (SSTE)			Sum of Squared Control Input (V^2) (SSCI)		
Commands and Cases	**PID**	**SMC**	**SMCRBFN**	**PID**	**SMC**	**SMCRBFN**
Sinusoidal trajectory at case 1	1.53×10^2	1.99×10^2	4.78×10^1	1.52×10^5	2.78×10^5	8.24×10^3
Trapezoidal trajectory at case 1	2.42×10^2	2.98×10^2	9.43×10^1	1.71×10^5	1.15×10^5	7.32×10^3
Sinusoidal trajectory at case 1	3.41×10^2	5.09×10^2	2.64×10^1	2.75×10^5	7.02×10^5	2.20×10^4
Trapezoidal trajectory at case 2	2.81×10^2	6.71×10^2	3.02×10^1	2.10×10^5	2.84×10^5	1.98×10^4

Source: Lin, F.J. et al., *IEEE Trans.*, MAG.-54(3), 1752, 2007.

The response is stable, the position tracking quality is good, and the control effort is reasonable for a large airgap.

The synthetic results in Table 20.1 [24] compare PID, SMC, and SMC-RBFN control systems and demonstrate that the SMC-RBFN performs best on an overall basis. An alternative adaptive back-stepping intelligent control system for the same prototype [7] has been shown to track complex position/time waveforms rigorously.

Still, the results in position tracking refer to small frequency bands, and further investigations into high-frequency band position control are needed for dynamically critical applications.

20.9 PM-PM REPULSIVE FORCE LEVITATION SYSTEM

In the search for high rigidity (stiffness), the PM-PM RFLS has been proposed for axial bearings. To operate with no (or minimum) control, the repulsive force is beneficial as it provides static stability; it does not provide positive damping in general. However, as discussed previously in this chapter, the passive small resistance electric damper may help to produce positive damping in such applications where the axial speed is small. So, either PED or AED may be used in a PM-PM RFLS (Figure 20.22).

The multiple PM disks (12, 24, etc.) are mounted in the aluminum plates (Figure 20.22). The number of PM disks per circle, their thickness, and the diameters of their placing on stator and rotor may be used for optimal design and stability. For a simplified analytical model, the magnets in air may be replaced by their intensities $F_1 = H_c A_2$ and $F_2 = H_c A_2$, where H_c is the coercive magnetic field (say $H_c = 1.78 \times 10^6$ A/m), residual flux density $B_r = 1.08$ T (approximately $\mu_{rec} = 1$ p.u.); A_1 and A_2 are the PM pole force areas. The PM pole intensity is considered as concentrated in the center of the PM pole. The force is

$$\overline{F} = \frac{\mu_0 F_1 F_2}{4\pi \cdot r_{12}^2} \overline{u}_{12} \tag{20.67}$$

where r_{12} is the distance between the PM pole centers.

The number of PM disks on rotor and stator per circle is the same; $\overline{u}_{12}$ is the direction of the force. There are quite a few components of this force (attraction and repulsive type) as the PMs may be "replaced" by air coils with an mmf of $I_0 = H_c \cdot h_{PM}$ (Figure 20.23). As it is easy to imagine, at small gap z_0, the resultant force will have an attraction character, while for larger gap it will become a repulsion force.

For a certain z/h_{PM} and PM radial length/z, the repulsive force will be maximum.

Concerning the radial force, it has to be negative (centralizing) for stable guidance, and this happens only for small radial displacements in combination with lower rotor than stator diameter of placing the PMs.

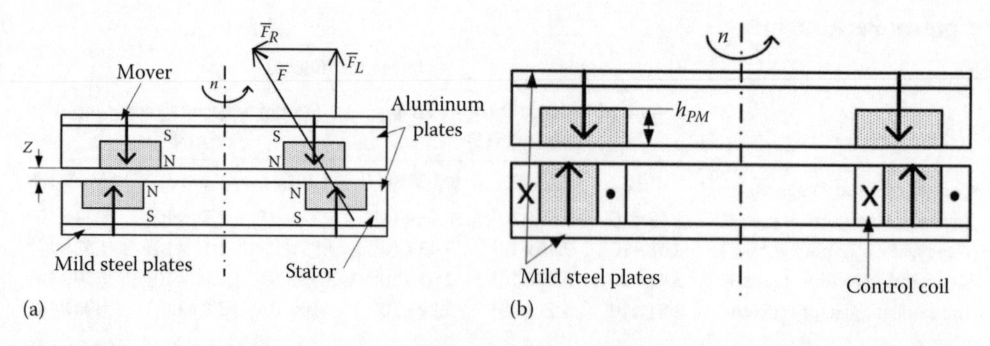

FIGURE 20.22 PM-PM repulsive force levitation system: (a) with passive (aluminum plate) damper and (b) with active electric damper.

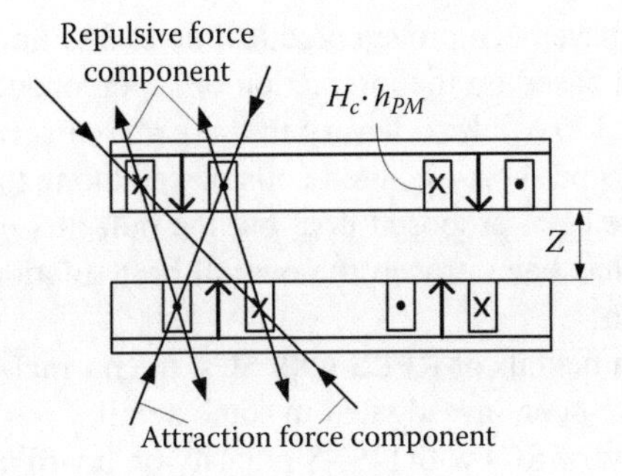

FIGURE 20.23 PMs as air coils.

For h_{PM} = 10 mm, at z = 2.5 mm, the repulsive force, for D_r = 47.5 mm and D_s = 50 mm, is maximum at 23 N [8], while the radial force is negative up to 1 mm radial displacement, but only at −0.5 N. When the rotor rotates, the 24 PMs on both rotor and stator assume different relative positions and thus the repulsive (vertical) levitation force pulsates heavily even with D_r = 74.5 mm and D_s = 50 mm (± 6 N).

However, much like for cogging force, if the number of PMs on stator and rotor is different (say 26/24), the levitation force pulsations during rotor rotations become small; alternatively, with two out of phase rows of PMs on the stator at D_{s1} = 50 mm and D_{s2} = 45 mm, with D_r = 47.5 mm, and 24 PMs on both rotor or stator, the repulsion force pulsations become small. While in principle the PM-PM RFLS works, there is still a lot of work to be done to

- Analyze it by 3D-FEM
- Elaborate an optimal design code to check the performance/cost
- Include the eddy currents in the aluminum plates
- Develop a state space circuit model for dynamics and control
- Investigate static and dynamic stability
- Introduce oscillation passive damping or closed-loop control via coils for refined levitation height tracking [25]

All the aforementioned issues are beyond our scope here, but seem worthy of further consideration by the interested reader.

20.10 SUMMARY

- Repulsive (electrodynamic) force may be developed in practical devices between
  - Conductors carrying currents of opposite polarity
  - A moving current carrying (or superconducting or a PM) coil raw and a fix secondary conducting sheet or ladder
  - Two rows of PMs of opposite polarity
  - One row of PMs and controlled current air coils
- The repulsive force tends to separate the two bodies with an intensity that increases when the distance (air gap) between the two decreases. It is thus statically stable. This is in contrast to attraction force levitation systems (AFLSs), which are statically unstable.
- Large repulsive forces may be produced with strong PMs or superconducting coils at air gaps 10 times larger than for AFLS (where 10–15 mm is typical); this way the track irregularities may be larger and thus lower costs are expected.
- As levitation means statically and dynamically stable hovering of a body (e.g., a MAGLEV), the dynamics and some control for repulsive force levitation system have to be investigated.

- For MAGLEVs, RFLS have been proven practical up to 585 km/h.
- The RFLS is in general based on the interaction of superconducting coils (SCs) on board of vehicle to produce 0.3–0.6 T flux density in the plane of secondary (track) conducting sheet, ladder, or eight-shape short-circuited coils placed along the guideway of MAGLEV.
- Normal-flux RFLS have been proposed first, but the null-flux eight-shape short-circuited vertical coil guideway has been proven the overall best solution for integrated levitation guidance and propulsion.
- To investigate the fundamentals of RFLS with SCs, the normal-flux topologies with sheet or ladder secondary have been investigated in some detail.
- As expected, when a row of SCs with NSNS polarity on board of MAGLEV moves above a sheet or ladder guideway, induced currents occur in the latter.
- As the speed increases, the frequency of these currents increases and thus their circuit becomes more and more inductive; at high speeds, the induced currents look almost opposite in polarity to those in the neighboring SC. This indicates a strong repulsive (levitation force) F_L. But a drag force F_d corresponds to the losses in the conducting guideway.
- While the levitation force grows monotonously with speed U, the drag force shows a peak value with speed, much like the torque in a dc stator–fed induction motor with cage rotor.
- Two criteria have been introduced to weigh the levitation "goodness": levitation force goodness $(F_L/F_d)_{umax}$ and levitation power goodness $F_L \cdot U\left(F_d \cdot U + P_{control}\right)$, where $P_{control}$ is the dynamic (control) power needed to replenish the SCs and for motion dynamic stabilization and control. The higher the two criteria values the better for some given initial system costs.
- Using a larger airgap than AFLS, the RFLS with SCs has been so far proven, at higher SCs length, capable to produce better performance at higher speeds (up to 585 km/h); the SCs volume for same levitation force tends to be smaller than that of dc-controlled iron-core coils.
- But the levitation force increases with speed, and the critical speed at which the drag force has its peak values leads to the conclusion that the vehicle has to start and accelerate on retractable wheels (much like an aircraft, while AFLSs work as helicopters do).
- For 3 m long, 0.5 m wide SC coils, RFLS have been built to produce $(F_L/F_d)_{130\ m/s}$ above 50/1 for sheet guideway, 120/1 for ladder guideways, and above 200/1 for null-flux (eight-shape track coils) systems.
- An analytical study of RFLS showed that by careful design SC flux in the track (sheet or ladder type) may vary almost sinusoidally in time. Also, the SC flux density in the guideway plane varies rather cosinusoidally along the transverse direction. Consequently, the vertical force flux density varies along the airgap exponentially. A practical field theory may be obtained to produce levitation and drag force dependence on speed and on system geometry. Curve fitting may be used to match these theoretical results to the image levitation force based (Neumann) formula and a proposed speed dependence.
- A similar (but based on circuit theory) model for the ladder and null-flux guideway system is obtained. In this case, however, additional pulsations in F_c and F_d occur; by a proper design, they may be reduced to negligible values.
- This way the aforementioned high levitation force goodness values have been obtained.
- Using these models in the motion equation (of a 1 DOF system), it is proven that the self-damping coefficient is negative at all speeds for the sheet guideway and above a small speed (15–25 m/s) for the ladder or null-flux guideway.
- Damping these natural (inherent) oscillations may be accomplished by
  - Passive electric damper (PED)
  - Active electric damper
  - Mechanical damper
  - Combined passive (on bogie) electric and mechanical damper

- PEDs are in general incapable of damping vehicle oscillations properly for the entire frequency spectrum or providing ride comfort up to maximum speed.
- The active electric damper with a room temperature controlled coil on board, placed between the SCs and the guideway, has been shown to be capable of producing satisfactory ride comfort up to maximum speed for a maximum of 5% of SC mmf in the control coil. An average of 5 kW/ton of control power is required at 550 km/h. This is only 40% more than the control power in AFLS at 360 km/h.
- The ride comfort is still defined by the Janeway curve, which limits acceleration versus frequency due to track irregularities (up to 20 Hz).
- Even the passive electric damper (two vertical aluminum sheets on the bogie, parallel to SC) with a secondary suspension system of SCs to bogie and cabin seems capable of providing good oscillation damping from 1 Hz upward.
- In the null-flux RFLS, when the eight-shape track coils are connected as a three-phase ac winding (with $q=1/2$ slots/pole/phase), and with connections of the coils on the two sides of the vehicle, integrated levitation and propulsion are obtained. This time vector control of the of the ac three-phase windings (as in a linear synchronous motor—Chapter 8) will control propulsion by iq control and stabilize the vertical vehicle oscillations by id control, instead of AED. Zero levitation power control may be attempted for RFLS as for AFLSs.
- The heteropolar SC or PM wheel on board rotating at $U_w > U$ peripheral speed above a track sheet or ladder can be designed to produce integrated levitation/propulsion/guidance at a smaller airgap (0.04–0.06 m). The upper half of so-called magnetic wheel is driven by a three-phase arch-shape stator much like in a rotary synchronous motor. A passive guideway MAGLEV is thus produced. Due to the smaller region of interaction, that is, weaker interaction of magnetic wheels with the guideway, the PMs (SCs) mmf has to be stronger than for flat RFLS of the same performance. Horizontal magnetic wheels with PMs and vertical axis with partial overlapping of guideway conductive sheets or ladders have also been proposed for RFLS [26].
- For standstill (bearings) or industrial MAGLEV platforms, repulsive force levitation may be produced by moving PMs and fix controlled current coils (much like for planar motion linear electric motors). Sliding mode control has proven very effective to track airgaps around 20 mm with good precision in such systems.
- PM-PM repulsive force levitation has also been proven adequate, with limited range lateral stabilization and passive (aluminum plate) damping, in axial bearings with rotors rotating at large speeds; active (close-loop) control is required to fully control such systems.
- It is estimated here that R&D on RFLS will develop quickly in the near future, both for super high-speed transport and for industrial applications (bearings and industrial MAGLEV platforms in clean rooms).

REFERENCES

1. G.T. Danby and J.P. Powell, Integrated systems for magnetic suspension and propulsion vehicles, *Record of Applied Superconductivity Conference*, Annapolis, MD, 1970.
2. J.R. Reitz, R.H. Borcherts et al., Preliminary design studies of magnetic suspensions for high speed transportation, *Final Report DOT-FRA-10026*, Washington, DC, March 1973.
3. T. Yamada and M. Iwamoto, Theoretical analysis of lift and drag forces on magnetically suspended high speed trains, *Electr. Eng. Jpn.*, 92(1), 1973, 53–61.
4. E. Ohno, M. Iwamoto, and T. Yamada, Characteristics of Superconducting magnetic suspensions and propulsion for high speed trains, *Proc. IEEE*, 61(5), 1973, 579–586.
5. S.A. Nasar and I. Boldea, *Linear Motion Electric Machines*, Chapter 7, John Wiley Interscience, New York, 1976.
6. I. Boldea and S.A. Nasar, *Linear Motion Electromagnetic Systems*, Chapter 11, John Wiley Interscience, New York, 1985.

7. F.J. Lin, L.T. Teng, and P.H. Shieh, Intelligent adaptive backstepping control for magnetic levitation apparatus, *IEEE Trans.*, MAG-54(5), 1973, 579–586.
8. S.C. Mukhopadhyay, J. Donaldson, G. Sengupta, S. Yamada, C. Chakraborthy, and D. Kacprzak, Fabrication of a repulsive-type magnetic bearing using novel arrangement of PM for vertical rotor suspension, *IEEE Trans.*, MAG-39(5), 2003, 3220–3222.
9. T. Yamada, M. Iwamoto, and T. Ito, Magnetic damping force in inductive magnetic levitation system for high speed trains, *Electr. Eng. Jpn.*, 94(1), 1974, 80–84.
10. M. Iwamoto, T. Yamada, and E. Ohno, Magnetic damping force in electrodynamically suspended trains, *IEEE Trans.*, MAG-10(3), 1974, 458–461.
11. R.H. Borcherts et al., Preliminary design studies of magnetic suspension for high speed ground transportation, *Record of FRA*, PB. 224 843, Washington, DC, June 1974.
12. S. Miikura, A. Kameari, M. Igarashi, and J.I. Kitano, Electromagnetic stiffness and damping effects in the secondary suspension of a superconducting MAGLEV vehicle, *Record of LDIA-1995*, Nagasaki, Japan, pp. 97–100, 1995.
13. D.M. Rote and Y. Cai, Review of dynamic stability of repulsive-force MAGLEV suspension systems, *IEEE Trans.*, MAG-38(2), 2002, 1383–1390.
14. A. Seki, Y. Osada, J.I. Kitano, and S. Miyamoto, Dynamics of the bogie a MAGLEV system with guideway irregularities, *IEEE Trans.*, MAG-32(5), 1996, 5043–5045.
15. J. de Boeij, M. Steinbach, and H.M. Gutierrez, Mathematical model of the 5 DOF sled dynamics of an electrodynamic MAGLEV system with a passive sled, *IEEE Trans.*, MAG-41(1), 2005, 460–465.
16. J. de Boeij, M. Steinbach, H.M. Gutierrez, Modeling the electromechanical interactions in a null-flux electrodynamic MAGLEV system, *IEEE Trans.*, MAG-41(1), 2005, 466–470.
17. R.D. Kent, Designing with null flux coils, *IEEE Trans.*, MAG-33(5), 1997, 4327–4334.
18. L.C. Davis and R.H. Borchers, Superconducting paddle wheels, screws and other propulsion units for high speed ground transportation, Scientific Research Staff, Ford Motor Co., Dearborn, MI, January 22, 1973.
19. N. Fujii, Y. Ito, and T. Yashihara, Characteristics of a moving magnet rotator over a conducting plate, *IEEE Trans.*, MAG-41(10), 2005, 3811–3813.
20. J. Bird and T.A. Lipo, An electrodynamic wheel: An integrated propulsion and levitation machine, *Record of IEEE IEMDC-2003*, pp. 1410–1416.
21. W. Ko and C. Ham, Novel approach to analyze the transient dynamics of an electrodynamics suspension MAGLEV, *IEEE Trans.*, MAG-43(6), 2007, 2603–2605.
22. Q. Han, C. Han, and R. Philips, Four and eight-piece Halbach array analysis and geometry optimization for MAGLEV, *IEE Proc.*, EPA-152(3), 2005, 535–542.
23. J.J.E. Slotine and W. Li, *Applied Nonlinear Control*, Prentice-Hall, Englewood Cliffs, NJ, 1991.
24. F.J. Lin, L.T. Teng, and P.H. Shieh, Intelligent sliding-mode control using RBFN for magnetic levitation system, *IEEE Trans.*, MAG-54(3), 2007, 1752–1762.
25. S.C. Mukhopadhyay, T. Ohji, M. Iwahara, and S. Yamada, Design analysis and control of a new repulsive type magnetic bearing, *IEE Proc.*, EPA-146(1), 1999, 33–40.
26. N. Fujii and S. Nonaka, Payload of revolving PM type magnet wheel, *Record of LDIA-1998*, Tokyo, Japan, pp. 351–354, 1998.

21 Active Guideway MAGLEVs

21.1 INTRODUCTION

By MAGLEVs we mean here magnetically levitated vehicles:

- For high-speed intercity movers
- For urban/suburban people movers
- For short-haul industrial platforms

Magnetic levitation means controlled suspension and guidance by means of electromagnetic (attraction) or electrodynamic (repulsion) forces. Magnetic levitation may be performed by dedicated dc fed controlled electromagnets—with or without PM assistance—in attraction force MAGLEVs (Chapter 19) or in dc superconducting magnets (or PMs) with air-core active damping coils (Chapter 20). The airgap between the vehicle (platform) and the active guideway has to be dynamically controlled, around rated airgap g_m, by ± 25%; g_m = 8–10 mm for attraction MAGLEVs and around 100 mm for repulsion force MAGLEVs vehicles. For industrial MAGLEV platforms, the airgap g_m could go as low as 1–2 mm.

The propulsion of MAGLEVs is produced with linear electric motors.

For active guideway MAGLEVS [1–5], in general, linear synchronous motors (LSMs) with iron-core (respectively, air-core) three-phase ac windings spread along the entire travel length are used.

In Chapters 7 and 8, the two main active guideway LSMs have been treated in detail:

- DC-excited iron-core LSM (active guideway)
- DC superconducting air-core LSM (active guideway)

In this chapter, the following items are dealt with:

- The structure, principles, and the control essentials of dc-excited iron-core active guideway LSM attraction force MAGLEV ("Transrapid")
- The structure, principles, and the control of dc-excited LSM air-core guideway repulsive force MAGLEV ("JRM Linear")
- The structure, principles, and the control of PM excited LSM iron-core guideway repulsive force MAGLEV ("General Atomics")

The various degrees of interaction of propulsive-levitation-guidance functions make the three mentioned active guideway MAGLEVs rather fully representative for people mover applications.

For active guideway industrial platforms, one more solution has been given notable attention: the doubly fed linear induction motor with long (active guideway) iron-core stator and wound mover. The wound mover is, in general, ac controlled at frequency f_2 by a partial ratings inverter on board or mover. For heavy starts, the wound mover may be short-circuited and a full-power dedicated on ground inverter (or soft starter) may be used to supply the stator. In view of its potential, this MAGLEV platform will also be offered due treatment in this chapter. Though the subject of this chapter is strongly related to specific industrial solutions, the presentation remains mostly scientific, coherent with the rest of the book. For the sake of self-sufficiency, some reminder knowledge on the linear motors for propulsion and on the magnetic suspension and guidance dc-controlled (or sc) electromagnets is given again in the chapter.

FIGURE 21.1 "Transrapid 08." (After http://upload.wikimedia.org/wikipedia/commons/d/d0/Shanghai_Transrapid_002.jpg)

The commercialization in 2002 of the first active guideway MAGLEV people mover (Transrapid, Figure 21.1) with the peak speed of 430 km/h (or more) and the advanced stage of the SC and PM active guideway MAGLEV people mover for 550 km/JRM and 160 km/h (General Atomics), respectively, are based on MAGLEV merits such as

- Lower vehicle weight/passenger at given speed
- Lower kwh/passenger/km (at given speed) than in high-speed wheel-suspension trains (such as TAGV, etc.)
- Notable lower track maintenance costs as the interaction forces with the guideway are distributed over 10^4 larger areas than in wheeled vehicles
- Lower noise and better ride comfort at super high speeds (400–550 km/h)

For active guideway MAGLEV systems, the cost of the active guideway and of the ground inverter stations is said to be compensated at high traffic density by the absence of full power transfer to the vehicle, which is hosting lower power equipment fed from linear generators on board, backed by a storage battery.

21.2 DC-EXCITED IRON-CORE LSM MAGLEV VEHICLES (TRANSRAPID)

The schematics of an active guideway dc-excited iron-core LSM MAGLEV vehicle is shown in Figure 21.2:

The "Transrapid like" system may be characterized as follows:

- The T shape guideway sustains, on the lower part, two long iron-core stators with three-phase ac cable windings fed, in synchronized manner, from on ground inverter power substations.
- On board, there are two rows of dc-excited poles in iron-core inductors, which make the LSM secondary.

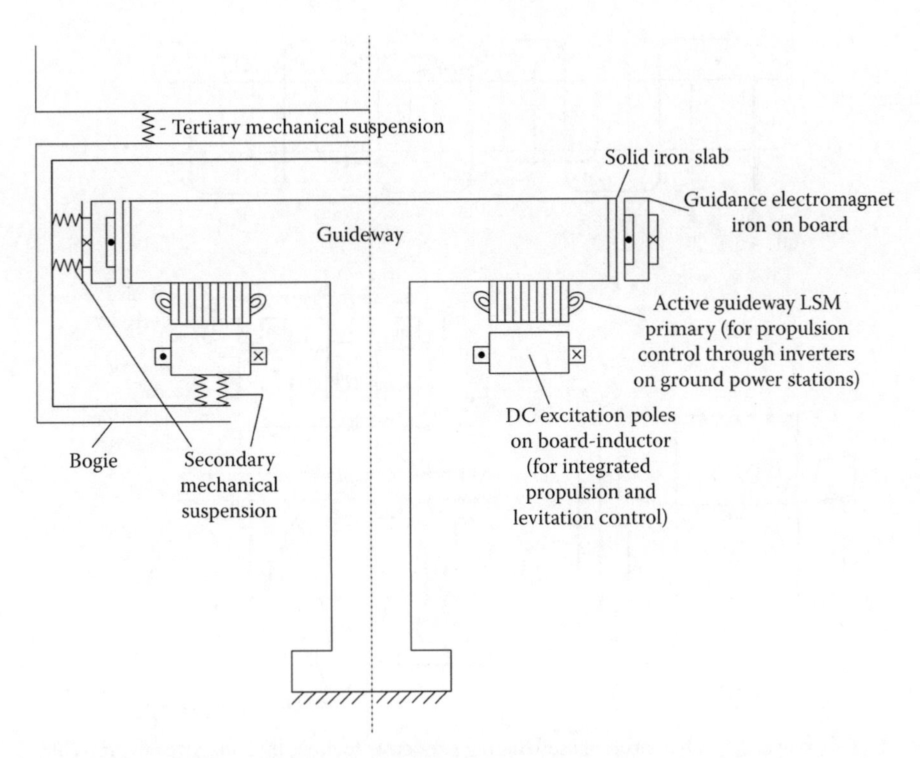

FIGURE 21.2 DC-excited iron-core LSM MAGLEV with active guideway.

- The magnetic flux density produced in the airgap by the dc coils (about 0.55–0.65 T) not only produces the stator ac emf, which is about in phase with controlled stator current (*iq* control), but also produces the fully stable magnetic suspension of the vehicle by advanced robust control of field coil currents (Chapter 7).
- To reduce the "unsprung mass," the dc excitation cores on board are attached to the bogie through a secondary mechanical suspension system; a tertiary (active) mechanical suspension system between the bogie and the passenger cabin provides the required comfort up to maximum speed (430 km/h, so far), for a reasonable dynamic energy consumption (3 W/kg) (Chapter 19).
- To reduce the cost of the laminated long stator, open slots (allowed by the large airgap ≈ 10 mm, rated) are filled with a single-layer diametrical aluminum-cable winding with $q=1$ slots/pole/phase (Figure 21.3); care is exercised for enough transposition of the elementary conductors in the cable to avoid large skin effect losses at around 200 Hz maximum fundamental frequency (at ≈ 400 km/h).
- The pole pitch τ is not much larger than stator stack width, to keep the end-connection winding length, weight, and losses small enough; but also, the stack width has to be limited to reduce the long stator costs (Figure 21.4).
- With vector control of LSM and $i_d \approx 0$, the interaction between propulsion and levitation is drastically reduced; a bit of around-zero i_d control may be used to damp some vehicle oscillations, etc.
- The guidance function is handled separately by lateral dc-controlled electromagnets acting "against" a solid vertical iron slab provided in the guideway structure.
- The electric energy on board of vehicle, required to supply the inductors and the guidance electromagnetics plus the auxiliaries, is produced by "electromagnetic induction,"—that is contact-less—by inserting small pole-pitch (half the stator slot pitch) multiphase coils in the inductor poles (Figure 21.4).

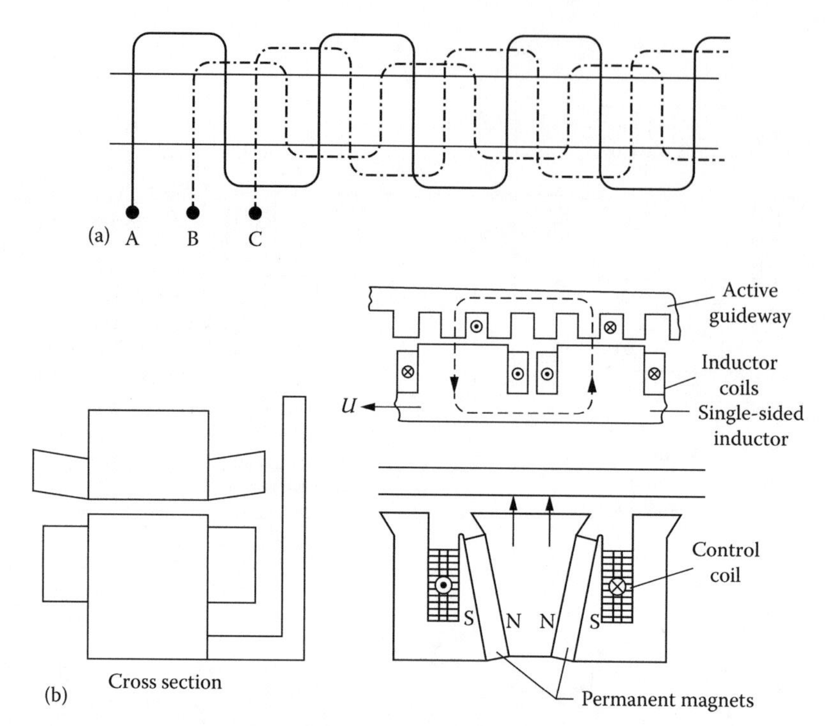

FIGURE 21.3 Long stator cable three-phase winding per active section, (a) and parts of LSM, (b).

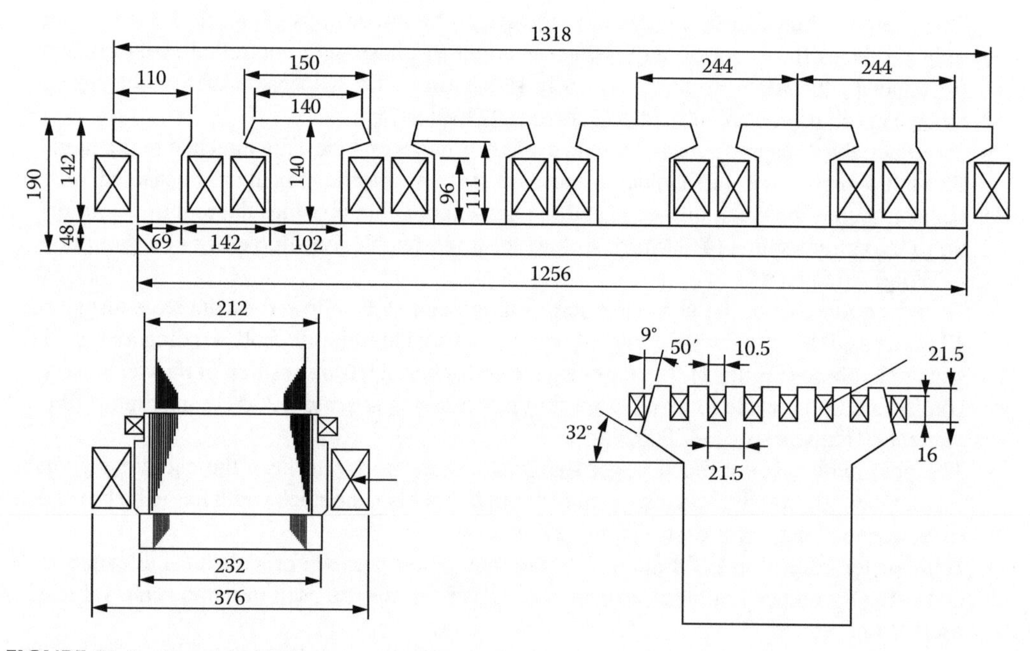

FIGURE 21.4 A practical inductor.

- To secure an acceptable propulsion efficiency and power factor (for reasonable inverter kVA ratings (and costs)), the long stator is fed section by section (Figure 21.5) from both ends (to reduce the number of power switches on ground that connect neighboring sections to a power substation) (Figure 21.6).

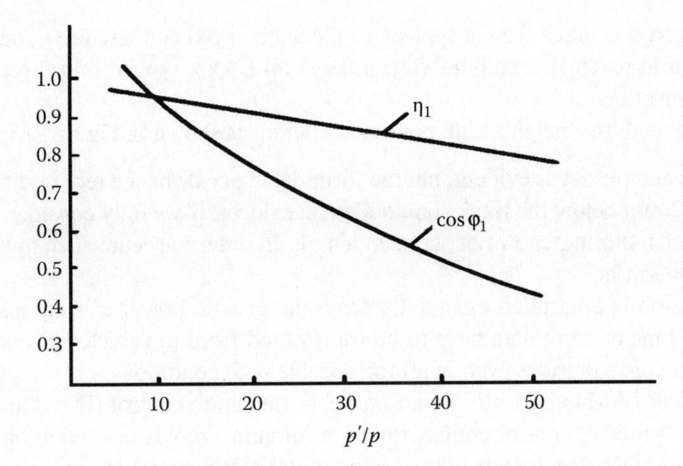

FIGURE 21.5 Typical propulsion efficiency and power factor versus active section length/inductor length (p'/p).

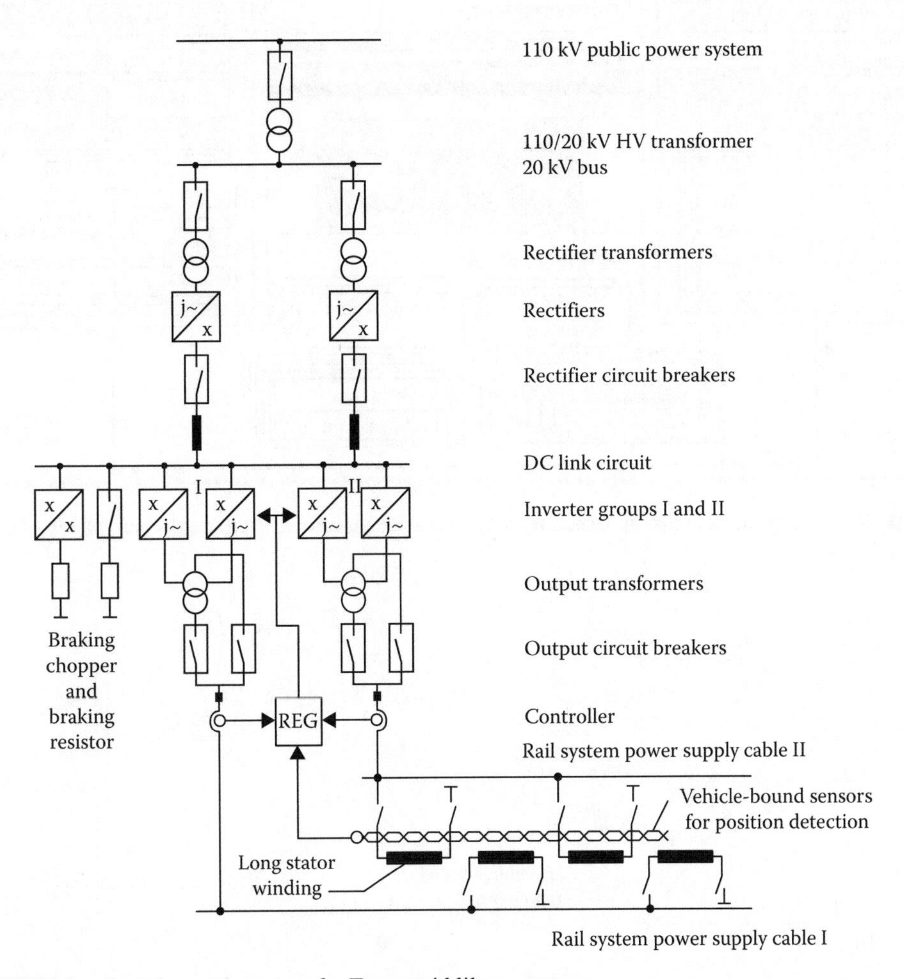

FIGURE 21.6 Generic supply system for Transrapid like systems.

As the electric energy is taken from a typical 10 kV public system, there are quite a few stages of energy conversion to reach the medium voltage level (4–6 kV), typical to a three-level (or more) medium voltage inverter.

A general view with two neighboring power substations is shown in Figure 21.7.

- The system complexity is evident; but the formidable performance required by a magnetic flight 8 ÷ 12 mm below the track should also be evident, if we only consider the precision alignment of 1–2 mm per 3 m of stator in length, in order to secure such low height flight at above 400 km/h.
- The propulsion is controlled essentially from the ground power substations and vehicle position and many more data have to be transmitted from/to vehicle by a complex radio system, to secure safe rides even in all foreseeable fault conditions.
- The control of LSM is basically based on field orientation control (FOC) and, as already mentioned, "pure" i_q control could provide minimum propulsion–levitation interference (Figure 21.8). FOC also provides easy regenerative breaking even at very low speeds. If the retrieved breaking energy cannot be sent back to the public power system, a resistor in the dc link may be controlled to dissipate it for a good cause.

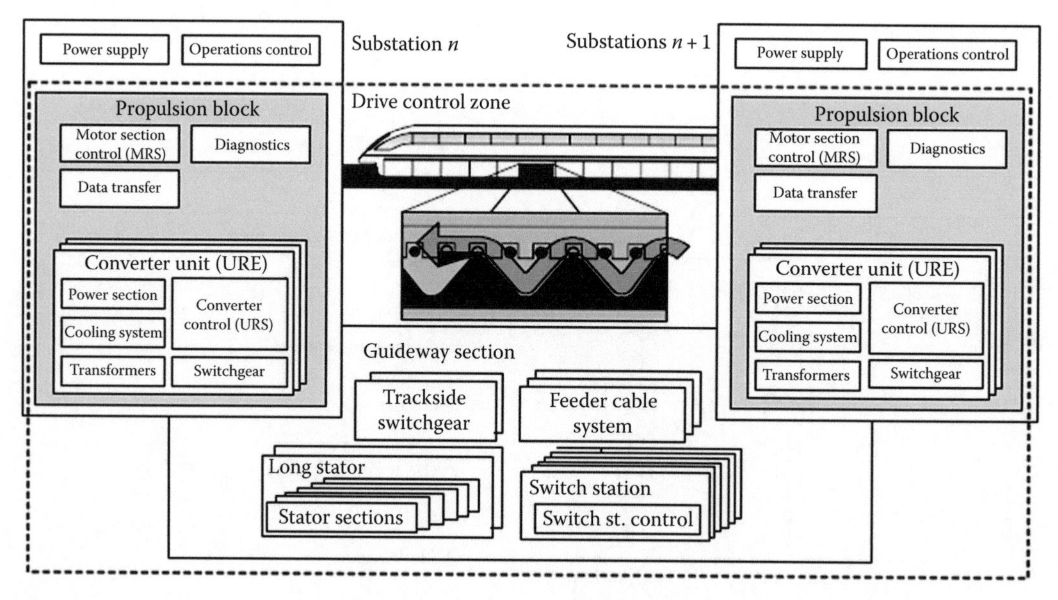

FIGURE 21.7 Typical propulsion structure with supply substations. (After Alscher, H. et al., *IEEE Spec.*, 8, 57, 1984.)

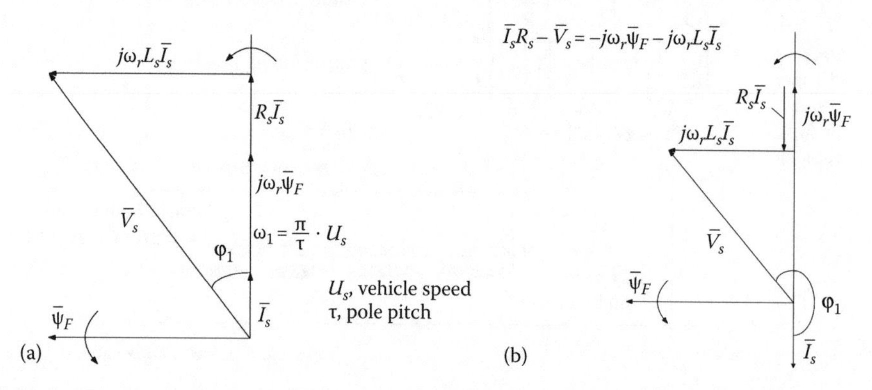

FIGURE 21.8 Pure i_q operation of LSM (a) motoring and (b) generating (regenerative braking).

- As the flight height is small, no safety wheels are required. But they may be needed to pull a faulty vehicle to the repair shop. Safety skids should be provided. Additional (safety) electric braking may be obtained through the guidance electromagnets, where a convenient ac additional current "insertion" will produce stronger eddy currents in the solid iron slabs in the track, working as an efficient eddy-current brake down to low speeds.
- An aerodynamic safety (parachute type) braking system may also be provided.
- As levitation robust (variable structure + PI) decentralized airgap dynamic control of each inductor units is used, the propulsion and the vehicle interference are handled as disturbances.
- For detailed control of attraction force suspension systems (see Chapter 19), it may be also feasible to control i_d also around zero in the sense of using it for damping some vertical inductor motion oscillations (not done so far, apparently).
- The "Transrapid like" system implies multidimensional motion damping and stabilization; such a complex objective is beyond our scope here.
- All in all, capable of less than 70 Wh/passenger/km at 400 km/h cruising speed, the Transrapid system has showed remarkable performance and its extension in very heavy traffic locations, throughout all continents, should be seriously considered.

21.3 SUPERCON MAGLEVs

By Supercon MAGLEVs, we mean the MAGLEVs with superconducting inductors on board of vehicles (Chapter 20), which handle both levitation and guidance (by, in general, repulsion forces) (Figure 21.9). In some special null-flux configuration with nonoverlapping stator three-phase

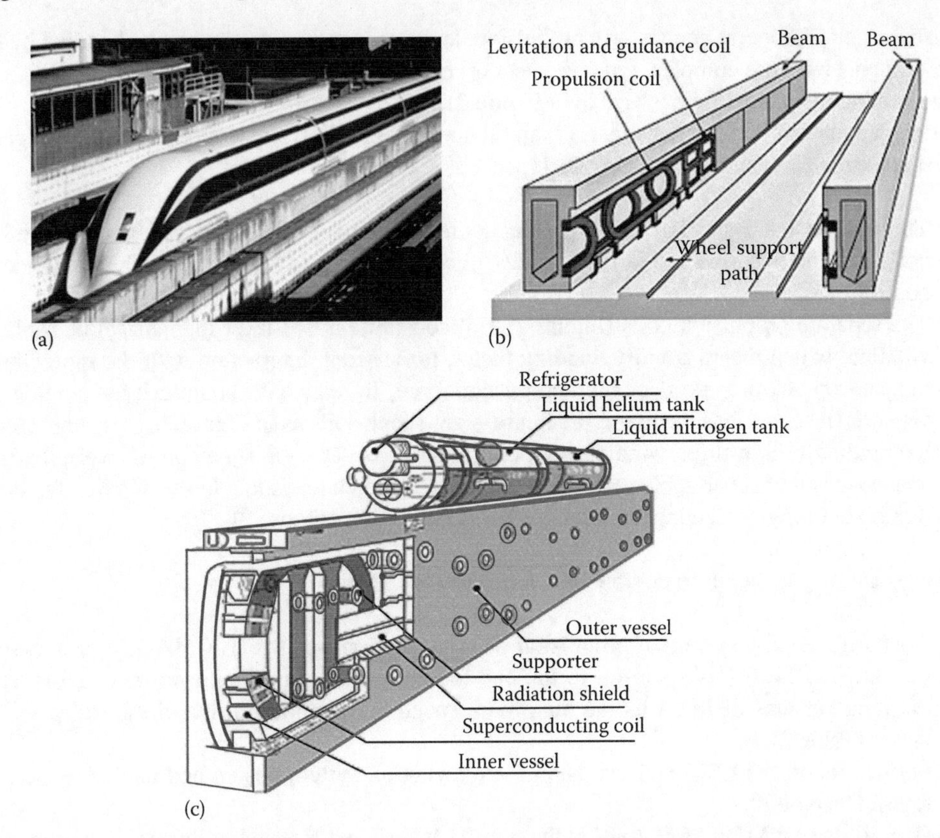

FIGURE 21.9 JR-MLX supercon MAGLEV, (a) general view (after http://ro.wikipedia.org/wiki/Fi%C8%99ier:JR-Maglev-MLX01-901_001.jpg), (b) stator coils distinct propulsion and levitation guidance, and (c) supercon magnet. (After Alscher, H. et al., *IEEE Spec.*, 8, 57, 1984.)

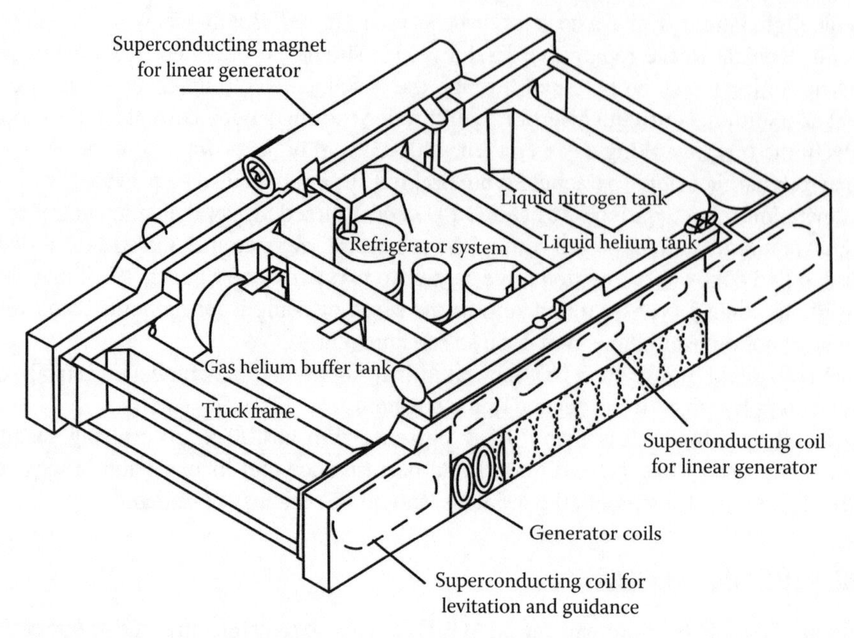

FIGURE 21.10 Generic magnetic bogie of "JL-MLX" MAGLEV. (After Alscher, H. et al., *IEEE Spec.*, 8, 57, 1984.)

eight-shape coils (properly connected), propulsion, levitation, and guidance may be handled by the same, though a bit more complex, guideway (stator) coil system.

The generic magnetic bogie is visible in Figure 21.10.

A detailed view of Supercon magnets (3) and stator propulsion (1) and levitation-guidance figure-eight-shape coils (2) is shown in Figure 21.11.

- The key design issue is related to the figure-eight-shape levitation–guidance ground coils, which are connected as in Figure 21.12, which provides for levitation and guidance control.
- The separate propulsion coils (Figure 21.11a) constitute a two-layer diametrical ac cable winding, which means a unity winding factor, for a strong interaction with the supercon magnets on board, to produce high propulsion force. To reduce the complexity of a guideway coil, it seems feasible to use the figure-eight-shape coils as in Figure 21.12, connected longitudinally in a three winding with coil pitch of $\tau/3$ (τ—the supercon magnet pitch). This means a nonoverlapping winding with $q=1/2$, which leads to a lower winding factor (0.867) but for great savings in total guideway coils costs (Figure 21.13).

In this case, it may be better to notably increase τ above 1.35 m.

- A generic feeding system of propulsion winding section for JR-MLX MAGLE is shown in Figure 21.14. It looks similar to the one for Transrapid, but the power involved is in the range of tens of MVA as the maximum speed is over 560 km/h and the trains are larger (Table 21.1).
- For details on the LSM and repulsion force levitation system design and control, please revisit Chapter 20.
- Overall, the JR-MLX MAGLEV is the only MAGLEV with tested performance, perhaps at 160 W/passenger/km above 450 km/h (up to 580 km/h). If such speeds are needed, this may be the first choice.

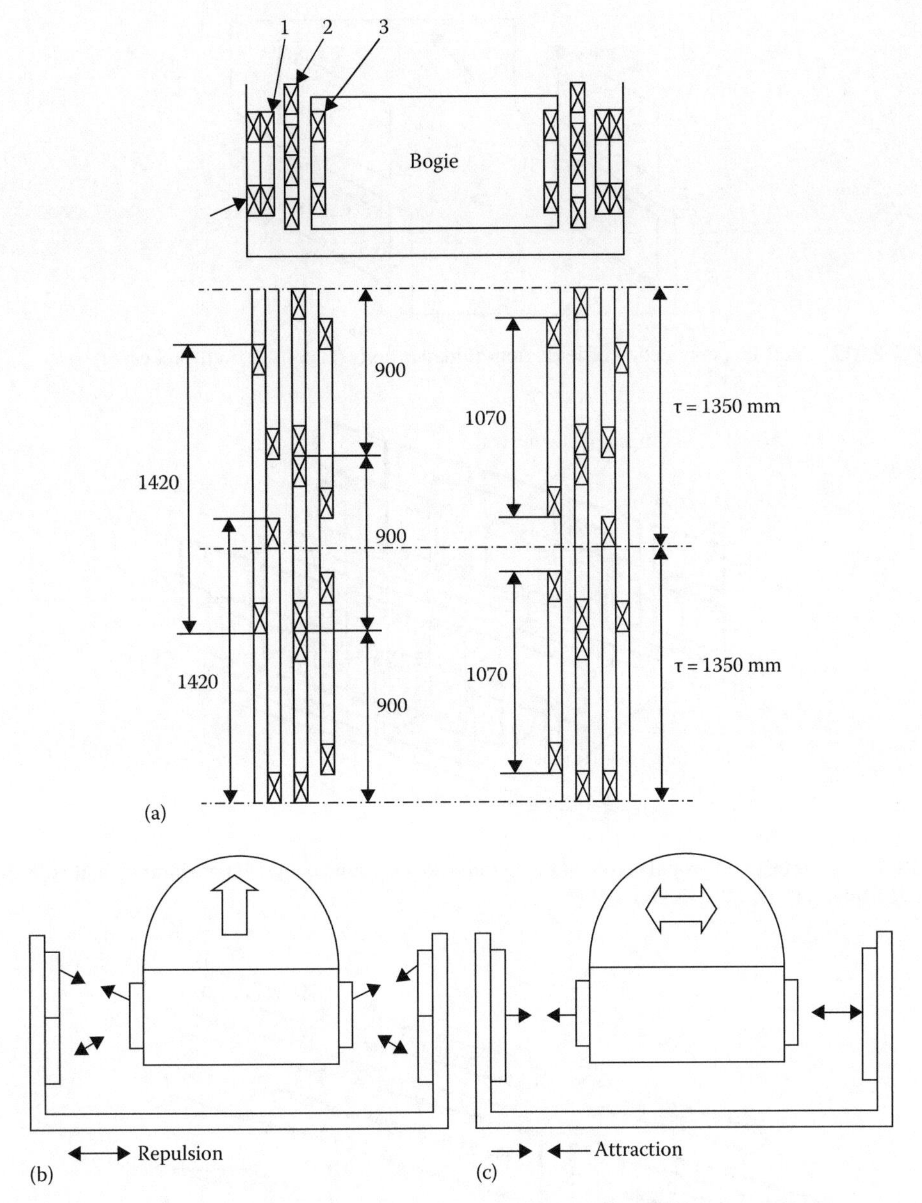

FIGURE 21.11 Supercon magnets on board (3) with propulsion coils (1) and levitation-guidance coils (2) (a) on ground, (b) levitation, and (c) guidance principles.

- In contrast with Transrapid, full levitation is reached above a critical speed (50 m/s or more) and thus acceleration on retractable wheels, up to a critical speed, is required.
- The large practical airgap of 100 mm allows for higher active track irregularities, for given ride comfort, at given speed. It also simplifies the stabilization of levitation-guidance functions; a control coil placed between the supercon magnets and the figure-eight-shape track coils provides active vertical motion damping at 0.3 W/kg (3 W/kg is required in all for Transrapid controlled levitation).
- Regenerative braking is feasible with supercon MAGLEVs also, from on ground inverters' control.

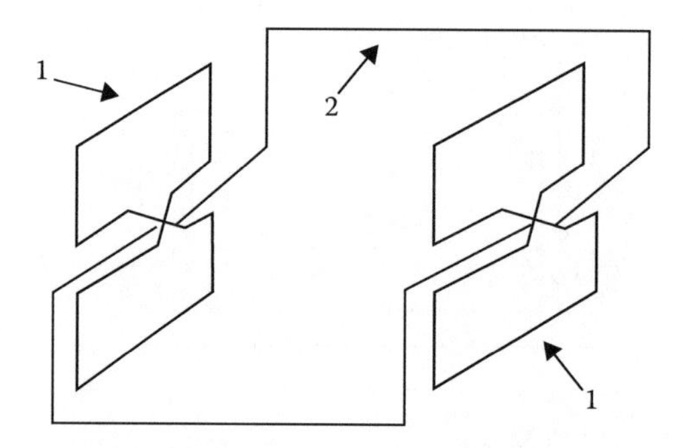

FIGURE 21.12 Null-flux connection of levitation-guidance coils (1, coils; 2, null-flux cable).

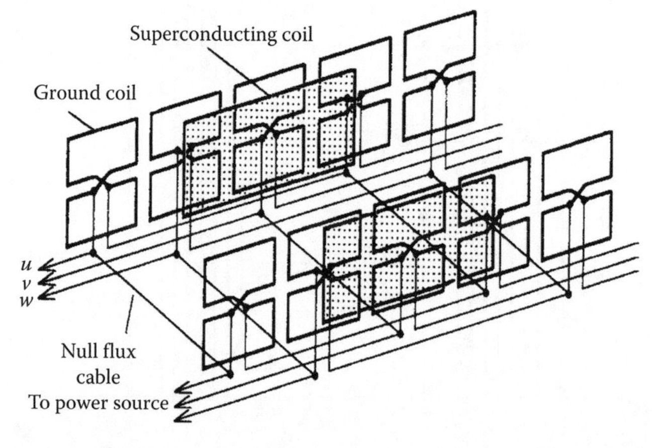

FIGURE 21.13 Integrated propulsion-levitation guidance on ground coils. (After Murai, T. and Fujiwara, S., *Trans. IEE Jpn.*, 116-D, 1289, 1996.)

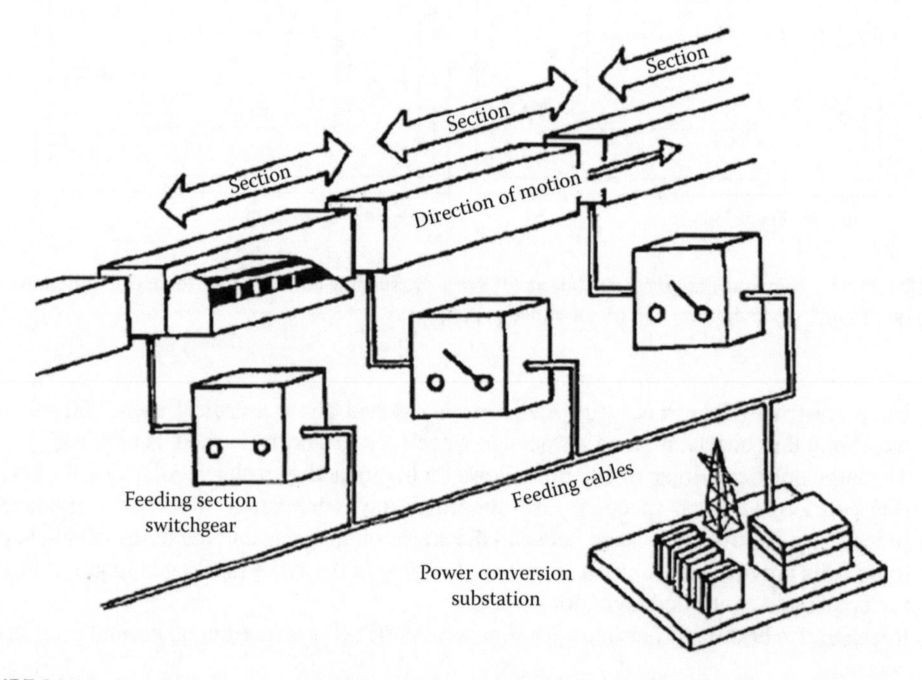

FIGURE 21.14 Feeding system of propulsion winding sections (JR-MLX).

TABLE 21.1
Specification Data of MLX 01 MAGLEV Trains

Specifications	MLX01 (First Train)	MLX01 (Second Train)
Maximum speed	550 km/h	
Number of cars	3	4
Pole pitch of electromagnets	1.35 m	
Vehicle configuration	Articulated bogie system with superconducting electromagnets	
Car body structure	Semi-monocoque structure using aluminum alloy	
Levitation height	0.1 m at 500 km/h	
Width of car	2.9 m car body, 3.22 bogie	
Height of car	3.28 m while levitating 3.22 on-gear running	
Length of end cars	28.0 m	
Length of intermediate car	21.6 m	24.3 m and 21.6 m
Cross section area of the car	8.9 m^2	
Maximum mass of fully loaded car	32 t end car 20 t intermediate car	33 t end car 22 t intermediate car
Length of the train	77.6 m	101.9 m

Sources: Courtesy of central Railway Company and Railway Technical Research Institute, Tokyo, Japan; After Alscher, H. et al., *IEEE Spec.*, 8, 57, 1984.

- The large drag force of levitation–guidance coils on ground at medium and low speeds provides "eddy current" braking but not down to zero speed. The same drag force impedes on vehicle acceleration efficiency, in contrast to Transrapid. Finally, the safety (and during acceleration) wheels may be equipped with disk brakes as done on aircraft, for safety and stop braking.

21.4 IRON-CORE ACTIVE GUIDEWAY URBAN PM-LSM MAGLEVs

In an effort to introduce active PM-LSM guideway urban MAGLEVs, with a simplified control, the system in Figure 21.15 was proposed in the last decade by "General Atomics" in the United States [7].

- It looks similar to Transrapid in terms of long iron-core stator with cable ac three-phase winding (Figure 21.16), but the stator core is split into two parts to provide practically lateral self-centralization (guidance). This solution may be acceptable for urban transportation purposes, if lateral (safety) rubber wheels are provided.
- For levitation, Litz-wire short-circuited coils in a nonmagnetic strong mold are placed along the track as in a repulsive force magnetic suspension system (Chapter 20).
- The on-ground levitation coils "see," in a null-flux configuration, the variation (due to speed) of the magnetic field (Figure 21.15) of dual-PM Halbach array system on board.
- The Litz-wire in the levitation coils is used to reduce the skin effect and thus, for given PM Halbach array pole pitching, reduce the critical speed (to 2.5 m/s!) when full levitation at about 25 mm height is obtained.

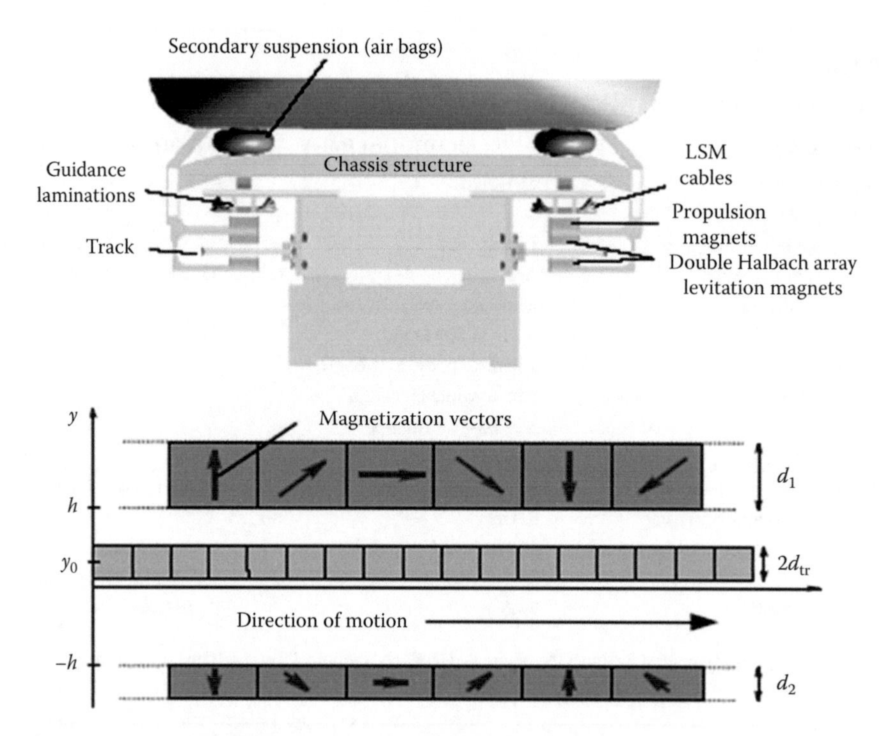

FIGURE 21.15 Active guideway PM-LSM urban MAGLEV. (After Borowy, B.S. et al., Controller design of linear synchronous motor for General Atomics urban MAGLEV, *Record of LDIA-2007*, 2007.)

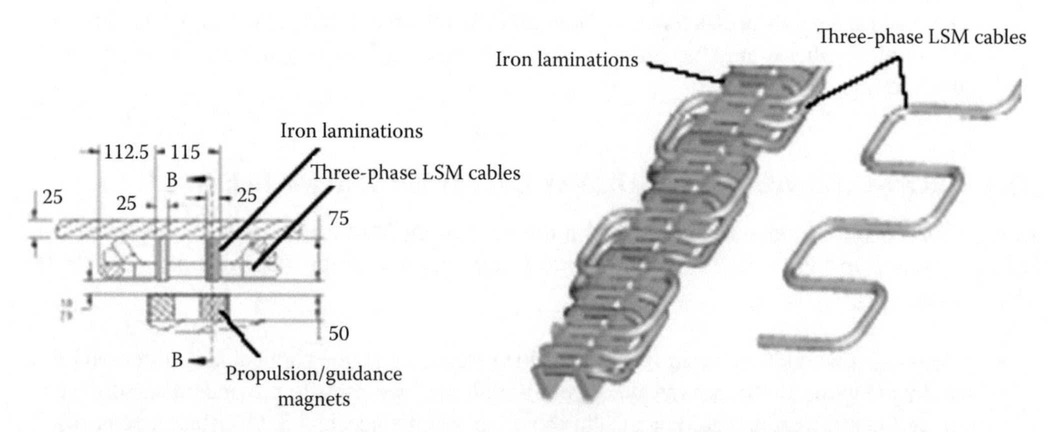

FIGURE 21.16 Three-phase stator LSM winding and propulsion-guidance PM inductor on board. (After Borowy, B.S. et al., Controller design of linear synchronous motor for General Atomics urban MAGLEV, *Record of LDIA-2007*, 2007.)

- The drag force, typical to repulsion force MAGLEVs, is also reduced; it is argued that no dedicated control is required for levitation; some levitation (vertical motion) damping may be performed by the FOC of the LSM, where i_d^* control is at disposal to modify to some extent the levitation force, produced by the PM inductor of the LSM.
- With full-size prototyping on the way (Table 21.2), the system feasibility—technically and commercially—is still to come.

TABLE 21.2
Key System Parameters

System Parameter	Value
Accessibility standards	Americans with Disabilities Act (ADA)
Weather	All-weather operation
Levitation/guidance	Permanent magnet Halbach arrays passive lift and guidance
Propulsion	Linear synchronous motor
Operation	Fully automated driverless train control
Safety	Automated train control, vehicle wraparound structure, restricted access
Speed, maximum operational	160 km/h (100 mph)
Speed, average	50 km/h (31 mph)
Vehicle size	12-m (39.4-ft) long × 2.6-m (8.5-ft) wide × 3-m (9.8-ft) tall
Average power consumption	50 kW
Grade, operating capability	>10%
Turn radius, design minimum	18.3 m (60 ft)
Passenger capacity	AW3 (crush load): 100 passengers total
Aesthetics philosophy	Guideway will blend with and enhance the environment

Source: After Borowy, B.S. et al., Controller design of linear synchronous motor for General Atomics urban MAGLEV, *Record of LDIA-2007*, 2007.

21.5 ACTIVE GUIDEWAY MULTIMOVER DOUBLY FED LIM MAGLEV INDUSTRIAL PLATFORMS

Though designed for high-speed people movers, the Transrapid MAGLEV may be scaled down to small powers for short-haul industrial platforms. However, because the dc inductor on board works at zero frequency, a long stator section can control just one vehicle. There are applications where, within one energized section (say without an inverter but with a soft starter supply) we should handle a few movers independently and/or in a synchronized convoy. As in such applications the airgap may be reduced to 1–3 mm for up to tens of meters of travel, the doubly fed LIM may be the way to go, to provide integrated propulsion–levitation field-oriented control.

An exemplary arrangement with two long stator sections and two movers (secondaries) is visible in Figure 21.17 [8], and it may be modeled as in Figure 21.18.

Now the two stator sections interact with both movers. This is a special situation; apparently, in such an extreme situation only convoy motion mode, same speed, of the movers is feasible. In most

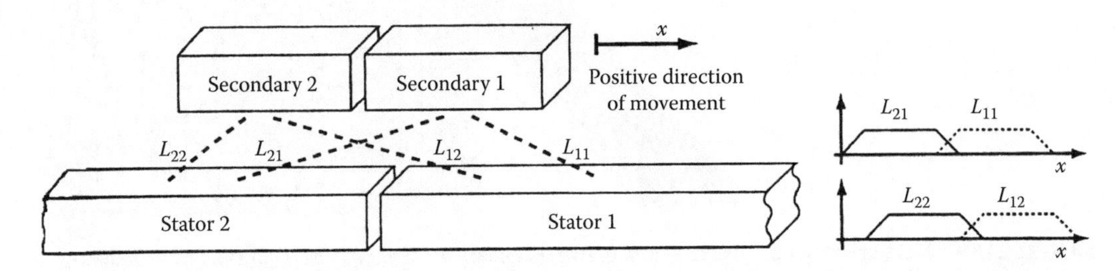

FIGURE 21.17 Coupling inductances with two three-phase stator sections and two three-phase movers in the neighborhood (doubly fed LIM). (After Henke, M. and Grotstollen, H., Control of linear drive test standard for MBP railway carriage, *Record of LDIA-2001*, Nagano, Japan, 2001.)

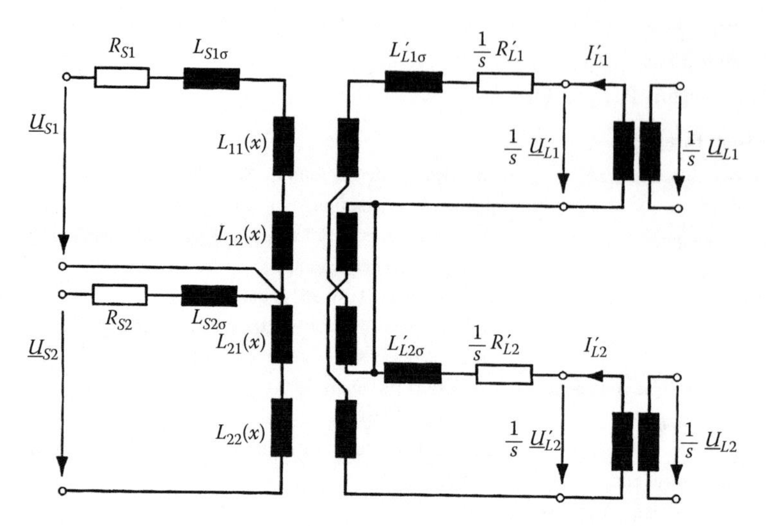

FIGURE 21.18 Equivalent circuit of the arrangement in Figure 21.17. (After Henke, M. and Grotstollen, H., Control of linear drive test standard for MBP railway carriage, *Record of LDIA-2001*, Nagano, Japan, 2001.)

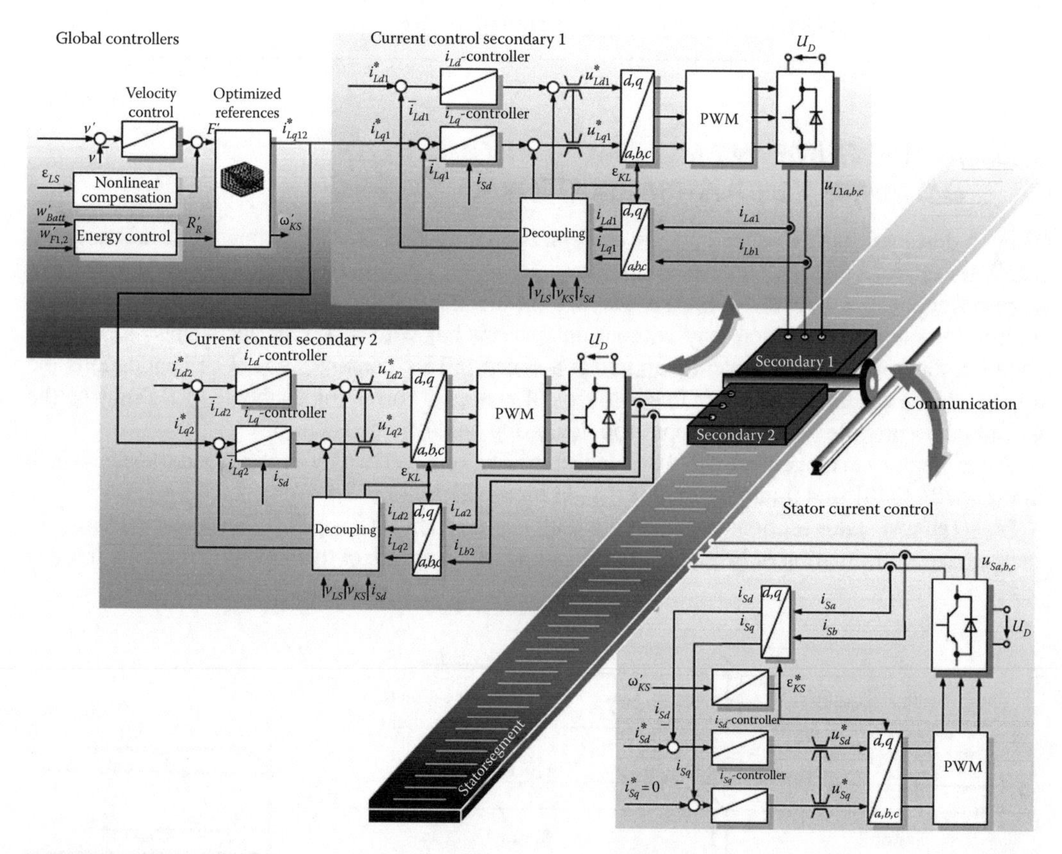

FIGURE 21.19 Field-oriented propulsion control of two secondary doubly fed LIM platform. (After Henke, M. and Grotstollen, H., Control of linear drive test standard for MBP railway carriage, *Record of LDIA-2001*, Nagano, Japan, 2001.)

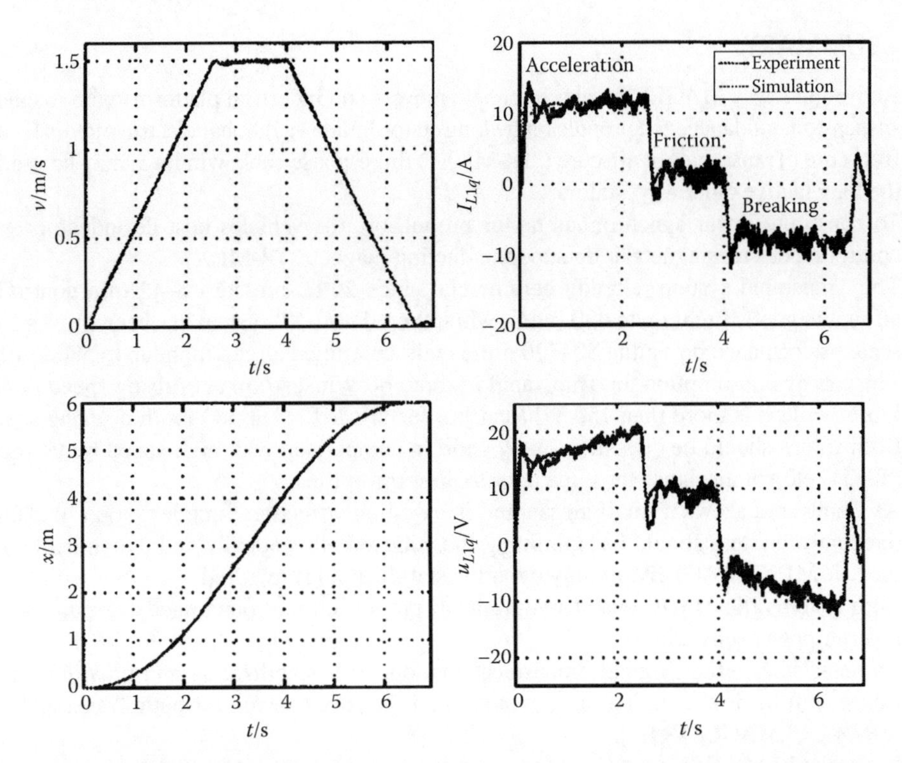

FIGURE 21.20 Speed (m/s), position (m), i_{L1q} (A), V_{L1q} (V) in acceleration–deceleration experiments. (After Henke, M. and Grotstollen, H., Control of linear drive test standard for MBP railway carriage, *Record of LDIA-2001*, Nagano, Japan, 2001.)

cases, however, we may consider two or more movers in the same stator section, which may travel at different speeds or in a convoy (same speed). The slip frequency of the two movers f_{2A} and f_{2B} are

$$f_{2A} = f_1 - \frac{U_A}{2\tau} = S_A f_1; \quad U_A, U_B \text{ speeds of movers A and B} \tag{21.1}$$

$$f_{2B} = f_1 - \frac{U_B}{2\tau} = S_B f_1 \tag{21.2}$$

Now distinct field-oriented propulsion control of the two three-phase wound movers A and B may be applied (Figure 22.19). A master and slave control system might be more appropriate [8].

Independent propulsion control of the two movers (not done yet) within a single stator section can be obtained by separate speed close-loop control in Figure 21.19 (after changes). For MAGLEV operation, the movers have to be placed below the stator section as there are attraction forces between the stator and movers. As the airgap is small, the airgap flux density may be 0.6–0.7 T, and thus enough attraction force can be produced. Thus suspension control—based on airgap and current robust control—can be produced via the i^*_{dL1} and i^*_{dL2} channels in Figure 21.19. Typical low-speed propulsion performance for convoy (same speed) lab operation is shown in Figure 21.20 [8].

Even position-sensorless controlled propulsion has been demonstrated. For lower cost, the stator inverter might be replaced by a soft starter, when, for heavy starts (if needed), the mover three-phase winding can be temporarily short-circuited. With a battery on board of mover, even its recharge (or s.o.c. control) can be orchestrated.

21.6 SUMMARY

- Active guideway MAGLEVs refer to people movers and industrial platforms with magnetic suspension, guidance, and propulsion by long stator (guideway) linear electric motors [9–11].
- Iron-core (Transrapid) or air-core ("JR-MLX") three-phase cable windings may be used in the long (active guideway) stators.
- To constitute linear synchronous motor propulsion, the vehicles host dc inductors with controlled electromagnets or dc-superconducting magnets (JR-MLX).
- The Transrapid system, already commercial since 2002, provides 8–12 mm controlled airgap magnetic flight up to 430 km/h, while the "JR-MLX" system has been proven full-scale performance providing 80–120 mm height controlled airgap flight up to 580 km/h.
- The energy consumption for Transrapid is about 60 Wh/seat/km at cruising speed of 400 km/h while it is more than 150 Wh/seat/km for "JR-MLX" at 580 km/h cruising speed. Both values should be considered very good in comparison with high-speed bullet trains (TAGV, etc.) or airplanes for same door to door travel time.
- As Transrapid shows a small airgap and is based on attraction force levitation guidance, the control system should be stabilizing the vehicle both statically and dynamically. For the "JR-MLX" MAGLEV, mainly dynamic stabilization is required.
- Both systems allow for regenerative braking by propulsion control initiated in the on-ground inverter power substations.
- As vehicles experience multidimensional motion, their stabilization and providing good riding comfort is a daunting task; apparently, it has been solved for both Transrapid and "JR-MLX" MAGLEVs [11].
- For urban MAGLEVs, the SC magnet inductors may be replaced by PMs on board; a null-flux repulsive force levitation system, with dual PM rows on board and short-circuited air-core coils on ground, has been added and thus the "General Atomics" system has been completed. It claims no need for levitation and guidance control, while the ride comfort is provided by mechanical suspension systems. The propulsion system and its control are as for Transrapid systems. The system is yet to be full-scale tested.
- For MAGLEV industrial platforms, the Transrapid system may be scaled down adding PMs in the inductor poles. "Zero power" levitation control may be applied by allowing the airgap to vary within some large (but safe) limits; beyond this airgap, safety (collision avoidance) control is performed.
- In addition, for MAGLEV platforms, but even for urban MAGLEV vehicles, the doubly fed LIM for propulsion and levitation can be used. The three-phase winding on board can handle, in interaction with a three-phase iron-core long stator, both propulsion and levitation control. The system may control even two vehicles independently, on a long enough active stator section, by controlling the on-board mover inverters at different slip frequencies f_{2A} and f_{2B}. For dc control in the mover windings, the system "degenerates" into Transrapid situation.
- Stator coil and PM-mover small MAGLEV platforms have been investigated in Chapter 18, dedicated to planar motor LOMs.

REFERENCES

1. H. Weh, Synchronous long stator motor with controlled normal force, ETZ.A-46, 1975 (in German), 409–413.
2. K. Meinrich and R. Kretzschmar, *Transrapid Maglev System*, Hestra Verlag, Darmstadt, Germany, 1989.
3. H. Kolm, R.D. Thornton, Y. Iwasa, and W. Brown, The magniplane system, *Cryogenics*, July 1975, 377–383.
4. E. Ohme, M. Iwamata, and T. Yamada, Characteristics of superconductive suspension and propulsion for high speed trains, *Proc. IEEE*, 61(405), 1973, 579–586.

5. M. Toshiaki and F. Shunsuke, Design of coil specifications in EDS Maglev using an optimization program, *Record of LDIA-1998*, Tokyo, Japan, pp. 343–346.
6. T. Murai and S. Fujiwara, Characteristics of combined propulsion, levitation and guidance system with asymmetric figure between upper and lower coils in EDS, *Trans. IEE Jpn.*, 116-D, 1996, 1289–1296.
7. B.S. Borowy, H. Gurol, and J.-K. Kim, Controller design of linear synchronous motor for General Atomics urban MAGLEV, *Record of LDIA-2007*.
8. M. Henke and H. Grotstollen, Control of linear drive test standard for MBP railway carriage, *Record of LDIA-2001*, Nagano, Japan.
9. M. Mihalachi and P. Mutschler, Capacitive sensors for position acquisition of linear drives between passive vehicles, *Record of OPTIM-2010*, 673–680.
10. H. Alscher, I. Boldea, A.R. Eastham, and M. Iguchi, Propelling passengers faster than speeding bullet, *IEEE Spectrum*, 8, 1984, 57–64.
11. D.M. Rate, Review of dynamic stability of repulsive force MAGLEV suspension system, *IEEE Trans.*, Mag-38(2), 2002, 1383–1390.
12. http://upload.wikimedia.org/wikipedia/commons/d/d0/Shanghai_Transrapid_002.jpg
13. http://ro.wikipedia.org/wiki/Fi%C8%99ier:JR-Maglev-MLX01-901_001.jpg

22 Passive Guideway MAGLEVs

22.1 INTRODUCTION

Passive guideway MAGLEVs are MAGLEV vehicles and industrial platforms whose active (electrically supplied) parts are all on board of the mover, and thus full power transfer to the vehicle is required. In counterbalance for a more costly vehicle, the guideway is passive and made of aluminum sheets and back solid iron slabs (for linear induction motors (LIM) propulsion) or of variable magnetic reluctance laminated or solid iron cores (for linear synchronous motor (LSM) propulsion).

Historically, MAGLEV research was revived in the 1960s in Germany and Japan by investigation of both active guideway and passive guideway solutions.

However, priority to active guideway for high-speed MAGLEVs has been given due to two main circumstantial technical limitations (or beliefs):

- Mega-Watt (MW)-level economical power transfer to MAGLEVs at 400–500 km/h was thought infeasible (this limitation has been removed by Tres Grande Vitesse in France in the 1990s, with above 550 km/h recently).
- The full-power converter weight/kVA was too large to leave enough volume and weight for load (passengers). The insulated gate bipolar transistor (IGBT) multilevel or MOSFT controlled transistor (MCT) inverters can today be built at less than 1 kg/kVA with forced cooling and small enough volume, so this limitation has also been eliminated. The removal of these two limitations explains why the passive guideway MAGLEVs are becoming popular again; for now only for urban MAGLEVs where they exhibit.
- Simple (constant voltage and frequency power substations).
- Independent vehicle control from on board.
- Simpler guideway switches.
- In case of vehicle faults, only that vehicle is removed quickly off the main guideway, which remains fully operational.

By now passive guideway urban MAGLEVs are getting solid ground, while in the near future their extension to intercity (400 km/h) transportation may be seriously considered.

An LIM propulsion passive guideway MAGLEV system (HSST in Japan) is commercially operational today, and one such system was operational for a few years at the airport of Birmingham (UK) on a 660 m long track; also another one is close to commercialization for an urban 8 km track (UMT in S. Korea). An experimental system is operational at Old Dominion University campus in the United States. In LIM propulsion MAGLEVs, the levitation and guidance are performed in general by dedicated controlled dc electromagnets that interact with solid iron track slabs. The drag force due to the eddy currents induced in the track slabs is rather small (10% of full propulsion force) for urban speeds (30 m/s) and the levitation (attractive) force per electromagnet weight is in general larger than 15/1, while (Chapter 19) the average energy consumption for levitation and guidance is around 2.0 kW/ton of vehicle.

In contrast, the LIM propulsion at 8–10 mm airgap (much smaller [2.5 mm] in the Old Dominion University experimental vehicle) leads to efficiency below 80% and a lagging power factor between 0.45 and 0.6. Even for such a moderate performance, the slip frequency in the LIM is kept such that a constant small total attraction normal force remains to be counteracted by the levitation system.

Integration of LIM propulsion and full levitation would mean a cage secondary with coarsely laminated core (track) to allow 0.6 T airgap flux density levels at 6–10 mm mechanical (now all magnetic) airgap.

In an effort to simplify the track topology (and reduce the costs), by propulsion-levitation integration, a homopolar linear synchronous motor (H-LSM) (Chapter 8) propulsion-levitation MAGLEV has been proposed and tested on a 4 ton prototype (Magibus.01, in Romania).

In a H-LSM (Chapter 8), both the dc and ac windings are placed on board of the "short primary," while the guideway is made of solid iron segments pole (τ) long, with 2τ periodicity, to produce sufficient magnetic anisotropy (4/1) at 6–10 mm airgap.

The control of dc winding produces fully controlled levitation while the inverter control of the ac winding currents (pure i_q control) performs propulsion control. This levitation-propulsion integration makes the rather lower thrust/weight capability of H-LSM acceptable; also, as the dc winding produces levitation and participates in propulsion, the total efficiency of levitation and propulsion is above 80% (in urban vehicles) and proves superior to LIM-MAGLEVs. Moreover, in urban vehicles, the power factor could be around 0.8 (lagging, always, for pure i_q control), and thus the inverter kVA ratings, weight, and cost are quite acceptable.

Guidance may be provided by distinct dc controlled electromagnets interacting with lateral solid iron slabs that may be part of the passive guideway mechanical structure itself.

This chapter introduces the principle and performance of existing LIM and H-LSM MAGLEVs with methods to further improve them. It discusses also new, potentially competitive, passive guideway systems such as

- Transverse-flux linear PMSM MAGLEVs
- Transverse-flux linear dc polarized switched reluctance motor MAGLEVs
- Multiphase two-level bipolar current reluctance motor MAGLEVs

22.2 LIM-MAGLEVs

A typical half cross section of an urban LIM-MAGLEV as shown in Figure 22.1 may be characterized as follows:

- The T-shape slender guideway contains a U-shape solid iron track, on the lower sides, which interacts with the dc controlled electromagnets on board the vehicle to produce controlled levitation and some guidance self-centralizing force; the damping of lateral motion may be handled by a mechanical damper, in urban applications.
- The LIM primary is placed on the vehicle also, but its secondary, placed along the track, is made of an aluminum slab (4–6 mm thick) on solid iron 20–25 mm thick plates (or track structure, itself). The LIM secondary is located on the upper part of track, for convenience. However, the normal total force of LIM, if of attraction character (for small slip frequency), opposes levitation and should be counteracted by the levitation system; the advantage of better LIM efficiency and power factor depends on LIM dynamic end effects' severity at cruising speed.

On the contrary, when the LIM is designed for zero normal force (attraction and repulsion normal forces cancel each other), the slip frequency tends to be large and thus the LIM efficiency and power factor are reduced notably.

- It may be feasible to use the field-oriented control (FOC) of LIM such that i_q provides propulsion control and i_d control helps in damping vertical vehicle oscillations, without inducing notable oscillation in the vehicle motion plane.

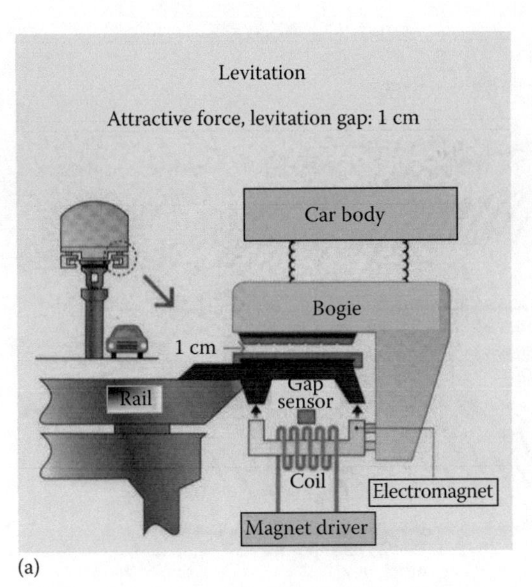

(a)

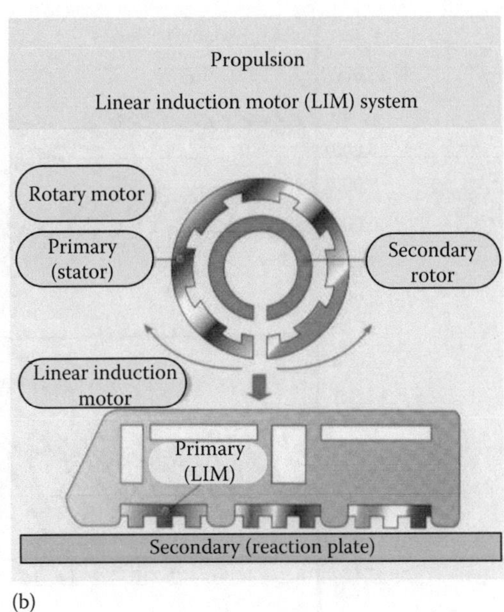

(b)

FIGURE 22.1 S. Korean LIM-MAGLEV: (a) cross section and (b) longitudinal view. (After Maglev (transport), Wikipedia, the free encyclopedia.)

- The commercial LIM-MAGLEV of HSST commenced operation in March 2005 in Aichi, Japan, on a nine-station 8.9 km track known as LINIMO (Figure 22.2). The trains go up to 100 km/h and the track shows a maximum gradient of 6% and a minimum curve radius of 75 m.
- Typical 8 mm airgap LIM propulsion performance when fed from PWM inverters for urban MAGLEVs is shown in Figure 22.3 (borrowed from Chapter 6).

FIGURE 22.2 “LINIMO” HSST (LIM-MAGLEV). (After http://en.wikipedia.org/wiki/File:Linimo_approaching_Banpaku_Kinen_Koen,_towards_Fujigaoka_Station.jpg)

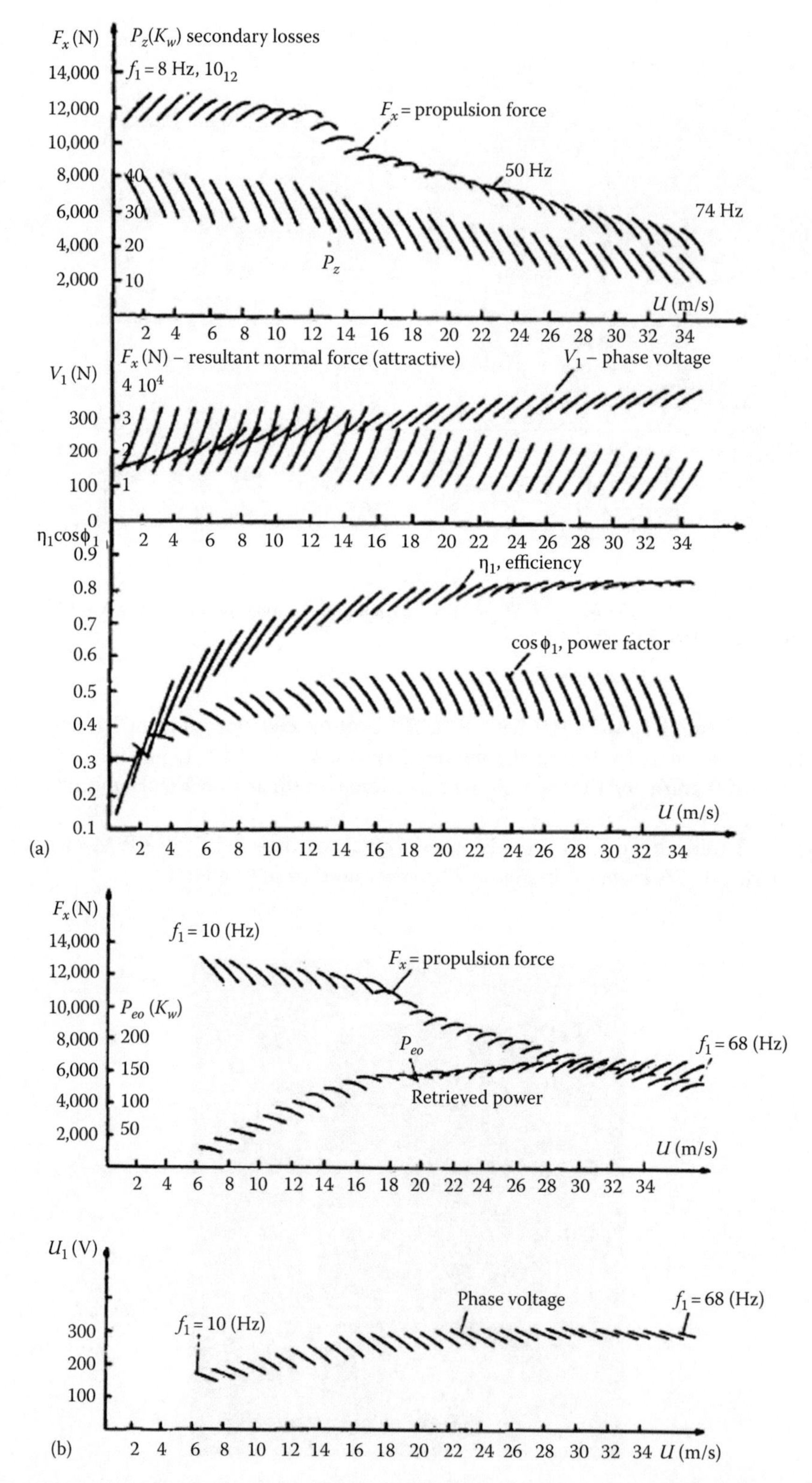

FIGURE 22.3 Typical performance with LIM in urban MAGLEVs: (a) propulsion and (b) regenerative braking.

- The efficiency may also be around 0.8 (1.5 W copper stator losses/*N* of thrust), but the power factor is 0.4–0.5, which puts a notable burden on the on-board inverter kVA (weight and cost).
- The (1.5–2) kW/ton of levitation/guidance is acceptably low. As the vehicle weight may be as low as mechanically feasible, reducing the weight of electric power equipment on board is a top priority to the point of minimum total net present value of the system (initial, plus energy, plus maintenance costs, for given commercial speed and traffic density).

22.2.1 Potential, Improved LIM-MAGLEV Concepts

- Scarce data are available on the commercial (or close to) LIM-MAGLEV systems as they are in their early days. But it seems evident that some ways to reduce the passive track separate components (for levitation and propulsion), with the normal force of LIM put to work, and/or the integration of full levitation and propulsion in the LIM with cage secondary, are ways worth following. Figure 22.4 shows a potential configuration in which both the LIM and the dc controlled electromagnets (now with PM assistance), placed below the track, act along the same track.

Now the LIM normal force "helps" levitation if the former has an attraction character and its control, in general, can assist the levitation stabilization.

The LIM primaries are interspaced with levitation electromagnets along the vehicle sides, on board. The LIM interaction with the guideway (Figure 22.4) may produce some centralizing guidance force.

Levitation control, with PM-assisted electromagnets, may be approached as zero current control (between, say, 70% and 120% rated airgap) and, then, safety airgap control takes it over (see Chapter 19).

The FOC of LIM through i_d current (Chapter 6) can provide levitation assistance control by damping vertical oscillations.

Decoupled control of LIM levitation (through i_d) and propulsion (through i_q) functions by FOC (or direct thrust and normal force control) is essential for good dynamic MAGLEV performance.

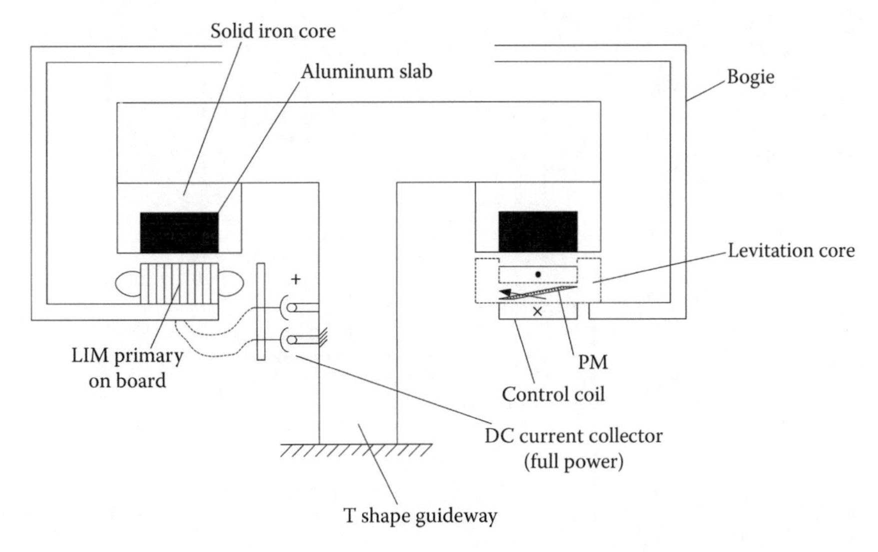

FIGURE 22.4 Integrated propulsion-levitation guideway with LIM + PM assisted dc control levitation electromagnets.

However, secondary-bogie and tertiary-cabin mechanical suspension systems are required to reduce "unsprung" mass and provide good ride comfort, even in urban LIM–MAGLEVs.

Another possibility is to use LIMs for full levitation and propulsion control, when higher airgap flux density is required (0.6 T or more). For a 6–10 mm mechanical airgap, this implies a cage secondary (placed in a coarse lamination iron core) along the track (Figure 22.5a).

The potential control system in Figure 22.5b shows that, with a secondary flux ($\hat{\overline{\Psi}}_r$) observer (which may integrate an airgap observer), the secondary flux position $\hat{\theta}_{\psi_r}$ is estimated (Chapter 6

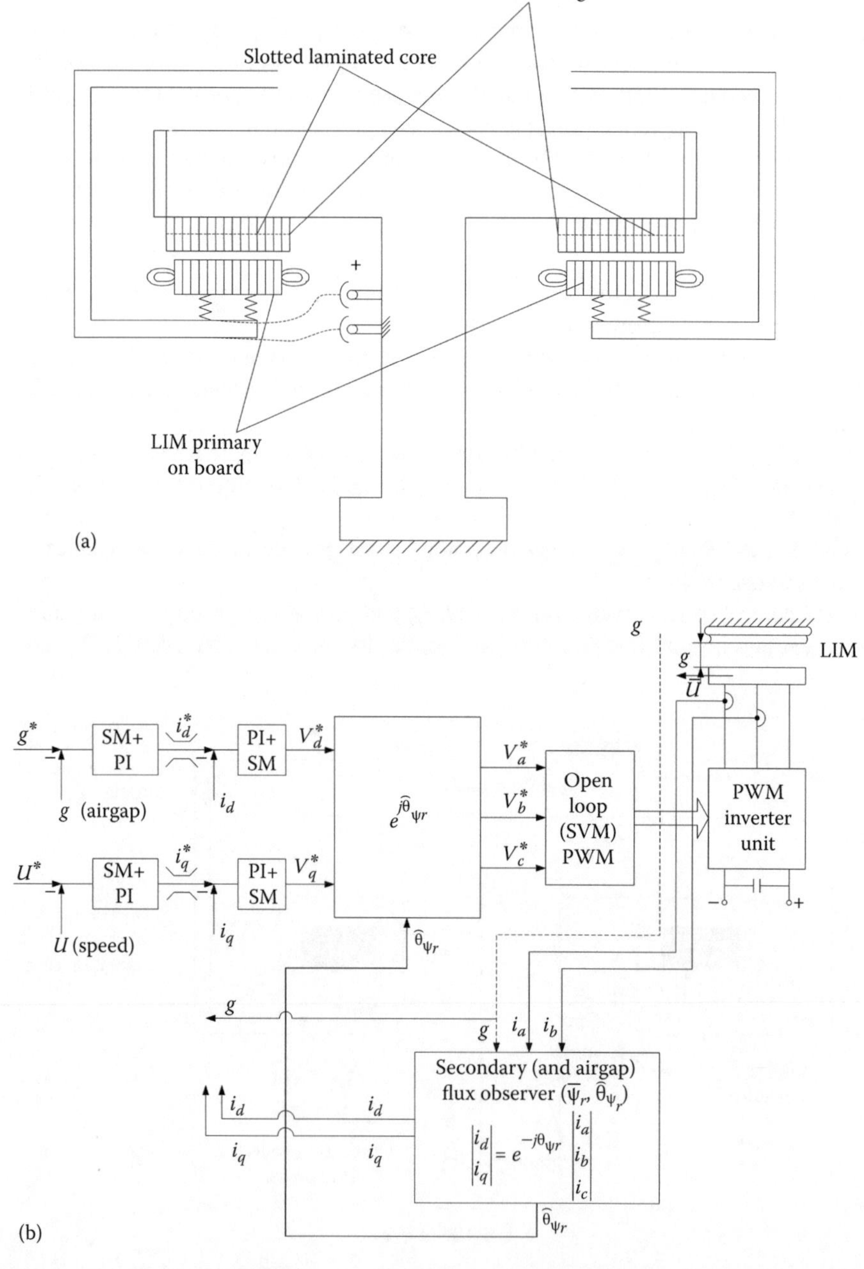

FIGURE 22.5 Generic integrated propulsion and full levitation LIM on urban MAGLEVs: (a) the system topology and (b) potential control system.

on LIMs). The robust SM + PI close loop regulators should handle both levitation and propulsion control with variable airgap: from start ($2g_{rated}$) to g_{rated} ±20% (for cruising). Needless to say that multiple LIM units (4 k units; $k \geq 1$) per cabin are required, each with its own PWM inverter and decentralized control, if the secondary and tertiary mechanical suspensions alleviate enough the interaction between LIM units.

The cage secondary of LIM may not cost too much to be practical for urban MAGLEVs; it is a mandatory solution to produce full levitation/propulsion by LIM for levitation force per LIM primary weight greater than 10–1. If in a vehicle the LIMs weigh 10% of all weight, then a 1.0 m/s² acceleration is available. This is enough in practice.

22.3 H-LSM MAGLEV (MAGNIBUS)

The Magnibus [2] system (Figure 22.6) uses linear homopolar (inductor) synchronous motors for integrated propulsion and full levitation. Guidance is provided by separated dc controlled electromagnets on board, interacting with lateral track solid iron guideway slabs (Figure 22.7).

For details on H-LSM, see Chapter 8.

FIGURE 22.6 Photograph of Magnibus. (After Boldea, I. et al., *IEEE Trans.*, VT-37(1), 213, 1988.)

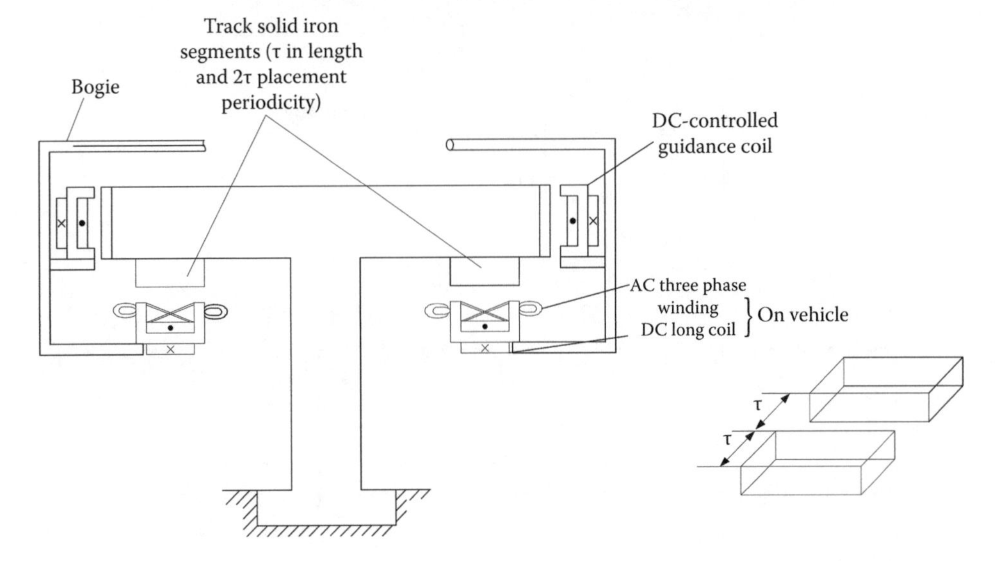

FIGURE 22.7 H-LSM MAGLEV system. (After Boldea, I. et al., *IEEE Trans.*, VT-37(1), 213, 1988.)

The main data of the experimental Magnibus-01 [2] are given in Table 22.1.

The levitation model, considered only for the dc winding of H-LSM in Figure 22.8a, has a structural diagram as in Figure 22.8b. The airgap δ and the absolute acceleration $\ddot{Z}_m$ are measured. Based on them an observer for airgap variation speed $\hat{Z}_\delta$ is built and integrated in a compensator that handles track irregularities accelerations indirectly (Figure 22.9).

The complete levitation control system (Figure 22.10) includes the compensator in Figure 22.9, but also adds the following:

- An outer PI airgap close loop regulator.
- An interior flux Ψ_T estimator signal (obtained with an emf search coil filter): this tends to "linearize" the system (Chapter 19) and thus yields good levitation control from 20 to 2 mm airgap. The flux derivative ($\dot{\Psi}^T$) and airgap sensor coils' positioning is shown in Figure 22.11. The airgap sensors operate around resonance at 100 kHz and thus produce reliable outputs.

TABLE 22.1
Magnibus-01 Data

Weight	4 ton	Passive guideway: solid iron segments 0.25 m × 0.1 m × 0.03 m
Length	4 m	Power collectors: 200 V dc (trolleybus double collectors)
Number of linear inductor motors	4	3 × 220/380 V ac (brush collectors) -to be eliminated
Motor weight	140 kg	
Number of poles per motor	$2p = 6$	
Rated airgap	$g_0 = 0.01$ m	Test track length: 150 m
Active stack width	$L_a = 0.11$ m	Levitation: attraction dc controlled
Pole pitch	$\tau = 0.1$ m	Propulsion: position ac controlled

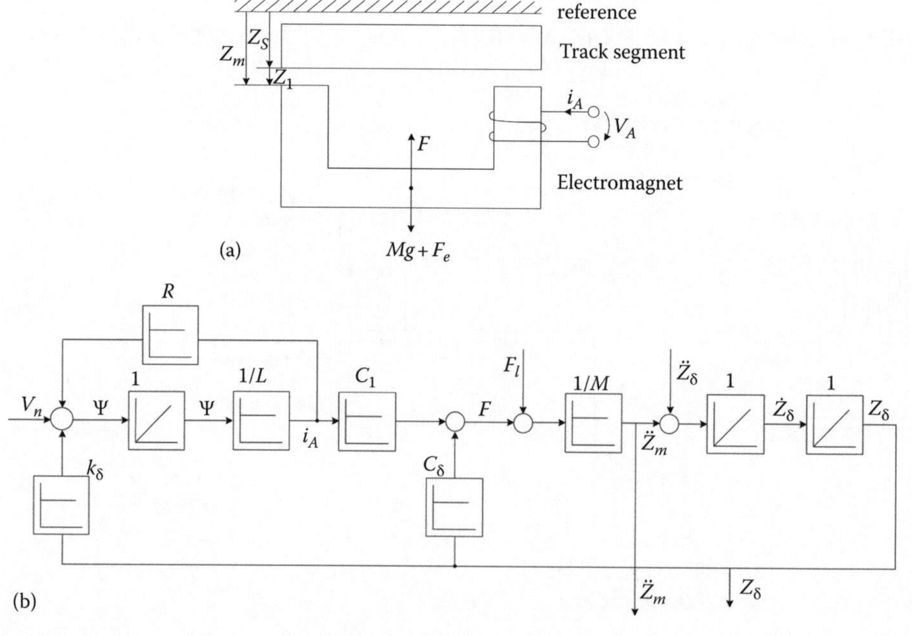

FIGURE 22.8 (a) Levitation model and (b) its structural diagram.

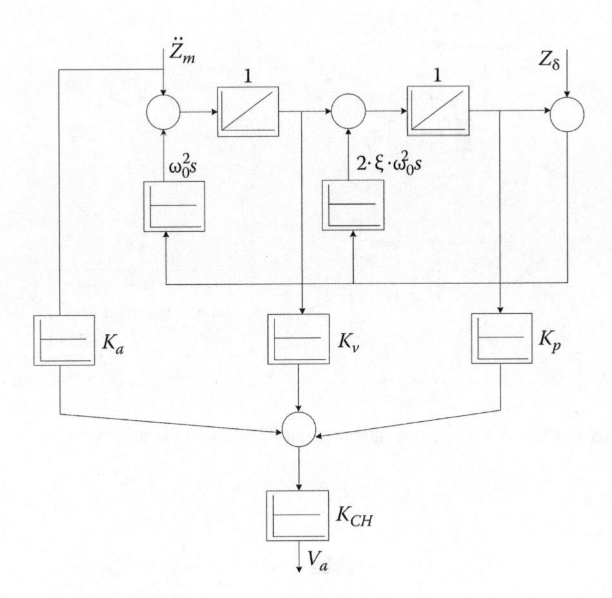

FIGURE 22.9 Compensator with $\dot{Z}_\delta$ observer.

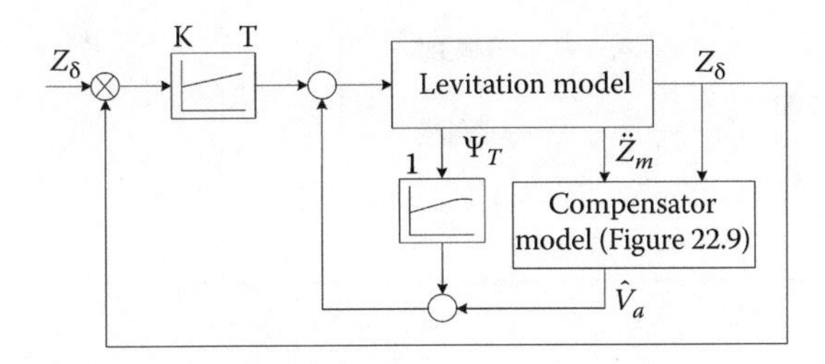

FIGURE 22.10 Complete levitation control system.

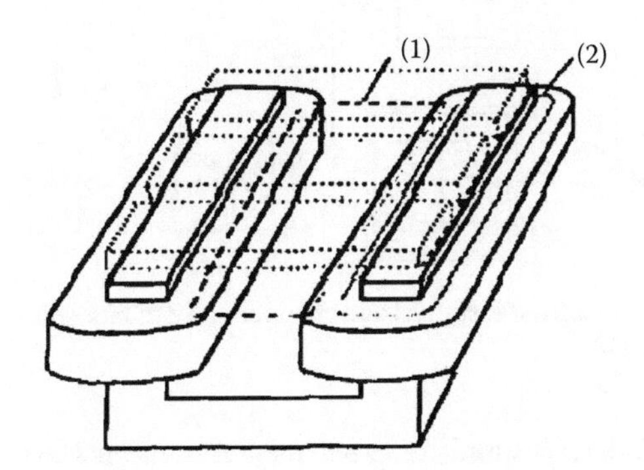

FIGURE 22.11 H-LSM with: (1) airgap sensor air-core coil and (2) flux derivative Ψ_T sensor coil.

The airgap and acceleration sensors' basic circuits are given in Figure 22.12.

In the Magnibus-01, a single quadrant fast-thyristor dc–dc converter (Figure 22.13a) with bipositional (fast) control (Figure 22.13b) was used. Today IGBT 2-quadrant dc–dc converters are standardly used for fast response in levitation systems (Transrapid, etc.)

As all 4 H-LSMs, placed at vehicle corners, are fixed to the bogie with secondary mechanical suspension systems, decentralized levitation control was successfully applied.

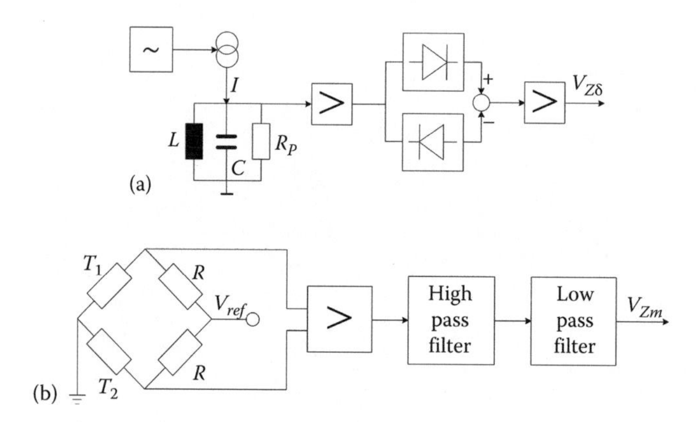

FIGURE 22.12 Airgap sensor (a) and acceleration sensor (b) basic circuits.

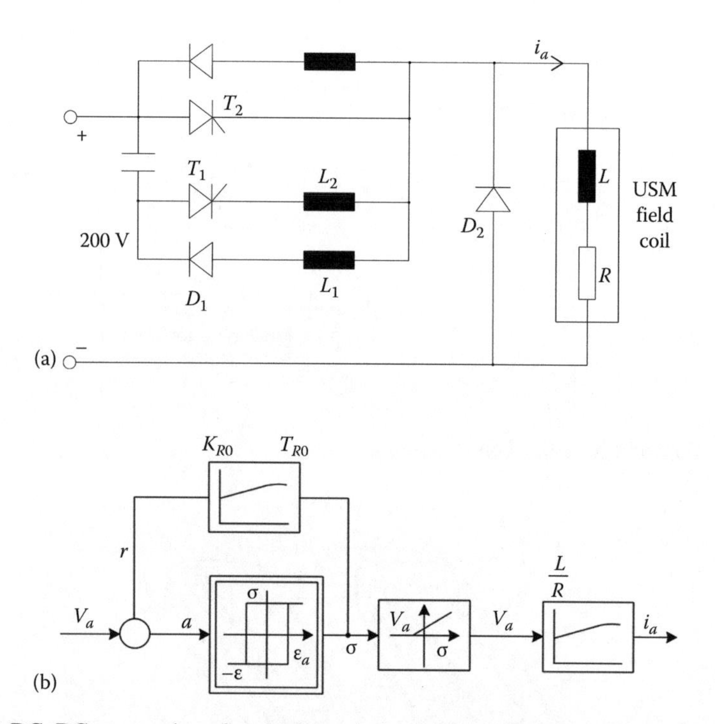

FIGURE 22.13 DC–DC one quadrant fast thyristor chopper (a) with bipositional control system (b).

Typical vehicle lifting (takeoff) standstill experimental transients from 20 to 12.5 mm and landing are shown in Figure 22.14 [2]; well-behaved responses are visible.

Responses to perturbations from neighboring H-LSMs are illustrated in Figure 22.15.

Finally, one airgap variation during vehicle acceleration and deceleration (with vehicle braking and speed reversal) is visible in Figure 22.16.

The position sensor with coils in air and its six signals per period are presented in Figure 22.17.

So rectangular current control in the current source inverter was used. Today, voltage-source PWM inverters could be used and motion-sensorless control for propulsion can be adapted.

It is feasible to use an inverter for each H-LSM primary and thus produce propulsion control so as to damp vehicle parasitic motions along the vertical axis.

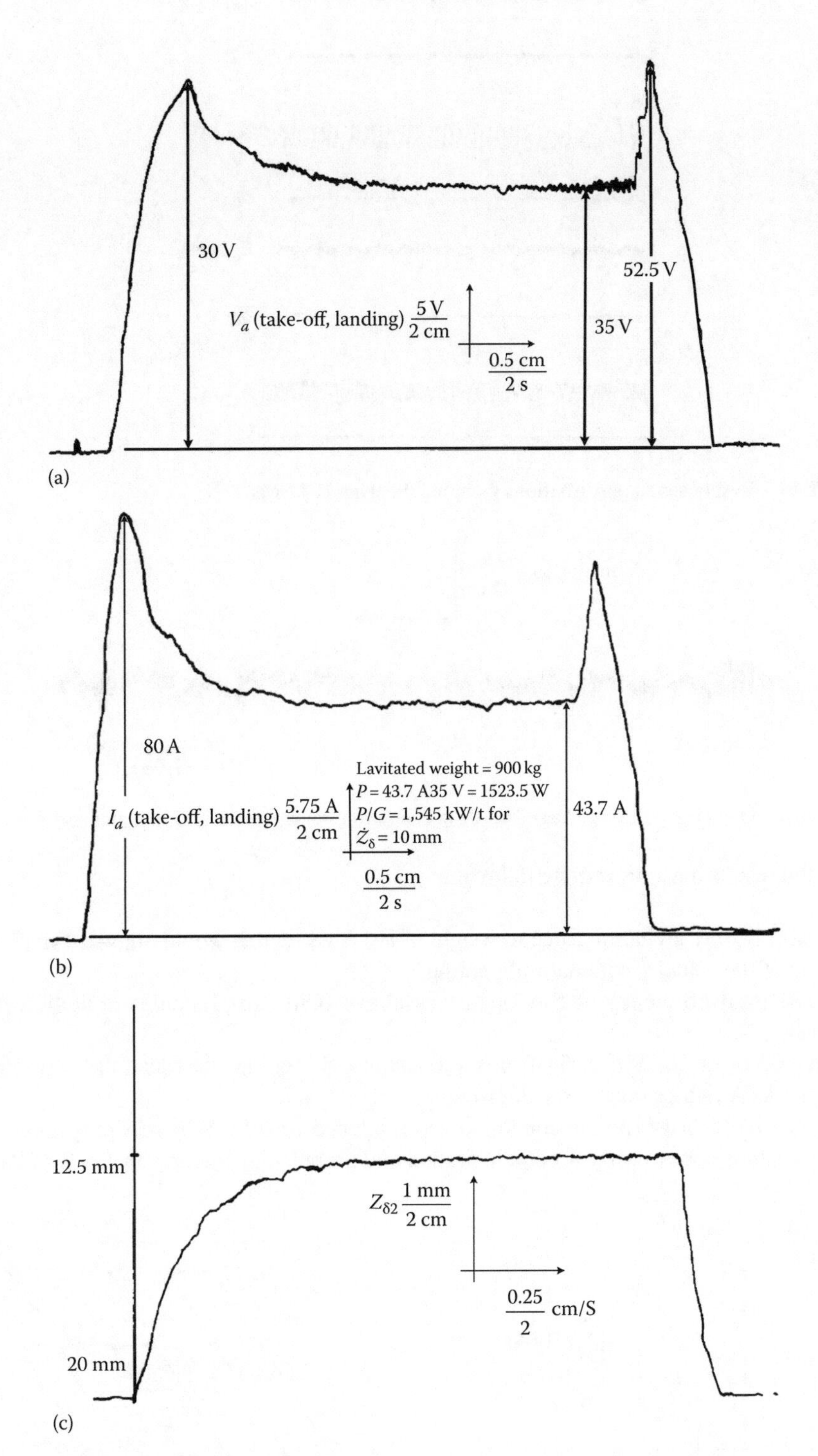

FIGURE 22.14 Experimental levitation takeoff (lifting) and landing at zero speed: (a) voltage, (b) dc current, and (c) airgap.

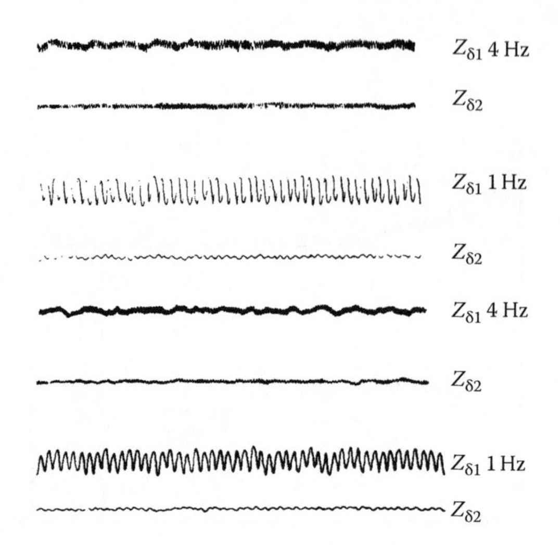

FIGURE 22.15 Responses to perturbations from neighboring H-LSMs.

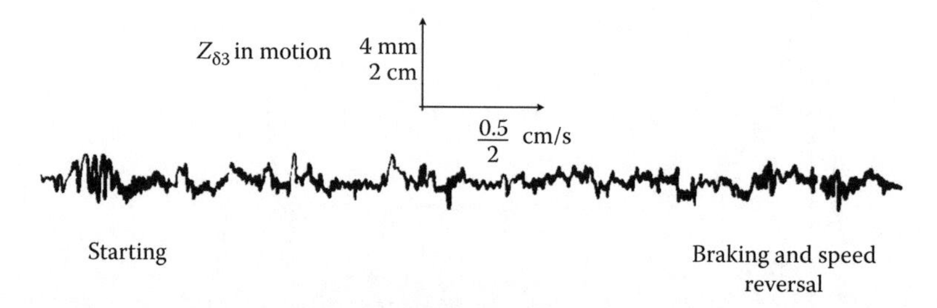

FIGURE 22.16 One airgap during Magnibus-01 acceleration, regenerative braking, and speed reversal.

Experimental results have proved the following:

- At least (7–10)/1 levitation force to weight of H-LSMs, which would allow a 1 m/s^2 acceleration of the vehicle, with natural cooling.
- H-LSM total efficiency at 5–10 m/s was above 0.80 (this includes levitation power losses).
- Power factor of H-LSM at 5–10 m/s was about 0.8 lagging, thus leading to reasonable inverter kVA ratings (and costs and weight).
- In general, the thrust (propulsion force) was produced with 1.5 W loss/N of thrust.
- DC windings power used for both, levitation and propulsion, amounts to 1.6–2 kW/ton.

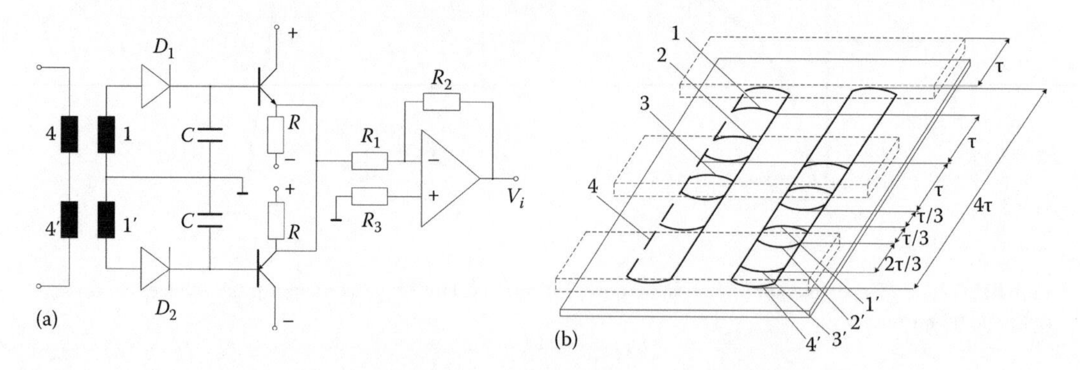

FIGURE 22.17 Position sensor for H-LSM control: (a) main circuits and (b) layout.

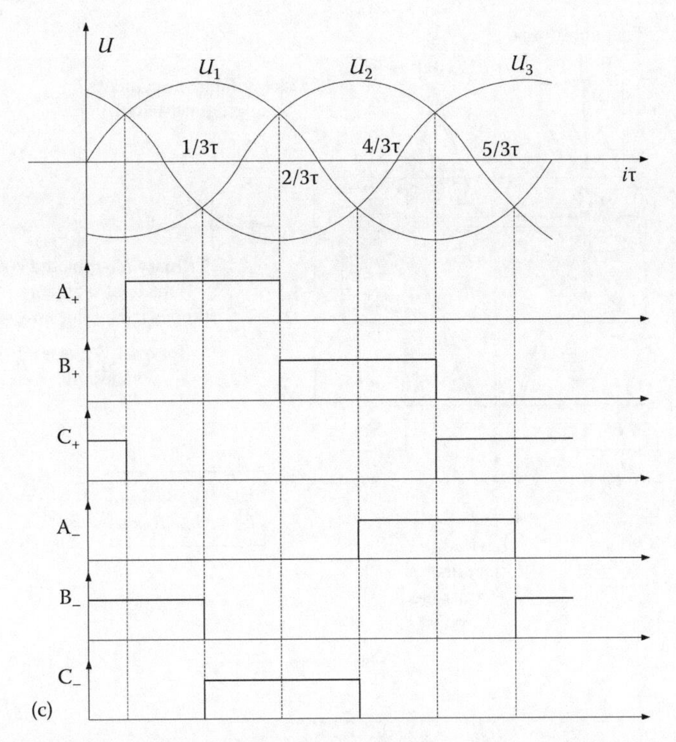

FIGURE 22.17 (continued) Position sensor for H-LSM control: (c) output signals.

- It is argued here that this superior performance might render Magnibus as feasible not only for urban but also for interurban use (up to 400 km/h or more). The vehicle independence in its full control, the passive track simplicity, and lower cost (even at mild traffic density), including simpler guideway switches, are the main merits of Magnibus, in addition to lower losses and better power factor. But, as in any passive guideway MAGLEV, full power transfer to the vehicle is required.
- For more on H-LSM and attraction-force levitation details, see Chapters 9 and 19.

22.4 POTENTIAL IMPROVEMENTS ON MAGNIBUS SYSTEM

Integrating not only propulsion and levitation by H-LSM but also controlled guidance (even for high speeds), while not increasing the segmented track weight, is illustrated in Figure 22.18.

The topology evidences:

- The permanent magnets (PMs) in a strong flux concentration configuration produce a rather large (larger than 1.2 T (1.4 T if possible)) flux density in the airgap between the track segments and the primary "teeth." The PMs alone should be able to suspend the vehicle at rated airgap for rated weight. When the weight increases, suspension is provided at a smaller gap; at lower weight the airgap is larger. In this case, the two coil currents i_{c1}, i_{c2} contain the components i_{lev}, which are controlled to zero (as in Chapter 19), unless the airgap goes out of the safe interval, $(0.6-1.2)g_{rated}$, when airgap control (with nonzero dynamic i_{lev}) is performed.
- On the other hand, the inclined airgaps (Figure 22.18a) produce additional levitation force and opposite guidance forces. The guidance force control is managed by the currents i_{guide} (same in both coils). The inclination angle may be designed to best suit the guidance and levitation force requirements.

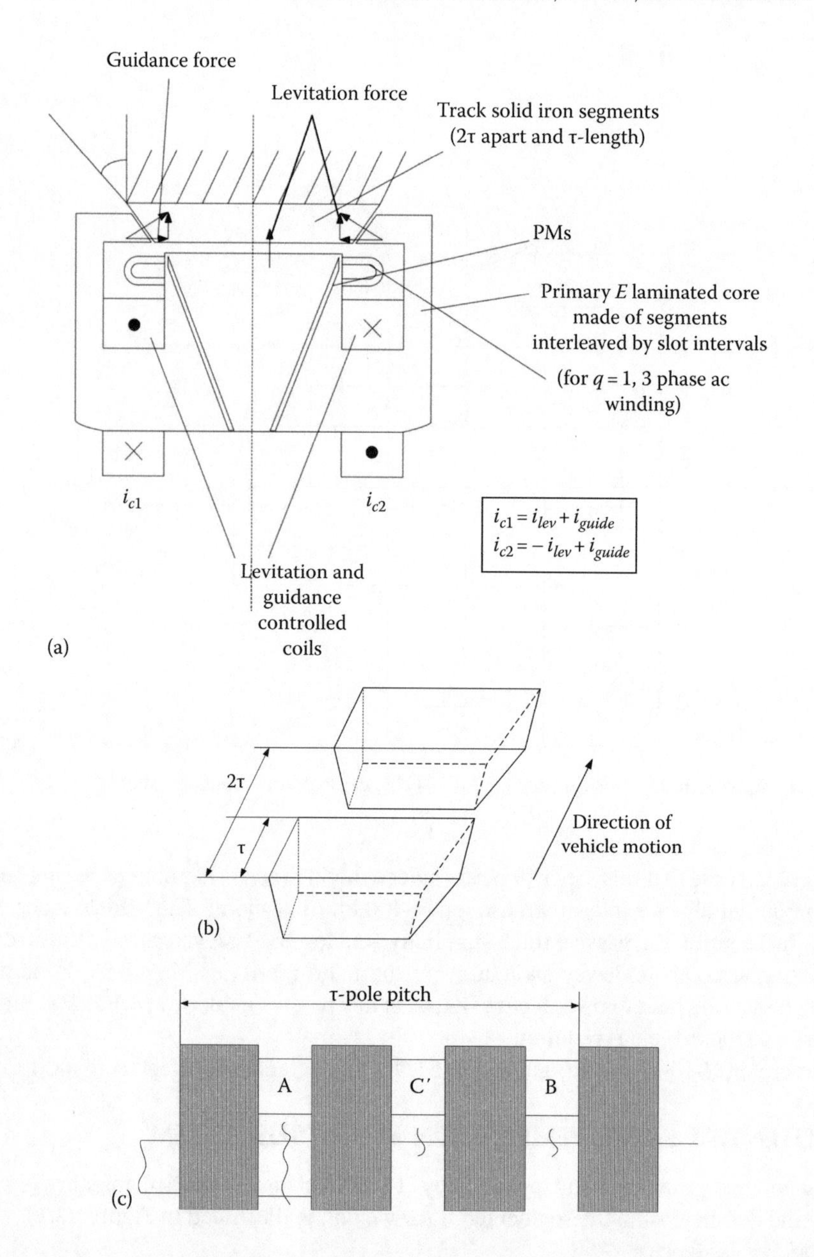

FIGURE 22.18 PM-assisted H-LSM MAGLEV (Magnibus 02): (a) cross section, (b) track segments, and (c) ac winding.

- Two dc–dc converters are required, and the frequency band of the levitation and guidance functions should be decoupled.
- If the PMs alone cannot lift the vehicle, then the i_{lev} will not be regulated around zero but around a nonzero value required for steady-state full levitation.
- In any case, the efficiency of the integrated propulsion-levitation guidance is improved above 90% due to PMs doing the main job, while the track weight remains remarkably small.
- It is almost needless to say that the same concept may be used by excluding the ac three-phase winding and making both the guideway and the primary core continuous, along the direction of travel, into a hybrid (PM assisted) levitation-guidance system suitable for LIM-MAGLEVs, etc.

22.5 TRANSVERSE-FLUX PM-LSM MAGLEVs

The transverse-flux PM-LSM (Chapter 14) has already been proposed (theoretically) for MAGLEVs in 1995, by its inventor [3] (Figure 22.19), despite the fact that the normal force/thrust ratio is only 3(4)/1, in general.

The TF-LSM has only PM excitation and long (multiple pole span) ac coils [4]. To reduce the track weight, TF-LSM uses single-phase blocks placed one after the other (shifted properly in the direction of motion) to produce, say, a three-phase machine (Figure 22.19b). It should be mentioned that the variable reluctance track (at 6–10 mm airgap) does show at best a 2/1 saliency ratio, and it has to be made of silicon laminations to avoid very large eddy current losses (a large drag force, also).

As the PM-produced maximum flux density in the airgap is basically ac and may hardly go over 1–1.2 T, due to very large flux fringing (leakage), and the iron to iron primary/secondary areas are at most 1/3 of total airgap primary area, there is not enough levitation force for full levitation of the vehicle as already mentioned (a 10/(15)/1 normal/thrust force is required for full levitation).

However, as in any linear synchronous machine, whose vector diagram is shown (in primary coordinates) in Figure 22.20, there is one more "tool" to produce additional levitation force: positive (magnetizing) i_d current:

$$\overline{I}_s R_s - \overline{V}_s = -j\omega_r \overline{\Psi}_s; \quad \overline{\Psi}_s = \Psi_{PM} + L_d I_d + jL_q I_q \tag{22.1}$$

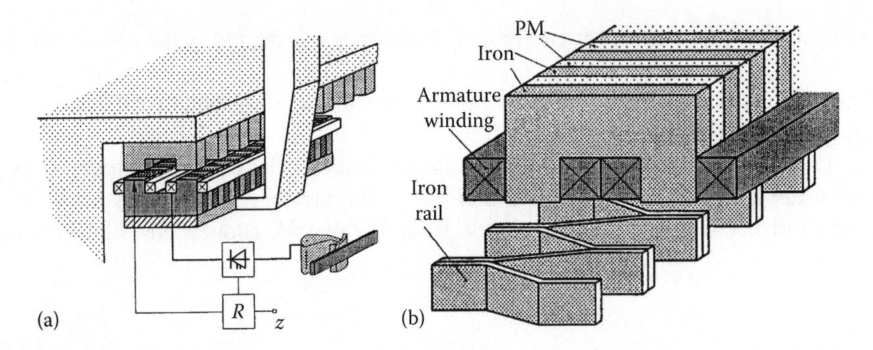

FIGURE 22.19 Proposed TF-LSM MAGLEV: (a) MAGLEV structure and (b) primary/secondary. (After Weh, M., Linear electromagnetic drives in traffic systems and industry, *Record of LDIA-1995*, Nagasaki, Japan, pp. 1–8, 1995.)

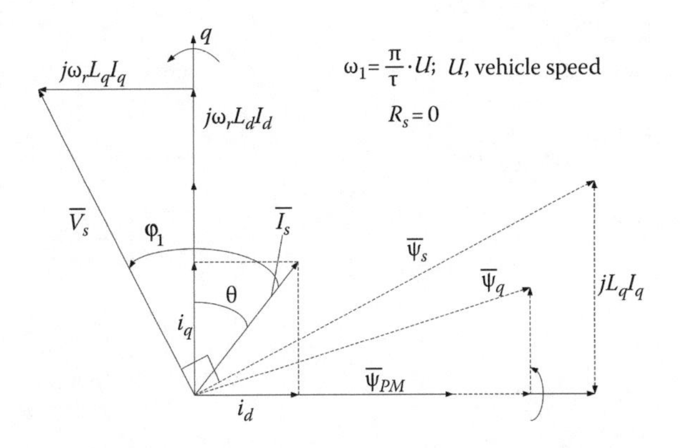

FIGURE 22.20 H-LSM vector diagram (steady state) in dq (primary coordinates).

It is a known fact that the normal (attraction) force is proportional to the airgap flux linkage Ψ_g amplitude squared:

$$F_n \sim K \times |\bar{\Psi}_g|^2 \tag{22.2}$$

Neglecting the leakage inductance, it may be said that $F_n \sim K' |\bar{\Psi}_s|^2$; but, with the large airgap of MAGLEVs, the error would be more than 50%.

The airgap flux Ψ_g is related to total airgap flux density in the airgap (B_g), which explains (22.2):

$$F_n \approx \frac{B_g^2}{2u_0} \text{Area} \tag{22.3}$$

- Both i_d and i_q currents produce an increase in levitation force but i_d more so, despite the fact that $L_q > L_d$ (by 20%–30% only, however)
- As the magnetic salience is very moderate $L_q/L_d < 1.3$, i_d does not produce a large braking force due to the reluctance effect:

$$F_x = \frac{3}{2}\frac{\pi}{\tau}\Big[\underset{\text{(positive)}}{\bar{\Psi}_{PM} I_q} + \underset{\text{(negative)}}{(L_d - L_q) I_d I_q}\Big]; \quad L_d < L_q; \quad I_d > 0 \tag{22.4}$$

However, the negative reluctance thrust (for $i_d > 0$) has to be considered in the propulsion design.

- Moreover, a positive i_d leads to a lower power factor (φ_1 lagging), and thus the inverter kVA ratings are slightly increased.
- But the fact that levitation and propulsion are handled only by the vector control (FOC) or direct thrust and normal force (airgap) control of PWM inverter is an important merit that may deserve full technical and economic investigation in full-scale applications (Figure 22.21).

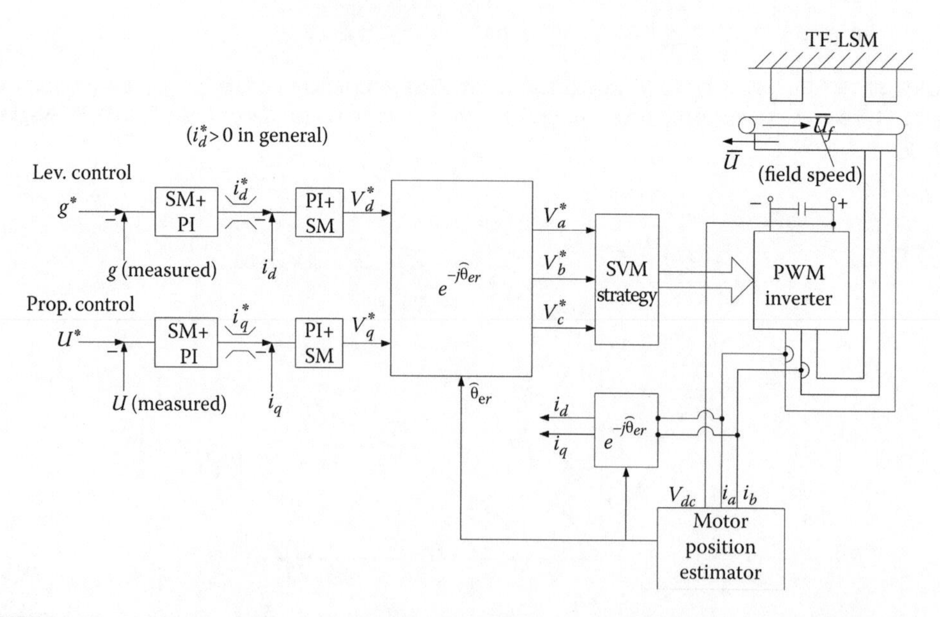

FIGURE 22.21 Generic levitation and propulsion control in a TF-PM-LSM potential MAGLEV.

- The thrust multiplier character of the ac multi-pole coil windings on TF-PM-LSM leads to lower copper losses/thrust, if the number of poles is large enough to cancel the fringing flux adverse effects, but at the cost of higher core losses due to higher fundamental frequency. An optimum problem is sensed here.
- The twisting (by pole pitch τ) of the guideway laminated cores (Figure 22.19b) remains a serious practical problem, but not so for industrial platforms (short travel).

22.6 DC-POLARIZED L-SRM MAGLEVs

A PM-less solution that maintains the transverse-flux topology but simplifies the guideway laminated segments stems from the dc-polarized linear switched reluctance motor (Figure 22.22).

- The number of primary and secondary poles (τ) is the same (as in TF-PM-LSM) and may be large; in principle $\tau > (5–7)g$; (g-the airgap).
- The dc coils are connected in series to cancel the total emf induced in them due to "full" coupling of each of them with one ac coil.
- Provided the homopolar flux dc coil airgap flux density $B_{dc} \approx 1.5T$ in the airgap has a rather sinusoidal strong ac component $B\sim$, the ac currents in the phases may be sinusoidal and thus vector control may be used in a standard PWM inverter. An additional leg for the controlling of i_{dc} for levitation control is added (Figure 22.22c).
- Basically, the normal force and thrust density are similar to those of H-LSM but with the additional advantage of moderate copper losses due to multiple (long) dc and ac coils (one ac coil per phase module), typical to transverse-flux machines. But, for g = 6–10 mm a saliency ratio of maximum 2/1 may be hoped for in a rather small pole pitch design ($\tau > (5–7)g$), required for high force densities.

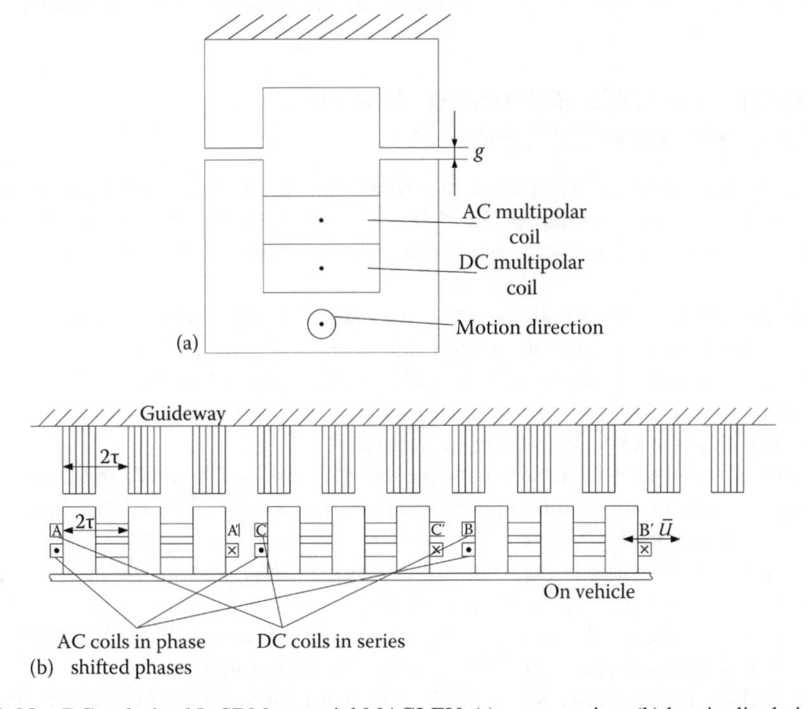

FIGURE 22.22 DC-polarized L-SRM potential MAGLEV: (a) cross section, (b) longitudinal view.

(continued)

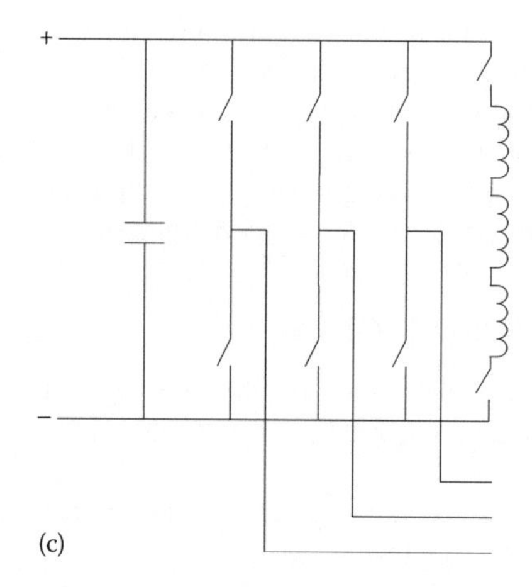

FIGURE 22.22 (continued) DC-polarized L-SRM potential MAGLEV: (c) inverter structure.

- The equivalent power factor of dc polarized L-SRM is basically lower than that of H-LSM, leading to a larger kVA inverter on board the vehicle.
- Moreover, the guideway segments have to be made of laminated steel as they experience notable ac magnetic fields. This leads to added guideway costs in comparison with the solid iron segments for H-LSM.
- It may be argued that dc + ac currents could be handled in the same coils of system in Figure 22.22. True, but then the whole inverter control has to be changed. This proposal is here to inspire new research, as comprehensive efforts—in theory and testing—are required to prove this solution, or other similar ones, practical or impractical for MAGLEV vehicles; for industrial platforms, however, its simplicity makes it very tempting.

22.7 MULTIPHASE (TRUE BRUSHLESS) LINEAR RELUCTANCE MACHINE MAGLEVs

Known also as "flux regulated reluctance machine" [5], the two-level bipolar current brushless reluctance machine [6] may be adapted, in its linear counterpart, for MAGLEVs to produce integrated propulsion and levitation with a good power factor (Figure 22.23a and b).

- The machine resembles the behavior of a nonexcited dc brush machine with brushes moved from the neutral axis to the high saliency pole corners (Figure 22.24).
- Out of the six phases (Figures 22.23 and 22.24), ideally two (A and B in this moment) are field phases and the other four (C, D, E, F) are thrust (torque) phases, as in the dc brush machine with shifted brushes and no excitation.
- Each phase "switches" roles based on mover (vehicle) position (Figure 22.23b). As expected, for given thrust, the minimum copper losses occur when $i_F^* = i_T^*$ (field and thrust current levels are equal to each other).
- The summation of currents may not always be zero, and thus an additional leg in the inverter handles the null current (Figure 22.23b).
- The higher the saliency the better, but for large airgap long travels (MAGLEV vehicles), even a segmented (one piece per pole) laminated guideway may do.
- As most of the primary area is active, a higher thrust and normal force density than most previous passive (or active) guideway MAGLEVs may be expected. So a 2.5/1 saliency ratio may already be acceptable.

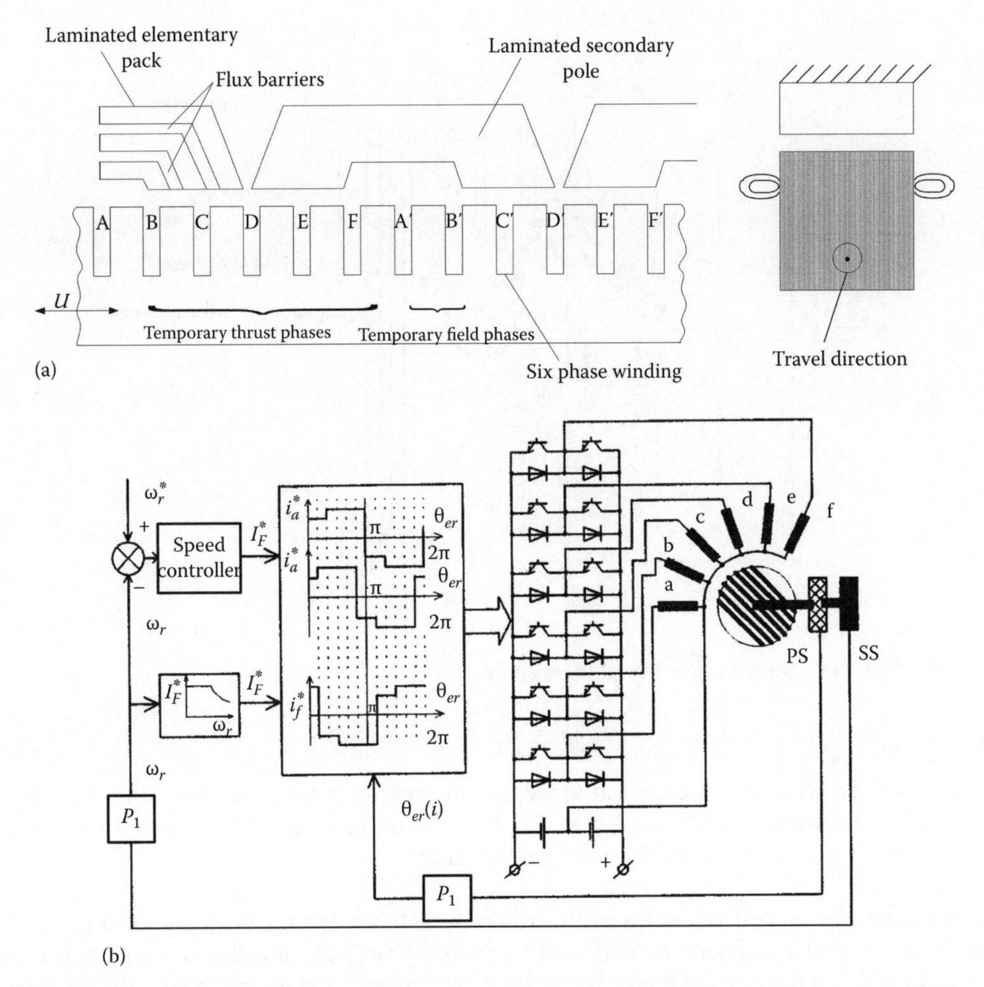

FIGURE 22.23 Generic bipolar two-level current L-RSM MAGLEV: (a) longitudinal cross section and (b) inverter dedicated topology. (After Boldea, I., *Reluctance Synchronous Machines and Drives*, Oxford University Press, 1996.)

- The flat top level of currents leads to a better inverter voltage (and kVA) utilization, with an ideal value closer to unity equivalent power factor.
- The levitation force is mainly controlled through the field phases (i_F^*) and thrust by the thrust phase currents (i_T^*). Six proximity (Hall type) position sensors (placed in slot tops) are required to "command" the bipolar two-level current waveforms (Figure 22.22b). Motion sensor-less propulsion control may also be feasible.
- There is some interference between levitation and propulsion, but it is rather small as the two currents' fields are basically orthogonal and always "tied" to the track poles position. (It may be inferred that even solid segments for a coarse lamination guideway may be used due to this characteristic).
- More than six phases may be used but, to limit the pole pitch in MAGLEV vehicles, six phases are considered a safe design start-up. A larger pole pitch leads to thicker guideway poles and thicker back iron in the primary.
- It has to be recognized that the pole-span ac coils in the discussed solution mean larger copper losses than for the long (multiple-pole span) coils of TF-RM-LSM. But the normal forces and thrust densities are at least larger, for a better kVA rating utilization

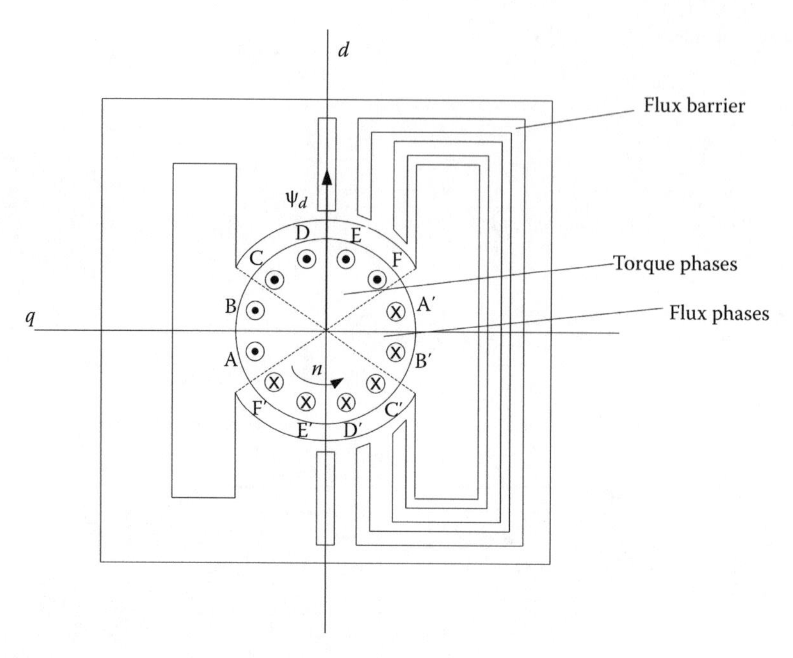

FIGURE 22.24 DC-brush unexcited machine analogy.

(power factor) in the inverter and lower iron losses (lower frequency due to larger pole pitch); moreover, no PMs are required in the solution in Figure 22.23, though low-cost (Ferrite) PMs may be added for better performance.

Tooth-wound dc + ac coil primaries, with variable reluctance secondary rating motors, may be "converted" to linear counterparts and tried for MAGLEVs [7,8]. However, it seems rather evident that the thrust density and power factor are notably lower than for the previously mentioned proposals.

All in all, these three competitive MAGLEV solutions look worthy of thorough quantitative analyses—technically and economically—for MAGLEV vehicles or for industrial platforms.

22.8 SUMMARY

- Passive guideway MAGLEVs refer to magnetically levitated "people movers" or industrial platforms with linear electric motor propulsion and passive (no PMs and no supplied windings) secondary (guideway).
- With all power equipment (for propulsion, levitation, guidance, and auxiliaries) on board of vehicle, full electric power transfer to the vehicle is required.
- Though the PWM inverters and full power collectors for propulsion are placed on board, the recent progress in power electronics with IGBT and IGCT PWM inverters has led to notably less than 1 kg/1 kVA with forced cooling; thus, the burden of power electronic weight on board is rather mild, allowing for enough load weight.
- Also, controlled mechanical-friction multi-MW power transmission to vehicles up to more than 500 km/h peak speed has been demonstrated with wheeled trains (TAGV, France); the inevitable wearing of the power collector due to accidental arching has been reduced to practically acceptable costs.

- This is how the passive guideway MAGLEV merits come into play:
  - Simpler (passive) guideway for electromagnetic propulsion, levitation, and guidance, at lower initial cost.
  - Standard power substations; even if single-phase ac 15 KV power is brought on board of such MAGLEV vehicles, the incurring step-down transformer and diode rectifier seem an acceptable burden.
  - Standard 800–1500 V dc power substations may be used for urban and suburban MAGLEVs.
  - The land intrusion and on ground stations' man supervision resources are much smaller than for active guideway MAGLEVs.
  - Each vehicle (train) is fully independent and thus, when faulty, it is simply taken off the track.
  - The guideway switches at intersections are much simpler than for active guideway MAGLEVs.
- Linear induction motor LIM propulsion with separate dc controlled electromagnets (without or with PM assistance) for levitation and guidance solutions [9] has led to the first commercial (and close to two) LIM-MAGLEVs for urban transportation.
- The ruggedness of LIM and its rather straightforward sensorless indirect field-oriented control from zero speed makes the solution attractive.
- However, the small but downward normal (attractive) force in LIM is opposite to the levitation force of dc controlled electromagnets, to preserve acceptable LIM efficiency (about 80%!) at still rather low, lagging, power factor (0.45–0.6) for 6–10 mm mechanical airgap.
- Moreover, the separate power electronics for full control of levitation and guidance may add notably (in peak kVA) to the vehicle full power rating and even weight (and cost).
- But, still, levitation and guidance may be provided at 2 kW/kg of vehicle for urban applications (20 m/s or so); to reduce this further, PMs are placed in electromagnets to cover full vehicle weight; and zero current control (at slightly variable airgap) for levitation (not done yet) may lead to 1 kW/kg for levitation and guidance (airgap control, instead of zero current control, is needed when the airgap exits the safe zone: of 0.6–1.2 rated value).
- A guideway structure, which serves both (by the solid iron core) the LIM and the levitation-guidance electromagnets, may use the LIM normal force for levitation control (or at least for vertical oscillations damping), when PMs are inserted in the levitation-guidance electromagnets.
- The full levitation control by LIMs—with no additional electromagnets—may be approached with cage (ladder) secondary in a coarsely laminated slotted iron guideway; at least for urban MAGLEVs and industrial platforms, this solution seems a way of the future.
- In order to reduce further the vehicle weight/passenger, mainly the electromagnetic and power electronics equipment peak KVA and weight, and further simplify the passive guideway, the homopolar linear synchronous motor (H-LSM) MAGLEV (Magnibus) concept was introduced and tested on a 150 m track with a 4 ton prototype.
- This time the propulsion and levitation passive track was made solely of solid mild iron segments with one pole pitch (out of two poles length) span. Separate lateral solid iron slabs in the guideway interact with additional guidance controlled electromagnets. As with forced cooling the H-LSM levitation force/weight may over pass 10/1 and levitation to propulsion force ratio is around 10(12)/1, the Magnibus system can provide $1-1.2\,\text{m/s}^2$ acceleration if the weight of H-LSMs represents 1/10(12) of vehicle total weight, which is quite reasonable.
- Also, because the eddy current drag force in the track solid iron segments due to the dynamic longitudinal effect is less than 10%–15% of rated thrust even at 300 km/h, the Magnibus system concept may be extended at 400 km/h interurban MAGLEVs.
- In urban MAGLEVs, the Magnibus system provided a total levitation-propulsion efficiency between 5 and 10 m/s of more than 80% at lagging power factor (i_q pure control) of

0.8 or slightly more, for an airgap of 10 mm. In addition, the ac copper losses per thrust was about 1.5 W/N.

- To further improve the total efficiency above 90%, a PM-assisted H-LSM capable of controlled propulsion, levitation, and guidance was introduced in this chapter. The levitation and guidance control are now handled mainly by two dc long coil currents with opposite ($\pm i_{lev}$) current components for levitation and guidance control $\left(\pm i_{guide}\right)$. Two dc–dc 2 quadrant converters control levitation and guidance. The segmented guideway weight (cost) is also notably reduced.
- One more transverse-flux PM-LSM passive guideway MAGLEV concept is illustrated, where both propulsion and levitation are controlled through a simple inverter at good efficiency (90%) but at lower power factor (0.6–0.7). Moreover, as the levitation/thrust ratio is only 4/1, to secure 1–1.2 m/s^2 vehicle acceleration, notably more flux density in the airgap should be provided by the ac winding d and q currents (in the orthogonal model). This will lower overall efficiency notably; also, the decoupled levitation-propulsion control seems a daunting job. A laminated segmented track is required, though. However, the solution seems to deserve full-scale vehicle experiments, especially for industrial MAGLEV platforms.
- Finally, two more PM-less passive guideway MAGLEV solutions have been proposed:
- The dc polarized transverse-flux linear dually salient machine MAGLEV
- The bipolar two-level multiphase (six phases) linear reluctance machine MAGLEV

They should provide full levitation and propulsion control (also some guidance (centralizing) force), with the first one providing better efficiency at lower power factor than the second one; both solutions require laminated variable reluctance secondary (track).

In conclusion, passive guideway MAGLEVs have revived and will do even better in the future for urban and then for interurban MAGLEV vehicles [10–12] and industrial platforms as power electronics weight is small enough to be accepted on board and full power transfer to vehicle up to 500 km/h in the MW range has become practical.

The passive guideway and of on ground standard power substations and its switches, the independent control of vehicle from on board at notably smaller Wh/passenger/km than for wheeled vehicles at comparative speed, and lower guideway maintenance (due to 10^4 smaller stress because of force distribution over a large area) make MAGLEVs, in our view, a viable part of the future of transportation. The technology is here to stay and progress.

REFERENCES

1. http://en.wikipedia.org/wiki/File:Linimo_approaching_Banpaku_Kinen_Koen,_towards_Fujigaoka_Station.jpg.
2. I. Boldea, A. Trica, G. Papusoiu, and S.A. Nasar, Field texts on a MAGLEV with passive guideway linear inductor motor transportation system, *IEEE Trans.*, VT-37(1), 1988, 213–219.
3. M. Weh, Linear electromagnetic drives in traffic systems and industry, *Record of LDIA-1995*, Nagasaki, Japan, 1995, pp. 1–8.
4. T.-K. Hoang, D.-H. Kang, and J.-Y. Lae, Comparisons between various designs of transverse flux linear motor in terms of thrust and normal force, *IEEE Trans.*, VT-46(10), 2010, 3795–3801.
5. J.D. Law, A. Chertox, and T.A. Lipo, Design performance of the field regulated reluctance machine, *Record of IEEE-IAS-1992 Meeting*, Houston, TX, Vol. 1, 1992, pp. 234–241.
6. I. Boldea, *Reluctance Synchronous Machines and Drives*, Chapter 6, Oxford University Press, 1996.
7. A. Zulu, B. Mecrow, and M. Armstrong, A wound-field three-phase flux-switching synchronous motor with all excitation sources on the stator, *IEEE-ECCE-2009*, 2009, pp. 1502–1509.
8. E. Sulaiman, T. Kosaka, and N. Matsui, A new structure of 24 slot/10 pole field excitation flux-switching synchronous machine for hybrid electric vehicles, *Record of EPE-2011*, Birmingham, U.K., 2011, pp. 1–10.

9. S. Kusagawa, K. Shutoh, and E. Masada, Electro-magnet suspension system with fuzzy control for a magnetically levitated railway system, *Record of LDIA-2003*, Birmingham, U.K., 2003, pp. 263–266.
10. B.S. Lee, Linear switched reluctance machine drive with electromagnetic levitation and guidance system, PhD Thesis, Virginia Tech, Blacksburg, VA, 2000.
11. I. Boldea and S.A. Nasar, *Linear Electric Actuators and Generators*, Cambridge University Press, London, U.K., 1997.
12. D.H. Kang, Design of PM excited transverse flux linear motor of inner mover type, *KIEE*, 5-B(2), 137–141, 2005.

Index

E

F

G

H

I

M

N

P

R

S

T

U

V

Z